Student's Solutions Manual

Precalculus
Graphical, Numerical, Algebraic
Seventh Edition, Media Update, and Florida Edition

Franklin D. Demana
The Ohio State University

Bert K. Waits
The Ohio State University

Gregory D. Foley
Ohio University

Daniel Kennedy
Baylor School

Addison-Wesley
is an imprint of

Reproduced by Pearson Addison-Wesley from electronic files supplied by the author.

Publishing as Pearson Addison-Wesley, 75 Arlington Street, Boston, MA 02116.

ISBN 0-321-36994-7

Contents

CHAPTER P	Prerequisites	1
CHAPTER 1	Functions and Graphs	29
	*Media Update and Florida Edition	233
CHAPTER 2	Polynomial, Power, and Rational Functions	38
CHAPTER 3	Exponential, Logistic, and Logarithmic Functions	73
CHAPTER 4	Trigonometric Functions	91
CHAPTER 5	Analytic Trigonometry	113
CHAPTER 6	Applications of Trigonometry	128
CHAPTER 7	Systems and Matrices	149
CHAPTER 8	Analytic Geometry in Two and Three Dimensions	172
CHAPTER 9	Discrete Mathematics	196
CHAPTER 10	An Introduction to Calculus: Limits, Derivatives, and Integrals	214
APPENDIX A	Algebra Review	228
APPENDIX C	Logic	231

Chapter P Prerequisites

Section P.1 Real Numbers

Quick Review P.1

1. $\{1, 2, 3, 4, 5, 6\}$

3. $\{-3, -2, -1\}$

5. **(a)** 1187.75 **(b)** -4.72

7. $(-2)^3 - 2(-2) + 1 = -3$; $(1.5)^3 - 2(1.5) + 1 = 1.375$

9. 0, 1, 2, 3, 4, 5, 6

Section P.1 Exercises

1. -4.625 (terminating)

3. $-2.1\overline{6}$ (repeating)

5.

−5 −4 −3 −2 −1 0 1 2 3 4 5

all real numbers less than or equal to 2 (to the left of and including 2)

7.

−2 −1 0 1 2 3 4 5 6 7 8

all real numbers less than 7 (to the left of 7)

9.

−5 −4 −3 −2 −1 0 1 2 3 4 5

all real numbers less than 0 (to the left of 0)

11. $-1 \leq x < 1$; all numbers between -1 and 1 including -1 and excluding 1

13. $-\infty < x < 5$, or $x < 5$; all numbers less then 5

15. $-1 < x < 2$; all numbers between -1 and 2, excluding both -1 and 2

17. $(-3, \infty)$; all numbers greater than -3

19. $(-2, 1)$; all numbers between -2 and 1, excluding both -2 and 1

21. $(-3, 4]$; all numbers between -3 and 4, excluding -3 and including 4

23. The real numbers greater than 4 and less than or equal to 9.

25. The real numbers greater than or equal to -3, or the real numbers which are at least -3.

27. The real numbers greater than -1.

29. $-3 < x \leq 4$; endpoints -3 and 4; bounded; half-open

31. $x < 5$; endpoint 5; unbounded; open

33. His age must be greater than or equal to 29: $x \geq 29$ or $[29, \infty)$; $x =$ Bill's age

35. The prices are between \$1.099 and \$1.399 (inclusive): $1.099 \leq x \leq 1.399$ or $[1.099, 1.399]$; $x =$ \$ per gallon of gasoline

37. $a(x^2 + b) = a \cdot x^2 + a \cdot b = ax^2 + ab$

39. $ax^2 + dx^2 = a \cdot x^2 + d \cdot x^2 = (a + d)x^2$

41. The opposite of $6 - \pi$, or $-(6 - \pi) = -6 + \pi = \pi - 6$

43. In -5^2, the base is 5.

45. **(a)** Associative property of multiplication

(b) Commutative property of multiplication

(c) Addition inverse property

(d) Addition identity property

(e) Distributive property of multiplication over addition

47. $\dfrac{x^2}{y^2}$

49. $\left(\dfrac{4}{x^2}\right)^2 = \dfrac{4^2}{(x^2)^2} = \dfrac{16}{x^4}$

51. $\dfrac{(x^{-3}y^2)^{-4}}{(y^6x^{-4})^{-2}} = \dfrac{x^{12}y^{-8}}{y^{-12}x^8} = \dfrac{x^4}{y^{-4}} = x^4y^4$

53. 3.6930338×10^{10}

55. $1.93175805 \times 10^{11}$

57. 4.839×10^8

59. 0.000 000 033 3

61. 5,870,000,000,000

63. $\dfrac{(1.35)(2.41) \times 10^{-7+8}}{1.25 \times 10^9} = \dfrac{3.2535 \times 10^1}{1.25 \times 10^9}$

$= \dfrac{3.2535}{1.25} \times 10^{1-9} = 2.6028 \times 10^{-8}$

65. **(a)** When $n = 0$, the equation $a^m a^n = a^{m+n}$ becomes $a^m a^0 = a^{m+0}$. That is, $a^m a^0 = a^m$. Since $a \neq 0$, we can divide both sides of the equation by a^m. Hence $a^0 = 1$.

(b) When $n = -m$, the equation $a^m a^n = a^{m+n}$ becomes $a^m a^{-m} = a^{m+(-m)}$. That is $a^{m-m} = a^0$. We know from part (a) that $a^0 = 1$. Since $a \neq 0$, we can divide both sides of the equation $a^m a^{-m} = 1$ by a^m. Hence $a^{-m} = \dfrac{1}{a^m}$.

67. False. If the real number is negative, the additive inverse is positive. For example, the additive inverse of -5 is 5.

69. $[-2, 1)$ corresponds to $-2 \leq x < 1$. The answer is E.

71. In $-7^2 = -(7^2)$, the base is 7. The answer is B.

73. The whole numbers are 0, 1, 2, 3, . . ., so the whole numbers with magnitude less than 7 are 0, 1, 2, 3, 4, 5, 6.

75. The integers are . . ., $-2, -1, 0, 1, 2,$ . . ., so the integers with magnitude less than 7 are $-6, -5, -4, -3, -2, -1, 0, 1, 2, 3, 4, 5, 6$.

Section P.2 Cartesian Coordinate System

Quick Review P.2

1.

0.5 1 1.5 2 2.5 3

Distance: $|\sqrt{7} - \sqrt{2}| = \sqrt{7} - \sqrt{2} \approx 1.232$

3.

−5 −4 −3 −2 −1 0 1 2 3 4 5

5.

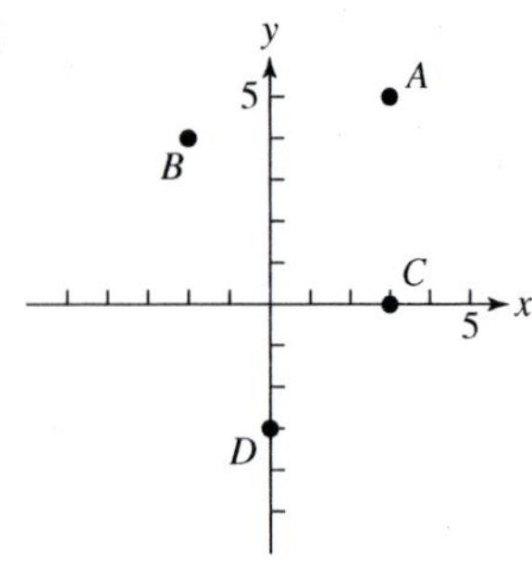

7. 5.5

9. 10

Section P.2 Exercises

1. $A(1, 0)$; $B(2, 4)$; $C(-3, -2)$; $D(0, -2)$

3. **(a)** First quadrant

(b) On the y-axis, between quadrants I and II

(c) Second quadrant

(d) Third quadrant

5. $3 + |-3| = 3 + 3 = 6$

7. $|(-2)3| = |-6| = 6$

9. Since $\pi \approx 3.14 < 4$, $|\pi - 4| = 4 - \pi$.

11. $|10.6 - (-9.3)| = |10.6 + 9.3| = 19.9$

13. $\sqrt{(-3 - 5)^2 + [-1 - (-1)]^2} = \sqrt{(-8)^2 + 0^2} = \sqrt{64}$
$= 8$

15. $\sqrt{(0 - 3)^2 + (0 - 4)^2} = \sqrt{3^2 + 4^2} = \sqrt{9 + 16}$
$= \sqrt{25} = 5$

17. $\sqrt{(-2 - 5)^2 + (0 - 0)^2} = \sqrt{(-7)^2 + 0^2} = \sqrt{49} = 7$

19. An isosceles triangle

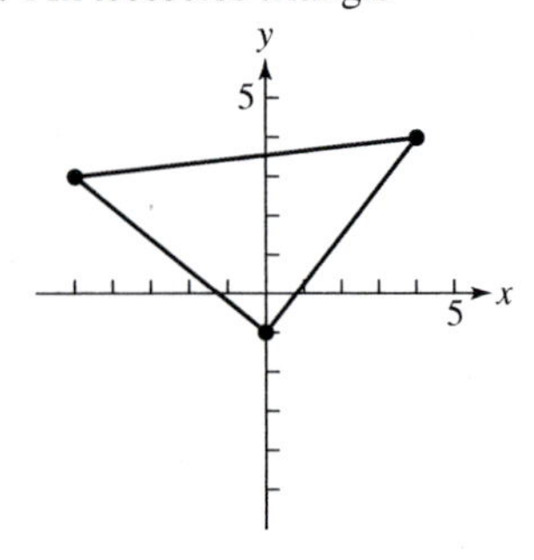

$\sqrt{(0 - (-5))^2 + (-1 - 3)^2} = \sqrt{5^2 + (-4)^2} = \sqrt{41}$

$\sqrt{(0 - (-4))^2 + (-1 - 4)^2} = \sqrt{(-4)^2 + (-5)^2}$
$= \sqrt{41}$

$\sqrt{(-5 - 4)^2 + (3 - 4)^2} = \sqrt{(-9)^2 + (-1)^2} = \sqrt{82}$

Perimeter $= 2\sqrt{41} + \sqrt{82} \approx 21.86$

Since $(\sqrt{41})^2 + (\sqrt{41})^2 = (\sqrt{82})^2$, this is a right triangle.

Area $= \frac{1}{2}(\sqrt{41})(\sqrt{41}) = 20.5$

21. A parallelogram

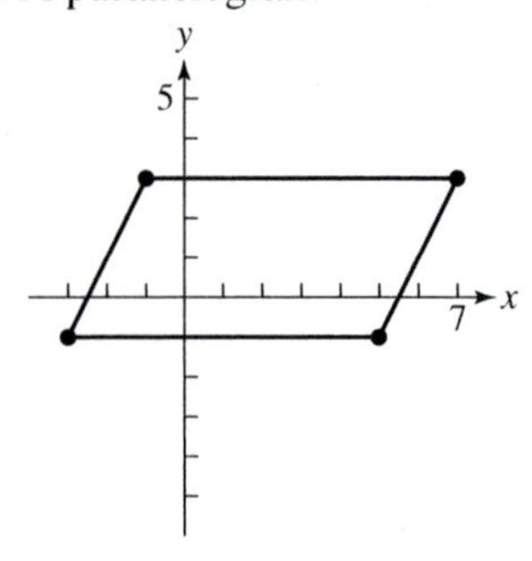

$\sqrt{[3 - (-1)]^2 + [-1 - (-3)]^2} = \sqrt{4^2 + 2^2} = \sqrt{20}$

This is a parallelogram with base 8 units and height 4 units. Perimeter $= 2\sqrt{20} + 16 \approx 24.94$;
Area $= 8 \cdot 4 = 32$

23. $\dfrac{10.6 + (-9.3)}{2} = \dfrac{1.3}{2} = 0.65$

25. $\left(\dfrac{-1 + 5}{2}, \dfrac{3 + 9}{2}\right) = \left(\dfrac{4}{2}, \dfrac{12}{2}\right) = (2, 6)$

27. $\left(\dfrac{-\frac{7}{3} + \frac{5}{3}}{2}, \dfrac{\frac{3}{4} + \left(-\frac{9}{4}\right)}{2}\right) = \left(\dfrac{-\frac{2}{3}}{2}, \dfrac{-\frac{6}{4}}{2}\right) = \left(-\dfrac{1}{3}, -\dfrac{3}{4}\right)$

29.

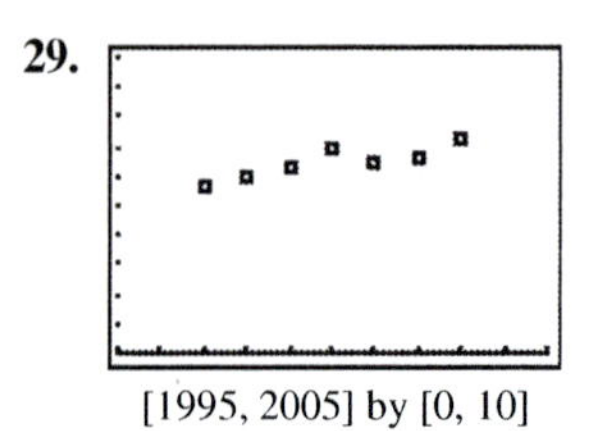

[1995, 2005] by [0, 10]

31.

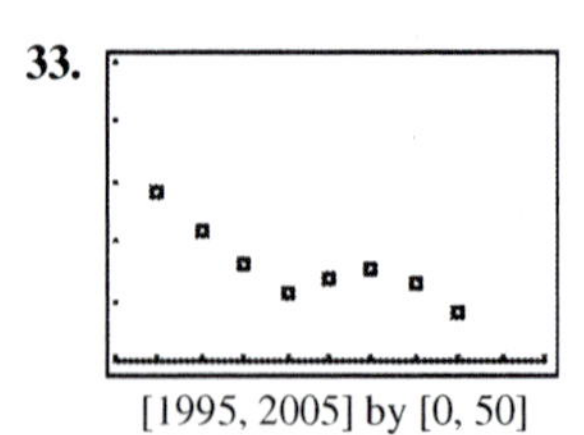

[1995, 2005] by [0, 150]

33.

[1995, 2005] by [0, 50]

35. **(a)** about \$183,000 **(b)** about \$277,000

37. The three side lengths (distances between pairs of points) are

$\sqrt{(4 - 1)^2 + (7 - 3)^2} = \sqrt{3^2 + 4^2} = \sqrt{9 + 16}$
$= \sqrt{25} = 5$

$\sqrt{(8 - 4)^2 + (4 - 7)^2} = \sqrt{4^2 + (-3)^2} = \sqrt{16 + 9}$
$= \sqrt{25} = 5$

$\sqrt{(8 - 1)^2 + (4 - 3)^2} = \sqrt{7^2 + 1^2} = \sqrt{49 + 1}$
$= \sqrt{50} = 5\sqrt{2}.$

Since two sides of the triangle formed have the same length, the triangle is isosceles.

39. **(a)** Vertical side: length $= 6 - (-2) = 8$; horizontal side: length $= 3 - (-2) = 5$; diagonal side:

$$\text{length} = \sqrt{[6-(-2)]^2 + [3-(-2)]^2} = \sqrt{8^2 + 5^2} = \sqrt{89}$$

(b) $8^2 + 5^2 = 64 + 25 = 89 = (\sqrt{89})^2$, so the Pythagorean Theorem implies the triangle is a right triangle.

41. $(x-1)^2 + (y-2)^2 = 5^2$, or $(x-1)^2 + (y-2)^2 = 25$

43. $[x-(-1)]^2 + [y-(-4)]^2 = 3^2$, or $(x+1)^2 + (y+4)^2 = 9$

45. $(x-3)^2 + (y-1)^2 = 6^2$, so the center is $(3, 1)$ and the radius is 6.

47. $(x-0)^2 + (y-0)^2 = (\sqrt{5})^2$, so the center is $(0, 0)$ and the radius is $\sqrt{5}$.

49. $|x - 4| = 3$

51. $|x - c| < d$

53. $\frac{1+a}{2} = 4$ and $\frac{2+b}{2} = 4$

$1 + a = 8$ $\quad$ $2 + b = 8$

$a = 7$ $\quad$ $b = 6$

55. The midpoint of the hypotenuse is $\left(\frac{5+0}{2}, \frac{0+7}{2}\right) = \left(\frac{5}{2}, \frac{7}{2}\right) = (2.5, 3.5)$. The distances from this point to the vertices are:

$$\sqrt{(2.5-0)^2 + (3.5-0)^2} = \sqrt{2.5^2 + 3.5^2} = \sqrt{6.25 + 12.25} = \sqrt{18.5}$$

$$\sqrt{(2.5-5)^2 + (3.5-0)^2} = \sqrt{(-2.5)^2 + 3.5^2} = \sqrt{6.25 + 12.25} = \sqrt{18.5}$$

$$\sqrt{(2.5-0)^2 + (3.5-7)^2} = \sqrt{2.5^2 + (-3.5)^2} = \sqrt{6.25 + 12.25} = \sqrt{18.5}.$$

57. $|x + 3| \geq 5$ means the distance from x to -3 must be 5 or more. So x can be 2 or more, or x can be -8 or less. That is, $x \leq -8$ or $x \geq 2$.

59. True. $\frac{\text{length of } AM}{\text{length of } AB} = \frac{1}{2}$ because M is the midpoint of AB.

By similar triangles, $\frac{\text{length of } AM'}{\text{length of } AC} = \frac{\text{length of } AM}{\text{length of } AB} = \frac{1}{2}$,

so M' is the midpoint of AC.

61. For a segment with endpoints at $a = -3$ and $b = 2$, the midpoint lies at $\frac{a+b}{2} = \frac{-3+2}{2} = \frac{-1}{2} = -\frac{1}{2}$.

The answer is C.

63. In the third quadrant, both coordinates are negative. The answer is E.

65. If the legs have lengths a and b, and the hypotenuse is c units long, then without loss of generality, we can assume the vertices are $(0, 0)$, $(a, 0)$, and $(0, b)$. Then the midpoint

of the hypotenuse is $\left(\frac{a+0}{2}, \frac{b+0}{2}\right) = \left(\frac{a}{2}, \frac{b}{2}\right)$.

The distance to the other vertices is

$$\sqrt{\left(\frac{a}{2}\right)^2 + \left(\frac{b}{2}\right)^2} = \sqrt{\frac{a^2}{4} + \frac{b^2}{4}} = \frac{c}{2} = \frac{1}{2}c.$$

For #67 and 69, note that since $P(a, b)$ is in the first quadrant, then a and b are positive. Hence, $-a$ and $-b$ are negative.

67. $Q(a, -b)$ is in the fourth quadrant and, since P and Q both have first coordinate a, PQ is perpendicular to the x-axis.

69. $Q(-a, -b)$ is in the third quadrant, and the midpoint of PQ is $\left(\frac{a+(-a)}{2}, \frac{b+(-b)}{2}\right) = (0, 0)$.

Section P.3 Linear Equations and Inequalities

Quick Review P.3

1. $2x + 5x + 7 + y - 3x + 4y + 2 = (2x + 5x - 3x) + (y + 4y) + (7 + 2) = 4x + 5y + 9$

3. $3(2x - y) + 4(y - x) + x + y = 6x - 3y + 4y - 4x + x + y = 3x + 2y$

5. $\frac{2}{y} + \frac{3}{y} = \frac{5}{y}$

7. $2 + \frac{1}{x} = \frac{2x}{x} + \frac{1}{x} = \frac{2x+1}{x}$

9. $\frac{x+4}{2} + \frac{3x-1}{5} = \frac{5(x+4)}{10} + \frac{2(3x-1)}{10} = \frac{5x + 20 + 6x - 2}{10} = \frac{11x + 18}{10}$

Section P.3 Exercises

1. **(a)** and **(c)**: $2(-3)^2 + 5(-3) = 2(9) - 15 = 18 - 15 = 3$, and $2\left(\frac{1}{2}\right)^2 + 5\left(\frac{1}{2}\right) = 2\left(\frac{1}{4}\right) + \frac{5}{2} = \frac{1}{2} + \frac{5}{2} = \frac{6}{2} = 3$. Meanwhile, substituting $x = -\frac{1}{2}$ gives -2 rather than 3.

3. **(b)**: $\sqrt{1 - 0^2} + 2 = \sqrt{1} + 2 = 1 + 2 = 3$. Meanwhile, substituting $x = -2$ or $x = 2$ gives $\sqrt{1-4} + 2 = \sqrt{-3} + 2$, which is undefined.

5. Yes: $-3x + 5 = 0$.

7. No: Subtracting x from both sides gives $3 = -5$, which is false and does not contain the variable x.

9. No: The equation has a root in it, so it is not linear.

11. $3x = 24$

$x = 8$

13. $3t = 12$

$t = 4$

15. $2x - 3 = 4x - 5$

$2x = 4x - 2$

$-2x = -2$

$x = 1$

17. $4 - 3y = 2y + 8$

$-3y = 2y + 4$

$-5y = 4$

$y = -\frac{4}{5} = -0.8$

19. $2\left(\frac{1}{2}x\right) = 2\left(\frac{7}{8}\right)$

$x = \frac{7}{4} = 1.75$

21. $2\left(\frac{1}{2}x + \frac{1}{3}\right) = 2(1)$

$x + \frac{2}{3} = 2$

$x = \frac{4}{3}$

23. $6 - 8z - 10z - 15 = z - 17$

$-18z - 9 = z - 17$

$-18z = z - 8$

$-19z = -8$

$z = \frac{8}{19}$

25. $4\left(\frac{2x - 3}{4} + 5\right) = 4(3x)$

$2x - 3 + 20 = 12x$

$2x + 17 = 12x$

$17 = 10x$

$x = \frac{17}{10} = 1.7$

27. $24\left(\frac{t + 5}{8} - \frac{t - 2}{2}\right) = 24\left(\frac{1}{3}\right)$

$3(t + 5) - 12(t - 2) = 8$

$3t + 15 - 12t + 24 = 8$

$-9t + 39 = 8$

$-9t = -31$

$t = \frac{31}{9}$

29. (a) The figure shows that $x = -2$ is a solution of the equation $2x^2 + x - 6 = 0$.

(b) The figure shows that $x = \frac{3}{2}$ is a solution of the equation $2x^2 + x - 6 = 0$.

31. (a): $2(0) - 3 = 0 - 3 = -3 < 7$. Meanwhile, substituting $x = 5$ gives 7 (which is not less than 7); substituting $x = 6$ gives 9.

33. (b) and (c): $4(2) - 1 = 8 - 1 = 7$ and $-1 < 7 \le 11$, and also $4(3) - 1 = 12 - 1 = 11$ and $-1 < 11 \le 11$. Meanwhile, substituting $x = 0$ gives -1 (which is not greater than -1).

35. 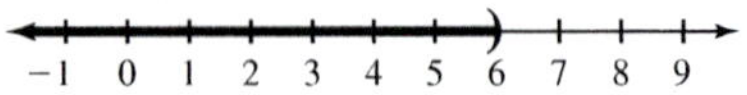 −1 0 1 2 3 4 5 6 7 8 9

37. 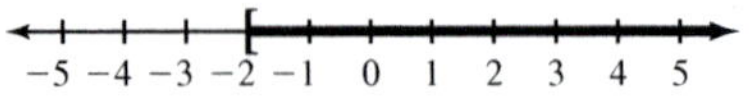 −5 −4 −3 −2 −1 0 1 2 3 4 5

$2x - 1 \le 4x + 3$

$2x \le 4x + 4$

$-2x \le 4$

$x \ge -2$

39. 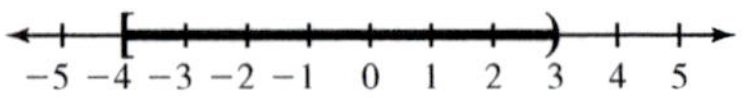 −5 −4 −3 −2 −1 0 1 2 3 4 5

$2 \le x + 6 < 9$

$-4 \le x < 3$

41. 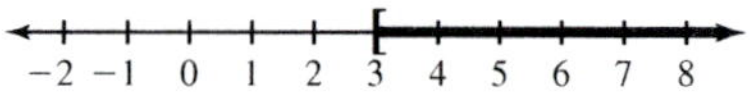 −2 −1 0 1 2 3 4 5 6 7 8

$10 - 6x + 6x - 3 \le 2x + 1$

$7 \le 2x + 1$

$6 \le 2x$

$3 \le x$

$x \ge 3$

43. $4\left(\frac{5x + 7}{4}\right) \le 4(-3)$

$5x + 7 \le -12$

$5x \le -19$

$x \le -\frac{19}{5}$

45. $3(4) \ge 3\left(\frac{2y - 5}{3}\right) \ge 3(-2)$

$12 \ge 2y - 5 \ge -6$

$17 \ge 2y \ge -1$

$\frac{17}{2} \ge y \ge -\frac{1}{2}$

$-\frac{1}{2} \le y \le \frac{17}{2}$

47. $0 \le 2z + 5 < 8$

$-5 \le 2z < 3$

$-\frac{5}{2} \le z < \frac{3}{2}$

49. $12\left(\frac{x - 5}{4} + \frac{3 - 2x}{3}\right) < 12(-2)$

$3(x - 5) + 4(3 - 2x) < -24$

$3x - 15 + 12 - 8x < -24$

$-5x - 3 < -24$

$-5x < -21$

$x > \frac{21}{5}$

51. $10\left(\frac{2y - 3}{2} + \frac{3y - 1}{5}\right) < 10(y - 1)$

$5(2y - 3) + 2(3y - 1) < 10y - 10$

$10y - 15 + 6y - 2 < 10y - 10$

$16y - 17 < 10y - 10$

$16y < 10y + 7$

$6y < 7$

$y < \frac{7}{6}$

53. $2\left[\frac{1}{2}(x - 4) - 2x\right] \le 2[5(3 - x)]$

$x - 4 - 4x \le 10(3 - x)$

$-3x - 4 \le 30 - 10x$

$-3x \le 34 - 10x$

$7x \le 34$

$x \le \frac{34}{7}$

55. $x^2 - 2x < 0$ for $x = 1$

57. $x^2 - 2x > 0$ for $x = 3, 4, 5, 6$

59. Multiply both sides of the first equation by 2.

61. (a) No: they have different solutions.

$3x = 6x + 9$	$x = 2x + 9$
$-3x = 9$	$-x = 9$
$x = -3$	$x = -9$

(b) Yes: the solution to both equations is $x = 4$.

$$\begin{aligned} 6x + 2 &= 4x + 10 \\ 6x &= 4x + 8 \\ 2x &= 8 \\ x &= 4 \end{aligned} \qquad \begin{aligned} 3x + 1 &= 2x + 5 \\ 3x &= 2x + 4 \\ x &= 4 \end{aligned}$$

63. False. $6 > 2$, but $-6 < -2$ because -6 lies to the left of -2 on the number line.

65. $3x + 5 = 2x + 1$
Subtracting 5 from each side gives $3x = 2x - 4$.
The answer is E.

67. $x(x + 1) = 0$
$x = 0$ or $x + 1 = 0$
$x = -1$
The answer is A.

69. (c) $\frac{800}{801} > \frac{799}{800}$

(d) $-\frac{103}{102} > -\frac{102}{101}$

(e) If your calculator returns 0 when you enter $2x + 1 < 4$, you can conclude that the value stored in x is not a solution of the inequality $2x + 1 < 4$.

71.
$$\begin{aligned} A &= \frac{1}{2}h(b_1 + b_2) \\ h(b_1 + b_2) &= 2A \\ b_1 + b_2 &= \frac{2A}{h} \\ b_1 &= \frac{2A}{h} - b_2 \end{aligned}$$

73.
$$\begin{aligned} C &= \frac{5}{9}(F - 32) \\ \frac{9}{5}C &= F - 32 \\ \frac{9}{5}C + 32 &= F \\ F &= \frac{9}{5}C + 32 \end{aligned}$$

Section P.4 Lines in the Plane

Exploration 1

1. The graphs of $y = mx + b$ and $y = mx + c$ have the same slope but different y-intercepts.

2.

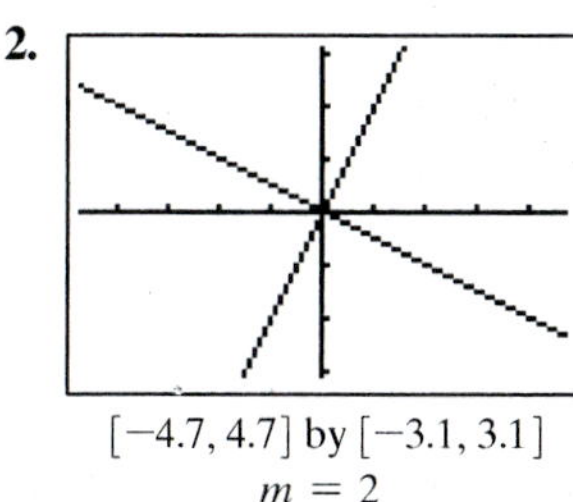

$[-4.7, 4.7]$ by $[-3.1, 3.1]$
$m = 2$

The angle between the two lines appears to be 90°.

3.

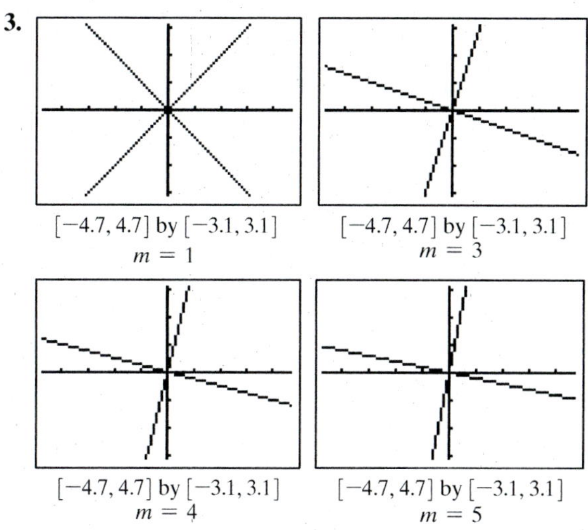

$[-4.7, 4.7]$ by $[-3.1, 3.1]$, $m = 1$ — $[-4.7, 4.7]$ by $[-3.1, 3.1]$, $m = 3$

$[-4.7, 4.7]$ by $[-3.1, 3.1]$, $m = 4$ — $[-4.7, 4.7]$ by $[-3.1, 3.1]$, $m = 5$

In each case, the two lines appear to be at right angles to one another.

Quick Review P.4

1.
$$\begin{aligned} -75x + 25 &= 200 \\ -75x &= 175 \\ x &= -\frac{7}{3} \end{aligned}$$

3.
$$\begin{aligned} 3(1 - 2x) + 4(2x - 5) &= 7 \\ 3 - 6x + 8x - 20 &= 7 \\ 2x - 17 &= 7 \\ 2x &= 24 \\ x &= 12 \end{aligned}$$

5.
$$\begin{aligned} 2x - 5y &= 21 \\ -5y &= -2x + 21 \\ y &= \frac{2}{5}x - \frac{21}{5} \end{aligned}$$

7.
$$\begin{aligned} 2x + y &= 17 + 2(x - 2y) \\ 2x + y &= 17 + 2x - 4y \\ y &= 17 - 4y \\ 5y &= 17 \\ y &= \frac{17}{5} \end{aligned}$$

9. $\frac{9 - 5}{-2 - (-8)} = \frac{4}{6} = \frac{2}{3}$

Section P.4 Exercises

1. $m = -2$

3. $m = \frac{9 - 5}{4 + 3} = \frac{4}{7}$

5. $m = \frac{3 + 5}{-1 + 2} = 8$

7. $2 = \frac{9 - 3}{5 - x} = \frac{6}{5 - x}$, so $x = 2$

9. $3 = \frac{y + 5}{4 + 3} = \frac{y + 5}{7}$, so $y = 16$

11. $y - 4 = 2(x - 1)$

13. $y + 4 = -2(x - 5)$

15. Since $m = 1$, we can choose $A = 1$ and $B = -1$. Since $x = -7$, $y = -2$ solves $x - y + C = 0$, C must equal 5: $x - y + 5 = 0$. Note that the coefficients can be multiplied by any non-zero number, e.g., another answer would be $2x - 2y + 10 = 0$. This comment also applies to the following problems.

17. Since $m = 0$, we can choose $A = 0$ and $B = 1$. Since $x = 1$, $y = -3$ solves $0x + y + C = 0$. C must equal 3: $0x + y + 3 = 0$, or $y + 3 = 0$. See comment in #15.

19. The slope is $m = 1 = -A/B$, so we can choose $A = 1$ and $B = -1$. Since $x = -1$, $y = 2$ solves $x - y + C = 0$, C must equal 3: $x - y + 3 = 0$. See comment in #15.

21. Begin with point-slope form: $y - 5 = -3(x - 0)$, so $y = -3x + 5$.

23. $m = -\frac{1}{4}$, so in point-slope form, $y - 5 = -\frac{1}{4}(x + 4)$, and therefore $y = -\frac{1}{4}x + 4$.

25. Solve for y: $y = -\frac{2}{5}x + \frac{12}{5}$.

27. Graph $y = 49 - 8x$; window should include $(6.125, 0)$ and $(0, 49)$, for example, $[-5, 10] \times [-10, 60]$.

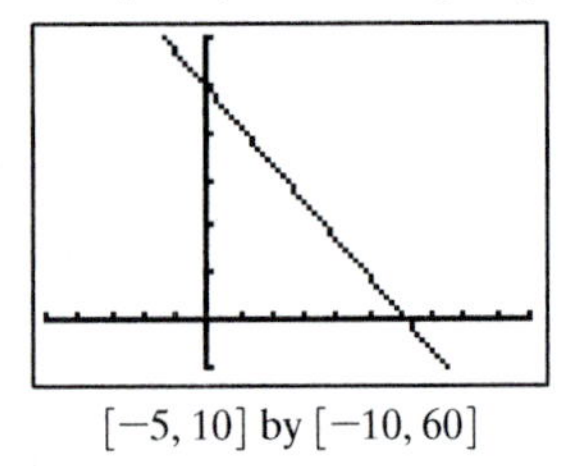

$[-5, 10]$ by $[-10, 60]$

29. Graph $y = (429 - 123x)/7$; window should include $(3.488, 0)$ and $(0, 61.29)$, for example, $[-1, 5] \times [-10, 80]$.

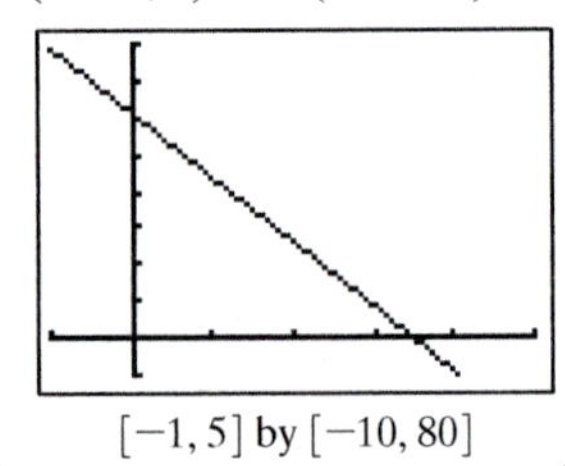

$[-1, 5]$ by $[-10, 80]$

31. (a): The slope is 1.5, compared to 1 in (b).

33. Substitute and solve: replacing y with 14 gives $x = 4$, and replacing x with 18 gives $y = 21$.

35. Substitute and solve: replacing y with 14 gives $x = -10$, and replacing x with 18 gives $y = -7$.

37. Ymin $= -30$, Ymax $= 30$, Yscl $= 3$

39. Ymin $= -20/3$, Ymax $= 20/3$, Yscl $= 2/3$

In #41 and 43, use the fact that parallel lines have the same slope; while the slopes of perpendicular lines multiply to give -1.

41. (a) Parallel: $y - 2 = 3(x - 1)$, or $y = 3x - 1$.

(b) Perpendicular: $y - 2 = -\frac{1}{3}(x - 1)$, or

$y = -\frac{1}{3}x + \frac{7}{3}$.

43. (a) Parallel: $2x + 3y = 9$, or $y = -\frac{2}{3}x + 3$.

(b) Perpendicular: $3x - 2y = 7$, or $y = \frac{3}{2}x - \frac{7}{2}$.

45. (a) $m = (67{,}500 - 42{,}000)/8 = 3187.5$, the y-intercept is $b = 42{,}000$ so $V = 3187.5t + 42{,}000$.

(b) The house is worth about \$72,500 after 9.57 years.

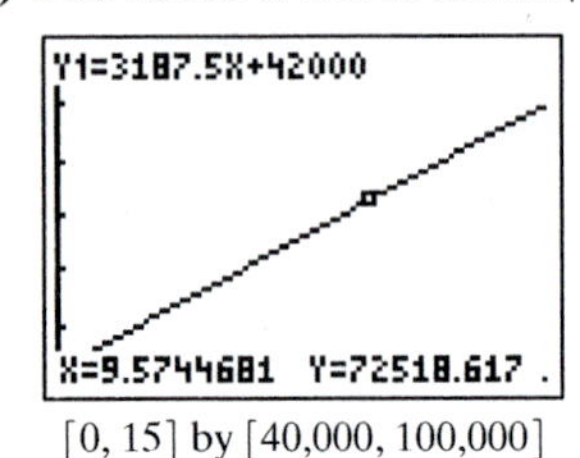

$[0, 15]$ by $[40{,}000, 100{,}000]$

(c) $3187.5t + 42{,}000 = 74{,}000$: $t = 10.04$.

(d) $t = 12$ years.

47. $y = \frac{3}{8}x$, where y is altitude and x is horizontal distance. The plane must travel $x = 32{,}000$ ft horizontally–just over 6 miles.

49. $m = \frac{3}{8} = 0.375 > \frac{4}{12} = 0.3\overline{3}$, so asphalt shingles are acceptable.

51. (a) Slope of the line between the points (1998, 5.9) and (1999, 6.3) is $m = \frac{6.3 - 5.9}{1999 - 1998} = \frac{0.4}{1} = 0.4$.
Using the *point-slope form* equation for the line, we have $y - 5.9 = 0.4(x - 1998)$, so $y = 0.4x - 793.3$.

(b) Using $y = 0.4x - 793.3$ and $x = 2002$, the model estimates Americans' expenditures in 2002 were \$7.5 trillion.

(c) Using $y = 0.4x - 793.3$ and $x = 2006$, the model predicts Americans' expenditures in 2006 will be \$9.1 trillion.

(d)

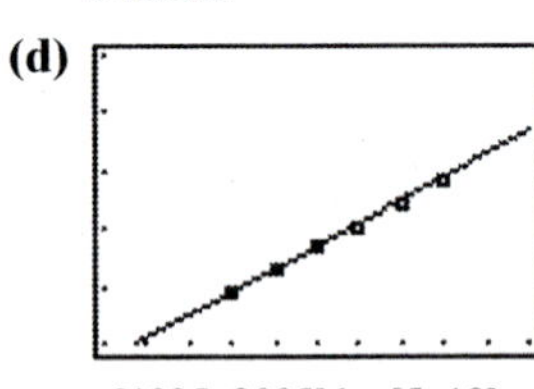

[1995, 2005] by [5, 10]

53. (a)

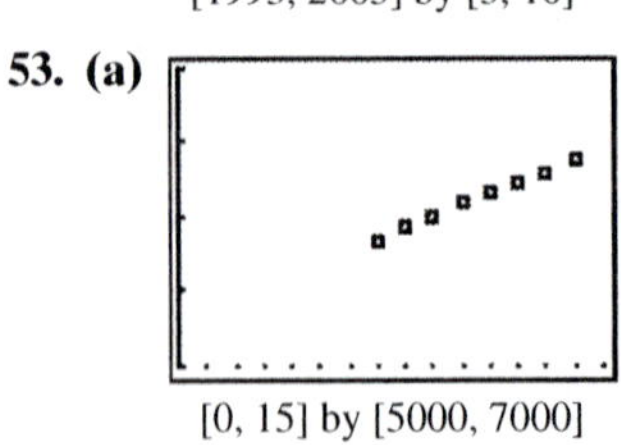

[0, 15] by [5000, 7000]

(b) Slope of the line between the points (7, 5852) and (14, 6377) is $m = \frac{6377 - 5852}{14 - 7} = \frac{525}{7} = 75$.

Using the *point-slope form* equation for the line, we have $y - 5852 = 75(x - 7)$, so $y = 75x + 5327$.

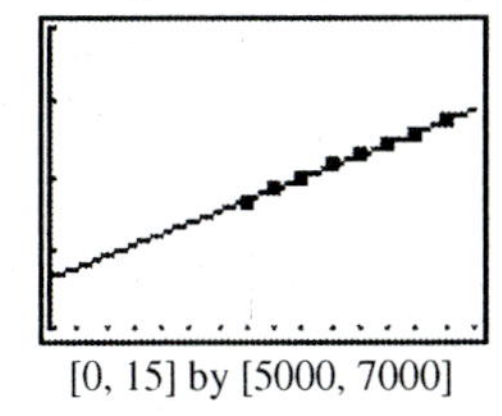

[0, 15] by [5000, 7000]

(c) The year 2006 is represented by $x = 16$. Using $y = 75x + 5327$ and $x = 16$, the model predicts the midyear world population in 2006 will be approximately 6527 million, which is a little larger than the Census Bureau estimate of 6525 million.

55. $\dfrac{8-0}{a-3} = \dfrac{4-0}{3-0}$

$\dfrac{8}{a-3} = \dfrac{4}{3}$

$24 = 4(a-3)$

$6 = a - 3$

$9 = a$

57. $\overline{AD} \parallel \overline{BC} \Rightarrow b = 5;$

$\overline{AB} \parallel \overline{DC} \Rightarrow \dfrac{5}{a-4} = \dfrac{5}{2} \Rightarrow a = 6$

59. (a) No, it is not possible for two lines with positive slopes to be perpendicular, because if both slopes are positive, they cannot multiply to -1.

(b) No, it is not possible for two lines with negative slopes to be perpendicular, because if both slopes are negative, they cannot multiply to -1.

61. False. The slope of a vertical line is undefined. For example, the vertical line through (3, 1) and (3, 6) would have a slope of $\dfrac{6-1}{3-3} = \dfrac{5}{0}$, which is undefined.

63. With $(x_1, y_1) = (-2, 3)$ and $m = 4$, the point-slope form equation $y - y_1 = m(x - x_1)$ becomes $y - 3 = 4[x - (-2)]$ or $y - 3 = 4(x + 2)$. The answer is A.

65. When a line has a slope of $m_1 = -2$, a perpendicular line must have a slope of $m_2 = -\dfrac{1}{m_1} = \dfrac{1}{2}$. The answer is E.

67. (a)

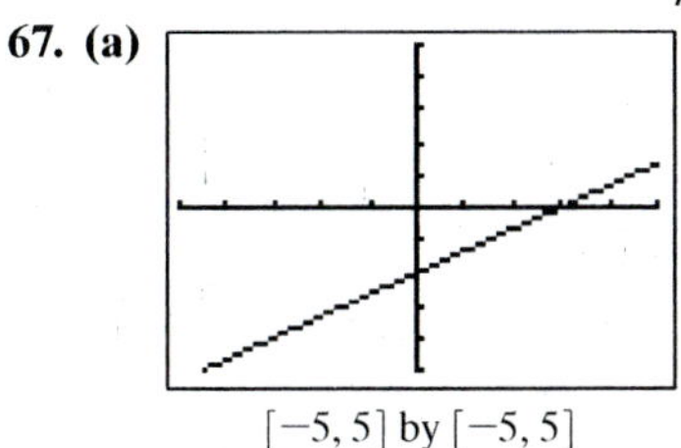

[−5, 5] by [−5, 5]

(b)

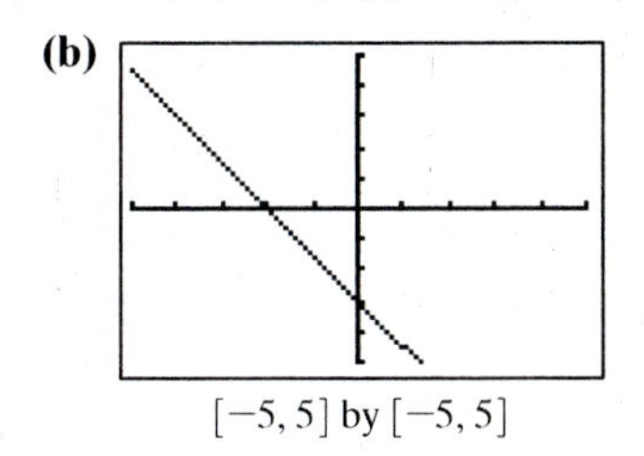

[−5, 5] by [−5, 5]

(c)

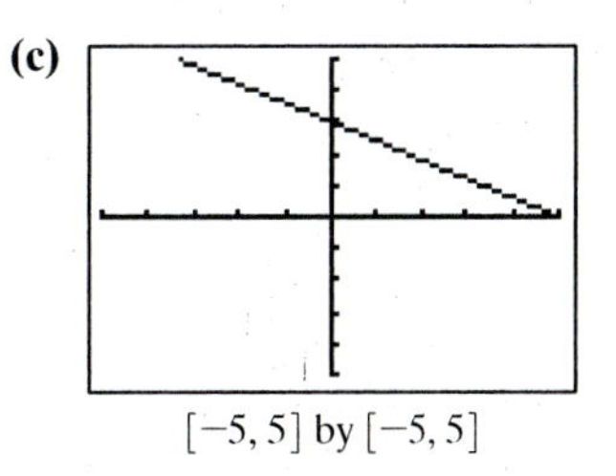

[−5, 5] by [−5, 5]

(d) From the graphs, it appears that a is the x-intercept and b is the y-intercept when $c = 1$.
Proof: The x-intercept is found by setting $y = 0$. When $c = 1$, we have $\dfrac{x}{a} + \dfrac{0}{b} = 1$. Hence $\dfrac{x}{a} = 1$ so $x = a$. The y-intercept is found by setting $x = 0$. When $c = 1$, we have $\dfrac{0}{a} + \dfrac{y}{b} = 1$. Hence $\dfrac{y}{b} = 1$, so $y = b$.

(e)

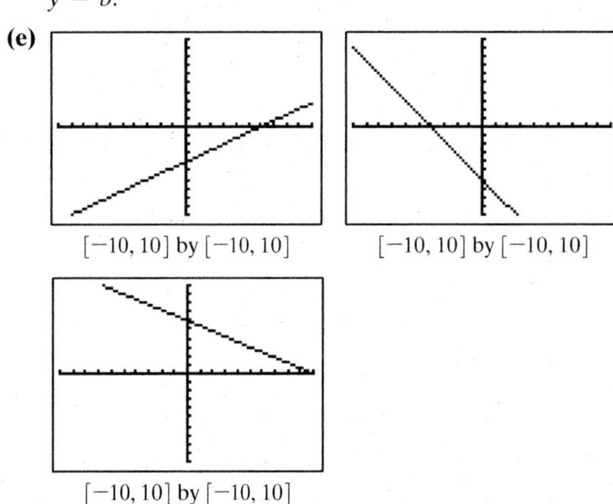

[−10, 10] by [−10, 10] [−10, 10] by [−10, 10]

[−10, 10] by [−10, 10]

From the graphs, it appears that a is half the x-intercept and b is half the y-intercept when $c = 2$.
Proof: When $c = 2$, we can divide both sides by 2 and we have $\dfrac{x}{2a} + \dfrac{y}{2b} = 1$. By part (d) the x-intercept is $2a$ and the y-intercept is $2b$.

(f) By a similar argument, when $c = -1$, a is the opposite of the x-intercept and b is the opposite of the y-intercept.

69. As in the diagram, we can choose one point to be the origin, and another to be on the x-axis. The midpoints of the sides, starting from the origin and working around counterclockwise in the diagram, are then $A\left(\dfrac{a}{2}, 0\right)$, $B\left(\dfrac{a+b}{2}, \dfrac{c}{2}\right)$, $C\left(\dfrac{b+d}{2}, \dfrac{c+e}{2}\right)$, and $D\left(\dfrac{d}{2}, \dfrac{e}{2}\right)$. The opposite sides are therefore parallel, since the slopes of the four lines connecting those points are: $m_{AB} = \dfrac{c}{b}$; $m_{BC} = \dfrac{e}{d-a}$; $m_{CD} = \dfrac{c}{b}$; $m_{DA} = \dfrac{e}{d-a}$.

71. A has coordinates $\left(\dfrac{b}{2}, \dfrac{c}{2}\right)$, while B is $\left(\dfrac{a+b}{2}, \dfrac{c}{2}\right)$, so the line containing A and B is the horizontal line $y = c/2$, and the distance from A to B is $\left|\dfrac{a+b}{2} - \dfrac{b}{2}\right| = \dfrac{a}{2}$.

Section P.5 Solving Equations Graphically, Numerically, and Algebraically

Exploration 1

1.

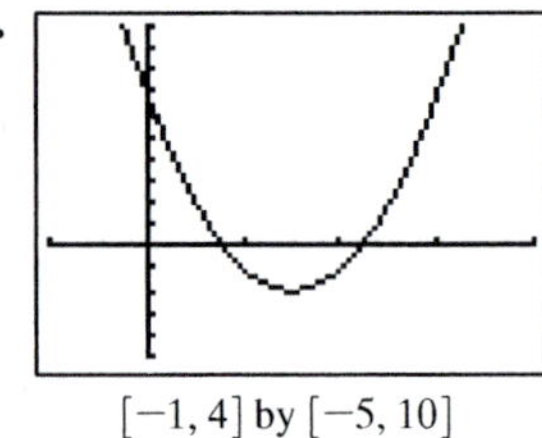

[−1, 4] by [−5, 10]

2. Using the numerical zoom, we find the zeros to be 0.79 and 2.21.

3.

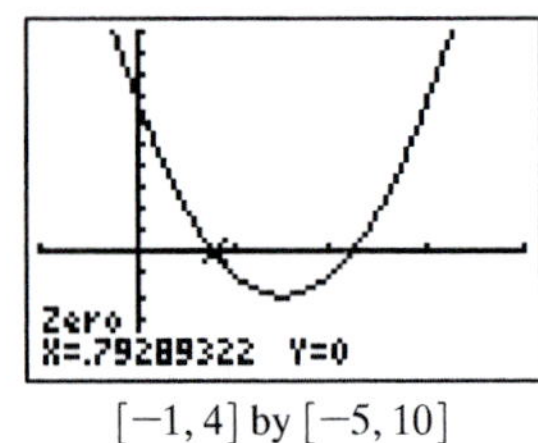

[−1, 4] by [−5, 10]

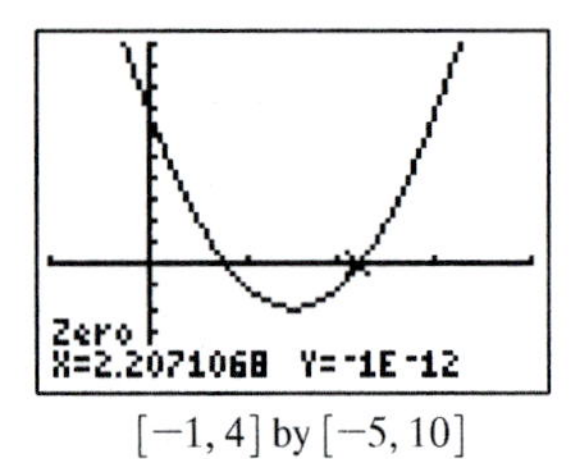

[−1, 4] by [−5, 10]

By this method we have zeros at 0.79 and 2.21.

4.

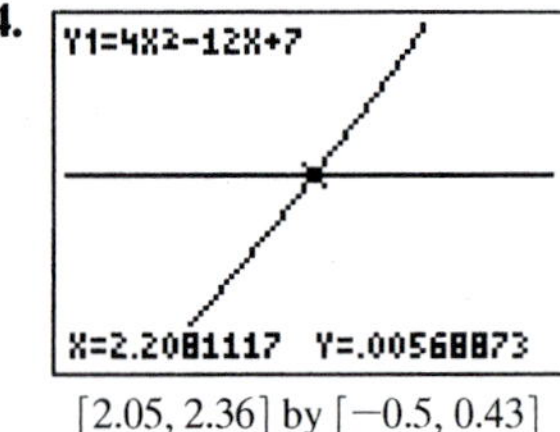

[2.05, 2.36] by [−0.5, 0.43]

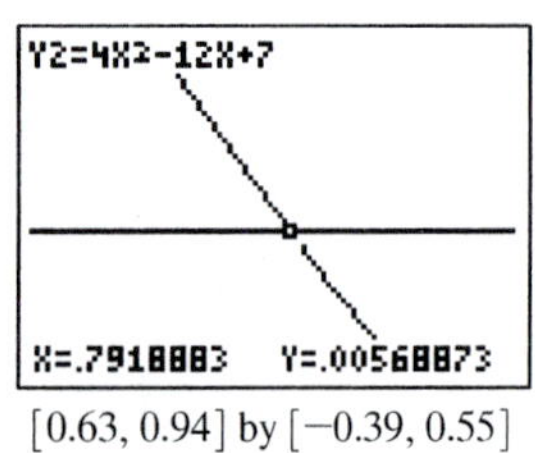

[0.63, 0.94] by [−0.39, 0.55]

Zooming in and tracing reveals the same zeros, correct to two decimal places.

5. The answers in parts 2, 3, and 4 are the same.

6. On a calculator, evaluating $4x^2 - 12x + 7$ when $x = 0.79$ gives $y = 0.0164$ and when $x = 2.21$ gives $y = 0.0164$, so the numbers 0.79 and 2.21 are approximate zeros.

7.

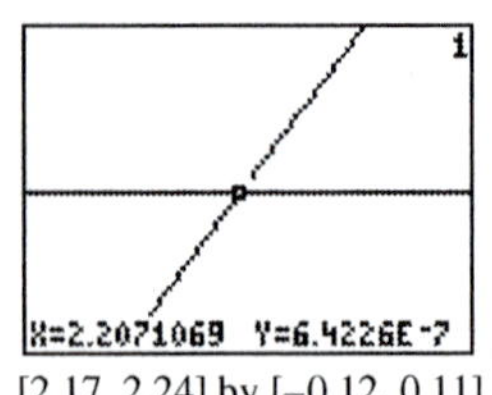

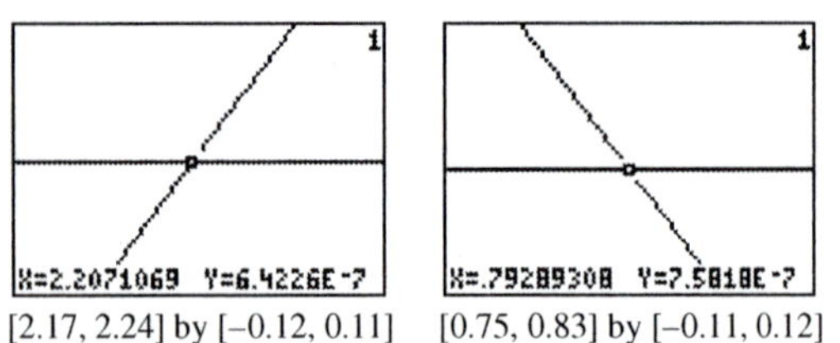

[2.17, 2.24] by [−0.12, 0.11] [0.75, 0.83] by [−0.11, 0.12]

Zooming in and tracing reveals zeros of 0.792893 and 2.207107 accurate to six decimal places. If rounded to two decimal places, these would be the same as the answers found in part 3.

Quick Review P.5

1. $(3x - 4)^2 = 9x^2 - 12x - 12x + 16 = 9x^2 - 24x + 16$

3. $(2x + 1)(3x - 5) = 6x^2 - 10x + 3x - 5$
$= 6x^2 - 7x - 5$

5. $25x^2 - 20x + 4 = (5x - 2)(5x - 2) = (5x - 2)^2$

7. $3x^3 + x^2 - 15x - 5 = x^2(3x + 1) - 5(3x + 1)$
$= (3x + 1)(x^2 - 5)$

9. $\dfrac{x}{2x + 1} - \dfrac{2}{x + 3}$

$= \dfrac{x(x + 3)}{(2x + 1)(x + 3)} - \dfrac{2(2x + 1)}{(2x + 1)(x + 3)}$

$= \dfrac{x^2 + 3x - 4x - 2}{(2x + 1)(x + 3)} = \dfrac{x^2 - x - 2}{(2x + 1)(x + 3)}$

$= \dfrac{(x - 2)(x + 1)}{(2x + 1)(x + 3)}$

Section P.5 Exercises

1.

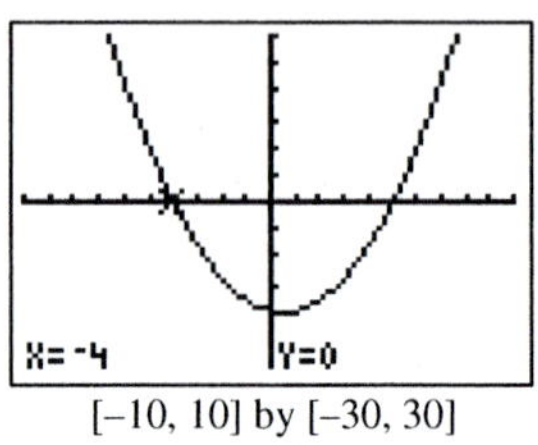

[−10, 10] by [−30, 30]

$x = -4$ or $x = 5$

The left side factors to $(x + 4)(x - 5) = 0$:

$x + 4 = 0$ or $x - 5 = 0$

$x = -4$ $\quad$ $x = 5$

3.

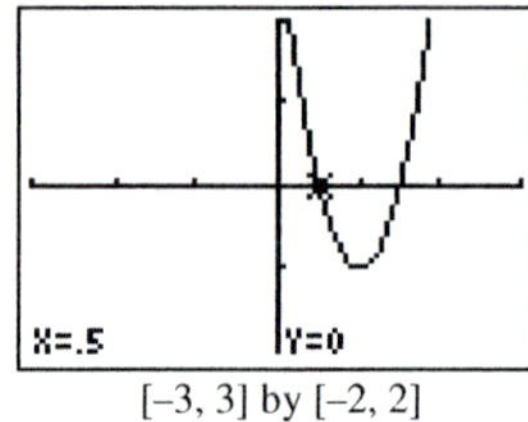

[−3, 3] by [−2, 2]

$x = 0.5$ or $x = 1.5$

The left side factors to $(2x - 1)(2x - 3) = 0$:

$2x - 1 = 0$ or $2x - 3 = 0$

$2x = 1$ $\quad$ $2x = 3$

$x = 0.5$ $\quad$ $x = 1.5$

5.

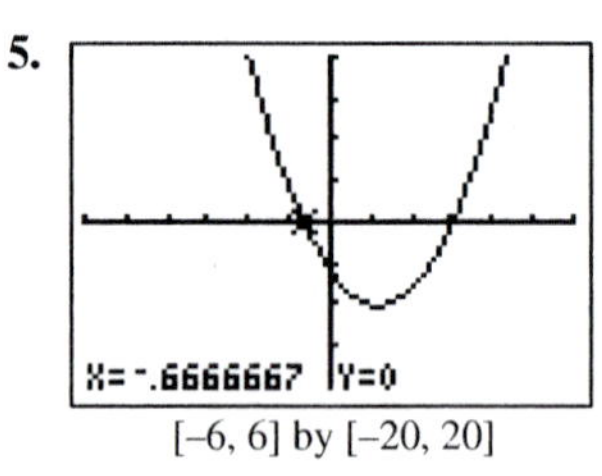

[−6, 6] by [−20, 20]

$x = -\dfrac{2}{3}$ or $x = 3$

Rewrite as $3x^2 - 7x - 6 = 0$; the left side factors to $(3x + 2)(x - 3) = 0$:

$3x + 2 = 0$ or $x - 3 = 0$

$x = -\dfrac{2}{3}$ $\quad$ $x = 3$

7. Rewrite as $(2x)^2 = 5^2$; then $2x = \pm 5$, or $x = \pm\dfrac{5}{2}$.

9. Divide both sides by 3 to get $(x + 4)^2 = \dfrac{8}{3}$. Then $x + 4 = \pm\sqrt{\dfrac{8}{3}}$ and $x = -4 \pm \sqrt{\dfrac{8}{3}}$.

11. Adding $2y^2 + 8$ to both sides gives $4y^2 = 14$. Divide both sides by 4 to get $y^2 = \dfrac{7}{2}$, so $y = \pm\sqrt{\dfrac{7}{2}}$.

13. $x^2 + 6x + 3^2 = 7 + 3^2$
$(x + 3)^2 = 16$
$x + 3 = \pm\sqrt{16}$
$x = -3 \pm 4$
$x = -7$ or $x = 1$

15. $x^2 - 7x = -\frac{5}{4}$
$x^2 - 7x + \left(-\frac{7}{2}\right)^2 = -\frac{5}{4} + \left(-\frac{7}{2}\right)^2$
$\left(x - \frac{7}{2}\right)^2 = 11$
$x - \frac{7}{2} = \pm\sqrt{11}$
$x = \frac{7}{2} \pm \sqrt{11}$
$x = \frac{7}{2} - \sqrt{11} \approx 0.18$ or $x = \frac{7}{2} + \sqrt{11} \approx 6.82$

17. $2x^2 - 7x + 9 = x^2 - 2x - 3 + 3x$
$2x^2 - 7x + 9 = x^2 + x - 3$
$x^2 - 8x = -12$
$x^2 - 8x + (-4)^2 = -12 + (-4)^2$
$(x - 4)^2 = 4$
$x - 4 = \pm 2$
$x = 4 \pm 2$
$x = 2$ or $x = 6$

19. $a = 1, b = 8$, and $c = -2$:
$$x = \frac{-8 \pm \sqrt{8^2 - 4(1)(-2)}}{2(1)} = \frac{-8 \pm \sqrt{72}}{2} = \frac{-8 \pm 6\sqrt{2}}{2} = -4 \pm 3\sqrt{2}$$
$x \approx -8.24$ or $x \approx 0.24$

21. $x^2 - 3x - 4 = 0$, so
$a = 1, b = -3$, and $c = -4$:
$$x = \frac{3 \pm \sqrt{(-3)^2 - 4(1)(-4)}}{2(1)} = \frac{3 \pm \sqrt{25}}{2} = \frac{3}{2} \pm \frac{5}{2}$$
$x = -1$ or $x = 4$

23. $x^2 + 5x - 12 = 0$, so
$a = 1, b = 5, c = -12$
$$x = \frac{-5 \pm \sqrt{(5)^2 - 4(1)(-12)}}{2(1)} = \frac{-5 \pm \sqrt{73}}{2} = -\frac{5}{2} \pm \frac{\sqrt{73}}{2}$$
$x \approx -6.77$ or $x \approx 1.77$

25. x-intercept: 3; y-intercept: -2

27. x-intercepts: $-2, 0, 2$; y-intercept: 0

29.

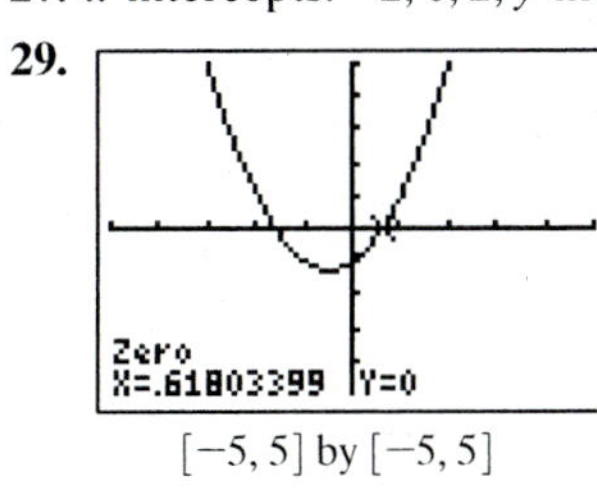

$[-5, 5]$ by $[-5, 5]$

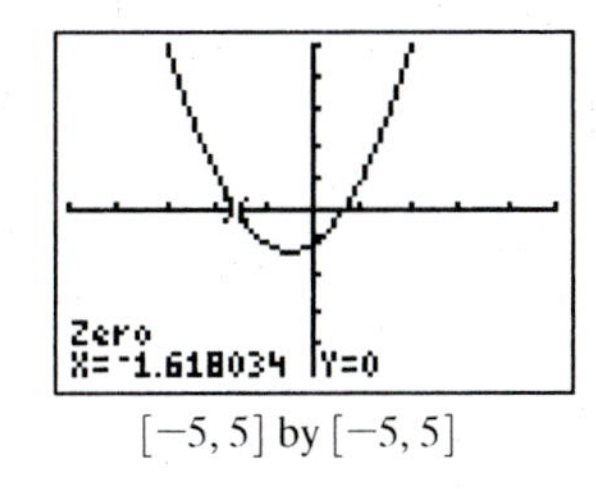

$[-5, 5]$ by $[-5, 5]$

31.

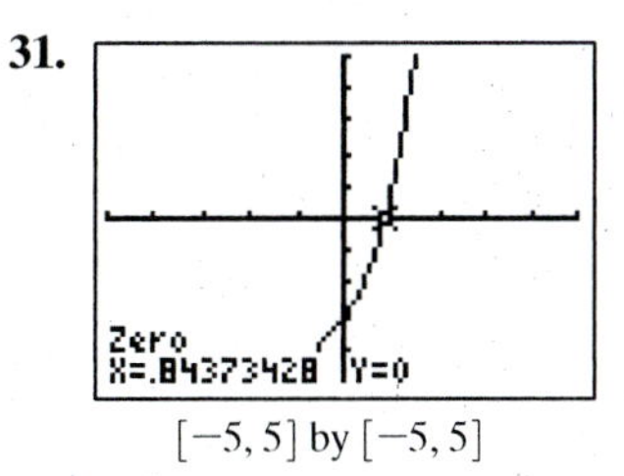

$[-5, 5]$ by $[-5, 5]$

33.

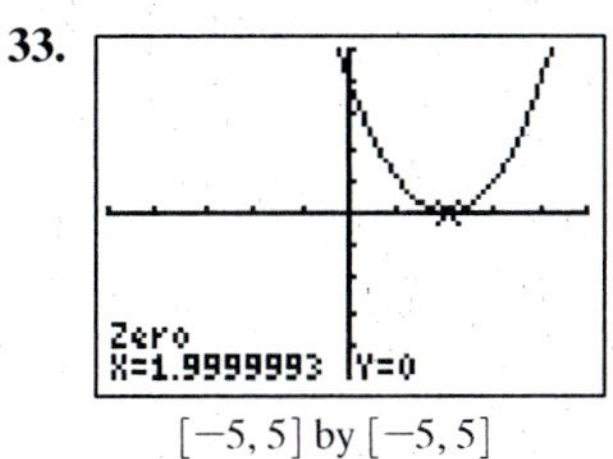

$[-5, 5]$ by $[-5, 5]$

35. $x^2 + 2x - 1 = 0$; $x \approx 0.4$

37. Using TblStart $= 1.61$ and ΔTbl $= 0.001$ gives a zero at 1.62.
Using TblStart $= -0.62$ and ΔTbl $= 0.001$ gives a zero at -0.62.

39. Graph $y = |x - 8|$ and $y = 2$: $t = 6$ or $t = 10$

41. Graph $y = |2x + 5|$ and $y = 7$: $x = 1$ or $x = -6$

43. Graph $y = |2x - 3|$ and $y = x^2$: $x = -3$ or $x = 1$

45. (a) The two functions are $y_1 = 3\sqrt{x + 4}$ (the one that begins on the x-axis) and $y_2 = x^2 - 1$.

(b) This is the graph of $y = 3\sqrt{x + 4} - x^2 + 1$.

(c) The x-coordinates of the intersections in the first picture are the same as the x-coordinates where the second graph crosses the x-axis.

47. The left side factors to $(x + 2)(x - 1) = 0$:
$x + 2 = 0$ or $x - 1 = 0$
$x = -2$ $\quad$ $x = 1$

49. $2x - 1 = 5$ or $2x - 1 = -5$
$2x = 6$ $\quad$ $2x = -4$
$x = 3$ $\quad$ $x = -2$

51. From the graph of $y = x^3 + 4x^2 - 3x - 2$ on $[-10, 10] \times [-10, 10]$, the solutions of the equation (x-intercepts of the graph) are $x \approx -4.56$, $x \approx -0.44$, $x = 1$.

53. $x^2 + 4x - 1 = 7$ or $x^2 + 4x - 1 = -7$
$x^2 + 4x - 8 = 0$ $\quad$ $x^2 + 4x + 6 = 0$
$x = \frac{-4 \pm \sqrt{16 + 32}}{2}$ $\quad$ $x = \frac{-4 \pm \sqrt{16 - 24}}{2}$
$x = -2 \pm 2\sqrt{3}$ $\quad$ — no real solutions to this equation.

55. Graph $y = |0.5x + 3|$ and $y = x^2 - 4$:
$x \approx -2.41$ or $x \approx 2.91$

57. (a) There must be 2 distinct real zeros, because $b^2 - 4ac > 0$ implies that $\pm\sqrt{b^2 - 4ac}$ are 2 distinct real numbers.

(b) There must be 1 real zero, because $b^2 - 4ac = 0$ implies that $\pm\sqrt{b^2 - 4ac} = 0$, so the root must be $x = -\frac{b}{a}$.

(c) There must be no real zeros, because $b^2 - 4ac < 0$ implies that $\pm\sqrt{b^2 - 4ac}$ are not real numbers.

59. Let x be the width of the field (in yd); the length is $x + 30$. Then the field is 80 yd wide and $80 + 30 = 110$ yd long.

$$8800 = x(x + 30)$$
$$0 = x^2 + 30x - 8800$$
$$0 = (x + 110)(x - 80)$$
$$0 = x + 110 \text{ or } 0 = x - 80$$
$$x = -110 \text{ or } x = 80$$

61. The area of the square is x^2. The area of the semicircle is $\frac{1}{2}\pi r^2 = \frac{1}{2}\pi\left(\frac{1}{2}x\right)^2$ since the radius of the semicircle is $\frac{1}{2}x$. Then $200 = x^2 + \frac{1}{2}\pi\left(\frac{1}{2}x\right)^2$. Solving this (graphically is easiest) gives $x \approx 11.98$ ft (since x must be positive).

63. False. Notice that for $x = -3$, $2x^2 = 2(-3)^2 = 18$. So x could also be -3.

65. For $x^2 - 5x + ?$ to be a perfect square, ? must be replaced by the square of half of -5, which is $\left(-\frac{5}{2}\right)^2$. The answer is B.

67. Since an absolute value cannot be negative, there are no solutions. The answer is E.

69. Graph $y = |x^2 - 4|$ and $y = c$ for several values of c.

(a) Let $c = 2$. The graph suggests $y = 2$ intersects $y = |x^2 - 4|$ four times.

$|x^2 - 4| = 2 \Rightarrow x^2 - 4 = 2$ or $x^2 - 4 = -2$

$x^2 = 6$ $\quad$ $x^2 = 2$

$x = \pm\sqrt{6}$ $\quad$ $x = \pm\sqrt{2}$

$|x^2 - 4| = 2$ has four solutions: $\{\pm\sqrt{2}, \pm\sqrt{6}\}$.

(b) Let $c = 4$. The graph suggests $y = 4$ intersects $y = |x^2 - 4|$ three times.

$|x^2 - 4| = 4 \Rightarrow x^2 - 4 = 4$ or $x^2 - 4 = -4$

$x^2 = 8$ $\quad$ $x^2 = 0$

$x = \pm\sqrt{8}$ $\quad$ $x = 0$

(c) Let $c = 5$. The graph suggest $y = 5$ intersects $y = |x^2 - 4|$ twice.

$|x^2 - 4| = 5 \Rightarrow x^2 - 4 = 5$ or $x^2 - 4 = -5$

$x^2 = 9$ $\quad$ $x^2 = -1$

$x = \pm 3$ $\quad$ no solution

$|x^2 - 4| = 5$ has two solutions: $\{\pm 3\}$.

(d) Let $c = -1$. The graph suggests $y = -1$ does not intersect $y = |x^2 - 4|$. Since absolute value is never negative, $|x^2 - 4| = -1$ has no solutions.

(e) There is no other possible number of solutions of this equation. For any c, the solution involves solving two quadratic equations, each of which can have 0, 1, or 2 solutions.

71. From #70(a), $x_1 + x_2 = -\frac{b}{a} = 5$. Since $a = 2$, this means $b = -10$. From #70(b), $x_1 \cdot x_2 = \frac{c}{a} = 3$; since $a = 2$, this means $c = 6$. The solutions are $\frac{10 \pm \sqrt{100 - 48}}{4}$; this reduces to $2.5 \pm \frac{1}{2}\sqrt{13}$, or approximately 0.697 and 4.303.

Section P.6 Complex Numbers

Quick Review P.6

1. $x + 9$

3. $a + 2d$

5. $x^2 - x - 6$

7. $x^2 - 2$

9. $x^2 - 2x - 1$

Section P.6 Exercises

In #1, 3, 5, and 7, add or subtract the real and imaginary parts separately.

1. $(2 - 3i) + (6 + 5i) = (2 + 6) + (-3 + 5)i = 8 + 2i$

3. $(7 - 3i) + (6 - i) = (7 + 6) + (-3 - 1)i = 13 - 4i$

5. $(2 - i) + (3 - \sqrt{-3}) = (2 + 3) + (-1 - \sqrt{3})i$
$= 5 - (1 + \sqrt{3})i$

7. $(i^2 + 3) - (7 + i^3) = (-1 + 3) - (7 - i)$
$= (2 - 7) + i = -5 + i$

In #9, 11, 13, and 15, multiply out and simplify, recalling that $i^2 = -1$.

9. $(2 + 3i)(2 - i) = 4 - 2i + 6i - 3i^2$
$= 4 + 4i + 3 = 7 + 4i$

11. $(1 - 4i)(3 - 2i) = 3 - 2i - 12i + 8i^2$
$= 3 - 14i - 8 = -5 - 14i$

13. $(7i - 3)(2 + 6i) = 14i + 42i^2 - 6 - 18i$
$= -42 - 6 - 4i = -48 - 4i$

15. $(-3 - 4i)(1 + 2i) = -3 - 6i - 4i - 8i^2$
$= -3 - 10i + 8 = 5 - 10i$

17. $\sqrt{-16} = 4i$

19. $\sqrt{-3} = \sqrt{3}i$

In #21 and 23, equate the real and imaginary parts.

21. $x = 2$, $y = 3$

23. $x = 1$, $y = 2$

In #25 and 27, multiply out and simplify, recalling that $i^2 = -1$.

25. $(3 + 2i)^2 = 9 + 12i + 4i^2 = 5 + 12i$

27. $\left(\frac{\sqrt{2}}{2} + \frac{\sqrt{2}}{2}i\right)^4 = \left(\frac{\sqrt{2}}{2}\right)^4(1 + i)^4$
$= \frac{1}{4}(1 + 2i + i^2)^2 = \frac{1}{4}(2i)^2 = \frac{1}{4}(-4) = -1$

In #29 and 31, recall that $(a + bi)(a - bi) = a^2 + b^2$.

29. $2^2 + 3^2 = 13$

31. $3^2 + 4^2 = 25$

In #33, 35, 37, and 39, multiply both the numerator and denominator by the complex conjugate of the denominator, recalling that $(a + bi)(a - bi) = a^2 + b^2$.

33. $\frac{1}{2 + i} \cdot \frac{2 - i}{2 - i} = \frac{2 - i}{5} = \frac{2}{5} - \frac{1}{5}i$

35. $\frac{2 + i}{2 - i} \cdot \frac{2 + i}{2 + i} = \frac{4 + 4i + i^2}{5} = \frac{3}{5} + \frac{4}{5}i$

37. $\frac{(2+i)^2(-i)}{1+i}\cdot\frac{1-i}{1-i} = \frac{(4+4i+i^2)(-i+i^2)}{2}$
$= \frac{(3+4i)(-1-i)}{2} = \frac{-3-3i-4i-4i^2}{2} = \frac{1}{2} - \frac{7}{2}i$

39. $\frac{(1-i)(2-i)}{1-2i}\cdot\frac{1+2i}{1+2i} = \frac{(2-i-2i+i^2)(1+2i)}{5}$
$= \frac{(1-3i)(1+2i)}{5} = \frac{1+2i-3i-6i^2}{5} = \frac{7}{5} - \frac{1}{5}i$

In #41 and 43, use the quadratic formula.

41. $x = -1 \pm 2i$

43. $x = \frac{7}{8} \pm \frac{\sqrt{15}}{8}i$

45. False. When $a = 0$, $z = a + bi$ becomes $z = bi$, and then $-\bar{z} = -(-bi) = bi = z$.

47. $(2 + 3i)(2 - 3i)$ is a product of conjugates and equals $2^2 + 3^2 = 13$. The answer is E.

49. Complex, nonreal solutions of polynomials with real coefficients always come in conjugate pairs. So another solution is $2 + 3i$, and the answer is A.

51. **(a)** $i = i$ $\quad i^5 = i\cdot i^4 = i$
$i^2 = -1$ $\quad i^6 = i^2\cdot i^4 = -1$
$i^3 = (-1)i = -i$ $\quad i^7 = i^3\cdot i^4 = -i$
$i^4 = (-1)^2 = 1$ $\quad i^8 = i^4\cdot i^4 = 1\cdot 1 = 1$

(b) $i^{-1} = \frac{1}{i} = \frac{1}{i}\cdot\frac{i}{i} = -i$ $\quad i^{-5} = \frac{1}{i}\cdot\frac{1}{i^4} = \frac{1}{i} = -i$
$i^{-2} = \frac{1}{i^2} = -1$ $\quad i^{-6} = \frac{1}{i^2}\cdot\frac{1}{i^4} = -1$
$i^{-3} = \frac{1}{i}\cdot\frac{1}{i^2} = -\frac{1}{i} = i$ $\quad i^{-7} = \frac{1}{i^3}\cdot\frac{1}{i^4} = -\frac{1}{i} = i$
$i^{-4} = \frac{1}{i^2}\cdot\frac{1}{i^2} = (-1)(-1) = 1$ $\quad i^{-8} = \frac{1}{i^4}\cdot\frac{1}{i^4} = 1\cdot 1 = 1$

(c) $i^0 = 1$

(d) Answers will vary.

53. Let a and b be any two real numbers. Then $(a + bi) - (a - bi) = (a - a) + (b + b)i = 0 + 2bi = 2bi$.

55. $\overline{(a + bi)\cdot(c + di)} = \overline{(ac - bd) + (ad + bc)i} = (ac - bd) - (ad + bc)i$ and $\overline{(a + bi)}\cdot\overline{(c + di)} = (a - bi)\cdot(c - di) = (ac - bd) - (ad + bc)i$ are equal.

57. $(-i)^2 - i(-i) + 2 = 0$ but $(i)^2 - i(i) + 2 \neq 0$. Because the coefficient of x in $x^2 - ix + 2 = 0$ is not a real number, the complex conjugate, i, of $-i$, need not be a solution.

■ Section P.7 Solving Inequalities Algebraically and Graphically

Quick Review P.7

1. $-7 < 2x - 3 < 7$
$-4 < 2x < 10$
$-2 < x < 5$

3. $|x + 2| = 3$
$x + 2 = 3$ or $x + 2 = -3$
$x = 1$ or $x = -5$

5. $x^3 - 4x = x(x^2 - 4) = x(x - 2)(x + 2)$

7. $\frac{z^2 - 25}{z^2 - 5z} = \frac{(z - 5)(z + 5)}{z(z - 5)} = \frac{z + 5}{z}$

9. $\frac{x}{x - 1} + \frac{x + 1}{3x - 4}$
$= \frac{x(3x - 4)}{(x - 1)(3x - 4)} + \frac{(x + 1)(x - 1)}{(x - 1)(3x - 4)}$
$= \frac{4x^2 - 4x - 1}{(x - 1)(3x - 4)}$

Section P.7 Exercises

1. $(-\infty, -9] \cup [1, \infty)$:
$x + 4 \geq 5$ or $x + 4 \leq -5$
$x \geq 1$ $\quad x \leq -9$

−12 −10 −8 −6 −4 −2 0 2 4 6 8

3. $(1, 5)$: $-2 < x - 3 < 2$
$1 < x < 5$

−2 −1 0 1 2 3 4 5 6 7 8

5. $\left(-\frac{2}{3}, \frac{10}{3}\right)$: $|4 - 3x| < 6$
$-6 < 4 - 3x < 6$
$-10 < -3x < 2$
$\frac{10}{3} > x > -\frac{2}{3}$

−5 −4 −3 −2 −1 0 1 2 3 4 5

7. $(-\infty, -11] \cup [7, \infty)$:
$\frac{x + 2}{3} \leq -3$ or $\frac{x + 2}{3} \geq 3$
$x + 2 \leq -9$ $\quad x + 2 \geq 9$
$x \leq -11$ $\quad x \geq 7$

−12 −10 −8 −6 −4 −2 0 2 4 6 8

9. $2x^2 + 17x + 21 = 0$
$(2x + 3)(x + 7) = 0$
$2x + 3 = 0$ or $x + 7 = 0$
$x = -\frac{3}{2}$ or $x = -7$

The graph of $y = 2x^2 + 17x + 21$ lies below the x-axis for $-7 < x < -\frac{3}{2}$. Hence $\left[-7, -\frac{3}{2}\right]$ is the solution since the endpoints are included.

11. $2x^2 + 7x - 15 = 0$
$(2x - 3)(x + 5) = 0$
$2x - 3 = 0$ or $x + 5 = 0$
$x = \frac{3}{2}$ or $x = -5$

The graph of $y = 2x^2 + 7x - 15$ lies above the x-axis for $x < -5$ and for $x > \frac{3}{2}$. Hence $(-\infty, -5) \cup \left(\frac{3}{2}, \infty\right)$ is the solution.

13.
$$\begin{aligned} 2 - 5x - 3x^2 &= 0 \\ (2 + x)(1 - 3x) &= 0 \\ 2 + x = 0 \quad &\text{or} \quad 1 - 3x = 0 \\ x = -2 \quad &\text{or} \quad x = \frac{1}{3} \end{aligned}$$

The graph of $y = 2 - 5x - 3x^2$ lies below the x-axis for $x < -2$ and for $x > \frac{1}{3}$. Hence $(-\infty, -2) \cup \left(\frac{1}{3}, \infty\right)$ is the solution.

15.
$$\begin{aligned} x^3 - x &= 0 \\ x(x^2 - 1) &= 0 \\ x(x + 1)(x - 1) &= 0 \\ x = 0 \quad \text{or} \quad x + 1 = 0 \quad &\text{or} \quad x - 1 = 0 \\ x = 0 \quad \text{or} \quad x = -1 \quad &\text{or} \quad x = 1 \end{aligned}$$

The graph of $y = x^3 - x$ lies above the x-axis for $x > 1$ and for $-1 < x < 0$. Hence $[-1, 0] \cup [1, \infty)$ is the solution.

17. The graph of $y = x^2 - 4x - 1$ is zero for $x \approx -0.24$ and $x \approx 4.24$, and lies below the x-axis for $-0.24 < x < 4.24$. Hence $(-0.24, 4.24)$ is the approximate solution.

19.
$$\begin{aligned} 6x^2 - 5x - 4 &= 0 \\ (3x - 4)(2x + 1) &= 0 \\ 3x - 4 = 0 \quad &\text{or} \quad 2x + 1 = 0 \\ x = \frac{4}{3} \quad &\text{or} \quad x = -\frac{1}{2} \end{aligned}$$

The graph of $y = 6x^2 - 5x - 4$ lies above the x-axis for $x < -\frac{1}{2}$ and for $x > \frac{4}{3}$. Hence $\left(-\infty, -\frac{1}{2}\right) \cup \left(\frac{4}{3}, \infty\right)$ is the solution.

21. The graph of $y = 9x^2 + 12x - 1$ appears to be zero for $x \approx -1.41$ and $x \approx 0.08$. and lies above the x-axis for $x < -1.41$ and $x > 0.08$. Hence $(-\infty, -1.41] \cup [0.08, \infty)$ is the approximate solution.

23.
$$\begin{aligned} 4x^2 - 4x + 1 &= 0 \\ (2x - 1)(2x - 1) &= 0 \\ (2x - 1)^2 &= 0 \\ 2x - 1 &= 0 \\ x &= \frac{1}{2} \end{aligned}$$

The graph of $y = 4x^2 - 4x + 1$ lies entirely above the x-axis, except at $x = \frac{1}{2}$. Hence $\left(-\infty, \frac{1}{2}\right) \cup \left(\frac{1}{2}, \infty\right)$ is the solution set.

25.
$$\begin{aligned} x^2 - 8x + 16 &= 0 \\ (x - 4)(x - 4) &= 0 \\ (x - 4)^2 &= 0 \\ x - 4 &= 0 \\ x &= 4 \end{aligned}$$

The graph of $y = x^2 - 8x + 16$ lies entirely above the x-axis, except at $x = 4$. Hence there is no solution.

27. The graph of $y = 3x^3 - 12x + 2$ is zero for $x \approx -2.08$, $x \approx 0.17$, and $x \approx 1.91$ and lies above the x-axis for $-2.08 < x < 0.17$ and $x > 1.91$. Hence, $[-2.08, 0.17] \cup [1.91, \infty)$ is the approximate solution.

29. $2x^3 + 2x > 5$ is equivalent to $2x^3 + 2x - 5 > 0$. The graph of $y = 2x^3 + 2x - 5$ is zero for $x \approx 1.11$ and lies above the x-axis for $x > 1.11$. So, $(1.11, \infty)$ is the approximate solution.

31. Answers may vary. Here are some possibilities.

(a) $x^2 + 1 > 0$

(b) $x^2 + 1 < 0$

(c) $x^2 \le 0$

(d) $(x + 2)(x - 5) \le 0$

(e) $(x + 1)(x - 4) > 0$

(f) $x(x - 4) \ge 0$

33. $s = -16t^2 + 256t$

(a)
$$\begin{aligned} -16t^2 + 256t &= 768 \\ -16t^2 + 256t - 768 &= 0 \\ t^2 - 16t + 48 &= 0 \\ (t - 12)(t - 4) &= 0 \\ t - 12 = 0 \quad &\text{or} \quad t - 4 = 0 \\ t = 12 \quad &\text{or} \quad t = 4 \end{aligned}$$

The projectile is 768 ft above ground twice: at $t = 4$ sec, on the way up, and $t = 12$ sec, on the way down.

(b) The graph of $s = -16t^2 + 256t$ lies above the graph of $s = 768$ for $4 < t < 12$. Hence the projectile's height will be at least 768 ft when t is in the interval $[4, 12]$.

(c) The graph of $s = -16t^2 + 256t$ lies below the graph of $s = 768$ for $0 < t < 4$ and $12 < t < 16$. Hence the projectile's height will be less than or equal to 768 ft when t is in the interval $(0, 4]$ or $[12, 16)$.

35. Solving the corresponding equation in the process of solving an inequality reveals the boundaries of the solution set. For example, to solve the inequality $x^2 - 4 \le 0$, we first solve the corresponding equation $x^2 - 4 = 0$ and find that $x = \pm 2$. The solution, $[-2, 2]$, of inequality has ± 2 as its boundaries.

37. (a) Let $x > 0$ be the width of a rectangle; then the length is $2x - 2$ and the perimeter is $P = 2[x + (2x - 2)]$. Solving $P < 200$ and $2x - 2 > 0$ gives 1 in. $< x <$ 34 in.

$$\begin{aligned} 2[x + (2x - 2)] &< 200 \quad \text{and} \quad 2x - 2 > 0 \\ 2(3x - 2) &< 200 \qquad\qquad\quad 2x > 2 \\ 6x - 4 &< 200 \qquad\qquad\quad\;\; x > 1 \\ 6x &< 204 \\ x &< 34 \end{aligned}$$

(b) The area is $A = x(2x - 2)$. We already know $x > 1$ from part (a). Solve $A \le 1200$.

$$\begin{aligned} x(2x - 2) &= 1200 \\ 2x^2 - 2x - 1200 &= 0 \\ x^2 - x - 600 &= 0 \\ (x - 25)(x + 24) &= 0 \\ x - 25 = 0 \quad &\text{or} \quad x + 24 = 0 \\ x = 25 \quad &\text{or} \quad x = -24 \end{aligned}$$

The graph of $y = 2x^2 - 2x - 1200$ lies below the x-axis for $1 < x < 25$, so $A \le 1200$ when x is in the interval $(1, 25]$.

39. Let x be the amount borrowed; then $\dfrac{200{,}000 + x}{50{,}000 + x} \ge 2$.

Solving for x reveals that the company can borrow no more than \$100,000.

41. True. The absolute value of any real number is always nonnegative, i.e., greater than or equal to zero.

43. The graph of $y = x^2 - 2x + 2$ lies entirely above the x-axis so $x^2 - 2x + 2 \ge 0$ for all real numbers x. The answer is D.

45. $x^2 \leq 1$ implies $-1 \leq x \leq 1$, so the solution is $[-1, 1]$. The answer is D.

47. $2x^2 + 7x - 15 = 10$ or $2x^2 + 7x - 15 = -10$
$2x^2 + 7x - 25 = 0$ $\quad$ $2x^2 + 7x - 5 = 0$
The graph of $y = 2x^2 + 7x - 25$ appears to be zero for $x \approx -5.69$ and $x \approx 2.19$. The graph of $y = 2x^2 + 7x - 5$ appears to be zero for $x \approx -4.11$ and $x \approx 0.61$.
Now look at the graphs of $y = |2x^2 + 7x - 15|$ and $y = 10$. The graph of $y = |2x^2 + 7x - 15|$ lies below the graph of $y = 10$ when $-5.69 < x < -4.11$ and when $0.61 < x < 2.19$. Hence $(-5.69, -4.11) \cup (0.61, 2.19)$ is the approximate solution.

Chapter P Review

1. Endpoints 0 and 5; bounded

3. $2(x^2 - x) = 2x^2 - 2x$

5. $\dfrac{(uv^2)^3}{v^2u^3} = \dfrac{u^3v^6}{u^3v^2} = v^4$

7. 3.68×10^9

9. 5,000,000,000

11. **(a)** 5.0711×10^{10}
(b) 4.63×10^9
(c) 5.0×10^8
(d) 3.995×10^9
(e) 1.4497×10^{10}

13. **(a)** Distance: $|14 - (-5)| = |19| = 19$
(b) Midpoint: $\dfrac{-5 + 14}{2} = \dfrac{9}{2} = 4.5$

15. The three side lengths (distances between pairs of points) are

$$\sqrt{[3 - (-2)]^2 + (11 - 1)^2} = \sqrt{5^2 + 10^2} = \sqrt{25 + 100} = \sqrt{125} = 5\sqrt{5}$$

$$\sqrt{(7 - 3)^2 + (9 - 11)^2} = \sqrt{4^2 + (-2)^2} = \sqrt{16 + 4} = \sqrt{20} = 2\sqrt{5}$$

$$\sqrt{[7 - (-2)]^2 + (9 - 1)^2} = \sqrt{9^2 + 8^2} = \sqrt{81 + 64} = \sqrt{145}.$$

Since $(2\sqrt{5})^2 + (5\sqrt{5})^2 = 20 + 125 = 145 = (\sqrt{145})^2$ —the sum of the squares of the two shorter side lengths equals the square of the long side length—the points determine a right triangle.

17. $(x - 0)^2 + (y - 0)^2 = 2^2$, or $x^2 + y^2 = 4$

19. $[x - (-5)]^2 + [y - (-4)]^2 = 3^2$, so the center is $(-5, -4)$ and the radius is 3.

21. **(a)** Distance between $(-3, 2)$ and $(-1, -2)$:

$$\sqrt{(-2 - 2)^2 + [-1 - (-3)]^2} = \sqrt{(-4)^2 + (2)^2} = \sqrt{16 + 4} = \sqrt{20} \approx 4.47$$

Distance between $(-3, 2)$ and $(5, 6)$:

$$\sqrt{(6 - 2)^2 + [5 - (-3)]^2} = \sqrt{4^2 + 8^2} = \sqrt{16 + 64} = \sqrt{80} \approx 8.94$$

Distance between $(5, 6)$ and $(-1, -2)$:

$$\sqrt{(-2 - 6)^2 + (-1 - 5)^2} = \sqrt{(-8)^2 + (-6)^2} = \sqrt{64 + 36} = \sqrt{100} = 10$$

(b) $(\sqrt{20})^2 + (\sqrt{80})^2 = 20 + 80 = 100 = 10^2$, so the Pythagorean Theorem guarantees the triangle is a right triangle.

23. $\dfrac{-1 + a}{2} = 3$ and $\dfrac{1 + b}{2} = 5$
$-1 + a = 6 \qquad 1 + b = 10$
$a = 7 \qquad b = 9$

25. $y + 1 = -\dfrac{2}{3}(x - 2)$

27. Beginning with point-slope form: $y + 2 = \dfrac{4}{5}(x - 3)$, so $y = \dfrac{4}{5}x - 4.4$.

29. $y = 4$

31. The slope of the given line is the same as the line we want: $m = -\dfrac{2}{5}$, so $y + 3 = -\dfrac{2}{5}(x - 2)$, and therefore $y = -\dfrac{2}{5}x - \dfrac{11}{5}$.

33. **(a)**

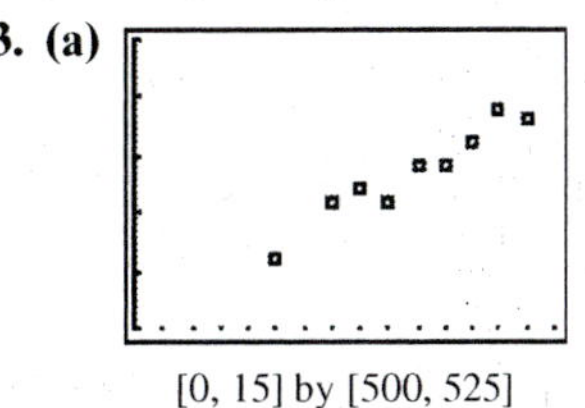

[0, 15] by [500, 525]

(b) Slope of the line between the points (5, 506) and (10, 514) is $m = \dfrac{514 - 506}{10 - 5} = \dfrac{8}{5} = 1.6$.
Using the *point-slope form* equation for the line, we have $y - 506 = 1.6(x - 5)$, so $y = 1.6x + 498$.

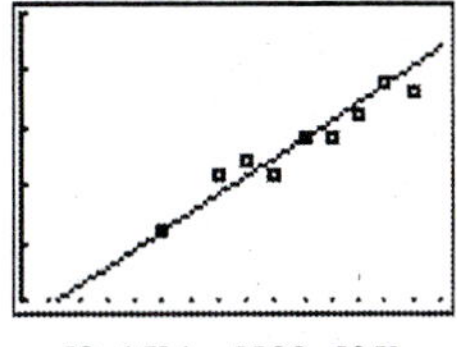

[0, 15] by [500, 525]

(c) The year 1996 is represented by $x = 6$. Using $y = 1.6x + 498$ and $x = 6$, we estimate the average SAT math score in 1996 to be 507.6, which is very close to the actual value 508.
(d) The year 2006 is represented by $x = 16$. Using $y = 1.6x + 498$ and $x = 16$, we predict the average SAT math score in 2006 will be 524.

35. $m = \dfrac{25}{10} = \dfrac{5}{2} = 2.5$

37. $3x - 4 = 6x + 5$
$-3x = 9$
$x = -3$

39. $2(5 - 2y) - 3(1 - y) = y + 1$
$10 - 4y - 3 + 3y = y + 1$
$7 - y = y + 1$
$-2y = -6$
$y = 3$

41.
$$\begin{aligned} x^2 - 4x - 3 &= 0 \\ x^2 - 4x &= 3 \\ x^2 - 4x + (2)^2 &= 3 + (2)^2 \\ (x - 2)^2 &= 7 \\ x - 2 &= \pm\sqrt{7} \end{aligned}$$
$$x - 2 = -\sqrt{7} \quad \text{or} \quad x - 2 = \sqrt{7}$$
$$x = 2 - \sqrt{7} \approx -0.65 \quad x = 2 + \sqrt{7} \approx 4.65$$

43.
$$\begin{aligned} 6x^2 + 7x &= 3 \\ 6x^2 + 7x - 3 &= 0 \\ (3x - 1)(2x + 3) &= 0 \end{aligned}$$
$$3x - 1 = 0 \quad \text{or} \quad 2x + 3 = 0$$
$$x = \frac{1}{3} \quad \text{or} \quad x = -\frac{3}{2}$$

45.
$$\begin{aligned} x(2x + 5) &= 4(x + 7) \\ 2x^2 + 5x &= 4x + 28 \\ 2x^2 + x - 28 &= 0 \\ (2x - 7)(x + 4) &= 0 \end{aligned}$$
$$2x - 7 = 0 \quad \text{or} \quad x + 4 = 0$$
$$x = \frac{7}{2} \quad \text{or} \quad x = -4$$

47.
$$\begin{aligned} 4x^2 - 20x + 25 &= 0 \\ (2x - 5)(2x - 5) &= 0 \\ (2x - 5)^2 &= 0 \\ 2x - 5 &= 0 \\ x &= \frac{5}{2} \end{aligned}$$

49.
$$\begin{aligned} x^2 &= 3x \\ x^2 - 3x &= 0 \\ x(x - 3) &= 0 \end{aligned}$$
$$x = 0 \quad \text{or} \quad x - 3 = 0$$
$$x = 0 \quad \text{or} \quad x = 3$$

51. Solving $x^2 - 6x + 13 = 0$ by using the quadratic formula with $a = 1$, $b = -6$, and $c = 13$ gives

$$\begin{aligned} x &= \frac{6 \pm \sqrt{(-6)^2 - 4(1)(13)}}{2(1)} = \frac{6 \pm \sqrt{-16}}{2} \\ &= \frac{6 \pm 4i}{2} = 3 \pm 2i \end{aligned}$$

53.
$$\begin{aligned} 2x^2 - 3x - 1 &= 0 \\ x^2 - \frac{3}{2}x - \frac{1}{2} &= 0 \\ x^2 - \frac{3}{2}x + \left(-\frac{3}{4}\right)^2 &= \frac{1}{2} + \left(-\frac{3}{4}\right)^2 \\ \left(x - \frac{3}{4}\right)^2 &= \frac{17}{16} \\ x - \frac{3}{4} &= \pm\frac{\sqrt{17}}{4} \\ x &= \frac{3}{4} \pm \frac{\sqrt{17}}{4} \end{aligned}$$
$$x = \frac{3}{4} - \frac{\sqrt{17}}{4} \approx -0.28 \quad \text{or} \quad x = \frac{3}{4} + \frac{\sqrt{17}}{4} \approx 1.78$$

55. The graph of $y = 3x^3 - 19x^2 - 14x$ is zero for $x = 0$, $x = -\frac{2}{3}$, and $x = 7$.

57. The graph of $y = x^3 - 2x^2 - 2$ is zero for $x \approx 2.36$.

59.
$$\begin{aligned} -2 < x + 4 &\le 7 \\ -6 < x &\le 3 \end{aligned}$$
Hence $(-6, 3]$ is the solution.

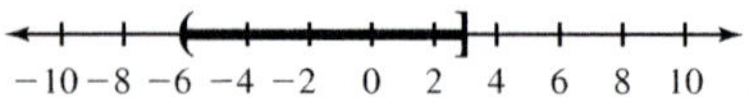

61.
$$\begin{aligned} \frac{3x - 5}{4} &\le -1 \\ 3x - 5 &\le -4 \\ 3x &\le 1 \\ x &\le \frac{1}{3} \end{aligned}$$
Hence $\left(-\infty, \frac{1}{3}\right]$ is the solution.

63. $3x + 4 \ge 2$ or $3x + 4 \le -2$
$$3x \ge -2 \quad \text{or} \quad 3x \le -6$$
$$x \ge -\frac{2}{3} \quad \text{or} \quad x \le -2$$
Hence $(-\infty, -2] \cup \left[-\frac{2}{3}, \infty\right)$ is the solution.

65. The graph of $y = 2x^2 - 2x - 1$ is zero for $x \approx -0.37$ and $x \approx 1.37$, and lies above the x-axis for $x < -0.37$ and for $x > 1.37$. Hence $(-\infty, -0.37) \cup (1.37, \infty)$ is the approximate solution.

67. $x^3 - 9x \le 3$ is equivalent to $x^3 - 9x - 3 \le 0$. The graph of $y = x^3 - 9x - 3$ is zero for $x \approx -2.82$, $x \approx -0.34$, and $x \approx 3.15$, and lies below the x-axis for $x < -2.82$ and for $-0.34 < x < 3.15$. Hence the approximate solution is $(-\infty, -2.82] \cup [-0.34, 3.15]$.

69.
$$\frac{x + 7}{5} > 2 \quad \text{or} \quad \frac{x + 7}{5} < -2$$
$$x + 7 > 10 \quad \text{or} \quad x + 7 < -10$$
$$x > 3 \quad \text{or} \quad x < -17$$
Hence $(-\infty, -17) \cup (3, \infty)$ is the solution.

71.
$$\begin{aligned} 4x^2 + 12x + 9 &= 0 \\ (2x + 3)(2x + 3) &= 0 \\ (2x + 3)^2 &= 0 \\ 2x + 3 &= 0 \\ x &= -\frac{3}{2} \end{aligned}$$
The graph of $y = 4x^2 + 12x + 9$ lies entirely above the x-axis except for $x = -\frac{3}{2}$. Hence all real numbers satisfy the inequality. So $(-\infty, \infty)$ is the solution.

73.
$$\begin{aligned} (3 - 2i) + (-2 + 5i) &= (3 - 2) + (-2 + 5)i \\ &= 1 + 3i \end{aligned}$$

75.
$$\begin{aligned} (1 + 2i)(3 - 2i) &= 3 - 2i + 6i - 4i^2 \\ &= 3 + 4i + 4 \\ &= 7 + 4i \end{aligned}$$

77.
$$\begin{aligned} (1 + 2i)^2(1 - 2i)^2 &= (1 + 4i + 4i^2)(1 - 4i + 4i^2) \\ &= (-3 + 4i)(-3 - 4i) \\ &= 9 - 12i + 12i - 16i^2 = 25 \end{aligned}$$

79. $\sqrt{-16} = \sqrt{(16)(-1)} = 4\sqrt{-1} = 4i$

81. $s = -16t^2 + 320t$

(a) $-16t^2 + 320t = 1538$

$-16t^2 + 320t - 1538 = 0$

The graph of $s = -16t^2 + 320t - 1538$ is zero at

$$t = \frac{-320 \pm \sqrt{320^2 - 4(-16)(-1538)}}{2(-16)}$$

$$= \frac{-320 \pm \sqrt{3968}}{-32} = \frac{40 \pm \sqrt{62}}{4}.$$

So $t = \dfrac{40 - \sqrt{62}}{4} \approx 8.03$ sec or

$t = \dfrac{40 + \sqrt{62}}{4} \approx 11.97.$

The projectile is 1538 ft above ground twice: at $t \approx 8$ sec, on the way up, and at $t \approx 12$ sec, on the way down.

(b) The graph of $s = -16t^2 + 320t$ lies below the graph of $s = 1538$ for $0 < t < 8$ and for $12 < t < 20$ (approximately). Hence the projectile's height will be at most 1538 ft when t is in the interval $(0, 8]$ or $[12, 20)$ (approximately).

(c) The graph of $s = -16t^2 + 320t$ lies above the graph of $s = 1538$ for $8 < t < 12$ (approximately). Hence the projectile's height will be greater than or equal to 1538 when t is in the interval $[8, 12]$ (approximately).

83. (a) Let $w > 0$ be the width of a rectangle; the length is $3w + 1$ and the perimeter is $P = 2[w + (3w + 1)]$. Solve $P \le 150$.

$2[w + (3w + 1)] \le 150$

$2(4w + 1) \le 150$

$8w + 2 \le 150$

$8w \le 148$

$w \le 18.5$

Thus $P \le 150$ cm when w is in the interval $(0, 18.5]$.

(b) The area is $A = w(3w + 1)$. Solve $A > 1500$.

$w(3w + 1) > 1500$

$3w^2 + w - 1550 > 0$

The graph of $A = 3w^2 + w - 1500$ appears to be zero for $w \approx 22.19$ when w is positive, and lies above the w-axis for $w > 22.19$. Hence, $A > 1500$ when w is in the interval $(22.19, \infty)$ (approximately).

Chapter 1
Functions and Graphs

■ Section 1.1 Modeling and Equation Solving

Exploration 1

1. $k = \dfrac{d}{m} = \dfrac{100 - 25}{100} = \dfrac{75}{100} = 0.75$
2. $t = 6.5\% + 0.5\% = 7\%$ or 0.07
3. $m = \dfrac{d}{k}$, $s = d + td$

   $s = pm$

   $$p = \frac{s}{m} = \frac{d + td}{\frac{d}{k}} = \frac{d + td}{1} \cdot \frac{k}{d} = \frac{d(1 + t)}{1} \cdot \frac{k}{d}$$

   $$= \frac{k(1 + t)}{1} = (0.75)(1.07) = 0.8025$$
4. Yes, because $\$36.99 \times 0.8025 = \$29.68$.
5. $\$100 \div 0.8025 = \$124.61$

Exploration 2

1. Because the linear model maintains a constant positive slope, it will eventually reach the point where 100% of the prisoners are female. It will then continue to rise, giving percentages above 100%, which are impossible.
2. Yes, because 2009 is still close to the data we are modeling. We would have much less confidence in the linear model for predicting the percentage 25 years from 2000.
3. One possible answer: Males are heavily dominant in violent crime statistics, while female crimes tend to be property crimes like burglary or shoplifting. Since property crimes rates are senstive to economic conditions, a statistician might look for adverse economic factors in 1990, especially those that would affect people near or below the poverty level.
4. Yes. Table 1.1 shows that the minimum wage worker had less purchasing power in 1990 than in any other year since 1955, which gives some evidence of adverse economic conditions among lower-income Americans that year. Nonetheless, a careful sociologist would certainly want to look at other data before claiming a connection between this statistic and the female crime rate.

Quick Review 1.1

1. $(x + 4)(x - 4)$
3. $(9y + 2)(9y - 2)$
5. $(4h^2 + 9)(4h^2 - 9) = (4h^2 + 9)(2h + 3)(2h - 3)$
7. $(x + 4)(x - 1)$
9. $(2x - 1)(x - 5)$

Section 1.1 Exercises

1. (d) (q)
3. (a) (p)
5. (e) (l)
7. (g) (t)
9. (i) (m)
11. **(a)** The percentage increased from 1954 to 1999 and then decreased slightly from 1999 to 2004.

    (b) The greatest increase occurred between 1974 and 1979.
13. Women (□), Men (+)

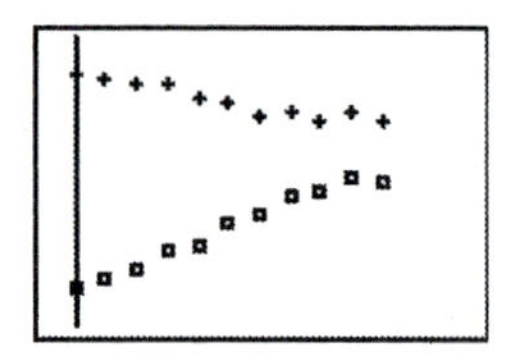

[–5, 55] by [23, 92]

15. To find the equation, first find the slope.

    Women: Slope $= \dfrac{\text{change in } y}{\text{change in } x} = \dfrac{58.5 - 32.3}{1999 - 1954} = \dfrac{26.2}{45}$ $= 0.582$. The y-intercept is 32.3, so the equation of the line is $y = 0.582x + 32.3$.

    Men: Slope $= \dfrac{74.0 - 83.5}{1999 - 1954} = \dfrac{-9.5}{45} = -0.211$. The y-intercept is 83.5, so the equation of the line is $y = -0.211x + 83.5$.

    In both cases, x represents the number of years after 1954.
17. For the percentages to be the same, we need to set the two equations equal to each other.

    $$0.582x + 32.3 = -0.211x + 83.5$$
    $$0.793x = 51.2$$
    $$x \approx 64.6$$

    So, approximately 65 years after 1954 (2018), the models predict that the percentages will be about the same. To check:

    Males: $y = (-0.211)(65) + 83.5 \approx 69.9\%$
    Females: $y = (0.582)(65) + 32.3 \approx 69.9\%$
19.

L1	L2	L3 3
316	12	3.7975
480	21	4.375
740	41	5.5405
1085	64	5.8986
1382	92	6.657
------	------	------

L3(1)=3.797468354...

21. Because all stepping stones have the same thickness, what matters is area.

    The area of a square stepping stone is
    $A_1 = 12 \cdot 12 = 144$ in.^{2}
    The area of a round stepping stone is

    $$A_2 = \pi\left(\frac{13}{2}\right)^2 \approx 3.14(6.5)^2 = 132.665 \text{ in.}^2$$

    The square stones give a greater amount of rock for the same price.

23. A scatter plot of the data suggests a parabola with its vertex at the origin.

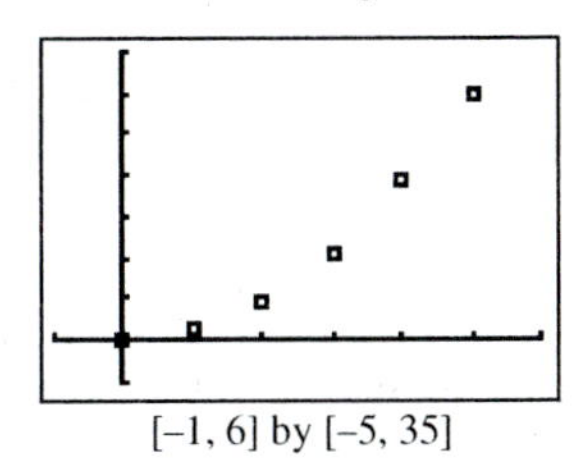

[–1, 6] by [–5, 35]

The model $y = 1.2t^2$ fits the data.

25. The lower line shows the minimum salaries, since they are lower than the average salaries.

27. The 1995 points are third from the right, Year 15, on both graphs. There is a clear drop in the average salary right after the 1994 strike.

29. Adding $2v^2 + 5$ to both sides gives $3v^2 = 13$. Divide both sides by 3 to get $v^2 = \frac{13}{3}$, so $v = \pm\sqrt{\frac{13}{3}}$.
$3v^2 = 13$ is equivalent to $3v^2 - 13 = 0$. The graph of $y = 3v^2 - 13$ is zero for $v \approx -2.08$ and for $v \approx 2.08$.

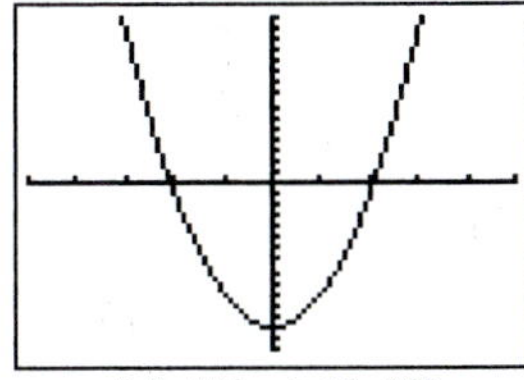

[–5, 5] by [–15, 15]

31.

$$2x^2 - 5x + 2 = x^2 - 5x + 6 + 3x$$
$$x^2 - 3x - 4 = 0$$
$$(x - 4)(x + 1) = 0$$
$$x - 4 = 0 \quad \text{or} \quad x + 1 = 0$$
$$x = 4 \quad \text{or} \quad x = -1$$

$2x^2 - 5x + 2 = (x - 3)(x - 2) + 3x$ is equivalent to $2x^2 - 8x + 2 - (x - 3)(x - 2) = 0$. The graph of $y = 2x^2 - 8x + 2 - (x - 3)(x - 2)$ is zero for $x = -1$ and for $x = 4$.

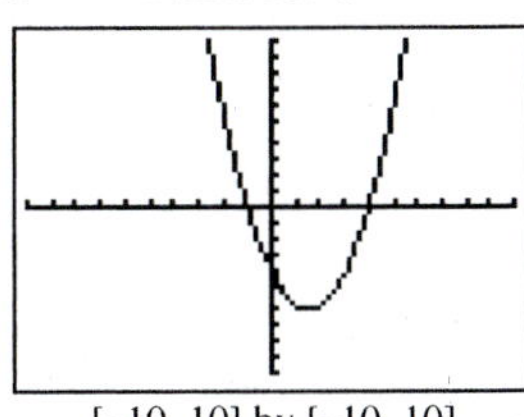

[–10, 10] by [–10, 10]

33. Rewrite as $2x^2 - 5x - 12 = 0$; the left side factors to $(2x + 3)(x - 4) = 0$:

$$2x + 3 = 0 \quad \text{or} \quad x - 4 = 0$$
$$2x = -3 \qquad x = 4$$
$$x = -1.5$$

The graph of $y = 2x^2 - 5x - 12$ is zero for $x = -1.5$ and for $x = 4$.

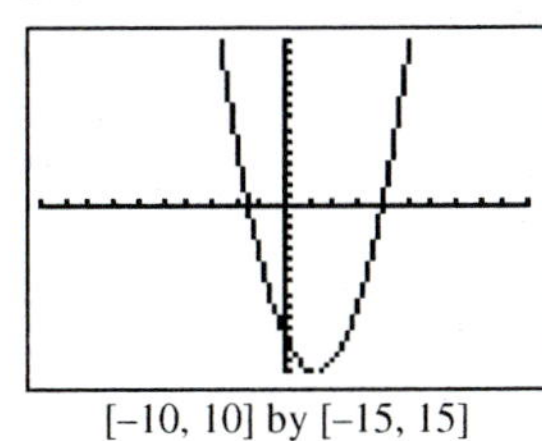

[–10, 10] by [–15, 15]

35. $x^2 + 7x - 14 = 0$, so $a = 1, b = 7$, and $c = -14$:

$$x = \frac{-7 \pm \sqrt{7^2 - 4(1)(-14)}}{2(1)} = \frac{-7 \pm \sqrt{105}}{2} = -\frac{7}{2} \pm \frac{1}{2}\sqrt{105}$$

The graph of $y = x^2 + 7x - 14$ is zero for $x \approx -8.62$ and for $x \approx 1.62$.

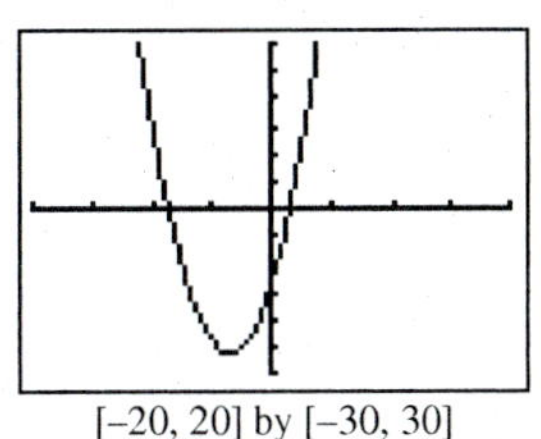

[–20, 20] by [–30, 30]

37. Change to $x^2 - 2x - 15 = 0$ (see below); this factors to $(x + 3)(x - 5) = 0$, so $x = -3$ or $x = 5$. Substituting the first of these shows that it is extraneous.

$$x + 1 = 2\sqrt{x + 4}$$
$$(x + 1)^2 = 2^2(\sqrt{x + 4})^2$$
$$x^2 + 2x + 1 = 4x + 16$$
$$x^2 - 2x - 15 = 0$$

The graph of $y = x + 1 - 2\sqrt{x + 4}$ is zero for $x = 5$.

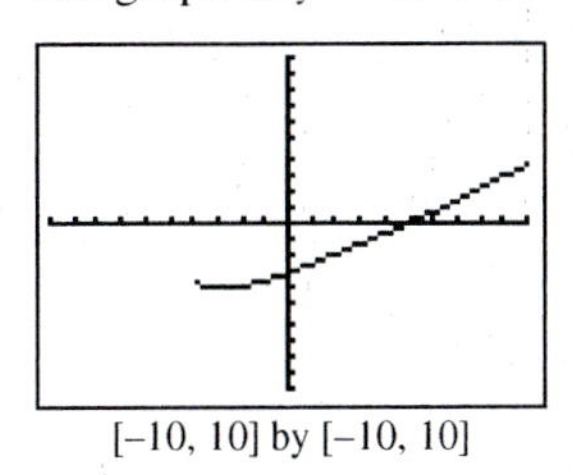

[–10, 10] by [–10, 10]

39. $x \approx 3.91$

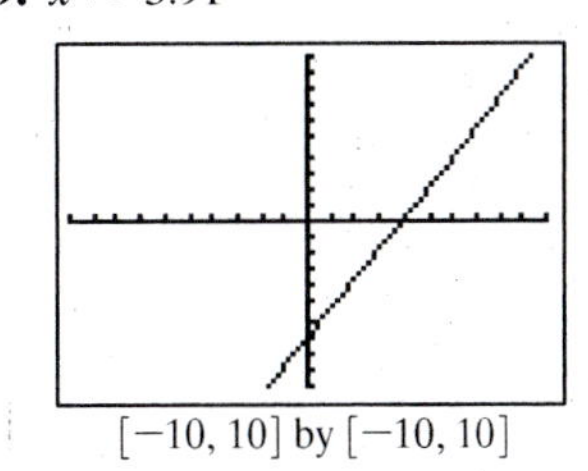

[–10, 10] by [–10, 10]

41. $x \approx 1.33$ or $x = 4$

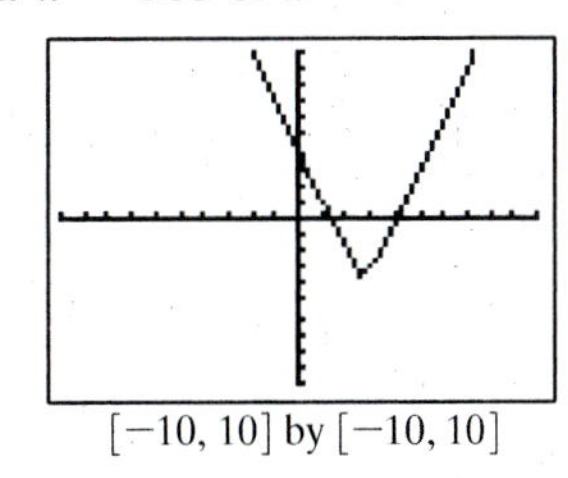

[–10, 10] by [–10, 10]

43. $x \approx 1.77$

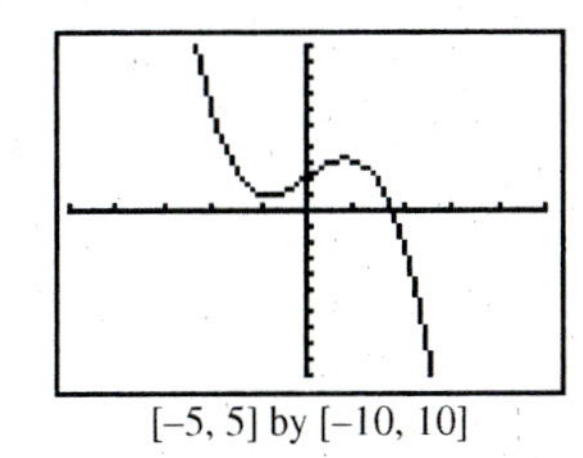

[–5, 5] by [–10, 10]

45. $x \approx -1.47$

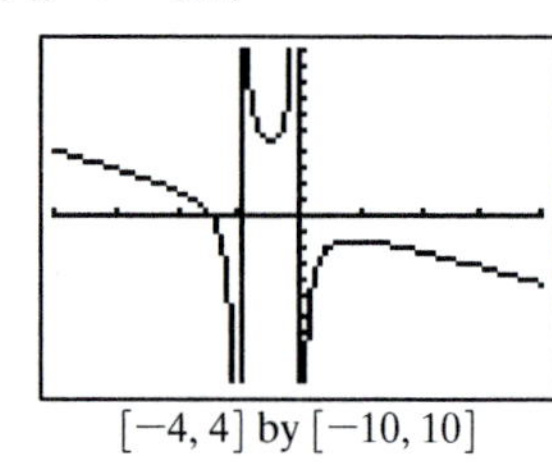

$[-4, 4]$ by $[-10, 10]$

47. Model the situation using $C = 0.18x + 32$, where x is the number of miles driven and C is the cost of a day's rental.

(a) Elaine's cost is $0.18(83) + 32 = \$46.94$.

(b) If for Ramon $C = \$69.80$, then

$$x = \frac{69.80 - 32}{0.18} = 210 \text{ miles.}$$

49. (a) $y = (x^{200})^{1/200} = x^{200/200} = x^1 = x$ for all $x \geq 0$.

(b) The graph looks like this:

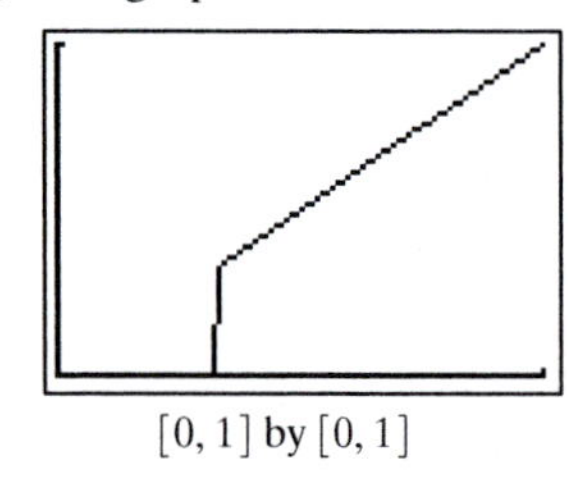

$[0, 1]$ by $[0, 1]$

(c) Yes, this is different from the graph of $y = x$.

(d) For values of x close to 0, x^{200} is so small that the calculator is unable to distinguish it from zero. It returns a value of $0^{1/200} = 0$ rather than x.

51. (a) $x = -3$ or $x = 1.1$ or $x = 1.15$.

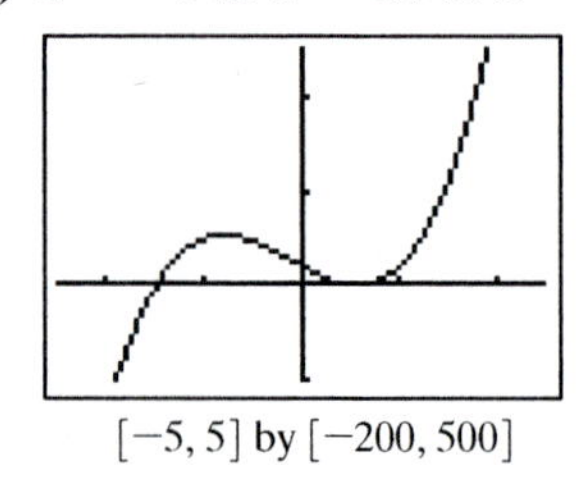

$[-5, 5]$ by $[-200, 500]$

(b) $x = -3$ only.

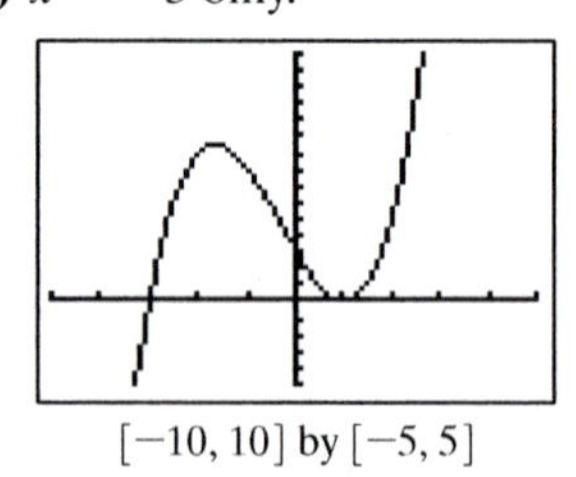

$[-10, 10]$ by $[-5, 5]$

53. Let n be any integer.
$n^2 + 2n = n(n + 2)$, which is either the product of two odd integers or the product of two even integers.

The product of two odd integers is odd.

The product of two even integers is a multiple of 4, since each even integer in the product contributes a factor of 2 to the product.

Therefore, $n^2 + 2n$ is either odd or a multiple of 4.

55. False. A product is zero if *any* factor is zero. That is, it takes only one zero factor to make the product zero.

57. This is a line with a negative slope and a y-intercept of 12. The answer is C. (The graph checks.)

59. As x increases by ones, the y-values get farther and farther apart, which implies an increasing slope and suggests a quadratic equation. The answer is B. (The equation checks.)

61. (a) March

(b) \$120

(c) June, after three months of poor performance

(d) Ahmad paid $(100)(\$120) = \$12{,}000$ for the stock and sold it for $(100)(\$100) = \$10{,}000$. He lost \$2,000 on the stock.

(e) After reaching a low in June, the stock climbed back to a price near \$140 by December. LaToya's shares had gained \$2000 by that point.

(f) One possible graph:

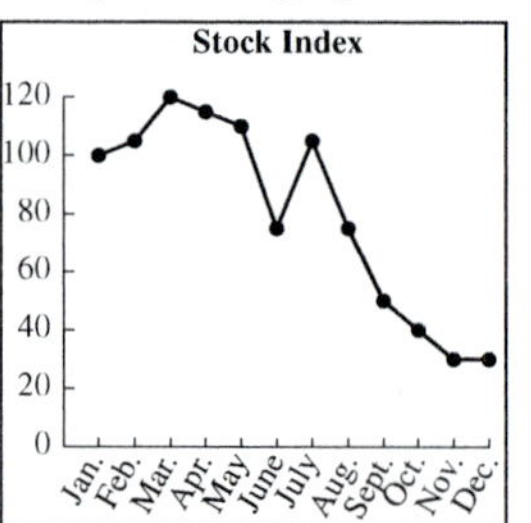

63. (a)

Subscribers

$[7, 15]$ by $[50, 200]$

Monthly Bills

$[7, 15]$ by $[35, 55]$

(b) The graph for subscribers appears to be linear. Since time t = the number of years after 1990, $t = 8$ for 1998 and $t = 14$ for 2004. The slope of the line is

$$\frac{180.4 - 69.2}{14 - 8} = \frac{111.2}{6} \approx 18.53.$$

Use the point-slope form to write the equation:
$y - 69.2 = 18.53(x - 8)$.
Solve for y: $y - 69.2 = 18.53x - 148.24$
$y = 18.53x - 79.04$
The linear model for subscribers as a function of years is $y = 18.53x - 79.04$.

(c) The fit is very good. The line goes through or is close to all the points.

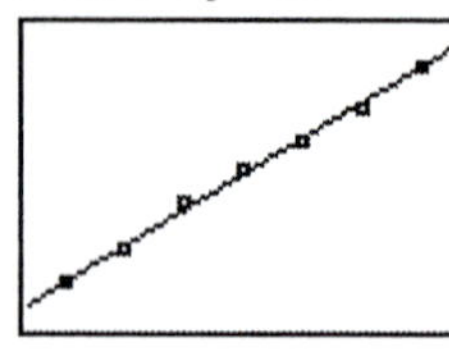

$[7, 15]$ by $[50, 200]$

(d) The monthly bill scatter plot has a curved shape that could be modeled more effectively by a function with a curved graph. Some possibilities include a quadratic function (parabola), a logarithmic function, a power function (e.g., square root), a logistic function, or a sine function.

(e)

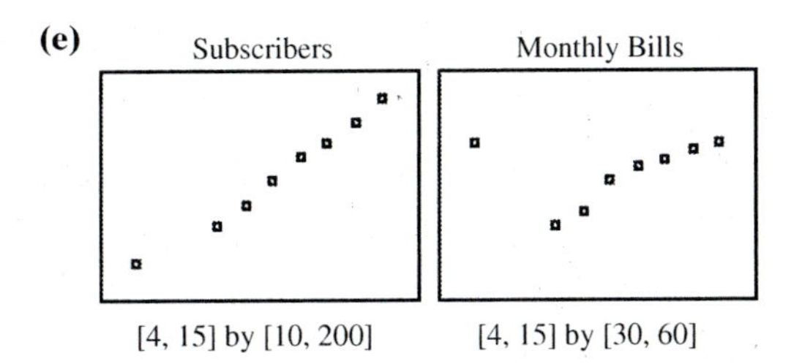

[4, 15] by [10, 200] [4, 15] by [30, 60]

(f) In 1995, cellular phone technology was still emerging, so the growth rate was not as fast as it was in more recent years. Thus, the slope from 1995 ($t = 5$) to 1998 ($t = 8$) is lower than the slope from 1998 to 2004. Cellular technology was more expensive before competition brought prices down. This explains the anomaly on the monthly bill scatter plot.

■ Section 1.2 Functions and Their Properties

Exploration 1

1. From left to right, the tables are (c) constant, (b) decreasing, and (a) increasing.

2.

X moves from	ΔX	ΔY1	X moves from	ΔX	ΔY2	X moves from	ΔX	ΔY3
−2 to −1	1	0	−2 to −1	1	−2	−2 to −1	1	2
−1 to 0	1	0	−1 to 0	1	−1	−1 to 0	1	2
0 to 1	1	0	0 to 1	1	−2	0 to 1	1	2
1 to 3	2	0	1 to 3	2	−4	1 to 3	2	3
3 to 7	4	0	3 to 7	4	−6	3 to 7	4	6

3. For an increasing function, $\Delta Y/\Delta X$ is positive. For a decreasing function, $\Delta Y/\Delta X$ is negative. For a constant function, $\Delta Y/\Delta X$ is 0.

4. For lines, $\Delta Y/\Delta X$ is the slope. Lines with positive slope are increasing, lines with negative slope are decreasing, and lines with 0 slope are constant, so this supports our answers to part 3.

Quick Review 1.2

1. $x^2 - 16 = 0$
$x^2 = 16$
$x = \pm 4$

3. $x - 10 < 0$
$x < 10$

5. As we have seen, the denominator of a function cannot be zero.
We need $x - 16 = 0$
$x = 16$

7. We need $x - 16 < 0$
$x < 16$

9. We need $3 - x \le 0$ and $x + 2 < 0$
$3 \le x$ $\quad x < -2$
$x < -2$ and $x \ge 3$

Section 1.2 Exercises

1. Yes, $y = \sqrt{x - 4}$ is a function of x, because when a number is substituted for x, there is at most one value produced for $\sqrt{x - 4}$.

3. No, $x = 2y^2$ does not determine y as a function of x, because when a positive number is substituted for x, y can be either $\sqrt{\frac{x}{2}}$ or $-\sqrt{\frac{x}{2}}$.

5. Yes

7. No

9. We need $x^2 + 4 \ge 0$; this is true for all real x.
Domain: $(-\infty, \infty)$.

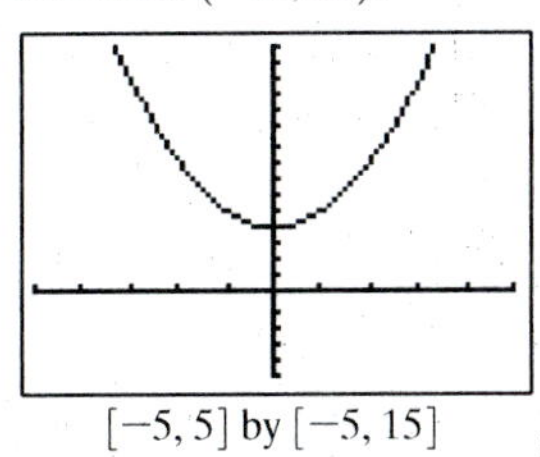
[−5, 5] by [−5, 15]

11. We need $x + 3 \neq 0$ and $x - 1 \neq 0$. Domain: $(-\infty, -3) \cup (-3, 1) \cup (1, \infty)$.

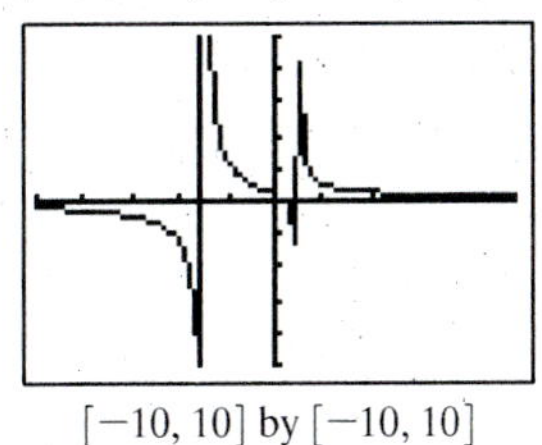
[−10, 10] by [−10, 10]

13. We notice that $g(x) = \dfrac{x}{x^2 - 5x} = \dfrac{x}{x(x - 5)}$.

As a result, $x - 5 \neq 0$ and $x \neq 0$.
Domain: $(-\infty, 0) \cup (0, 5) \cup (5, \infty)$.

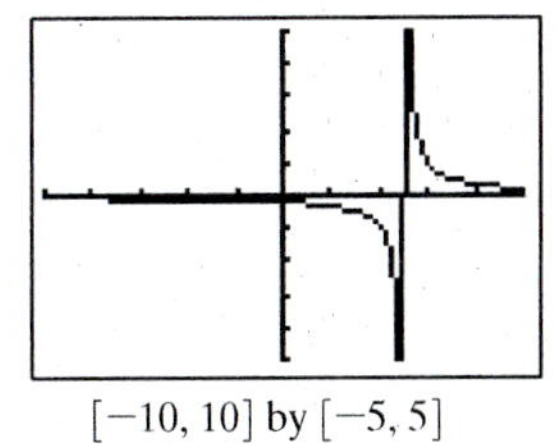
[−10, 10] by [−5, 5]

15. We need $x + 1 \neq 0$, $x^2 + 1 \neq 0$, and $4 - x \ge 0$. The first requirement means $x \neq -1$, the second is true for all x, and the last means $x \le 4$. The domain is therefore $(-\infty, -1) \cup (-1, 4]$.

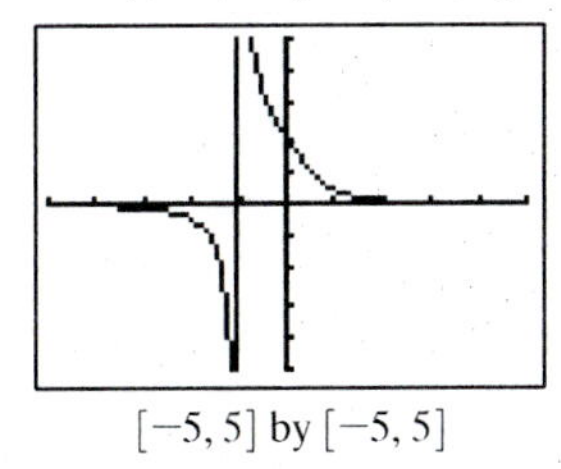
[−5, 5] by [−5, 5]

17. $f(x) = 10 - x^2$ can take on any negative value. Because x^2 is nonnegative, $f(x)$ cannot be greater than 10. The range is $(-\infty, 10]$.

19. The range of a function is most simply found by graphing it. As our graph shows, the range of $f(x)$ is $(-\infty, -1) \cup [0, \infty)$.

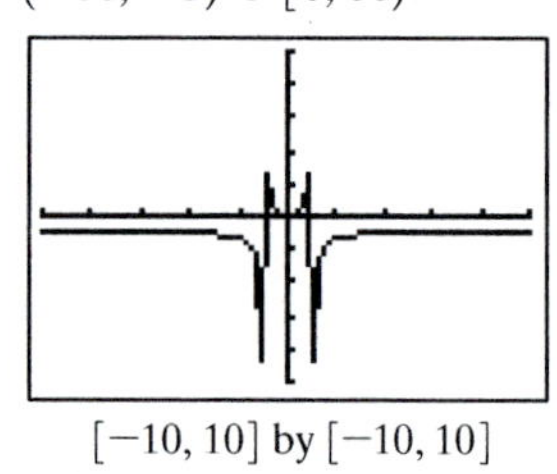

$[-10, 10]$ by $[-10, 10]$

21. Yes, non-removable

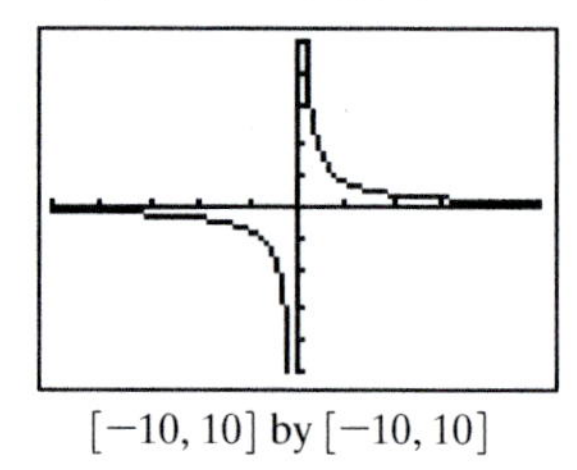

$[-10, 10]$ by $[-10, 10]$

23. Yes, non-removable

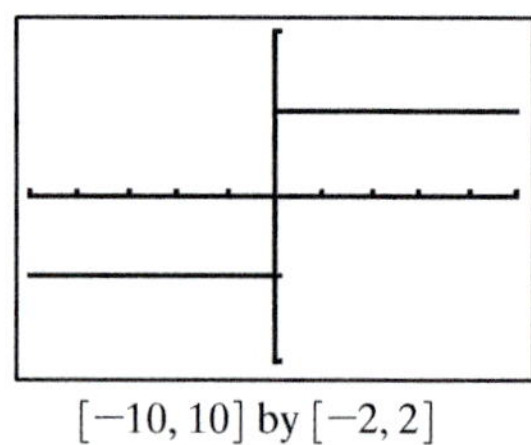

$[-10, 10]$ by $[-2, 2]$

25. Local maxima at $(-1, 4)$ and $(5, 5)$, local minimum at $(2, 2)$. The function increases on $(-\infty, -1]$, decreases on $[-1, 2]$, increases on $[2, 5]$, and decreases on $[5, \infty)$.

27. $(-1, 3)$ and $(3, 3)$ are neither. $(1, 5)$ is a local maximum, and $(5, 1)$ is a local minimum. The function increases on $(-\infty, 1]$, decreases on $[1, 5]$, and increases on $[5, \infty)$.

29. Decreasing on $(-\infty, -2]$; increasing on $[-2, \infty)$

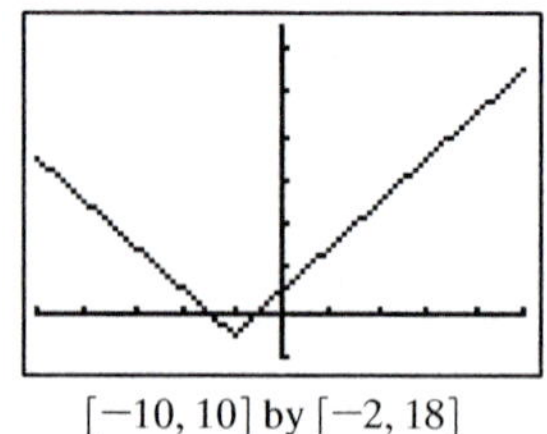

$[-10, 10]$ by $[-2, 18]$

31. Decreasing on $(-\infty, -2]$; constant on $[-2, 1]$; increasing on $[1, \infty)$

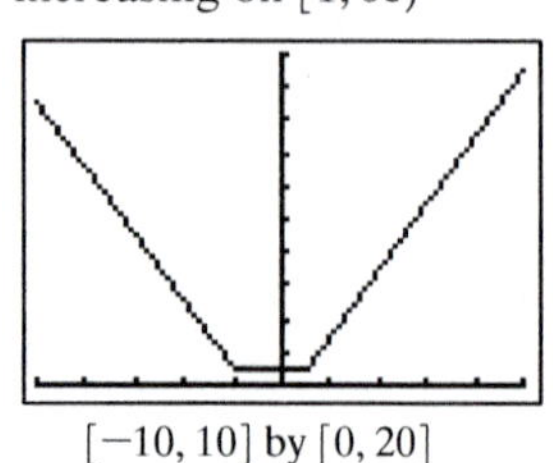

$[-10, 10]$ by $[0, 20]$

33. Increasing on $(-\infty, 1]$; decreasing on $[1, \infty)$

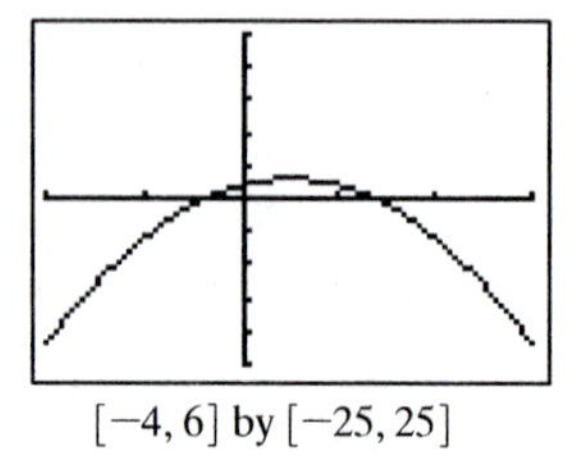

$[-4, 6]$ by $[-25, 25]$

35. Constant functions are always bounded.

37. $2^x > 0$ for all x, so y is bounded below by $y = 0$.

39. Since $y = \sqrt{1 - x^2}$ is always positive, we know that $y \geq 0$ for all x. We must also check for an upper bound:

$$x^2 > 0$$
$$-x^2 < 0$$
$$1 - x^2 < 1$$
$$\sqrt{1 - x^2} < \sqrt{1}$$
$$\sqrt{1 - x^2} < 1$$

Thus, y is bounded.

41. f has a local minimum when $x = 0.5$, where $y = 3.75$. It has no maximum.

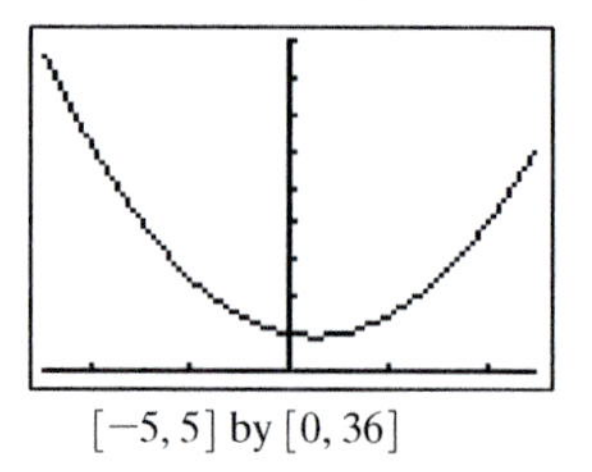

$[-5, 5]$ by $[0, 36]$

43. Local minimum: $y \approx -4.09$ at $x \approx -0.82$.
Local maximum: $y \approx -1.91$ at $x \approx 0.82$.

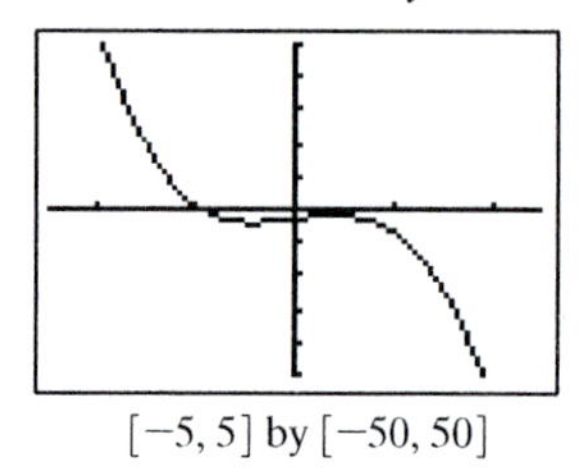

$[-5, 5]$ by $[-50, 50]$

45. Local maximum: $y \approx 9.16$ at $x \approx -3.20$.
Local minima: $y = 0$ at $x = 0$ and $y = 0$ at $x = -4$.

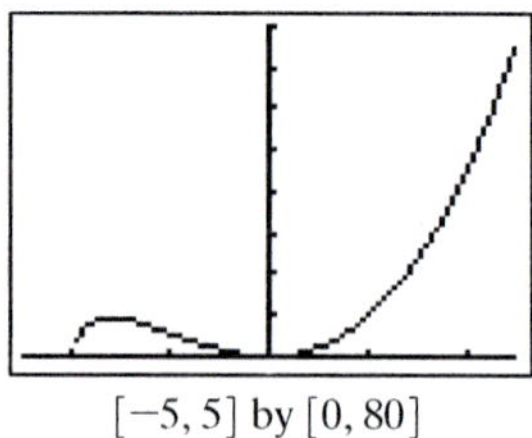

$[-5, 5]$ by $[0, 80]$

47. Even: $f(-x) = 2(-x)^4 = 2x^4 = f(x)$

49. Even: $f(-x) = \sqrt{(-x)^2 + 2} = \sqrt{x^2 + 2} = f(x)$

51. Neither: $f(-x) = -(-x)^2 + 0.03(-x) + 5 = -x^2 - 0.03x + 5$, which is neither $f(x)$ nor $-f(x)$.

53. Odd: $g(-x) = 2(-x)^3 - 3(-x)$
$= -2x^3 + 3x = -g(x)$

55. The quotient $\dfrac{x}{x - 1}$ is undefined at $x = 1$, indicating that $x = 1$ is a vertical asymptote. Similarly, $\lim_{x \to \infty} \dfrac{x}{x - 1} = 1$, indicating a horizontal asymptote at $y = 1$. The graph confirms these.

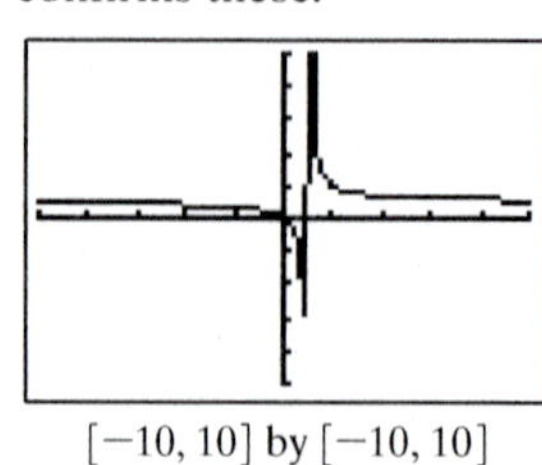

$[-10, 10]$ by $[-10, 10]$

57. The quotient $\frac{x+2}{3-x}$ is undefined at $x = 3$, indicating a possible vertical asymptote at $x = 3$. Similarly, $\lim_{x\to\infty} \frac{x+2}{3-x} = -1$, indicating a possible horizontal asymptote at $y = -1$. The graph confirms these asymptotes.

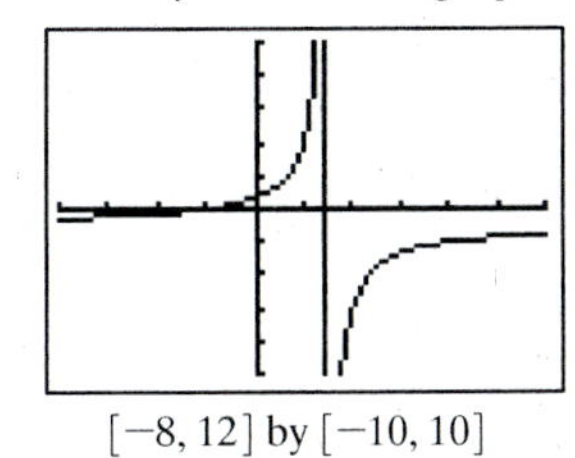

$[-8, 12]$ by $[-10, 10]$

59. The quotient $\frac{x^2+2}{x^2-1}$ is undefined at $x = 1$ and $x = -1$. So we expect two vertical asymptotes. Similarly, the $\lim_{x\to\infty} \frac{x^2+2}{x^2-1} = 1$, so we expect a horizontal asymptote at $y = 1$. The graph confirms these asymptotes.

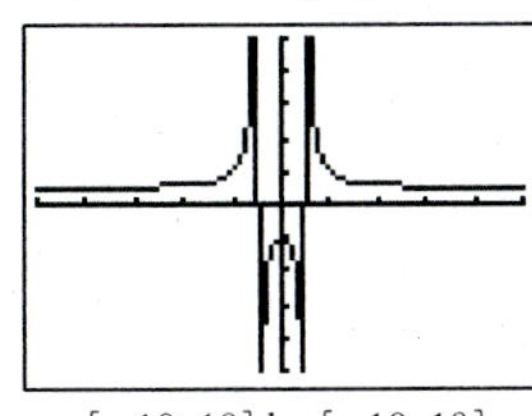

$[-10, 10]$ by $[-10, 10]$

61. The quotient $\frac{4x-4}{x^3-8}$ does not exist at $x = 2$, so we expect a vertical asymptote there. Similarly, $\lim_{x\to\infty} \frac{4x-4}{x^3+8} = 0$, so we expect a horizontal asymptote at $y = 0$. The graph confirms these asymptotes.

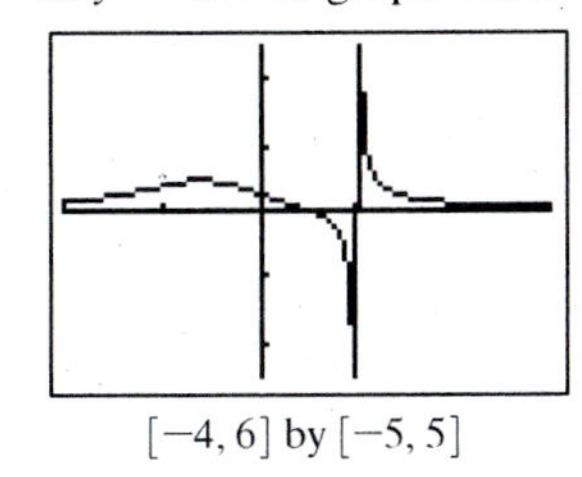

$[-4, 6]$ by $[-5, 5]$

63. The denominator is zero when $x = -\frac{1}{2}$, so there is a vertical asymptote at $x = -\frac{1}{2}$. When x is very large, $\frac{x+2}{2x+1}$ behaves much like $\frac{x}{2x} = \frac{1}{2}$, so there is a horizontal asymptote at $y = \frac{1}{2}$. The graph matching this description is (b).

65. The denominator cannot equal zero, so there is no vertical asymptote. When x is very large, $\frac{x+2}{2x^2+1}$ behaves much like $\frac{x}{2x^2} = \frac{1}{2x}$, which for large x is close to zero. So there is a horizontal asymptote at $y = 0$. The graph matching this description is (a).

67. **(a)** Since, $\lim_{x\to\infty} \frac{x}{x^2-1} = 0$, we expect a horizontal asymptote at $y = 0$. To find where our function crosses $y = 0$, we solve the equation

$$\frac{x}{x^2-1} = 0$$
$$x = 0 \cdot (x^2 - 1)$$
$$x = 0$$

The graph confirms that $f(x)$ crosses the horizontal asymptote at $(0, 0)$.

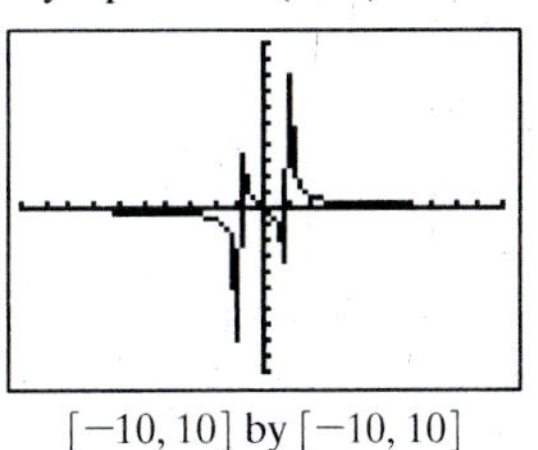

$[-10, 10]$ by $[-10, 10]$

(b) Since $\lim_{x\to\infty} \frac{x}{x^2+1} = 0$, we expect a horizontal asymptote at $y = 0$. To find where our function crosses $y = 0$, we solve the equation:

$$\frac{x}{x^2+1} = 0$$
$$x = 0 \cdot (x^2 + 1)$$
$$x = 0$$

The graph confirms that $g(x)$ crosses the horizontal asymptote at $(0, 0)$.

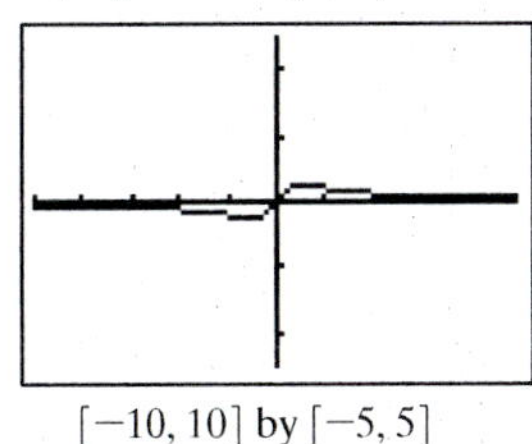

$[-10, 10]$ by $[-5, 5]$

(c) Since $\lim_{x\to\infty} \frac{x^2}{x^3+1} = 0$, we expect a horizontal asymptote at $y = 0$. To find where $h(x)$ crosses $y = 0$, we solve the equation

$$\frac{x^2}{x^3+1} = 0$$
$$x^2 = 0 \cdot (x^3 + 1)$$
$$x^2 = 0$$
$$x = 0$$

The graph confirms that $h(x)$ intersects the horizontal asymptote at $(0, 0)$.

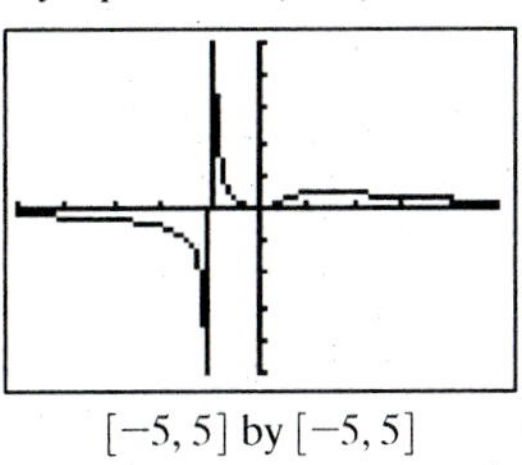

$[-5, 5]$ by $[-5, 5]$

69. **(a)** The vertical asymptote is $x = 0$, and this function is undefined at $x = 0$ (because a denominator can't be zero).

(b)

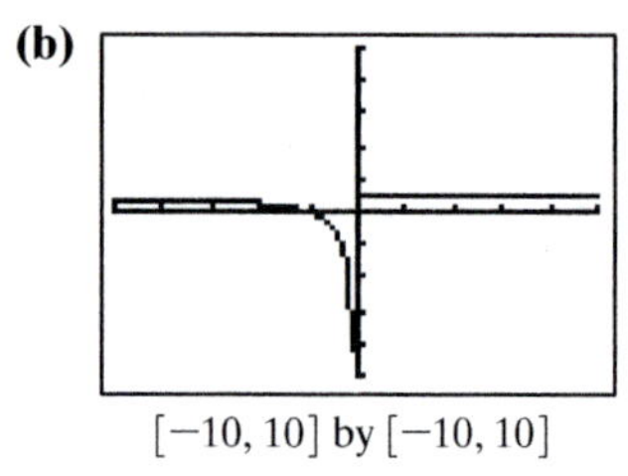

$[-10, 10]$ by $[-10, 10]$

Add the point $(0, 0)$.

(c) Yes. It passes the vertical line test.

71. True. This is what it means for a set of points to be the graph of a function.

73. Temperature is a continuous variable, whereas the other quantities all vary in steps. The answer is B.

75. Air pressure drops with increasing height. All the other functions either steadily increase or else go both up and down. The answer is C.

77. (a)

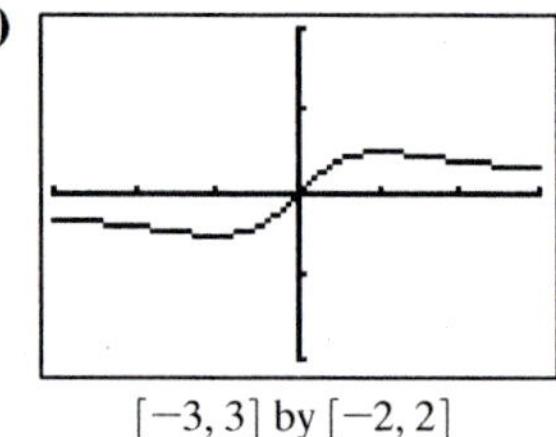

$[-3, 3]$ by $[-2, 2]$

$k = 1$

(b) $\frac{x}{1 + x^2} < 1 \Leftrightarrow x < 1 + x^2 \Leftrightarrow x^2 - x + 1 > 0$

But the discriminant of $x^2 - x + 1$ is negative (-3), so the graph never crosses the x-axis on the interval $(0, \infty)$.

(c) $k = -1$

(d) $\frac{x}{1 + x^2} > -1 \Leftrightarrow x > -1 - x^2 \Leftrightarrow x^2 + x + 1 > 0$

But the discriminant of $x^2 + x + 1$ is negative (-3), so the graph never crosses the x-axis on the interval $(-\infty, 0)$.

79. One possible graph:

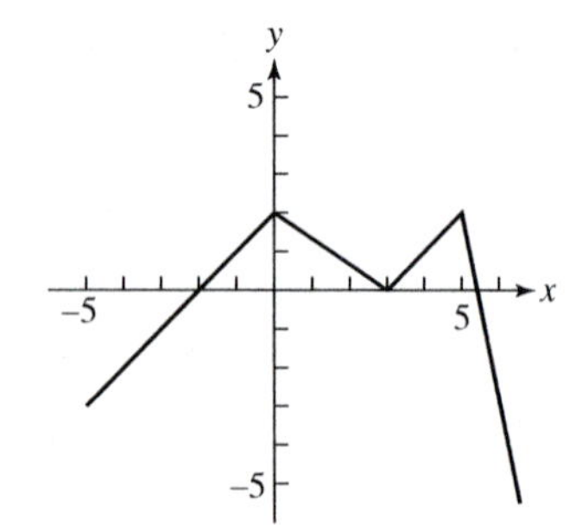

81. One possible graph:

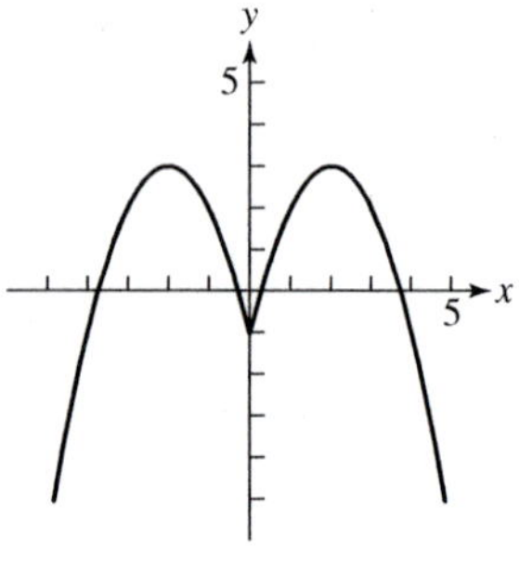

83. (a)

$$x^2 > 0$$
$$-0.8x^2 < 0$$
$$2 - 0.8x^2 < 2$$

$f(x)$ is bounded above by $y = 2$. To determine if $y = 2$ is in the range, we must solve the equation for x: $2 = 2 - 0.8x^2$

$$0 = -0.8x^2$$
$$0 = x^2$$
$$0 = x$$

Since $f(x)$ exists at $x = 0$, $y = 2$ is in the range.

(b) $\lim_{x\to\infty} \frac{3x^2}{3 + x^2} = \lim_{x\to\infty} \frac{3x^2}{x^2} = \lim_{x\to\infty} 3 = 3$. Thus, $g(x)$ is bounded by $y = 3$. However, when we solve for x, we get

$$3 = \frac{3x^2}{3 + x^2}$$
$$3(3 + x^2) = 3x^2$$
$$9 + 3x^2 = 3x^2$$
$$9 = 0$$

Since $9 \neq 0$, $y = 3$ is not in the range of $g(x)$.

(c) $h(x)$ is not bounded above.

(d) For all values of x, we know that $\sin(x)$ is bounded above by $y = 1$. Similarly, $2\sin(x)$ is bounded above by $y = 2 \cdot 1 = 2$. It is in the range.

(e) $\lim_{x\to\infty} \frac{4x}{x^2 + 2x + 1} = \lim_{x\to\infty} \frac{4x}{(x + 1)^2} =$

$\lim_{x\to\infty} 4\left(\frac{x}{x + 1}\right)\left(\frac{1}{x + 1}\right) = \lim_{x\to\infty} \frac{4}{x + 1}$

(since $x + 1 \approx x$ for very large x) $= 0$.

[Similarly, $\lim_{x\to-\infty} \frac{4x}{x^2 + x + 1} = 0$] As a result, we know that $g(x)$ is bounded by $y = 0$ as x goes to ∞ and $-\infty$.

However, $g(x) > 0$ for all $x > 0$ (since $(x + 1)^2 > 0$ always and $4x > 0$ when $x > 0$), so we must check points near $x = 0$ to determine where the function is at its maximum. [Since $g(x) < 0$ for all $x < 0$ (since $(x + 1)^2 > 0$ always and $4x < 0$ when $x < 0$) we can ignore those values of x since we are concerned only with the upper bound of $g(x)$.] Examining our graph, we see that $g(x)$ has an upper bound at $y = 1$, which occurs when $x = 1$. The least upper bound of $g(x) = 1$, and it is in the range of $g(x)$.

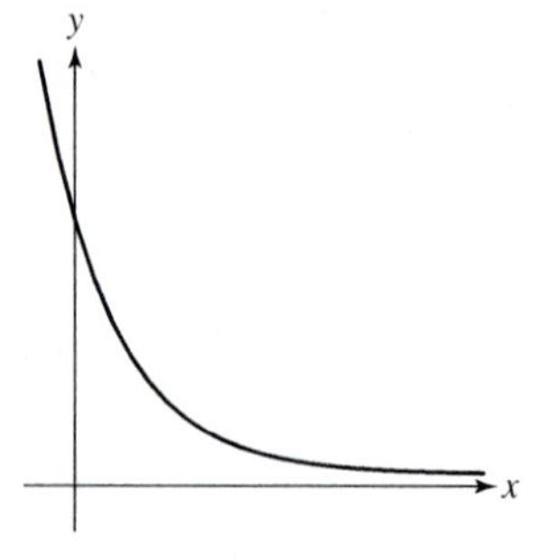

85. Since f is odd, $f(-x) = -f(x)$ for all x. In particular, $f(-0) = -f(0)$. This is equivalent to saying that $f(0) = -f(0)$, and the only number which equals its opposite is 0. Therefore, $f(0) = 0$, which means the graph must pass through the origin.

Section 1.3 Twelve Basic Functions

Exploration 1

1. The graphs of $f(x) = \frac{1}{x}$ and $f(x) = \ln x$ have vertical asymptotes at $x = 0$.
2. The graph of $g(x) = \frac{1}{x} + \ln x$ (shown below) does have a vertical asymptote at $x = 0$.

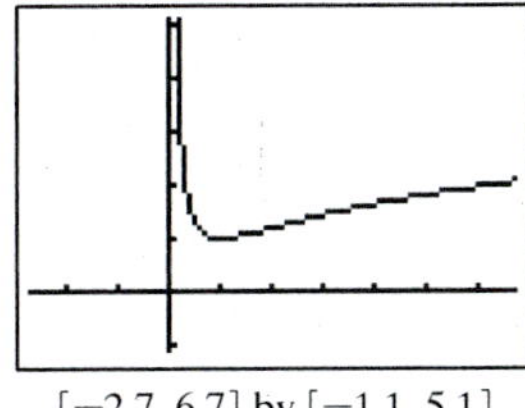

$[-2.7, 6.7]$ by $[-1.1, 5.1]$

3. The graphs of $f(x) = \frac{1}{x}$, $f(x) = e^x$, and $f(x) = \frac{1}{1 + e^{-x}}$ have horizontal asymptotes at $y = 0$.
4. The graph of $g(x) = \frac{1}{x} + e^x$ (shown below) does have a horizontal asymptote at $y = 0$.

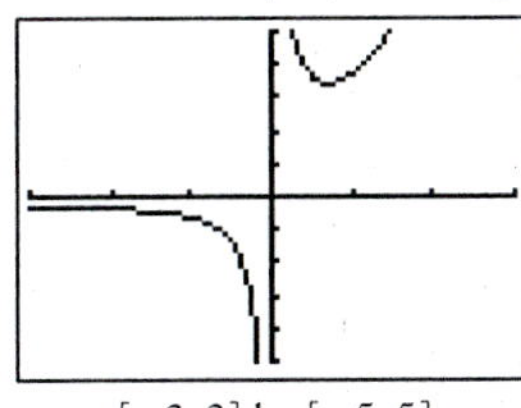

$[-3, 3]$ by $[-5, 5]$

5.

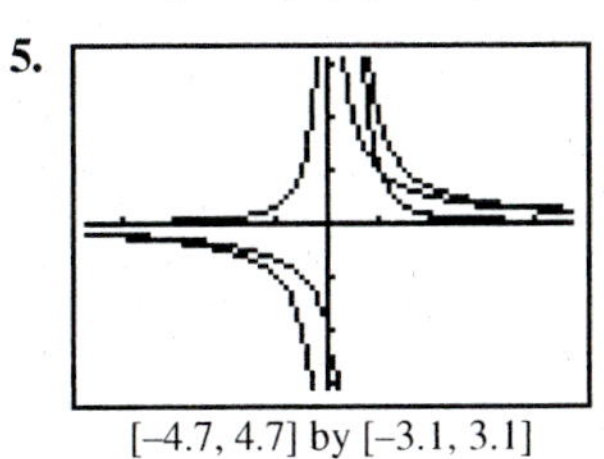

$[-4.7, 4.7]$ by $[-3.1, 3.1]$

Both $f(x) = \frac{1}{x}$ and $g(x) = \frac{1}{2x^2 - x} = \frac{1}{x(2x - 1)}$ have vertical asymptotes at $x = 0$, but $h(x) = f(x) + g(x)$ does not; it has a removable discontinuity.

Quick Review 1.3

1. 59.34
3. $7 - \pi$
5. 0
7. 3
9. -4

Section 1.3 Exercises

1. $y = x^3 + 1$; (e)
3. $y = -\sqrt{x}$; (j)
5. $y = -x$; (i)
7. $y = \text{int}(x + 1)$; (k)
9. $y = (x + 2)^3$; (d)
11. $2 - \frac{4}{1 + e^{-x}}$; (l)
13. Exercise 8
15. Exercises 7, 8
17. Exercises 2, 4, 6, 10, 11, 12
19. $y = x$, $y = x^3$, $y = \frac{1}{x}$, $y = \sin x$
21. $y = x^2$, $y = \frac{1}{x}$, $y = |x|$
23. $y = \frac{1}{x}$, $y = e^x$, $y = \frac{1}{1 + e^{-x}}$
25. $y = \frac{1}{x}$, $y = \sin x$, $y = \cos x$, $y = \frac{1}{1 + e^{-x}}$
27. $y = x$, $y = x^3$, $y = \frac{1}{x}$, $y = \sin x$
29. Domain: All reals
    Range: $[-5, \infty)$

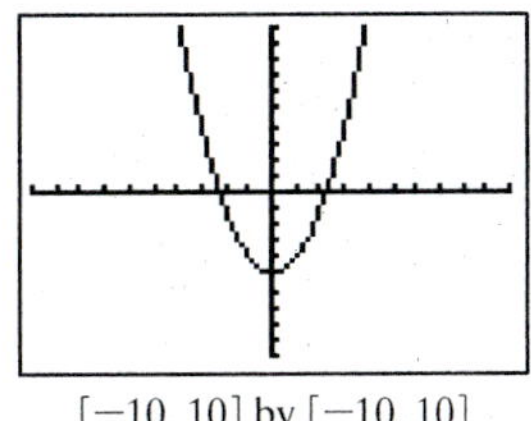

$[-10, 10]$ by $[-10, 10]$

31. Domain: $(-6, \infty)$
    Range: All reals

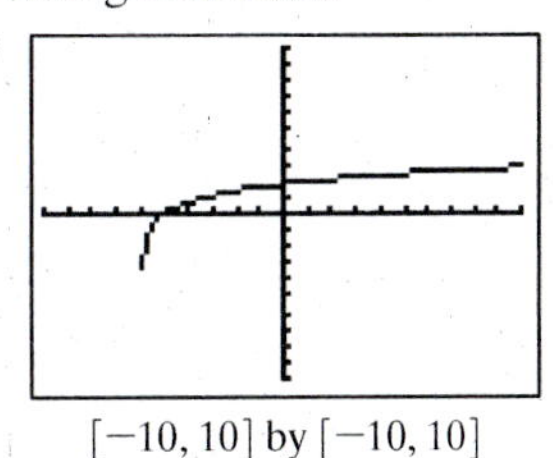

$[-10, 10]$ by $[-10, 10]$

33. Domain: All reals
    Range: All integers

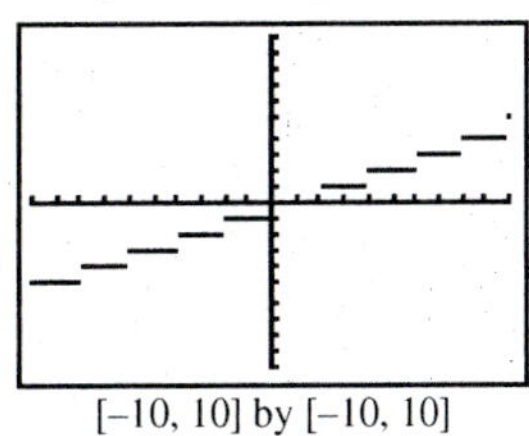

$[-10, 10]$ by $[-10, 10]$

35.

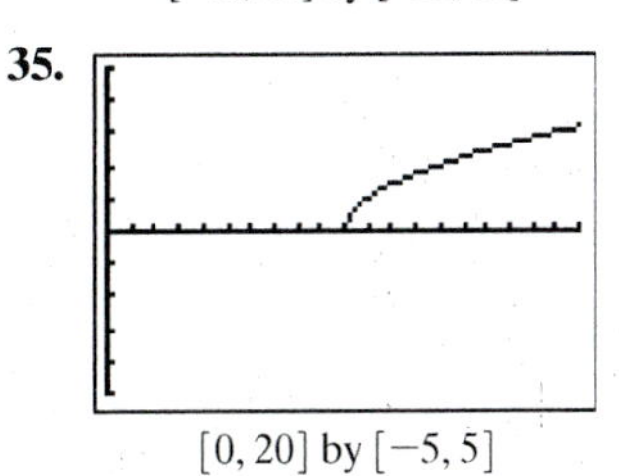

$[0, 20]$ by $[-5, 5]$

(a) $r(x)$ is increasing on $[10, \infty)$.
(b) $r(x)$ is neither odd nor even.
(c) The one extreme is a minimum value of 0 at $x = 10$.
(d) $r(x) = \sqrt{x - 10}$ is the square root function, shifted 10 units right.

37.

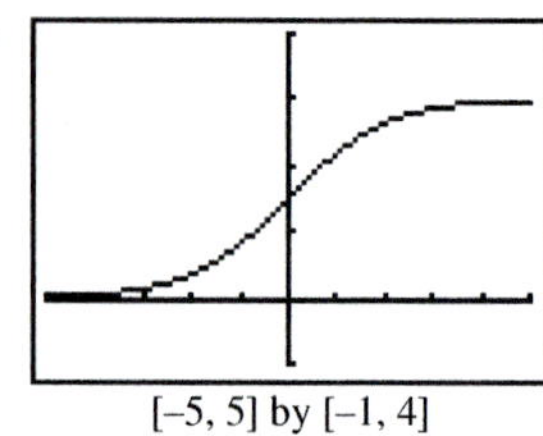

$[-5, 5]$ by $[-1, 4]$

(a) $f(x)$ is increasing on $(-\infty, \infty)$.

(b) $f(x)$ is neither odd nor even.

(c) There are no extrema.

(d) $f(x) = \dfrac{3}{1 + e^{-x}}$ is the logistic function, $\dfrac{1}{1 + e^{-x}}$, stretched vertically by a factor of 3.

39.

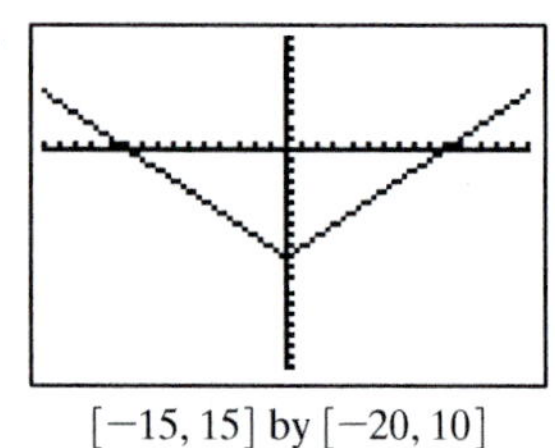

$[-15, 15]$ by $[-20, 10]$

(a) $h(x)$ is increasing on $[0, \infty)$ and decreasing on $(-\infty, 0]$.

(b) $h(x)$ is even, because it is symmetric about the y-axis.

(c) The one extremum is a minimum value of -10 at $x = 0$.

(d) $h(x) = |x| - 10$ is the absolute value function, $|x|$, shifted 10 units down.

41.

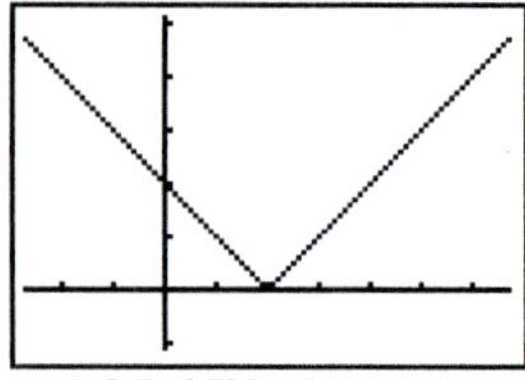

$[-2.7, 6.7]$ by $[-1.1, 5.1]$

(a) $s(x)$ is increasing on $[2, \infty)$ and decreasing on $(-\infty, 2]$.

(b) $s(x)$ is neither odd nor even.

(c) The one extremum is a minimum value of 0 at $x = 2$.

(d) $s(x) = |x - 2|$ is the absolute value function, $|x|$, shifted 2 units to the right.

43. The end behavior approaches the horizontal asymptotes $y = 2$ and $y = -2$.

45.

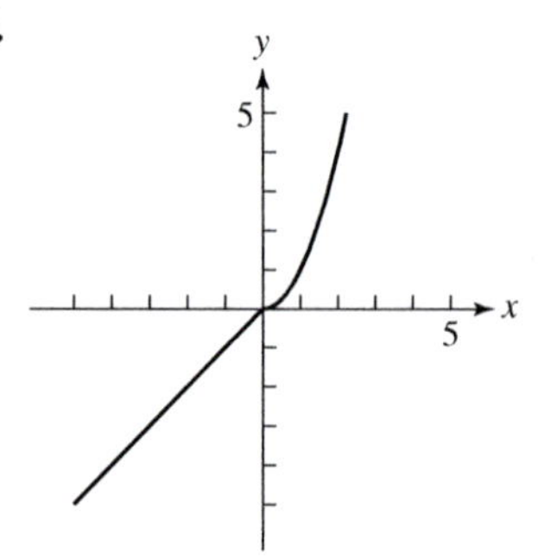

There are no points of discontinuity.

47.

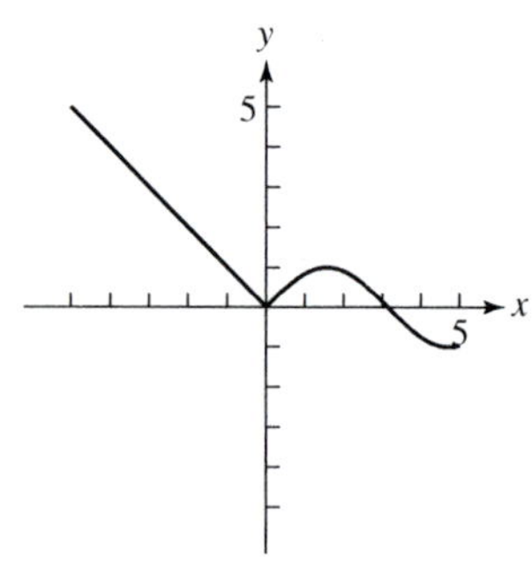

There are no points of discontinuity.

49.

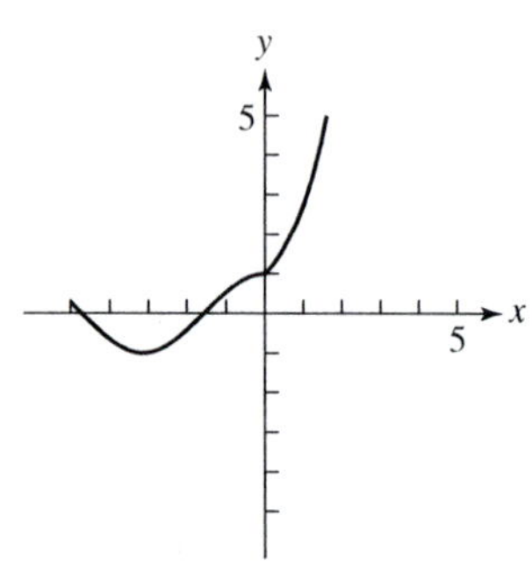

There are no points of discontinuity.

51.

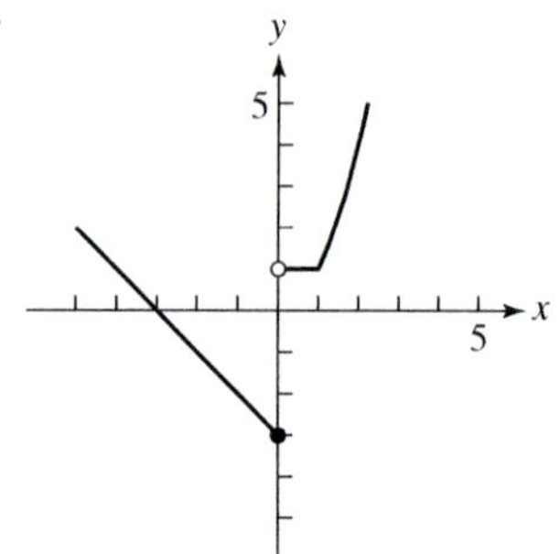

There is a point of discontinuity at $x = 0$.

53. (a)

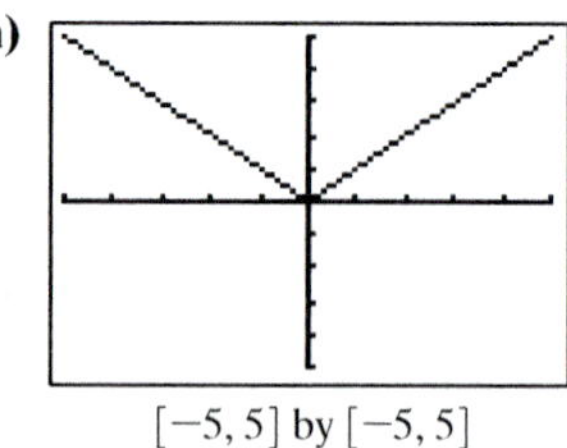

$[-5, 5]$ by $[-5, 5]$

This is $g(x) = |x|$.

(b) Squaring x and taking the (positive) square root has the same effect as the absolute value function.

$f(x) = \sqrt{x^2} = \sqrt{|x|^2} = |x| = g(x)$

55. (a)

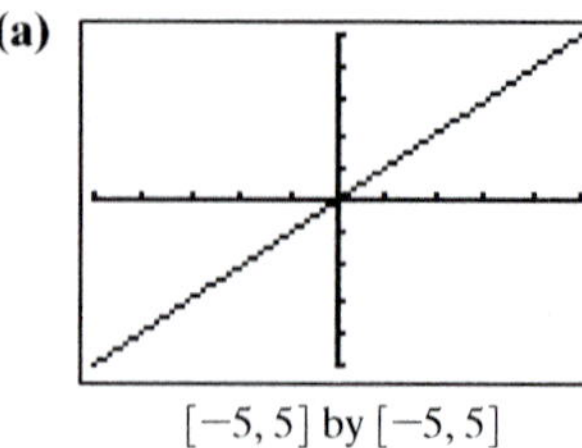

$[-5, 5]$ by $[-5, 5]$

This is the function $f(x) = x$.

(b) The fact that $\ln(e^x) = x$ shows that the natural logarithm function takes on arbitrarily large values. In particular, it takes on the value L when $x = e^L$.

57. The Greatest Integer Function $f(x) = \text{int}(x)$

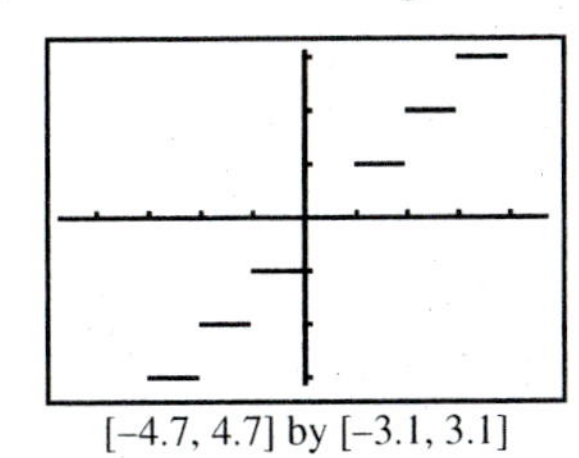

[–4.7, 4.7] by [–3.1, 3.1]

Domain: all real numbers
Range: all integers
Continuity: There is a discontinuity at each integer value of x.
Increasing/decreasing behavior: constant on intervals of the form $[k, k + 1)$, where k is an integer
Symmetry: none
Boundedness: not bounded
Local extrema: every non-integer is both a local minimum and local maximum
Horizontal asymptotes: none
Vertical asymptotes: none
End behavior: $\text{int}(x) \to -\infty$ as $x \to -\infty$ and $\text{int}(x) \to \infty$ as $x \to \infty$.

59. True. The asymptotes are $x = 0$ and $x = 1$.

61. $3 < 3 + \dfrac{1}{1 + e^{-x}} < 4$. The answer is D.

63. The answer is E. The others all have either a restricted domain or intervals where the function is decreasing or constant.

65. **(a)** A product of two odd functions is even.

(b) A product of two even functions is even.

(c) A product of an odd function and an even function is odd.

67. **(a)** Pepperoni count ought to be proportional to the area of the pizza, which is proportional to the square of the radius.

(b) $12 = k(4)^2$

$k = \dfrac{12}{16} = \dfrac{3}{4} = 0.75$

(c) Yes, very well.

(d) The fact that the pepperoni count fits the expected quadratic model so perfectly suggests that the pizzeria uses such a chart. If repeated observations produced the same results, there would be little doubt.

69. **(a)** At $x = 0$, $\dfrac{1}{x}$ does not exist, $e^x = 1$, $\ln x$ is not defined, $\cos x = 1$, and $\dfrac{1}{1 + e^{-x}} = 1$.

(b) for $f(x) = x$, $f(x + y) = x + y = f(x) + f(y)$

(c) for $f(x) = e^x$, $f(xy) = e^{xy} = e^x e^y = f(x) \cdot f(y)$

(d) for $f(x) = \ln x$, $f(x + y) = \ln(xy) = \ln(x) + \ln(y) = f(x) + f(y)$

(e) The odd functions: x, x^3, $\dfrac{1}{x}$, $\sin x$

■ Section 1.4 Building Functions from Functions

Exploration 1

If $f = 2x - 3$ and $g = \dfrac{x + 3}{2}$, then

$$f \circ g = 2\left(\frac{x+3}{2}\right) - 3 = x + 3 - 3 = x.$$

If $f = |2x + 4|$ and $g = \dfrac{(x - 2)(x + 2)}{2}$,

then $f \circ g = 2\left(\dfrac{(x-2)(x+2)}{2}\right) + 4$
$= (x - 2)(x + 2) + 4 = x^2 - 4 + 4 = x^2$.

If $f = \sqrt{x}$ and $g = x^2$, then $f \circ g = \sqrt{x^2} = |x|$. Note, we use the absolute value of x because g is defined for $-\infty < x < +\infty$, while f is defined only for positive values of x. The absolute value function is always positive.

If $f = x^5$ and $g = x^{0.6}$, then $f \circ g = (x^{0.6})^5 = x^3$.

If $f = x - 3$ and $g = \ln(e^3 x)$, then $f \circ g = \ln(e^3 x) - 3 = \ln(e^3) + \ln x - 3 = 3 \ln e + \ln x - 3 = 3 + \ln x - 3 = \ln x$.

If $f = 2 \sin x \cos x$ and $g = \dfrac{x}{2}$, then $f \circ g = 2 \sin \dfrac{x}{2} \cos \dfrac{x}{2} = \sin\left(2\left(\dfrac{x}{2}\right)\right) = \sin x$. This is the double angle formula (*see Section 5.4*). You can see this graphically.

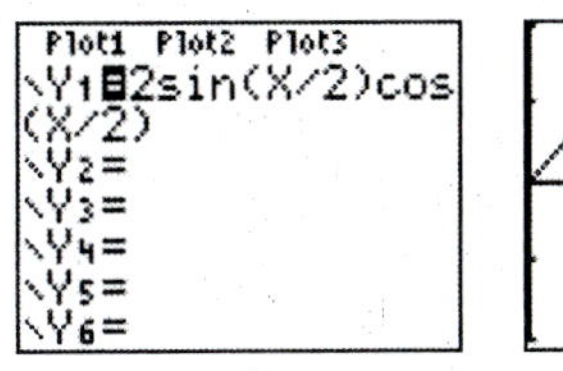

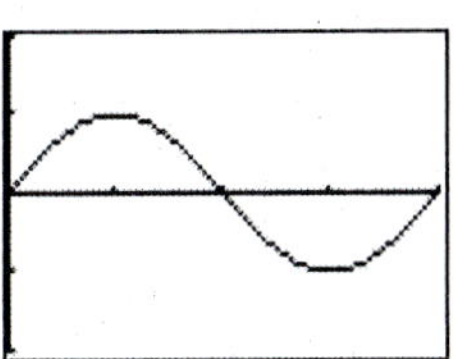

[0, 2π] by [–2, 2]

If $f = 1 - 2x^2$ and $g = \sin\left(\dfrac{x}{2}\right)$,

then $f \circ g = 1 - 2\left(\sin^2\left(\dfrac{x}{2}\right)\right) = \cos\left(2\left(\dfrac{x}{2}\right)\right) = \cos x$.

(The double angle formula for $\cos 2x$ is $\cos 2x = \cos^2 x - \sin^2 x = (1 - \sin^2 x) - \sin^2 x = 1 - 2\sin^2 x$. *See Section 5.3.*) This can be seen graphically:

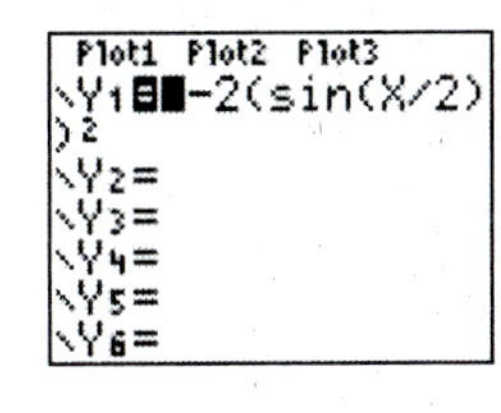

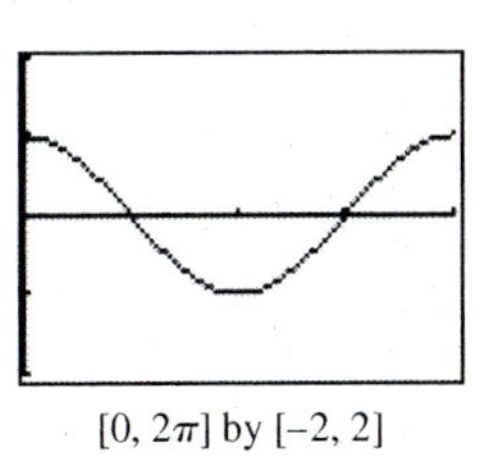

[0, 2π] by [–2, 2]

f	g	$f \circ g$
$2x - 3$	$\frac{x+3}{2}$	x
$\|2x + 4\|$	$\frac{(x-2)(x+2)}{2}$	x^2
$\sqrt{x}$	x^2	$\|x\|$
x^5	$x^{0.6}$	x^3
$x - 3$	$\ln(e^3 x)$	$\ln x$
$2 \sin x \cos x$	$\frac{x}{2}$	$\sin x$
$1 - 2x^2$	$\sin\left(\frac{x}{2}\right)$	$\cos x$

Quick Review 1.4

1. $(-\infty, -3) \cup (-3, \infty)$

3. $(-\infty, 5]$

5. $[1, \infty)$

7. $(-\infty, \infty)$

9. $(-1, 1)$

Section 1.4 Exercises

1. $(f + g)(x) = 2x - 1 + x^2$; $(f - g)(x) = 2x - 1 - x^2$; $(fg)(x) = (2x - 1)(x^2) = 2x^3 - x^2$.
There are no restrictions on any of the domains, so all three domains are $(-\infty, \infty)$.

3. $(f + g)(x) = \sqrt{x} + \sin x$; $(f - g)(x) = \sqrt{x} - \sin x$; $(fg)(x) = \sqrt{x} \sin x$.
Domain in each case is $[0, \infty)$. For $\sqrt{x}$, $x \geq 0$. For $\sin x$, $-\infty < x < \infty$.

5. $(f/g)(x) = \dfrac{\sqrt{x+3}}{x^2}$; $x + 3 \geq 0$ and $x \neq 0$, so the domain is $[-3, 0) \cup (0, \infty)$.
$(g/f)(x) = \dfrac{x^2}{\sqrt{x+3}}$; $x + 3 > 0$, so the domain is $(-3, \infty)$.

7. $(f/g)(x) = \dfrac{x^2}{\sqrt{1 - x^2}}$. The denominator cannot be zero and the term under the square root must be positive, so $1 - x^2 > 0$. Therefore, $x^2 < 1$, which means that $-1 < x < 1$. The domain is $(-1, 1)$.
$(g/f)(x) = \dfrac{\sqrt{1 - x^2}}{x^2}$. The term under the square root must be nonnegative, so $1 - x^2 \geq 0$ (or $x^2 \leq 1$). The denominator cannot be zero, so $x \neq 0$. Therefore, $-1 \leq x < 0$ or $0 < x \leq 1$. The domain is $[-1, 0) \cup (0, 1]$.

9.

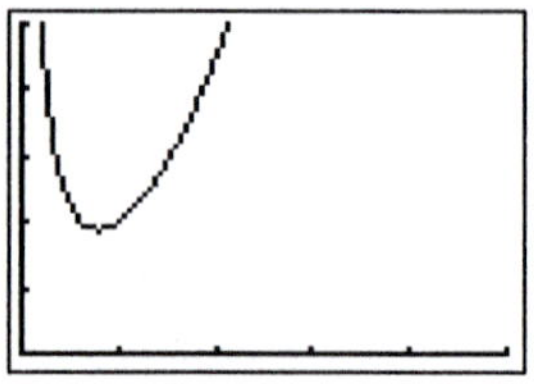

[0, 5] by [0, 5]

11. $(f \circ g)(3) = f(g(3)) = f(4) = 5$;
$(g \circ f)(-2) = g(f(-2)) = g(-7) = -6$

13. $(f \circ g)(3) = f(g(3)) = f(\sqrt{3 + 1}) = f(2) = 2^2 + 4 = 8$;
$(g \circ f)(-2) = g(f(-2)) = g((-2)^2 + 4) = g(8) = \sqrt{8 + 1} = 3$

15. $f(g(x)) = 3(x - 1) + 2 = 3x - 3 + 2 = 3x - 1$. Because both f and g have domain $(-\infty, \infty)$, the domain of $f(g(x))$ is $(-\infty, \infty)$.
$g(f(x)) = (3x + 2) - 1 = 3x + 1$; again, the domain is $(-\infty, \infty)$.

17. $f(g(x)) = (\sqrt{x + 1})^2 - 2 = x + 1 - 2 = x - 1$. The domain of g is $x \geq -1$, while the domain of f is $(-\infty, \infty)$, so the domain of $f(g(x))$ is $x \geq -1$, or $[-1, \infty)$.
$g(f(x)) = \sqrt{(x^2 - 2) + 1} = \sqrt{x^2 - 1}$. The domain of f is $(-\infty, \infty)$, while the domain of g is $[-1, \infty)$, so $g(f(x))$ requires that $f(x) \geq -1$.
This means $x^2 - 2 \geq -1$, or $x^2 \geq 1$, which means $x \leq -1$ or $x \geq 1$. Therefore the domain of $g(f(x))$ is $(-\infty, -1] \cup [1, \infty)$.

19. $f(g(x)) = f(\sqrt{1 - x^2}) = (\sqrt{1 - x^2})^2 = 1 - x^2$; the domain is $[-1, 1]$.
$g(f(x)) = g(x^2) = \sqrt{1 - (x^2)^2} = \sqrt{1 - x^4}$; the domain is $[-1, 1]$.

21. $f(g(x)) = f\left(\dfrac{1}{3x}\right) = \dfrac{1}{2(1/3x)} = \dfrac{1}{2/3x} = \dfrac{3x}{2}$; the domain is $(-\infty, 0) \cup (0, \infty)$.
$g(f(x)) = g\left(\dfrac{1}{2x}\right) = \dfrac{1}{3(1/2x)} = \dfrac{1}{3/2x} = \dfrac{2x}{3}$; the domain is $(-\infty, 0) \cup (0, \infty)$.

23. One possibility: $f(x) = \sqrt{x}$ and $g(x) = x^2 - 5x$

25. One possibility: $f(x) = |x|$ and $g(x) = 3x - 2$

27. One possibility: $f(x) = x^5 - 2$ and $g(x) = x - 3$

29. One possibility: $f(x) = \cos x$ and $g(x) = \sqrt{x}$.

31. $r = 48 + 0.03t$ in., so $V = \frac{4}{3}\pi r^3 = \frac{4}{3}\pi(48 + 0.03t)^3$; when $t = 300$,
$V = \frac{4}{3}\pi(48 + 9)^3 = 246{,}924\pi \approx 775{,}734.6 \text{ in}^3$.

33. The initial area is $(5)(7) = 35 \text{ km}^2$. The new length and width are $l = 5 + 2t$ and $w = 7 + 2t$, so $A = lw = (5 + 2t)(7 + 2t)$. Solve $(7 + 2t)(5 + 2t) = 175$ (5 times its original size), either graphically or algebraically: the positive solution is $t \approx 3.63$ seconds.

35. $3(1) + 4(1) = 3 + 4 = 7 \neq 5$
$3(4) + 4(-2) = 12 - 8 = 4 \neq 5$
$3(3) + 4(-1) = 9 - 4 = 5$
The answer is $(3, -1)$.

37. $y^2 = 25 - x^2$, $y = \sqrt{25 - x^2}$ and $y = -\sqrt{25 - x^2}$

39. $y^2 = x^2 - 25$, $y = \sqrt{x^2 - 25}$ and $y = -\sqrt{x^2 - 25}$

41. $x + |y| = 1 \Rightarrow |y| = -x + 1 \Rightarrow y = -x + 1$ or $y = -(-x + 1)$. $y = 1 - x$ and $y = x - 1$

43. $y^2 = x^2 \Rightarrow y = x$ and $y = -x$ or $y = |x|$ and $y = -|x|$

45. False. If $g(x) = 0$, then $\left(\frac{f}{g}\right)(x)$ is not defined and 0 is not in the domain of $\left(\frac{f}{g}\right)(x)$, even though 0 may be in the domains of both $f(x)$ and $g(x)$.

47. Composition of functions isn't necessarily commutative. The answer is C.

49. $(f \circ f)(x) = f(x^2 + 1) = (x^2 + 1)^2 + 1 = (x^4 + 2x^2 + 1) + 1 = x^4 + 2x^2 + 2$. The answer is E.

51. If $f(x) = e^x$ and $g(x) = 2 \ln x$, then $f(g(x)) = f(2 \ln x) = e^{2 \ln x} = (e^{\ln x})^2 = x^2$. The domain is $(0, \infty)$.

If $f(x) = (x^2 + 2)^2$ and $g(x) = \sqrt{x - 2}$, then $f(g(x)) = f(\sqrt{x - 2}) = ((\sqrt{x - 2})^2 + 2)^2 = (x - 2 + 2)^2 = x^2$. The domain is $[2, \infty)$.

If $f(x) = (x^2 - 2)^2$ and $g(x) = \sqrt{2 - x}$, then $f(g(x)) = f(\sqrt{2 - x}) = ((\sqrt{2 - x})^2 - 2)^2 = (2 - x - 2)^2 = x^2$. The domain is $(-\infty, 2]$.

If $f(x) = \frac{1}{(x - 1)^2}$ and $g(x) = \frac{x + 1}{x}$, then

$$f(g(x)) = f\left(\frac{x + 1}{x}\right) = \frac{1}{\left(\frac{x + 1}{x} - 1\right)^2} = \frac{1}{\left(\frac{x + 1 - x}{x}\right)^2} = \frac{1}{\frac{1}{x^2}} = x^2.$$

The domain is $x \neq 0$.

If $f(x) = x^2 - 2x + 1$ and $g(x) = x + 1$, then $f(g(x)) = f(x + 1) = (x + 1)^2 - 2(x + 1) + 1 = ((x + 1) - 1)^2 = x^2$. The domain is $(-\infty, \infty)$.

If $f(x) = \left(\frac{x + 1}{x}\right)^2$ and $g(x) = \frac{1}{x - 1}$, then

$$f(g(x)) = f\left(\frac{1}{x - 1}\right) = \left(\frac{\frac{1}{x - 1} + 1}{\frac{1}{x - 1}}\right)^2 = \left(\frac{\frac{1 + x - 1}{x - 1}}{\frac{1}{x - 1}}\right)^2 = x^2.$$

The domain is $x \neq 1$.

f	g	D
e^x	$2 \ln x$	$(0, \infty)$
$(x^2 + 2)^2$	$\sqrt{x - 2}$	$[2, \infty)$
$(x^2 - 2)^2$	$\sqrt{2 - x}$	$(-\infty, 2]$
$\frac{1}{(x - 1)^2}$	$\frac{x + 1}{x}$	$x \neq 0$
$x^2 - 2x + 1$	$x + 1$	$(-\infty, \infty)$
$\left(\frac{x + 1}{x}\right)^2$	$\frac{1}{x - 1}$	$x \neq 1$

53. (a) $(f + g)(x) = (g + f)(x) = f(x)$ if $g(x) = 0$.

(b) $(fg)(x) = (gf)(x) = f(x)$ if $g(x) = 1$.

(c) $(f \circ g)(x) = (g \circ f)(x) = f(x)$ if $g(x) = x$.

55. $y^2 + x^2y - 5 = 0$. Using the quadratic formula,

$$y = \frac{-x^2 \pm \sqrt{(x^2)^2 - 4(1)(-5)}}{2} = \frac{-x^2 \pm \sqrt{x^4 + 20}}{2}$$

so, $y_1 = \frac{-x^2 + \sqrt{x^4 + 20}}{2}$

and $y_2 = \frac{-x^2 - \sqrt{x^4 + 20}}{2}$.

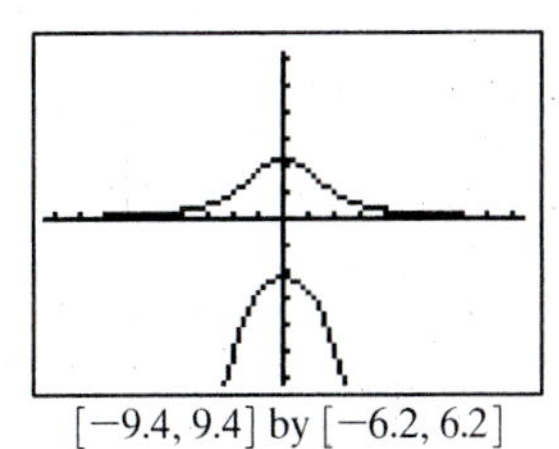

$[-9.4, 9.4]$ by $[-6.2, 6.2]$

■ Section 1.5 Parametric Relations and Inverses

Exploration 1

1. T starts at -4, at the point $(8, -3)$. It stops at $T = 2$, at the point $(8, 3)$. 61 points are computed.
2. The graph is smoother because the plotted points are closer together.
3. The graph is less smooth because the plotted points are further apart. In CONNECT mode, they are connected by straight lines.
4. The smaller the Tstep, the slower the graphing proceeds. This is because the calculator has to compute more X and Y values.
5. The grapher skips directly from the point $(0, -1)$ to the point $(0, 1)$, corresponding to the T-values $T = -2$ and $T = 0$. The two points are connected by a straight line, hidden by the Y-axis.
6. With the Tmin set at -1, the grapher begins at the point $(-1, 0)$, missing the bottom of the curve entirely.
7. Leave everything else the same, but change Tmin back to -4 and Tmax to -1.

Quick Review 1.5

1. $3y = x + 6$, so $y = \frac{x + 6}{3} = \frac{1}{3}x + 2$

3. $y^2 = x - 4$, so $y = \pm\sqrt{x - 4}$

5.
$$\begin{aligned} x(y + 3) &= y - 2 \\ xy + 3x &= y - 2 \\ xy - y &= -3x - 2 \\ y(x - 1) &= -(3x + 2) \\ y &= -\frac{3x + 2}{x - 1} = \frac{3x + 2}{1 - x} \end{aligned}$$

7.
$$\begin{aligned} x(y - 4) &= 2y + 1 \\ xy - 4x &= 2y + 1 \\ xy - 2y &= 4x + 1 \\ y(x - 2) &= 4x + 1 \\ y &= \frac{4x + 1}{x - 2} \end{aligned}$$

9. $x = \sqrt{y+3},\ y \ge -3$ [and $x \ge 0$]

$x^2 = y + 3,\ y \ge -3$, and $x \ge 0$

$y = x^2 - 3,\ y \ge -3$, and $x \ge 0$

Section 1.5 Exercises

1. $x = 3(2) = 6,\ y = 2^2 + 5 = 9$. The answer is $(6, 9)$.

3. $x = 3^3 - 4(3) = 15,\ y = \sqrt{3+1} = 2$. The answer is $(15, 2)$.

5. (a)

t	$(x, y) = (2t, 3t - 1)$
-3	$(-6, -10)$
-2	$(-4, -7)$
-1	$(-2, -4)$
0	$(0, -1)$
1	$(2, 2)$
2	$(4, 5)$
3	$(6, 8)$

(b) $t = \dfrac{x}{2},\ y = 3\left(\dfrac{x}{2}\right) - 1 = 1.5x - 1$. This is a function.

(c)

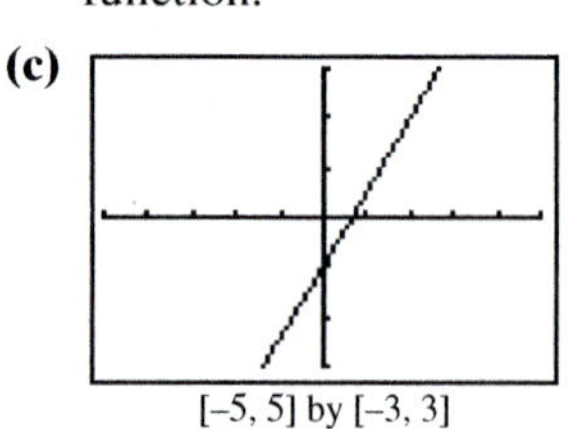

[–5, 5] by [–3, 3]

7. (a)

t	$(x, y) = (t^2, t - 2)$
-3	$(9, -5)$
-2	$(4, -4)$
-1	$(1, -3)$
0	$(0, -2)$
1	$(1, -1)$
2	$(4, 0)$
3	$(9, 1)$

(b) $t = y + 2,\ x = (y + 2)^2$. This is not a function.

(c)

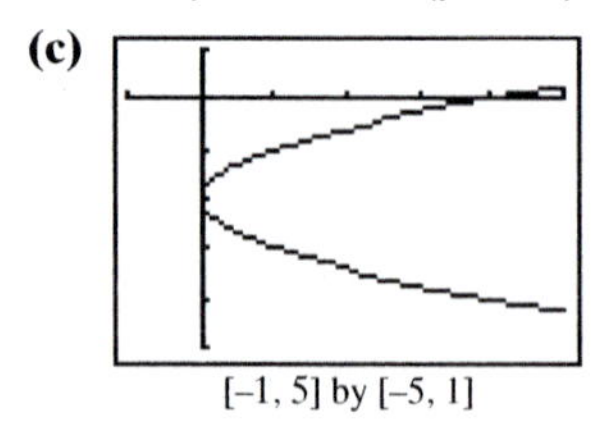

[–1, 5] by [–5, 1]

9. (a) By the vertical line test, the relation is not a function.

(b) By the horizontal line test, the relation's inverse is a function.

11. (a) By the vertical line test, the relation is a function.

(b) By the horizontal line test, the relation's inverse is a function.

13. $y = 3x - 6 \Rightarrow$

$$x = 3y - 6$$
$$3y = x + 6$$
$$f^{-1}(x) = y = \frac{x+6}{3} = \frac{1}{3}x + 2;\ (-\infty, \infty)$$

15. $y = \dfrac{2x-3}{x+1} \Rightarrow$

$$x = \frac{2y-3}{y+1}$$
$$x(y+1) = 2y - 3$$
$$xy + x = 2y - 3$$
$$xy - 2y = -x - 3$$
$$y(x-2) = -(x+3)$$
$$f^{-1}(x) = y = -\frac{x+3}{x-2} = \frac{x+3}{2-x};$$
$$(-\infty, 2) \cup (2, \infty)$$

17. $y = \sqrt{x-3},\ x \ge 3,\ y \ge 0 \Rightarrow$

$$x = \sqrt{y-3},\quad x \ge 0,\ y \ge 3$$
$$x^2 = y - 3,\quad x \ge 0,\ y \ge 3$$
$$f^{-1}(x) = y = x^2 + 3,\quad x \ge 0$$

19. $y = x^3 \Rightarrow$

$$x = y^3$$
$$f^{-1}(x) = y = \sqrt[3]{x};\ (-\infty, \infty)$$

21. $y = \sqrt[3]{x+5} \Rightarrow$

$$x = \sqrt[3]{y+5}$$
$$x^3 = y + 5$$
$$f^{-1}(x) = y = x^3 - 5;\ (-\infty, \infty)$$

23. One-to-one

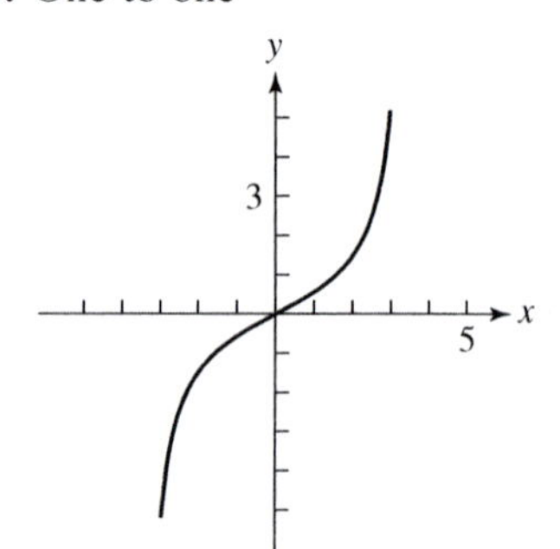

25. One-to-one

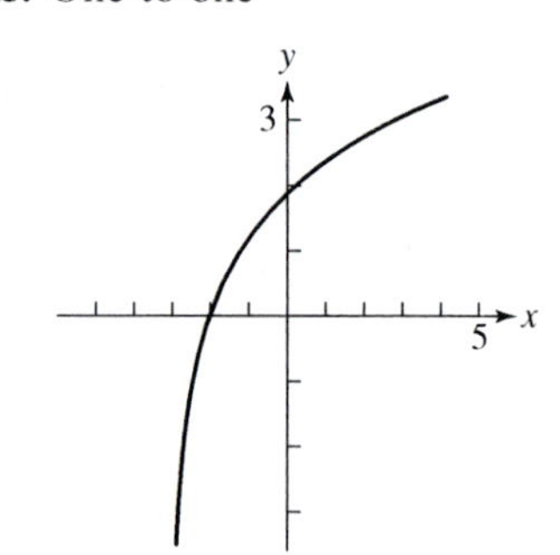

27. $f(g(x)) = 3\left[\dfrac{1}{3}(x+2)\right] - 2 = x + 2 - 2 = x;$

$g(f(x)) = \dfrac{1}{3}[(3x-2)+2] = \dfrac{1}{3}(3x) = x$

29. $f(g(x)) = [(x-1)^{1/3}]^3 + 1 = (x-1)^1 + 1$

$= x - 1 + 1 = x;$

$g(f(x)) = [(x^3+1)-1]^{1/3} = (x^3)^{1/3} = x^1 = x$

31. $f(g(x)) = \dfrac{\dfrac{1}{x-1}+1}{\dfrac{1}{x-1}} = (x-1)\left(\dfrac{1}{x-1}+1\right)$

$= 1 + x - 1 = x;$

$g(f(x)) = \dfrac{1}{\dfrac{x+1}{x}-1} = \left(\dfrac{1}{\dfrac{x+1}{x}-1}\right)\cdot\dfrac{x}{x}$

$= \dfrac{x}{x+1-x} = \dfrac{x}{1} = x$

33. (a) $y = (1.08)(100) = 108$ euros

(b) $x = \dfrac{y}{1.08} = \dfrac{25}{27}y$. This converts euros ($x$) to dollars ($y$).

(c) $x = (0.9259)(48) = \$44.44$

35. $y = e^x$ and $y = \ln x$ are inverses. If we restrict the domain of the function $y = x^2$ to the interval $[0, \infty)$, then the restricted function and $y = \sqrt{x}$ are inverses.

37. $y = |x|$

39. True. All the ordered pairs swap domain and range values.

41. The inverse of the relation given by $x^2y + 5y = 9$ is the relation given by $y^2x + 5x = 9$.

$(1)^2(2) + 5(2) = 2 + 10 = 12 \neq 9$
$(1)^2(-2) + 5(-2) = -2 - 10 = -12 \neq 9$
$(2)^2(-1) + 5(-1) = -4 - 5 = -9 \neq 9$
$(-1)^2(2) + 5(2) = 2 + 10 = 12 \neq 9$
$(-2)^2(1) + 5(1) = 4 + 5 = 9$

The answer is E.

43. $f(x) = 3x - 2$
$y = 3x - 2$

The inverse relation is

$x = 3y - 2$
$x + 2 = 3y$
$\dfrac{x+2}{3} = y$
$f^{-1}(x) = \dfrac{x+2}{3}$

The answer is C.

45. (Answers may vary.)

(a) If the graph of f is unbroken, its reflection in the line $y = x$ will be also.

(b) Both f and its inverse must be one-to-one in order to be inverse functions.

(c) Since f is odd, $(-x, -y)$ is on the graph whenever (x, y) is. This implies that $(-y, -x)$ is on the graph of f^{-1} whenever (x, y) is. That implies that f^{-1} is odd.

(d) Let $y = f(x)$. Since the ratio of Δy to Δx is positive, the ratio of Δx to Δy is positive. Any ratio of Δy to Δx on the graph of f^{-1} is the same as some ratio of Δx to Δy on the graph of f, hence positive. This implies that f^{-1} is increasing.

47. (a) $\dfrac{\Delta y}{\Delta x} = \dfrac{97-70}{88-52} = \dfrac{27}{36} = 0.75$, which gives us the slope of the equation. To find the rest of the equation, we use one of the initial points

$y - 70 = 0.75(x - 52)$
$y = 0.75x - 39 + 70$
$y = 0.75x + 31$

(b) To find the inverse, we substitute y for x and x for y, and then solve for y:

$x = 0.75y + 31$
$x - 31 = 0.75y$
$y = \dfrac{4}{3}(x - 31)$

The inverse function converts scaled scores to raw scores.

49. (a) It does not clear the fence.

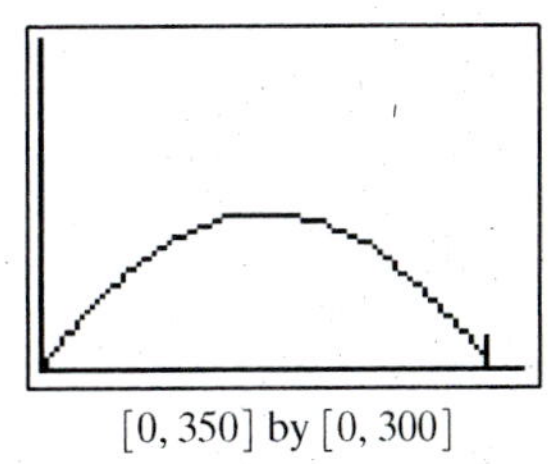

[0, 350] by [0, 300]

(b) It still does not clear the fence.

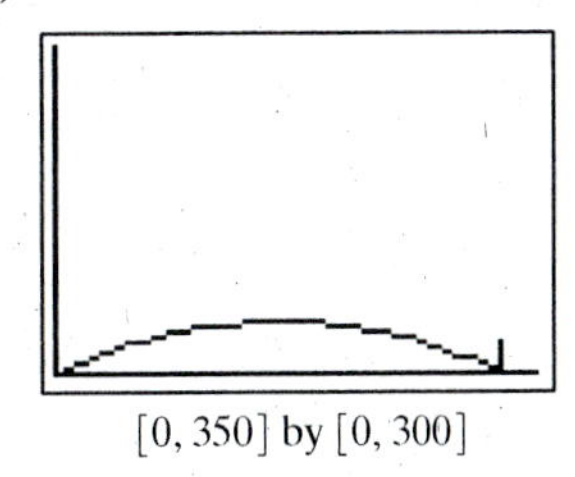

[0, 350] by [0, 300]

(c) Optimal angle is 45°. It clears the fence.

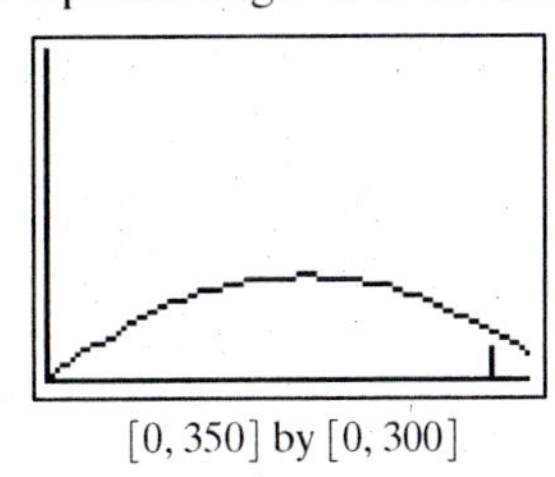

[0, 350] by [0, 300]

51. When $k = 1$, the scaling function is linear. Opinions will vary as to which is the best value of k.

■ Section 1.6 Graphical Transformations

Exploration 1

1.

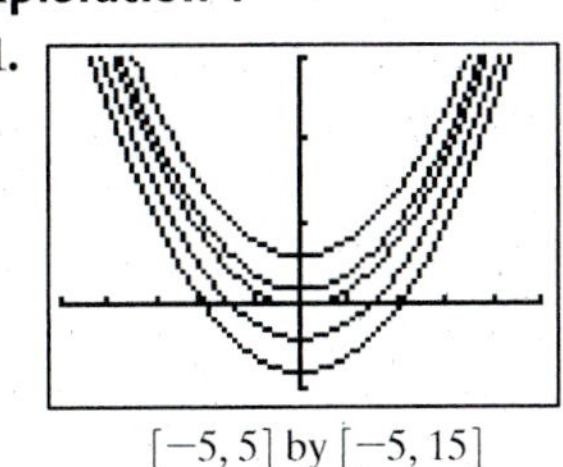

[−5, 5] by [−5, 15]

They raise or lower the parabola along the y-axis.

2.

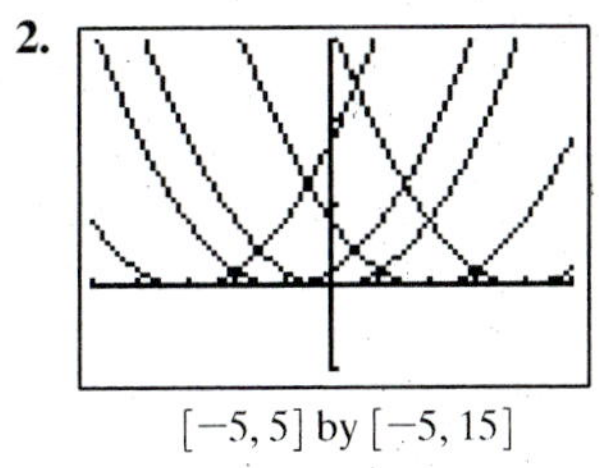

[−5, 5] by [−5, 15]

They move the parabola left or right along the x-axis.

3.

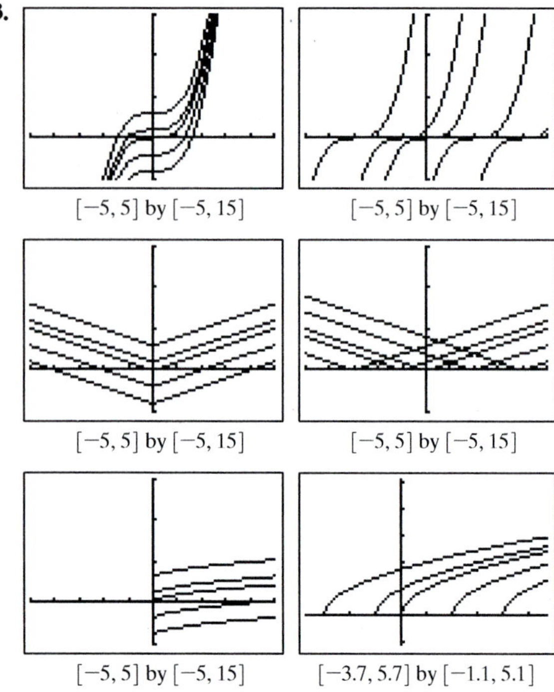

[−5, 5] by [−5, 15] [−5, 5] by [−5, 15]

[−5, 5] by [−5, 15] [−5, 5] by [−5, 15]

[−5, 5] by [−5, 15] [−3.7, 5.7] by [−1.1, 5.1]

Yes

Exploration 2

1.

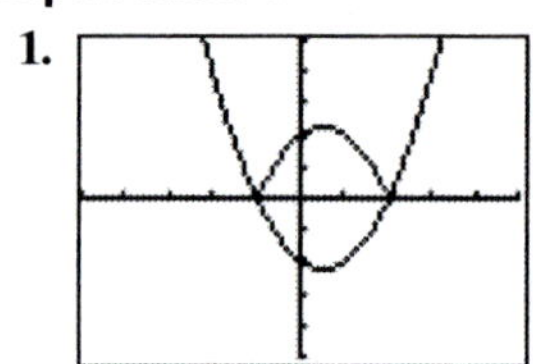

Graph C. Points with positive y-coordinates remain unchanged, while points with negative y-coordinates are reflected across the x-axis.

2.

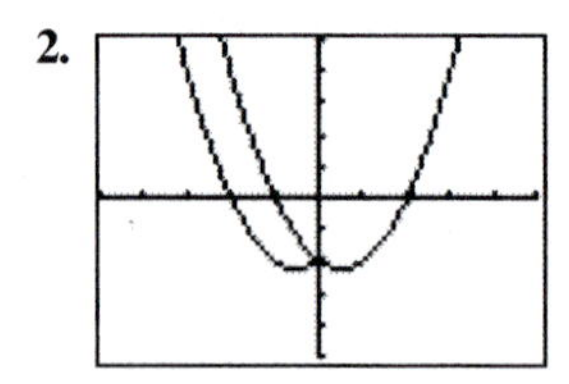

Graph A. Points with positive x-coordinates remain unchanged. Since the new function is even, the graph for negative x-values will be a reflection of the graph for positive x-values.

3.

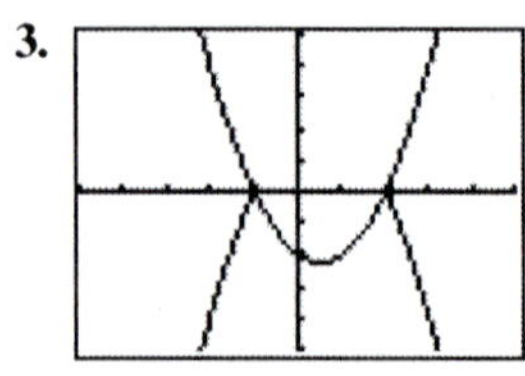

Graph F. The graph will be a reflection across the x-axis of graph C.

4.

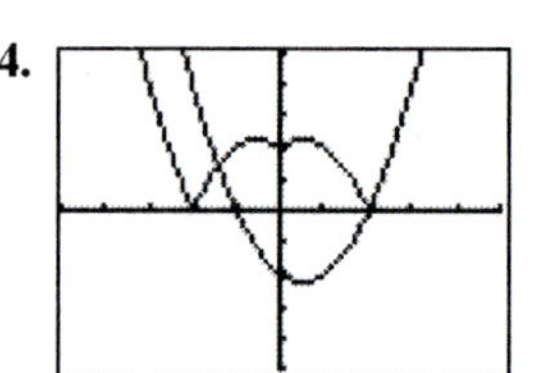

Graph D. The points with negative y-coordinates in graph A are reflected across the x-axis.

Exploration 3

1.

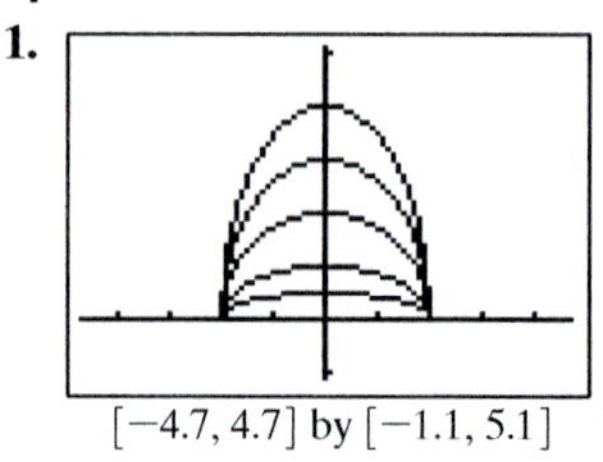

[−4.7, 4.7] by [−1.1, 5.1]

The 1.5 and the 2 stretch the graph vertically; the 0.5 and the 0.25 shrink the graph vertically.

2.

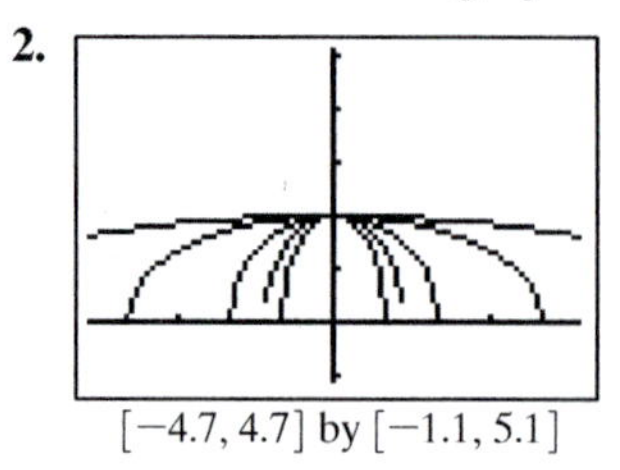

[−4.7, 4.7] by [−1.1, 5.1]

The 1.5 and the 2 shrink the graph horizontally; the 0.5 and the 0.25 stretch the graph horizontally.

Quick Review 1.6

1. $(x + 1)^2$
3. $(x + 6)^2$
5. $(x - 5/2)^2$
7. $x^2 - 4x + 4 + 3x - 6 + 4 = x^2 - x + 2$
9. $(x^3 - 3x^2 + 3x - 1) + 3(x^2 - 2x + 1) - 3x + 3$
   $= x^3 - 3x^2 + 2 + 3x^2 - 6x + 3 = x^3 - 6x + 5$

Section 1.6 Exercises

1. Vertical translation down 3 units
3. Horizontal translation left 4 units
5. Horizontal translation to the right 100 units
7. Horizontal translation to the right 1 unit, and vertical translation up 3 units
9. Reflection across x-axis
11. Reflection across y-axis

For #13–19, recognize $y = c \cdot x^3$ $(c > 0)$ as a vertical stretch (if $c > 1$) or shrink (if $0 < c < 1$) of factor c, and $y = (c \cdot x)^3$ as a horizontal shrink (if $c > 1$) or stretch (if $0 < c < 1$) of factor $1/c$. Note also that $y = (c \cdot x)^3 = c^3x^3$, so that for this function, any horizontal stretch/shrink can be interpreted as an equivalent vertical shrink/stretch (and vice versa).

13. Vertically stretch by 2
15. Horizontally stretch by $1/0.2 = 5$, or vertically shrink by $0.2^3 = 0.008$

17. $g(x) = \sqrt{x - 6} + 2 = f(x - 6)$; starting with f, translate right 6 units to get g.

19. $g(x) = -(x + 4 - 2)^3 = -f(x + 4)$; starting with f, translate left 4 units, and reflect across the x-axis to get g.

21.

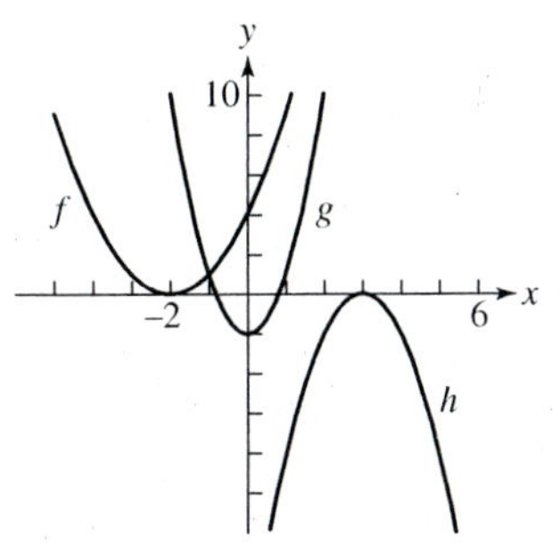

23.

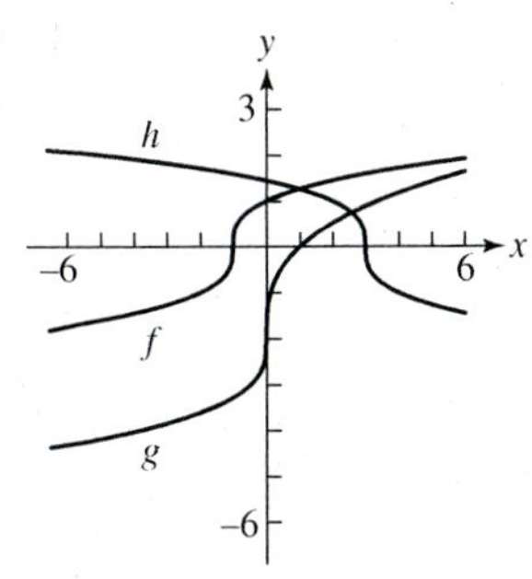

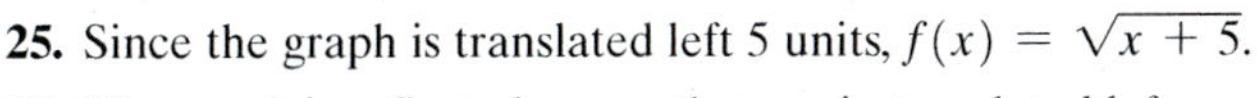

25. Since the graph is translated left 5 units, $f(x) = \sqrt{x + 5}$.

27. The graph is reflected across the x-axis, translated left 2 units, and translated up 3 units. $y = -\sqrt{x}$ would be reflected across the x-axis, $y = -\sqrt{x + 2}$ adds the horizontal translation, and finally, the vertical translation gives $f(x) = -\sqrt{x + 2} + 3 = 3 - \sqrt{x + 2}$.

29. **(a)** $y = -f(x) = -(x^3 - 5x^2 - 3x + 2)$
$= -x^3 + 5x^2 + 3x - 2$

(b) $y = f(-x) = (-x)^3 - 5(-x)^2 - 3(-x) + 2$
$= -x^3 - 5x^2 + 3x + 2$

31. **(a)** $y = -f(x) = -(\sqrt[3]{8x}) = -2\sqrt[3]{x}$

(b) $y = f(-x) = \sqrt[3]{8(-x)} = \sqrt[3]{-8x} = -2\sqrt[3]{x}$

33. Let f be an odd function; that is, $f(-x) = -f(x)$ for all x in the domain of f. To reflect the graph of $y = f(x)$ across the y-axis, we make the transformation $y = f(-x)$. But $f(-x) = -f(x)$ for all x in the domain of f, so this transformation results in $y = -f(x)$. That is exactly the translation that reflects the graph of f across the x-axis, so the two reflections yield the same graph.

35. **37.**

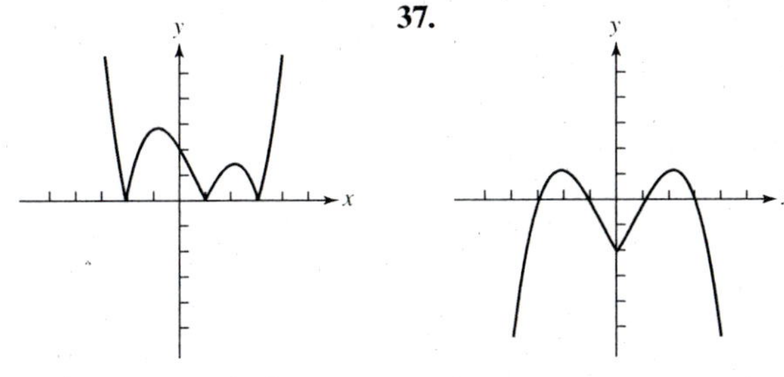

39. **(a)** $y_1 = 2y = 2(x^3 - 4x) = 2x^3 - 8x$

(b) $y_2 = f\left(\frac{x}{\frac{1}{3}}\right) = f(3x) = (3x)^3 - 4(3x) = 27x^3 - 12x$

41. **(a)** $y_1 = 2y = 2(x^2 + x - 2) = 2x^2 + 2x - 4$

(b) $y_2 = f(3x) = (3x)^2 + 3x - 2 = 9x^2 + 3x - 2$

43. Starting with $y = x^2$, translate right 3 units, vertically stretch by 2, and translate down 4 units.

45. Starting with $y = x^2$, horizontally shrink by $\frac{1}{3}$ and translate down 4 units.

47. First stretch (multiply right side by 3): $y = 3x^2$, then translate (replace x with $x - 4$): $y = 3(x - 4)^2$.

49. First translate left (replace x with $x + 2$): $y = |x + 2|$, then stretch (multiply right side by 2): $y = 2|x + 2|$, then translate down (subtract 4 from the right side): $y = 2|x + 2| - 4$.

To make the sketches for #51–53, it is useful to apply the described transformations to several selected points on the graph. The original graph here has vertices $(-2, -4)$, $(0, 0)$, $(2, 2)$, and $(4, 0)$; in the solutions below, the images of these four points are listed.

51. Translate left 1 unit, then vertically stretch by 3, and finally translate up 2 units. The four vertices are transformed to $(-3, -10)$, $(-1, 2)$, $(1, 8)$, and $(3, 2)$.

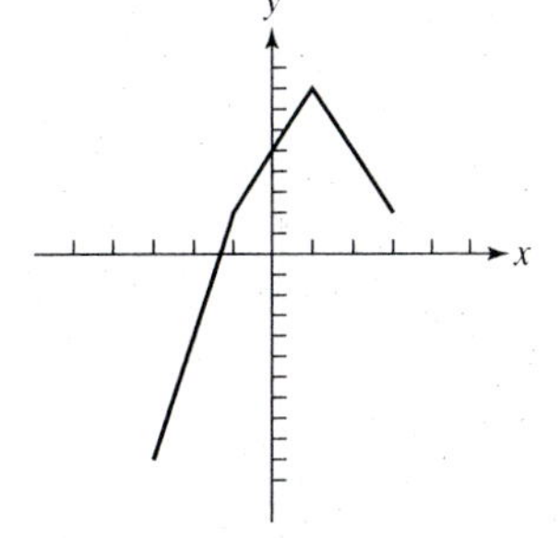

53. Horizontally shrink by $\frac{1}{2}$. The four vertices are transformed to $(-1, -4)$, $(0, 0)$, $(1, 2)$, $(2, 0)$.

55. Reflections have more effect on points that are farther away from the line of reflection. Translations affect the distance of points from the axes, and hence change the effect of the reflections.

57. First vertically stretch by $\frac{9}{5}$, then translate up 32 units.

59. False. $y = f(x + 3)$ is $y = f(x)$ translated 3 units to the *left*.

61. To vertically stretch $y = f(x)$ by a factor of 3, multiply the $f(x)$ by 3. The answer is C.

63. To translate $y = f(x)$ 2 units up, add 2 to $f(x)$: $y = f(x) + 2$. To reflect the result across the y-axis, replace x with $-x$. The answer is A.

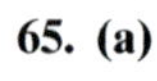

65. (a)

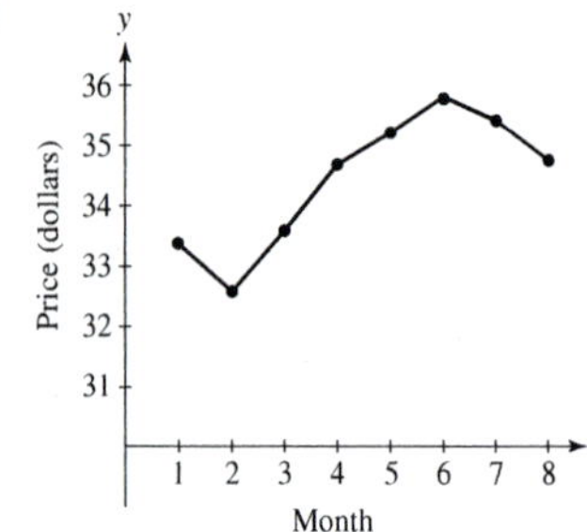

(b) Change the y-value by multiplying by the conversion rate from dollars to yen, a number that changes according to international market conditions. This results in a vertical stretch by the conversion rate.

67. (a) The original graph is on the left; the graph of $y = |f(x)|$ is on the right.

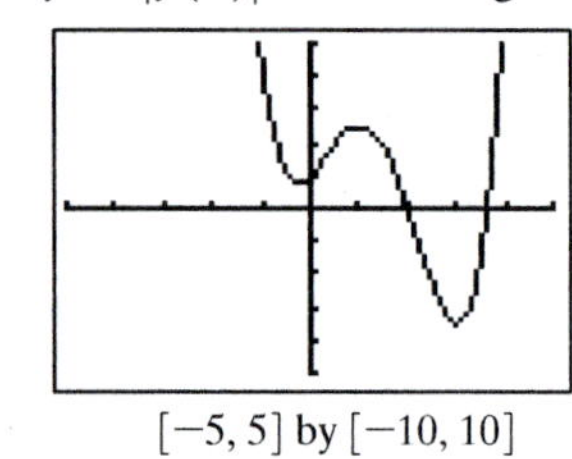

$[-5, 5]$ by $[-10, 10]$

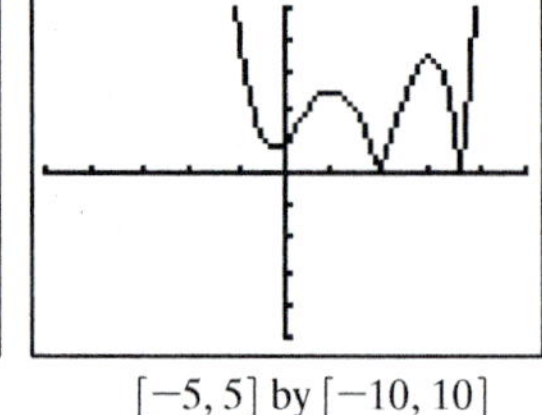

$[-5, 5]$ by $[-10, 10]$

(b) The original graph is on the left; the graph of $y = f(|x|)$ is on the right.

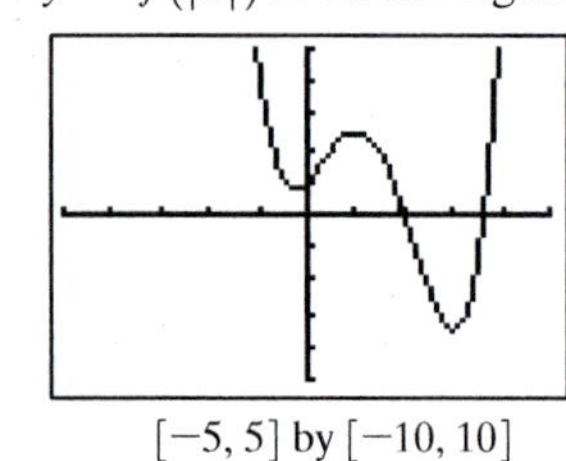

$[-5, 5]$ by $[-10, 10]$

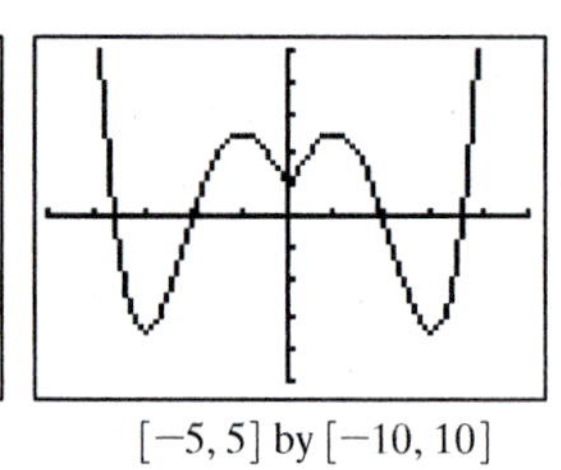

$[-5, 5]$ by $[-10, 10]$

(c)

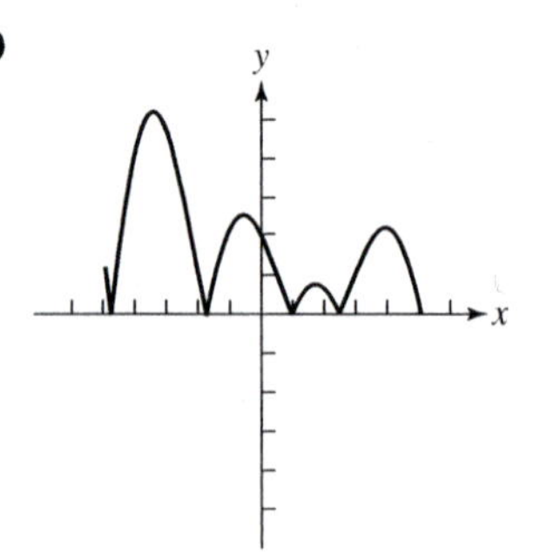

(d)

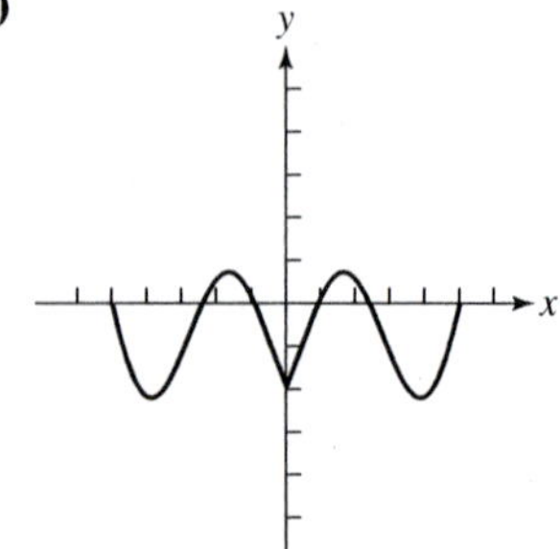

■ Section 1.7 Modeling with Functions

Exploration 1

1.

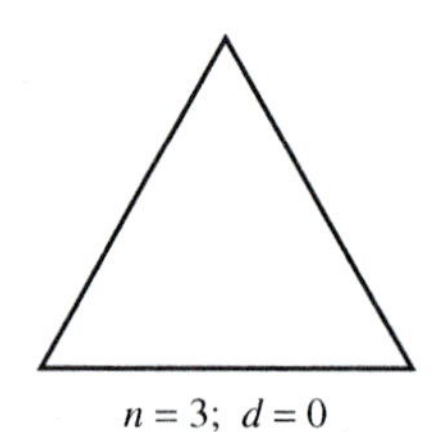

$n = 3;\ d = 0$

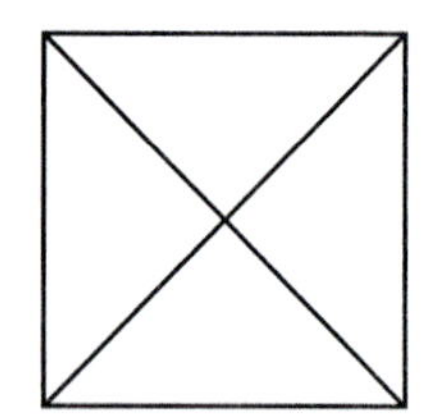

$n = 4;\ d = 2$

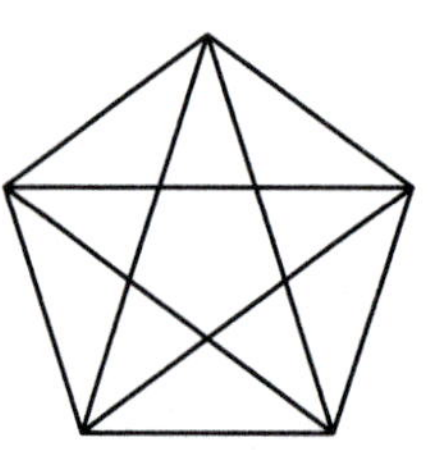

$n = 5;\ d = 5$

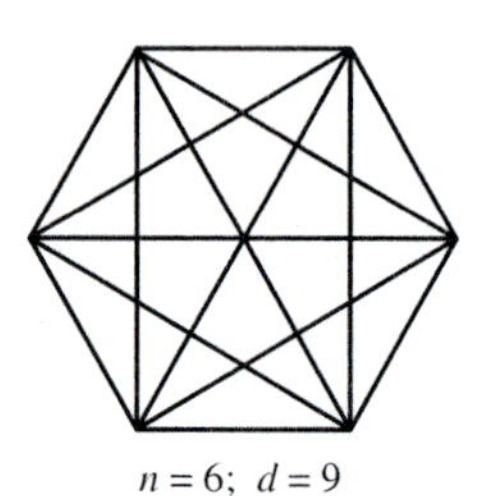

$n = 6;\ d = 9$

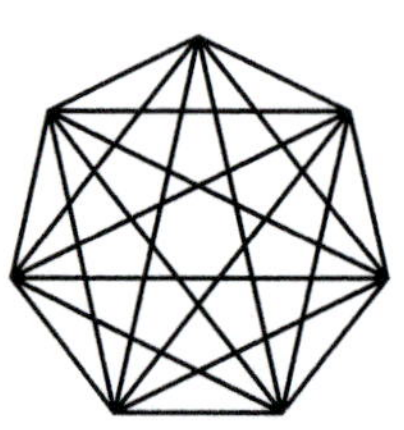

$n = 7;\ d = 14$

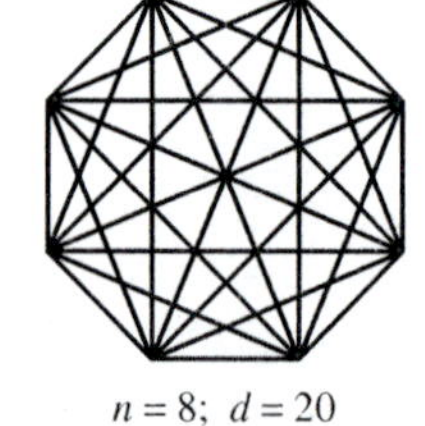

$n = 8;\ d = 20$

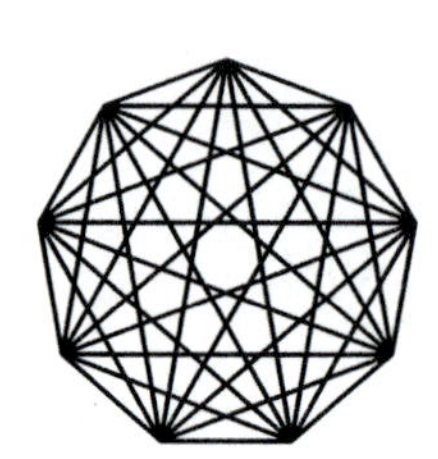

$n = 9;\ d = 27$

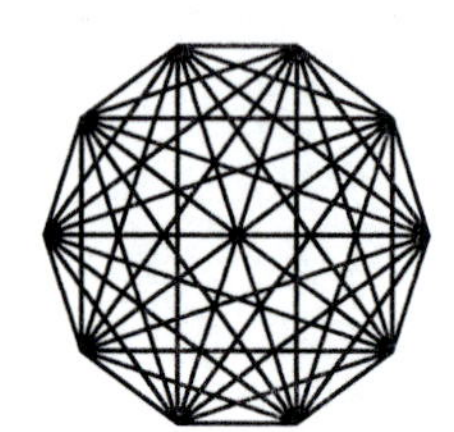

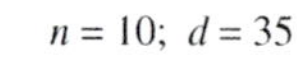

$n = 10;\ d = 35$

2.

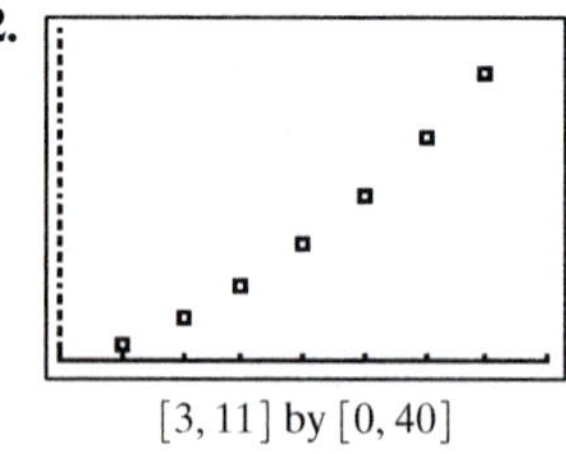

$[3, 11]$ by $[0, 40]$

3. Linear: $r^2 = 0.9758$
Power: $r^2 = 0.9903$
Quadratic: $R^2 = 1$
Cubic: $R^2 = 1$
Quartic: $R^2 = 1$

4. The best-fit curve is quadratic: $y = 0.5x^2 - 1.5x$. The cubic and quartic regressions give this same curve.

5. Since the quadratic curve fits the points perfectly, there is nothing to be gained by adding a cubic term or a quartic term. The coefficients of these terms in the regressions are zero.

6. $y = 0.5x^2 - 1.5x$. At $x = 128$,
$y = 0.5(128)^2 - 1.5(128) = 8000$

Quick Review 1.7

1. $h = 2(A/b)$

3. $h = V/(\pi r^2)$

5. $r = \sqrt[3]{\frac{3V}{4\pi}}$

7. $h = \frac{A - 2\pi r^2}{2\pi r} = \frac{A}{2\pi r} - r$

9. $P = \frac{A}{(1 + r/n)^{nt}} = A\left(1 + \frac{r}{n}\right)^{-nt}$

Section 1.7 Exercises

1. $3x + 5$

3. $0.17x$

5. $A = \ell w = (x + 12)(x)$

7. $x + 0.045x = (1 + 0.045)x = 1.045x$

9. $x - 0.40x = 0.60x$

11. Let C be the total cost and n be the number of items produced; $C = 34{,}500 + 5.75n$.

13. Let R be the revenue and n be the number of items sold; $R = 3.75n$.

15. The basic formula for the volume of a right circular cylinder is $V = \pi r^2 h$, where r is the radius and h is height. Since height equals diameter ($h = d$) and the diameter is two times r ($d = 2r$), we know $h = 2r$. Then, $V = \pi r^2(2r) = 2\pi r^3$.

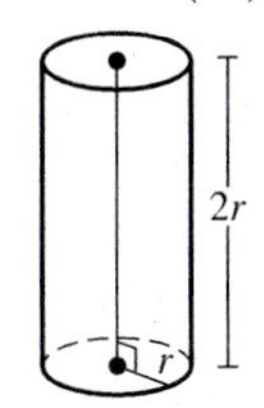

17. Let a be the length of the base. Then the other two sides of the triangle have length two times the base, or $2a$. Since the triangle is isoceles, a perpendicular dropped from the "top" vertex to the base is perpendicular. As a result, $h^2 + \left(\frac{a}{2}\right)^2 = (2a)^2$, or $h^2 = 4a^2 - \frac{a^2}{4} = \frac{16a^2 - a^2}{4}$ $= \frac{15a^2}{4}$, so $h = \sqrt{\frac{15a^2}{4}} = \frac{a\sqrt{15}}{2}$. The triangle's area is $A = \frac{1}{2}bh = \frac{1}{2}(a)\left(\frac{a\sqrt{15}}{2}\right) = \frac{a^2\sqrt{15}}{4}$.

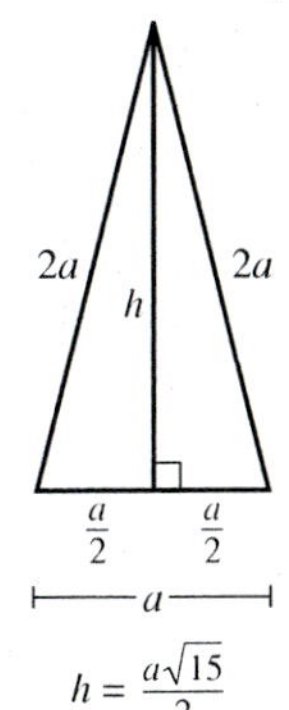

19. Let r be the radius of the sphere. Since the sphere is tangent to all six faces of the cube, we know that the height (and width, and depth) of the cube is equal to the sphere's diameter, which is two times r ($2r$). The surface area of the cube is the sum of the area of all six faces, which equals $2r \cdot 2r = 4r^2$. Thus, $A = 6 \cdot 4r^2 = 24r^2$.

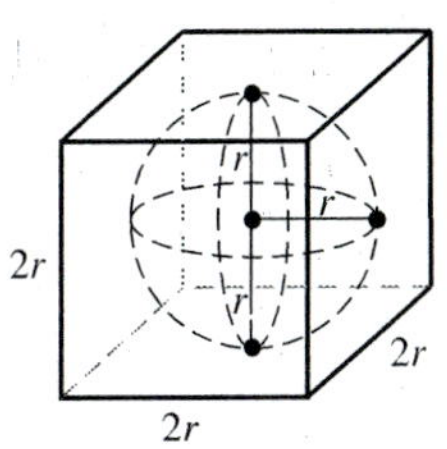

21. Solving $x + 4x = 620$ gives $x = 124$, so $4x = 496$. The two numbers are 124 and 496.

23. $1.035x = 36{,}432$, so $x = 35{,}200$

25. $182 = 52t$, so $t = 3.5$ hr.

27. $0.60(33) = 19.8$; $0.75(27) = 20.25$. The \$33 shirt sells for \$19.80. The \$27 shirt sells for \$20.25. The \$33 shirt is a better bargain, because the sale price is cheaper.

29.
$$71\,065\,000(1 + x) = 82\,400\,000$$
$$71\,065\,000x = 82\,400\,000 - 71\,065\,000$$
$$x = \frac{82\,400\,000 - 71\,065\,000}{71\,065\,000} \approx 0.1595$$

There was a 15.95% increase in sales.

31. **(a)** $0.10x + 0.45(100 - x) = 0.25(100)$.

(b) Graph $y_1 = 0.1x + 0.45(100 - x)$ and $y_2 = 25$. Use $x \approx 57.14$ gallons of the 10% solution and about 42.86 gal of the 45% solution.

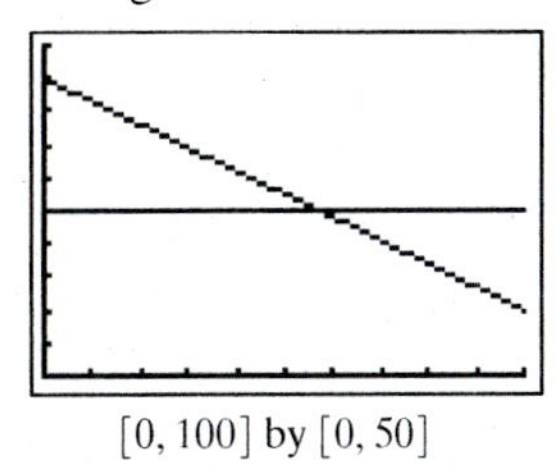

[0, 100] by [0, 50]

33. **(a)** The height of the box is x, and the base measures $10 - 2x$ by $18 - 2x$. $V(x) = x(10 - 2x)(18 - 2x)$

(b) Because one side of the original piece of cardboard measures 10 in., $2x$ must be greater than 0 but less than 10, so that $0 < x < 5$. The domain of $V(x)$ is $(0, 5)$.

(c) Graphing $V(x)$ produces a cubic-function curve that between $x = 0$ and $x = 5$ has a maximum at approximately $(2.06, 168.1)$. The cut-out squares should measure approximately 2.06 in. by 2.06 in.

35. Equation of the parabola, to pass through $(-16, 8)$ and $(16, 8)$:

$y = kx^2$

$8 = k\,(\pm 16)^2$

$k = \frac{8}{256} = \frac{1}{32}$

$y = \frac{1}{32}x^2$

y-coordinate of parabola 8 in. from center:

$y = \frac{1}{32}(8)^2 = 2$

From that point to the top of the dish is $8 - 2 = 6$ in.

37. Original volume of water:
$V_0 = \frac{1}{3}\pi r^2 h = \frac{1}{3}\pi(9)^2(24) \approx 2035.75$ in.^{3}
Volume lost through faucet:
$V_1 =$ time $\times$ rate $= (120$ sec$)(5$ in.^{3}/sec$) = 600$ in.^{3}
Find volume:
$V_f = V_0 - V_1 = 2035.75 - 600 = 1435.75$
Since the final cone-shaped volume of water has radius and height in a 9-to-24 ratio, or $r = \frac{3}{8}h$:

$$V_f = \frac{1}{3}\pi\left(\frac{3}{8}h\right)^2 h = \frac{3}{64}\pi h^3 = 1435.75$$

Solving, we obtain $h \approx 21.36$ in.

39. Bicycle's speed in feet per second:
$(2 \times \pi \times 16$ in./rot$)(2$ rot/sec$) = 64\pi$ in./sec
Unit conversion:
$(64\pi \text{ in./sec})\left(\frac{1}{12}\text{ ft/in.}\right)\left(\frac{1}{5280}\text{ mi/ft}\right)(3600 \text{ sec/hr})$
≈ 11.42 mi/hr

41. True. The correlation coefficient is close to 1 (or -1) if there is a good fit. A correlation coefficient near 0 indicates a very poor fit.

43. The pattern of points is S-shaped, which suggests a cubic model. The answer is C.

45. The points appear to lie along an upward-opening parabola. The answer is B.

47. **(a)** $C = 100{,}000 + 30x$

(b) $R = 50x$

(c) $100{,}000 + 30x = 50x$
$100{,}000 = 20x$
$x = 5000$ pairs of shoes

(d) Graph $y_1 = 100{,}000 + 30x$ and $y_2 = 50x$; these graphs cross when $x = 5000$ pairs of shoes. The point of intersection corresponds to the break-even point, where $C = R$.

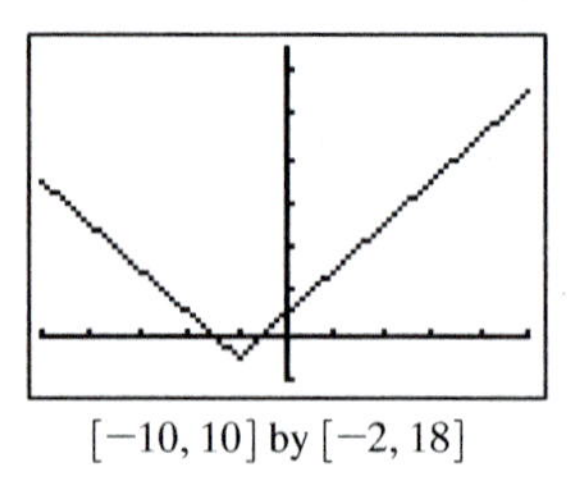

$[-10, 10]$ by $[-2, 18]$

49. **(a)** $y_1 = u(x) = 125{,}000 + 23x$.

(b) $y_2 = s(x) = 125{,}000 + 23x + 8x = 125{,}000 + 31x$.

(c) $y_3 = r_u(x) = 56x$.

(d) $y_4 = R_s(x) = 79x$.

(e)

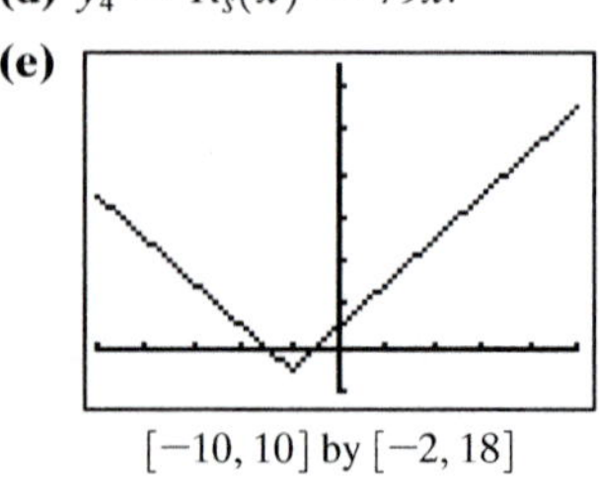

$[-10, 10]$ by $[-2, 18]$

(f) You should recommend stringing the rackets; fewer strung rackets need to be sold to begin making a profit (since the intersection of y_2 and y_4 occurs for smaller x than the intersection of y_1 and y_3).

51. **(a)**

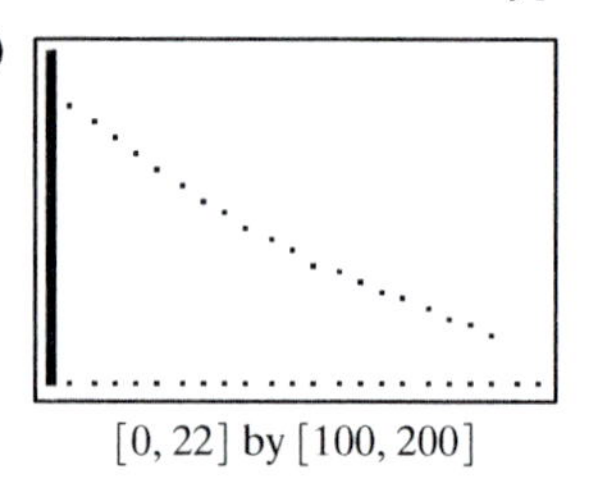

$[0, 22]$ by $[100, 200]$

(b) List L3 = {112.3, 106.5, 101.5, 96.6, 92.0, 87.2, 83.1, 79.8, 75.0, 71.7, 68, 64.1, 61.5, 58.5, 55.9, 53.0, 50.8, 47.9, 45.2, 43.2}

(c) The regression equation is $y = 118.07 \times 0.951^x$. It fits the data extremely well.

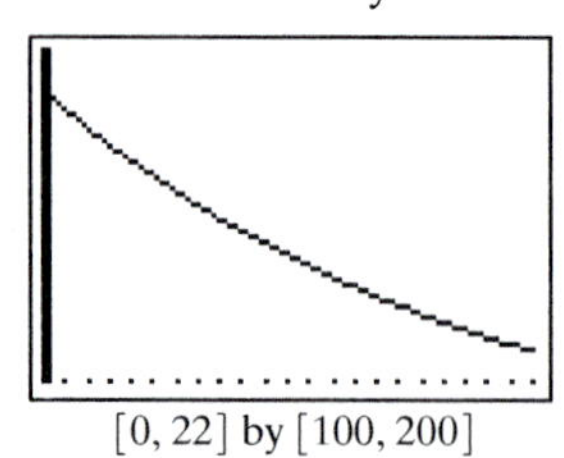

$[0, 22]$ by $[100, 200]$

■ Chapter 1 Review

1. (d)

3. (i)

5. (b)

7. (g)

9. (a)

11. **(a)** All reals **(b)** All reals

13. **(a)** All reals

(b) $g(x) = x^2 + 2x + 1 = (x + 1)^2$.
At $x = -1$, $g(x) = 0$, the function's minimum.
The range is $[0, \infty)$.

15. **(a)** All reals

(b) $|x| \geq 0$ for all x, so $3|x| \geq 0$ and $3|x| + 8 \geq 8$ for all x. The range is $[8, \infty)$.

17. **(a)** $f(x) = \frac{x}{x^2 - 2x} = \frac{x}{x(x-2)}$. $x \neq 0$ and $x - 2 \neq 0$, $x \neq 2$. The domain is all reals except 0 and 2.

(b) For $x > 2$, $f(x) > 0$ and for $x < 2$, $f(x) < 0$. $f(x)$ does not cross $y = 0$, so the range is all reals except $f(x) = 0$.

19. Continuous

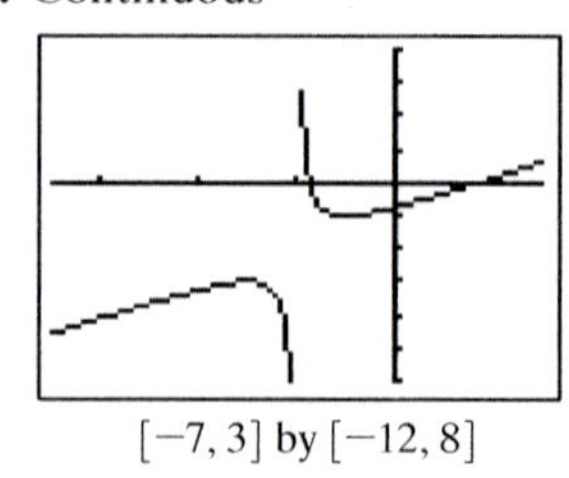

$[-7, 3]$ by $[-12, 8]$

21. **(a)** $x^2 - 5x \neq 0$, $x(x - 5) \neq 0$, so $x \neq 0$ and $x \neq 5$. We expect vertical asymptotes at $x = 0$ and $x = 5$.

(b) $y = 0$

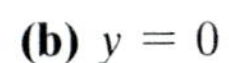

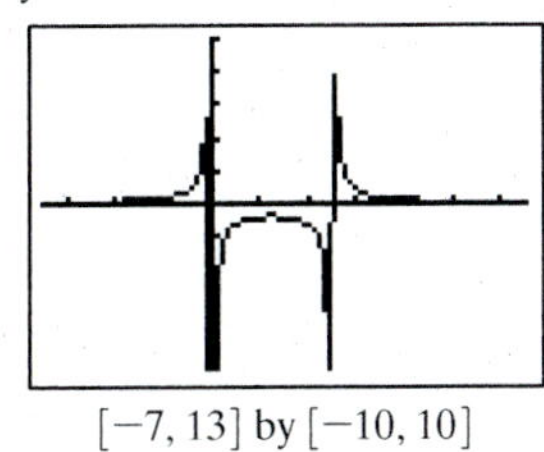

$[-7, 13]$ by $[-10, 10]$

23. **(a)** None

(b) Since $\lim\limits_{x\to\infty} \dfrac{7x}{\sqrt{x^2 + 10}} = 7$ and $\lim\limits_{x\to-\infty} \dfrac{7x}{\sqrt{x^2 + 10}} = -7$, we expect horizontal asymptotes at $y = 7$ and $y = -7$.

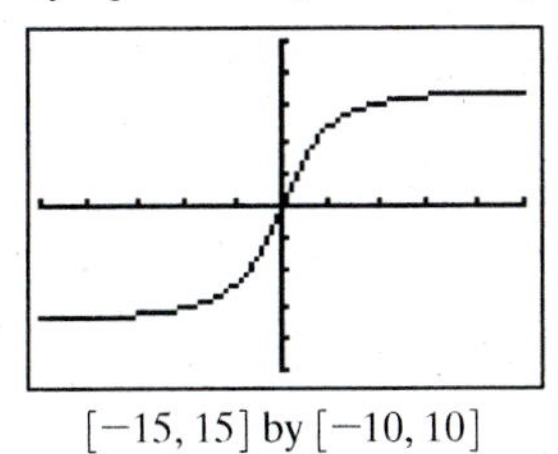

$[-15, 15]$ by $[-10, 10]$

25. $(-\infty, \infty)$

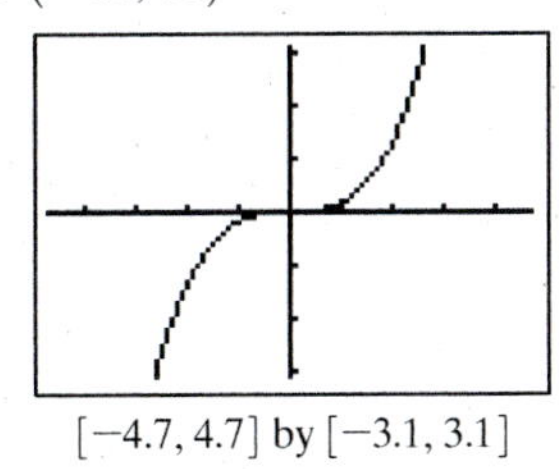

$[-4.7, 4.7]$ by $[-3.1, 3.1]$

27. As the graph illustrates, y is increasing over the intervals $(-\infty, -1)$, $(-1, 1)$, and $(1, \infty)$.

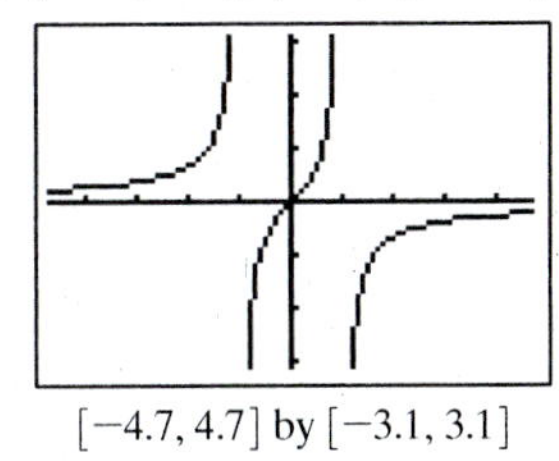

$[-4.7, 4.7]$ by $[-3.1, 3.1]$

29. $-1 \le \sin x \le 1$, but $-\infty < x < \infty$, so $f(x)$ is not bounded.

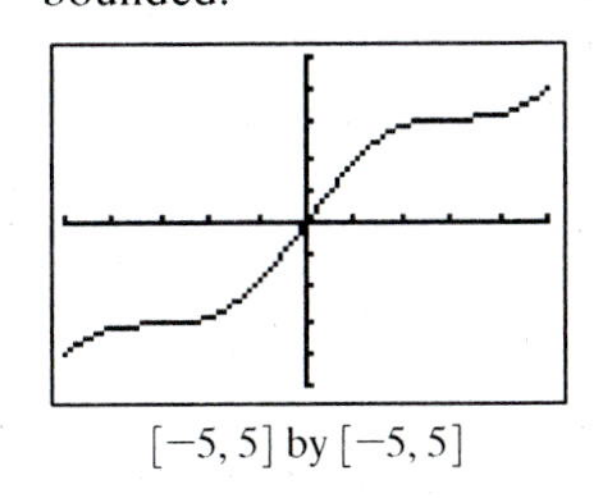

$[-5, 5]$ by $[-5, 5]$

31. $e^x > 0$ for all x, so $-e^x < 0$ and $5 - e^x < 5$ for all x. $h(x)$ is bounded above.

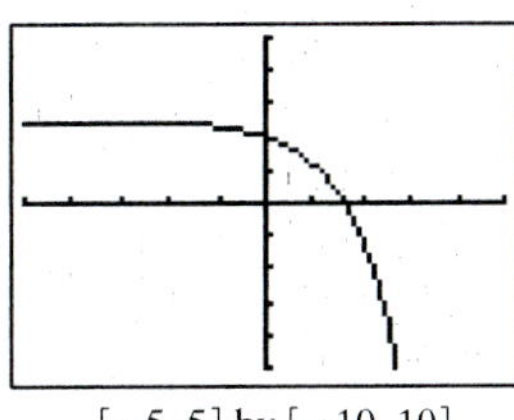

$[-5, 5]$ by $[-10, 10]$

33. **(a)** None **(b)** -7, at $x = -1$

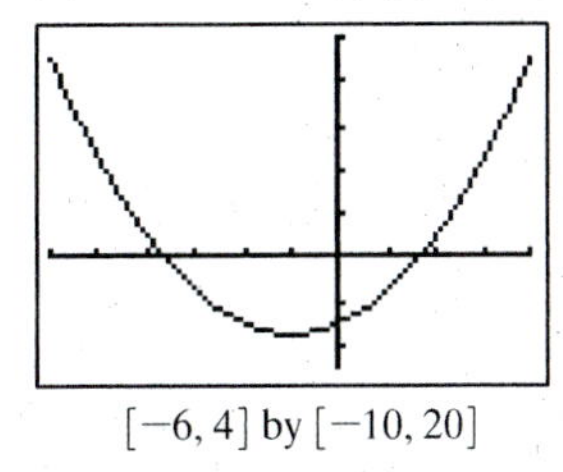

$[-6, 4]$ by $[-10, 20]$

35. **(a)** -1, at $x = 0$ **(b)** None

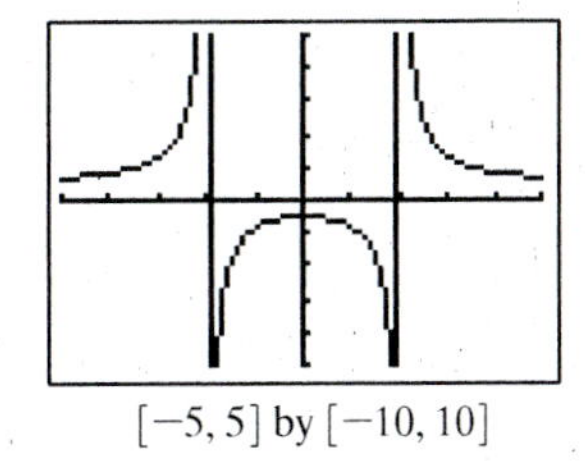

$[-5, 5]$ by $[-10, 10]$

37. The function is even since it is symmetrical about the y-axis.

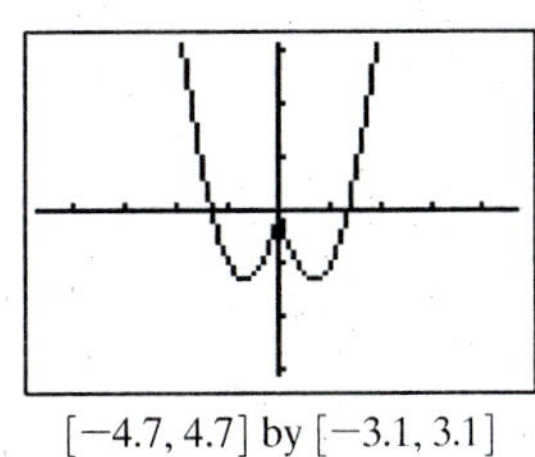

$[-4.7, 4.7]$ by $[-3.1, 3.1]$

39. Since no symmetry is exhibited, the function is neither.

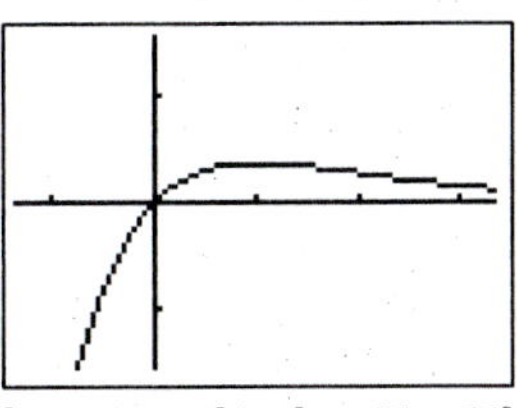

$[-1.35, 3.35]$ by $[-1.55, 1.55]$

41. $x = 2y + 3$, $2y = x - 3$, $y = \dfrac{x-3}{2}$, so $f^{-1}(x) = \dfrac{x-3}{2}$.

43. $x = \dfrac{2}{y}$, $xy = 2$, $y = \dfrac{2}{x}$, so $f^{-1}(x) = \dfrac{2}{x}$.

45.

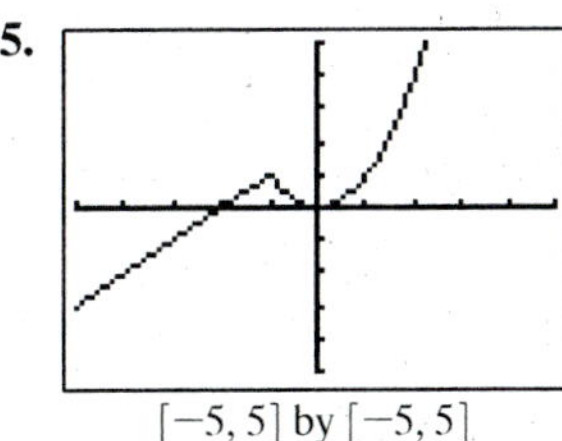

$[-5, 5]$ by $[-5, 5]$

47.

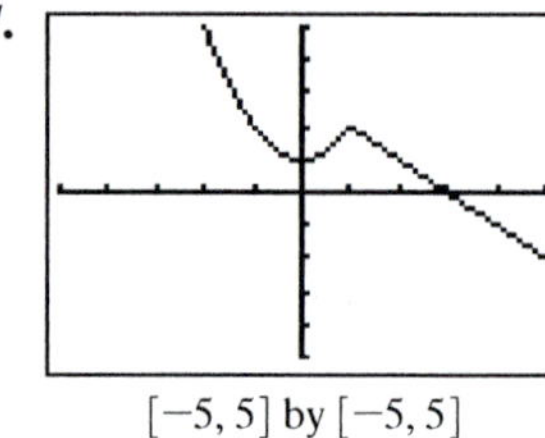

$[-5, 5]$ by $[-5, 5]$

49.

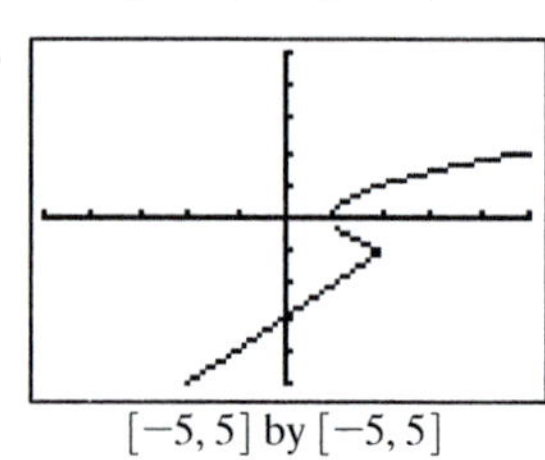

$[-5, 5]$ by $[-5, 5]$

51.

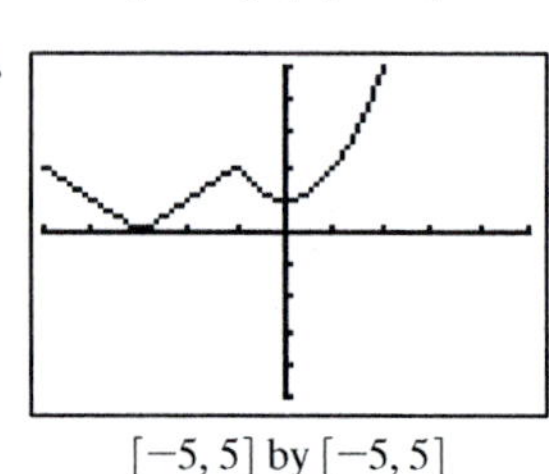

$[-5, 5]$ by $[-5, 5]$

53. $(f \circ g)(x) = f(g(x)) = f(x^2 - 4) = \sqrt{x^2 - 4}$.
Since $x^2 - 4 \geq 0$, $x^2 \geq 4$, $x \leq -2$ or $x \geq 2$.
The domain is $(-\infty, -2] \cup [2, \infty)$.

55. $(f \cdot g)(x) = f(x) \cdot g(x) = \sqrt{x} \cdot (x^2 - 4)$.
Since $\sqrt{x} \geq 0$, the domain is $[0, \infty)$.

57. $\lim_{x \to \infty} \sqrt{x} = \infty$. (Large negative values are not in the domain.)

59. $r^2 = \left(\frac{s}{2}\right)^2 + \left(\frac{s}{2}\right)^2 = \frac{2s^2}{4}$, $r = \sqrt{\frac{2s^2}{4}} = \frac{s\sqrt{2}}{2}$.

The area of the circle is

$$A = \pi r^2 = \pi\left(\frac{s\sqrt{2}}{2}\right)^2 = \frac{2\pi s^2}{4} = \frac{\pi s^2}{2}$$

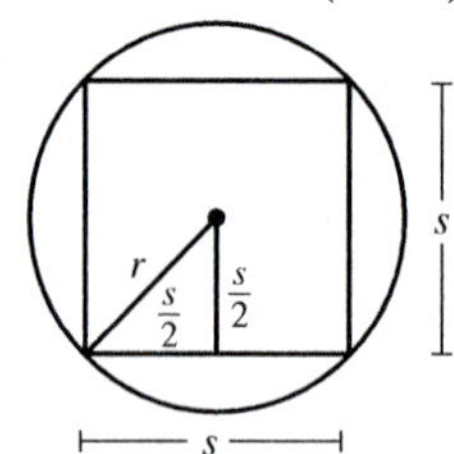

$r = \frac{s\sqrt{2}}{2}$

61. $d = 2r$, $r = \frac{d}{2}$, so the radius of. the tank is 10 feet.
Volume is $V = \pi r^2 \cdot h = \pi(10)^2 \cdot h = 100\pi h$

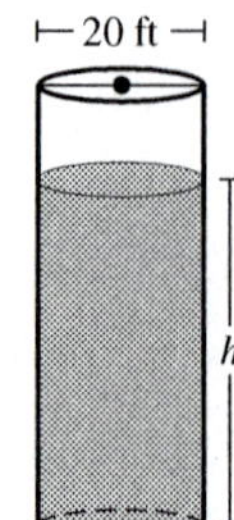

63. Since $V = 4000\pi - 2t$, we know that $\pi r^2 h = 4000\pi - 2t$. In this case, $r = 10'$, so $100\pi h = 4000\pi - 2t$, $h = \frac{4000\pi - 2t}{100\pi} = 40 - \frac{t}{50\pi}$

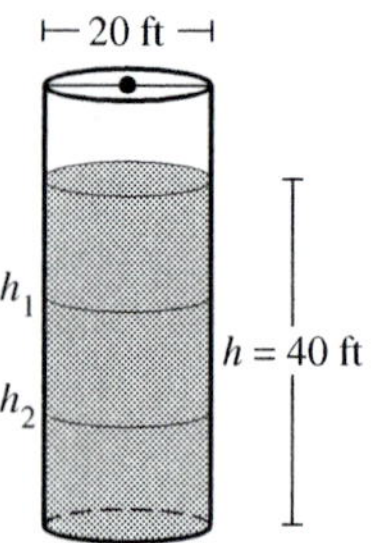

65. **(a)**

[4, 15] by [940, 1700]

(b) The regression line is $y = 61.133x + 725.333$.

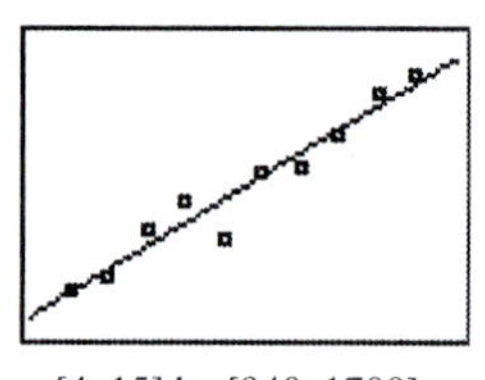

[4, 15] by [940, 1700]

(c) $61.133(20) + 725.333 \approx 1948$ (thousands of barrels)

67. **(a)** $r^2 + \left(\frac{h}{2}\right)^2 = (\sqrt{3})^2$,

$\frac{h^2}{4} = 3 - r^2$, $h^2 = 12 - 4r^2$, $h = \sqrt{12 - 4r^2}$,

$h = 2\sqrt{3 - r^2}$

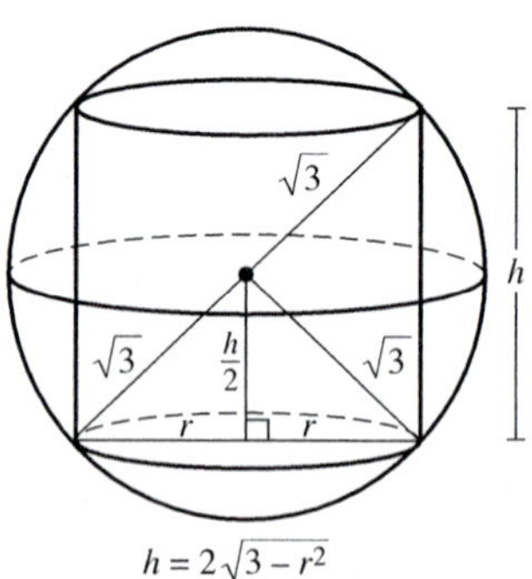

$h = 2\sqrt{3 - r^2}$

(b) $V = \pi r^2 h = (\pi r^2)(2\sqrt{3 - r^2}) = 2\pi r^2\sqrt{3 - r^2}$

(c) Since $\sqrt{3 - r^2} \geq 0$, $3 - r^2 \geq 0$
$3 \geq r^2$, $-\sqrt{3} \leq r \leq \sqrt{3}$.
However, $r < 0$ are invalid values, so the domain is $[0, \sqrt{3}]$.

(d)

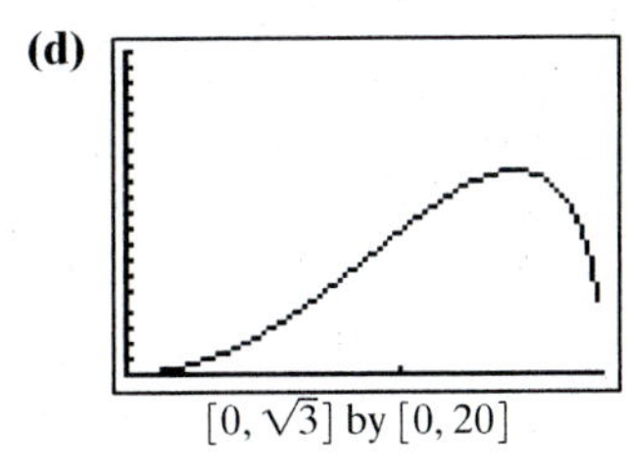

$[0, \sqrt{3}]$ by $[0, 20]$

(e) 12.57 in^3

Chapter 1 Project

1.

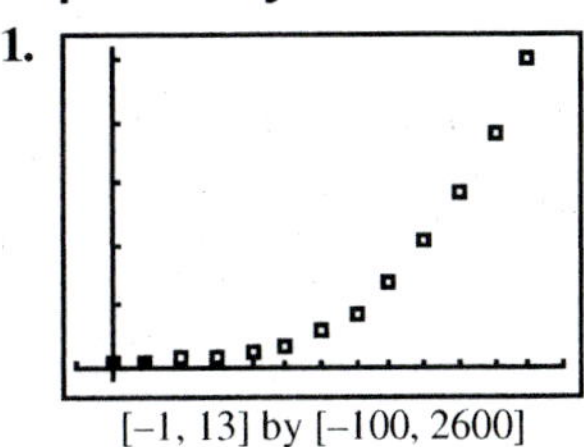

$[-1, 13]$ by $[-100, 2600]$

2. The exponential regression produces $y \approx 21.956(1.511)^x$.

3. 2000: For $x = 13$, $y \approx 4690$
2001: For $x = 14$, $y \approx 7085$

4. The model, which is based on data from the early, high-growth period of Starbucks Coffee's company history, does not account for the effects of gradual market saturation by Starbucks and its competitors. The actual growth in the number of locations is slowing while the model increases more rapidly.

5. The logistic regression produces

$$y \approx \frac{4914.198}{1 + 269.459\,e^{-0.486x}}.$$

6. 2000: For $x = 13$, $y \approx 3048$
2001: For $x = 14$, $y \approx 3553$

These predictions are less than the actual numbers, but are not off by as much as the numbers derived from the exponential model were. For the year 2020 ($x = 33$), the logistic model predicts about 4914 locations. (This prediction is probably too conservative.)

Chapter 2
Polynomial, Power, and Rational Functions

Section 2.1 Linear and Quadratic Functions and Modeling

Exploration 1

1. $-\$2000$ per year
2. The equation will have the form $v(t) = mt + b$. The value of the building after 0 year is $v(0) = m(0) + b = b = 50{,}000$.

   The slope m is the rate of change, which is -2000 (dollars per year). So an equation for the value of the building (in dollars) as a function of the time (in years) is $v(t) = -2000t + 50{,}000$.
3. $v(0) = 50{,}000$ and $v(16) = -2000(16) + 50{,}000 = 18{,}000$ dollars
4. The equation $v(t) = 39{,}000$ becomes
   $$-2000t + 50{,}000 = 39{,}000$$
   $$-2000t = -11{,}000$$
   $$t = 5.5 \text{ years}$$

Quick Review 2.1

1. $y = 8x + 3.6$

3. $y - 4 = -\dfrac{3}{5}(x + 2)$, or $y = -0.6x + 2.8$

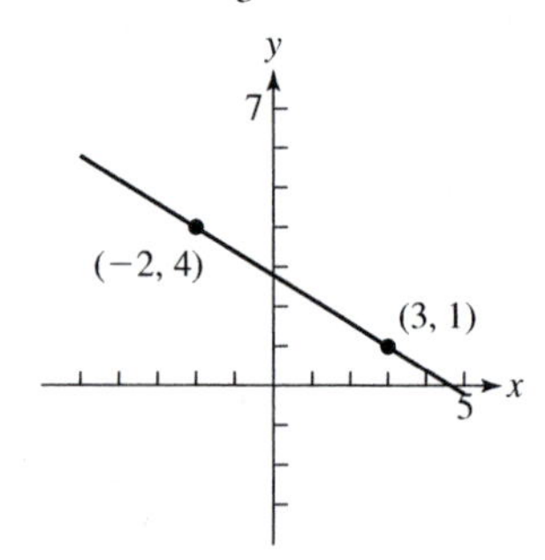

5. $(x + 3)^2 = (x + 3)(x + 3) = x^2 + 3x + 3x + 9 = x^2 + 6x + 9$

7. $3(x - 6)^2 = 3(x - 6)(x - 6) = (3x - 18)(x - 6) = 3x^2 - 18x - 18x + 108 = 3x^2 - 36x + 108$

9. $2x^2 - 4x + 2 = 2(x^2 - 2x + 1) = 2(x - 1)(x - 1) = 2(x - 1)^2$

Section 2.1 Exercises

1. Not a polynomial function because of the exponent -5
3. Polynomial of degree 5 with leading coefficient 2
5. Not a polynomial function because of cube root

7. $m = \dfrac{5}{7}$ so $y - 4 = \dfrac{5}{7}(x - 2) \Rightarrow f(x) = \dfrac{5}{7}x + \dfrac{18}{7}$

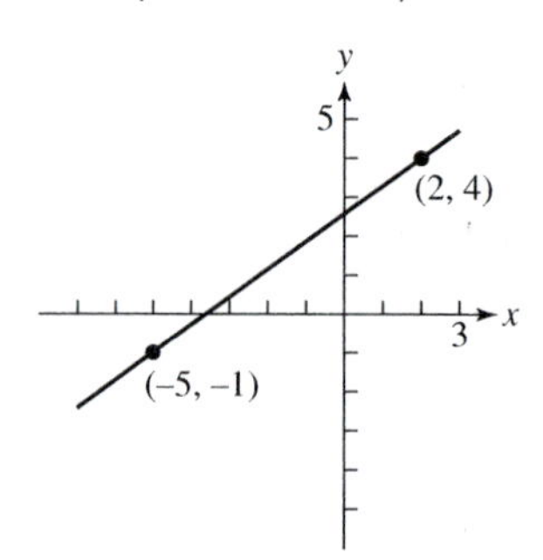

9. $m = -\dfrac{4}{3}$ so $y - 6 = -\dfrac{4}{3}(x + 4) \Rightarrow f(x) = -\dfrac{4}{3}x + \dfrac{2}{3}$

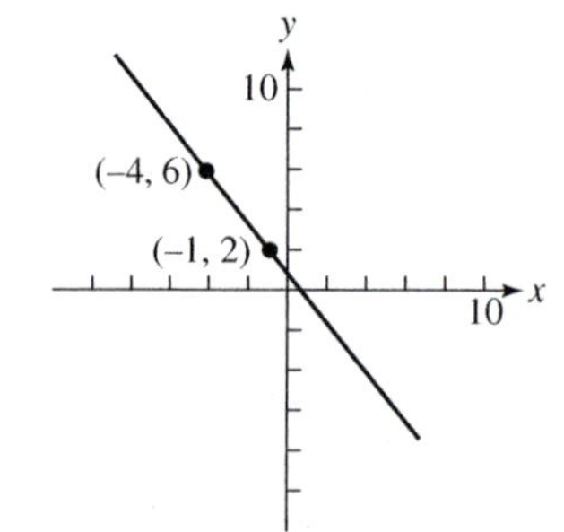

11. $m = -1$ so $y - 3 = -1(x - 0) \Rightarrow f(x) = -x + 3$

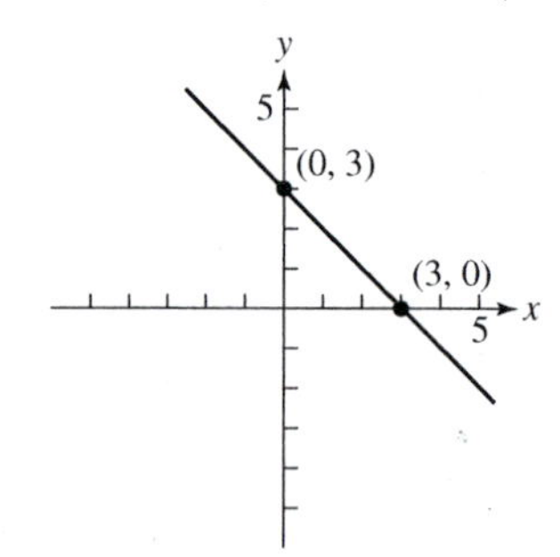

13. (a)—the vertex is at $(-1, -3)$, in Quadrant III, eliminating all but (a) and (d). Since $f(0) = -1$, it must be (a).
15. (b)—the vertex is in Quadrant I, at $(1, 4)$, meaning it must be either (b) or (f). Since $f(0) = 1$, it cannot be (f): if the vertex in (f) is $(1, 4)$, then the intersection with the y-axis would be about $(0, 3)$. It must be (b).
17. (e)—the vertex is at $(1, -3)$ in Quadrant IV, so it must be (e).

19. Translate the graph of $f(x) = x^2$ 3 units right to obtain the graph of $h(x) = (x - 3)^2$, and translate this graph 2 units down to obtain the graph of $g(x) = (x - 3)^2 - 2$.

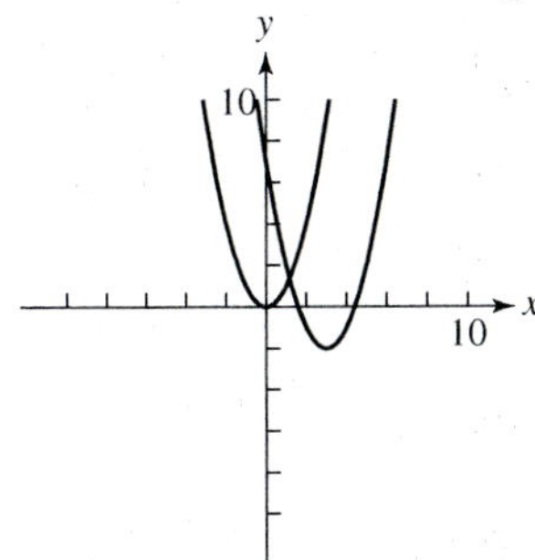

21. Translate the graph of $f(x) = x^2$ 2 units left to obtain the graph of $h(x) = (x + 2)^2$, vertically shrink this graph by a factor of $\frac{1}{2}$ to obtain the graph of $k(x) = \frac{1}{2}(x + 2)^2$, and translate this graph 3 units down to obtain the graph of $g(x) = \frac{1}{2}(x + 2)^2 - 3$.

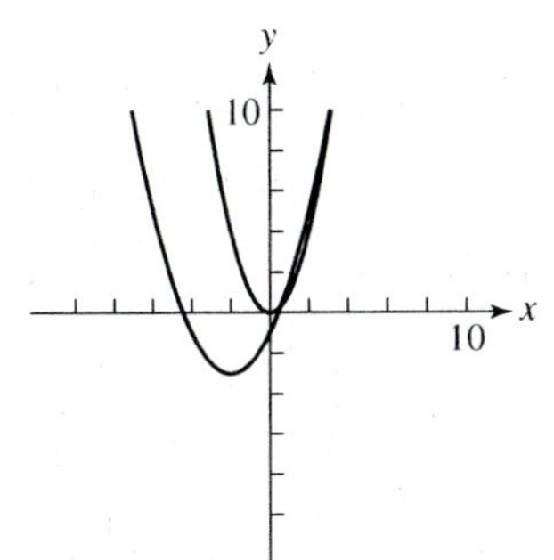

For #23–31, with an equation of the form $f(x) = a(x - h)^2 + k$, the vertex is (h, k) and the axis is $x = h$.

23. Vertex: $(1, 5)$; axis: $x = 1$

25. Vertex: $(1, -7)$; axis: $x = 1$

27. $f(x) = 3\left(x^2 + \frac{5}{3}x\right) - 4$

$= 3\left(x^2 + 2\cdot\frac{5}{6}x + \frac{25}{36}\right) - 4 - \frac{25}{12} = 3\left(x + \frac{5}{6}\right)^2 - \frac{73}{12}$

Vertex: $\left(-\frac{5}{6}, -\frac{73}{12}\right)$; axis: $x = -\frac{5}{6}$

29. $f(x) = -(x^2 - 8x) + 3$

$= -(x^2 - 2\cdot 4x + 16) + 3 + 16 = -(x - 4)^2 + 19$

Vertex: $(4, 19)$; axis: $x = 4$

31. $g(x) = 5\left(x^2 - \frac{6}{5}x\right) + 4$

$= 5\left(x^2 - 2\cdot\frac{3}{5}x + \frac{9}{25}\right) + 4 - \frac{9}{5} = 5\left(x - \frac{3}{5}\right)^2 + \frac{11}{5}$

Vertex: $\left(\frac{3}{5}, \frac{11}{5}\right)$; axis: $x = \frac{3}{5}$

33. $f(x) = (x^2 - 4x + 4) + 6 - 4 = (x - 2)^2 + 2$.
Vertex: $(2, 2)$; axis: $x = 2$; opens upward; does not intersect x–axis.

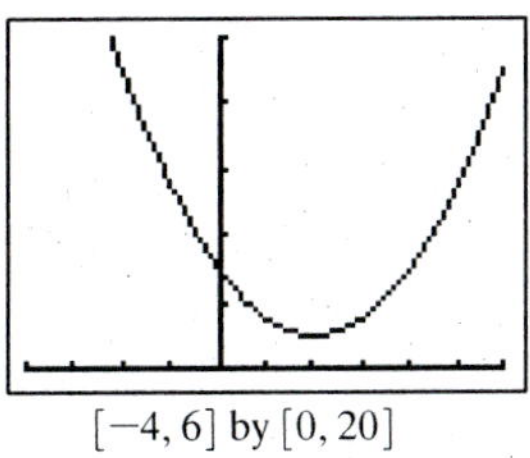

$[-4, 6]$ by $[0, 20]$

35. $f(x) = -(x^2 + 16x) + 10$

$= -(x^2 + 16x + 64) + 10 + 64 = -(x + 8)^2 + 74$.

Vertex: $(-8, 74)$; axis: $x = -8$; opens downward; intersects x–axis at about -16.602 and $0.602 (-8 \pm \sqrt{74})$.

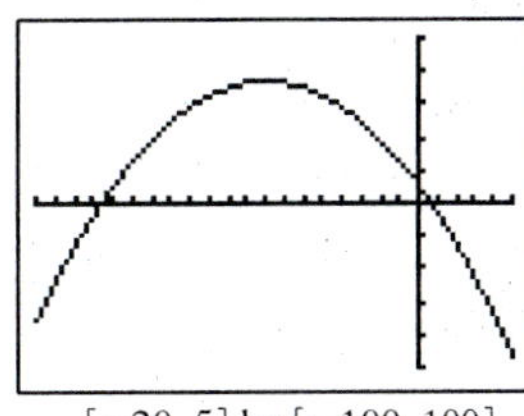

$[-20, 5]$ by $[-100, 100]$

37. $f(x) = 2(x^2 + 3x) + 7$

$= 2\left(x^2 + 3x + \frac{9}{4}\right) + 7 - \frac{9}{2} = 2\left(x + \frac{3}{2}\right)^2 + \frac{5}{2}$

Vertex: $\left(-\frac{3}{2}, \frac{5}{2}\right)$; axis: $x = -\frac{3}{2}$; opens upward; does not intersect the x-axis; vertically stretched by 2.

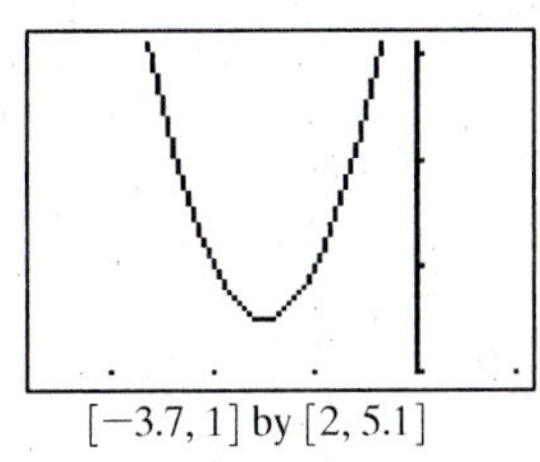

$[-3.7, 1]$ by $[2, 5.1]$

For #39–43, use the form $y = a(x - h)^2 + k$, taking the vertex (h, k) from the graph or other given information.

39. $h = -1$ and $k = -3$, so $y = a(x + 1)^2 - 3$. Now substitute $x = 1$, $y = 5$ to obtain $5 = 4a - 3$, so $a = 2$: $y = 2(x + 1)^2 - 3$.

41. $h = 1$ and $k = 11$, so $y = a(x - 1)^2 + 11$. Now substitute $x = 4$, $y = -7$ to obtain $-7 = 9a + 11$, so $a = -2$: $y = -2(x - 1)^2 + 11$.

43. $h = 1$ and $k = 3$, so $y = a(x - 1)^2 + 3$. Now substitute $x = 0$, $y = 5$ to obtain $5 = a + 3$, so $a = 2$: $y = 2(x - 1)^2 + 3$.

45. Strong positive

47. Weak positive

49. (a)

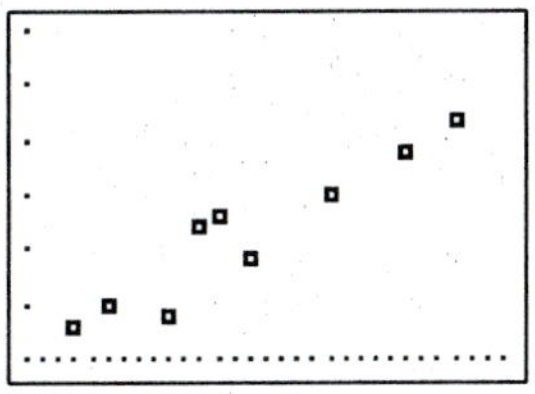

$[15, 45]$ by $[20, 50]$

(b) Strong positive

51. $m = -\frac{2350}{5} = -470$ and $b = 2350$,
so $v(t) = -470t + 2350$.
At $t = 3$, $v(3) = (-470)(3) + 2350 = \$940$.

53. **(a)** $y \approx 0.541x + 4.072$. The slope, $m \approx 0.541$, represents the average annual increase in hourly compensation for production workers, about \$0.54 per year.

(b) Setting $x = 40$ in the regression equation leads to $y \approx \$25.71$.

55. **(a)** [0, 100] by [0, 1000] is one possibility.

(b) When $x \approx 107.335$ or $x \approx 372.665$ — either 107, 335 units or 372, 665 units.

57. If the strip is x feet wide, the area of the strip is $A(x) = (25 + 2x)(40 + 2x) - 1000 \text{ ft}^2$.
This equals 504 ft^2 when $x = 3.5$ ft.

59. **(a)** $R(x) = (26{,}000 - 1000x)(0.50 + 0.05x)$.

(b) Many choices of Xmax and Ymin are reasonable. Shown is [0, 15] by [10,000, 17,000].

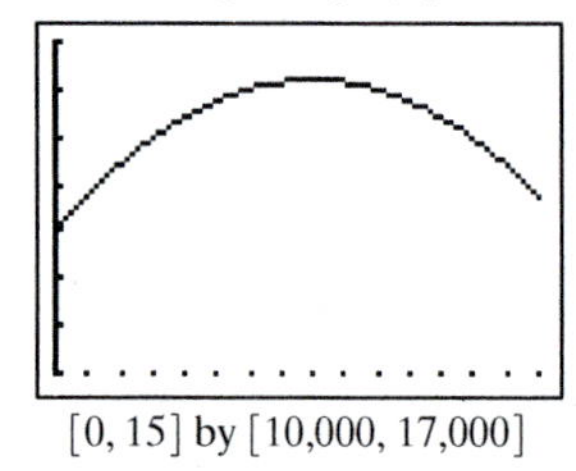

[0, 15] by [10,000, 17,000]

(c) The maximum revenue — \$16,200 — is achieved when $x = 8$; that is, charging 90 cents per can.

61. **(a)** $g \approx 32 \text{ ft/sec}^2$. $s_0 = 83$ ft and $v_0 = 92$ ft/sec. So the models are height $= s(t) = -16t^2 + 92t + 83$ and vertical velocity $= v(t) = -32t + 92$. The maximum height occurs at the vertex of $s(t)$.
$h = -\frac{b}{2a} = -\frac{92}{2(-16)} = 2.875$, and
$k = s(2.875) = 215.25$. The maximum height of the baseball is about 215 ft above the field.

(b) The amount of time the ball is in the air is a zero of $s(t)$. Using the quadratic formula, we obtain

$$t = \frac{-92 \pm \sqrt{92^2 - 4(-16)(83)}}{2(-16)}$$

$$= \frac{-92 \pm \sqrt{13{,}776}}{-32} \approx -0.79 \text{ or } 6.54.$$ Time is not negative, so the ball is in the air about 6.54 seconds.

(c) To determine the ball's vertical velocity when it hits the ground, use $v(t) = -32t + 92$, and solve for $t = 6.54$. $v(6.54) = -32(6.54) + 92 \approx -117$ ft/sec when it hits the ground.

63. **(a)** $h = -16t^2 + 80t - 10$. The graph is shown in the window [0, 5] by [−10, 100].

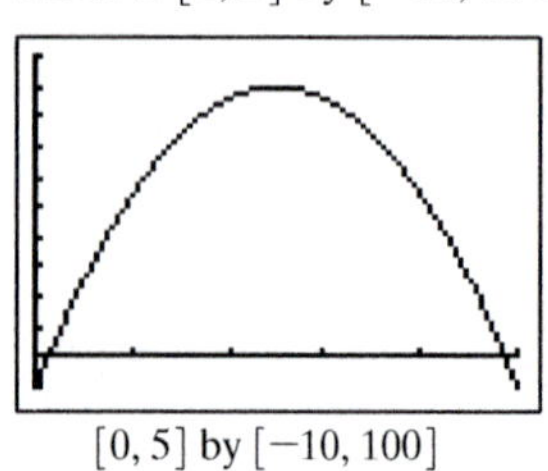

[0, 5] by [−10, 100]

(b) The maximum height is 90 ft, 2.5 sec after it is shot.

65. The quadratic regression is $y \approx 0.449x^2 + 0.934x + 114.658$. Plot this curve together with the curve $y = 450$, and then find the intersection to find when the number of patent applications will reach 450,000. Note that we use $y = 450$ because the data were given as a number of thousands. The intersection occurs at $x \approx 26.3$, so the number of applications will reach 450,000 approximately 26 years after 1980—in 2006.

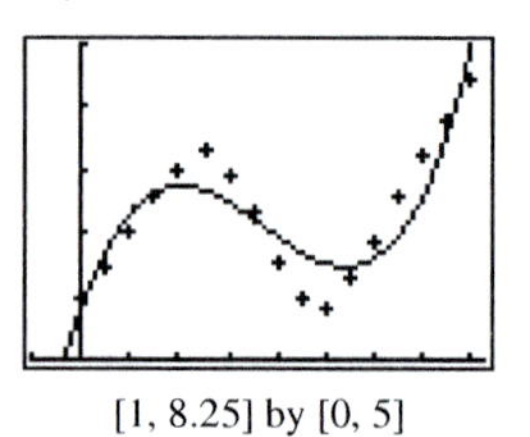

[1, 8.25] by [0, 5]

67. **(a)**

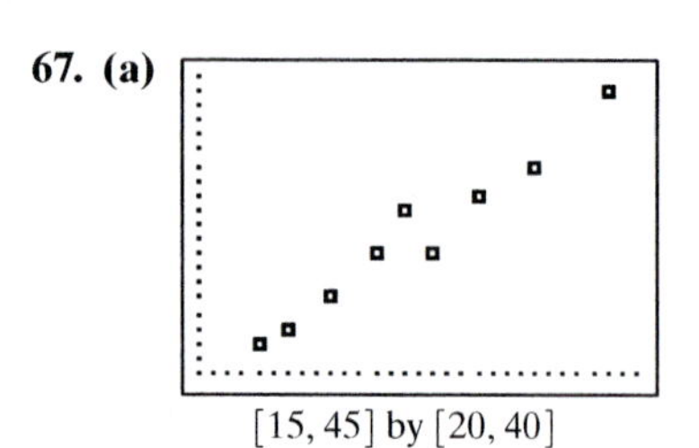

[15, 45] by [20, 40]

(b) $y \approx 0.68x + 9.01$

(c) On average, the children gained 0.68 pound per month.

(d)

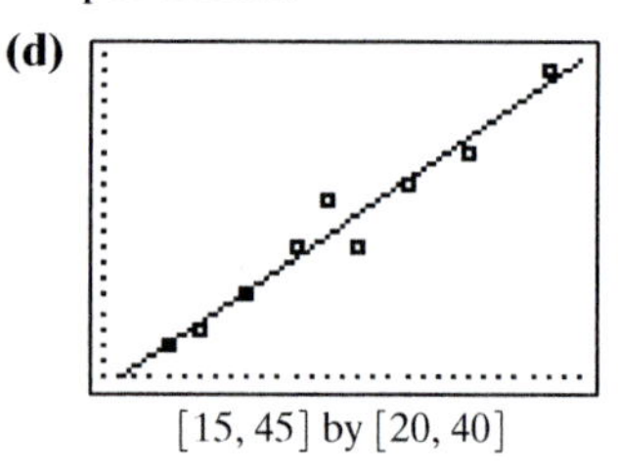

[15, 45] by [20, 40]

(e) ≈ 29.41 lbs

69. The Identity Function $f(x) = x$

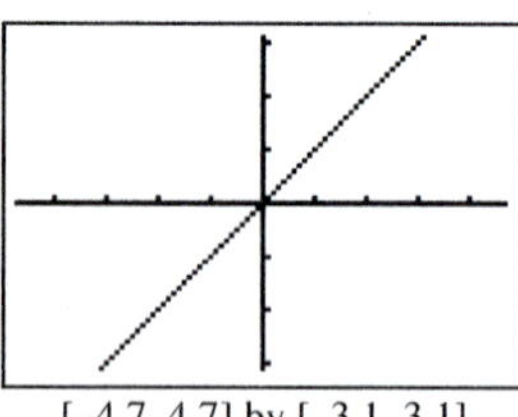

[−4.7, 4.7] by [−3.1, 3.1]

Domain: $(-\infty, \infty)$
Range: $(-\infty, \infty)$
Continuity: The function is continuous on its domain.
Increasing–decreasing behavior: Increasing for all x
Symmetry: Symmetric about the origin
Boundedness: Not bounded
Local extrema: None
Horizontal asymptotes: None
Vertical asymptotes: None
End behavior: $\lim_{x\to-\infty} f(x) = -\infty$ and $\lim_{x\to\infty} f(x) = \infty$

71. False. For $f(x) = 3x^2 + 2x - 3$, the initial value is $f(0) = -3$.

73. $m = \frac{1-3}{4-(-2)} = \frac{-2}{6} = -\frac{1}{3}$. The answer is E.

For #75, $f(x) = 2(x + 3)^2 - 5$ corresponds to $f(x) = a(x - h)^2 + k$ with $a = 2$ and $(h, k) = (-3, -5)$.

75. The axis of symmetry runs vertically through the vertex: $x = -3$. The answer is B.

77. (a) Graphs (i), (iii), and (v) are linear functions. They can all be represented by an equation $y = ax + b$, where $a \neq 0$.

(b) In addition to graphs (i), (iii), and (v), graphs (iv) and (vi) are also functions, the difference is that (iv) and (vi) are *constant* functions, represented by $y = b, b \neq 0$.

(c) (ii) is not a function because a single value x (i.e., $x = -2$) results in a multiple number of y-values. In fact, there are infinitely many y-values that are valid for the equation $x = -2$.

79. Answers will vary. One possibility: When using the least-squares method, mathematicians try to minimize the residual $y_i - (ax_i + b)$, i.e., place the "predicted" y-values as close as possible to the actual y-values. If mathematicians reversed the ordered pairs, the objective would change to minimizing the residual $x_i - (cy_i + d)$, i.e., placing the "predicted" x-values as close as possible to the actual x-values. In order to obtain an exact inverse, the x- and y-values for each xy pair would have to be almost exactly the same distance from the regression line—which is statistically impossible in practice.

81. (a) If $ax^2 + bx^2 + c = 0$, then

$$x = \frac{-b \pm \sqrt{b^2 - 4ac}}{2a}$$ by the quadratic formula. Thus, $x_1 = \frac{-b + \sqrt{b^2 - 4ac}}{2a}$ and

$$x_2 = \frac{-b - \sqrt{b^2 - 4ac}}{2a}$$ and

$$x_1 + x_2 = \frac{-b + \sqrt{b^2 - 4ac} - b - \sqrt{b^2 - 4ac}}{2a}$$

$$= \frac{-2b}{2a} = \frac{-b}{a} = -\frac{b}{a}.$$

(b) Similarly,

$$x_1 \cdot x_2 = \left(\frac{-b + \sqrt{b^2 - 4ac}}{2a}\right)\left(\frac{-b - \sqrt{b^2 - 4ac}}{2a}\right)$$

$$= \frac{b^2 - (b^2 - 4ac)}{4a^2} = \frac{4ac}{4a^2} = \frac{c}{a}.$$

83. Multiply out $f(x)$ to get $x^2 - (a + b)x + ab$. Complete the square to get $\left(x - \frac{a + b}{2}\right)^2 + ab - \frac{(a + b)^2}{4}$. The vertex is then (h, k) where $h = \frac{a + b}{2}$ and

$$k = ab - \frac{(a + b)^2}{4} = -\frac{(a - b)^2}{4}.$$

85. The Constant Rate of Change Theorem states that a function defined on all real numbers is a linear function if and only if it has a constant nonzero average rate of change between any two points on its graph. To prove this, suppose $f(x) = mx + b$ with m and b constants and $m \neq 0$. Let x_1 and x_2 be real numbers with $x_1 \neq x_2$. Then the average rate of change is

$$\frac{f(x_2) - f(x_1)}{x_2 - x_1} = \frac{(mx_2 + b) - (mx_1 + b)}{x_2 - x_1} = \frac{mx_2 - mx_1}{x_2 - x_1} = \frac{m(x_2 - x_1)}{x_2 - x_1} = m,$$ a nonzero constant.

Now suppose that m and x_1 are constants, with $m \neq 0$. Let x be a real number such that $x \neq x_1$, and let f be a function defined on all real numbers such that $\frac{f(x) - f(x_1)}{x - x_1} = m$. Then $f(x) - f(x_1) = m(x - x_1) = mx - mx_1$, and $f(x) = mx + (f(x_1) - mx_1)$. $f(x_1) - mx_1$ is a constant; call it b. Then $f(x_1) - mx_1 = b$; so, $f(x_1) = b + mx_1$ and $f(x) = b + mx$ for all $x \neq x_1$. Thus, f is a linear function.

Section 2.2 Power Functions with Modeling

Exploration 1

1.

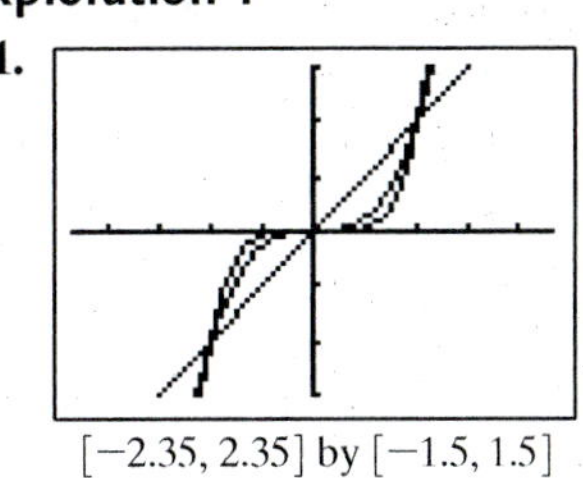

$[-2.35, 2.35]$ by $[-1.5, 1.5]$

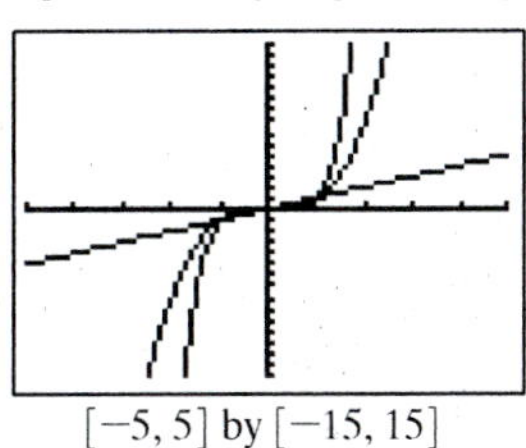

$[-5, 5]$ by $[-15, 15]$

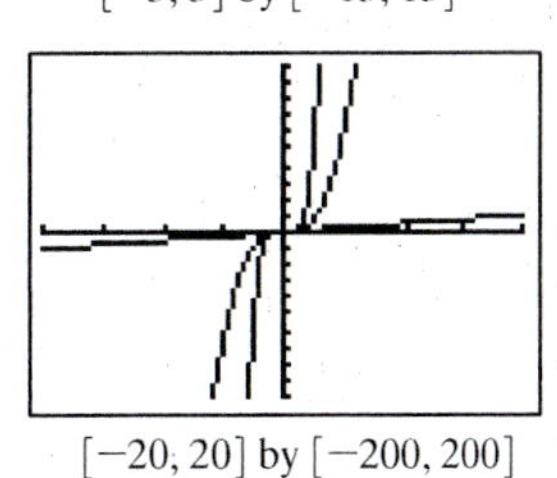

$[-20, 20]$ by $[-200, 200]$

The pairs $(0, 0)$, $(1, 1)$, and $(-1, -1)$ are common to all three graphs. The graphs are similar in that if $x < 0$, $f(x)$, $g(x)$, and $h(x) < 0$ and if $x > 0$, $f(x)$, $g(x)$, and $h(x) > 0$. They are different in that if $|x| < 1$, $f(x)$, $g(x)$, and $h(x) \to 0$ at dramatically different rates, and if $|x| > 1$, $f(x)$, $g(x)$, and $h(x) \to \infty$ at dramatically different rates.

2.

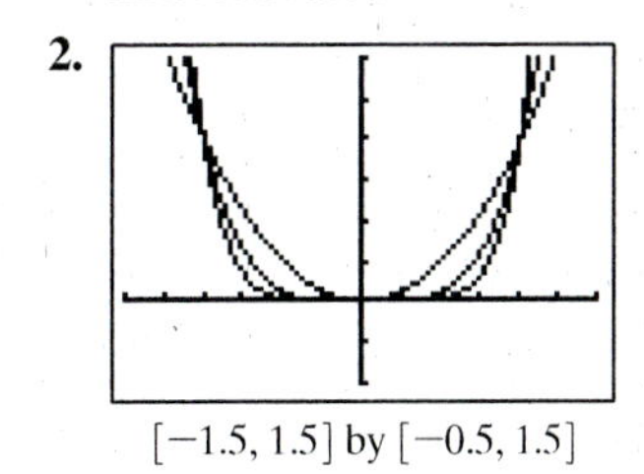

$[-1.5, 1.5]$ by $[-0.5, 1.5]$

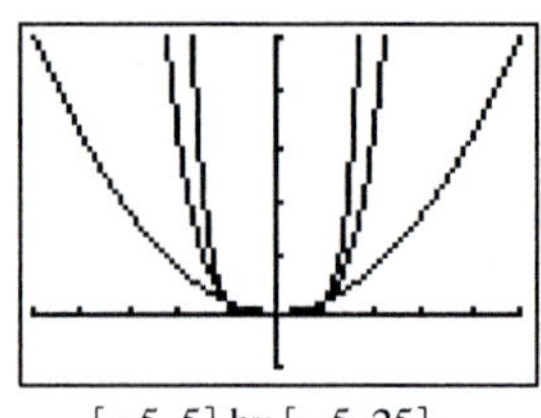
$[-5, 5]$ by $[-5, 25]$

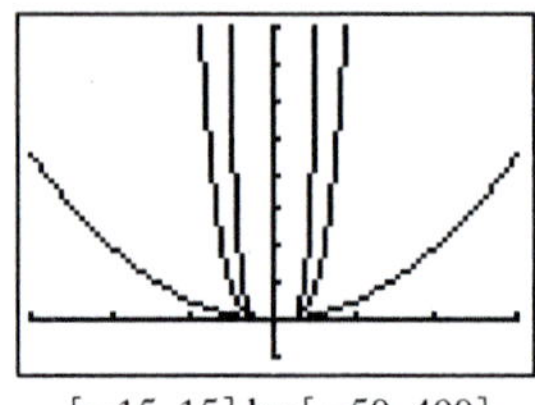
$[-15, 15]$ by $[-50, 400]$

The pairs $(0, 0)$, $(1, 1)$, and $(-1, 1)$ are common to all three graphs. The graphs are similar in that for $x \neq 0$, $f(x)$, $g(x)$, and $h(x) > 0$. They are diffferent in that if $|x| < 1$, $f(x)$, $g(x)$, and $h(x) \to 0$ at dramatically different rates, and if $|x| > 1$, $f(x)$, $g(x)$, and $h(x) \to \infty$ at dramatically different rates.

Quick Review 2.2

1. $\sqrt[3]{x^2}$
3. $\dfrac{1}{d^2}$
5. $\dfrac{1}{\sqrt[5]{q^4}}$
7. $3x^{3/2}$
9. $\approx 1.71x^{-4/3}$

Section 2.2 Exercises

1. power $= 5$, constant $= -\dfrac{1}{2}$
3. not a power function
5. power $= 1$, constant $= c^2$
7. power $= 2$, constant $= \dfrac{g}{2}$
9. power $= -2$, constant $= k$
11. degree $= 0$, coefficient $= -4$
13. degree $= 7$, coefficient $= -6$
15. degree $= 2$, coefficient $= 4\pi$
17. $A = ks^2$
19. $I = V/R$
21. $E = mc^2$
23. The weight w of an object varies directly with its mass m, with the constant of variation g.
25. The refractive index n of a medium is inversely proportional to v, the velocity of light in the medium, with constant of variation c, the constant velocity of light in free space.
27. power $= 4$, constant $= 2$
 Domain: $(-\infty, \infty)$
 Range: $[0, \infty)$
 Continuous
 Decreasing on $(-\infty, 0)$. Increasing on $(0, \infty)$.
 Even. Symmetric with respect to y-axis.
 Bounded below, but not above
 Local minimum at $x = 0$.
 Asymptotes: None
 End Behavior: $\lim_{x\to-\infty} 2x^4 = \infty$, $\lim_{x\to\infty} 2x^4 = \infty$

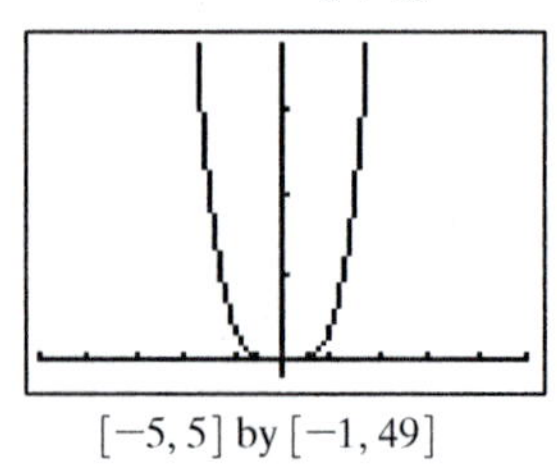
$[-5, 5]$ by $[-1, 49]$

29. power $= \dfrac{1}{4}$, constant $= \dfrac{1}{2}$
 Domain: $[0, \infty)$
 Range: $[0, \infty)$
 Continuous
 Increasing on $[0, \infty)$
 Bounded below
 Neither even nor odd
 Local minimum at $(0, 0)$
 Asymptotes: None
 End Behavior: $\lim_{x\to\infty} \dfrac{1}{2}\sqrt[4]{x} = \infty$

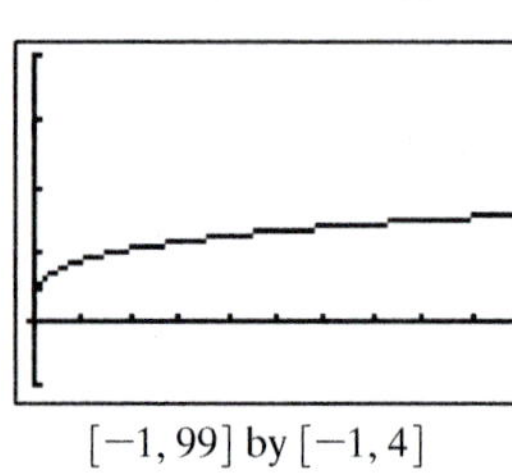
$[-1, 99]$ by $[-1, 4]$

31. Start with $y = x^4$ and shrink vertically by $\dfrac{2}{3}$. Since $f(-x) = \dfrac{2}{3}(-x)^4 = \dfrac{2}{3}x^4$, f is even.

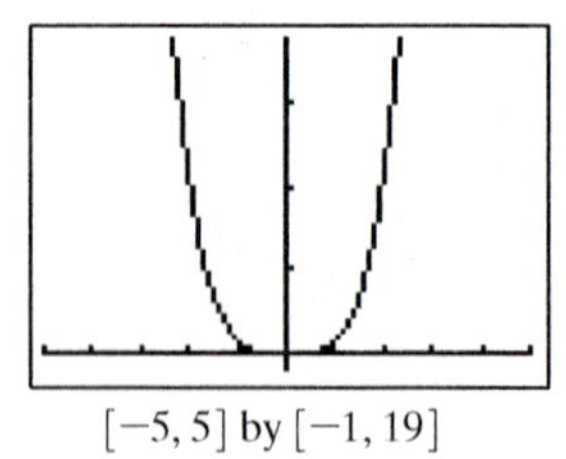
$[-5, 5]$ by $[-1, 19]$

33. Start with $y = x^5$, then stretch vertically by 1.5 and reflect over the x-axis. Since $f(-x) = -1.5(-x)^5 = 1.5x^5 = -f(x)$, f is odd.

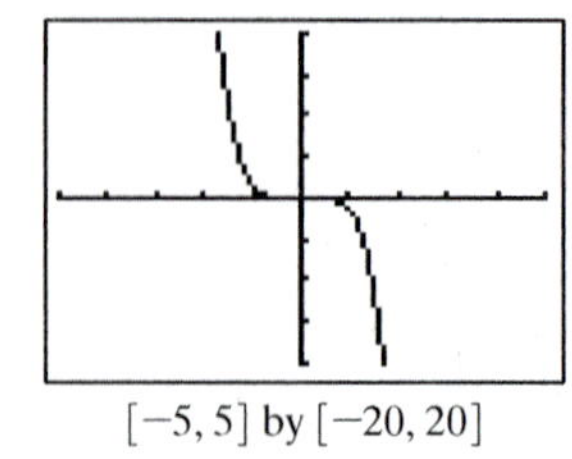
$[-5, 5]$ by $[-20, 20]$

35. Start with $y = x^8$, then shrink vertically by $\frac{1}{4}$. Since $f(-x) = \frac{1}{4}(-x)^8 = \frac{1}{4}x^8 = f(x)$, f is even.

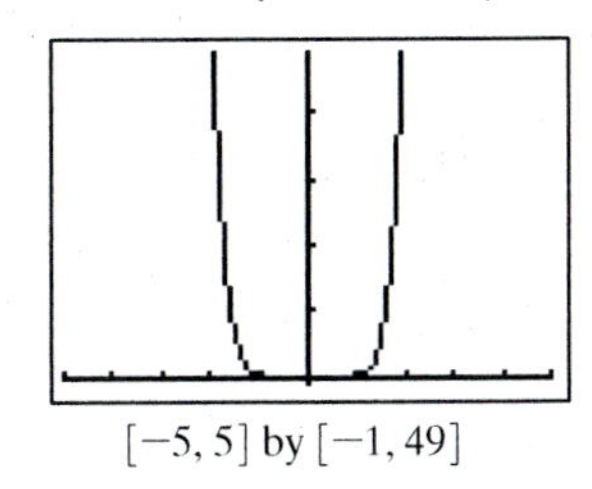

$[-5, 5]$ by $[-1, 49]$

37. (g)

39. (d)

41. (h)

43. $k = 3, a = \frac{1}{4}$. In the first quadrant, the function is increasing and concave down. f is undefined for $x < 0$.

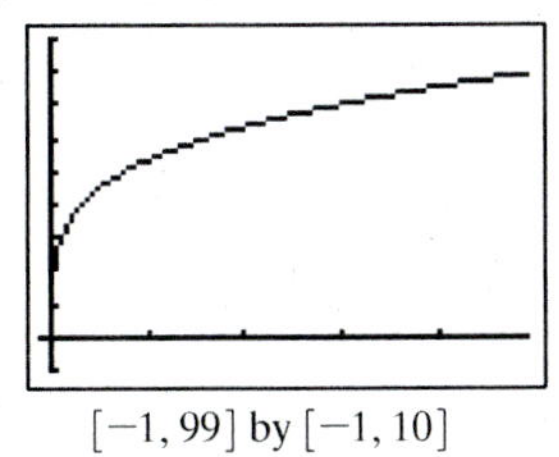

$[-1, 99]$ by $[-1, 10]$

45. $k = -2, a = \frac{4}{3}$. In the fourth quadrant, f is decreasing and concave down. $f(-x) = -2(\sqrt[3]{(-x)^4})$ $= -2(\sqrt[3]{x^4}) = -2x^{4/3} = f(x)$, so f is even.

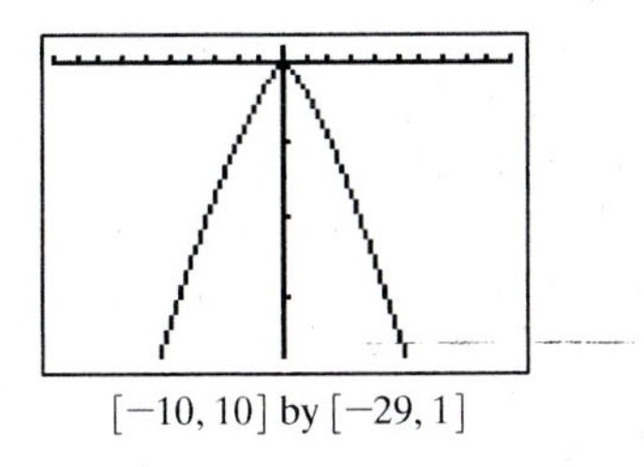

$[-10, 10]$ by $[-29, 1]$

47. $k = \frac{1}{2}, a = -3$. In the first quadrant, f is decreasing and concave up. $f(-x) = \frac{1}{2}(-x)^{-3} = \frac{1}{2(-x)^3} = -\frac{1}{2}x^{-3}$ $= -f(x)$, so f is odd.

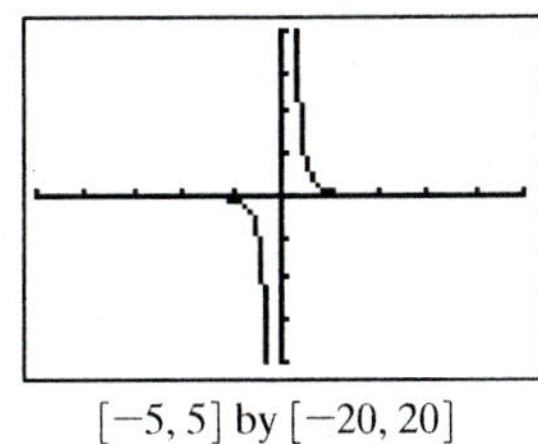

$[-5, 5]$ by $[-20, 20]$

49. $y = \frac{8}{x^2}$, power $= -2$, constant $= 8$

51. $V = \frac{kT}{P}$, so $k = \frac{PV}{T} = \frac{(0.926 \text{ atm})(3.46 \text{ L})}{302°\text{K}}$ $= 0.0106 \frac{\text{atm–L}}{\text{K}}$

At $P = 1.452$ atm, $V = \dfrac{\left(\frac{0.0106 \text{ atm–L}}{\text{K}}\right)(302°\text{K})}{1.452 \text{ atm}}$ $= 2.21$ L

53. $n = \frac{c}{v}$, so $v = \frac{c}{n} = \dfrac{\left(\frac{3.00 \times 10^8 \text{ m}}{\text{sec}}\right)}{2.42} = 1.24 \times 10^8 \frac{\text{m}}{\text{sec}}$

55. (a)

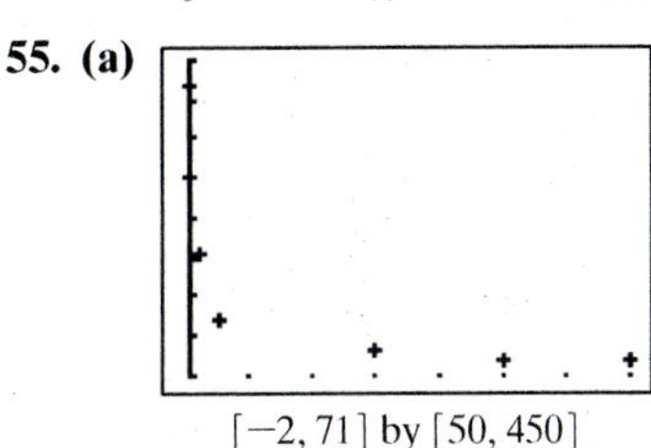

$[-2, 71]$ by $[50, 450]$

(b) $r \approx 231.204 \cdot w^{-0.297}$

(c)

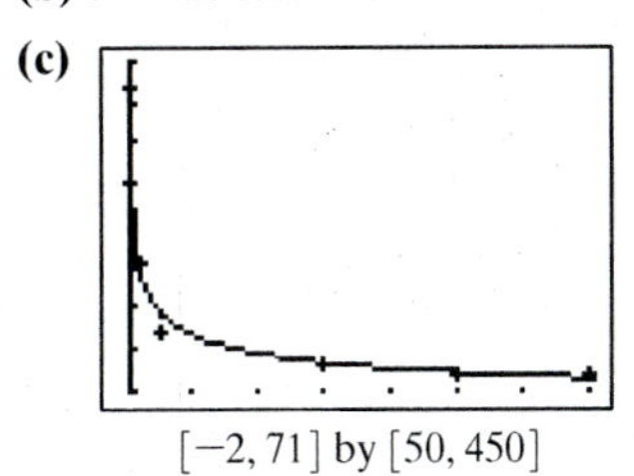

$[-2, 71]$ by $[50, 450]$

(d) Approximately 37.67 beats/min, which is very close to Clark's observed value.

57. (a)

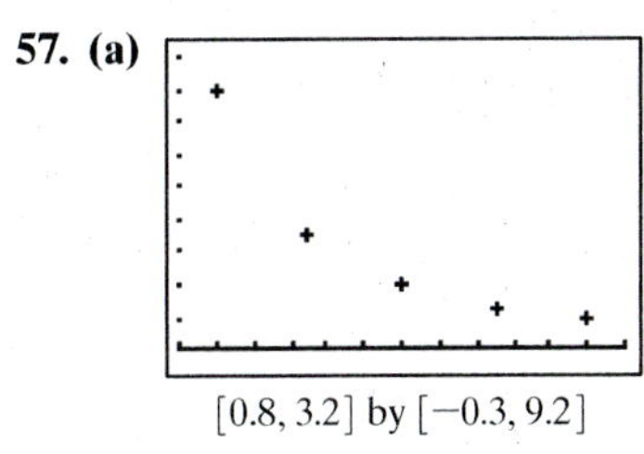

$[0.8, 3.2]$ by $[-0.3, 9.2]$

(b) $y \approx 7.932 \cdot x^{-1.987}$; yes

(c)

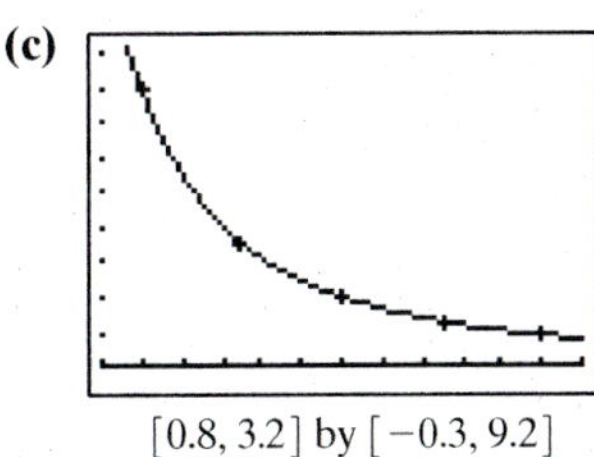

$[0.8, 3.2]$ by $[-0.3, 9.2]$

(d) Approximately $2.76 \frac{\text{W}}{\text{m}^2}$ and $0.697 \frac{\text{W}}{\text{m}^2}$, respectively.

59. False. $f(-x) = (-x)^{1/3} = -(x^{1/3}) = -f(x)$ and so the function is odd. It is symmetric about the origin, not the y-axis.

61. $f(0) = -3(0)^{-1/3} = -3 \cdot \frac{1}{0^{1/3}} = -3 \cdot \frac{1}{0}$ is undefined.

Also, $f(-1) = -3(-1)^{-1/3} = -3(-1) = 3$,
$f(1) = -3(1)^{-1/3} = -3(1) = -3$, and
$f(3) = -3(3)^{-1/3} \approx -2.08$. The answer is E.

63. $f(x) = x^{3/2} = (x^{1/2})^3 = (\sqrt{x})^3$ is defined for $x \geq 0$. The answer is B.

65. (a)

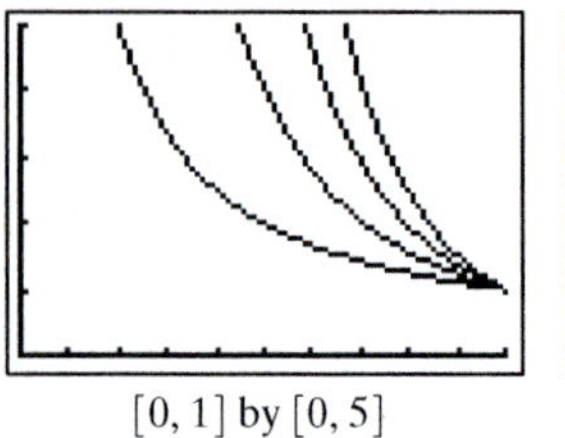
[0, 1] by [0, 5]

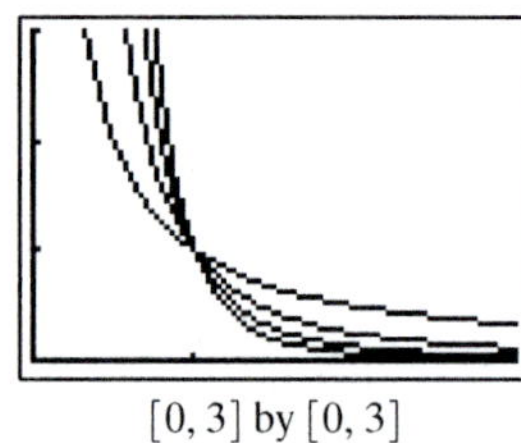
[0, 3] by [0, 3]

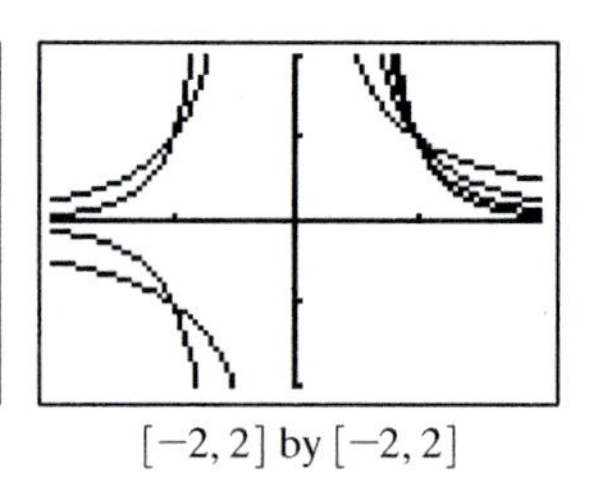
[−2, 2] by [−2, 2]

The graphs of $f(x) = x^{-1}$ and $h(x) = x^{-3}$ are similar and appear in the 1st and 3rd quadrants only. The graphs of $g(x) = x^{-2}$ and $k(x) = x^{-4}$ are similar and appear in the 1st and 2nd quadrants only. The pair (1, 1) is common to all four functions.

	f	g	h	k
Domain	$x \neq 0$	$x \neq 0$	$x \neq 0$	$x \neq 0$
Range	$y \neq 0$	$y > 0$	$y \neq 0$	$y > 0$
Continuous	yes	yes	yes	yes
Increasing		$(-\infty, 0)$		$(-\infty, 0)$
Decreasing	$(-\infty, 0), (0, \infty)$	$(0, \infty)$	$(-\infty, 0), (0, \infty)$	$(0, \infty)$
Symmetry	w.r.t. origin	w.r.t. y-axis	w.r.t. origin	w.r.t. y-axis
Bounded	not	below	not	below
Extrema	none	none	none	none
Asymptotes	x-axis, y-axis	x-axis, y-axis	x-axis, y-axis	x-axis, y-axis
End Behavior	$\lim_{x \to \pm\infty} f(x) = 0$	$\lim_{x \to \pm\infty} g(x) = 0$	$\lim_{x \to \pm\infty} h(x) = 0$	$\lim_{x \to \pm\infty} k(x) = 0$

(b)

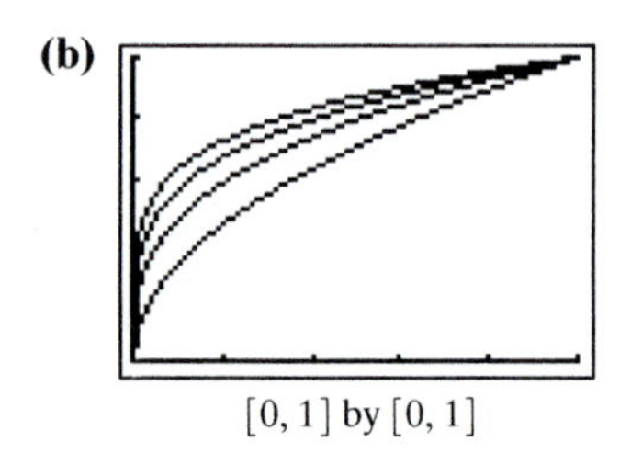
[0, 1] by [0, 1]

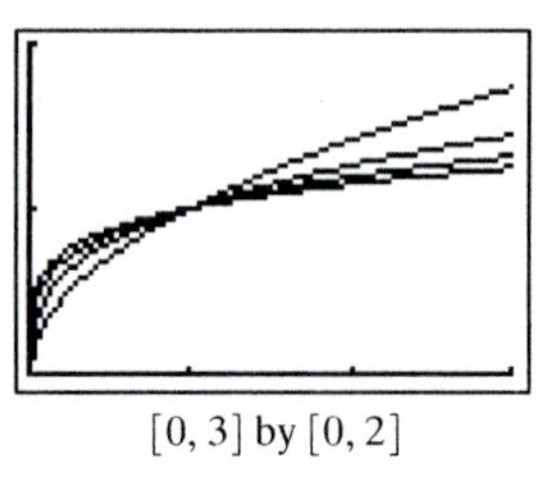
[0, 3] by [0, 2]

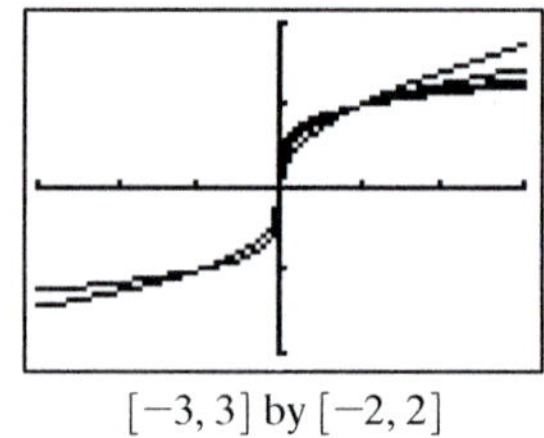
[−3, 3] by [−2, 2]

The graphs of $f(x) = x^{1/2}$ and $h(x) = x^{1/4}$ are similar and appear in the 1st quadrant only. The graphs of $g(x) = x^{1/3}$ and $k(x) = x^{1/5}$ are similar and appear in the 1st and 3rd quadrants only. The pairs (0, 0), (1, 1) are common to all four functions.

	f	g	h	k
Domain	$[0, \infty)$	$(-\infty, \infty)$	$[0, \infty)$	$(-\infty, \infty)$
Range	$y \geq 0$	$(-\infty, \infty)$	$y \geq 0$	$(-\infty, \infty)$
Continuous	yes	yes	yes	yes
Increasing	$[0, \infty)$	$(-\infty, \infty)$	$[0, \infty)$	$(-\infty, \infty)$
Decreasing				
Symmetry	none	w.r.t. origin	none	w.r.t. origin
Bounded	below	not	below	not
Extrema	min at (0, 0)	none	min at (0, 0)	none
Asymptotes	none	none	none	none
End behavior	$\lim_{x \to \infty} f(x) = \infty$	$\lim_{x \to \infty} g(x) = \infty$ $\lim_{x \to -\infty} g(x) = -\infty$	$\lim_{x \to \infty} h(x) = \infty$	$\lim_{x \to \infty} k(x) = \infty$ $\lim_{x \to -\infty} k(x) = -\infty$

67. Our new table looks like:

Table 2.10 (revised) Average Distances and Orbit Periods for the Six Innermost Planets

Planet	Average Distance from Sun (Au)	Period of Orbit (yrs)
Mercury	0.39	0.24
Venus	0.72	0.62
Earth	1	1
Mars	1.52	1.88
Jupiter	5.20	11.86
Saturn	9.54	29.46

Source: Shupe, Dorr, Payne, Hunsiker, et al., National Geographic Atlas of the World (rev. 6th ed.). Washington, DC: National Geographic Society, 1992, plate 116.

Using these new data, we find a power function model of: $y \approx 0.99995 \cdot x^{1.50115} \approx x^{1.5}$. Since y represents years, we set $y = T$ and since x represents distance, we set $x = a$ then, $y = x^{1.5} \Rightarrow T = a^{3/2} \Rightarrow (T)^2 = (a^{3/2})^2 \Rightarrow T^2 = a^3$.

69. If f is even,

$f(x) = f(-x)$, so $\dfrac{1}{f(x)} = \dfrac{1}{f(-x)}$, $(f(x) \neq 0)$.

Since $g(x) = \dfrac{1}{f(x)} = \dfrac{1}{f(-x)} = g(-x)$, g is also even.

If g is even,

$g(x) = g(-x)$, so $g(-x) = \dfrac{1}{f(-x)} = g(x) = \dfrac{1}{f(x)}$.

Since $\dfrac{1}{f(-x)} = \dfrac{1}{f(x)}$, $f(-x) = f(x)$, and f is even.

If f is odd,

$f(x) = -f(x)$, so $\dfrac{1}{f(x)} = -\dfrac{1}{f(x)}$, $f(x) \neq 0$.

Since $g(x) = \dfrac{1}{f(x)} = -\dfrac{1}{f(x)} = -g(x)$, g is also odd.

If g is odd,

$g(x) = g(-x)$, so $g(-x) = \dfrac{1}{f(-x)} = -g(x) = -\dfrac{1}{f(x)}$.

Since $\dfrac{1}{f(-x)} = -\dfrac{1}{f(x)}$, $f(-x) = -f(x)$, and f is odd.

71. (a) The force F acting on an object varies jointly as the mass m of the object and the acceleration a of the object.

(b) The kinetic energy KE of an object varies jointly as the mass m of the object and the square of the velocity v of the object.

(c) The force of gravity F acting on two objects varies jointly as their masses m_1 and m_2 and inversely as the square of the distance r between their centers, with the constant of variation G, the universal gravitational constant.

Section 2.3 Polynomial Functions of Higher Degree with Modeling

Exploration 1

1. (a) $\lim_{x\to\infty} 2x^3 = \infty$, $\lim_{x\to-\infty} 2x^3 = -\infty$

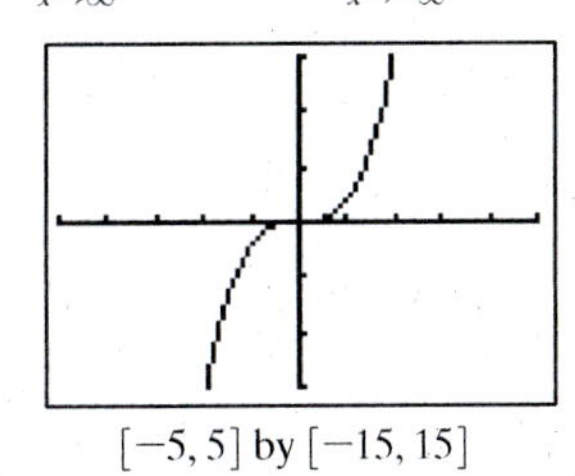

$[-5, 5]$ by $[-15, 15]$

(b) $\lim_{x\to\infty}(-x^3) = -\infty$, $\lim_{x\to-\infty}(-x^3) = \infty$

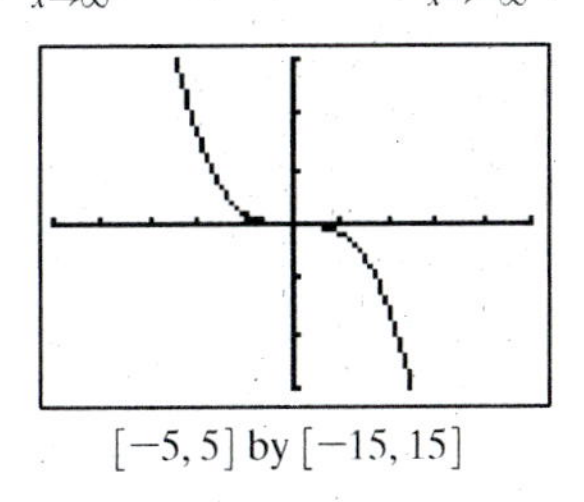

$[-5, 5]$ by $[-15, 15]$

(c) $\lim_{x\to\infty} x^5 = \infty$, $\lim_{x\to-\infty} x^5 = -\infty$

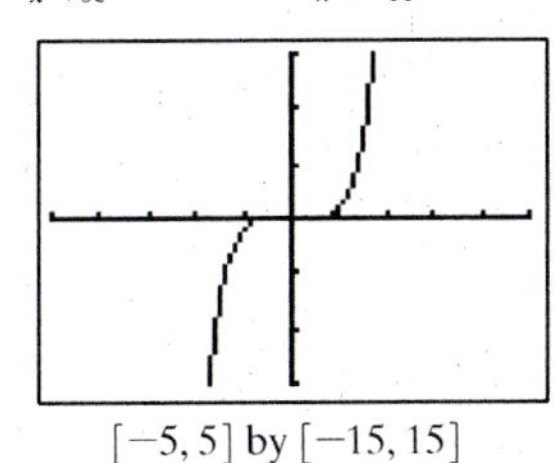

$[-5, 5]$ by $[-15, 15]$

(d) $\lim_{x\to\infty}(-0.5x^7) = -\infty$, $\lim_{x\to-\infty}(-0.5x^7) = \infty$

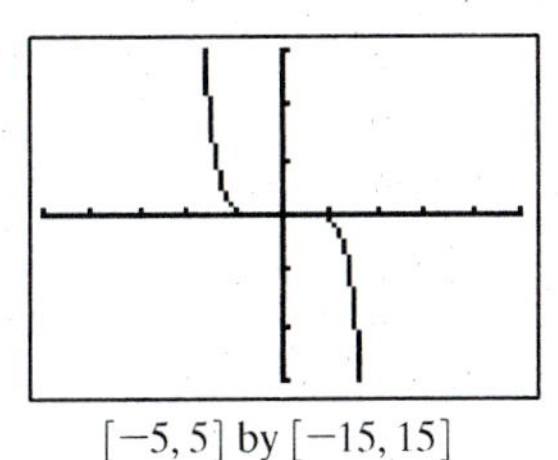

$[-5, 5]$ by $[-15, 15]$

2. (a) $\lim_{x\to\infty}(-3x^4) = -\infty$, $\lim_{x\to-\infty}(-3x^4) = -\infty$

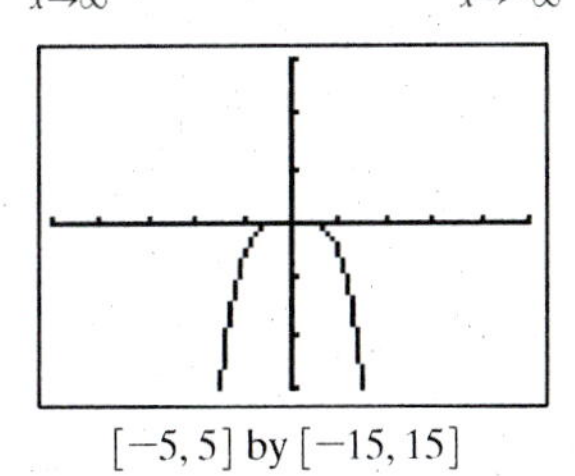

$[-5, 5]$ by $[-15, 15]$

(b) $\lim_{x\to\infty} 0.6x^4 = \infty$, $\lim_{x\to-\infty} 0.6x^4 = \infty$

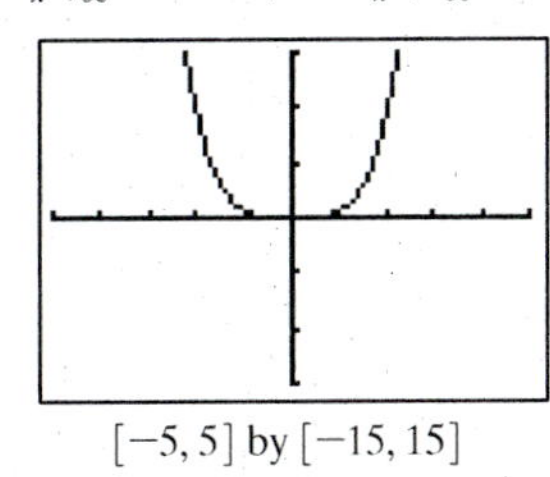

$[-5, 5]$ by $[-15, 15]$

(c) $\lim_{x\to\infty} 2x^6 = \infty$, $\lim_{x\to-\infty} 2x^6 = \infty$

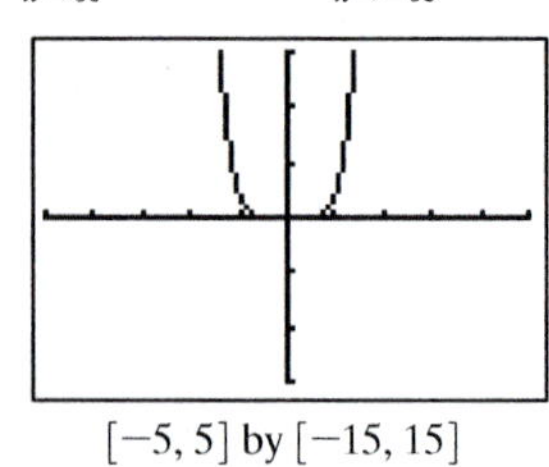

[−5, 5] by [−15, 15]

(d) $\lim_{x\to\infty} (-0.5x^2) = -\infty$, $\lim_{x\to-\infty} (-0.5x^2) = -\infty$

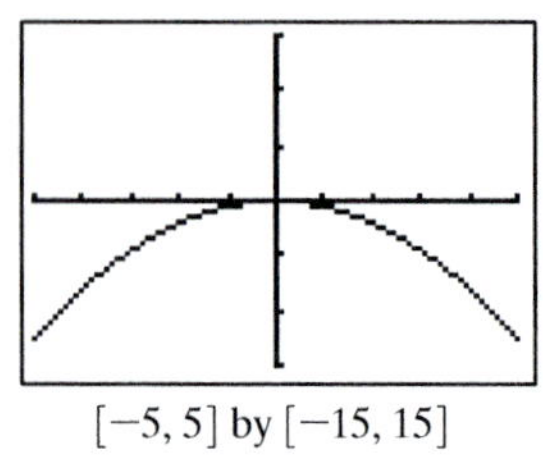

[−5, 5] by [−15, 15]

3. (a) $\lim_{x\to\infty} (-0.3x^5) = -\infty$, $\lim_{x\to-\infty} (-0.3x^5) = \infty$

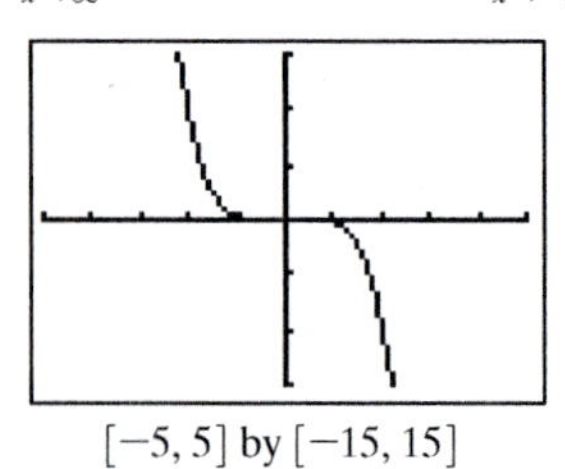

[−5, 5] by [−15, 15]

(b) $\lim_{x\to\infty} (-2x^2) = -\infty$, $\lim_{x\to-\infty} (-2x^2) = -\infty$

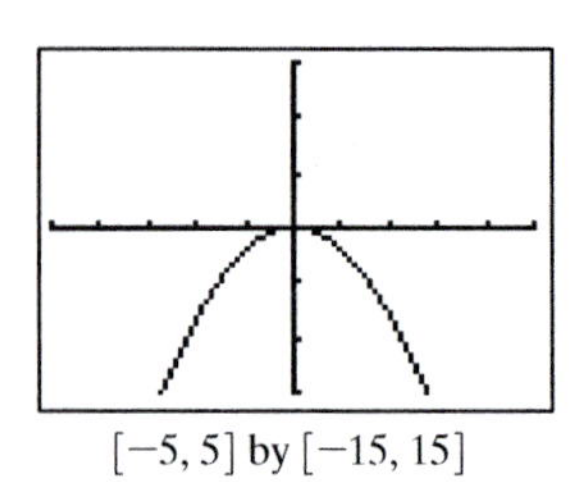

[−5, 5] by [−15, 15]

(c) $\lim_{x\to\infty} 3x^4 = \infty$, $\lim_{x\to-\infty} 3x^4 = \infty$

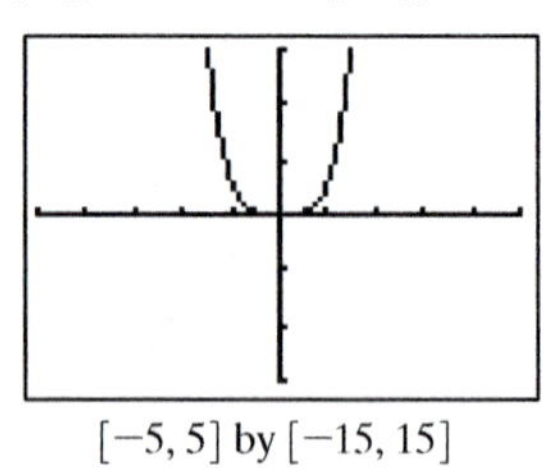

[−5, 5] by [−15, 15]

(d) $\lim_{x\to\infty} 2.5x^3 = \infty$, $\lim_{x\to-\infty} 2.5x^3 = -\infty$

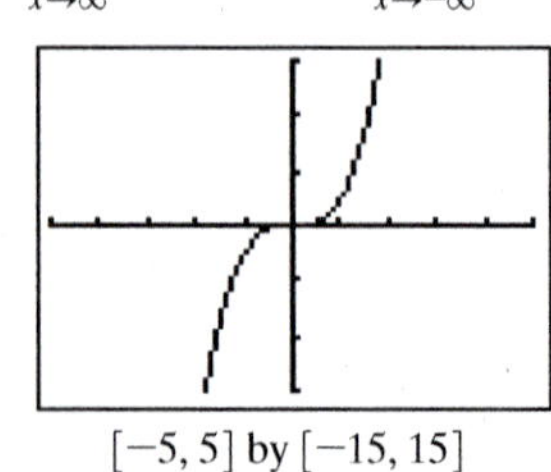

[−5, 5] by [−15, 15]

If $a_n > 0$ and $n > 0$, $\lim_{x\to\infty} f(x) = \infty$ and $\lim_{x\to-\infty} f(x) = -\infty$. If $a_n < 0$ and $n > 0$, $\lim_{x\to\infty} f(x) = -\infty$ and $\lim_{x\to-\infty} f(x) = \infty$.

Exploration 2

1. $y = 0.0061x^3 + 0.0177x^2 - 0.5007x + 0.9769$ It is an exact fit, which we expect with only 4 data points!

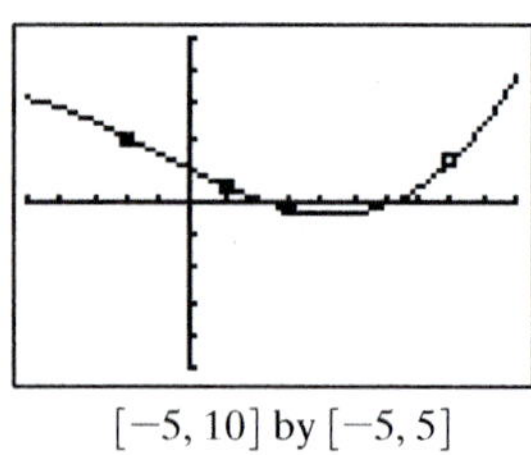

[−5, 10] by [−5, 5]

2. $y = -0.375x^4 + 6.917x^3 - 44.125x^2 + 116.583x - 111$ It is an exact fit, exactly what we expect with only 5 data points!

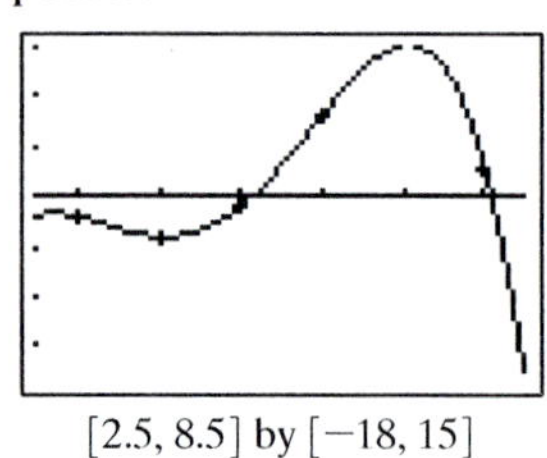

[2.5, 8.5] by [−18, 15]

Quick Review 2.3

1. $(x - 4)(x + 3)$

3. $(3x - 2)(x - 3)$

5. $x(3x - 2)(x - 1)$

7. $x = 0, x = 1$

9. $x = -6, x = -3, x = 1.5$

Section 2.3 Exercises

1. Start with $y = x^3$, shift to the right by 3 units, and then stretch vertically by 2. y-intercept: $(0, -54)$

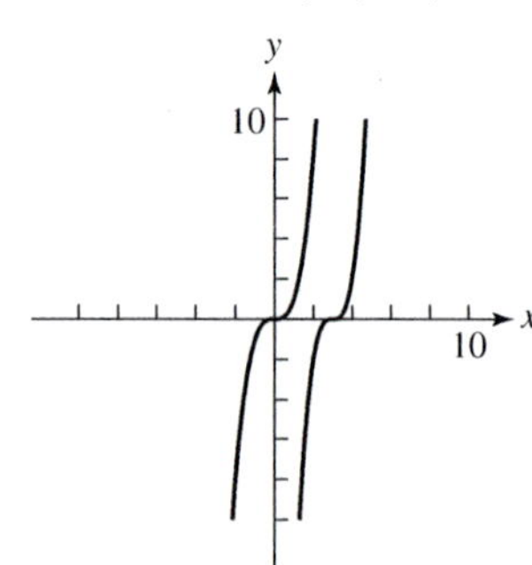

3. Start with $y = x^3$, shift to the left by 1 unit, vertically shrink by $\frac{1}{2}$, reflect over the x-axis, and then vertically shift up 2 units. y-intercept: $\left(0, \frac{3}{2}\right)$

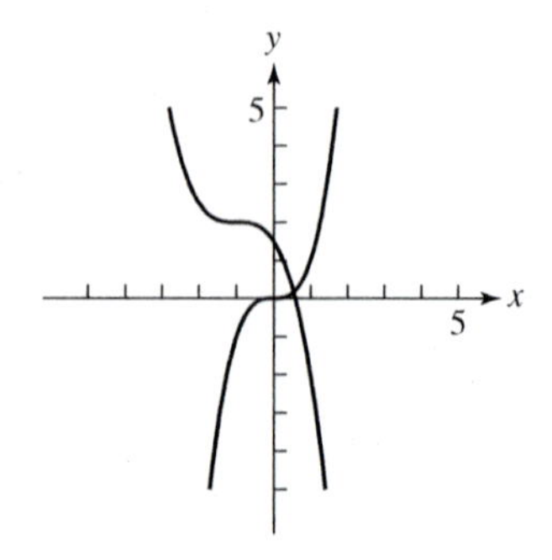

5. Start with $y = x^4$, shift to the left 2 units, vertically stretch by 2, reflect over the x-axis, and vertically shift down 3 units. y-intercept: $(0, -35)$

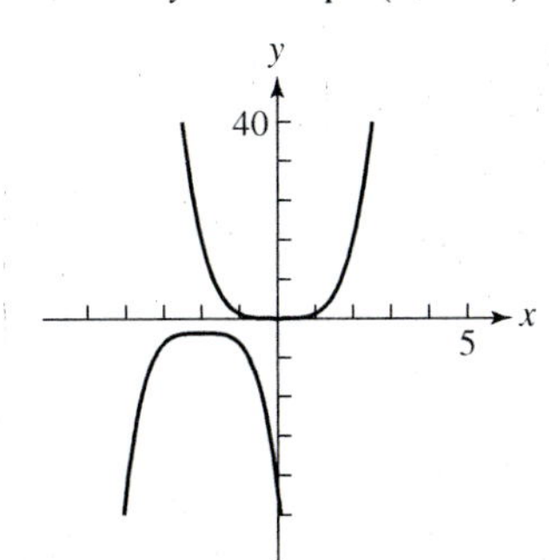

7. local maximum: $\approx (0.79, 1.19)$, zeros: $x = 0$ and $x \approx 1.26$. The general shape of f is like $y = -x^4$, but near the origin, f behaves a lot like its other term, $2x$. f is neither even nor odd.

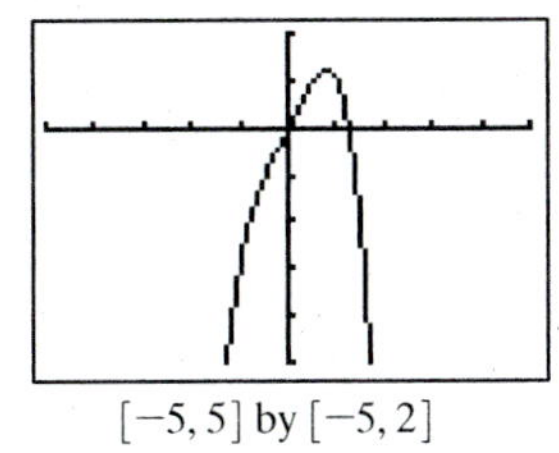

$[-5, 5]$ by $[-5, 2]$

9. Cubic function, positive leading coefficient. The answer is (c).

11. Higher than cubic, positive leading coefficient. The answer is (a).

13. One possibility:

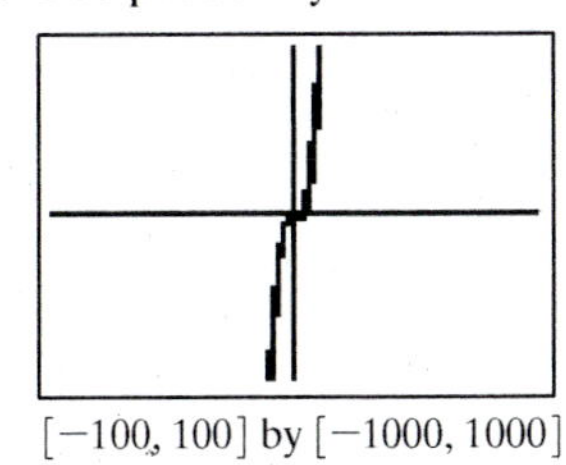

$[-100, 100]$ by $[-1000, 1000]$

15. One possibility:

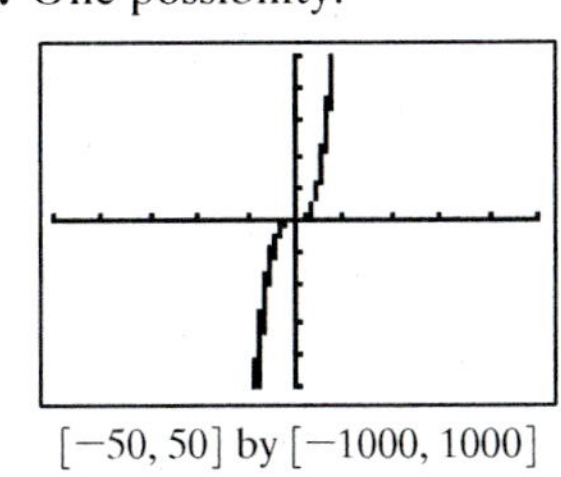

$[-50, 50]$ by $[-1000, 1000]$

For #17–23, when one end of a polynomial function's graph curves up into Quadrant I or II, this indicates a limit at ∞. And when an end curves down into Quadrant III or IV, this indicates a limit at $-\infty$.

17.

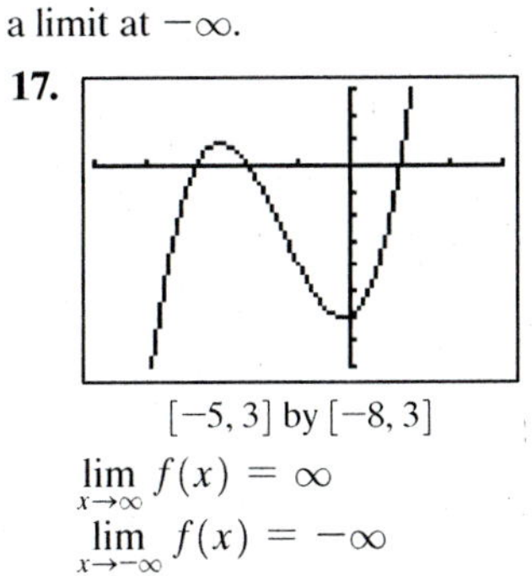

$[-5, 3]$ by $[-8, 3]$

$\lim_{x\to\infty} f(x) = \infty$
$\lim_{x\to-\infty} f(x) = -\infty$

19.

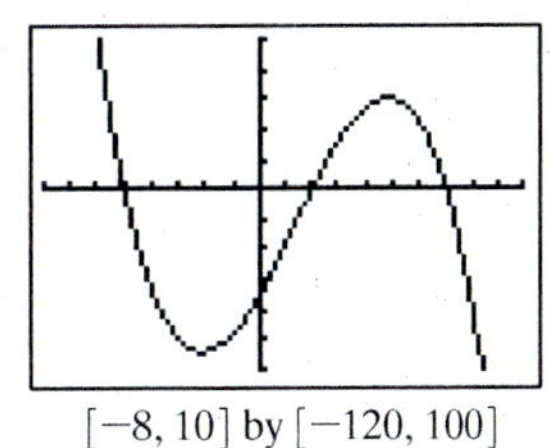

$[-8, 10]$ by $[-120, 100]$

$\lim_{x\to\infty} f(x) = -\infty$
$\lim_{x\to-\infty} f(x) = \infty$

21.

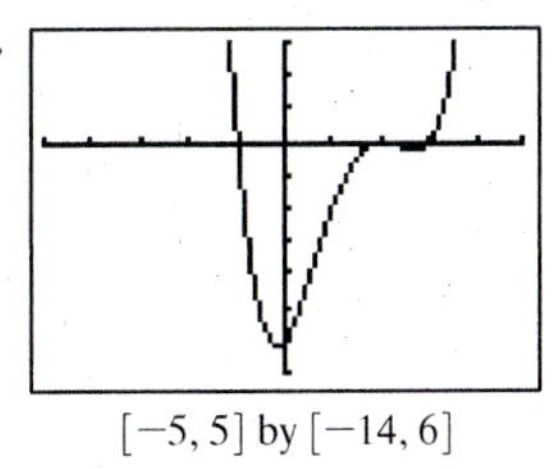

$[-5, 5]$ by $[-14, 6]$

$\lim_{x\to\infty} f(x) = \infty$
$\lim_{x\to-\infty} f(x) = \infty$

23.

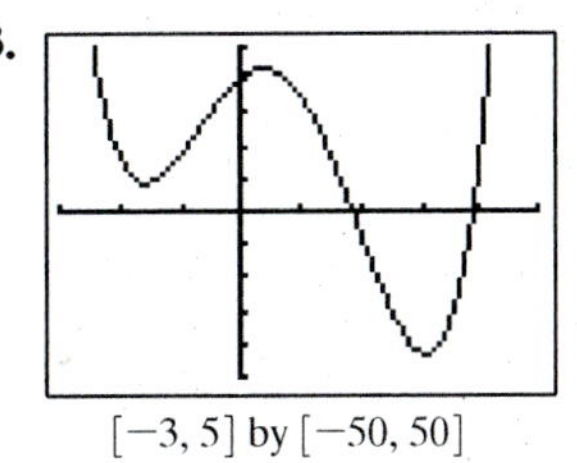

$[-3, 5]$ by $[-50, 50]$

$\lim_{x\to\infty} f(x) = \infty$
$\lim_{x\to-\infty} f(x) = \infty$

For #25–27, the end behavior of a polynomial is governed by the highest-degree term.

25. $\lim_{x\to\infty} f(x) = \infty$, $\lim_{x\to-\infty} f(x) = \infty$

27. $\lim_{x\to\infty} f(x) = -\infty$, $\lim_{x\to-\infty} f(x) = \infty$

29. (a); There are 3 zeros: they are -2.5, 1, and 1.1.

31. (c); There are 3 zeros: approximately -0.273 (actually $-3/11$), -0.25, and 1.

For #33–35, factor or apply the quadratic formula.

33. -4 and 2

35. $2/3$ and $-1/3$

For #37–39, factor out x, then factor or apply the quadratic formula.

37. 0, $-2/3$, and 1

39. Degree 3; zeros: $x = 0$ (multiplicity 1, graph crosses x-axis), $x = 3$ (multiplicity 2, graph is tangent)

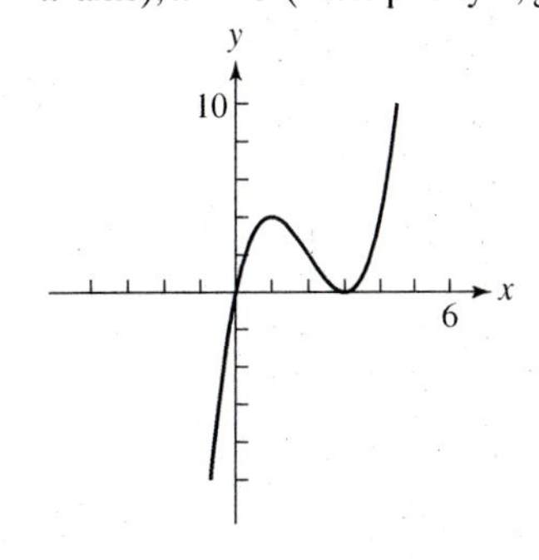

41. Degree 5; zeros: $x = 1$ (multiplicity 3, graph crosses x-axis), $x = -2$ (multiplicity 2, graph is tangent)

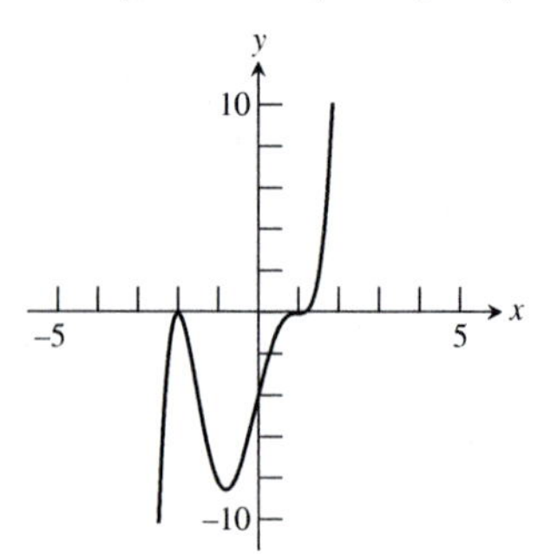

43. Zeros: $-2.43, -0.74, 1.67$

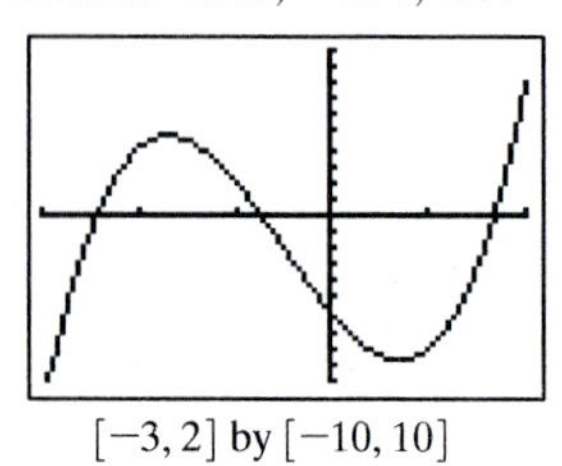

$[-3, 2]$ by $[-10, 10]$

45. Zeros: $-2.47, -1.46, 1.94$

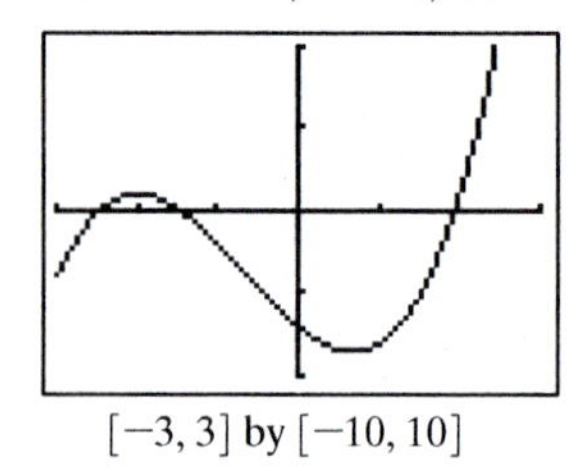

$[-3, 3]$ by $[-10, 10]$

47. Zeros: $-4.90, -0.45, 1, 1.35$

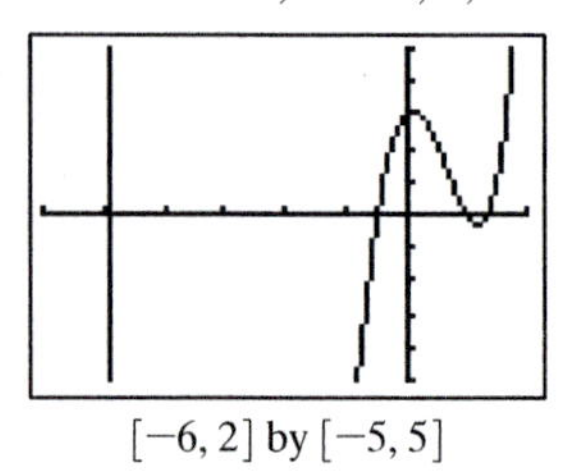

$[-6, 2]$ by $[-5, 5]$

49. $0, -6$, and 6. Algebraically — factor out x first.

51. $-5, 1$, and 11. Graphically.

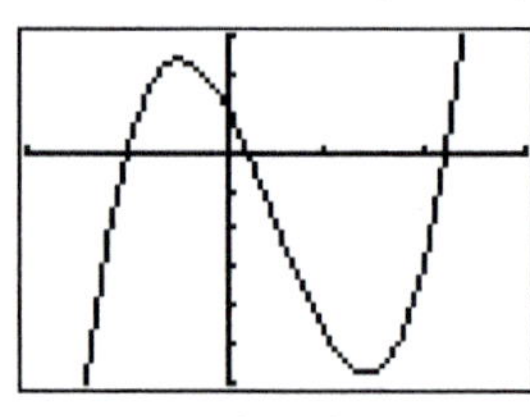

$[-10, 15]$ by $[-300, 150]$

For #53–55, the "minimal" polynomials are given; any constant (or any other polynomial) can be multiplied by the answer given to give another answer.

53. $f(x) = (x - 3)(x + 4)(x - 6)$
$= x^3 - 5x^2 - 18x + 72$

55. $f(x) = (x - \sqrt{3})(x + \sqrt{3})(x - 4)$
$= (x^2 - 3)(x - 4) = x^3 - 4x^2 - 3x + 12$

57. $y = 0.25x^3 - 1.25x^2 - 6.75x + 19.75$

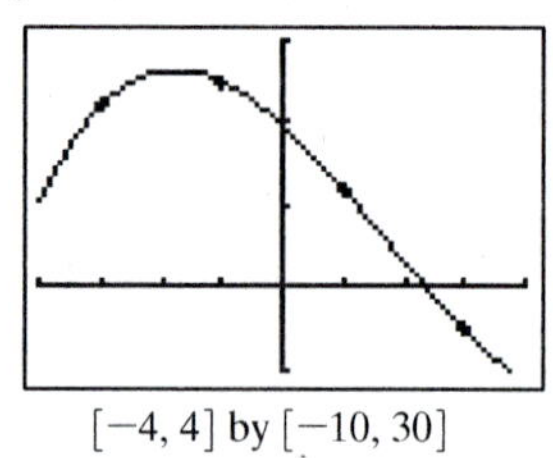

$[-4, 4]$ by $[-10, 30]$

59. $y = -2.21x^4 + 45.75x^3 - 339.79x^2 + 1075.25x - 1231$

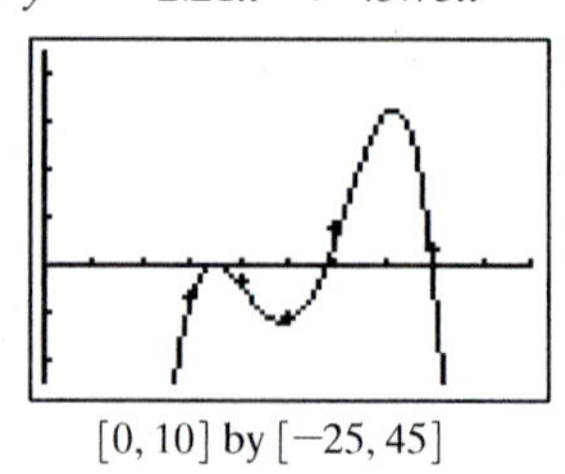

$[0, 10]$ by $[-25, 45]$

61. $f(x) = x^7 + x + 100$ has an odd-degree leading term, which means that in its end behavior it will go toward $-\infty$ at one end and toward ∞ at the other. Thus the graph must cross the x-axis at least once. That is to say, $f(x)$ takes on both positive and negative values, and thus by the Intermediate Value Theorem, $f(x) = 0$ for some x.

63. (a)

$[0, 60]$ by $[-10, 210]$

(b) $y = 0.051x^2 + 0.97x + 0.26$

(c)

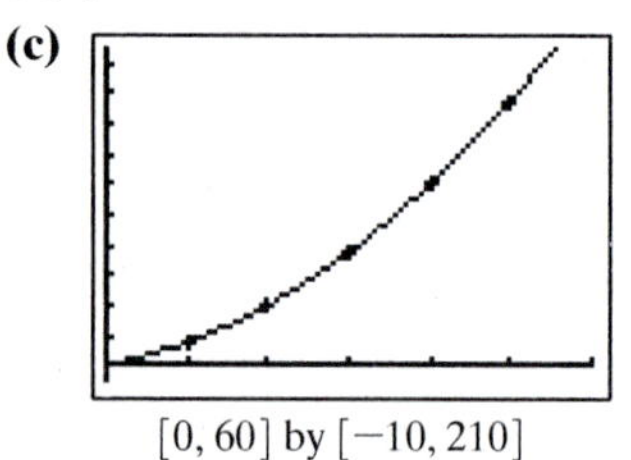

$[0, 60]$ by $[-10, 210]$

(d) $y(25) \approx 56.39$ ft

(e) Using the quadratic equation to solve $0 = 0.051x^2 + 0.97x + (0.26 - 300)$, we find two answers: $x = 67.74$ mph and $x = -86.76$ mph. Clearly the negative value is extraneous.

65. (a)

$[0, 0.8]$ by $[0, 1.20]$

(b) 0.3391 cm from the center of the artery

67. The volume is $V(x) = x(10 - 2x)(25 - 2x)$; use any x with $0 < x \leq 0.929$ or $3.644 \leq x < 5$.

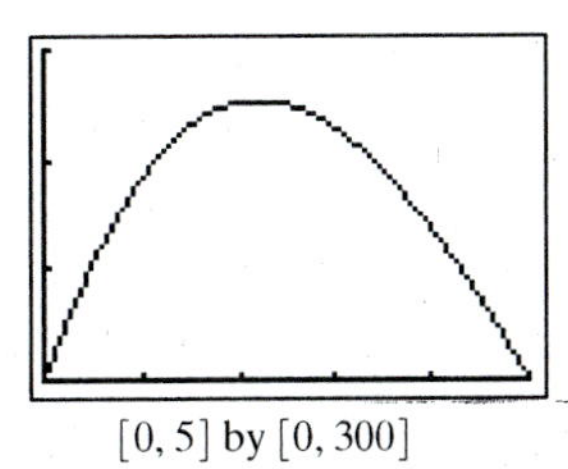

$[0, 5]$ by $[0, 300]$

69. True. Because f is continuous and
$f(1) = (1)^3 - (1)^2 - 2 = -2 < 0$
while $f(2) = (2)^3 - (2)^2 - 2 = 2 > 0$,
the Intermediate Value Theorem assures us that the graph of f crosses the x-axis $(f(x) = 0)$ somewhere between $x = 1$ and $x = 2$.

71. When $x = 0$, $f(x) = 2(x - 1)^3 + 5 = 2(-1)^3 + 5 = 3$. The answer is C.

73. The graph indicates three zeros, each of multiplicity 1: $x = -2$, $x = 0$, and $x = 2$. The end behavior indicates a negative leading coefficient. So $f(x) = -x(x + 2)(x - 2)$, and the answer is B.

75. The first view shows the end behavior of the function, but obscures the fact that there are two local maxima and a local minimum (and 4 x-axis intersections) between -3 and 4. These are visible in the second view, but missing is the minimum near $x = 7$ and the x-axis intersection near $x = 9$. The second view suggests a degree 4 polynomial rather than degree 5.

77. The exact behavior near $x = 1$ is hard to see. A zoomed–in view around the point $(1, 0)$ suggests that the graph just touches the x-axis at 0 without actually crossing it — that is, $(1, 0)$ is a local maximum. One possible window is $[0.9999, 1.0001]$ by $[-1 \times 10^{-7}, 1 \times 10^{-7}]$.

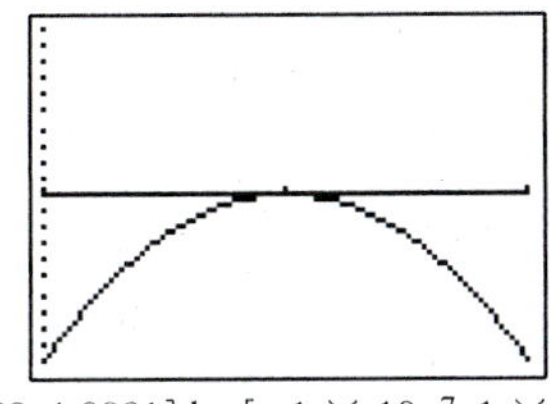

$[0.9999, 1.0001]$ by $[-1 \times 10^{-7}, 1 \times 10^{-7}]$

79. A maximum and minimum are not visible in the standard window, but can be seen on the window $[0.2, 0.4]$ by $[5.29, 5.3]$.

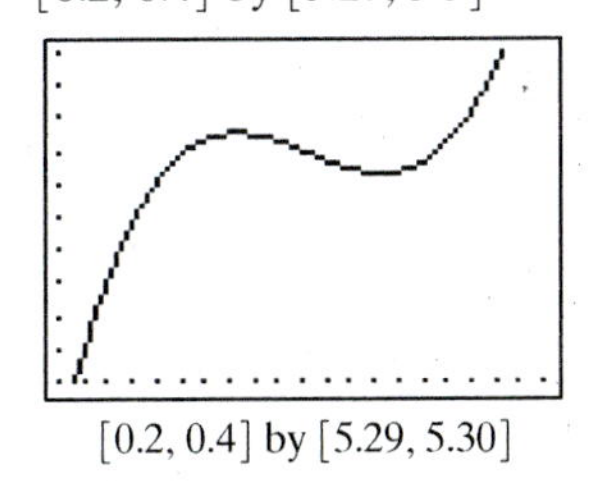

$[0.2, 0.4]$ by $[5.29, 5.30]$

81. The graph of $y = 3(x^3 - x)$ (shown on the window $[-2, 2]$ by $[-5, 5]$) increases, then decreases, then increases; the graph of $y = x^3$ only increases. Therefore, this graph cannot be obtained from the graph of $y = x^3$ by the transformations studied in Chapter 1 (translations, reflections, and stretching/shrinking). Since the right side includes only these transformations, there can be no solution.

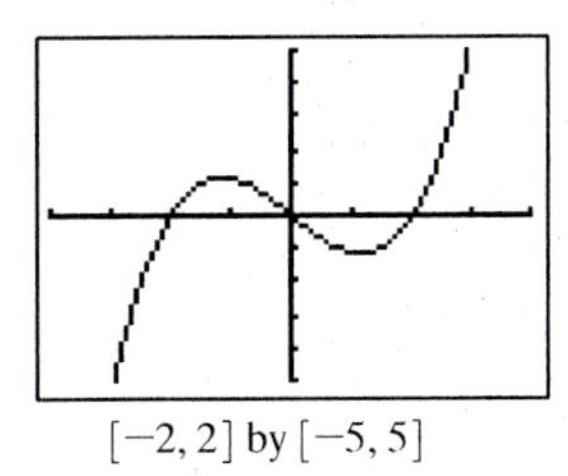

$[-2, 2]$ by $[-5, 5]$

83. **(a)** Substituting $x = 2$, $y = 7$, we find that $7 = 5(2 - 2) + 7$, so Q is on line L, and also $f(2) = -8 + 8 + 18 - 11 = 7$, so Q is on the graph of $f(x)$.

(b) Window $[1.8, 2.2]$ by $[6, 8]$. Calculator output will not show the detail seen here.

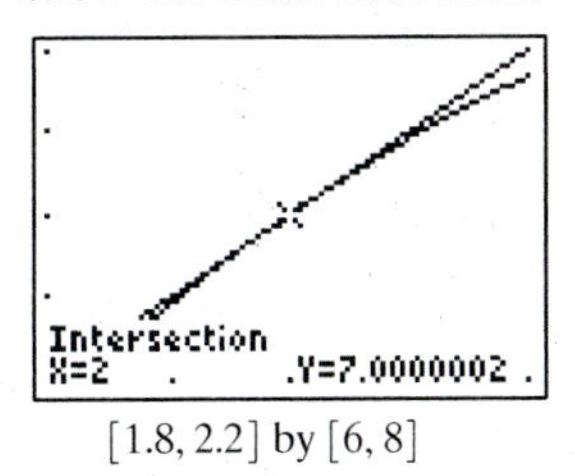

$[1.8, 2.2]$ by $[6, 8]$

(c) The line L also crosses the graph of $f(x)$ at $(-2, -13)$.

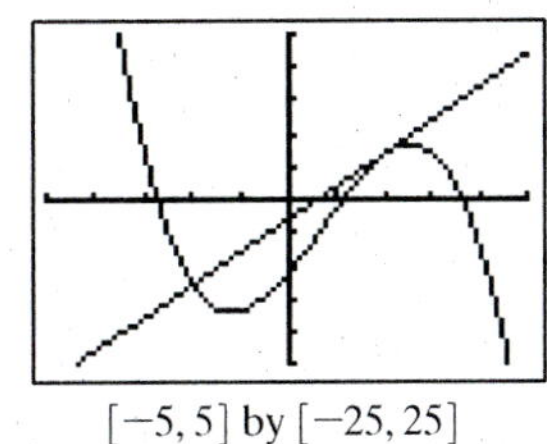

$[-5, 5]$ by $[-25, 25]$

85. **(a)** Label the points of the diagram as shown, adding the horizontal segment $\overline{FH}$. Therefore, ΔECB is similar (in the geometric sense) to ΔHGB, and also ΔABC is similar to ΔAFH. Therefore:

$$\frac{HG}{EC} = \frac{BG}{BC}, \text{ or } \frac{8}{x} = \frac{D - u}{D}, \text{ and also } \frac{AF}{AB} = \frac{FH}{BC},$$
$$\text{or } \frac{y - 8}{y} = \frac{D - u}{D}. \text{ Then } \frac{8}{x} = \frac{y - 8}{y}.$$

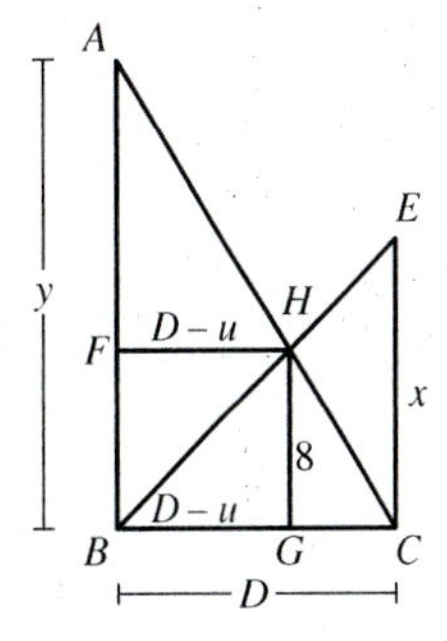

(b) Equation (a) says $\frac{8}{x} = 1 - \frac{8}{y}$. Multiply both sides by xy: $8y = xy - 8x$. Subtract xy from both sides and factor: $y(8 - x) = -8$. Divide both sides by $8 - x$: $y = \frac{-8}{8 - x}$. Factor out -1 from numerator and denominator: $y = \frac{8}{x - 8}$.

(c) Applying the Pythagorean Theorem to ΔEBC and ΔABC, we have $x^2 + D^2 = 20^2$ and $y^2 + D^2 = 30^2$, which combine to give $D^2 = 400 - x^2 = 900 - y^2$, or $y^2 - x^2 = 500$. Substituting $y = 8x/(x - 8)$, we get $\left(\dfrac{8x}{x-8}\right)^2 - x^2 = 500$, so that $\dfrac{64x^2}{(x-8)^2} - x^2 = 500$, or $64x^2 - x^2(x-8)^2 = 500(x-8)^2$. Expanding this gives $500x^2 - 8000x + 32{,}000 = 64x^2 - x^4 + 16x^3 - 64x^2$. This is equivalent to $x^4 - 16x^3 + 500x^2 - 8000x + 32{,}000 = 0$.

(d) The two solutions are $x \approx 5.9446$ and $x \approx 11.7118$. Based on the figure, x must be between 8 and 20 for this problem, so $x \approx 11.7118$. Then $D = \sqrt{20^2 - x^2} \approx 16.2121$ ft.

Section 2.4 Real Zeros of Polynomial Functions

Quick Review 2.4

1. $x^2 - 4x + 7$

3. $7x^3 + x^2 - 3$

5. $x(x^2 - 4) = x(x^2 - 2^2) = x(x+2)(x-2)$

7. $4(x^2 + 2x - 15) = 4(x+5)(x-3)$

9. $(x^3 + 2x^2) - (x+2) = x^2(x+2) - 1(x+2)$
$= (x+2)(x^2-1) = (x+2)(x+1)(x-1)$

Section 2.4 Exercises

1.

$$\begin{array}{r|l} & x - 1 \\ \hline x - 1 & x^2 - 2x + 3 \\ & \underline{x^2 - x\phantom{{}+3}} \\ & -x + 3 \\ & \underline{-x + 1} \\ & 2 \end{array}$$

$f(x) = (x-1)^2 + 2$; $\dfrac{f(x)}{x-1} = x - 1 + \dfrac{2}{x-1}$

3.

$$\begin{array}{r|l} & x^2 + x + 4 \\ \hline x + 3 & x^3 + 4x^2 + 7x - 9 \\ & \underline{x^3 + 3x^2} \\ & x^2 + 7x \\ & \underline{x^2 + 3x} \\ & 4x - 9 \\ & \underline{4x + 12} \\ & -21 \end{array}$$

$f(x) = (x^2 + x + 4)(x+3) - 21$;
$\dfrac{f(x)}{x+3} = x^2 + x + 4 - \dfrac{21}{x+3}$

5.

$$\begin{array}{r|l} & x^2 - 4x + 12 \\ \hline x^2 + 2x - 1 & x^4 - 2x^3 + 3x^2 - 4x + 6 \\ & \underline{x^4 + 2x^3 - x^2} \\ & -4x^3 + 4x^2 - 4x \\ & \underline{-4x^3 - 8x^2 + 4x} \\ & 12x^2 - 8x + 6 \\ & \underline{12x^2 + 24x - 12} \\ & -32x + 18 \end{array}$$

$f(x) = (x^2 - 4x + 12)(x^2 + 2x - 1) - 32x + 18$;
$\dfrac{f(x)}{x^2 + 2x - 1} = x^2 - 4x + 12 + \dfrac{-32x + 18}{x^2 + 2x - 1}$

7. $\dfrac{x^3 - 5x^2 + 3x - 2}{x+1} = x^2 - 6x + 9 + \dfrac{-11}{x+1}$

$$\begin{array}{r|rrrr} -1 & 1 & -5 & 3 & -2 \\ & & -1 & 6 & -9 \\ \hline & 1 & -6 & 9 & -11 \end{array}$$

9. $\dfrac{9x^3 + 7x^2 - 3x}{x - 10} = 9x^2 + 97x + 967 + \dfrac{9670}{x-10}$

$$\begin{array}{r|rrrr} 10 & 9 & 7 & -3 & 0 \\ & & 90 & 970 & 9670 \\ \hline & 9 & 97 & 967 & 9670 \end{array}$$

11. $\dfrac{5x^4 - 3x + 1}{4 - x}$

$= -5x^3 - 20x^2 - 80x - 317 + \dfrac{-1269}{4-x}$

$$\begin{array}{r|rrrrr} 4 & -5 & 0 & 0 & 3 & -1 \\ & & -20 & -80 & -320 & -1268 \\ \hline & -5 & -20 & -80 & -317 & -1269 \end{array}$$

13. The remainder is $f(2) = 3$.

15. The remainder is $f(-3) = -43$.

17. The remainder is $f(2) = 5$.

19. Yes: 1 is a zero of the second polynomial.

21. No: when $x = 2$, the second polynomial evaluates to 10.

23. Yes: -2 is a zero of the second polynomial.

25. From the graph it appears that $(x+3)$ and $(x-1)$ are factors.

$$\begin{array}{r|rrrr} -3 & 5 & -7 & -49 & 51 \\ & & -15 & 66 & -51 \\ \hline 1 & 5 & -22 & -17 & 0 \\ & & 5 & -17 & \\ \hline & 5 & -17 & 0 & \end{array}$$

$f(x) = (x+3)(x-1)(5x-17)$

27. $2(x+2)(x-1)(x-4) = 2x^3 - 6x^2 - 12x + 16$

29. $2(x-2)\left(x - \dfrac{1}{2}\right)\left(x - \dfrac{3}{2}\right)$
$= \dfrac{1}{2}(x-2)(2x-1)(2x-3)$
$= 2x^3 - 8x^2 + \dfrac{19}{2}x - 3$

31. Since $f(-4) = f(3) = f(5) = 0$, it must be that $(x + 4)$, $(x - 3)$, and $(x - 5)$ are factors of f. So $f(x) = k(x + 4)(x - 3)(x - 5)$ for some constant k.
Since $f(0) = 180$, we must have $k = 3$. So $f(x) = 3(x + 4)(x - 3)(x - 5)$

33. Possible rational zeros: $\dfrac{\pm 1}{\pm 1, \pm 2, \pm 3, \pm 6}$, or $\pm 1, \pm\frac{1}{2}, \pm\frac{1}{3}, \pm\frac{1}{6}$; 1 is a zero.

35. Possible rational zeros: $\dfrac{\pm 1, \pm 3, \pm 9}{\pm 1, \pm 2}$, or $\pm 1, \pm 3, \pm 9, \pm\frac{1}{2}, \pm\frac{3}{2}, \pm\frac{9}{2}$; $\frac{3}{2}$ is a zero.

37.

3	2	−4	1	−2
		6	6	21
	2	2	7	19

Since all numbers in the last line are ≥ 0, 3 is an upper bound for the zeros of f.

39.

2	1	−1	1	1	−12
		2	2	6	14
	1	1	3	7	2

Since all values in the last line are ≥ 0, 2 is an upper bound for the zeros of $f(x)$.

41.

−1	3	−4	1	3
		−3	7	−8
	3	−7	8	−5

Since the values in the last line alternate signs, −1 is a lower bound for the zeros of $f(x)$.

43.

0	1	−4	7	−2
		0	0	0
	1	−4	7	−2

Since the values in the last line alternate signs, 0 is a lower bound for the zeros of $f(x)$.

45. By the Upper and Lower Bound Tests, −5 is a lower bound and 5 is an upper bound. No zeros outside window.

−5	6	−11	−7	8	−34
		−30	205	−990	4910
	6	−41	198	−982	4876

5	6	−11	−7	8	−34
		30	95	440	2240
	6	19	88	448	2206

47. Synthetic division shows that the Upper and Lower Bound Tests were not met. There *are* zeros not shown (approx. −11.002 and 12.003), because −5 and 5 are not bounds for zeros of $f(x)$.

−5	1	−4	−129	396	−8	3
		−5	45	420	−4080	20,440
	1	−9	−84	816	−4088	−20,443

5	1	−4	−129	396	−8	3
		5	5	−620	−1120	−5640
	1	1	−124	−224	−1128	−5637

For #49–55, determine the rational zeros using a grapher (and the Rational Zeros Test as necessary). Use synthetic division to reduce the function to a quadratic polynomial, which can be solved with the quadratic formula (or otherwise). The first two are done in detail; for the rest, we show only the synthetic division step(s).

49. Possible rational zeros: $\dfrac{\pm 1, \pm 2, \pm 3, \pm 6}{\pm 1, \pm 2}$, or $\pm 1, \pm 2, \pm 3, \pm 6, \pm\frac{1}{2}, \pm\frac{3}{2}$. The only rational zero is $\frac{3}{2}$. Synthetic division (below) leaves $2x^2 - 4$, so the irrational zeros are $\pm\sqrt{2}$.

3/2	2	−3	−4	6
		3	0	−6
	2	0	−4	0

51. Rational: −3; irrational: $1 \pm \sqrt{3}$

−3	1	1	−8	−6
		−3	6	6
	1	−2	−2	0

53. Rational: −1 and 4; irrational: $\pm\sqrt{2}$

−1	1	−3	−6	6	8
		−1	4	2	−8
	1	−4	−2	8	0

4	1	−4	−2	8
		4	0	−8
	1	0	−2	0

55. Rational: $-\frac{1}{2}$ and 4; irrational: none

4	2	−7	−2	−7	−4
		8	4	8	4
	2	1	2	1	0

−1/2	2	1	2	1
		−1	0	−1
	2	0	2	0

57. The supply and demand graphs are shown on the window [0, 50] by [0, 100]. They intersect when $p = \$36.27$, at which point the supply and demand equal 53.7.

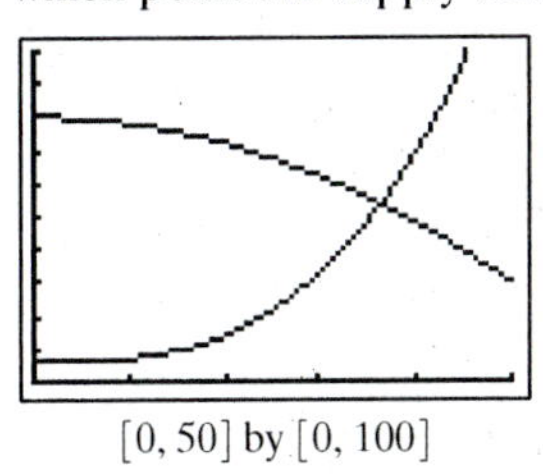

[0, 50] by [0, 100]

59. Using the Remainder Theorem, the remainder is $(-1)^{40} - 3 = -2$.

61. (a) Lower bound:

$\underline{-5}\rfloor$	1	2	−11	−13	38
		−5	15	−20	165
	1	−3	4	−33	203

Upper bound:

$\underline{4}\rfloor$	1	2	−11	−13	38
		4	24	52	156
	1	6	13	39	194

The Upper and Lower Bound Tests are met, so all real zeros of f lie on the interval $[-5, 4]$.

(b) Potential rational zeros:

$$\frac{\textit{Factors of } 38}{\textit{Factors of } 1} : \frac{\pm 1, \pm 2, \pm 19, \pm 38}{\pm 1}$$

A graph shows that 2 is most promising, so we verify with synthetic division:

$\underline{2}\rfloor$	1	2	−11	−13	38
		2	8	−6	−38
	1	4	−3	−19	0

Use the Remainder Theorem:

$f(-2) = 20 \neq 0$ $\quad f(-38) = 1{,}960{,}040$

$f(-1) = 39 \neq 0$ $\quad f(38) = 2{,}178{,}540$

$f(1) = 17 \neq 0$ $\quad f(-19) = 112{,}917$

$f(19) = 139{,}859$

Since all possible rational roots besides 2 yield non-zero function values, there are no other rational roots.

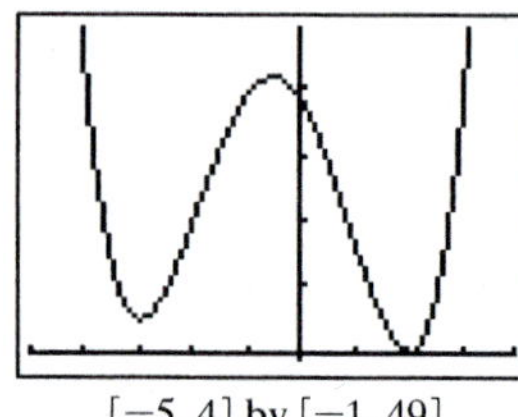

$[-5, 4]$ by $[-1, 49]$

(c) $f(x) = (x - 2)(x^3 + 4x^2 - 3x - 19)$

(d) From our graph, we find that one irrational zero of x is $x \approx 2.04$.

(e) $f(x) \approx (x - 2)(x - 2.04)(x^2 + 6.04x + 9.3216)$

63. False. $x - a$ is a factor if and only if $f(a) = 0$. So $(x + 2)$ is a factor if and only if $f(-2) = 0$.

65. The statement $f(3) = 0$ means that $x = 3$ is a zero of $f(x)$ and that 3 is an x-intercept of the graph of $f(x)$. And it follows that $x - 3$ is a factor of $f(x)$ and thus that the remainder when $f(x)$ is divided by $x - 3$ is zero. So the answer is A.

67. $f(x) = (x + 2)(x^2 + x - 1) - 3$ yields a remainder of -3 when divided by either $x + 2$ or $x^2 + x - 1$, from which it follows that $x + 2$ is not a factor of $f(x)$ and that $f(x)$ is not evenly divisible by $x + 2$. The answer is B.

69. (a) The volume of a sphere is $V = \frac{4}{3}\pi r^3$. In this case, the radius of the buoy is 1, so the buoy's volume is $\frac{4}{3}\pi$.

(b) Total weight $= \text{volume} \cdot \frac{\text{weight}}{\text{unit volume}}$ $= \text{volume} \cdot \text{density}$. In this case, the density of the buoy is $\frac{1}{4}d$, so, the weight W_b of the buoy is $W_b = \frac{4\pi}{3} \cdot \frac{1}{4}d = \frac{d\pi}{3}$.

(c) The weight of the displaced water is $W_{H_2O} = \text{volume} \cdot \text{density}$. We know from geometry that the volume of a spherical cap is

$V = \frac{\pi}{6}(3r^2 + h^2)h$, so,

$W_{H_2O} = \frac{\pi}{6}(3r^2 + x^2)x \cdot d = \frac{\pi d}{6}x(3r^2 + x^2)$

(d) Setting the two weights equal, we have:

$W_b = W_{H_2O}$

$\frac{\pi d}{3} = \frac{\pi d}{6}(3r^2 + x^2)x$

$2 = (3r^2 + x^2)x$

$0 = (6x - 3x^2 + x^2)x - 2$

$0 = -2x^3 + 6x^2 - 2$

$0 = x^3 - 3x^2 + 1$

Solving graphically, we find that $x \approx 0.6527$ m, the depth that the buoy will sink.

71. (a) Shown is one possible view, on the window $[0, 600]$ by $[0, 500]$.

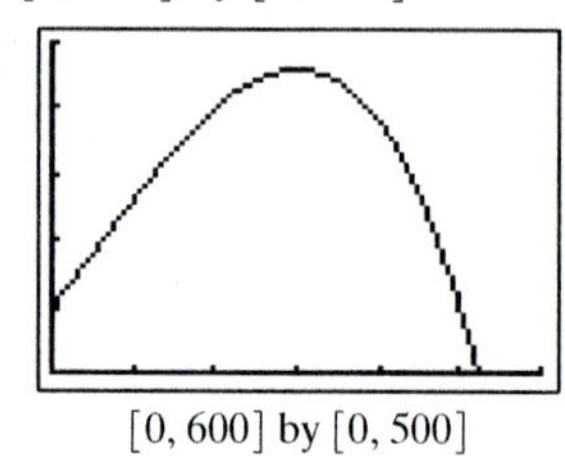

$[0, 600]$ by $[0, 500]$

(b) The maximum population, after 300 days, is 460 turkeys.

(c) $P = 0$ when $t \approx 523.22$ — about 523 days after release.

(d) Answers will vary. One possibility: After the population increases to a certain point, they begin to compete for food and eventually die of starvation.

73. (a) 2 sign changes in $f(x)$, 1 sign change in $f(-x) = -x^3 + x^2 + x + 1$; 0 or 2 positive zeros, 1 negative zero.

(b) No positive zeros, 1 or 3 negative zeros.

(c) 1 positive zero, no negative zeros.

(d) 1 positive zero, 1 negative zero.

75. $\dfrac{4x^3 - 5x^2 + 3x + 1}{2x - 1}$

$= \dfrac{2x^3 - \frac{5}{2}x^2 + \frac{3}{2}x + \frac{1}{2}}{x - \frac{1}{2}}$ Divide numerator and denominator by 2.

zero of divisor	$\frac{1}{2}$ |	2	$-\frac{5}{2}$	$\frac{3}{2}$	$\frac{1}{2}$ Write coefficients of dividend.
line for products			1	$-\frac{3}{4}$	$\frac{3}{8}$
line for sums		2	$-\frac{3}{2}$	$\frac{3}{4}$	$\frac{7}{8}$ Quotient, remainder

Copy 2 into the first quotient position. Multiply $2 \cdot \frac{1}{2} = 1$ and add this to $-\frac{5}{2}$. Multiply $-\frac{3}{2} \cdot \frac{1}{2} = -\frac{3}{4}$ and add this to $\frac{3}{2}$. Multiply $\frac{3}{4} \cdot \frac{1}{2} = \frac{3}{8}$ and add this to $\frac{1}{2}$. The last line tells us $\left(x - \frac{1}{2}\right)\left(2x^2 - \frac{3}{2}x + \frac{3}{4}\right) + \frac{7}{8}$

$= 2x^3 - \frac{5}{4}x^2 + \frac{3}{2}x + \frac{1}{2}$.

77. (a) $g(x) = 3f(x)$, so the zeros of f and the zeros of g are identical. If the coefficients of a polynomial are rational, we may multiply that polynomial by the least common multiple (LCM) of the denominators of the coefficients to obtain a polynomial, with integer coefficients, that has the same zeros as the original.

(b) The zeros of $f(x)$ are the same as the zeros of $6f(x) = 6x^3 - 7x^2 - 40x + 21$. Possible rational zeros: $\dfrac{\pm1, \pm3, \pm7, \pm21}{\pm1, \pm2, \pm3, \pm6}$, or $\pm1, \pm3, \pm7, \pm21, \pm\frac{1}{2}, \pm\frac{3}{2}, \pm\frac{7}{2}, \pm\frac{21}{2}, \pm\frac{1}{3}, \pm\frac{7}{3}, \pm\frac{1}{6}, \pm\frac{7}{6}$. The actual zeros are $-7/3$, $1/2$, and 3.

(c) The zeros of $f(x)$ are the same as the zeros of $12f(x) = 12x^3 - 30x^2 - 37x + 30$.

Possible rational zeros:
$\dfrac{\pm1, \pm2, \pm3, \pm5, \pm6, \pm10, \pm15, \pm30}{\pm1, \pm2, \pm3, \pm4, \pm6, \pm12}$, or
$\pm1, \pm2, \pm3, \pm5, \pm6, \pm10, \pm15, \pm30, \pm\frac{1}{2}, \pm\frac{3}{2}, \pm\frac{5}{2}, \pm\frac{15}{2}, \pm\frac{1}{3}, \pm\frac{2}{3}, \pm\frac{5}{3}, \pm\frac{10}{3}, \pm\frac{1}{4}, \pm\frac{3}{4}, \pm\frac{5}{4}, \pm\frac{15}{4}, \pm\frac{1}{6}, \pm\frac{5}{6}, \pm\frac{1}{12}, \pm\frac{5}{12}$.

There are no rational zeros.

79. (a) Approximate zeros: $-3.126, -1.075, 0.910, 2.291$

(b) $f(x) \approx g(x)$
$= (x + 3.126)(x + 1.075)(x - 0.910)(x - 2.291)$

(c) Graphically: Graph the original function and the approximate factorization on a variety of windows and observe their similarity. Numerically: Compute $f(c)$ and $g(c)$ for several values of c.

Section 2.5 Complex Zeros and the Fundamental Theorem of Algebra

Exploration 1

1. $f(2i) = (2i)^2 - i(2i) + 2 = -4 + 2 + 2 = 0$;
$f(-i) = (-i)^2 - i(-i) + 2 = -1 - 1 + 2 = 0$; no.
2. $g(i) = i^2 - i + (1 + i) = -1 - i + 1 + i = 0$;
$g(1 - i) = (1 - i)^2 - (1 - i) + (1 + i) = -2i + 2i = 0$; no.
3. The Complex Conjugate Zeros Theorem does not necessarily hold true for a polynomial function with *complex* coefficients.

Quick Review 2.5

1. $(3 - 2i) + (-2 + 5i) = (3 - 2) + (-2 + 5)i = 1 + 3i$
3. $(1 + 2i)(3 - 2i) = 1(3 - 2i) + 2i(3 - 2i) = 3 - 2i + 6i - 4i^2 = 7 + 4i$
5. $(2x - 3)(x + 1)$
7. $x = \dfrac{5 \pm \sqrt{25 - 4(1)(11)}}{2} = \dfrac{5 \pm \sqrt{-19}}{2} = \dfrac{5}{2} \pm \dfrac{\sqrt{19}}{2}i$
9. $\dfrac{\pm1, \pm2}{\pm1, \pm3}$, or $\pm1, \pm2, \pm\frac{1}{3}, \pm\frac{2}{3}$

Section 2.5 Exercises

1. $(x - 3i)(x + 3i) = x^2 - (3i)^2 = x^2 + 9$. The factored form shows the zeros to be $x = \pm3i$. The absence of real zeros means that the graph has no x-intercepts.
3. $(x - 1)(x - 1)(x + 2i)(x - 2i)$
$= (x^2 - 2x + 1)(x^2 + 4)$
$= x^4 - 2x^3 + 5x^2 - 8x + 4$. The factored form shows the zeros to be $x = 1$ (multiplicity 2) and $x = \pm2i$. The real zero $x = 1$ is the x-intercept of the graph.

In #5–15, any constant multiple of the given polynomial is also an answer.

5. $(x - i)(x + i) = x^2 + 1$
7. $(x - 1)(x - 3i)(x + 3i) = (x - 1)(x^2 + 9) = x^3 - x^2 + 9x - 9$
9. $(x - 2)(x - 3)(x - i)(x + i) = (x - 2)(x - 3)(x^2 + 1) = x^4 - 5x^3 + 7x^2 - 5x + 6$
11. $(x - 5)(x - 3 - 2i)(x - 3 + 2i) = (x - 5)(x^2 - 6x + 13) = x^3 - 11x^2 + 43x - 65$
13. $(x - 1)^2(x + 2)^3 = x^5 + 4x^4 + x^3 - 10x^2 - 4x + 8$
15. $(x - 2)^2(x - 3 - i)(x - 3 + i)$
$= (x - 2)^2(x^2 - 6x + 10)$
$= (x^2 - 4x + 4)(x^2 - 6x + 10)$
$= x^4 - 10x^3 + 38x^2 - 64x + 40$

In #17–19, note that the graph crosses the x-axis at odd-multiplicity zeros, and "kisses" (touches but does not cross) the x-axis where the multiplicity is even.

17. (b)

19. (d)

In #21–25, the number of complex zeros is the same as the degree of the polynomial; the number of real zeros can be determined from a graph. The latter always differs from the former by an even number (when the coefficients of the polynomial are real).

21. 2 complex zeros; none real.

23. 3 complex zeros; 1 real.

25. 4 complex zeros; 2 real.

In #27–31, look for real zeros using a graph (and perhaps the Rational Zeros Test). Use synthetic division to factor the polynomial into one or more linear factors and a quadratic factor. Then use the quadratic formula to find complex zeros.

27. Inspection of the graph reveals that $x = 1$ is the only real zero. Dividing $f(x)$ by $x - 1$ leaves $x^2 + x + 5$ (below). The quadratic formula gives the remaining zeros of $f(x)$.

$$\begin{array}{r|rrrr} 1 & 1 & 0 & 4 & -5 \\ & & 1 & 1 & 5 \\ \hline & 1 & 1 & 5 & 0 \end{array}$$

Zeros: $x = 1, x = -\frac{1}{2} \pm \frac{\sqrt{19}}{2} i$

$f(x)$

$= (x - 1)\left[x - \left(-\frac{1}{2} - \frac{\sqrt{19}}{2} i\right)\right]\left[x - \left(-\frac{1}{2} + \frac{\sqrt{19}}{2} i\right)\right]$

$= \frac{1}{4}(x - 1)(2x + 1 + \sqrt{19}\,i)(2x + 1 - \sqrt{19}\,i)$

29. Zeros: $x = \pm 1$(graphically) and $x = -\frac{1}{2} \pm \frac{\sqrt{23}}{2} i$ (applying the quadratic formula to $x^2 + x + 6$).

$$\begin{array}{r|rrrrr} 1 & 1 & 1 & 5 & -1 & -6 \\ & & 1 & 2 & 7 & 6 \\ \hline & 1 & 2 & 7 & 6 & 0 \end{array}$$

$$\begin{array}{r|rrrr} -1 & 1 & 2 & 7 & 6 \\ & & -1 & -1 & -6 \\ \hline & 1 & 1 & 6 & 0 \end{array}$$

$f(x) = (x - 1)(x + 1)\left[x - \left(-\frac{1}{2} - \frac{\sqrt{23}}{2} i\right)\right]$
$\left[x - \left(-\frac{1}{2} + \frac{\sqrt{23}}{2} i\right)\right]$

$= \frac{1}{4}(x - 1)(x + 1)(2x + 1 + \sqrt{23}\,i)\,(2x + 1 - \sqrt{23}\,i)$

31. Zeros: $x = -\frac{7}{3}$ and $x = \frac{3}{2}$ (graphically) and $x = 1 \pm 2i$ (applying the quadratic formula to $6x^2 - 12x + 30 = 6(x^2 - 2x + 5)$).

$$\begin{array}{r|rrrrr} -7/3 & 6 & -7 & -1 & 67 & -105 \\ & & -14 & 49 & -112 & 105 \\ \hline & 6 & -21 & 48 & -45 & 0 \end{array}$$

$$\begin{array}{r|rrrr} 3/2 & 6 & -21 & 48 & -45 \\ & & 9 & -18 & 45 \\ \hline & 6 & -12 & 30 & 0 \end{array}$$

$f(x) = (3x + 7)(2x - 3)[x - (1 - 2i)]$
$[x - (1 + 2i)]$
$= (3x + 7)(2x - 3)(x - 1 + 2i)(x - 1 - 2i)$

In #33–35, since the polynomials' coefficients are real, for the given zero $z = a + bi$, the complex conjugate $\bar{z} = a - bi$ must also be a zero. Divide $f(x)$ by $x - z$ and $x - \bar{z}$ to reduce to a quadratic.

33. First divide $f(x)$ by $x - (1 + i)$ (synthetically). Then divide the result, $x^3 + (-1 + i)x^2 - 3x + (3 - 3i)$, by $x - (1 - i)$. This leaves the polynominal $x^2 - 3$. Zeros: $x = \pm\sqrt{3}, x = 1 \pm i$

$$\begin{array}{r|rrrrr} 1+i & 1 & -2 & -1 & 6 & -6 \\ & & 1+i & -2 & -3-3i & 6 \\ \hline & 1 & -1+i & -3 & 3-3i & 0 \end{array}$$

$$\begin{array}{r|rrrr} 1-i & 1 & -1+i & -3 & 3-3i \\ & & 1-i & 0 & -3+3i \\ \hline & 1 & 0 & -3 & 0 \end{array}$$

$f(x) = (x - \sqrt{3})(x + \sqrt{3})[x - (1 - i)][x - (1 + i)]$
$= (x - \sqrt{3})(x + \sqrt{3})(x - 1 + i)(x - 1 - i)$

35. First divide $f(x)$ by $x - (3 - 2i)$. Then divide the result, $x^3 + (-3 - 2i)x^2 - 2x + 6 + 4i$, by $x - (3 + 2i)$. This leaves $x^2 - 2$. Zeros: $x = \pm\sqrt{2}, x = 3 \pm 2i$

$$\begin{array}{r|rrrrr} 3-2i & 1 & -6 & 11 & 12 & -26 \\ & & 3-2i & -13 & -6+4i & 26 \\ \hline & 1 & -3-2i & -2 & 6+4i & 0 \end{array}$$

$$\begin{array}{r|rrrr} 3+2i & 1 & -3-2i & -2 & 6+4i \\ & & 3+2i & 0 & -6-4i \\ \hline & 1 & 0 & -2 & 0 \end{array}$$

$f(x) = (x - \sqrt{2})(x + \sqrt{2})[x - (3 - 2i)]$
$[x - (3 + 2i)]$
$= (x - \sqrt{2})(x + \sqrt{2})(x - 3 + 2i)(x - 3 - 2i)$

For #37–41, find real zeros graphically, then use synthetic division to find the quadratic factors. Only the synthetic divison step is shown.

37. $f(x) = (x - 2)(x^2 + x + 1)$

$$\begin{array}{r|rrrr} 2 & 1 & -1 & -1 & -2 \\ & & 2 & 2 & 2 \\ \hline & 1 & 1 & 1 & 0 \end{array}$$

39. $f(x) = (x - 1)(2x^2 + x + 3)$

$$\begin{array}{r|rrrr} 1 & 2 & -1 & 2 & -3 \\ & & 2 & 1 & 3 \\ \hline & 2 & 1 & 3 & 0 \end{array}$$

41. $f(x) = (x - 1)(x + 4)(x^2 + 1)$

$$\begin{array}{r|rrrrr} 1 & 1 & 3 & -3 & 3 & -4 \\ & & 1 & 4 & 1 & 4 \\ \hline & 1 & 4 & 1 & 4 & 0 \end{array}$$

$$\begin{array}{r|rrrr} -4 & 1 & 4 & 1 & 4 \\ & & -4 & 0 & -4 \\ \hline & 1 & 0 & 1 & 0 \end{array}$$

43. Solve for h: $\frac{\pi}{3}(15h^2 - h^3)(62.5) = \frac{4}{3}\pi\,(125)(20)$, so that $15h^2 - h^3 = 160$. Of the three solutions (found graphically), only $h \approx 3.776$ ft makes sense in this setting.

45. Yes: $(x + 2)(x^2 + 1) = x^3 + 2x^2 + x + 2$ is one such polynomial. Other examples can be obtained by multiplying any other quadratic with no real zeros by $(x + 2)$.

47. No: if all coefficients are real, $1 - 2i$ and $1 + i$ must also be zeros, giving 5 zeros for a degree 4 polynomial.

49. $f(x)$ must have the form $a(x - 3)(x + 1)(x - 2 + i)\,(x - 2 - i)$; since $f(0) = a(-3)(1)(-2 + i)(-2 - i) = -15a = 30$, we know that $a = -2$. Multiplied out, this gives $f(x) = -2x^4 + 12x^3 - 20x^2 - 4x + 30$.

51. **(a)** The model is $D \approx -0.0820t^3 + 0.9162t^2 - 2.5126t + 3.3779$.

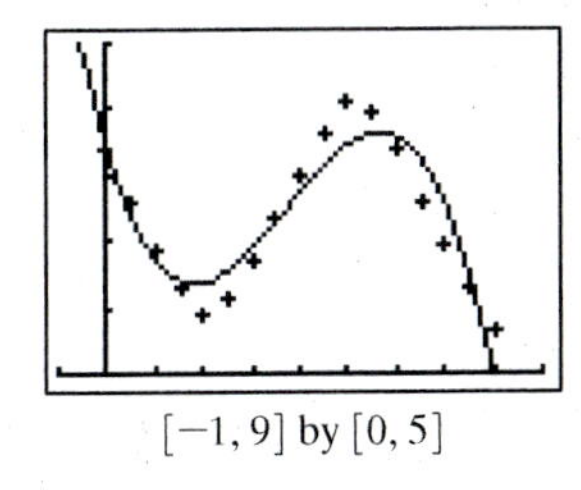

$[-1, 9]$ by $[0, 5]$

(b) Sally walks toward the detector, turns and walks away (or walks backward), then walks toward the detector again.

(c) The model "changes direction" at $t \approx 1.81$ sec ($D \approx 1.35$ m) and $t \approx 5.64$ sec (when $D \approx 3.65$ m).

53. False. Complex, nonreal solutions always come in conjugate pairs, so that if $1 - 2i$ is a zero, then $1 + 2i$ must also be a zero.

55. Both the sum and the product of two complex conjugates are real numbers, and the absolute value of a complex number is always real. The square of a complex number, on the other hand, need not be real. The answer is E.

57. Because the complex, non-real zeros of a real-coefficient polynomial always come in conjugate pairs, a polynomial of degree 5 can have either 0, 2, or 4 non-real zeros. The answer is C.

59. **(a)**

Power	Real Part	Imaginary Part
7	8	−8
8	16	0
9	16	16
10	0	32

(b) $(1 + i)^7 = 8 - 8i$
$(1 + i)^8 = 16$
$(1 + i)^9 = 16 + 16i$
$(1 + i)^{10} = 32i$

(c) Reconcile as needed.

61. $f(i) = i^3 - i(i)^2 + 2i(i) + 2 = -i + i - 2 + 2 = 0$. One can also take the last number of the bottom row from synthetic division.

63. Synthetic division shows that $f(i) = 0$ (the remainder), and at the same time gives $f(x) \div (x - i) = x^2 + 3x - i = h(x)$, so $f(x) = (x - i)(x^2 + 3x - i)$.

$$\begin{array}{r|rrrr} i & 1 & 3 - i & -4i & -1 \\ & & i & 3i & 1 \\ \hline & 1 & 3 & -i & 0 \end{array}$$

65. From graphing (or the Rational Zeros Test), we expect $x = 2$ to be a zero of $f(x) = x^3 - 8$. Indeed, $f(2) = 8 - 8 = 0$. So, $x = 2$ *is* a zero of $f(x)$. Using synthetic division we obtain:

$$\begin{array}{r|rrrr} 2 & 1 & 0 & 0 & -8 \\ & & 2 & 4 & 8 \\ \hline & 1 & 2 & 4 & 0 \end{array}$$

$f(x) = (x - 2)(x^2 + 2x + 4)$. We then apply the quadratic formula to find that the cube roots of $x^3 - 8$ are $2, -1 + \sqrt{3}i$, and $-1 - \sqrt{3}i$.

■ Section 2.6 Graphs of Rational Functions

Exploration 1

1. $g(x) = \dfrac{1}{x - 2}$

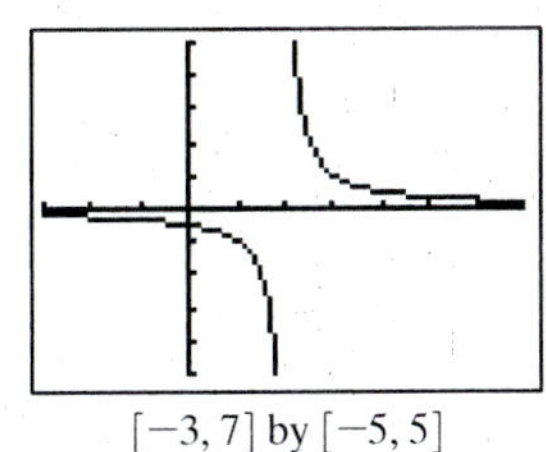

$[-3, 7]$ by $[-5, 5]$

2. $h(x) = -\dfrac{1}{x - 5}$

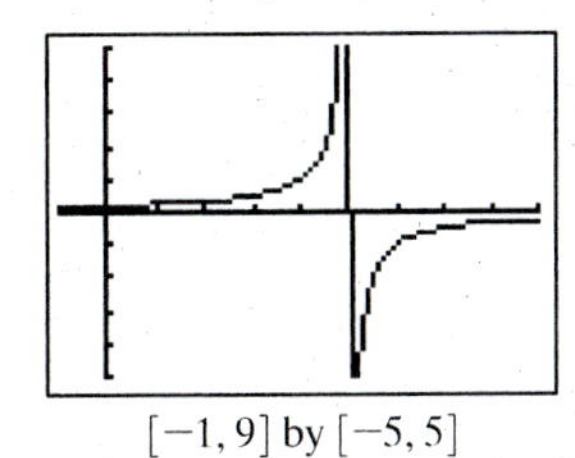

$[-1, 9]$ by $[-5, 5]$

3. $k(x) = \dfrac{3}{x + 4} - 2$

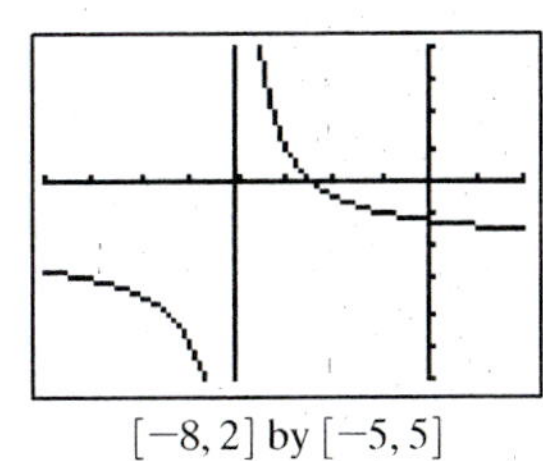

$[-8, 2]$ by $[-5, 5]$

Quick Review 2.6

1. $f(x) = (2x - 1)(x + 3) \Rightarrow x = -3$ or $x = \frac{1}{2}$

3. $g(x) = (x + 2)(x - 2) \Rightarrow x = \pm 2$

5. $h(x) = (x - 1)(x^2 + x + 1) \Rightarrow x = 1$

7.
$$\begin{array}{r} 2 \\ x - 3 \overline{)\, 2x + 1} \\ \underline{2x - 6} \\ 7 \end{array}$$
Quotient: 2, Remainder: 7

9.
$$\begin{array}{r} 3 \\ x \overline{)\, 3x - 5} \\ \underline{3x \quad\;\;} \\ -5 \end{array}$$
Quotient: 3, Remainder: −5

Section 2.6 Exercises

1. The domain of $f(x) = 1/(x + 3)$ is all real numbers $x \neq -3$. The graph suggests that $f(x)$ has a vertical asymptote at $x = -3$.

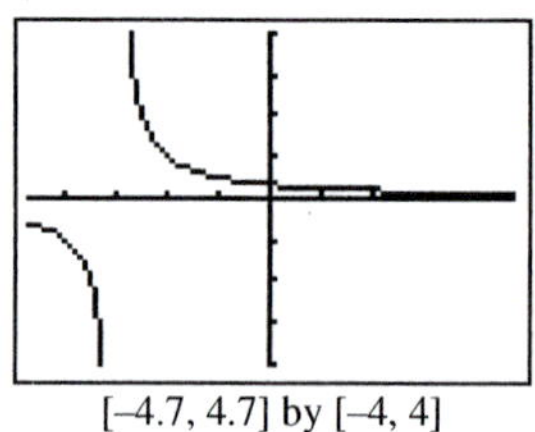

[−4.7, 4.7] by [−4, 4]

As x approaches −3 from the left, the values of $f(x)$ decrease without bound. As x approaches −3 from the right, the values of $f(x)$ increase without bound. That is, $\lim_{x \to -3^-} f(x) = -\infty$ and $\lim_{x \to -3^+} f(x) = \infty$.

3. The domain of $f(x) = -1/(x^2 - 4)$ is all real numbers $x \neq -2, 2$. The graph suggests that $f(x)$ has vertical asymptotes at $x = -2$ and $x = 2$.

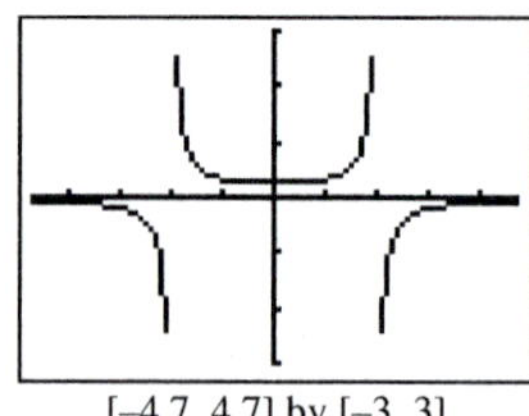

[−4.7, 4.7] by [−3, 3]

As x approaches −2 from the left, the values of $f(x)$ decrease without bound, and as x approaches −2 from the right, the values of $f(x)$ increase without bound. As x approaches 2 from the left, the values of $f(x)$ increase without bound, and as x approaches 2 from the right, the values of $f(x)$ decrease without bound. That is, $\lim_{x \to -2^-} f(x) = -\infty$, $\lim_{x \to -2^+} f(x) = \infty$, $\lim_{x \to 2^-} f(x) = \infty$, and $\lim_{x \to 2^+} f(x) = -\infty$.

5. Translate right 3 units. Asymptotes: $x = 3$, $y = 0$.

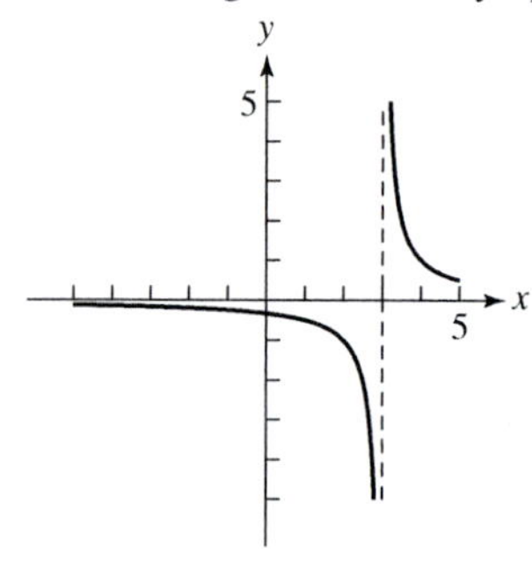

7. Translate left 3 units, reflect across x-axis, vertically stretch by 7, translate up 2 units.
Asymptotes: $x = -3$, $y = 2$.

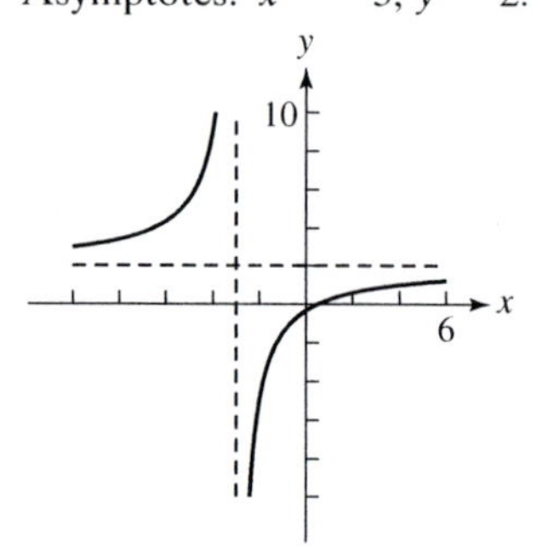

9. Translate left 4 units, vertically stretch by 13, translate down 2 units. Asymptotes: $x = -4$, $y = -2$.

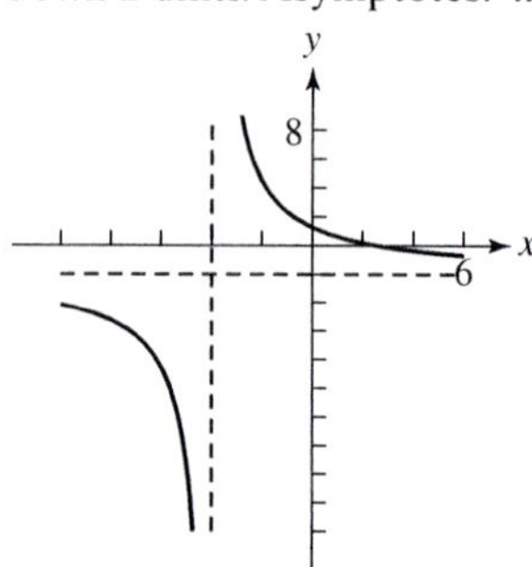

11. $\lim_{x \to 3^-} f(x) = \infty$

13. $\lim_{x \to \infty} f(x) = 0$

15. $\lim_{x \to -3^+} f(x) = \infty$

17. $\lim_{x \to -\infty} f(x) = 5$

19. The graph of $f(x) = (2x^2 - 1)/(x^2 + 3)$ suggests that there are no vertical asymptotes and that the horizontal asymptote is $y = 2$.

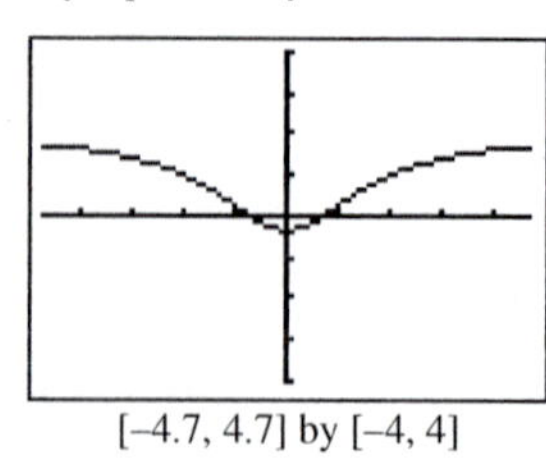

[−4.7, 4.7] by [−4, 4]

The domain of $f(x)$ is all real numbers, so there are indeed no vertical asymptotes. Using polynomial long division, we find that

$$f(x) = \frac{2x^2 - 1}{x^2 + 3} = 2 - \frac{7}{x^2 + 3}$$

When the value of $|x|$ is large, the denominator $x^2 + 3$ is a large positive number, and $7/(x^2 + 3)$ is a small positive number, getting closer to zero as $|x|$ increases. Therefore,

$$\lim_{x\to\infty} f(x) = \lim_{x\to-\infty} f(x) = 2,$$

so $y = 2$ is indeed a horizontal asymptote.

21. The graph of $f(x) = (2x + 1)/(x^2 - x)$ suggests that there are vertical asymptotes at $x = 0$ and $x = 1$, with $\lim_{x\to0^-} f(x) = \infty$, $\lim_{x\to0^+} f(x) = -\infty$, $\lim_{x\to1^-} f(x) = -\infty$, and $\lim_{x\to1^+} f(x) = \infty$, and that the horizontal asymptote is $y = 0$.

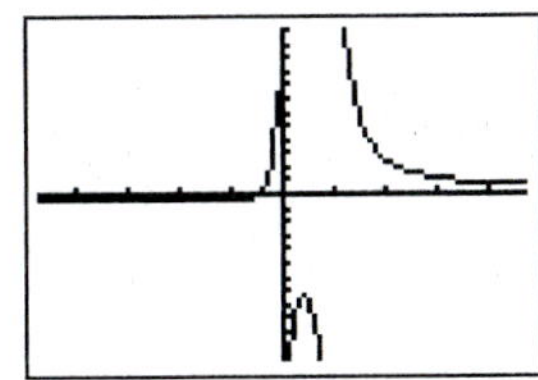

[–4.7, 4.7] by [–12, 12]

The domain of $f(x) = (2x + 1)/(x^2 - x) = (2x + 1)/[x(x - 1)]$ is all real numbers $x \neq 0, 1$, so there are indeed vertical asymptotes at $x = 0$ and $x = 1$. Rewriting one rational expression as two, we find that

$$f(x) = \frac{2x + 1}{x^2 - x} = \frac{2x}{x^2 - x} + \frac{1}{x^2 - x}$$
$$= \frac{2}{x - 1} + \frac{1}{x^2 - x}$$

When the value of $|x|$ is large, both terms get close to zero. Therefore,

$$\lim_{x\to\infty} f(x) = \lim_{x\to-\infty} f(x) = 0,$$

so $y = 0$ is indeed a horizontal asymptote.

23. Intercepts: $\left(0, \frac{2}{3}\right)$ and $(2, 0)$. Asymptotes: $x = -1$, $x = 3$, and $y = 0$.

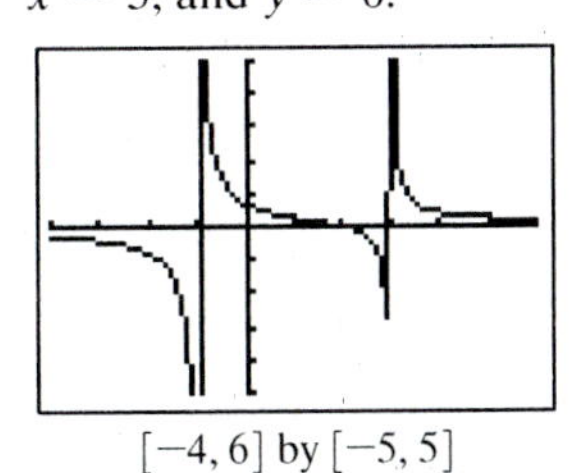

[−4, 6] by [−5, 5]

25. No intercepts. Asymptotes: $x = -1$, $x = 0$, $x = 1$, and $y = 0$.

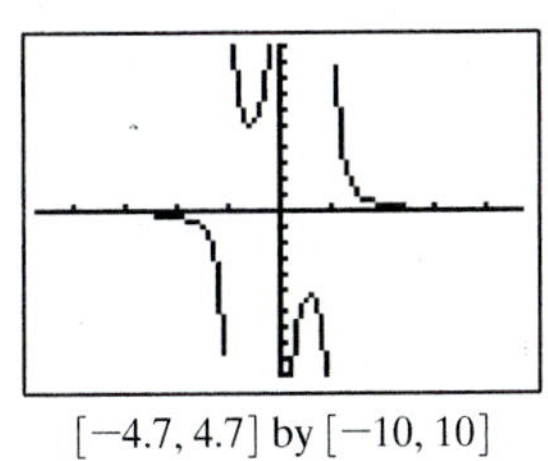

[−4.7, 4.7] by [−10, 10]

27. Intercepts: $(0, 2)$, $(-1.28, 0)$, and $(0.78, 0)$. Asymptotes: $x = 1$, $x = -1$, and $y = 2$.

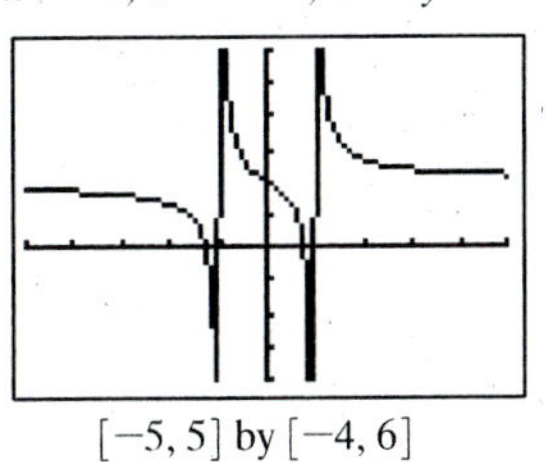

[−5, 5] by [−4, 6]

29. Intercept: $\left(0, \frac{3}{2}\right)$. Asymptotes: $x = -2$, $y = x - 4$.

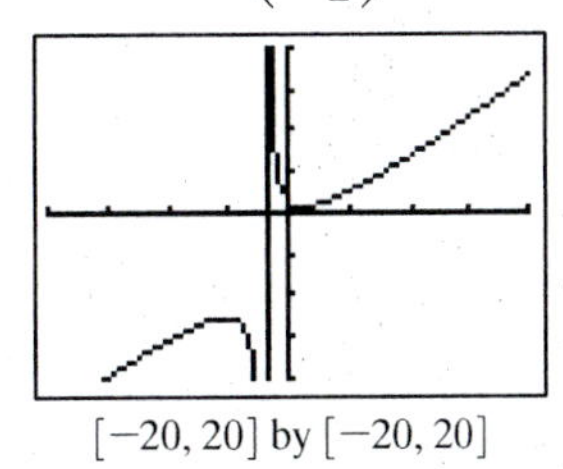

[−20, 20] by [−20, 20]

31. (d); Xmin = −2, Xmax = 8, Xscl = 1, and Ymin = −3, Ymax = 3, Yscl = 1.

33. (a); Xmin = −3, Xmax = 5, Xscl = 1, and Ymin = −5, Ymax = 10, Yscl = 1.

35. (e); Xmin = −2, Xmax = 8, Xscl = 1, and Ymin = −3, Ymax = 3, Yscl = 1.

37. For $f(x) = 2/(2x^2 - x - 3)$, the numerator is never zero, and so $f(x)$ never equals zero and the graph has no x-intercepts. Because $f(0) = -2/3$, the y-intercept is $-2/3$. The denominator factors as $2x^2 - x - 3 = (2x - 3)(x + 1)$, so there are vertical asymptotes at $x = -1$ and $x = 3/2$. And because the degree of the numerator is less than the degree of the denominator, the horizontal asymptote is $y = 0$. The graph supports this information and allows us to conclude that

$$\lim_{x\to-1^-} f(x) = \infty, \lim_{x\to-1^+} f(x) = -\infty, \lim_{x\to(3/2)^-} f(x) = -\infty,$$
and $\lim_{x\to(3/2)^+} f(x) = \infty$.

The graph also shows a local maximum of $-16/25$ at $x = 1/4$.

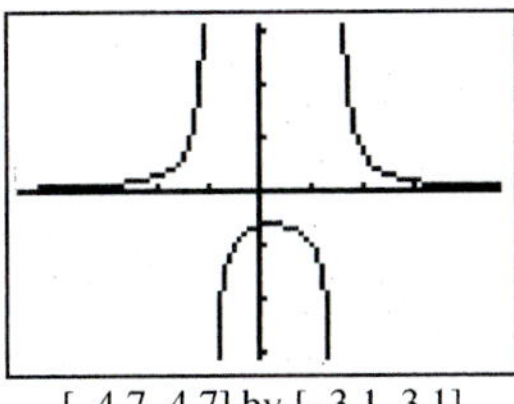

[–4.7, 4.7] by [–3.1, 3.1]

Intercept: $\left(0, -\frac{2}{3}\right)$

Domain: $(-\infty, -1) \cup \left(-1, \frac{3}{2}\right) \cup \left(\frac{3}{2}, \infty\right)$

Range: $\left(-\infty, -\frac{16}{25}\right] \cup (0, \infty)$

Continuity: All $x \neq -1, \frac{3}{2}$

Increasing on $(-\infty, -1)$ and $\left(-1, \frac{1}{4}\right]$

Decreasing on $\left[\frac{1}{4}, \frac{3}{2}\right)$ and $\left(\frac{3}{2}, \infty\right)$
Not symmetric.
Unbounded.
Local maximum at $\left(\frac{1}{4}, -\frac{16}{25}\right)$
Horizontal asymptote: $y = 0$
Vertical asymptotes: $x = -1$ and $x = 3/2$
End behavior: $\lim_{x\to-\infty} f(x) = \lim_{x\to\infty} f(x) = 0$

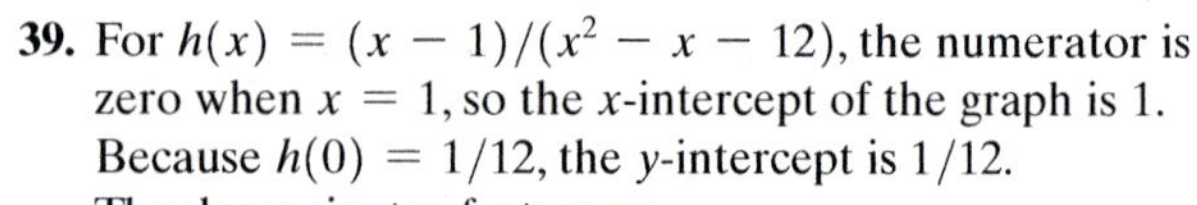

39. For $h(x) = (x - 1)/(x^2 - x - 12)$, the numerator is zero when $x = 1$, so the x-intercept of the graph is 1. Because $h(0) = 1/12$, the y-intercept is $1/12$. The denominator factors as
$x^2 - x - 12 = (x + 3)(x - 4)$,
so there are vertical asymptotes at $x = -3$ and $x = 4$. And because the degree of the numerator is less than the degree of the denominator, the horizontal asymptote is $y = 0$. The graph supports this information and allows us to conclude that
$\lim_{x\to-3^-} h(x) = -\infty$, $\lim_{x\to-3^+} h(x) = \infty$, $\lim_{x\to4^-} h(x) = -\infty$,
and $\lim_{x\to4^+} h(x) = \infty$.

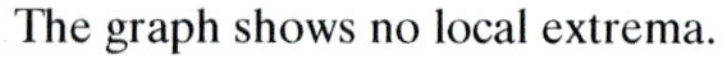

The graph shows no local extrema.

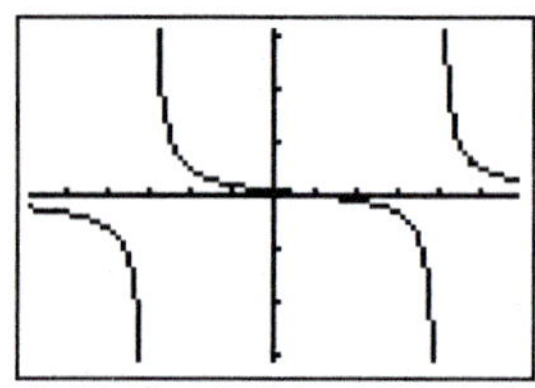

[–5.875, 5.875] by [–3.1, 3.1]

Intercepts: $\left(0, \frac{1}{12}\right)$, $(1, 0)$
Domain: $(-\infty, -3) \cup (-3, 4) \cup (4, \infty)$
Range: $(-\infty, \infty)$
Continuity: All $x \neq -3, 4$
Decreasing on $(-\infty, -3)$, $(-3, 4)$, and $(4, \infty)$
Not symmetric.
Unbounded.
No local extrema.
Horizontal asymptote: $y = 0$
Vertical asymptotes: $x = -3$ and $x = 4$
End behavior: $\lim_{x\to-\infty} h(x) = \lim_{x\to\infty} h(x) = 0$

41. For $f(x) = (x^2 + x - 2)/(x^2 - 9)$, the numerator factors as
$x^2 + x - 2 = (x + 2)(x - 1)$,
so the x-intercepts of the graph are -2 and 1. Because $f(0) = 2/9$, the y-intercept is $2/9$. The denominator factors as
$x^2 - 9 = (x + 3)(x - 3)$,
so there are vertical asymptotes at $x = -3$ and $x = 3$. And because the degree of the numerator equals the degree of the denominator with a ratio of leading terms that equals 1, the horizontal asymptote is $y = 1$. The graph supports this information and allows us to conclude that
$\lim_{x\to-3^-} f(x) = \infty$, $\lim_{x\to-3^+} f(x) = -\infty$, $\lim_{x\to3^-} f(x) = -\infty$,
and $\lim_{x\to3^+} f(x) = \infty$.
The graph also shows a local maximum of about 0.260 at about $x = -0.675$.

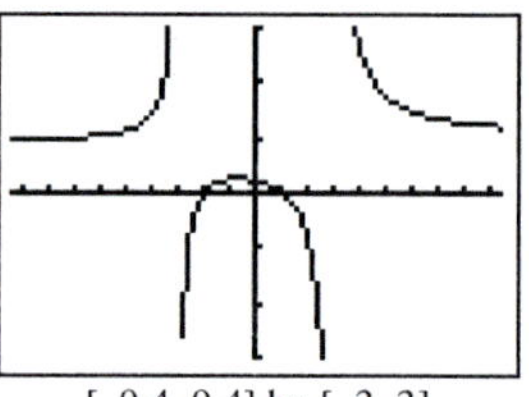

[–9.4, 9.4] by [–3, 3]

Intercepts: $(-2, 0)$, $(1, 0)$, $\left(0, \frac{2}{9}\right)$
Domain: $(-\infty, -3) \cup (-3, 3) \cup (3, \infty)$
Range: $(-\infty, 0.260] \cup (1, \infty)$
Continuity: All $x \neq -3, 3$
Increasing on $(-\infty, -3)$ and $(-3, -0.675]$
Decreasing on $[-0.675, 3)$ and $(3, \infty)$
Not symmetric.
Unbounded.
Local maximum at about $(-0.675, 0.260)$
Horizontal asymptote: $y = 1$
Vertical asymptotes: $x = -3$ and $x = 3$
End behavior: $\lim_{x\to-\infty} f(x) = \lim_{x\to\infty} f(x) = 1$.

43. For $h(x) = (x^2 + 2x - 3)/(x + 2)$, the numerator factors as
$x^2 + 2x - 3 = (x + 3)(x - 1)$,
so the x-intercepts of the graph are -3 and 1. Because $h(0) = -3/2$, the y-intercept is $-3/2$. The denominator is zero when $x = -2$, so there is a vertical asymptote at $x = -2$. Using long division, we rewrite $h(x)$ as

$$h(x) = \frac{x^2 + 2x - 2}{x + 2} = x - \frac{2}{x + 2},$$

so the end-behavior asymptote of $h(x)$ is $y = x$. The graph supports this information and allows us to conclude that
$\lim_{x\to-2^-} h(x) = \infty$ and $\lim_{x\to-2^+} h(x) = -\infty$.
The graph shows no local extrema.

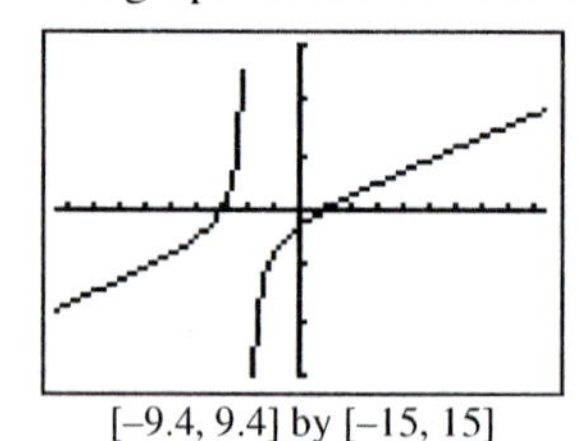

[–9.4, 9.4] by [–15, 15]

Intercepts: $(-3, 0)$, $(1, 0)$, $\left(0, -\frac{3}{2}\right)$
Domain: $(-\infty, -2) \cup (-2, \infty)$
Range: $(-\infty, \infty)$
Continuity: All $x \neq -2$
Increasing on $(-\infty, -2)$ and $(-2, \infty)$
Not symmetric.
Unbounded.
No local extrema.
Horizontal asymptote: None
Vertical asymptote: $x = -2$
Slant asymptote: $y = x$
End behavior: $\lim_{x\to-\infty} h(x) = -\infty$ and $\lim_{x\to\infty} h(x) = \infty$.

45. Divide $x^2 - 2x - 3$ by $x - 5$ to show that
$$f(x) = \frac{x^2 - 2x + 3}{x - 5} = x + 3 + \frac{18}{x - 5}.$$
The end-behavior asymptote of $f(x)$ is $y = x + 3$.

(a)

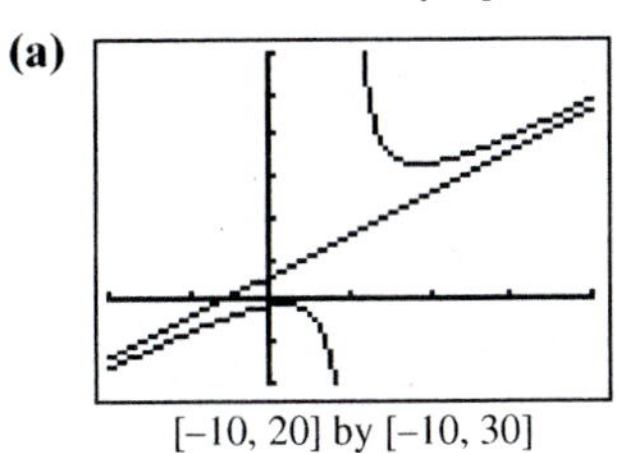

[–10, 20] by [–10, 30]

(b)

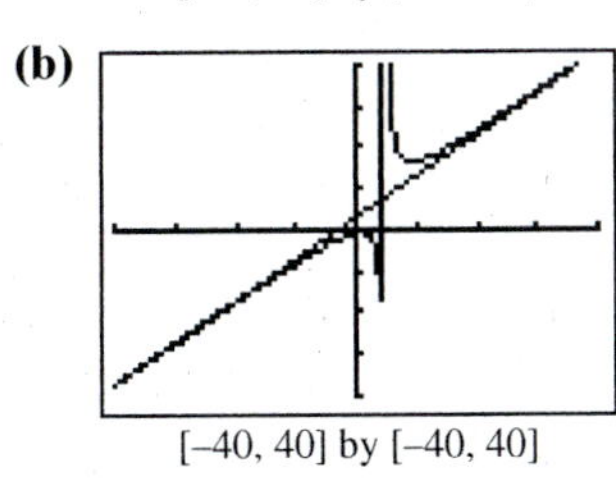

[–40, 40] by [–40, 40]

47. Divide $x^3 - x^2 + 1$ by $x + 2$ to show that
$$f(x) = \frac{x^3 - x^2 + 1}{x + 2} = x^2 - 3x + 6 - \frac{11}{x + 2}.$$
The end-behavior asymptote of $f(x)$ is $y = x^2 - 3x + 6$.

(a)

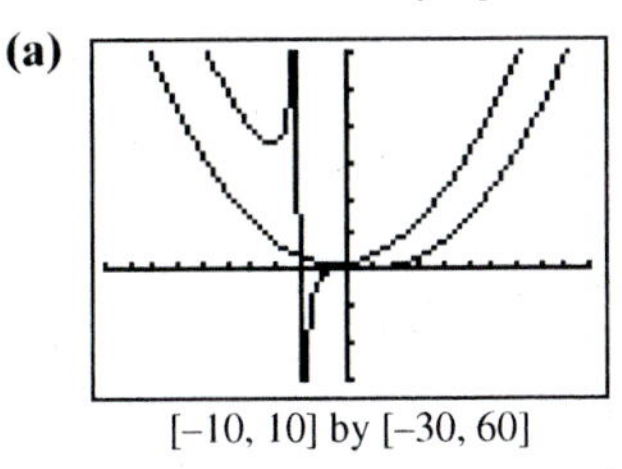

[–10, 10] by [–30, 60]

(b)

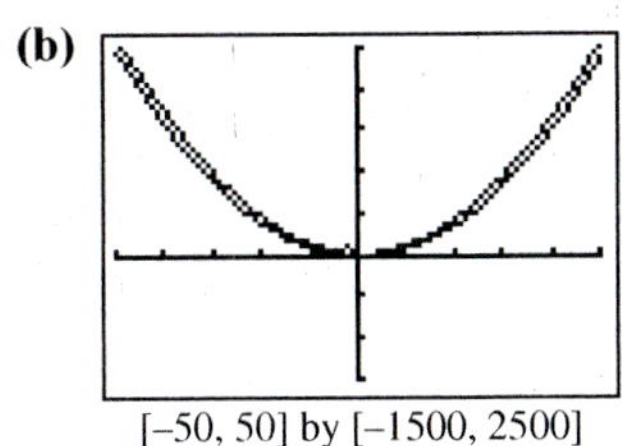

[–50, 50] by [–1500, 2500]

49. Divide $x^4 - 2x + 1$ by $x - 2$ to show that
$$f(x) = \frac{x^4 - 2x + 1}{x - 2} = x^3 + 2x^2 + 4x + 6 + \frac{13}{x - 2}.$$
The end-behavior asymptote of $f(x)$ is
$y = x^3 + 2x^2 + 4x + 6$.

(a)

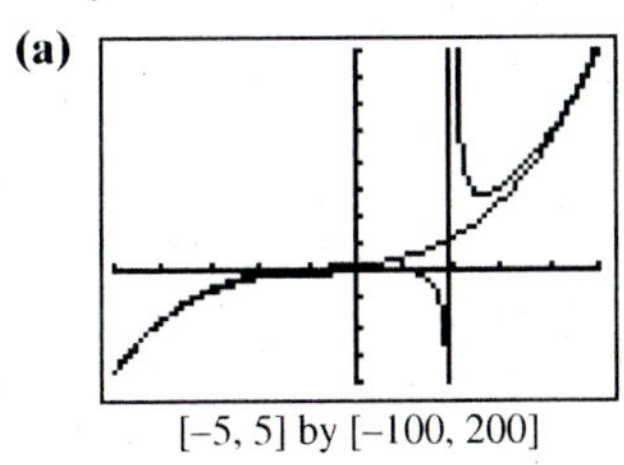

[–5, 5] by [–100, 200]

(b)

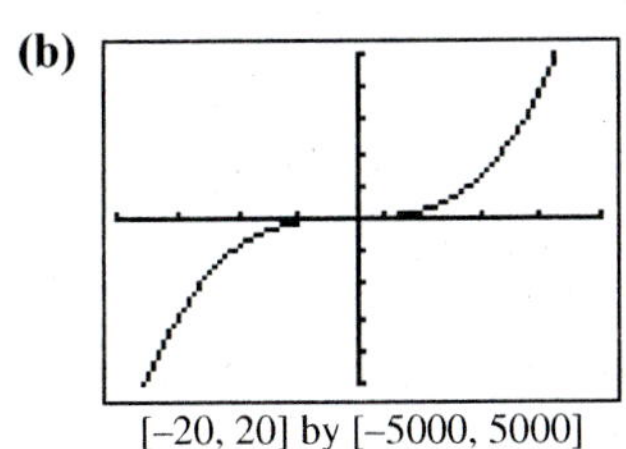

[–20, 20] by [–5000, 5000]

51. For $f(x) = (3x^2 - 2x + 4)/(x^2 - 4x + 5)$, the numerator is never zero, and so $f(x)$ never equals zero and the graph has no x-intercepts. Because $f(0) = 4/5$, the y-intercept is $4/5$. The denominator is never zero, and so there are no vertical asymptotes. And because the degree of the numerator equals the degree of the denominator with a ratio of leading terms that equals 3, the horizontal asymptote is $y = 3$. The graph supports this information. The graph also shows a local maximum of about 14.227 at about $x = 2.445$ and a local minimum of about 0.773 at about $x = -0.245$.

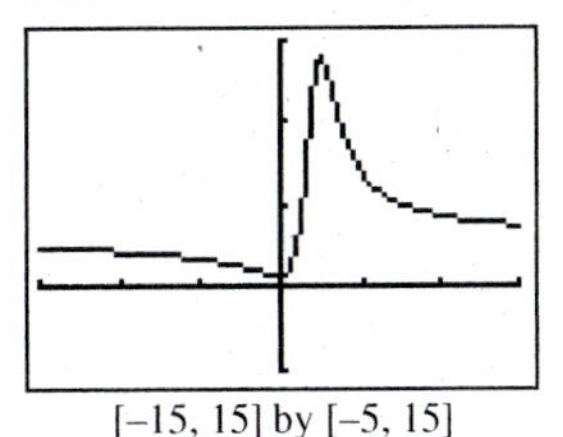

[–15, 15] by [–5, 15]

Intercept: $\left(0, \frac{4}{5}\right)$
Domain: $(-\infty, \infty)$
Range: $[0.773, 14.227]$
Continuity: $(-\infty, \infty)$
Increasing on $[-0.245, 2.445]$
Decreasing on $(-\infty, -0.245]$, $[2.445, \infty)$
Not symmetric.
Bounded.
Local maximum at $(2.445, 14.227)$; local minimum at $(-0.245, 0.773)$
Horizontal asymptote: $y = 3$
No vertical asymptotes.
End behavior: $\lim_{x \to -\infty} f(x) = \lim_{x \to \infty} f(x) = 3$

53. For $h(x) = (x^3 - 1)/(x - 2)$, the numerator factors as $x^3 - 1 = (x - 1)(x^2 + x + 1)$,
so the x-intercept of the graph is 1. The y-intercept is $h(0) = 1/2$. The denominator is zero when $x = 2$, so the vertical asymptote is $x = 2$. Because we can rewrite $h(x)$ as
$$h(x) = \frac{x^3 - 1}{x - 2} = x^2 + 2x + 4 + \frac{7}{x - 2},$$
we know that the end-behavior asymptote is $y = x^2 + 2x + 4$. The graph supports this information and allows us to conclude that
$\lim_{x \to 2^-} h(x) = -\infty$, $\lim_{x \to 2^+} h(x) = \infty$.
The graph also shows a local maximum of about 0.586 at about $x = 0.442$, a local minimum of about 0.443 at about $x = -0.384$, and another local minimum of about 25.970 at about $x = 2.942$.

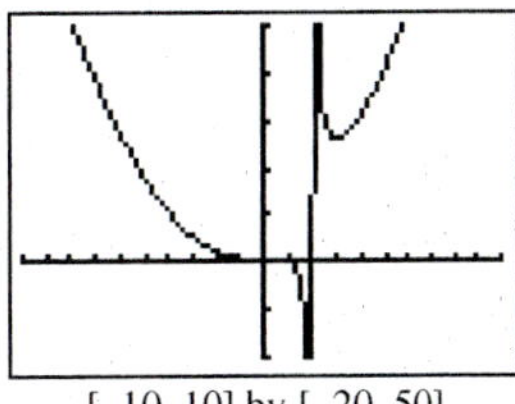

[–10, 10] by [–20, 50]

Intercepts: $(1, 0)$, $\left(0, \frac{1}{2}\right)$
Domain: $(-\infty, 2) \cup (2, \infty)$
Range: $(-\infty, \infty)$
Continuity: All real $x \neq 2$
Increasing on $[-0.384, 0.442]$, $[2.942, \infty)$

Decreasing on $(-\infty, -0.384]$, $[0.442, 2)$, $(2, 2.942]$
Not symmetric.
Unbounded.
Local maximum at $(0.442, 0.586)$; local minimum at $(-0.384, 0.443)$ and $(2.942, 25.970)$
No horizontal asymptote. End-behavior asymptote: $y = x^2 + 2x + 4$
Vertical asymptote: $x = 2$
End behavior: $\lim_{x\to-\infty} h(x) = \lim_{x\to\infty} h(x) = \infty$

55. $f(x) = (x^3 - 2x^2 + x - 1)/(2x - 1)$ has only one x-intercept, and we can use the graph to show that it is about 1.755. The y-intercept is $f(0) = 1$. The denominator is zero when $x = 1/2$, so the vertical asymptote is $x = 1/2$. Because we can rewrite $f(x)$ as

$$f(x) = \frac{x^3 - 2x^2 + x - 1}{2x - 1} = \frac{1}{2}x^2 - \frac{3}{4}x + \frac{1}{8} - \frac{7}{16(2x - 1)},$$

we know that the end-behavior asymptote is $y = \frac{1}{2}x^2 - \frac{3}{4}x + \frac{1}{8}$. The graph supports this information and allows us to conclude that $\lim_{x\to1/2^-} f(x) = \infty$, $\lim_{x\to1/2^+} f(x) = -\infty$.
The graph also shows a local minimum of about 0.920 at about $x = -0.184$.

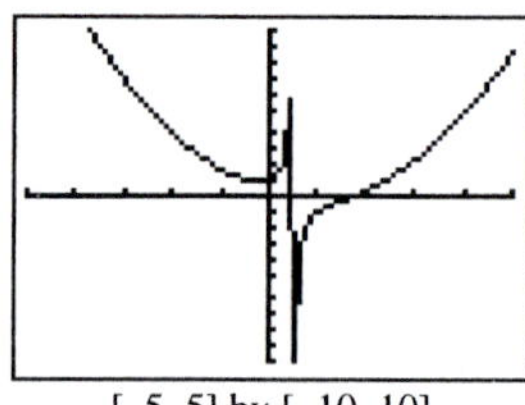
[–5, 5] by [–10, 10]

Intercepts: $(1.755, 0)$, $(0, 1)$
Domain: All $x \neq \frac{1}{2}$
Range: $(-\infty, \infty)$
Continuity: All $x \neq \frac{1}{2}$
Increasing on $[-0.184, 0.5)$, $(0.5, \infty)$
Decreasing on $(-\infty, -0.184]$
Not symmetric.
Unbounded.
Local minimum at $(-0.184, 0.920)$
No horizontal asymptote. End-behavior asymptote: $y = \frac{1}{2}x^2 - \frac{3}{4}x + \frac{1}{8}$
Vertical asymptote: $x = \frac{1}{2}$
End behavior: $\lim_{x\to-\infty} f(x) = \lim_{x\to\infty} f(x) = \infty$

57. For $h(x) = (x^4 + 1)/(x + 1)$, the numerator is never zero, and so $h(x)$ never equals zero and the graph has no x-intercepts. Because $h(0) = 1$, the y-intercept is 1. So the one intercept is the point $(0, 1)$. The denominator is zero when $x = -1$, so $x = -1$ is a vertical asymptote. Divide $x^4 + 1$ by $x + 1$ to show that

$$h(x) = \frac{x^4 + 1}{x + 1} = x^3 - x^2 + x - 1 + \frac{2}{x + 1}.$$

The end-behavior asymptote of $h(x)$ is $y = x^3 - x^2 + x - 1$.

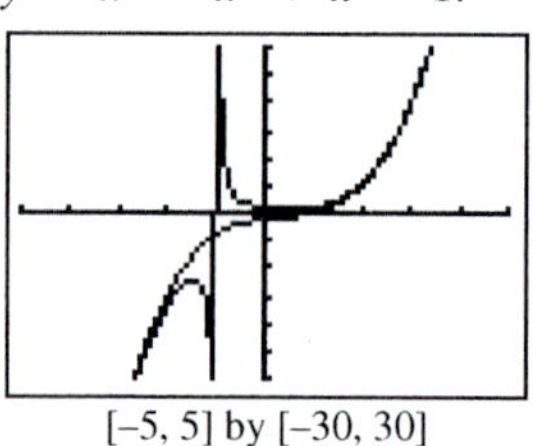
[–5, 5] by [–30, 30]

59. For $f(x) = (x^5 - 1)/(x + 2)$, the numerator factors as $x^5 - 1 = (x - 1)(x^4 + x^3 + x^2 + x + 1)$, and since the second factor is never zero (as can be verified by Descartes' Rule of Signs or by graphing), the x-intercept of the graph is 1. Because $f(0) = -1/2$, the y-intercept is $-1/2$. So the intercepts are $(1, 0)$ and $(0, -1/2)$.
The denominator is zero when $x = -2$, so $x = -2$ is a vertical asymptote. Divide $x^5 - 1$ by $x + 2$ to show that

$$f(x) = \frac{x^5 - 1}{x + 2} = x^4 - 2x^3 + 4x^2 - 8x + 16 - \frac{33}{x + 2}$$

The end-behavior asymptote of $f(x)$ is $y = x^4 - 2x^3 + 4x^2 - 8x + 16$.

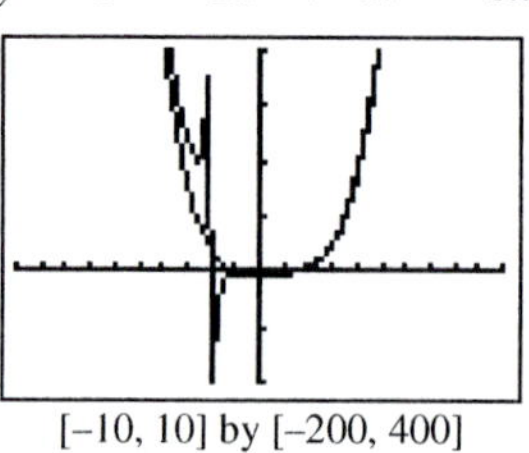
[–10, 10] by [–200, 400]

61. $h(x) = (2x^3 - 3x + 2)/(x^3 - 1)$ has only one x-intercept, and we can use the graph to show that it is about -1.476. Because $h(0) = -2$, the y-intercept is -2. So the intercepts are $(-1.476, 0)$ and $(0, -2)$. The denominator is zero when $x = 1$, so $x = 1$ is a vertical asymptote. Divide $2x^3 - 3x + 2$ by $x^3 - 1$ to show that

$$h(x) = \frac{2x^3 - 3x + 2}{x^3 - 1} = 2 - \frac{3x - 4}{x^3 - 1}.$$

The end-behavior asymptote of $h(x)$ is $y = 2$, a horizontal line.

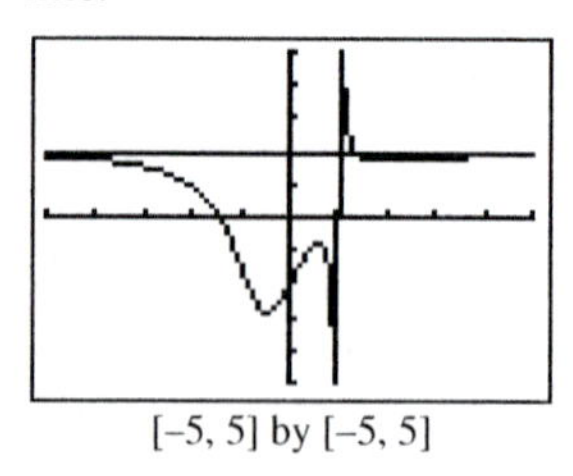
[–5, 5] by [–5, 5]

63. False. If the denominator is never zero, there will be no vertical asymptote. For example, $f(x) = 1/(x^2 + 1)$ is a rational function and has no vertical asymptotes.

65. The excluded values are those for which $x^3 + 3x = 0$, namely 0 and -3. The answer is E.

67. Since $x + 5 = 0$ when $x = -5$, there is a vertical asymptote. And because $x^2/(x + 5) = x - 5 + 25/(x + 5)$, the end behavior is characterized by the slant asymptote $y = x - 5$. The answer is D.

69. **(a)** No: the domain of f is $(-\infty, 3) \cup (3, \infty)$; the domain of g is all real numbers.

(b) No: while it is not defined at 3, it does not tend toward $\pm\infty$ on either side.

(c) Most grapher viewing windows do not reveal that f is undefined at 3.

(d) Almost—but not quite; they are equal for all $x \neq 3$.

71. (a) The volume is $f(x) = k/x$, where x is pressure and k is a constant. $f(x)$ is a quotient of polynomials and hence is rational, but $f(x) = k \cdot x^{-1}$, so is a power function with constant of variation k and pressure -1.

(b) If $f(x) = kx^a$, where a is a negative integer, then the power function f is also a rational function.

(c) $V = \dfrac{k}{P}$, so $k = (2.59)(0.866) = 2.24294$.

If $P = 0.532$, then $V = \dfrac{2.24294}{0.532} \approx 4.22$ L.

73. Horizontal asymptotes: $y = -2$ and $y = 2$.

Intercepts: $\left(0, -\dfrac{3}{2}\right), \left(\dfrac{3}{2}, 0\right)$

$$h(x) = \begin{cases} \dfrac{2x - 3}{x + 2} & x \geq 0 \\ \dfrac{2x - 3}{-x + 2} & x < 0 \end{cases}$$

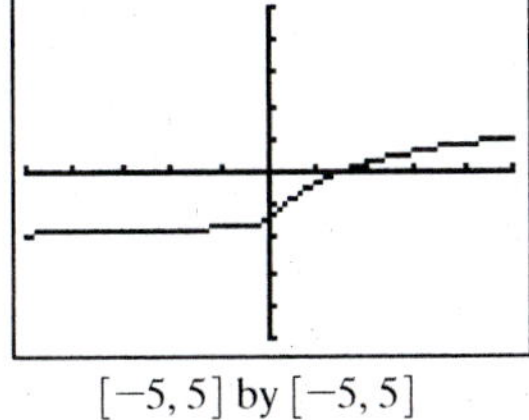

$[-5, 5]$ by $[-5, 5]$

75. Horizontal asymptotes: $y = \pm 3$.

Intercepts: $\left(0, \dfrac{5}{4}\right), \left(\dfrac{5}{3}, 0\right)$

$$f(x) = \begin{cases} \dfrac{5 - 3x}{x + 4} & x \geq 0 \\ \dfrac{5 - 3x}{-x + 4} & x < 0 \end{cases}$$

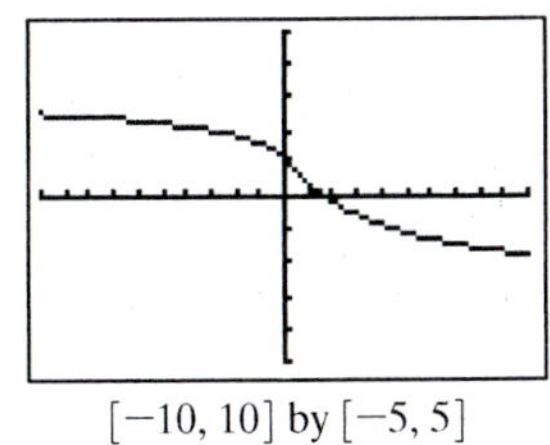

$[-10, 10]$ by $[-5, 5]$

77. The graph of f is the graph of $y = \dfrac{1}{x}$ shifted horizontally $-d/c$ units, stretched vertically by a factor of $|bc - ad|/c^2$, reflected across the x-axis if and only if $bc - ad < 0$, and then shifted vertically by a/c.

Section 2.7 Solving Equations in One Variable

Quick Review 2.7

1. The denominator is $x^2 + x - 12 = (x - 3)(x + 4)$, so the new numerator is $2x(x + 4) = 2x^2 + 8x$.

3. The LCD is the LCM of 12, 18, and 6, namely 36.

$$\frac{5}{12} + \frac{7}{18} - \frac{5}{6} = \frac{15}{36} + \frac{14}{36} - \frac{30}{36} = -\frac{1}{36}$$

5. The LCD is $(2x + 1)(x - 3)$.

$$\frac{x}{2x+1} - \frac{2}{x-3} = \frac{x(x-3)}{(2x+1)(x-3)} - \frac{2(2x+1)}{(2x+1)(x-3)}$$
$$= \frac{x^2 - 3x - 4x - 2}{(2x+1)(x-3)}$$
$$= \frac{x^2 - 7x - 2}{(2x+1)(x-3)}$$

7. For $2x^2 - 3x - 1 = 0$: $a = 2$, $b = -3$, and $c = -1$.

$$x = \frac{-b \pm \sqrt{b^2 - 4ac}}{2a}$$
$$= \frac{-(-3) \pm \sqrt{(-3)^2 - 4(2)(-1)}}{2(2)}$$
$$= \frac{3 \pm \sqrt{9 - (-8)}}{4} = \frac{3 \pm \sqrt{17}}{4}$$

9. For $3x^2 + 2x - 2 = 0$: $a = 3$, $b = 2$, and $c = -2$.

$$x = \frac{-b \pm \sqrt{b^2 - 4ac}}{2a}$$
$$= \frac{-2 \pm \sqrt{(2)^2 - 4(3)(-2)}}{2(3)}$$
$$= \frac{-2 \pm \sqrt{4 - (-24)}}{6} = \frac{-2 \pm \sqrt{28}}{6}$$
$$= \frac{-2 \pm 2\sqrt{7}}{6} = \frac{-1 \pm \sqrt{7}}{3}$$

Section 2.7 Exercises

1. Algebraically: $\dfrac{x-2}{3} + \dfrac{x+5}{3} = \dfrac{1}{3}$

$$(x - 2) + (x + 5) = 1$$
$$2x + 3 = 1$$
$$2x = -2$$
$$x = -1$$

Numerically: For $x = -1$,

$$\frac{x-2}{3} + \frac{x+5}{3} = \frac{-1-2}{3} + \frac{-1+5}{3}$$
$$= \frac{-3}{3} + \frac{4}{3}$$
$$= \frac{1}{3}$$

3. Algebraically: $x + 5 = \dfrac{14}{x}$

$$x^2 + 5x = 14 \quad (x \neq 0)$$
$$x^2 + 5x - 14 = 0$$
$$(x - 2)(x + 7) = 0$$
$$x - 2 = 0 \quad \text{or} \quad x + 7 = 0$$
$$x = 2 \quad \text{or} \quad x = -7$$

Numerically: For $x = 2$,

$x + 5 = 2 + 5 = 7$ and

$\dfrac{14}{x} = \dfrac{14}{2} = 7$.

For $x = -7$,

$x + 5 = -7 + 5 = -2$ and

$\dfrac{14}{x} = \dfrac{14}{-7} = -2$.

5. Algebraically: $x + \dfrac{4x}{x-3} = \dfrac{12}{x-3}$

$x(x-3) + 4x = 12 \; (x \neq 3)$
$x^2 - 3x + 4x = 12$
$x^2 + x - 12 = 0$
$(x+4)(x-3) = 0$
$x + 4 = 0$ or $x - 3 = 0$
$x = -4$ or $x = 3$ — but $x = 3$ is extraneous.

Numerically: For $x = -4$,

$x + \dfrac{4x}{x-3} = -4 + \dfrac{4(-4)}{-4-3} = -4 + \dfrac{16}{7} = -\dfrac{12}{7}$ and

$\dfrac{12}{x-3} = \dfrac{12}{-4-3} = -\dfrac{12}{7}$.

7. Algebraically: $x + \dfrac{10}{x} = 7$

$x^2 + 10 = 7x \quad (x \neq 0)$
$x^2 - 7x + 10 = 0$
$(x-2)(x-5) = 0$
$x - 2 = 0$ or $x - 5 = 0$
$x = 2$ or $x = 5$

Graphically: The graph of $f(x) = x + \dfrac{10}{x} - 7$ suggests that the x-intercepts are 2 and 5.

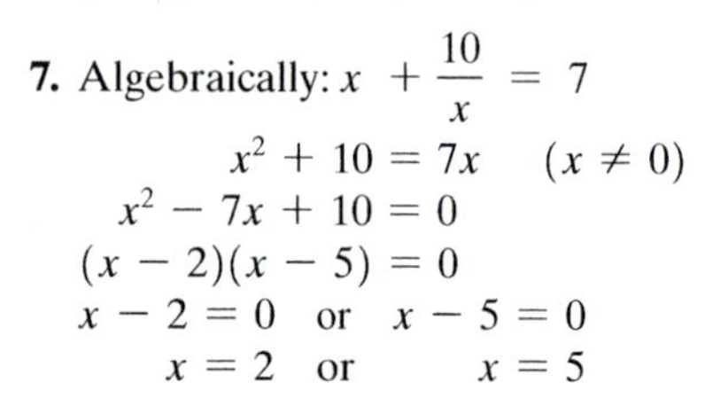
[–9.4, 9.4] by [–15, 5]

Then the solutions are $x = 2$ and $x = 5$.

9. Algebraically: $x + \dfrac{12}{x} = 7$

$x^2 + 12 = 7x \quad (x \neq 0)$
$x^2 - 7x + 12 = 0$
$(x-3)(x-4) = 0$
$x - 3 = 0$ or $x - 4 = 0$
$x = 3$ or $x = 4$

Graphically: The graph of $f(x) = x + \dfrac{12}{x} - 7$ suggests that the x-intercepts are 3 and 4.

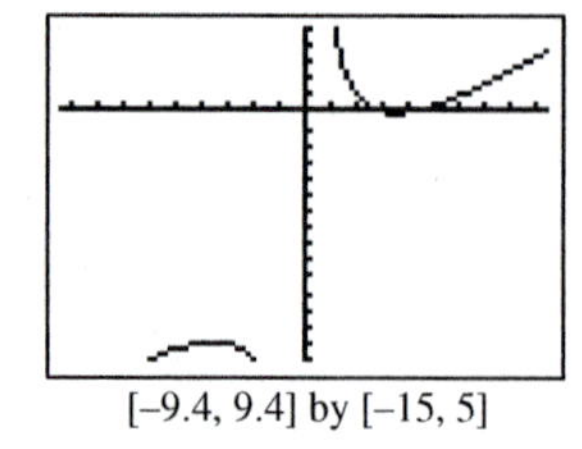
[–9.4, 9.4] by [–15, 5]

Then the solutions are $x = 3$ and $x = 4$.

11. Algebraically: $2 - \dfrac{1}{x+1} = \dfrac{1}{x^2+x}$

[and $x^2 + x = x(x+1)$]
$2(x^2 + x) - x = 1 \quad (x \neq 0, -1)$
$2x^2 + x - 1 = 0$
$(2x-1)(x+1) = 0$
$2x - 1 = 0$ or $x + 1 = 0$
$x = \dfrac{1}{2}$ or $x = -1$
— but $x = -1$ is extraneous.

Graphically: The graph of $f(x) = 2 - \dfrac{1}{x+1} - \dfrac{1}{x^2+x}$ suggests that the x-intercept is $\dfrac{1}{2}$. There is a hole at $x = -1$.

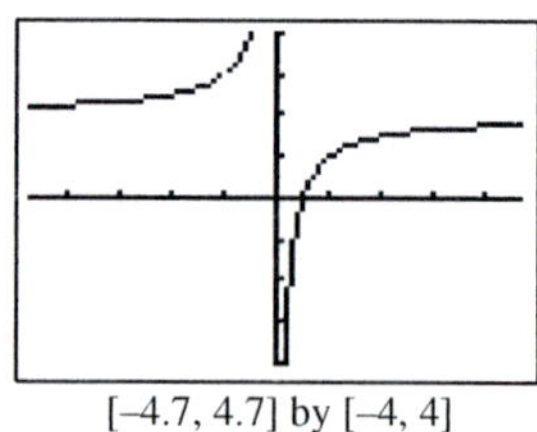
[–4.7, 4.7] by [–4, 4]

Then the solution is $x = \dfrac{1}{2}$.

13. Algebraically: $\dfrac{3x}{x+5} + \dfrac{1}{x-2} = \dfrac{7}{x^2+3x-10}$

[and $x^2 + 3x - 10 = (x+5)(x-2)$]
$3x(x-2) + (x+5) = 7 \quad (x \neq -5, 2)$
$3x^2 - 5x - 2 = 0$
$(3x+1)(x-2) = 0$
$3x + 1 = 0$ or $x - 2 = 0$
$x = -\dfrac{1}{3}$ or $x = 2$
— but $x = 2$ is extraneous.

Graphically: The graph of

$f(x) = \dfrac{3x}{x+5} + \dfrac{1}{x-2} - \dfrac{7}{x^2+3x-10}$ suggests that the x-intercept is $-\dfrac{1}{3}$. There is a hole at $x = 2$.

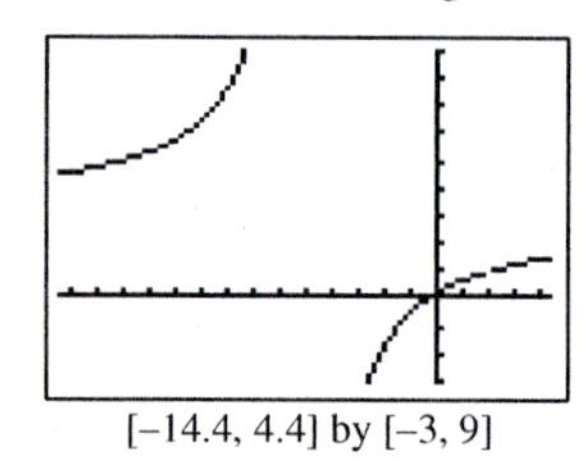
[–14.4, 4.4] by [–3, 9]

Then the solution is $x = -\dfrac{1}{3}$.

15. Algebraically: $\dfrac{x-3}{x} - \dfrac{3}{x+1} + \dfrac{3}{x^2+x} = 0$

[and $x^2 + x = x(x+1)$]
$(x-3)(x+1) - 3x + 3 = 0 \quad (x \neq 0, -1)$
$x^2 - 5x = 0$
$x(x-5) = 0$
$x = 0$ or $x - 5 = 0$
$x = 0$ or $x = 5$ — but $x = 0$ is extraneous.

Graphically: The graph of

$f(x) = \dfrac{x-3}{x} - \dfrac{3}{x+1} + \dfrac{3}{x^2+x}$ suggests that the x-intercept is 5. The x-axis hides a hole at $x = 0$.

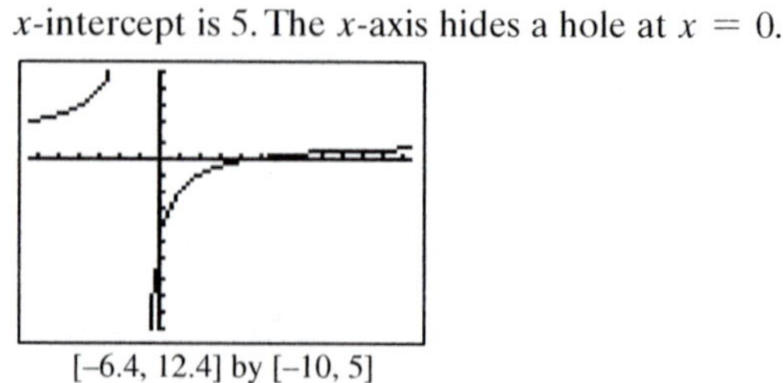
[–6.4, 12.4] by [–10, 5]

Then the solution is $x = 5$.

17. Algebraically: $\frac{3}{x+2} + \frac{6}{x^2+2x} = \frac{3-x}{x}$

[and $x^2 + 2x = x(x+2)$]

$3x + 6 = (3 - x)(x + 2) \quad (x \neq -2, 0)$

$3x + 6 = -x^2 + x + 6$

$x^2 + 2x = 0$

$x(x + 2) = 0$

$x = 0$ or $x + 2 = 0$

$x = 0$ or $x = -2$

— but both solutions are extraneous.

No real solutions.

Graphically: The graph of

$f(x) = \frac{3}{x+2} + \frac{6}{x^2+2x} - \frac{3-x}{x}$ suggests that there are no x-intercepts. There is a hole at $x = -2$, and the x-axis hides a "hole" at $x = 0$.

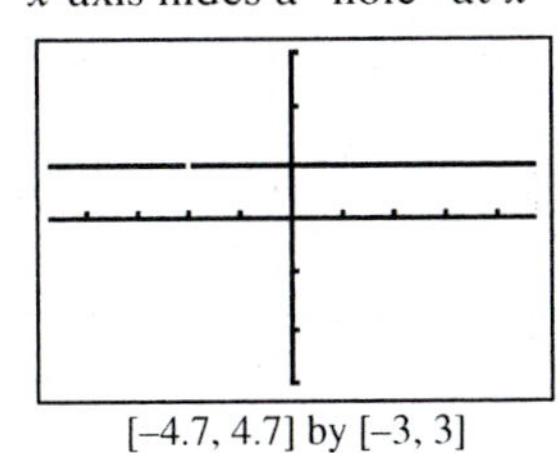

[–4.7, 4.7] by [–3, 3]

Then there are no real solutions.

19. There is no x-intercept at $x = -2$. That is the extraneous solution.

21. Neither possible solution corresponds to an x-intercept of the graph. Both are extraneous.

23. $\frac{2}{x-1} + x = 5$

$2 + x(x - 1) = 5(x - 1) \quad (x \neq 1)$

$x^2 - x + 2 = 5x - 5$

$x^2 - 6x + 7 = 0$

$$x = \frac{-(-6) \pm \sqrt{(-6)^2 - 4(1)(7)}}{2(1)}$$

$$x = \frac{6 \pm \sqrt{8}}{2} = 3 \pm \sqrt{2}$$

$x = 3 + \sqrt{2} \approx 4.414$ or

$x = 3 - \sqrt{2} \approx 1.586$

25. $\frac{x^2 - 2x + 1}{x + 5} = 0$

$x^2 - 2x + 1 = 0 \quad (x \neq 5)$

$(x - 1)^2 = 0$

$x - 1 = 0$

$x = 1$

27. $\frac{4x}{x+4} + \frac{5}{x-1} = \frac{15}{x^2 + 3x - 4}$

[and $x^2 + 3x - 4 = (x + 4)(x - 1)$]

$4x(x - 1) + 5(x + 4) = 15 \quad (x \neq -4, 1)$

$4x^2 + x + 5 = 0$

The discriminant is $b^2 - 4ac = 1^2 - 4(4)(5) = -79 < 0$. There are no real solutions.

29. $x^2 + \frac{5}{x} = 8$

$x^3 + 5 = 8x \quad (x \neq 0)$

Using a graphing calculator to find the x-intercepts of $f(x) = x^3 - 8x + 5$ yields the solutions $x \approx -3.100$, $x \approx 0.661$, and $x \approx 2.439$.

31. (a) The total amount of solution is $(125 + x)$ mL; of this, the amount of acid is x plus 60% of the original amount, or $x + 0.6(125)$.

(b) $y = 0.83$

(c) $C(x) = \frac{x + 75}{x + 125} = 0.83$. Multiply both sides by $x + 125$, then rearrange to get $0.17x = 28.75$, so that $x \approx 169.12$ mL.

33. (a) $C(x) = \frac{3000 + 2.12x}{x}$

(b) A profit is realized if $C(x) < 2.75$, or $3000 + 2.12x < 2.75x$. Then $3000 < 0.63x$, so that $x > 4761.9$—4762 hats per week.

(c) They must have $2.75x - (3000 + 2.12x) > 1000$ or $0.63x > 4000$: 6350 hats per week.

35. (a) If x is the length, then $182/x$ is the width.

$$P(x) = 2x + 2\left(\frac{182}{x}\right) = 2x + \frac{364}{x}$$

(b) The graph of $P(x) = 2x + 364/x$ has a minimum when $x \approx 13.49$, so that the rectangle is square. Then $P(13.49) = 2(13.49) + 364/13.49 \approx 53.96$ ft.

37. (a) Since $V = \pi r^2 h$, the height here is $V/(\pi r^2)$. And since in general, $S = 2\pi r^2 + 2\pi rh = 2\pi r^2 + 2V/r$, here $S(x) = 2\pi x^2 + 1000/x$ ($0.5\ L = 500\ \text{cm}^3$).

(b) Solving $2\pi x^2 + 1000/x = 900$ graphically by finding the zeros of $f(x) = 2\pi x^2 + 1000/x - 900$ yields two solutions: either $x \approx 1.12$ cm, in which case $h \approx 126.88$ cm, or $x \approx 11.37$ cm, in which case $h \approx 1.23$ cm.

39. (a) $\frac{1}{R} = \frac{1}{R_1} + \frac{1}{R_2}$

$\frac{1}{R} = \frac{1}{2.3} + \frac{1}{x}$

$2.3x = xR + 2.3R$

$R(x) = \frac{2.3x}{x + 2.3}$

(b) $2.3x = xR + 2.3R$

$x = \frac{2.3R}{2.3 - R}$

For $R = 1.7$, $x \approx 6.52$ ohms.

41. (a) Drain A can drain $1/4.75$ of the pool per hour, while drain B can drain $1/t$ of the pool per hour. Together, they can drain a fraction

$$D(t) = \frac{1}{4.75} + \frac{1}{t} = \frac{t + 4.75}{4.75t}$$

of the pool in 1 hour.

(b) The information implies that $D(t) = 1/2.6$, so we solve $\frac{1}{2.6} = \frac{1}{4.75} + \frac{1}{t}$.

Graphically: The function $f(t) = \frac{1}{2.6} - \frac{1}{4.75} - \frac{1}{t}$ has a zero at $t \approx 5.74$ h, so that is the solution.

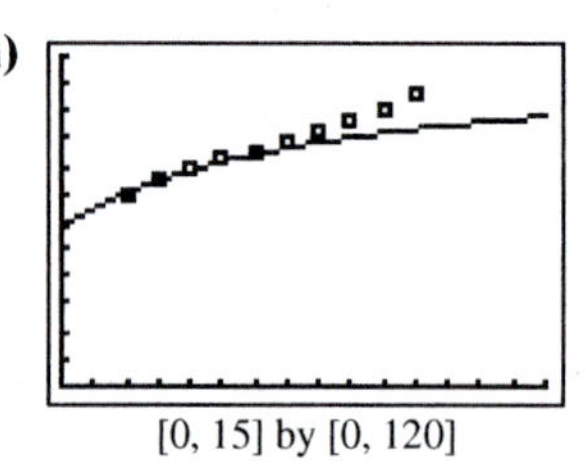

[0, 10] by [−0.25, 0.25]

Algebraically:
$$\frac{1}{2.6} = \frac{1}{4.75} + \frac{1}{t}$$
$$4.75t = 2.6t + 2.6(4.75)$$
$$t = \frac{2.6\,(4.75)}{4.75 - 2.6}$$
$$\approx 5.74$$

43. (a)

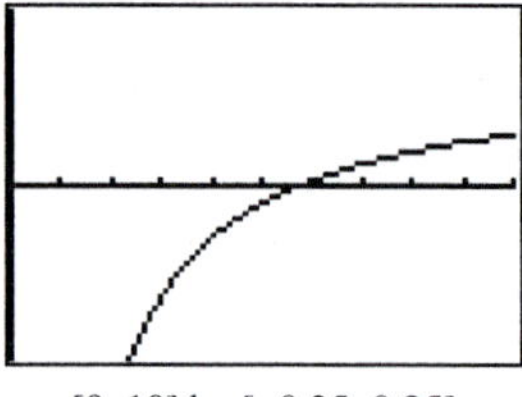

[0, 15] by [0, 120]

(b) When $x = 15$, $y = 120 - 500/(15 + 8) \approx 98.3$. In 2005, sales are estimated at about $98.3 billion.

45. False. An extraneous solution is a value that, though generated by the solution-finding process, does not work in the original equation. In an equation containing rational expressions, an extraneous solution is typically a solution to the version of the equation that has been cleared of fractions but not to the original version.

47.
$$x - \frac{3x}{x+2} = \frac{6}{x+2}$$
$$x(x+2) - 3x = 6 \quad (x \neq -2)$$
$$x^2 - x - 6 = 0$$
$$(x-3)(x+2) = 0$$
$x = 3$ or $x = -2$ — but $x = -2$ is extraneous.
The answer is D.

49.
$$\frac{x}{x+2} + \frac{2}{x-5} = \frac{14}{x^2 - 3x - 10}$$
[and $x^2 - 3x - 10 = (x+2)(x-5)$]
$$x(x-5) + 2(x+2) = 14 \quad (x \neq -2, 5)$$
$$x^2 - 3x - 10 = 0$$
$$(x+2)(x-5) = 0$$
$x = -2$ or $x = 5$ — but both solutions are extraneous.
The answer is E.

51. (a) The LCD is $x^2 + 2x = x(x+2)$.
$$f(x) = \frac{x-3}{x} + \frac{3}{x+2} + \frac{6}{x^2+2x}$$
$$= \frac{(x-3)(x+2)}{x^2+2x} + \frac{3x}{x^2+2x} + \frac{6}{x^2+2x}$$
$$= \frac{x^2 - x - 6 + 3x + 6}{x^2 + 2x}$$
$$= \frac{x^2 + 2x}{x^2 + 2x}$$

(b) All $x \neq 0, -2$

(c) $f(x) = \begin{cases} 1 & x \neq -2, 0 \\ \text{undefined} & x = -2 \text{ or } x = 0 \end{cases}$

(d) The graph appears to be the horizontal line $y = 1$ with holes at $x = -2$ and $x = 0$.

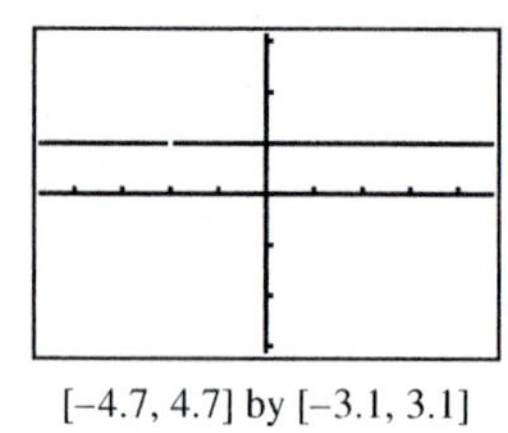

[−4.7, 4.7] by [−3.1, 3.1]

This matches the definition in part (c).

53.
$$y = 1 - \frac{1}{1-x}$$
$$y(1-x) = (1-x) - 1$$
$$y - xy = -x$$
$$y = xy - x$$
$$x = \frac{y}{y-1}$$

55.
$$y = 1 + \cfrac{1}{1 + \cfrac{1}{1-x}}$$
$$y = 1 + \frac{1-x}{1-x+1}$$
$$y = 1 + \frac{1-x}{2-x}$$
$$y(2-x) = (2-x) + (1-x)$$
$$2y - xy = 3 - 2x$$
$$2y - 3 = xy - 2x$$
$$x = \frac{2y-3}{y-2}$$

Section 2.8 Solving Inequalities in One Variable

Exploration 1

1. (a)

(+)(−)(+)	(+)(−)(+)	(+)(+)(+)
Negative	Negative	Positive

Boundary points on the x-axis: −3, 2

(b)

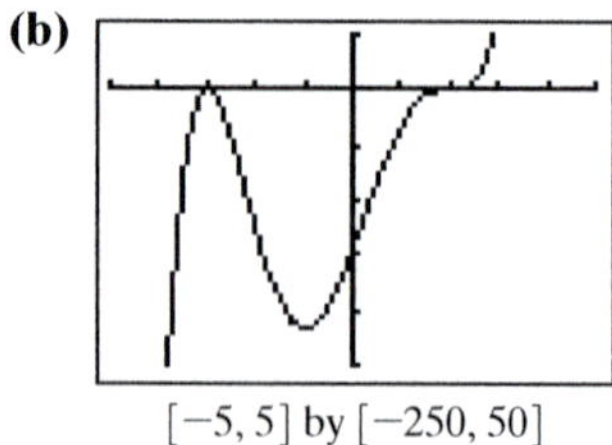

[−5, 5] by [−250, 50]

2. (a)

(−)(+)(−)(+)	(−)(+)(−)(+)	(−)(+)(+)(+)
Positive	Positive	Negative

Boundary points on the x-axis: −2, −1

(b)

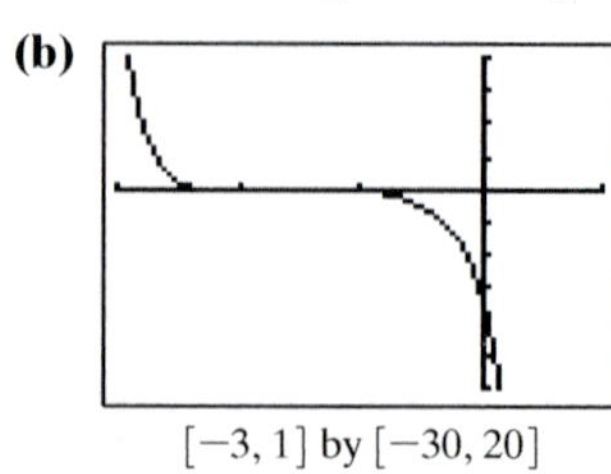

[−3, 1] by [−30, 20]

3. (a)

(+)(+)(−)(−)	(+)(+)(+)(−)	(+)(+)(+)(−)	
Positive	Negative	Negative	x
	−4	2	

(b)

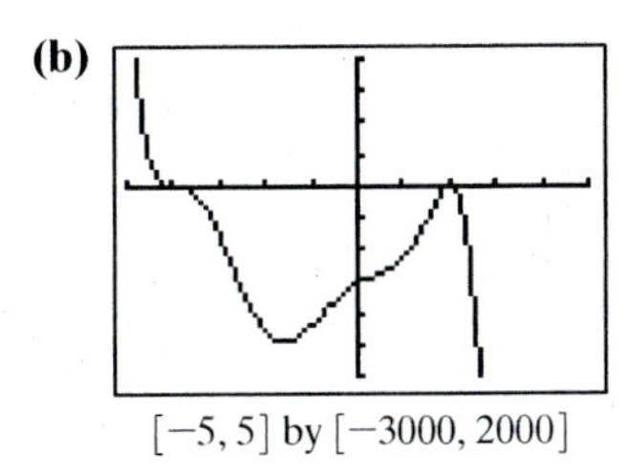

$[-5, 5]$ by $[-3000, 2000]$

Quick Review 2.8

1. $\lim_{x\to\infty} f(x) = \infty,\ \lim_{x\to-\infty} f(x) = -\infty$

3. $\lim_{x\to\infty} g(x) = \infty,\ \lim_{x\to-\infty} g(x) = \infty$

5. $\dfrac{x^3+5}{x}$

7. $\dfrac{x(x-3)-2(2x+1)}{(2x+1)(x-3)} = \dfrac{x^2-3x-4x-2}{(2x+1)(x-3)}$
$= \dfrac{x^2-7x-2}{(2x+1)(x-3)} = \dfrac{x^2-7x-2}{2x^2-5x-3}$

9. (a) $\dfrac{\pm1, \pm3}{\pm1, \pm2}$ or $\pm1, \pm\dfrac{1}{2}, \pm3, \pm\dfrac{3}{2}$

(b) A graph suggests that -1 and $\dfrac{3}{2}$ are good candidates for zeros.

$$\begin{array}{r|rrrr} -1 & 2 & 1 & -4 & -3 \\ & & -2 & 1 & 3 \\ \hline 3/2 & 2 & -1 & -3 & 0 \\ & & 3 & 3 & \\ \hline & 2 & 2 & 0 & \end{array}$$

$2x^3 + x^2 - 4x - 3 = (x+1)\left(x-\dfrac{3}{2}\right)(2x+2)$
$= (x+1)(2x-3)(x+1)$

Section 2.8 Exercises

1. (a) $f(x) = 0$ when $x = -2, -1, 5$

(b) $f(x) > 0$ when $-2 < x < -1$ or $x > 5$

(c) $f(x) < 0$ when $x < -2$ or $-1 < x < 5$

(−)(−)(−)	(+)(−)(−)	(+)(+)(−)	(+)(+)(+)	
Negative	Positive	Negative	Positive	x
	−2	−1	5	

3. (a) $f(x) = 0$ when $x = -7, -4, 6$

(b) $f(x) > 0$ when $x < -7$ or $-4 < x < 6$ or $x > 6$

(c) $f(x) < 0$ when $-7 < x < -4$

$(-)(-)(-)^2$	$(+)(-)(-)^2$	$(+)(+)(-)^2$	$(+)(+)(+)^2$	
Positive	Negative	Positive	Positive	x
	−7	−4	6	

5. (a) $f(x) = 0$ when $x = 8, -1$

(b) $f(x) > 0$ when $-1 < x < 8$ or $x > 8$

(c) $f(x) < 0$ when $x < -1$

$(+)(-)^2(-)^3$	$(+)(-)^2(+)^3$	$(+)(+)^2(+)^3$	
Negative	Positive	Positive	x
	−1	8	

7. $(x+1)(x-3)^2 = 0$ when $x = -1, 3$

$(-)(-)^2$	$(+)(-)^2$	$(+)(+)^2$	
Negative	Positive	Positive	x
	−1	3	

By the sign chart, the solution of $(x+1)(x-3)^2 > 0$ is $(-1, 3) \cup (3, \infty)$.

9. $(x+1)(x^2-3x+2) = (x+1)(x-1)(x-2) = 0$ when $x = -1, 1, 2$

(−)(−)(−)	(+)(−)(−)	(+)(+)(−)	(+)(+)(+)	
Negative	Positive	Negative	Positive	x
	−1	1	2	

By the sign chart, the solution of $(x+1)(x-1)(x-2) < 0$ is $(-\infty, -1) \cup (1, 2)$.

11. By the Rational Zeros Theorem, the possible rational zeros are $\pm1, \pm\dfrac{1}{2}, \pm2, \pm3, \pm\dfrac{3}{2}, \pm6$. A graph suggests that $-2, \dfrac{1}{2}$, and 3 are good candidates to be zeros.

$$\begin{array}{r|rrrr} -2 & 2 & -3 & -11 & 6 \\ & & -4 & 14 & -6 \\ \hline 3 & 2 & -7 & 3 & 0 \\ & & 6 & -3 & \\ \hline & 2 & -1 & 0 & \end{array}$$

$2x^3 - 3x^2 - 11x + 6 = (x+2)(x-3)(2x-1) = 0$ when $x = -2, 3, \dfrac{1}{2}$

(−)(−)(−)	(+)(−)(−)	(+)(−)(+)	(+)(+)(+)	
Negative	Positive	Negative	Positive	x
	−2	$\frac{1}{2}$	3	

By the sign chart, the solution of $(x+2)(x-3)(2x-1) \ge 0$ is $\left[-2, \dfrac{1}{2}\right] \cup [3, \infty)$.

13. The zeros of $f(x) = x^3 - x^2 - 2x$ appear to be $-1, 0$, and 2. Substituting these values into f confirms this. The graph shows that the solution of $x^3 - x^2 - 2x \ge 0$ is $[-1, 0] \cup [2, \infty)$.

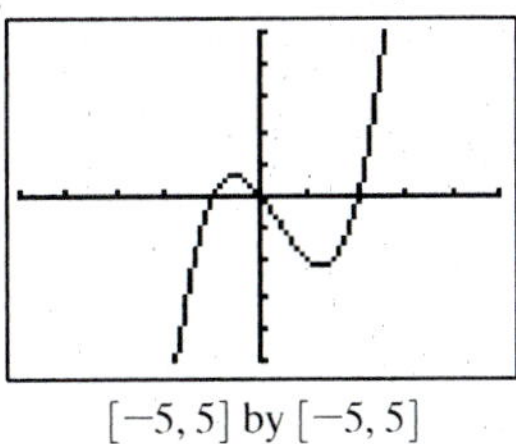

$[-5, 5]$ by $[-5, 5]$

15. The zeros of $f(x) = 2x^3 - 5x^2 - x + 6$ appear to be $-1, \dfrac{3}{2}$ and 2. Substituting these values into f confirms this. The graph shows that the solution of $2x^3 - 5x^2 - x + 6 > 0$ is $\left(-1, \dfrac{3}{2}\right) \cup (2, \infty)$.

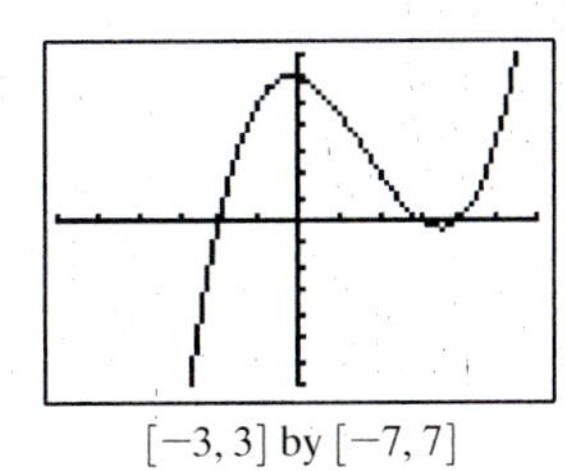

$[-3, 3]$ by $[-7, 7]$

17. The only zero of $f(x) = 3x^3 - 2x^2 - x + 6$ is found graphically to be $x \approx -1.15$. The graph shows that the solution of $3x^3 - 2x^2 - x + 6 \geq 0$ is approximately $[-1.15, \infty)$.

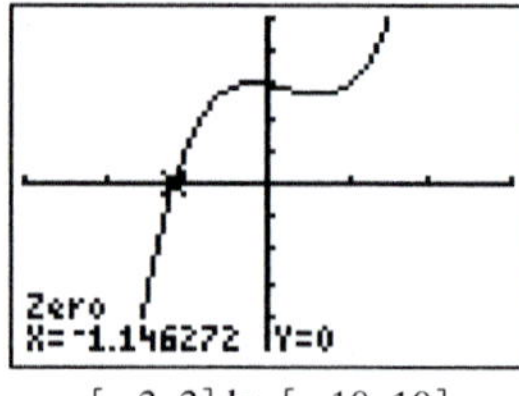

$[-3, 3]$ by $[-10, 10]$

19. The zeros of $f(x) = 2x^4 - 3x^3 - 6x^2 + 5x + 6$ appear to be $-1, \frac{3}{2}$, and 2. Substituting these into f confirms this. The graph shows that the solution of $2x^4 - 3x^3 - 6x^2 + 5x + 6 < 0$ is $\left(\frac{3}{2}, 2\right)$.

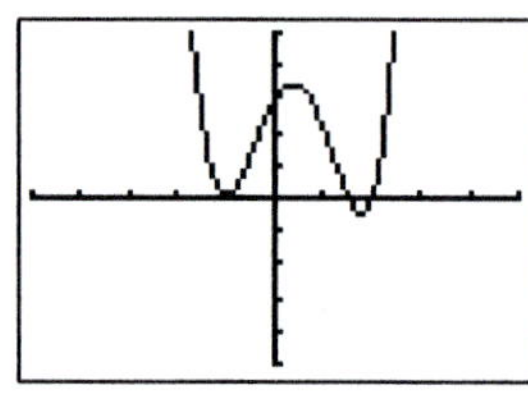

$[-5, 5]$ by $[-10, 10]$

21. $f(x) = (x^2 + 4)(2x^2 + 3)$

(a) The solution is $(-\infty, \infty)$, because both factors of $f(x)$ are always positive.

(b) $(-\infty, \infty)$, for the same reason as in part (a).

(c) There are no solutions, because both factors of $f(x)$ are always positive.

(d) There are no solutions, for the same reason as in part (c).

23. $f(x) = (2x^2 - 2x + 5)(3x - 4)^2$
The first factor is always positive because the leading term has a positive coefficient and the discriminant $(-2)^2 - 4(2)(5) = -36$ is negative. The only zero is $x = 4/3$, with multiplicity two, since that is the solution for $3x - 4 = 0$.

(a) True for all $x \neq \frac{4}{3}$

(b) $(-\infty, \infty)$

(c) There are no solutions.

(d) $x = \frac{4}{3}$

25. (a) $f(x) = 0$ when $x = 1$

(b) $f(x)$ is undefined when $x = -\frac{3}{2}, 4$

(c) $f(x) > 0$ when $-\frac{3}{2} < x < 1$ or $x > 4$

(d) $f(x) < 0$ when $x < -\frac{3}{2}$ or $1 < x < 4$

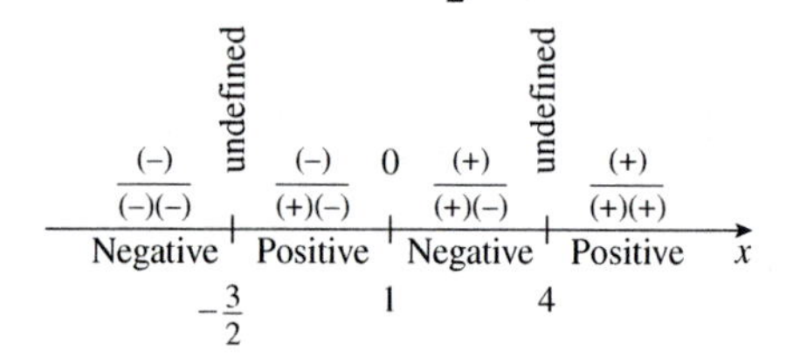

27. (a) $f(x) = 0$ when $x = 0, -3$

(b) $f(x)$ is undefined when $x < -3$

(c) $f(x) > 0$ when $x > 0$

(d) $f(x) < 0$ when $-3 < x < 0$

0 0
(–)(+) (+)(+)
Undefined | Negative | Positive x
–3 0

29. (a) $f(x) = 0$ when $x = -5$

(b) $f(x)$ is undefined when $x = -\frac{1}{2}, x = 1, x < -5$

(c) $f(x) > 0$ when $-5 < x < -\frac{1}{2}$ or $x > 1$

(d) $f(x) < 0$ when $-\frac{1}{2} < x < 1$

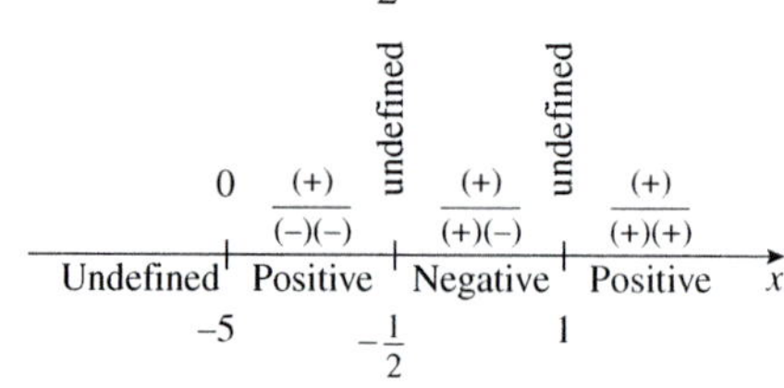

31. (a) $f(x) = 0$ when $x = 3$

(b) $f(x)$ is undefined when $x = 4, x < 3$

(c) $f(x) > 0$ when $3 < x < 4$ or $x > 4$

(d) None. $f(x)$ is never negative

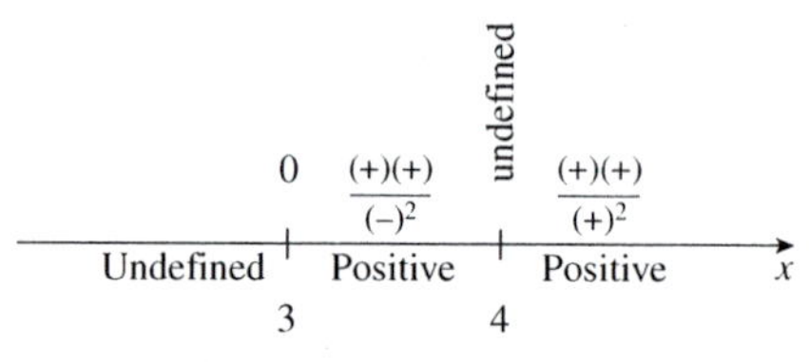

33. $f(x) = \frac{x - 1}{x^2 - 4} = \frac{x - 1}{(x + 2)(x - 2)}$ has points of potential sign change at $x = -2, 1, 2$.

undefined undefined
(–)/(–)(–) (–)/(+)(–) 0 (+)/(+)(–) (+)/(+)(+)
Negative | Positive | Negative | Positive x
–2 1 2

By the sign chart, the solution of $\frac{x - 1}{x^2 - 4} < 0$ is $(-\infty, -2) \cup (1, 2)$.

35. $f(x) = \frac{x^2 - 1}{x^2 + 1} = \frac{(x + 1)(x - 1)}{(x^2 + 1)}$ has points of potential sign change at $x = -1, 1$.

(–)(–)/(+) 0 (+)(–)/(+) 0 (+)(+)/(+)
Positive | Negative | Positive x
–1 1

By the sign chart, the solution of $\frac{x^2 - 1}{x^2 + 1} \leq 0$ is $[-1, 1]$.

37. $f(x) = \dfrac{x^2 + x - 12}{x^2 - 4x + 4} = \dfrac{(x+4)(x-3)}{(x-2)^2}$ has points of potential sign change at $x = -4, 2, 3$.

$\frac{(-)(-)}{(-)^2}$	0	$\frac{(+)(-)}{(-)^2}$	undefined	$\frac{(+)(-)}{(+)^2}$	0	$\frac{(+)(+)}{(+)^2}$	
Positive	-4	Negative	2	Negative	3	Positive	x

By the sign chart, the solution of $\dfrac{x^2 + x - 12}{x^2 - 4x + 4} > 0$ is $(-\infty, -4) \cup (3, \infty)$.

39. $f(x) = \dfrac{x^3 - x}{x^2 + 1} = \dfrac{x(x+1)(x-1)}{x^2 + 1}$ has points of potential sign change at $x = -1, 0, 1$.

$\frac{(-)(-)(-)}{(+)}$	0	$\frac{(-)(+)(-)}{(+)}$	0	$\frac{(+)(+)(-)}{(+)}$	0	$\frac{(+)(+)(+)}{(+)}$	
Negative	-1	Positive	0	Negative	1	Positive	x

By the sign chart, the solution of $\dfrac{x^3 - x}{x^2 + 1} \geq 0$ is $[-1, 0] \cup [1, \infty)$.

41. $f(x) = x|x - 2|$ has points of potential sign change at $x = 0, 2$.

$(-)\lvert-\rvert$	0	$(+)\lvert-\rvert$	0	$(+)\lvert+\rvert$	
Negative	0	Positive	2	Positive	x

By the sign chart, the solution of $x|x - 2| > 0$ is $(0, 2) \cup (2, \infty)$.

43. $f(x) = (2x - 1)\sqrt{x + 4}$ has a point of potential sign change at $x = \dfrac{1}{2}$. Note that the domain of f is $[-4, \infty)$.

	0	$(-)(+)$	0	$(+)(+)$	
Undefined	-4	Negative	$\frac{1}{2}$	Positive	x

By the sign chart, the solution of $(2x - 1)\sqrt{x + 4} < 0$ is $\left(-4, \dfrac{1}{2}\right)$.

45. $f(x) = \dfrac{x^3(x - 2)}{(x + 3)^2}$ has points of potential sign change at $x = -3, 0, 2$.

$\frac{(-)^3(-)}{(-)^2}$	undefined	$\frac{(-)^3(-)}{(+)^2}$	0	$\frac{(+)^3(-)}{(+)^2}$	0	$\frac{(+)^3(+)}{(+)^2}$	
Positive	-3	Positive	0	Negative	2	Positive	x

By the sign chart, the solution of $\dfrac{x^3(x - 2)}{(x + 3)^2} < 0$ is $(0, 2)$.

47. $f(x) = x^2 - \dfrac{2}{x} = \dfrac{x^3 - 2}{x}$ has points of potential sign change at $x = 0, \sqrt[3]{2}$.

$\frac{(-)}{(-)}$	undefined	$\frac{(-)}{(+)}$	0	$\frac{(+)}{(+)}$	
Positive	0	Negative	$\sqrt[3]{2}$	Positive	x

By the sign chart, the solution of $x^2 - \dfrac{2}{x} > 0$ is $(-\infty, 0) \cup (\sqrt[3]{2}, \infty)$.

49. $f(x) = \dfrac{1}{x + 1} + \dfrac{1}{x - 3} = \dfrac{2(x - 1)}{(x + 1)(x - 3)}$ has points of potential sign change at $x = -1, 1, 3$.

$\frac{(-)}{(-)(-)}$	undefined	$\frac{(-)}{(+)(-)}$	0	$\frac{(+)}{(+)(-)}$	undefined	$\frac{(+)}{(+)(+)}$	
Negative	-1	Positive	1	Negative	3	Positive	x

By the sign chart, the solution of $\dfrac{1}{x + 1} + \dfrac{1}{x - 3} \leq 0$ is $(-\infty, -1) \cup [1, 3)$.

51. $f(x) = (x + 3)|x - 1|$ has points of potential sign change at $x = -3, 1$.

$(-)\lvert-\rvert$	0	$(+)\lvert-\rvert$	0	$(+)\lvert+\rvert$	
Negative	-3	Positive	1	Positive	x

By the sign chart, the solution of $(x + 3)|x - 1| \geq 0$ is $[-3, \infty)$.

53. $f(x) = \dfrac{(x - 5)|x - 2|}{\sqrt{2x - 2}}$ has points of potential sign change at $x = 2, 5$. Note that the domain of f is $(1, \infty)$.

	undefined	$\frac{(-)\lvert-\rvert}{(+)}$	0	$\frac{(-)\lvert+\rvert}{(+)}$	0	$\frac{(+)\lvert+\rvert}{(+)}$	
Undefined	1	Negative	2	Negative	5	Positive	x

By the sign chart, the solution of $\dfrac{(x - 5)|x - 2|}{\sqrt{2x - 2}} \geq 0$ is $[5, \infty)$.

55. One way to solve the inequality is to graph $y = 3(x - 1) + 2$ and $y = 5x + 6$ together, then find the interval along the x-axis where the first graph is below or intersects the second graph. Another way is to solve for x algebraically.

57. Let $x > 0$ be the width of a rectangle; then the length is $2x - 2$ and the perimeter is $P = 2[x + (2x - 2)]$. Solving $P < 200$ and $2x - 2 > 0$ (below) gives 1 in. $< x <$ 34 in.

$$\begin{aligned} 2[x + (2x - 2)] &< 200 \\ 2(3x - 2) &< 200 \\ 6x - 4 &< 200 \\ 6x &< 204 \\ x &< 34 \end{aligned} \quad \text{and} \quad \begin{aligned} 2x - 2 &> 0 \\ 2x &> 2 \\ x &> 1 \end{aligned}$$

59. The lengths of the sides of the box are x, $12 - 2x$, and $15 - 2x$, so the volume is $x(12 - 2x)(15 - 2x)$. To solve $x(12 - 2x)(15 - 2x) \le 100$, graph $f(x) = x(12 - 2x)(15 - 2x) - 100$ and find the zeros: $x \approx 0.69$ and $x \approx 4.20$.

[0, 6] by [−100, 100]

From the graph, the solution of $f(x) \le 0$ is approximately $[0, 0.69] \cup [4.20, 6]$. The squares should be such that either 0 in. $\le x \le$ 0.69 in. or 4.20 in. $\le x \le$ 6 in.

61. **(a)** $\frac{1}{2}L = 500 \text{ cm}^3$

$$V = \pi x^2 h = 500 \Rightarrow h = \frac{500}{\pi x^2}$$

$$S = 2\pi x h + 2\pi x^2 = 2\pi x\left(\frac{500}{\pi x^2}\right) + 2\pi x^2$$

$$= \frac{1000}{x} + 2\pi x^2 = \frac{1000 + 2\pi x^3}{x}$$

(b) Solve $S < 900$ by graphing $\frac{1000 + 2\pi x^3}{x} - 900$ and finding its zeros: $x \approx 1.12$ and $x \approx 11.37$

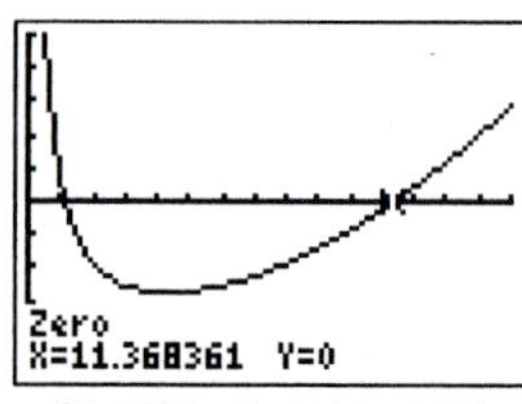

[0, 15] by [−1000, 1000]

From the graph, the solution of $S - 900 < 0$ is approximately $(1.12, 11.37)$. So the radius is between 1.12 cm and 11.37 cm. The corresponding height must be between 1.23 cm and 126.88 cm.

(c) Graph S and find the minimum graphically.

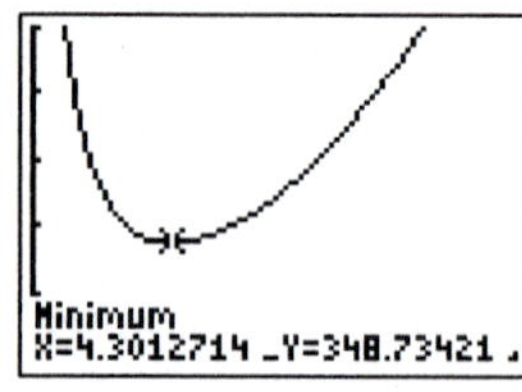

[0, 15] by [0, 1000]

The minimum surface area is about 348.73 cm^2.

63. **(a)** $y \approx 993.870x + 19{,}025.768$

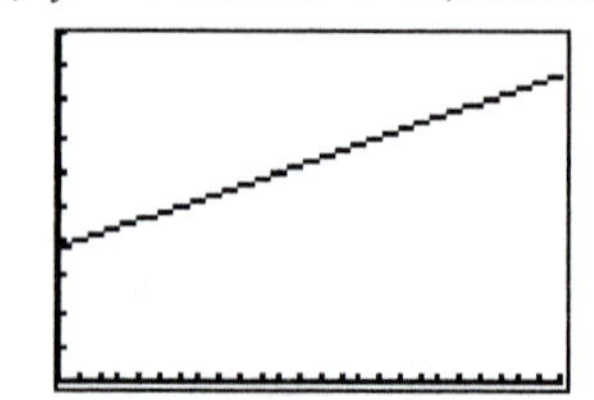

(b) From the graph of $y = 993.870x + 19{,}025.768$, we find that $y = 40{,}000$ when $x \approx 21.2$. The per capita income will exceed $40,000 in the year 2011.

65. False. Because the factor x^4 has an even power, it does not change sign at $x = 0$.

67. x must be positive but less than 1. The answer is C.

69. The statement is true so long as the denominator is negative and the numerator is nonzero. Thus x must be less than 3 but nonzero. The answer is D.

71. $f(x) = \frac{(x - 1)(x + 2)^2}{(x - 3)(x + 1)}$

Vertical asymptotes: $x = -1$, $x = 3$

x-intercepts: $(-2, 0)$, $(1, 0)$

y-intercept: $\left(0, \frac{4}{3}\right)$

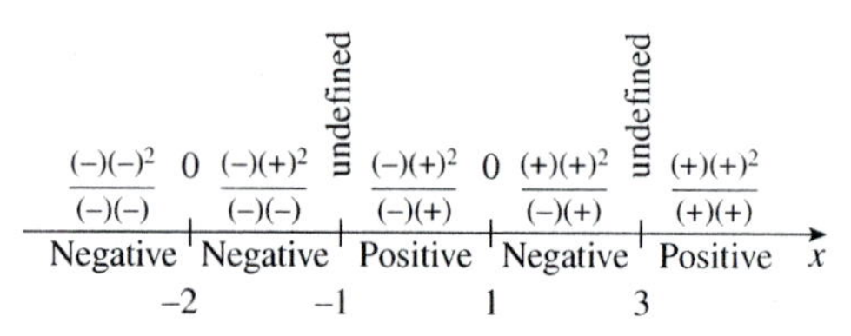

By hand:

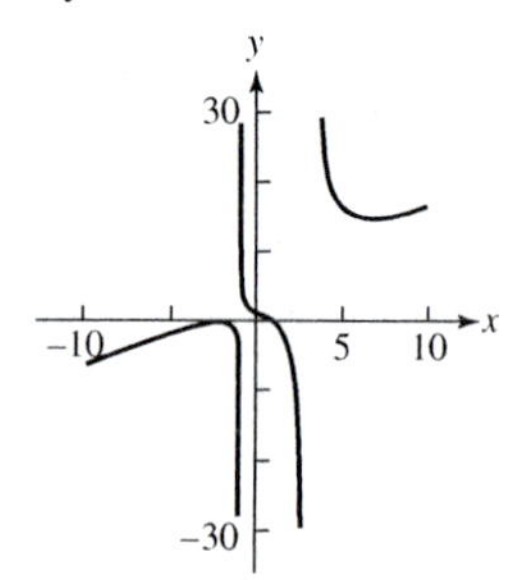

Grapher:

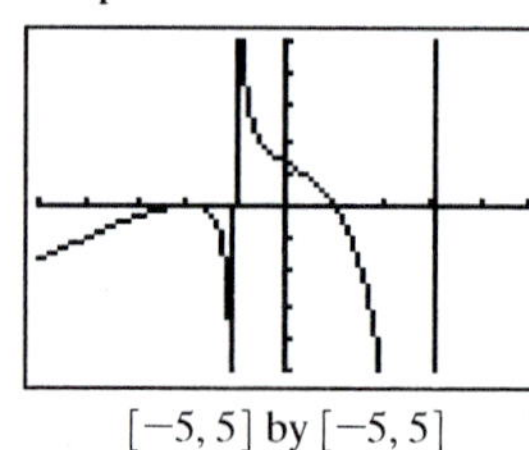

[−5, 5] by [−5, 5]

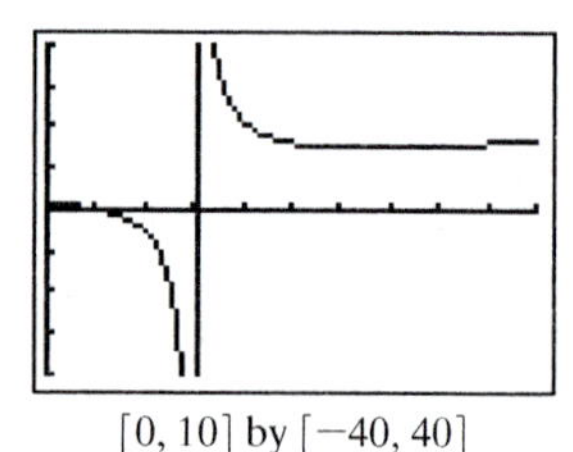

[0, 10] by [−40, 40]

73. **(a)** $|x - 3| < 1/3 \Rightarrow |3x - 9| < 1 \Rightarrow |3x - 5 - 4| < 1 \Rightarrow |f(x) - 4| < 1$.

For example:

$$|f(x) - 4| = |(3x - 5) - 4| = |3x - 9|$$

$$= 3|x - 3| < 3\left(\frac{1}{3}\right) = 1$$

(b) If x stays within the dashed vertical lines, $f(x)$ will stay within the dashed horizontal lines. For the example in part (a), the graph shows that for $\frac{8}{3} < x < \frac{10}{3}$ $\left(\text{that is, } |x - 3| < \frac{1}{3}\right)$, we have $3 < f(x) < 5$ (that is, $|f(x) - 4| < 1$).

(c) $|x - 3| < 0.01 \Rightarrow |3x - 9| < 0.03 \Rightarrow |3x - 5 - 4| < 0.03 \Rightarrow |f(x) - 4| < 0.03$. The dashed lines would be closer to $x = 3$ and $y = 4$.

75. One possible answer: Given $0 < a < b$, multiplying both sides of $a < b$ by a gives $a^2 < ab$; multiplying by b gives $ab < b^2$. Then, by the transitive property of inequality, we have $a^2 < b^2$.

■ Chapter 2 Review

For #1, first find the slope of the line. Then use algebra to put into $y = mx + b$ format.

1. $m = \dfrac{-9 - (-2)}{4 - (-3)} = \dfrac{-7}{7} = -1, (y + 9) = -1(x - 4),$
$y = -x - 5$

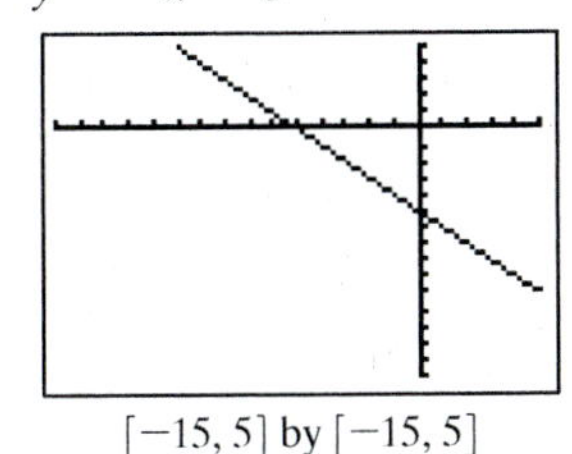

$[-15, 5]$ by $[-15, 5]$

3. Starting from $y = x^2$, translate right 2 units and vertically stretch by 3 (either order), then translate up 4 units.

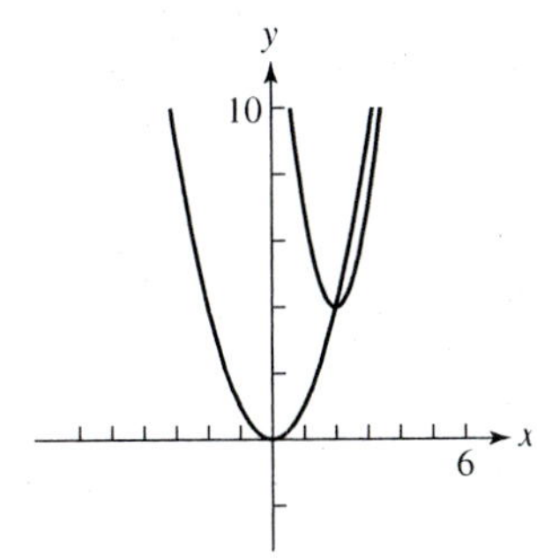

5. Vertex: $(-3, 5)$; axis: $x = -3$

7. $f(x) = -2(x^2 + 8x) - 31$
$= -2(x^2 + 8x + 16) + 32 - 31 = -2(x + 4)^2 + 1;$
Vertex: $(-4, 1)$; axis: $x = -4$

For #9–11, use the form $y = a(x - h)^2 + k$, where (h, k), the vertex, is given.

9. $h = -2$ and $k = -3$ are given, so $y = a(x + 2)^2 - 3$.
Using the point $(1, 2)$, we have $2 = 9a - 3$, so $a = \dfrac{5}{9}$:
$y = \dfrac{5}{9}(x + 2)^2 - 3.$

11. $h = 3$ and $k = -2$ are given, so $y = a(x - 3)^2 - 2$.
Using the point $(5, 0)$, we have $0 = 4a - 2$, so $a = \dfrac{1}{2}$:
$y = \dfrac{1}{2}(x - 3)^2 - 2.$

13.

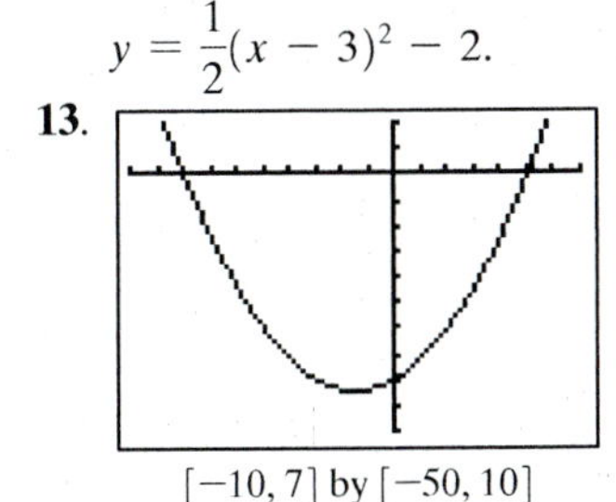

$[-10, 7]$ by $[-50, 10]$

15.

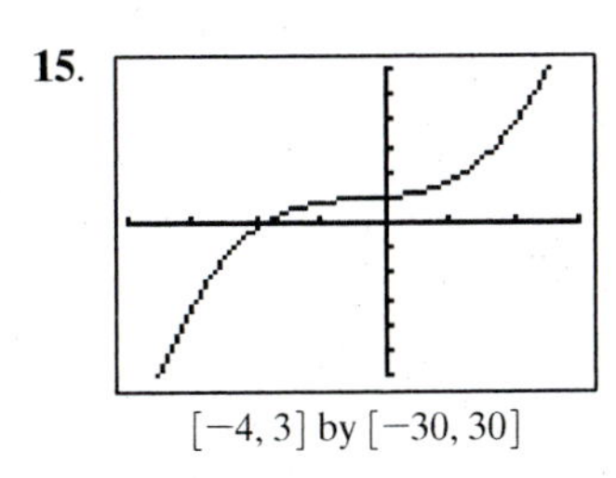

$[-4, 3]$ by $[-30, 30]$

17. $S = kr^2 \quad (k = 4\pi)$

19. The force F needed varies directly with the distance x from its resting position, with constant of variation k.

21. $k = 4, a = \dfrac{1}{3}$. In Quadrant I, $f(x)$ is increasing and concave down since $0 < a < 1$.

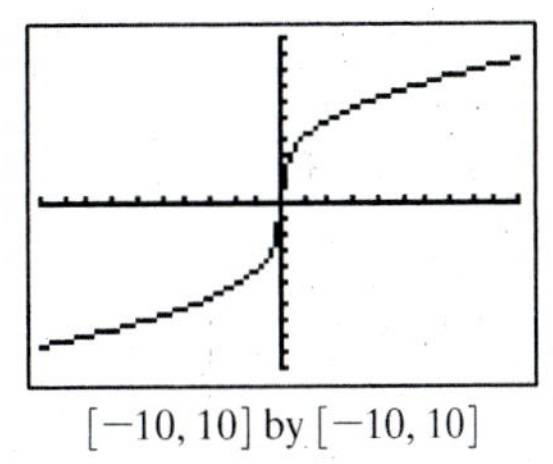

$[-10, 10]$ by $[-10, 10]$

$f(-x) = 4(-x)^{1/3} = -4x^{1/3} = -f(x)$, so f is odd.

23. $k = -2, a = -3$. In Quadrant IV, f is increasing and concave down. $f(-x) = -2(-x)^{-3} = \dfrac{-2}{(-x)^3} = \dfrac{-2}{-x^3}$
$= \dfrac{2}{x^3} = 2x^{-3} = -f(x)$, so f is odd.

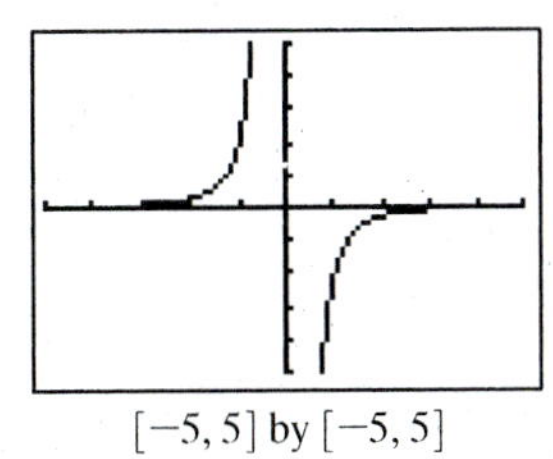

$[-5, 5]$ by $[-5, 5]$

25. $\dfrac{2x^3 - 7x^2 + 4x - 5}{x - 3} = 2x^2 - x + 1 - \dfrac{2}{x - 3}$

$$
\begin{array}{r|l}
 & 2x^2 - x + 1 \\ \hline
x - 3 & 2x^3 - 7x^2 + 4x - 5 \\
 & 2x^3 - 6x^2 \\ \hline
 & -x^2 + 4x \\
 & -x^2 + 3x \\ \hline
 & x - 5 \\
 & x - 3 \\ \hline
 & -2
\end{array}
$$

27. $\dfrac{2x^4 - 3x^3 + 9x^2 - 14x + 7}{x^2 + 4}$
$= 2x^2 - 3x + 1 + \dfrac{-2x + 3}{x^2 + 4}$

$$
\begin{array}{r|l}
 & 2x^2 - 3x + 1 \\ \hline
x^2 + 4 & 2x^4 - 3x^3 + 9x^2 - 14x + 7 \\
 & 2x^4 + 8x^2 \\ \hline
 & -3x^3 + x^2 - 14x + 7 \\
 & -3x^3 - 12x \\ \hline
 & x^2 - 2x + 7 \\
 & x^2 + 4 \\ \hline
 & -2x + 3
\end{array}
$$

29. Remainder: $f(-2) = -39$

31. Yes: 2 is a zero of the second polynomial.

33. $$\begin{array}{r|rrrr} 5 & 1 & -5 & 3 & 4 \\ & & 5 & 0 & 15 \\ \hline & 1 & 0 & 3 & 19 \end{array}$$

Yes, $x = 5$ is an upper bound for the zeros of $f(x)$ because all entries on the bottom row are nonnegative.

35. $$\begin{array}{r|rrrrr} -3 & 4 & 4 & -15 & -17 & -2 \\ & & -12 & 24 & -27 & 132 \\ \hline & 4 & -8 & 9 & -44 & 130 \end{array}$$

Yes, $x = -3$ is a lower bound for the zeros of $f(x)$ because all entries on the bottom row alternate signs.

37. Possible rational zeros: $\dfrac{\pm 1, \pm 2, \pm 3, \pm 6}{\pm 1, \pm 2}$,

or $\pm 1, \pm 2, \pm 3, \pm 6, \pm\frac{1}{2}, \pm\frac{3}{2}$; $-\frac{3}{2}$ and 2 are zeros.

39. $(1 + i)^3 = (1 + 2i + i^2)(1 + i) = (2i)(1 + i)$
$= -2 + 2i$

41. $i^{29} = i$

For #43, use the quadratic formula.

43. $x = \dfrac{6 \pm \sqrt{36 - 52}}{2} = \dfrac{6 \pm 4i}{2} = 3 \pm 2i$

45. (c) $f(x) = (x - 2)^2$ is a quadratic polynomial that has vertex $(2, 0)$ and y-intercept $(0, 4)$, so its graph must be graph (c).

47. (b) $f(x) = (x - 2)^4$ is a quartic polynomial that passes through $(2, 0)$ and $(0, 16)$, so its graph must be graph (b).

In #49–51, use a graph and the Rational Zeros Test to determine zeros.

49. Rational: 0 (multiplicity 2) — easily seen by inspection. Irrational: $5 \pm \sqrt{2}$ (using the quadratic formula, after taking out a factor of x^2). No non-real zeros.

51. Rational: none. Irrational: approximately -2.34, 0.57, 3.77. No non-real zeros.

53. The only rational zero is $-\frac{3}{2}$. Dividing by $x + \frac{3}{2}$ (below) leaves $2x^2 - 12x + 20$, which has zeros $\dfrac{12 \pm \sqrt{144 - 160}}{4} = 3 \pm i$. Therefore
$f(x) = (2x + 3)[x - (3 - i)][x - (3 + i)]$
$= (2x + 3)(x - 3 + i)(x - 3 - i)$.

$$\begin{array}{r|rrrr} -3/2 & 2 & -9 & 2 & 30 \\ & & -3 & 18 & -30 \\ \hline & 2 & -12 & 20 & 0 \end{array}$$

55. All zeros are rational: $1, -1, \frac{2}{3}$, and $-\frac{5}{2}$. Therefore $f(x) = (3x - 2)(2x + 5)(x - 1)(x + 1)$; this can be confirmed by multiplying out the terms or graphing the original function and the factored form of the function.

In #57–59, determine rational zeros (graphically or otherwise) and divide synthetically until a quadratic remains. If more real zeros remain, use the quadratic formula.

57. The only real zero is 2; dividing by $x - 2$ leaves the quadratic factor $x^2 + x + 1$, so
$f(x) = (x - 2)(x^2 + x + 1)$.

$$\begin{array}{r|rrrr} 2 & 1 & -1 & -1 & -2 \\ & & 2 & 2 & 2 \\ \hline & 1 & 1 & 1 & 0 \end{array}$$

59. The two real zeros are 1 and $\frac{3}{2}$; dividing by $x - 1$ and $x - \frac{3}{2}$ leaves the quadratic factor $2x^2 - 4x + 10$, so
$f(x) = (2x - 3)(x - 1)(x^2 - 2x + 5)$.

$$\begin{array}{r|rrrrr} 1 & 2 & -9 & 23 & -31 & 15 \\ & & 2 & -7 & 16 & -15 \\ \hline & 2 & -7 & 16 & -15 & 0 \end{array} \qquad \begin{array}{r|rrrr} 3/2 & 2 & -7 & 16 & -15 \\ & & 3 & -6 & 15 \\ \hline & 2 & -4 & 10 & 0 \end{array}$$

61. $(x - \sqrt{5})(x + \sqrt{5})(x - 3) = x^3 - 3x^2 - 5x + 15$. Other answers may be found by multiplying this polynomial by any real number.

63. $(x - 3)(x + 2)(3x - 1)(2x + 1)$
$= 6x^4 - 5x^3 - 38x^2 - 5x + 6$ (This may be multiplied by any real number.)

65. $(x + 2)^2(x - 4)^2 = x^4 - 4x^3 - 12x^2 + 32x + 64$ (This may be multiplied by any real number.)

67. $f(x) = -1 + \dfrac{2}{x - 5}$; translate right 5 units and vertically stretch by 2 (either order), then translate down 1 unit. Horizontal asymptote: $y = -1$; vertical asymptote: $x = 5$.

69. Asymptotes: $y = 1$, $x = -1$, and $x = 1$.
Intercept: $(0, -1)$.

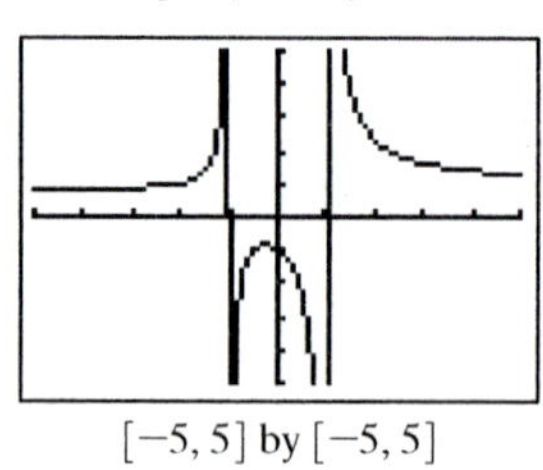

$[-5, 5]$ by $[-5, 5]$

71. End-behavior asymptote: $y = x - 7$.
Vertical asymptote: $x = -3$. Intercept: $\left(0, \frac{5}{3}\right)$.

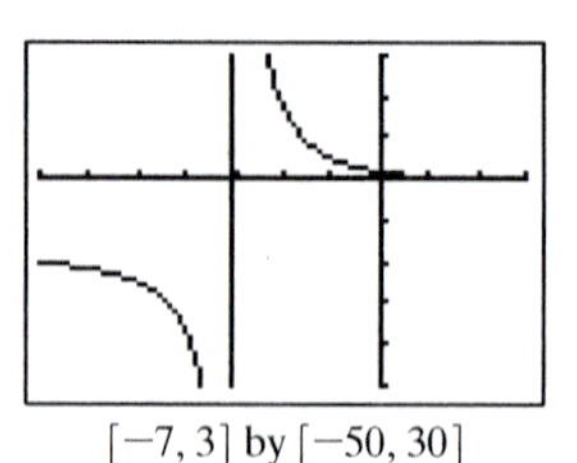

$[-7, 3]$ by $[-50, 30]$

73. $f(x) = \dfrac{x^3 + x^2 - 2x + 5}{x + 2}$ has only one x-intercept, and we can use the graph to show that it is about -2.552. The y-intercept is $f(0) = 5/2$. The denominator is zero when $x = -2$, so the vertical asymptote is $x = -2$. Because we can rewrite $f(x)$ as

$$f(x) = \frac{x^3 + x^2 - 2x + 5}{x + 2} = x^2 - x + \frac{5}{x + 2},$$

we know that the end-behavior asymptote is $y = x^2 - x$. The graph supports this information and allows us to conclude that

$\lim_{x \to -2^-} = -\infty$, $\lim_{x \to -2^+} = \infty$.

The graph also shows a local minimum of about 1.63 at about $x = 0.82$.

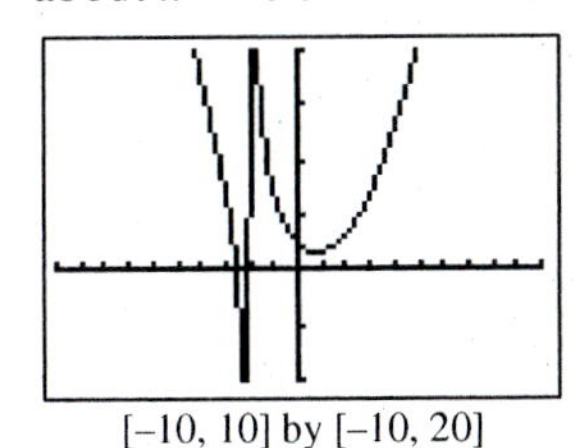

[–10, 10] by [–10, 20]

y-intercept: $\left(0, \frac{5}{2}\right)$

x-intercept: $(-2.55, 0)$

Domain: All $x \neq -2$

Range: $(-\infty, \infty)$

Continuity: All $x \neq -2$

Increasing on $[0.82, \infty)$

Decreasing on $(-\infty, -2), (-2, 0.82]$

Not symmetric.

Unbounded.

Local minimum: (0.82, 1.63)

No horizontal asymptote. End-behavior asymptote: $y = x^2 - x$

Vertical asymptote: $x = -2$.

End behavior: $\lim_{x \to -\infty} f(x) = \lim_{x \to \infty} f(x) = \infty$

75. Multiply by x: $2x^2 - 11x + 12 = 0$, so $x = \frac{3}{2}$ or $x = 4$.

For #77, find the zeros of $f(x)$ and then determine where the function is positive or negative by creating a sign chart.

77. $f(x) = (x - 3)(2x + 5)(x + 2)$, so the zeros of $f(x)$ are $x = \left\{-\frac{5}{2}, -2, 3\right\}$.

(–)(–)(–)	(–)(+)(–)	(–)(+)(+)	(+)(+)(+)
Negative	Positive	Negative	Positive

Boundaries on the x-axis: $-\frac{5}{2}$, -2, 3

As our sign chart indicates, $f(x) < 0$ on the interval $\left(-\infty, -\frac{5}{2}\right) \cup (-2, 3)$.

79. Zeros of numerator and denominator: -3, -2, and 2.

Choose -4, -2.5, 0, and 3; $\dfrac{x + 3}{x^2 - 4}$ is positive at -2.5 and 3, and equals 0 at -3, so the solution is $[-3, -2) \cup (2, \infty)$.

81. Since the function is always positive, we need only worry about the equality $(2x - 1)^2|x + 3| = 0$. By inspection, we see this holds true only when $x = \left\{-3, \frac{1}{2}\right\}$.

83. Synthetic division reveals that we *cannot* conclude that 5 is an upper bound (since there are both positive and negative numbers on the bottom row), while -5 *is* a lower bound (because all numbers on the bottom row alternate signs). Yes, there is another zero (at $x \approx 10.0002$).

5⌋	1	–10	–3	28	20	–2
		5	–25	–140	–560	–2700
	1	–5	–28	–112	–540	–2702

–5⌋	1	–10	–3	28	20	–2
		–5	75	–360	1660	–8400
	1	–15	72	–332	1680	–8402

85. (a) $V = $ (height)(width)(length) $= x(30 - 2x)(70 - 2x)$ in.^{3}

(b) Either $x \approx 4.57$ or $x \approx 8.63$ in.

87. (a) The tank is made up of a cylinder, with volume $\pi x^2(140 - 2x)$, and a sphere, with volume $\frac{4}{3}\pi x^3$. Thus, $V = \frac{4}{3}\pi x^3 + \pi x^2(140 - 2x)$.

(b)

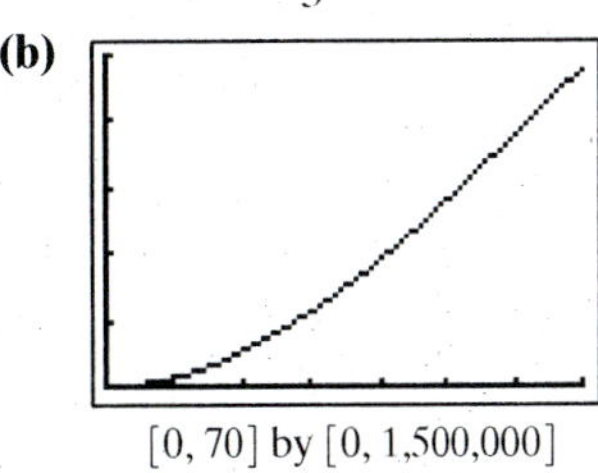

[0, 70] by [0, 1,500,000]

(c) The largest volume occurs when $x = 70$ (so it is actually a sphere). This volume is $\frac{4}{3}\pi(70)^3 \approx 1{,}436{,}755$ ft^3.

89. (a) $y = 1.401x + 4.331$

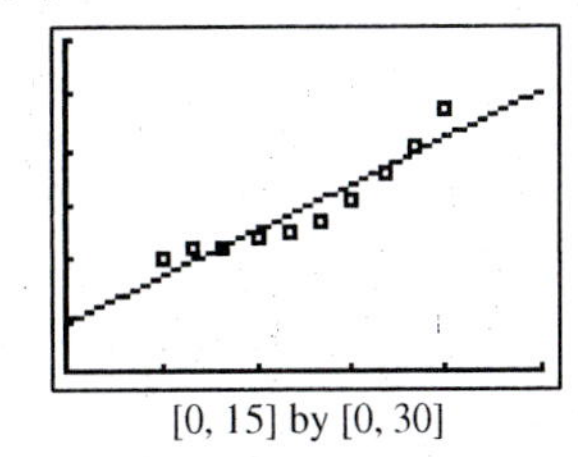

[0, 15] by [0, 30]

(b) $y = 0.188x^2 - 1.411x + 13.331$

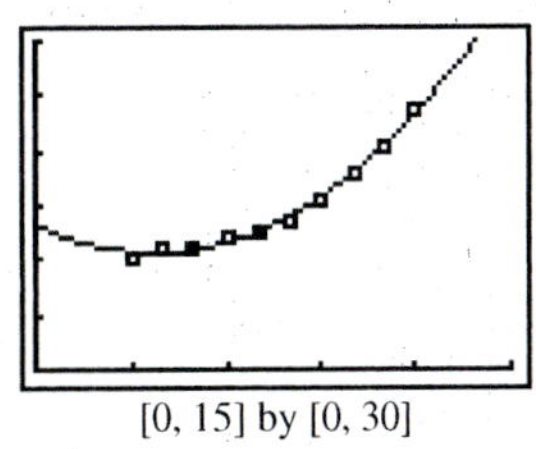

[0, 15] by [0, 30]

(c) Linear: Solving $1.401x + 4.331 = 30$ graphically, we find that $y = 30$ when $x \approx 18.32$. The spending will exceed \$30 million in the year 2008.

Quadratic: Solving $0.188x^2 - 1.411x + 13.331 = 30$ graphically, we find that $y = 30$ when $x \approx 13.89$. The spending will exceed \$30 million in the year 2003.

91. **(a)** $P(15) = 325$, $P(70) = 600$, $P(100) = 648$

(b) $y = \frac{640}{0.8} = 800$

(c) The deer population approaches (but never equals) 800.

93. **(a)** $C(x) = \frac{50}{50 + x}$

(b) Shown is the window $[0, 50]$ by $[0, 1]$, with the graphs of $y = C(x)$ and $y = 0.6$. The two graphs cross when $x \approx 33.33$ ounces of distilled water.

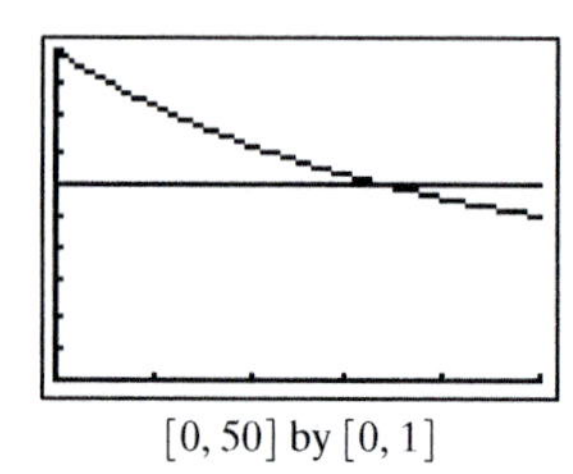

$[0, 50]$ by $[0, 1]$

(c) Algebraic solution of $\frac{50}{50 + x} = 0.6$ leads to $50 = 0.6(50 + x)$, so that $0.6x = 20$, or $x = \frac{100}{3} \approx 33.33$.

95. **(a)** Let y be the height of the tank; $1000 = x^2y$, so $y = 1000/x^2$. The surface area equals the area of the base plus 4 times the area of one side. Each side is a rectangle with dimensions $x \times y$, so $S = x^2 + 4xy = x^2 + 4000/x$.

(b) Solve $x^2 + 4000/x = 600$, or $x^3 - 600x + 4000 = 0$ (a graphical solution is easiest): Either $x = 20$, giving dimensions 20 ft by 20 ft by 2.5 ft or $x \approx 7.32$, giving approximate dimensions 7.32 by 7.32 by 18.66.

(c) $7.32 < x < 20$ (lower bound approximate), so y must be between 2.5 ft and about 18.66 ft.

Chapter 2 Project

Answers are based on the sample data shown in the table.

1.

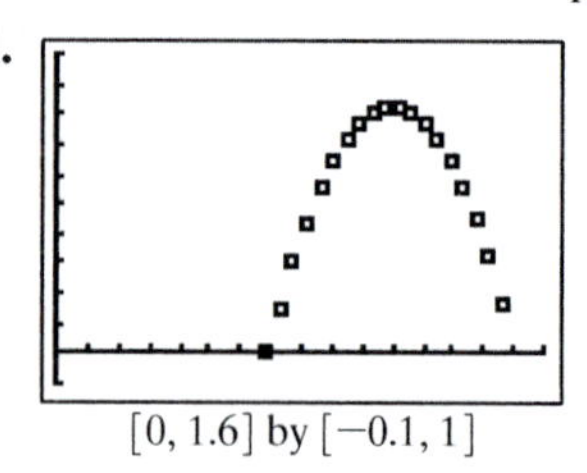

$[0, 1.6]$ by $[-0.1, 1]$

2. We estimate the vertex to lie halfway between the two data points with the greatest height, so that h is the average of 1.075 and 1.118, or about 1.097. We estimate k to be 0.830, which is slightly greater than the greatest height in the data, 0.828.
Noting that $y = 0$ when $x = 0.688$, we solve $0 = a(0.688 - 1.097)^2 + 0.830$ to find $a \approx -4.962$. So the estimated quadratic model is $y = -4.962(x - 1.097)^2 + 0.830$.

3. The sign of a affects the direction the parabola opens. The magnitude of a affects the vertical stretch of the graph. Changes to h cause horizontal shifts to the graph, while changes to k cause vertical shifts.

4. $y \approx -4.962x^2 - 10.887x - 5.141$

5. $y \approx -4.968x^2 + 10.913x - 5.160$

6. $y \approx -4.968x^2 + 10.913x - 5.160$

$$\approx -4.968\,(x^2 - 2.1967x + 1.0386)$$

$$= -4.968\left[x^2 - 2.1967x + \left(\frac{2.1967}{2}\right)^2 - \left(\frac{2.1967}{2}\right)^2 + 1.0386\right]$$

$$= -4.968\left[\left(x - \frac{2.1967}{2}\right)^2 - 0.1678\right]$$

$$\approx -4.968\,(x - 1.098)^2 + 0.833$$

Chapter 3
Exponential, Logistic, and Logarithmic Functions

■ Section 3.1 Exponential and Logistic Functions

Exploration 1

1. The point (0, 1) is common to all four graphs, and all four functions can be described as follows:

Domain: $(-\infty, \infty)$
Range: $(0, \infty)$
Continuous
Always increasing
Not symmetric
No local extrema
Bounded below by $y = 0$, which is also the only asymptote
$\lim_{x\to\infty} f(x) = \infty$, $\lim_{x\to-\infty} f(x) = 0$

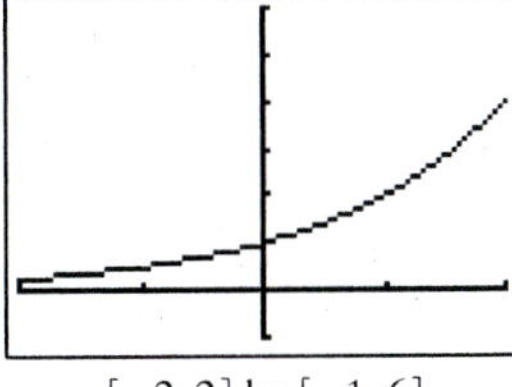

$y_1 = 2^x$

$[-2, 2]$ by $[-1, 6]$

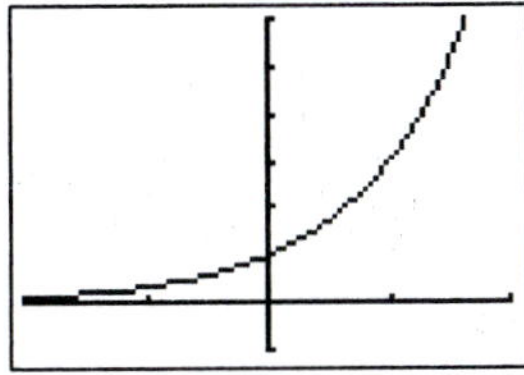

$y_2 = 3^x$

$[-2, 2]$ by $[-1, 6]$

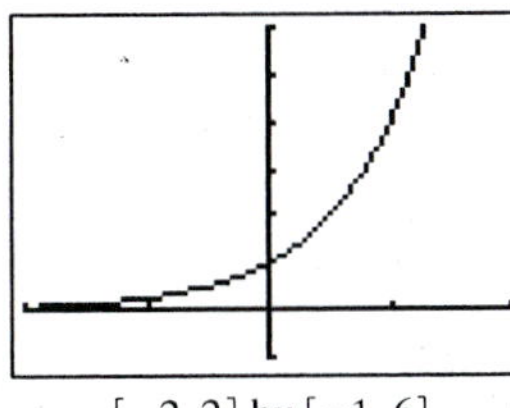

$y_3 = 4^x$

$[-2, 2]$ by $[-1, 6]$

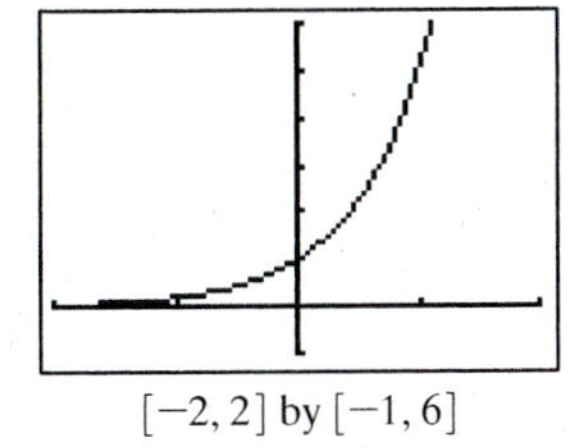

$y_4 = 5^x$

$[-2, 2]$ by $[-1, 6]$

2. The point (0, 1) is common to all four graphs, and all four functions can be described as follows:

Domain: $(-\infty, \infty)$
Range: $(0, \infty)$
Continuous
Always decreasing
Not symmetric
No local extrema
Bounded below by $y = 0$, which is also the only asymptote
$\lim_{x\to\infty} g(x) = 0$, $\lim_{x\to-\infty} g(x) = \infty$

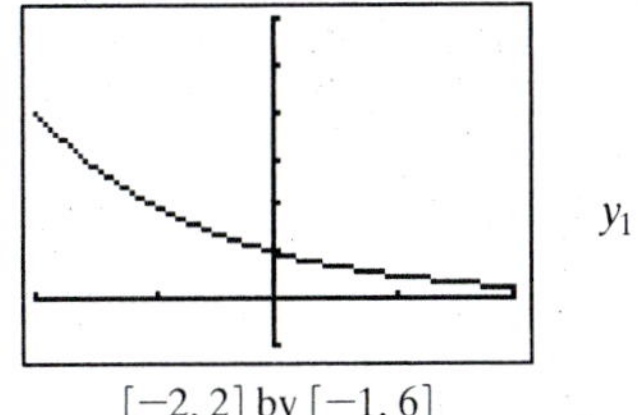

$y_1 = \left(\frac{1}{2}\right)^x$

$[-2, 2]$ by $[-1, 6]$

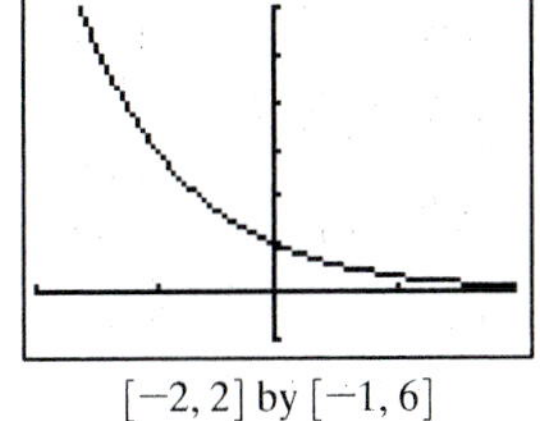

$y_2 = \left(\frac{1}{3}\right)^x$

$[-2, 2]$ by $[-1, 6]$

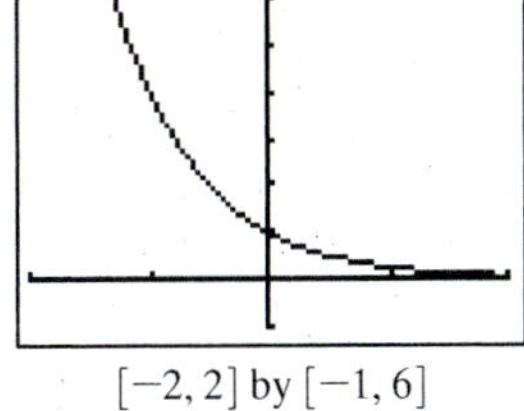

$y_3 = \left(\frac{1}{4}\right)^x$

$[-2, 2]$ by $[-1, 6]$

$y_4 = \left(\frac{1}{5}\right)^x$

$[-2, 2]$ by $[-1, 6]$

Exploration 2

1.

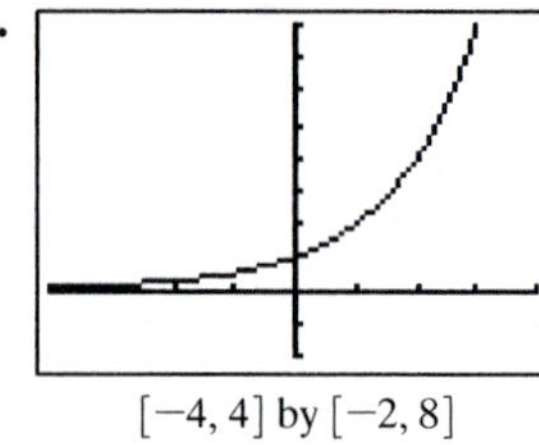

$[-4, 4]$ by $[-2, 8]$

$f(x) = 2^x$

2.

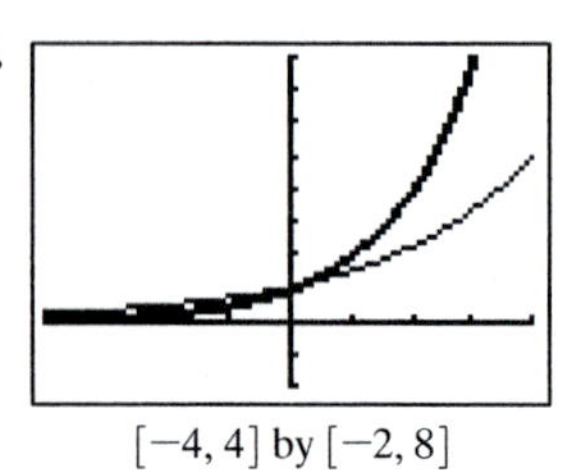

$[-4, 4]$ by $[-2, 8]$

$f(x) = 2^x$
$g(x) = e^{0.4x}$

$[-4, 4]$ by $[-2, 8]$

$f(x) = 2^x$
$g(x) = e^{0.5x}$

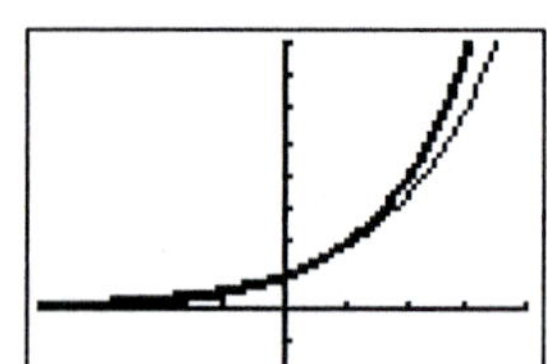

$[-4, 4]$ by $[-2, 8]$

$f(x) = 2^x$
$g(x) = e^{0.6x}$

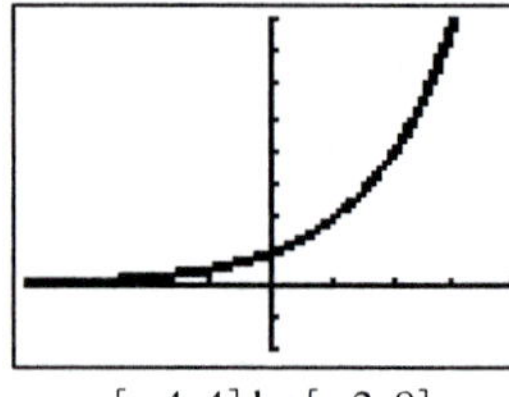

$[-4, 4]$ by $[-2, 8]$

$f(x) = 2^x$
$g(x) = e^{0.7x}$

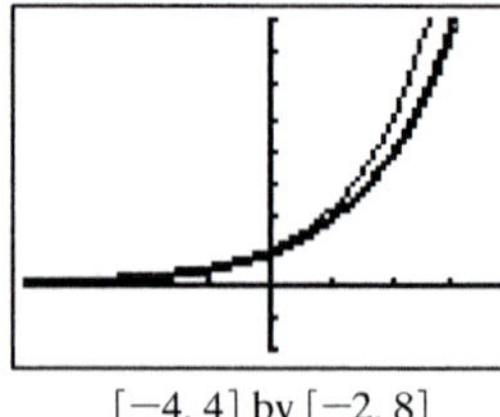

$[-4, 4]$ by $[-2, 8]$

$f(x) = 2^x$
$g(x) = e^{0.8x}$

$k = 0.7$ most closely matches the graph of $f(x)$.

3. $k \approx 0.693$

Quick Review 3.1

1. $\sqrt[3]{-216} = -6$ since $(-6)^3 = -216$
3. $27^{2/3} = (3^3)^{2/3} = 3^2 = 9$
5. $\dfrac{1}{2^{12}}$
7. $\dfrac{1}{a^6}$
9. -1.4 since $(-1.4)^5 = -5.37824$

Section 3.1 Exercises

1. Not an exponential function because the base is variable and the exponent is constant. It is a power function.
3. Exponential function, with an initial value of 1 and base of 5.
5. Not an exponential function because the base is variable.
7. $f(0) = 3 \cdot 5^0 = 3 \cdot 1 = 3$
9. $f\left(\dfrac{1}{3}\right) = -2 \cdot 3^{1/3} = -2\sqrt[3]{3}$
11. $f(x) = \dfrac{3}{2} \cdot \left(\dfrac{1}{2}\right)^x$
13. $f(x) = 3 \cdot (\sqrt{2})^x = 3 \cdot 2^{x/2}$
15. Translate $f(x) = 2^x$ by 3 units to the right. Alternatively, $g(x) = 2^{x-3} = 2^{-3} \cdot 2^x = \dfrac{1}{8} \cdot 2^x = \dfrac{1}{8} \cdot f(x)$, so it can be obtained from $f(x)$ using a vertical shrink by a factor of $\dfrac{1}{8}$.

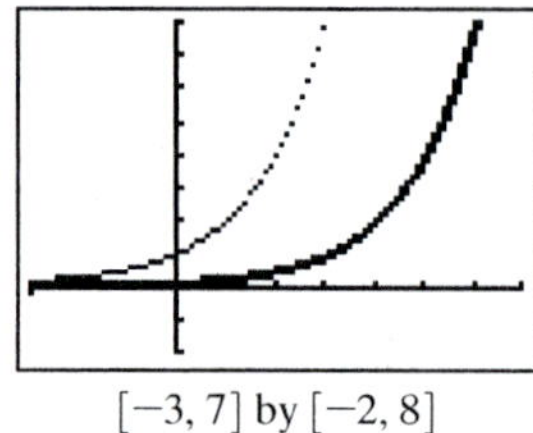

$[-3, 7]$ by $[-2, 8]$

17. Reflect $f(x) = 4^x$ over the y-axis.

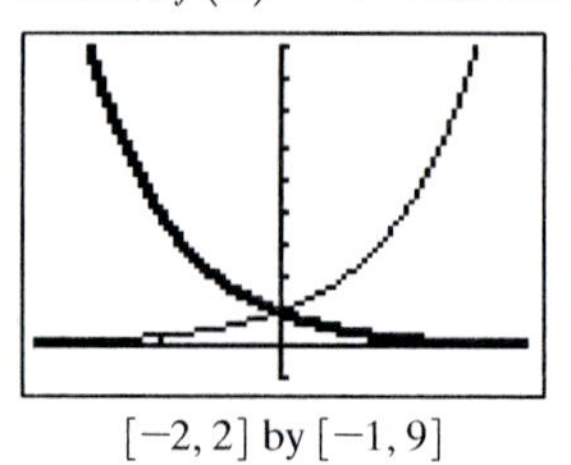

$[-2, 2]$ by $[-1, 9]$

19. Vertically stretch $f(x) = 0.5^x$ by a factor of 3 and then shift 4 units up.

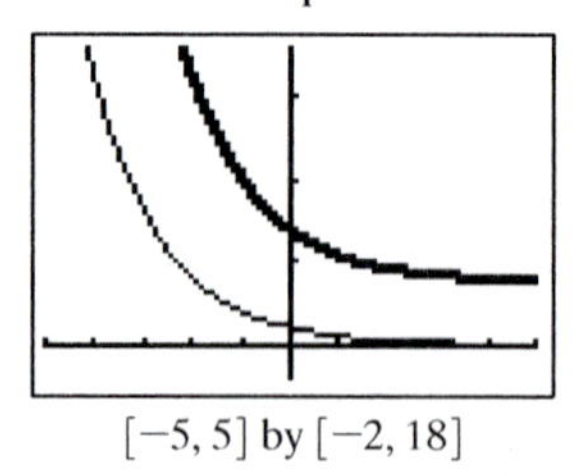

$[-5, 5]$ by $[-2, 18]$

21. Reflect $f(x) = e^x$ across the y-axis and horizontally shrink by a factor of 2.

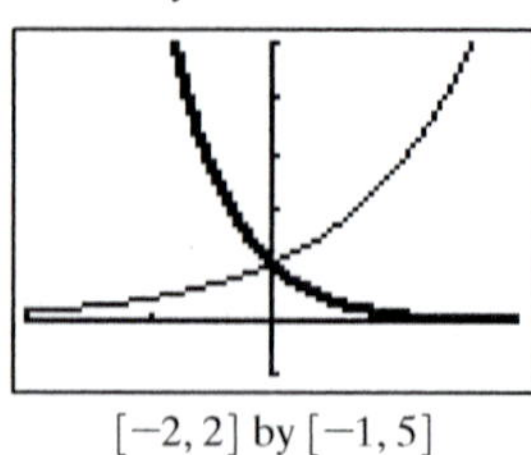

$[-2, 2]$ by $[-1, 5]$

23. Reflect $f(x) = e^x$ across the y-axis, horizontally shrink by a factor of 3, translate 1 unit to the right, and vertically stretch by a factor of 2.

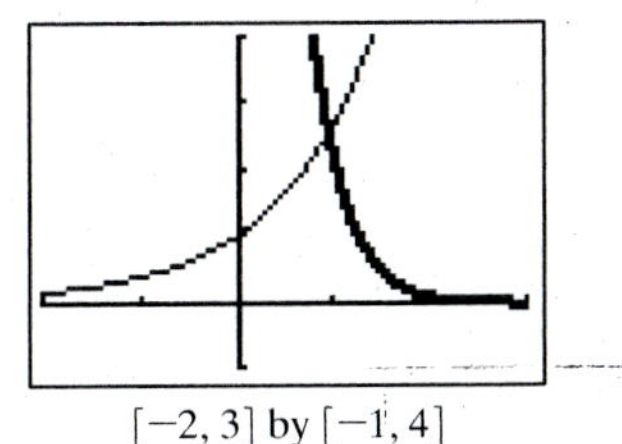

$[-2, 3]$ by $[-1, 4]$

25. Graph (a) is the only graph shaped and positioned like the graph of $y = b^x, b > 1$.

27. Graph (c) is the reflection of $y = 2^x$ across the x-axis.

29. Graph (b) is the graph of $y = 3^{-x}$ translated down 2 units.

31. Exponential decay; $\lim_{x\to\infty} f(x) = 0$; $\lim_{x\to-\infty} f(x) = \infty$

33. Exponential decay: $\lim_{x\to\infty} f(x) = 0$; $\lim_{x\to-\infty} f(x) = \infty$

35. $x < 0$

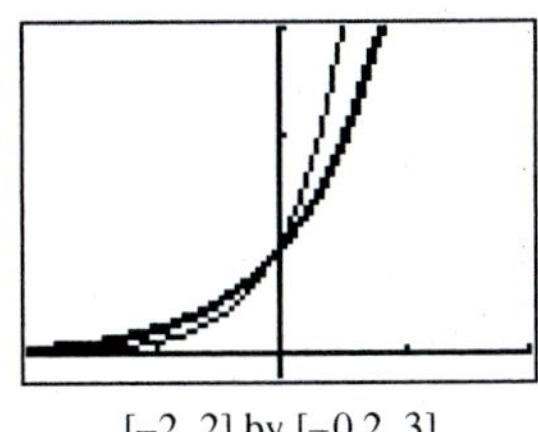

$[-2, 2]$ by $[-0.2, 3]$

37. $x < 0$

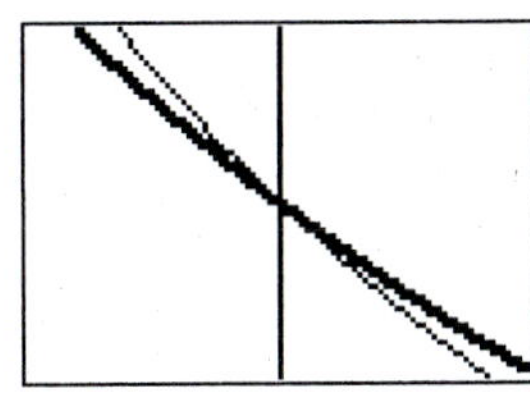

$[-0.25, 0.25]$ by $[0.75, 1.25]$

39. $y_1 = y_3$, since $3^{2x+4} = 3^{2(x+2)} = (3^2)^{x+2} = 9^{x+2}$.

41. y-intercept: $(0, 4)$. Horizontal asymptotes: $y = 0$, $y = 12$.

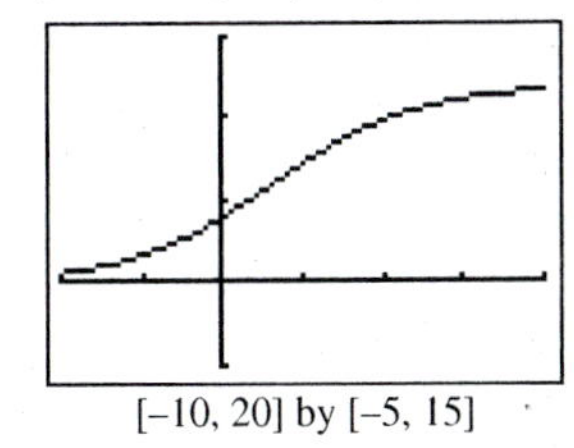

$[-10, 20]$ by $[-5, 15]$

43. y-intercept: $(0, 4)$. Horizontal asymptotes: $y = 0$, $y = 16$.

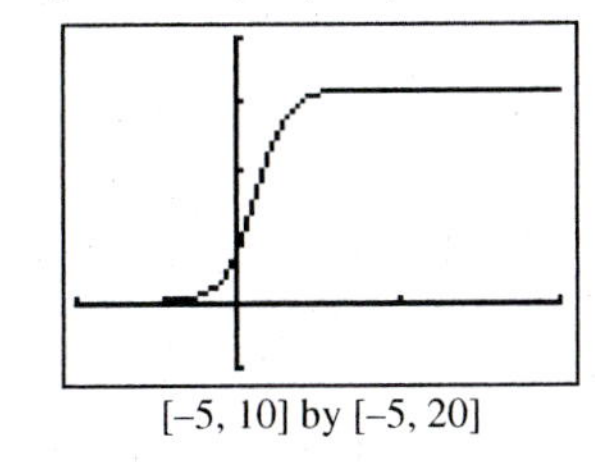

$[-5, 10]$ by $[-5, 20]$

45.

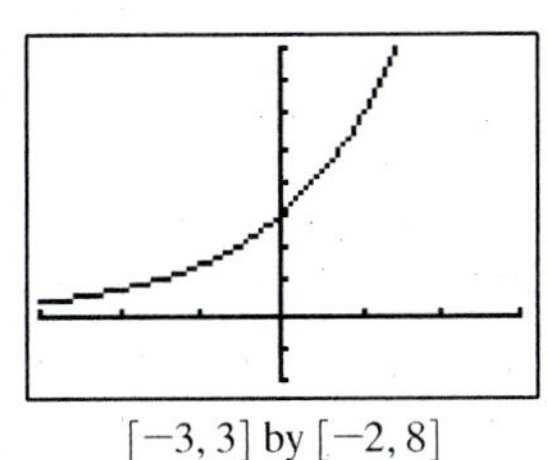

$[-3, 3]$ by $[-2, 8]$

Domain: $(-\infty, \infty)$
Range: $(0, \infty)$
Continuous
Always increasing
Not symmetric
Bounded below by $y = 0$, which is also the only asymptote
No local extrema
$\lim_{x\to\infty} f(x) = \infty$, $\lim_{x\to-\infty} f(x) = 0$

47.

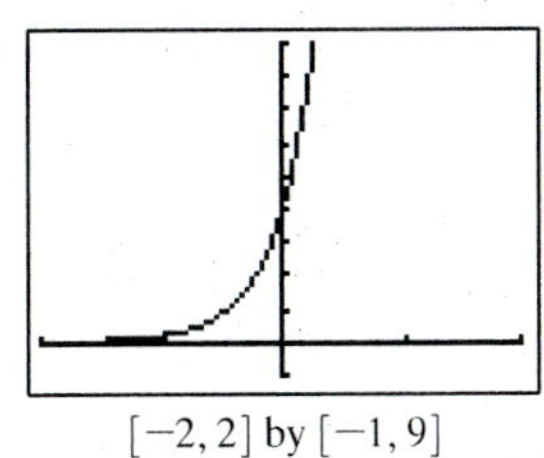

$[-2, 2]$ by $[-1, 9]$

Domain: $(-\infty, \infty)$
Range: $(0, \infty)$
Continuous
Always increasing
Not symmetric
Bounded below by $y = 0$, which is the only asymptote
No local extrema
$\lim_{x\to\infty} f(x) = \infty$, $\lim_{x\to-\infty} f(x) = 0$

49.

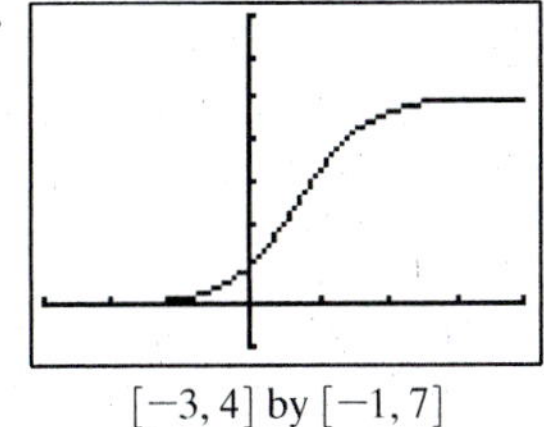

$[-3, 4]$ by $[-1, 7]$

Domain: $(-\infty, \infty)$
Range: $(0, 5)$
Continuous
Always increasing
Symmetric about $(0.69, 2.5)$
Bounded below by $y = 0$ and above by $y = 5$; both are asymptotes
No local extrema
$\lim_{x\to\infty} f(x) = 5$, $\lim_{x\to-\infty} f(x) = 0$

For #51, refer to Example 7 on page 285 in the text.

51. Let $P(t)$ be Austin's population t years after 1990. Then with exponential growth, $P(t) = P_0 b^t$ where $P_0 = 465{,}622$. From Table 3.7, $P(10) = 465{,}622\, b^{10} = 656{,}562$. So,

$$b = \sqrt[10]{\frac{656{,}562}{465{,}622}} \approx 1.0350.$$

Solving graphically, we find that the curve $y = 465{,}622(1.0350)^t$ intersects the line $y = 800{,}000$ at $t \approx 15.75$. Austin's population will pass 800,000 in 2006.

53. Using the results from Exercises 51 and 52, we represent Austin's population as $y = 465{,}622(1.0350)^t$ and Columbus's population as $y = 632{,}910(1.0118)^t$. Solving graphically, we find that the curves intersect at $t \approx 13.54$. The two populations will be equal, at 741,862, in 2003.

55. Solving graphically, we find that the curve $y = \dfrac{12.79}{(1 + 2.402e^{-0.0309x})}$ intersects the line $y = 10$ when $t \approx 69.67$. Ohio's population stood at 10 million in 1969.

57. **(a)** When $t = 0$, $B = 100$.

(b) When $t = 6$, $B \approx 6394$.

59. False. If $a > 0$ and $0 < b < 1$, or if $a < 0$ and $b > 1$, then $f(x) = a \cdot b^x$ is decreasing.

61. Only 8^x has the form $a \cdot b^x$ with a nonzero and b positive but not equal to 1. The answer is E.

63. The growth factor of $f(x) = a \cdot b^x$ is the base b. The answer is A.

65. **(a)**

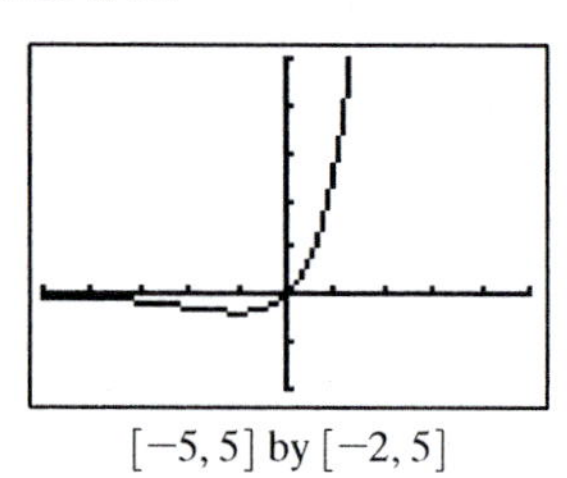

$[-5, 5]$ by $[-2, 5]$

Domain: $(-\infty, \infty)$

Range: $\left[-\dfrac{1}{e}, \infty\right)$

Intercept: $(0, 0)$

Decreasing on $(-\infty, -1]$: Increasing on $[-1, \infty)$

Bounded below by $y = -\dfrac{1}{e}$

Local minimum at $\left(-1, -\dfrac{1}{e}\right)$

Asymptote $y = 0$.

$\lim_{x\to\infty} f(x) = \infty$, $\lim_{x\to-\infty} f(x) = 0$

(b)

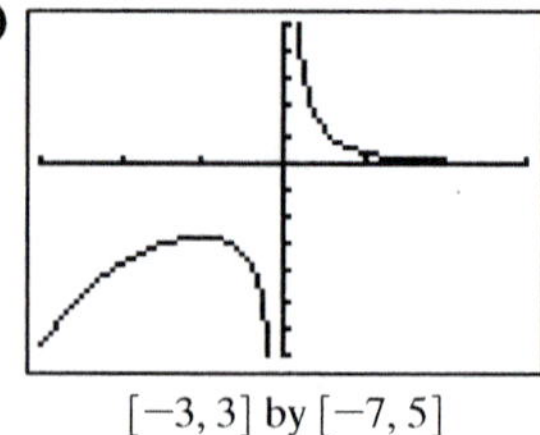

$[-3, 3]$ by $[-7, 5]$

Domain: $(-\infty, 0) \cup (0, \infty)$

Range: $(-\infty, -e] \cup (0, \infty)$

No intercepts

Increasing on $(-\infty, -1]$;

Decreasing on $[-1, 0) \cup (0, \infty)$

Not bounded

Local maxima at $(-1, -e)$

Asymptotes: $x = 0$, $y = 0$.

$\lim_{x\to\infty} g(x) = 0$, $\lim_{x\to-\infty} g(x) = -\infty$

67. **(a)** y_1—$f(x)$ decreases less rapidly as x decreases.

(b) y_3—as x increases, $g(x)$ decreases ever more rapidly.

69. $a \neq 0$, $c = 2$.

71. $a > 0$ and $b > 1$, or $a < 0$ and $0 < b < 1$.

73. Since $0 < b < 1$, $\lim_{x\to-\infty}(1 + a \cdot b^x) = \infty$ and $\lim_{x\to\infty}(1 + a \cdot b^x) = 1$. Thus, $\lim_{x\to-\infty}\dfrac{c}{1 + a \cdot b^x} = 0$ and $\lim_{x\to\infty}\dfrac{c}{1 + a \cdot b^x} = c$.

■ Section 3.2 Exponential and Logistic Modeling

Quick Review 3.2

1. 0.15

3. $(1.07)(23)$

5. $b^2 = \dfrac{160}{40} = 4$, so $b = \pm\sqrt{4} = \pm 2$.

7. $b = \sqrt[6]{\dfrac{838}{782}} \approx 1.01$

9. $b = \sqrt[4]{\dfrac{91}{672}} \approx 0.61$

Section 3.2 Exercises

For #1–19, use the model $P(t) = P_0(1 + r)^t$.

1. $r = 0.09$, so $P(t)$ is an exponential growth function of 9%.

3. $r = -0.032$, so $f(x)$ is an exponential decay function of 3.2%.

5. $r = 1$, so $g(t)$ is an exponential growth function of 100%.

7. $f(x) = 5 \cdot (1 + 0.17)^x = 5 \cdot 1.17^x$ (x = years)

9. $f(x) = 16 \cdot (1 - 0.5)^x = 16 \cdot 0.5^x$ (x = months)

11. $f(x) = 28{,}900 \cdot (1 - 0.026)^x = 28{,}900 \cdot 0.974^x$ (x = years)

13. $f(x) = 18 \cdot (1 + 0.052)^x = 18 \cdot 1.052^x$ (x = weeks)

15. $f(x) = 0.6 \cdot 2^{x/3}$ (x = days)

17. $f(x) = 592 \cdot 2^{-x/6}$ (x = years)

19. $f_0 = 2.3$, $\dfrac{2.875}{2.3} = 1.25 = r + 1$, so $f(x) = 2.3 \cdot 1.25^x$ (Growth Model)

For #21, use $f(x) = f_0 \cdot b^x$

21. $f_0 = 4$, so $f(x) = 4 \cdot b^x$. Since $f(5) = 4 \cdot b^5 = 8.05$, $b^5 = \dfrac{8.05}{4}$, $b = \sqrt[5]{\dfrac{8.05}{4}} \approx 1.15$. $f(x) \approx 4 \cdot 1.15^x$

For #23–27, use the model $f(x) = \dfrac{c}{1 + a \cdot b^x}$.

23. $c = 40$, $a = 3$, so $f(1) = \dfrac{40}{1 + 3b} = 20$, $20 + 60b = 40$, $60b = 20$, $b = \dfrac{1}{3}$, thus $f(x) = \dfrac{40}{1 + 3 \cdot \left(\dfrac{1}{3}\right)^x}$.

25. $c = 128, a = 7$, so $f(5) = \dfrac{128}{1 + 7b^5} = 32$,

$128 = 32 + 224b^5, 224b^5 = 96, b^5 = \dfrac{96}{224}$,

$b = \sqrt[5]{\dfrac{96}{224}} \approx 0.844$, thus $f(x) \approx \dfrac{128}{1 + 7 \cdot 0.844^x}$.

27. $c = 20, a = 3$, so $f(2) = \dfrac{20}{1 + 3b^2} = 10, 20 = 10 + 30b^2$,

$30b^2 = 10, b^2 = \dfrac{1}{3}, b = \sqrt{\dfrac{1}{3}} \approx 0.58$,

thus $f(x) = \dfrac{20}{1 + 3 \cdot 0.58^x}$.

29. $P(t) = 736{,}000(1.0149)^t$; $P(t) = 1{,}000{,}000$ when $t \approx 20.73$ years, or the year 2020.

31. The model is $P(t) = 6250(1.0275)^t$.

(a) In 1915: about $P(25) \approx 12{,}315$. In 1940: about $P(50) \approx 24{,}265$.

(b) $P(t) = 50{,}000$ when $t \approx 76.65$ years after 1890 — in 1966.

33. (a) $y = 6.6\left(\dfrac{1}{2}\right)^{t/14}$, where t is time in days.

(b) After 38.11 days.

35. One possible answer: Exponential and linear functions are similar in that they are always increasing or always decreasing. However, the two functions vary in how *quickly* they increase or decrease. While a linear function will increase or decrease at a steady rate over a given interval, the rate at which exponential functions increase or decrease over a given interval will vary.

37. One possible answer: From the graph we see that the doubling time for this model is 4 years. This is the time required to double from 50,000 to 100,000, from 100,000 to 200,000, or from any population size to twice that size. Regardless of the population size, it takes 4 years for it to double.

39. When $t = 1, B \approx 200$—the population doubles every hour.

For #41, use the formula $P(h) = 14.7 \cdot 0.5^{h/3.6}$, where h is miles above sea level.

41. $P(10) = 14.7 \cdot 0.5^{10/3.6} = 2.14$ lb/in^2

43. The exponential regression model is $P(t) = 1149.61904(1.012133)^t$, where $P(t)$ is measured in thousands of people and t is years since 1900. The predicted population for Los Angeles for 2003 is $P(103) \approx 3981$, or 3,981,000 people. This is an overestimate of 161,000 people, an error of $\dfrac{161{,}000}{3{,}820{,}000} \approx 0.04 = 4\%$.

The equation in #45 can be solved either algebraically or graphically; the latter approach is generally faster.

45. (a) $P(0) = 16$ students.

(b) $P(t) = 200$ when $t \approx 13.97$ — about 14 days.

(c) $P(t) = 300$ when $t \approx 16.90$ — about 17 days.

47. The logistic regression model is $P(x) = \dfrac{837.7707752}{1 + 9.668309563e^{-.015855579x}}$, where x is the number of years since 1900 and $P(x)$ is measured in millions of people. In the year 2010, $x = 110$, so the model predicts a population of

$P(110) = \dfrac{837.7707752}{1 + 9.668309563e^{(-.015855579)(110)}} = \dfrac{837.7707752}{1 + 9.668309563e^{(-1.74411369)}} = \dfrac{837.7707752}{2.690019034} \approx 311.4$
$\approx 311{,}400{,}000$ people.

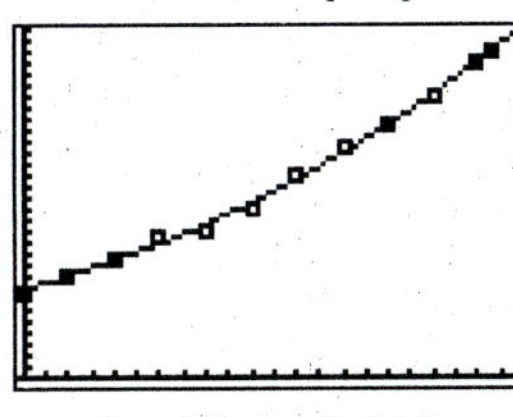

[−1, 109] by [0, 310]

49. $P(x) \approx \dfrac{19.875}{1 + 57.993e^{-0.035005x}}$ where x is the number of years after 1800 and P is measured in millions. Our model is the same as the model in Exercise 56 of Section 3.1.

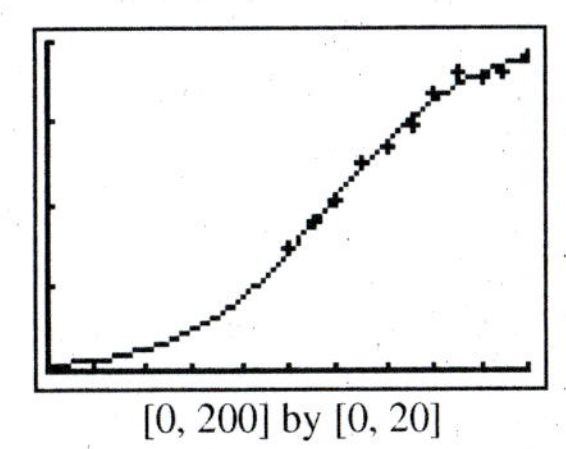

[0, 200] by [0, 20]

51. False. This is true for *logistic* growth, not for exponential growth.

53. The base is $1.049 = 1 + 0.049$, so the constant percentage growth rate is $0.049 = 4.9\%$. The answer is C.

55. The growth can be modeled as $P(t) = 1 \cdot 2^{t/4}$. Solve $P(t) = 1000$ to find $t \approx 39.86$. The answer is D.

57. (a) $P(x) \approx \dfrac{694.27}{1 + 7.90e^{-0.017x}}$, where x is the number of years since 1900 and P is measured in millions. $P(100) \approx 277.9$, or 277,900,000 people.

(b) The logistic model underestimates the 2000 population by about 3.5 million, an error of around 1.2%.

(c) The logistic model predicted a value closer to the actual value than the exponential model, perhaps indicating a better fit.

59. $\sinh(-x) = \dfrac{e^{-x} - e^{-(-x)}}{2} = \dfrac{e^{-x} - e^x}{2}$

$= -\left(\dfrac{e^x - e^{-x}}{2}\right) = -\sinh(x)$, so the function is odd.

61. (a) $\dfrac{\sinh(x)}{\cosh(x)} = \dfrac{\dfrac{e^x - e^{-x}}{2}}{\dfrac{e^x + e^{-x}}{2}}$

$= \dfrac{e^x - e^{-x}}{2} \cdot \dfrac{2}{e^x + e^{-x}} = \dfrac{e^x - e^{-x}}{e^x + e^{-x}} = \tanh(x)$.

(b) $\tanh(-x) = \dfrac{e^{-x} - e^{-(-x)}}{e^{-x} + e^{-(-x)}} = \dfrac{e^{-x} - e^x}{e^{-x} + e^x}$

$= -\dfrac{e^x - e^{-x}}{e^x + e^{-x}} = -\tanh(x)$, so the function is odd.

(c) $f(x) = 1 + \tanh(x) = 1 + \dfrac{e^x - e^{-x}}{e^x + e^{-x}}$

$= \dfrac{e^x + e^{-x} + e^x - e^{-x}}{e^x + e^{-x}} = \dfrac{2e^x}{e^x + e^{-x}}$

$= \dfrac{e^x}{e^x}\left(\dfrac{2}{1 + e^{-x}e^{-x}}\right) = \dfrac{2}{1 + e^{-2x}}$,

which is a logistic function of $c = 2, a = 1$, and $k = 2$.

Section 3.3 Logarithmic Functions and Their Graphs

Exploration 1

1.

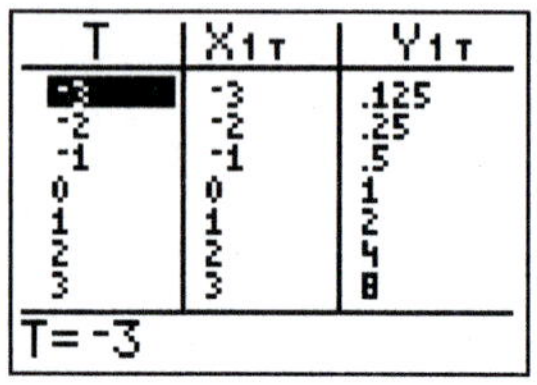

T	X1T	Y1T
-3	-3	.125
-2	-2	.25
-1	-1	.5
0	0	1
1	1	2
2	2	4
3	3	8

T=-3

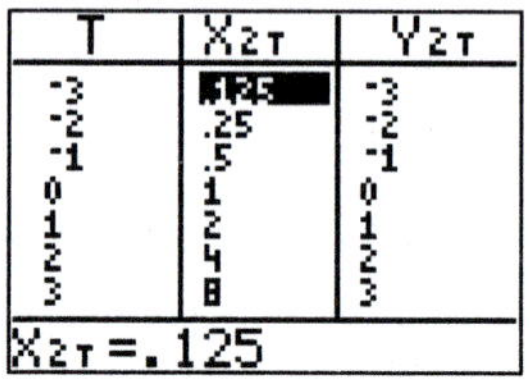

T	X2T	Y2T
-3	.125	-3
-2	.25	-2
-1	.5	-1
0	1	0
1	2	1
2	4	2
3	8	3

X2T=.125

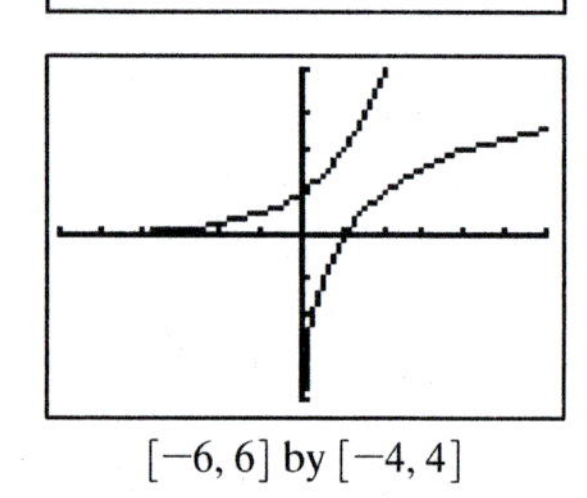

$[-6, 6]$ by $[-4, 4]$

2. Same graph as part 1.

Quick Review 3.3

1. $\dfrac{1}{25} = 0.04$

3. $\dfrac{1}{5} = 0.2$

5. $\dfrac{2^{33}}{2^{28}} = 2^5 = 32$

7. $5^{1/2}$

9. $\left(\dfrac{1}{e}\right)^{1/2} = e^{-1/2}$

Section 3.3 Exercises

1. $\log_4 4 = 1$ because $4^1 = 4$

3. $\log_2 32 = 5$ because $2^5 = 32$

5. $\log_5 \sqrt[3]{25} = \dfrac{2}{3}$ because $5^{2/3} = \sqrt[3]{25}$

7. $\log 10^3 = 3$

9. $\log 100{,}000 = \log 10^5 = 5$

11. $\log \sqrt[3]{10} = \log 10^{1/3} = \dfrac{1}{3}$

13. $\ln e^3 = 3$

15. $\ln \dfrac{1}{e} = \ln e^{-1} = -1$

17. $\ln \sqrt[4]{e} = \ln e^{1/4} = \dfrac{1}{4}$

19. 3, because $b^{\log_b 3} = 3$ for any $b > 0$.

21. $10^{\log(0.5)} = 10^{\log_{10}(0.5)} = 0.5$

23. $e^{\ln 6} = e^{\log_e 6} = 6$

25. $\log 9.43 \approx 0.9745 \approx 0.975$ and $10^{0.9745} \approx 9.43$

27. $\log(-14)$ is undefined because $-14 < 0$.

29. $\ln 4.05 \approx 1.399$ and $e^{1.399} \approx 4.05$

31. $\ln(-0.49)$ is undefined because $-0.49 < 0$.

33. $x = 10^2 = 100$

35. $x = 10^{-1} = \dfrac{1}{10} = 0.1$

37. $f(x)$ is undefined for $x > 1$. The answer is (d).

39. $f(x)$ is undefined for $x < 3$. The answer is (a).

41. Starting from $y = \ln x$: translate left 3 units.

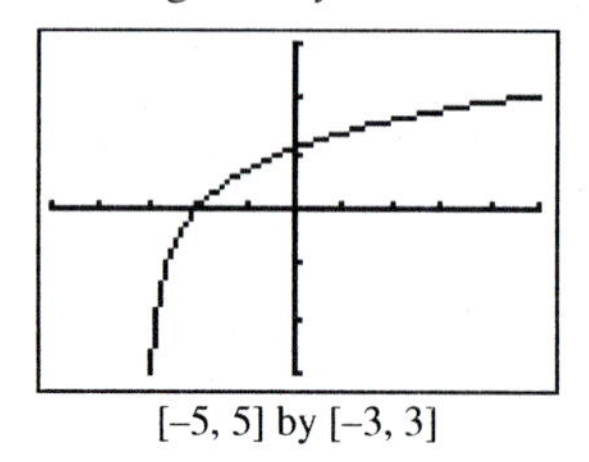

[–5, 5] by [–3, 3]

43. Starting from $y = \ln x$: reflect across the y-axis and translate up 3 units.

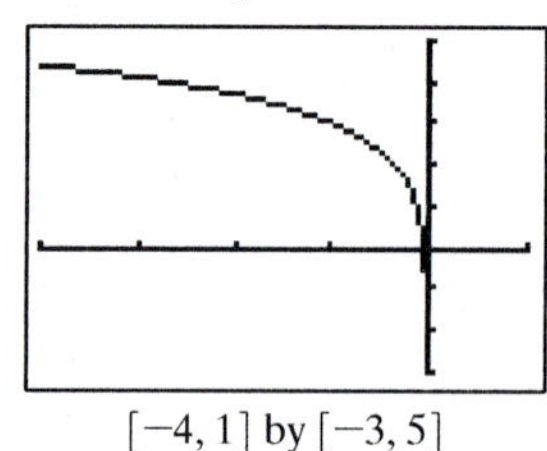

$[-4, 1]$ by $[-3, 5]$

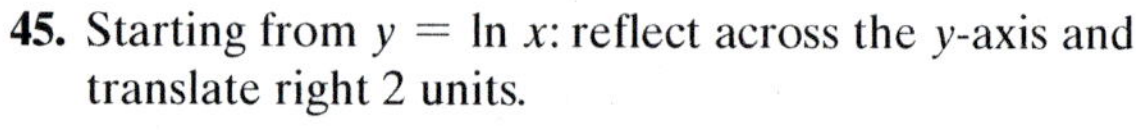

45. Starting from $y = \ln x$: reflect across the y-axis and translate right 2 units.

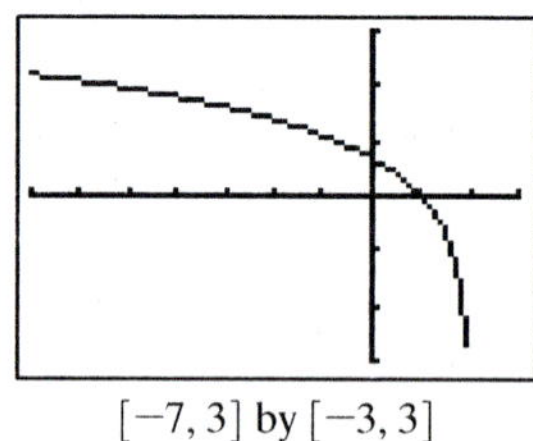

$[-7, 3]$ by $[-3, 3]$

47. Starting from $y = \log x$: translate down 1 unit.

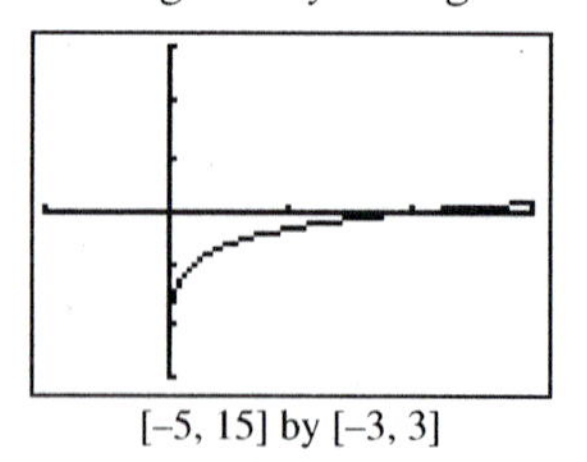

[–5, 15] by [–3, 3]

49. Starting from $y = \log x$: reflect across both axes and vertically stretch by 2.

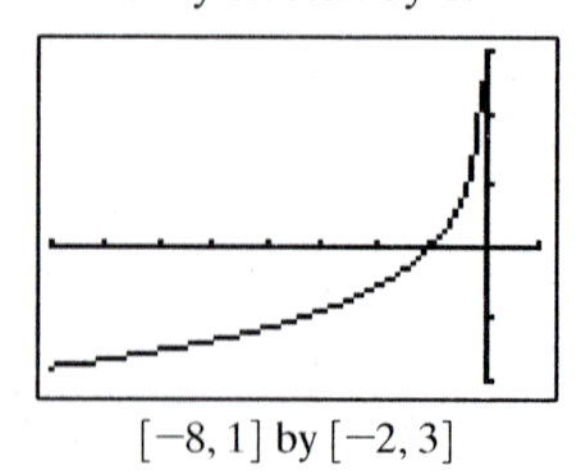

$[-8, 1]$ by $[-2, 3]$

51. Starting from $y = \log x$: reflect across the y-axis, translate right 3 units, vertically stretch by 2, translate down 1 unit.

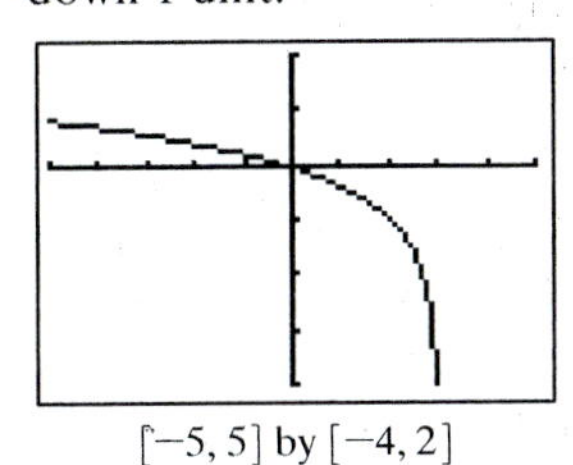

$[-5, 5]$ by $[-4, 2]$

53.

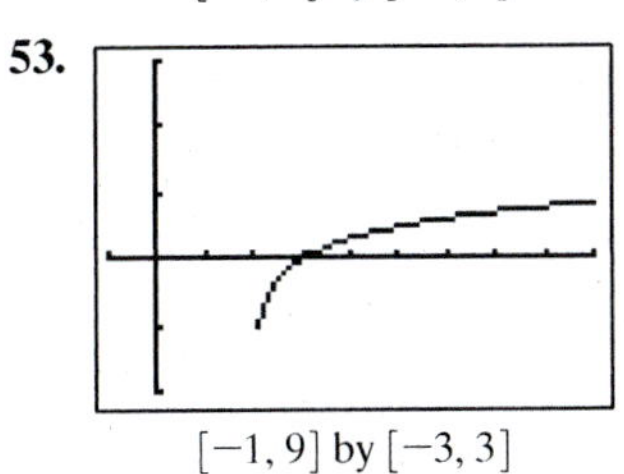

$[-1, 9]$ by $[-3, 3]$

Domain: $(2, \infty)$
Range: $(-\infty, \infty)$
Continuous
Always increasing
Not symmetric
Not bounded
No local extrema
Asymptote at $x = 2$
$\lim_{x\to\infty} f(x) = \infty$

55.

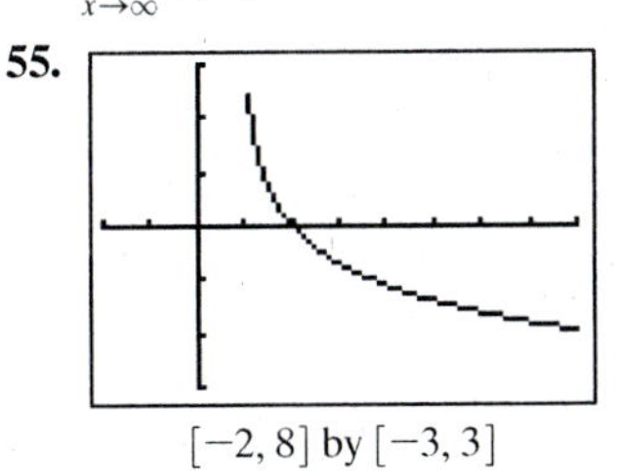

$[-2, 8]$ by $[-3, 3]$

Domain: $(1, \infty)$
Range: $(-\infty, \infty)$
Continuous
Always decreasing
Not symmetric
Not bounded
No local extrema
Asymptotes: $x = 1$
$\lim_{x\to\infty} f(x) = -\infty$

57.

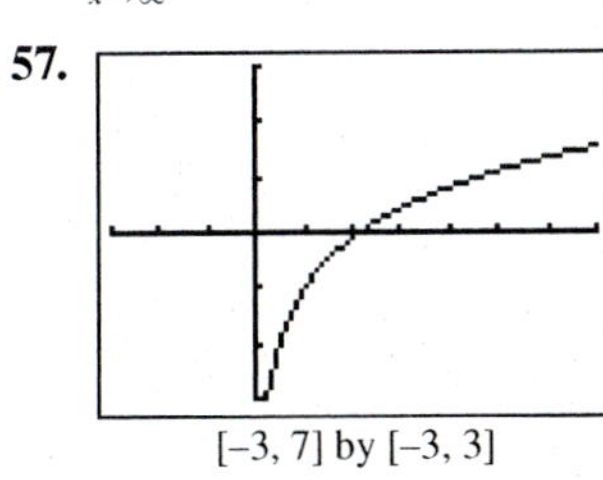

$[-3, 7]$ by $[-3, 3]$

Domain: $(0, \infty)$
Range: $(-\infty, \infty)$
Continuous
Increasing on its domain
No symmetry
Not bounded
No local extrema
Asymptote at $x = 0$
$\lim_{x\to\infty} f(x) = \infty$

59. (a) $\beta = 10 \log\left(\frac{10^{-11}}{10^{-12}}\right) = 10 \log 10 = 10(1) = 10$ dB

(b) $\beta = 10 \log\left(\frac{10^{-5}}{10^{-12}}\right) = 10 \log 10^{7} = 10(7) = 70$ dB

(c) $\beta = 10 \log\left(\frac{10^{3}}{10^{-12}}\right) = 10 \log 10^{15} = 10(15) = 150$ dB

61. The logarithmic regression model is $y = -246461780.3 + 32573678.51 \ln x$, where x is the year and y is the population. Graph the function and use TRACE to find that $x \approx 2023$ when $y \approx 150{,}000{,}000$. The population of San Antonio will reach 1,500,000 people in the year 2023.

L1	L2	L3
1970		------
1980	785940	
1990	935933	
2000	1.15E6	
------	------	

L2(1)=654153

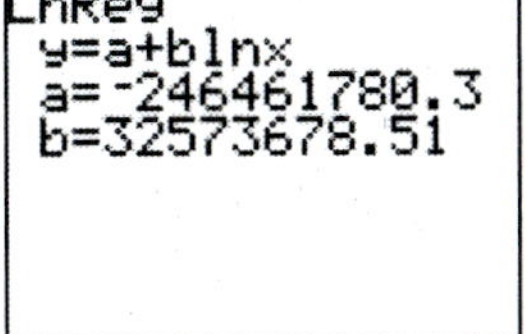

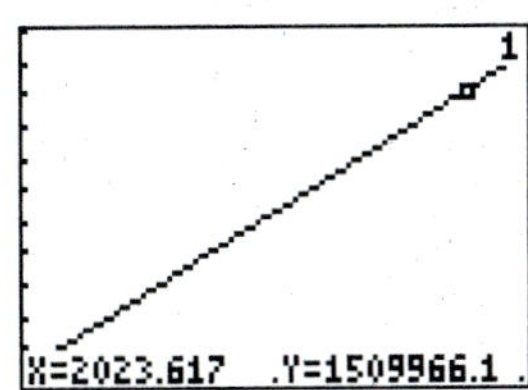

[1970, 2030] by [600,000, 1,700,000]

63. True, by the definition of a logarithmic function.

65. $\log 2 \approx 0.30103$. The answer is C.

67. The graph of $f(x) = \ln x$ lies entirely to the right of the origin. The answer is B.

69.

$f(x)$	3^x	$\log_3 x$
Domain	$(-\infty, \infty)$	$(0, \infty)$
Range	$(0, \infty)$	$(-\infty, \infty)$
Intercepts	$(0, 1)$	$(1, 0)$
Asymptotes	$y = 0$	$x = 0$

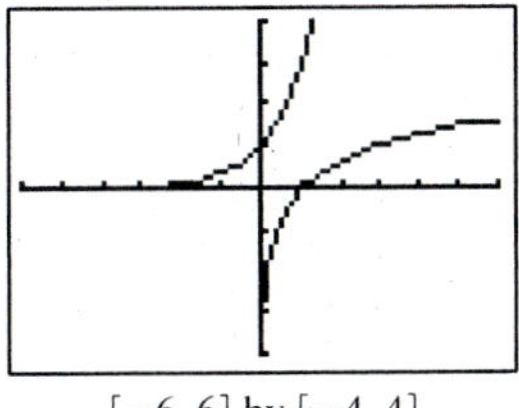

$[-6, 6]$ by $[-4, 4]$

71. $b = \sqrt[e]{e}$. The point that is common to both graphs is (e, e).

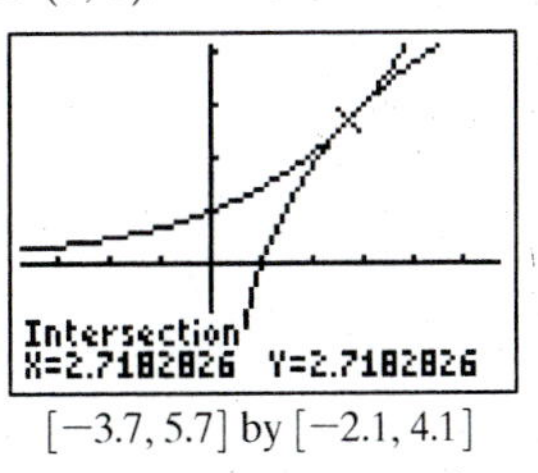

$[-3.7, 5.7]$ by $[-2.1, 4.1]$

73. Reflect across the x-axis.

Section 3.4 Properties of Logarithmic Functions

Exploration 1

1. $\log(2 \cdot 4) \approx 0.90309$,
$\log 2 + \log 4 \approx 0.30103 + 0.60206 \approx 0.90309$
2. $\log\left(\frac{8}{2}\right) \approx 0.60206$, $\log 8 - \log 2 \approx 0.90309 - 0.30103 \approx 0.60206$
3. $\log 2^3 \approx 0.90309$, $3 \log 2 \approx 3(0.30103) \approx 0.90309$
4. $\log 5 = \log\left(\frac{10}{2}\right) = \log 10 - \log 2 \approx 1 - 0.30103 = 0.69897$
5. $\log 16 = \log 2^4 = 4 \log 2 \approx 1.20412$
$\log 32 = \log 2^5 = 5 \log 2 \approx 1.50515$
$\log 64 = \log 2^6 = 6 \log 2 \approx 1.80618$
6. $\log 25 = \log 5^2 = 2 \log 5 = 2 \log\left(\frac{10}{2}\right) = 2(\log 10 - \log 2) \approx 1.39794$
$\log 40 = \log(4 \cdot 10) = \log 4 + \log 10 \approx 1.60206$
$\log 50 = \log\left(\frac{100}{2}\right) = \log 100 - \log 2 \approx 1.69897$
The list consists of 1, 2, 4, 5, 8, 16, 20, 25, 32, 40, 50, 64, and 80.

Exploration 2

1. False
2. False; $\log_3(7x) = \log_3 7 + \log_3 x$
3. True
4. True
5. False; $\log \frac{x}{4} = \log x - \log 4$
6. True
7. False; $\log_5 x^2 = \log_5 x + \log_5 x = 2 \log_5 x$
8. True

Quick Review 3.4

1. $\log 10^2 = 2$
3. $\ln e^{-2} = -2$
5. $\frac{x^5y^{-2}}{x^2y^{-4}} = x^{5-2}y^{-2-(-4)} = x^3y^2$
7. $(x^6y^{-2})^{1/2} = (x^6)^{1/2}(y^{-2})^{1/2} = \frac{|x|^3}{|y|}$
9. $\frac{(u^2v^{-4})^{1/2}}{(27u^6v^{-6})^{1/3}} = \frac{|u||v|^{-2}}{3u^2v^{-2}} = \frac{1}{3|u|}$

Section 3.4 Exercises

1. $\ln 8x = \ln 8 + \ln x = 3 \ln 2 + \ln x$
3. $\log \frac{3}{x} = \log 3 - \log x$
5. $\log_2 y^5 = 5 \log_2 y$
7. $\log x^3y^2 = \log x^3 + \log y^2 = 3 \log x + 2 \log y$
9. $\ln \frac{x^2}{y^3} = \ln x^2 - \ln y^3 = 2 \ln x - 3 \ln y$
11. $\log \sqrt[4]{\frac{x}{y}} = \frac{1}{4}(\log x - \log y) = \frac{1}{4}\log x - \frac{1}{4}\log y$
13. $\log x + \log y = \log xy$
15. $\ln y - \ln 3 = \ln(y/3)$
17. $\frac{1}{3}\log x = \log x^{1/3} = \log \sqrt[3]{x}$
19. $2 \ln x + 3 \ln y = \ln x^2 + \ln y^3 = \ln(x^2y^3)$
21. $4 \log(xy) - 3 \log(yz) = \log(x^4y^4) - \log(y^3z^3) = \log\left(\frac{x^4y^4}{y^3z^3}\right) = \log\left(\frac{x^4y}{z^3}\right)$

In #23–27, natural logarithms are shown, but common (base-10) logarithms would produce the same results.

23. $\frac{\ln 7}{\ln 2} \approx 2.8074$
25. $\frac{\ln 175}{\ln 8} \approx 2.4837$
27. $\frac{\ln 12}{\ln 0.5} = -\frac{\ln 12}{\ln 2} \approx -3.5850$
29. $\log_3 x = \frac{\ln x}{\ln 3}$
31. $\log_2(a + b) = \frac{\ln(a + b)}{\ln 2}$
33. $\log_2 x = \frac{\log x}{\log 2}$
35. $\log_{1/2}(x + y) = \frac{\log(x + y)}{\log(1/2)} = -\frac{\log(x + y)}{\log 2}$
37. Let $x = \log_b R$ and $y = \log_b S$.
Then $b^x = R$ and $b^y = S$, so that
$\frac{R}{S} = \frac{b^x}{b^y} = b^{x-y}$
$\log_b\left(\frac{R}{S}\right) = \log_b b^{x-y} = x - y = \log_b R - \log_b S$
39. Starting from $g(x) = \ln x$: vertically shrink by a factor $1/\ln 4 \approx 0.72$.

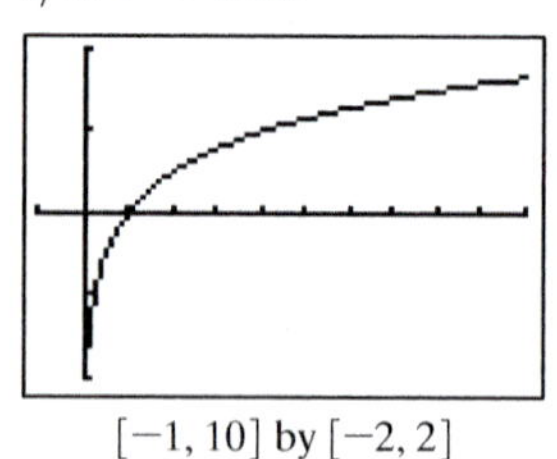

$[-1, 10]$ by $[-2, 2]$

41. Starting from $g(x) = \ln x$: reflect across the x-axis, then vertically shrink by a factor $1/\ln 3 \approx 0.91$.

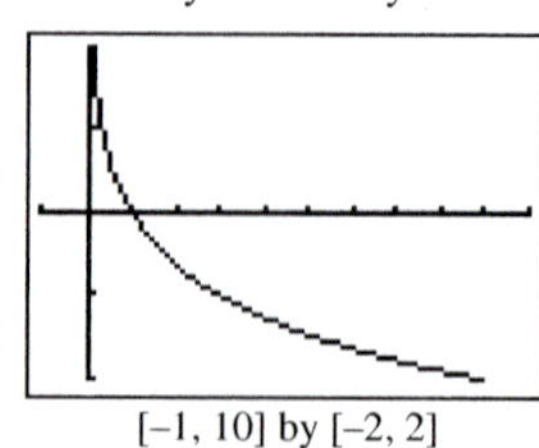

$[-1, 10]$ by $[-2, 2]$

43. (b): $[-5, 5]$ by $[-3, 3]$, with Xscl = 1 and Yscl = 1 (graph $y = \ln(2 - x)/\ln 4$).

45. (d): $[-2, 8]$ by $[-3, 3]$, with Xscl $= 1$ and Yscl $= 1$ (graph $y = \ln(x-2)/\ln 0.5$).

47.

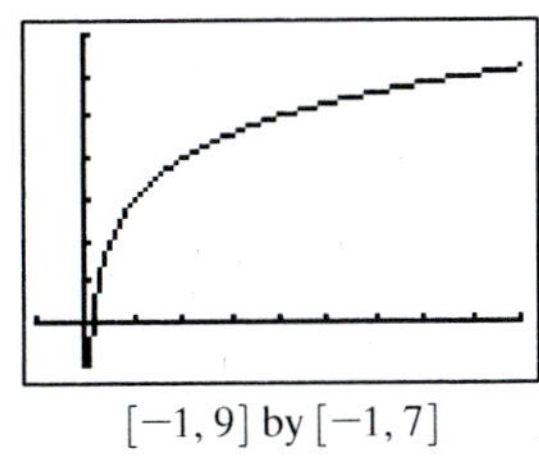

$[-1, 9]$ by $[-1, 7]$

Domain: $(0, \infty)$
Range: $(-\infty, \infty)$
Continuous
Always increasing
Asymptote: $x = 0$
$\lim_{x \to \infty} f(x) = \infty$

$f(x) = \log_2(8x) = \dfrac{\ln(8x)}{\ln(2)}$

49.

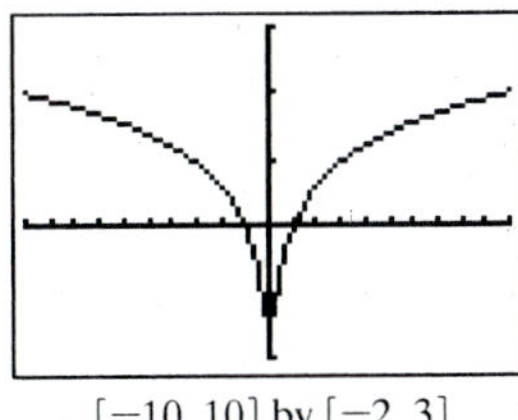

$[-10, 10]$ by $[-2, 3]$

Domain: $(-\infty, 0) \cup (0, \infty)$
Range: $(-\infty, \infty)$
Discontinuous at $x = 0$
Decreasing on interval $(-\infty, 0)$; increasing on interval $(0, \infty)$
Asymptote: $x = 0$
$\lim_{x \to \infty} f(x) = \infty$, $\lim_{x \to -\infty} f(x) = \infty$,

51. In each case, take the exponent of 10, add 12, and multiply the result by 10.

(a) 0
(b) 10
(c) 60
(d) 80
(e) 100
(f) 120 $(1 = 10^0)$

53. $\log \dfrac{I}{12} = -0.00235(40) = -0.094$, so
$I = 12 \cdot 10^{-0.094} \approx 9.6645$ lumens.

55. From the change-of-base formula, we know that
$f(x) = \log_3 x = \dfrac{\ln x}{\ln 3} = \dfrac{1}{\ln 3} \cdot \ln x \approx 0.9102 \ln x$.
$f(x)$ can be obtained from $g(x) = \ln x$ by vertically stretching by a factor of approximately 0.9102.

57. True. This is the product rule for logarithms.

59. $\log 12 = \log(3 \cdot 4) = \log 3 + \log 4$ by the product rule. The answer is B.

61. $\ln x^5 = 5 \ln x$ by the power rule. The answer is A.

63. (a) $f(x) = 2.75 \cdot x^{5.0}$

(b) $f(7.1) \approx 49{,}616$

(c)

$\ln(x)$	1.39	1.87	2.14	2.30
$\ln(y)$	7.94	10.37	11.71	12.53

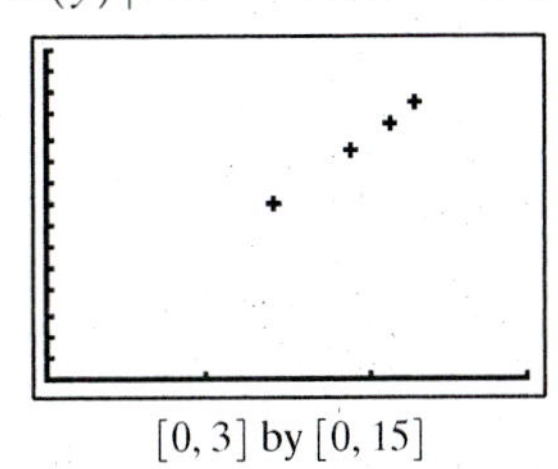

$[0, 3]$ by $[0, 15]$

(d) $\ln(y) = 5.00 \ln x + 1.01$

(e) $a \approx 5, b \approx 1$ so $f(x) = e^1 x^5 = ex^5 \approx 2.72x^5$. The two equations are the same.

65. (a)

$\log(w)$	−0.70	−0.52	0.30	0.70	1.48	1.70	1.85
$\log(r)$	2.62	2.48	2.31	2.08	1.93	1.85	1.86

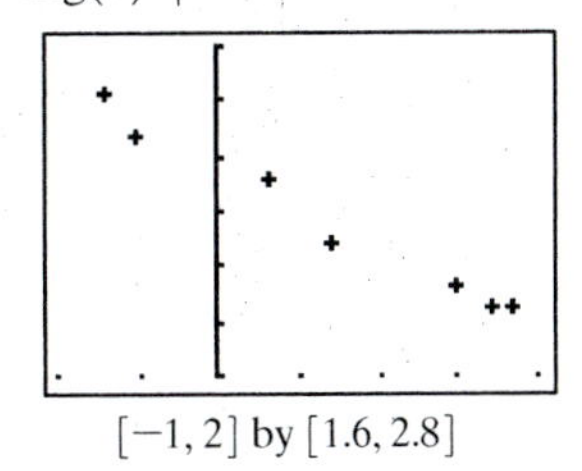

$[-1, 2]$ by $[1.6, 2.8]$

(b) $\log r = (-0.30) \log w + 2.36$

(c)

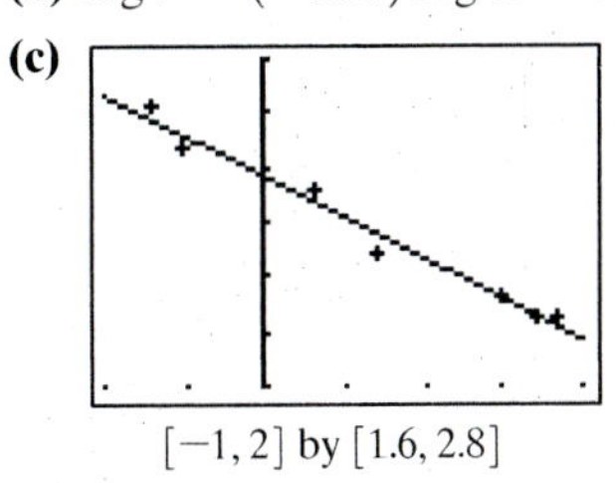

$[-1, 2]$ by $[1.6, 2.8]$

(d) $\log r = (-0.30) \log(450) + 2.36 \approx 1.58, r \approx 37.69$, very close

(e) One possible answer: Consider the power function $y = a \cdot x^b$ then:

$$\begin{aligned} \log y &= \log(a \cdot x^b) \\ &= \log a + \log x^b \\ &= \log a + b \log x \\ &= b(\log x) + \log a \end{aligned}$$

which is clearly a linear function of the form $f(t) = mt + c$ where $m = b, c = \log a, f(t) = \log y$ and $t = \log x$. As a result, there is a linear relationship between $\log y$ and $\log x$.

For #67, solve graphically.

67. $\approx 6.41 < x < 93.35$

69. (a)

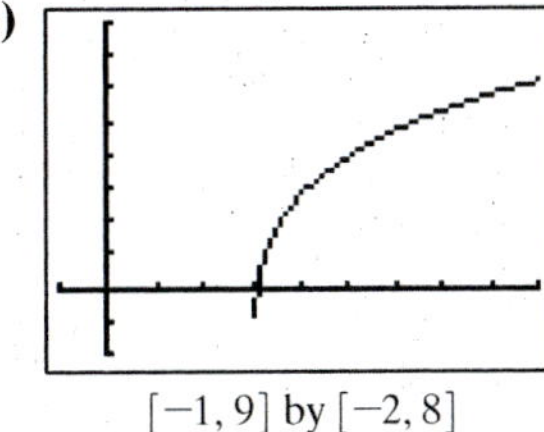

$[-1, 9]$ by $[-2, 8]$

Domain of f and g: $(3, \infty)$

(b)

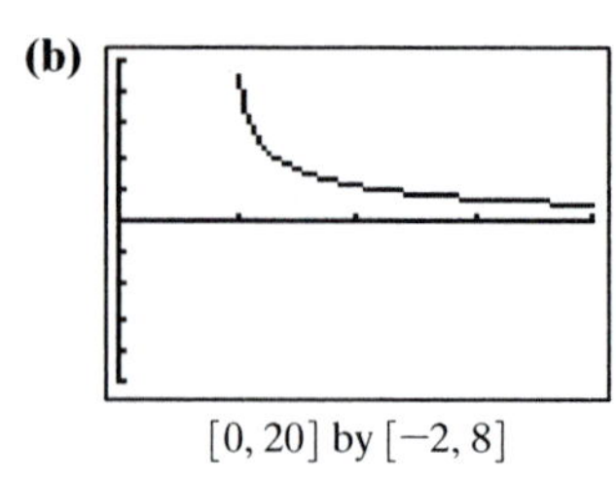

[0, 20] by [−2, 8]

Domain of f and g: $(5, \infty)$

(c)

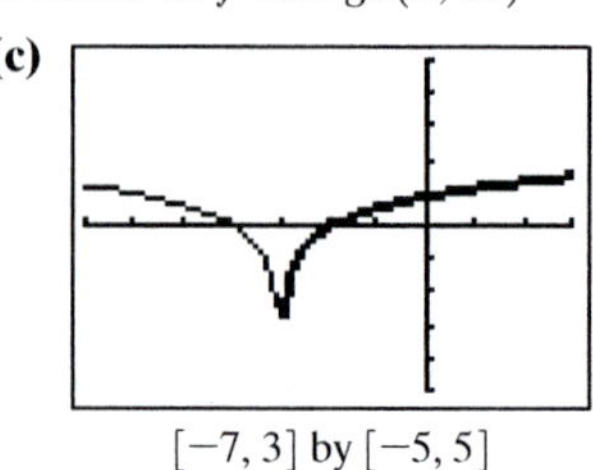

[−7, 3] by [−5, 5]

Domain of f: $(-\infty, -3) \cup (-3, \infty)$
Domain of g: $(-3, \infty)$
Answers will vary.

71. Let $y = \dfrac{\log x}{\ln x}$. By the change-of-base formula,

$$y = \frac{\log x}{\dfrac{\log x}{\log e}} = \log x \cdot \frac{\log e}{\log x} = \log e \approx 0.43$$

Thus, y is a constant function.

Section 3.5 Equation Solving and Modeling

Exploration 1

1. $\log(4 \cdot 10) \approx 1.60206$
$\log(4 \cdot 10^2) \approx 2.60206$
$\log(4 \cdot 10^3) \approx 3.60206$
$\log(4 \cdot 10^4) \approx 4.60206$
$\log(4 \cdot 10^5) \approx 5.60206$
$\log(4 \cdot 10^6) \approx 6.60206$
$\log(4 \cdot 10^7) \approx 7.60206$
$\log(4 \cdot 10^8) \approx 8.60206$
$\log(4 \cdot 10^9) \approx 9.60206$
$\log(4 \cdot 10^{10}) \approx 10.60206$

2. The integers increase by 1 for every increase in a power of 10.

3. The decimal parts are exactly equal.

4. $4 \cdot 10^{10}$ is nine orders of magnitude greater than $4 \cdot 10$.

Quick Review 3.5

In #1–3, graphical support (i.e., graphing both functions on a square window) is also useful.

1. $f(g(x)) = e^{2\ln(x^{1/2})} = e^{\ln x} = x$ and $g(f(x)) = \ln(e^{2x})^{1/2} = \ln(e^x) = x$.

3. $f(g(x)) = \dfrac{1}{3}\ln(e^{3x}) = \dfrac{1}{3}(3x) = x$ and $g(f(x)) = e^{3(1/3\ln x)} = e^{\ln x} = x$.

5. 7.783×10^8 km

7. 602,000,000,000,000,000,000,000

9. $(1.86 \times 10^5)(3.1 \times 10^7) = (1.86)(3.1) \times 10^{5+7} = 5.766 \times 10^{12}$

Section 3.5 Exercises

For #1–17, take a logarithm of both sides of the equation, when appropriate.

1. $36\left(\frac{1}{3}\right)^{x/5} = 4$
$\left(\frac{1}{3}\right)^{x/5} = \frac{1}{9}$
$\left(\frac{1}{3}\right)^{x/5} = \left(\frac{1}{3}\right)^2$
$\frac{x}{5} = 2$
$x = 10$

3. $2 \cdot 5^{x/4} = 250$
$5^{x/4} = 125$
$5^{x/4} = 5^3$
$\frac{x}{4} = 3$
$x = 12$

5. $10^{-x/3} = 10$, so $-x/3 = 1$, and therefore $x = -3$.

7. $x = 10^4 = 10{,}000$

9. $x - 5 = 4^{-1}$, so $x = 5 + 4^{-1} = 5.25$.

11. $x = \dfrac{\ln 4.1}{\ln 1.06} = \log_{1.06} 4.1 \approx 24.2151$

13. $e^{0.035x} = 4$, so $0.035x = \ln 4$, and therefore $x = \dfrac{1}{0.035}\ln 4 \approx 39.6084$.

15. $e^{-x} = \dfrac{3}{2}$, so $-x = \ln\dfrac{3}{2}$, and therefore $x = -\ln\dfrac{3}{2} \approx -0.4055$.

17. $\ln(x - 3) = \dfrac{1}{3}$, so $x - 3 = e^{1/3}$, and therefore $x = 3 + e^{1/3} \approx 4.3956$.

19. We must have $x(x + 1) > 0$, so $x < -1$ or $x > 0$. Domain: $(-\infty, -1) \cup (0, \infty)$; graph (e).

21. We must have $\dfrac{x}{x+1} > 0$, so $x < -1$ or $x > 0$. Domain: $(-\infty, -1) \cup (0, \infty)$; graph (d).

23. We must have $x > 0$. Domain: $(0, \infty)$; graph (a).

For #25–37, algebraic solutions are shown (and are generally the only way to get *exact* answers). In many cases solving graphically would be faster; graphical support is also useful.

25. Write both sides as powers of 10, leaving $10^{\log x^2} = 10^6$, or $x^2 = 1{,}000{,}000$. Then $x = 1000$ or $x = -1000$.

27. Write both sides as powers of 10, leaving $10^{\log x^4} = 10^2$, or $x^4 = 100$. Then $x^2 = 10$, and $x = \pm\sqrt{10}$.

29. Multiply both sides by $3 \cdot 2^x$, leaving $(2^x)^2 - 1 = 12 \cdot 2^x$, or $(2^x)^2 - 12 \cdot 2^x - 1 = 0$. This is quadratic in 2^x, leading to $2^x = \dfrac{12 \pm \sqrt{144 + 4}}{2} = 6 \pm \sqrt{37}$. Only $6 + \sqrt{37}$ is positive, so the only answer is $x = \dfrac{\ln(6 + \sqrt{37})}{\ln 2} = \log_2(6 + \sqrt{37}) \approx 3.5949$.

31. Multiply both sides by $2e^x$, leaving $(e^x)^2 + 1 = 8e^x$, or $(e^x)^2 - 8e^x + 1 = 0$. This is quadratic in e^x, leading to

$$e^x = \frac{8 \pm \sqrt{64 - 4}}{2} = 4 \pm \sqrt{15}. \text{ Then}$$

$$x = \ln(4 \pm \sqrt{15}) \approx \pm 2.0634.$$

33. $\frac{500}{200} = 1 + 25e^{0.3x}$, so $e^{0.3x} = \frac{3}{50} = 0.06$, and therefore

$$x = \frac{1}{0.3} \ln 0.06 \approx -9.3780.$$

35. Multiply by 2, then combine the logarithms to obtain $\ln \frac{x+3}{x^2} = 0$. Then $\frac{x+3}{x^2} = e^0 = 1$, so $x + 3 = x^2$.

The solutions to this quadratic equation are

$$x = \frac{1 \pm \sqrt{1 + 12}}{2} = \frac{1}{2} \pm \frac{1}{2}\sqrt{13} \approx 2.3028.$$

37. $\ln[(x - 3)(x + 4)] = 3 \ln 2$, so $(x - 3)(x + 4) = 8$, or $x^2 + x - 20 = 0$. This factors to $(x - 4)(x + 5) = 0$, so $x = 4$ (an actual solution) or $x = -5$ (extraneous, since $x - 3$ and $x + 4$ must be positive).

39. A \$100 bill has the value of 1000, or 10^3, dimes so they differ by an order of magnitude of 3.

41. $7 - 5.5 = 1.5$. They differ by an order of magnitude of 1.5.

43. Given

$$\beta_1 = 10 \log \frac{I_1}{I_0} = 95$$

$$\beta_2 = 10 \log \frac{I_2}{I_0} = 65,$$

we seek the logarithm of the ratio I_1/I_2.

$$10 \log \frac{I_1}{I_0} - 10 \log \frac{I_2}{I_0} = \beta_1 - \beta_2$$

$$10\left(\log \frac{I_1}{I_0} - \log \frac{I_2}{I_0}\right) = 95 - 65$$

$$10 \log \frac{I_1}{I_2} = 30$$

$$\log \frac{I_1}{I_2} = 3$$

The two intensities differ by 3 orders of magnitude.

45. Assuming that T and B are the same for the two quakes, we have $7.9 = \log a_1 - \log T + B$ and $6.6 = \log a_2 - \log T + B$, so $7.9 - 6.6 = 1.3 = \log(a_1/a_2)$. Then $a_1/a_2 = 10^{1.3}$, so $a_1 \approx 19.95a_2$—the Mexico City amplitude was about 20 times greater.

47. (a) Carbonated water: $-\log [H^+] = 3.9$

$$\log [H^+] = -3.9$$

$$[H^+] = 10^{-3.9} \approx 1.26 \times 10^{-4}$$

Household ammonia: $-\log [H^+] = 11.9$

$$\log [H^+] = -11.9$$

$$[H^+] = 10^{-11.9} \approx 1.26 \times 10^{-12}$$

(b) $\frac{[H^+] \text{ of carbonated water}}{[H^+] \text{ of household ammonia}} = \frac{10^{-3.9}}{10^{-11.9}} = 10^8$

(c) They differ by an order of magnitude of 8.

The equation in #49 can be solved either algebraically or graphically; the latter approach is generally faster.

49. Substituting known information into $T(t) = T_m + (T_0 - T_m)e^{-kt}$ leaves $T(t) = 22 + 70e^{-kt}$. Using $T(12) = 50 = 22 + 70e^{-12k}$, we have $e^{-12k} = \frac{2}{5}$, so

$k = -\frac{1}{12} \ln \frac{2}{5} \approx 0.0764$. Solving $T(t) = 30$ yields $t \approx 28.41$ minutes.

51. (a)

[0, 40] by [0, 80]

(b)

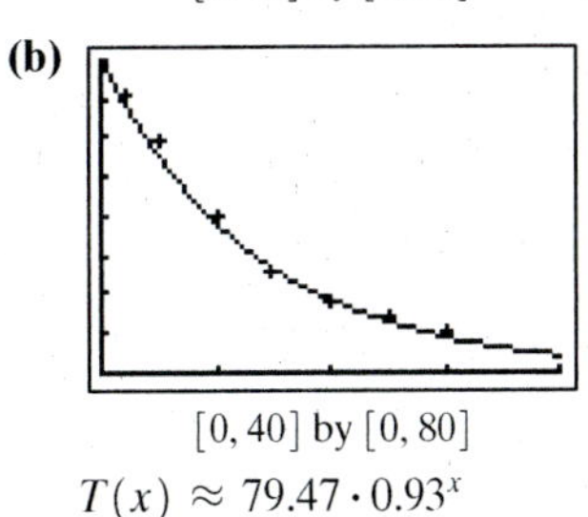

[0, 40] by [0, 80]

$$T(x) \approx 79.47 \cdot 0.93^x$$

(c) $T(0) + 10 = 79.47 + 10 \approx 89.47$°C

53. (a)

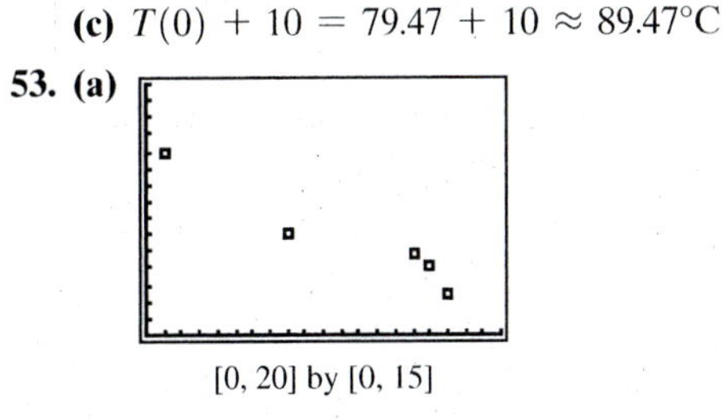

[0, 20] by [0, 15]

(b) The scatter plot is better because it accurately represents the times between the measurements. The equal spacing on the bar graph suggests that the measurements were taken at equally spaced intervals, which distorts our perceptions of how the consumption has changed over time.

55. Logarithmic seems best — the scatterplot of (x, y) looks most logarithmic. (The data can be modeled by $y = 3 + 2 \ln x$.)

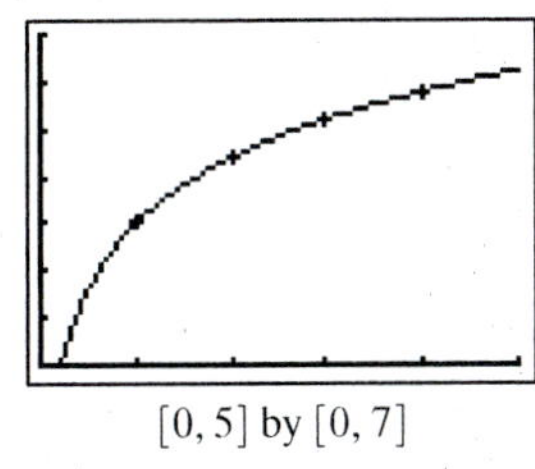

[0, 5] by [0, 7]

57. Exponential — the scatterplot of (x, y) is *exactly* exponential. (The data can be modeled by $y = \frac{3}{2} \cdot 2^x$.)

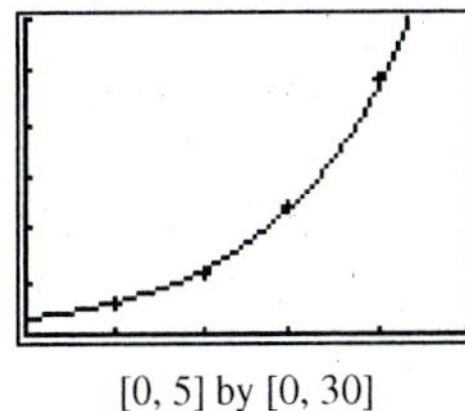

[0, 5] by [0, 30]

59. False. The order of magnitude of a positive number is its *common* logarithm.

61. $2^{3x-1} = 32$
$2^{3x-1} = 2^5$
$3x - 1 = 5$
$x = 2$
The answer is B.

63. Given

$$R_1 = \log \frac{a_1}{T} + B = 8.1$$

$$R_2 = \log \frac{a_2}{T} + B = 6.1,$$

we seek the ratio of amplitudes (severities) a_1/a_2.

$$\left(\log \frac{a_1}{T} + B\right) - \left(\log \frac{a_2}{T} + B\right) = R_1 - R_2$$

$$\log \frac{a_1}{T} - \log \frac{a_2}{T} = 8.1 - 6.1$$

$$\log \frac{a_1}{a_2} = 2$$

$$\frac{a_1}{a_2} = 10^2 = 100$$

The answer is E.

65. A logistic regression $\left(f(x) = \dfrac{4443}{1 + 169.96e^{-0.0354x}}\right)$ most closely matches the data, and would provide a natural "cap" to the population growth at approx. 4.4 million people. (Note: x = number of years since 1900.)

67. (a)

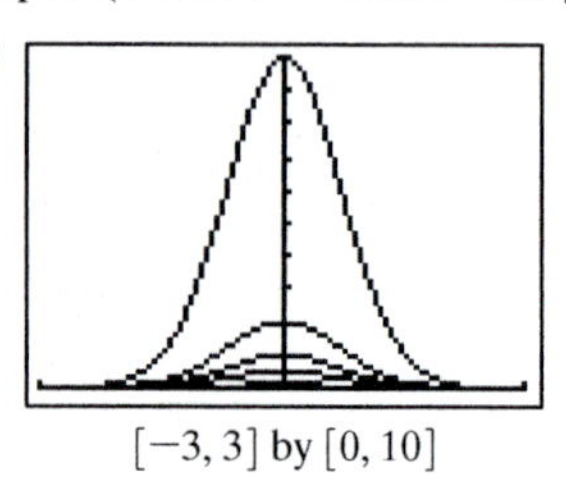

$[-3, 3]$ by $[0, 10]$

As k increases, the bell curve stretches vertically. Its height increases and the slope of the curve seems to steepen.

(b)

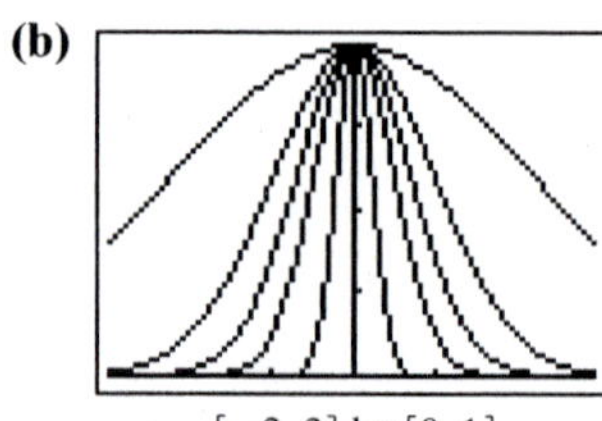

$[-3, 3]$ by $[0, 1]$

As c increases, the bell curve compresses horizontally. Its slope seems to steepen, increasing more rapidly to $(0, 1)$ and decreasing more rapidly from $(0, 1)$.

69. (a) r cannot be negative since it is a distance.

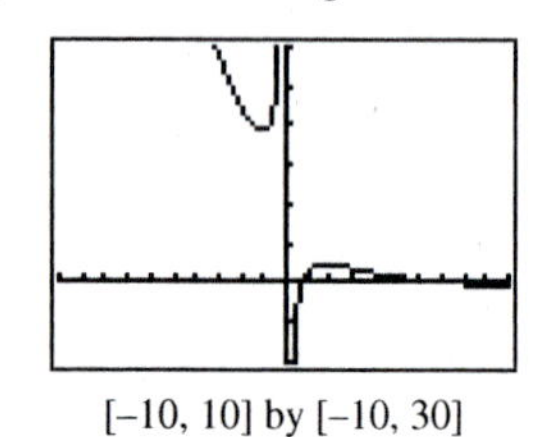

$[-10, 10]$ by $[-10, 30]$

(b) $[0, 10]$ by $[-5, 3]$ is a good choice. The maximum energy, approximately 2.3807, occurs when $r \approx 1.729$.

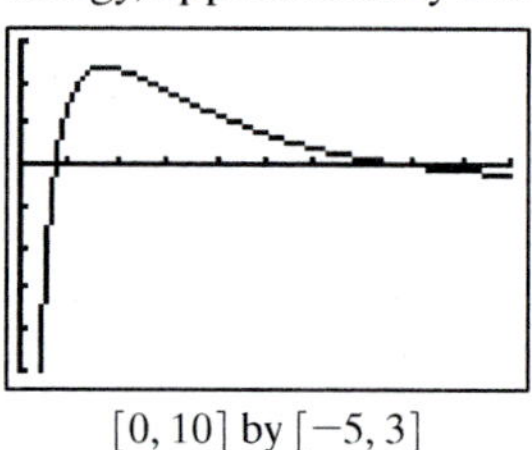

$[0, 10]$ by $[-5, 3]$

71. One possible answer: We "map" our data so that all points (x, y) are plotted as $(\ln x, y)$. If these "new" points are linear—and thus can be represented by some standard linear regression $y = ax + b$—we make the same substitution $(x \to \ln x)$ and find $y = a \ln x + b$, a logarithmic regression.

The equations and inequalities in #73–75 must be solved graphically—they cannot be solved algebraically. For #77, algebraic solution is possible, although a graphical approach may be easier.

73. $x \approx 1.3066$

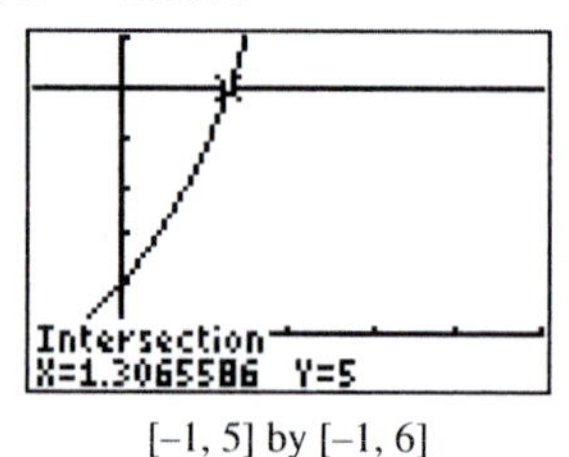

$[-1, 5]$ by $[-1, 6]$

75. $0 < x < 1.7115$ (approx.)

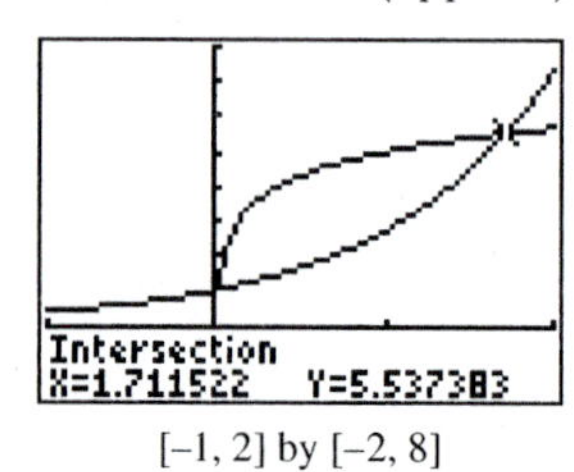

$[-1, 2]$ by $[-2, 8]$

77. $\log x - 2 \log 3 > 0$, so $\log(x/9) > 0$. Then $\frac{x}{9} > 10^0 = 1$, so $x > 9$.

■ Section 3.6 Mathematics of Finance

Exploration 1

1.

k	A
10	1104.6
20	1104.9
30	1105
40	1105
50	1105.1
60	1105.1
70	1105.1
80	1105.1
90	1105.1
100	1105.1

A approaches a limit of about 1105.1.

2. $y = 1000e^{0.1} \approx 1105.171$ is an upper bound (and asymptote) for $A(x)$. $A(x)$ approaches, but never equals, this bound.

Quick Review 3.6

1. $200 \cdot 0.035 = 7$

3. $\frac{1}{4} \cdot 7.25\% = 1.8125\%$

5. $\frac{78}{120} = 0.65 = 65\%$

7. $0.32x = 48$ gives $x = 150$

9. $300\,(1 + 0.05) = 315$ dollars

Section 3.6 Exercises

1. $A = 1500(1 + 0.07)^6 \approx \$2251.10$

3. $A = 12{,}000(1 + 0.075)^7 \approx \$19{,}908.59$

5. $A = 1500\left(1 + \frac{0.07}{4}\right)^{20} \approx \$2122.17$

7. $A = 40{,}500\left(1 + \frac{0.038}{12}\right)^{240} \approx \$86{,}496.26$

9. $A = 1250e^{(0.054)(6)} \approx \$1728.31$

11. $A = 21{,}000e^{(0.037)(10)} \approx \$30{,}402.43$

13. $FV = 500 \cdot \dfrac{\left(1 + \frac{0.07}{4}\right)^{24} - 1}{\frac{0.07}{4}} \approx \$14{,}755.51$

15. $FV = 450 \cdot \dfrac{\left(1 + \frac{0.0525}{12}\right)^{120} - 1}{\frac{0.0525}{12}} \approx \$70{,}819.63$

17. $PV = 815.37 \cdot \dfrac{1 - \left(1 + \frac{0.047}{12}\right)^{-60}}{\frac{0.047}{12}} \approx \$43{,}523.31$

19. $R = \dfrac{PV \cdot i}{1 - (1 + i)^{-n}} = \dfrac{(18{,}000)\left(\frac{0.054}{12}\right)}{1 - \left(1 + \frac{0.054}{12}\right)^{-72}} \approx \$293.24$

In #21–23, the time must be rounded up to the end of the next compounding period.

21. Solve $2300\left(1 + \frac{0.09}{4}\right)^{4t} = 4150: (1.0225)^{4t} = \frac{83}{46}$, so

$t = \frac{1}{4}\frac{\ln(83/46)}{\ln 1.0225} \approx 6.63$ years — round to 6 years 9 months (the next full compounding period).

23. Solve $15{,}000\left(1 + \frac{0.08}{12}\right)^{12t} = 45{,}000: (1.0067)^{12t} = 3$, so

$t = \frac{1}{12}\frac{\ln 3}{\ln 1.0067} \approx 13.71$ years — round to 13 years 9 months (the next full compounding period). Note: A graphical solution provides $t \approx 13.78$ years—round to 13 years 10 months.

25. Solve $22{,}000\left(1 + \frac{r}{365}\right)^{(365)(5)} = 36{,}500$:

$1 + \frac{r}{365} = \left(\frac{73}{44}\right)^{1/1825}$, so $r \approx 10.13\%$

27. Solve $14.6\,(1 + r)^6 = 22$: $1 + r = \left(\frac{110}{73}\right)^{1/6}$, so

$r \approx 7.07\%$.

In #29, the time must be rounded up to the end of the next compounding period.

29. Solve $\left(1 + \frac{0.0575}{4}\right)^{4t} = 2$: $t = \frac{1}{4}\frac{\ln 2}{\ln 1.014375} \approx 12.14$ — round to 12 years 3 months.

For #31–33, use the formula $S = Pe^{rt}$.

31. Time to double: solve $2 = e^{0.09t}$, leading to

$t = \frac{1}{0.09}\ln 2 \approx 7.7016$ years. After 15 years:

$S = 12{,}500e^{(0.09)(15)} \approx \$48{,}217.82$

33. APR: solve $2 = e^{4r}$, leading to $r = \frac{1}{4}\ln 2 \approx 17.33\%$.

After 15 years: $S = 9500e^{(0.1733)(15)} \approx \$127{,}816.26$ (using the "exact" value of r).

In #35–39, the time must be rounded up to the end of the next compounding period (except in the case of continuous compounding).

35. Solve $\left(1 + \frac{0.04}{4}\right)^{4t} = 2$: $t = \frac{1}{4}\frac{\ln 2}{\ln 1.01} \approx 17.42$ — round to 17 years 6 months.

37. Solve $1.07^t = 2$: $t = \frac{\ln 2}{\ln 1.07} \approx 10.24$ — round to 11 years.

39. Solve $\left(1 + \frac{0.07}{12}\right)^{12t} = 2$: $t = \frac{1}{12}\frac{\ln 2}{\ln(1 + 0.07/12)}$ ≈ 9.93 — round to 10 years.

For #41–43, observe that the initial balance has no effect on the APY.

41. $\text{APY} = \left(1 + \frac{0.06}{4}\right)^4 - 1 \approx 6.14\%$

43. $\text{APY} = e^{0.063} - 1 \approx 6.50\%$

45. The APYs are $\left(1 + \frac{0.05}{12}\right)^{12} - 1 \approx 5.1162\%$ and $\left(1 + \frac{0.051}{4}\right)^4 - 1 \approx 5.1984\%$. So, the better investment is 5.1% compounded quarterly.

For #47–49, use the formula $S = R\dfrac{(1 + i)^n - 1}{i}$.

47. $i = \frac{0.0726}{12} = 0.00605$ and $R = 50$, so

$S = 50\dfrac{(1.00605)^{(12)(25)} - 1}{0.00605} \approx \$42{,}211.46.$

49. $i = \frac{0.124}{12} = 0.0103$; solve

$250{,}000 = R\dfrac{(1.0103)^{(12)(20)} - 1}{0.0103}$ to obtain $R \approx \$239.42$ per month (round up, since \$239.41 will not be adequate).

For #51–53, use the formula $A = R\dfrac{1-(1+i)^{-n}}{i}$.

51. $i = \dfrac{0.0795}{12} = 0.006625$; solve

$9000 = R\dfrac{1-(1.006625)^{-(12)(4)}}{0.006625}$ to obtain $R \approx \$219.51$ per month.

53. $i = \dfrac{0.0875}{12} = 0.0072917$; solve

$86{,}000 = R\dfrac{1-(1.0072917)^{-(12)(30)}}{0.0072917}$ to obtain $R \approx \$676.57$ per month (roundup, since \$676.56 will not be adequate).

55. (a) With $i = \dfrac{0.12}{12} = 0.01$, solve

$86{,}000 = 1050\dfrac{1-(1.01)^{-n}}{0.01}$; this leads to

$(1.01)^{-n} = 1 - \dfrac{860}{1050} = \dfrac{19}{105}$, so $n \approx 171.81$

months, or about 14.32 years. The mortgage will be paid off after 172 months (14 years, 4 months). The last payment will be less than \$1050. A reasonable estimate of the final payment can be found by taking the fractional part of the computed value of n above, 0.81, and multiplying by \$1050, giving about \$850.50. To figure the exact amount of the final payment,

solve $86{,}000 = 1050\dfrac{1-(1.01)^{-171}}{0.01} + R(1.01)^{-172}$

(the present value of the first 171 payments, plus the present value of a payment of R dollars 172 months from now). This gives a final payment of $R \approx \$846.57$.

(b) The total amount of the payments under the original plan is $360 \cdot \$884.61 = \$318{,}459.60$. The total using the higher payments is $172 \cdot \$1050 = \$180{,}660$ (or $171 \cdot \$1050 + \$846.57 = \$180{,}396.57$ if we use the correct amount of the final payment)—a difference of \$137,859.60 (or \$138,063.03 using the correct final payment).

57. One possible answer: The APY is the percentage increase from the initial balance $S(0)$ to the end-of-year balance $S(1)$; specifically, it is $S(1)/S(0) - 1$. Multiplying the initial balance by P results in the end-of-year balance being multiplied by the same amount, so that the ratio remains unchanged. Whether we start with a \$1 investment, or a \$1000 investment, $\text{APY} = \left(1 + \dfrac{r}{k}\right)^k - 1$.

59. One possible answer: Some of these situations involve counting things (e.g., populations), so that they can only take on whole number values — exponential models which predict, e.g., 439.72 fish, have to be interpreted in light of this fact.

Technically, bacterial growth, radioactive decay, and compounding interest also are "counting problems" — for example, we cannot have fractional bacteria, or fractional atoms of radioactive material, or fractions of pennies. However, because these are generally very large numbers, it is easier to ignore the fractional parts. (This might also apply when one is talking about, e.g., the population of the whole world.)

Another distinction: while we often use an exponential model for all these situations, it generally fits better (over long periods of time) for radioactive decay than for most of the others. Rates of growth in populations (esp. human populations) tend to fluctuate more than exponential models suggest. Of course, an exponential model also fits well in compound interest situations where the interest rate is held constant, but there are many cases where interest rates change over time.

61. False. The limit, with continuous compounding, is $A = Pe^{rt} = 100\,e^{0.05} \approx \$105.13$.

63. $A = P(1 + r/k)^{kt} = 2250(1 + 0.07/4)^{4(6)} \approx \$3412.00$. The answer is B.

65. $FV = R((1+i)^n - 1)/i = 300((1 + 0.00375)^{240} - 1)/0.00375 \approx \$116{,}437.31$. The answer is E.

67. The last payment will be \$364.38.

69. (a) Matching up with the formula $S = R\dfrac{(1+i)^n - 1}{i}$, where $i = r/k$, with r being the rate and k being the number of payments per year, we find $r = 8\%$.

(b) $k = 12$ payments per year.

(c) Each payment is $R = \$100$.

Chapter 3 Review

1. $f\left(\dfrac{1}{3}\right) = -3 \cdot 4^{1/3} = -3\sqrt[3]{4}$

For #3, recall that exponential functions have the form $f(x) = a \cdot b^x$.

3. $a = 3$, so $f(2) = 3 \cdot b^2 = 6$, $b^2 = 2$, $b = \sqrt{2}$, $f(x) = 3 \cdot 2^{x/2}$

5. $f(x) = 2^{-2x} + 3$ — starting from 2^x, horizontally shrink by $\frac{1}{2}$, reflect across y-axis, and translate up 3 units.

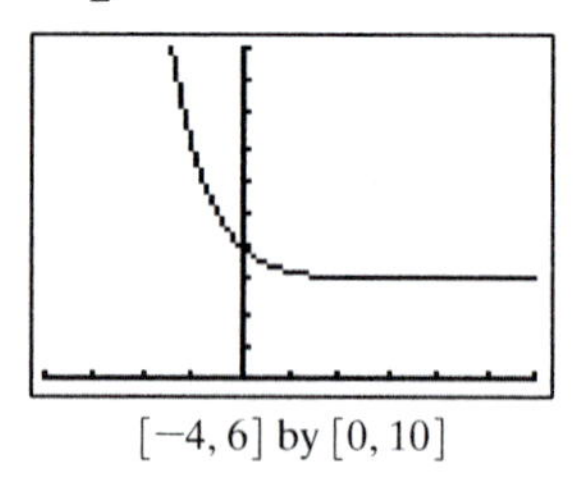

[−4, 6] by [0, 10]

7. $f(x) = -2^{-3x} - 3$ — starting from 2^x, horizontally shrink by $\frac{1}{3}$, reflect across the y-axis, reflect across x-axis, translate down 3 units.

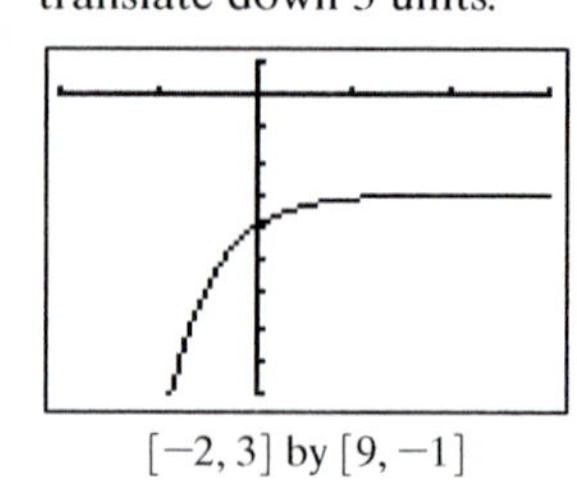

[−2, 3] by [9, −1]

9. Starting from e^x, horizontally shrink by $\frac{1}{2}$, then translate right $\frac{3}{2}$ units — or translate right 3 units, then horizontally shrink by $\frac{1}{2}$.

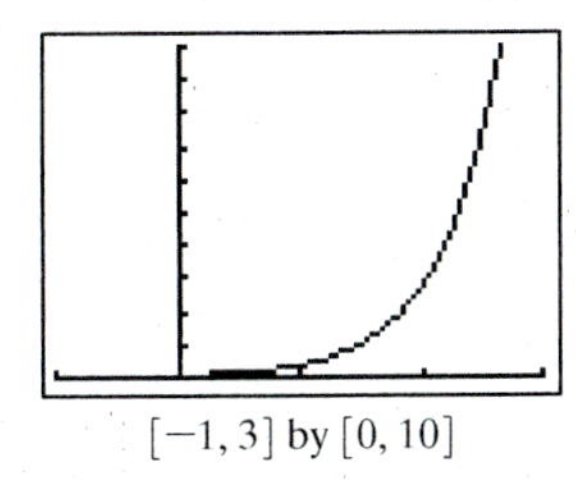

$[-1, 3]$ by $[0, 10]$

11. $f(0) = \frac{100}{5+3} = 12.5$, $\lim_{x\to-\infty} f(x) = 0$, $\lim_{x\to\infty} f(x) = 20$
y-intercept: $(0, 12.5)$; Asymptotes: $y = 0$ and $y = 20$

13. It is an exponential decay function.
$\lim_{x\to\infty} f(x) = 2$, $\lim_{x\to-\infty} f(x) = \infty$

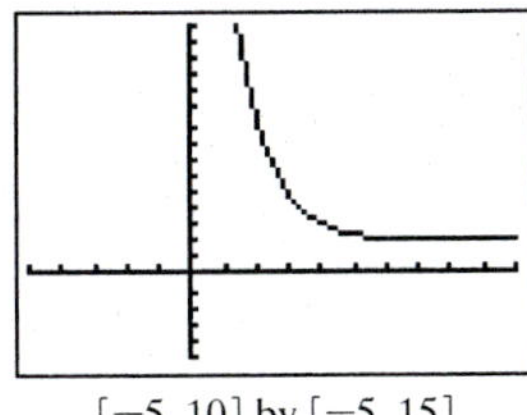

$[-5, 10]$ by $[-5, 15]$

15.

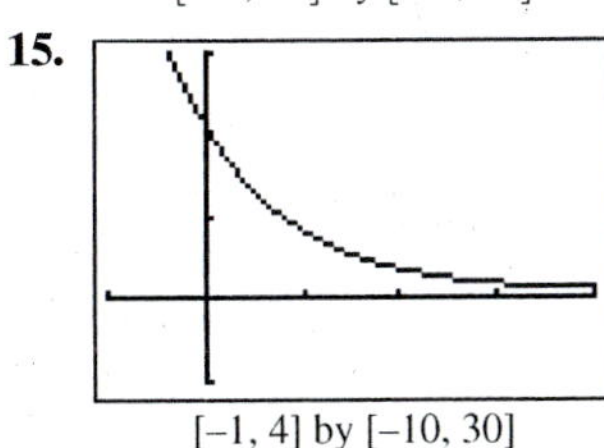

$[-1, 4]$ by $[-10, 30]$

Domain: $(-\infty, \infty)$
Range: $(1, \infty)$
Continuous
Always decreasing
Not symmetric
Bounded below by $y = 1$, which is also the only asymptote
No local extrema
$\lim_{x\to\infty} f(x) = 1$, $\lim_{x\to-\infty} f(x) = \infty$

17.

$[-5, 10]$ by $[-2, 8]$

Domain: $(-\infty, \infty)$
Range: $(0, 6)$
Continuous
Increasing
Symmetric about $(1.20, 3)$
Bounded above by $y = 6$ and below by $y = 0$, the two asymptotes
No extrema
$\lim_{x\to\infty} f(x) = 6$, $\lim_{x\to-\infty} f(x) = 0$

For #19–21, recall that exponential functions are of the form $f(x) = a \cdot (1 + r)^{kx}$.

19. $a = 24$, $r = 0.053$, $k = 1$; so $f(x) = 24 \cdot 1.053^x$, where $x =$ days.

21. $a = 18$, $r = 1$, $k = \frac{1}{21}$, so $f(x) = 18 \cdot 2^{x/21}$, where $x =$ days.

For #23–25, recall that logistic functions are expressed in $f(x) = \frac{c}{1 + ae^{-bx}}$.

23. $c = 30$, $a = 1.5$, so $f(2) = \frac{30}{1 + 1.5e^{-2b}} = 20$,
$30 = 20 + 30e^{-2b}$, $30e^{-2b} = 10$, $e^{-2b} = \frac{1}{3}$,
$-2b \ln e = \ln \frac{1}{3} \approx -1.0986$, so $b \approx 0.55$.
Thus, $f(x) = \frac{30}{1 + 1.5e^{-0.55x}}$

25. $c = 20$, $a = 3$, so $f(3) = \frac{20}{1 + 3e^{-3b}} = 10$,
$20 = 10 + 30e^{-3b}$, $30e^{-3b} = 10$, $e^{-3b} = \frac{10}{30} = \frac{1}{3}$,
$-3b \ln e = \ln \frac{1}{3} \approx -1.0986$, so $b \approx 0.37$.
Thus, $f(x) \approx \frac{20}{1 + 3e^{-0.37x}}$

27. $\log_2 32 = \log_2 2^5 = 5 \log_2 2 = 5$

29. $\log \sqrt[3]{10} = \log 10^{\frac{1}{3}} = \frac{1}{3} \log 10 = \frac{1}{3}$

31. $x = 3^5 = 243$

33. $\left(\frac{x}{y}\right) = e^{-2}$
$x = \frac{y}{e^2}$

35. Translate left 4 units.

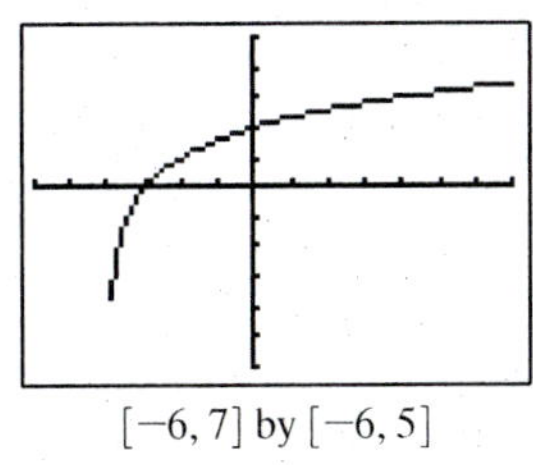

$[-6, 7]$ by $[-6, 5]$

37. Translate right 1 unit, reflect across x-axis, and translate up 2 units.

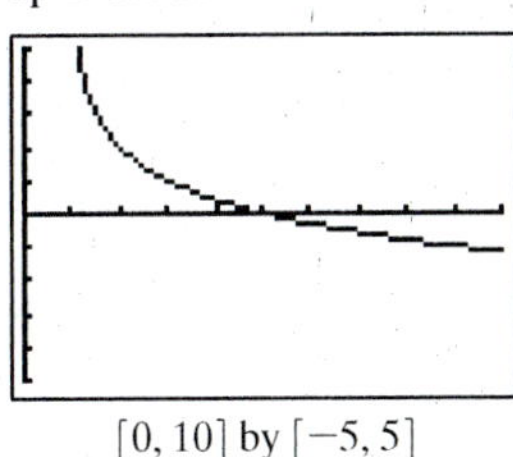

$[0, 10]$ by $[-5, 5]$

39. 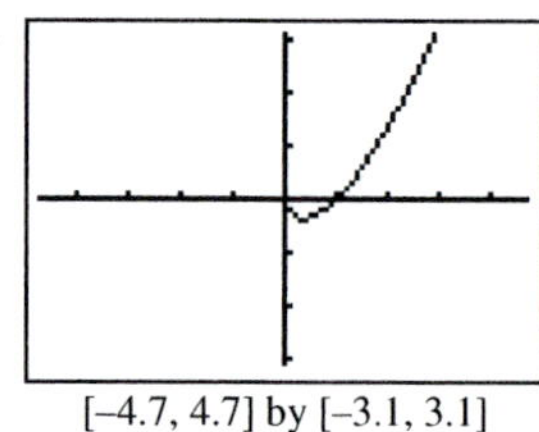

[–4.7, 4.7] by [–3.1, 3.1]

Domain: $(0, \infty)$

Range: $\left[-\frac{1}{e}, \infty\right) \approx [-0.37, \infty)$

Continuous

Decreasing on $(0, 0.37]$; increasing on $[0.37, \infty)$

Not symmetric

Bounded below

Local minimum at $\left(\frac{1}{e}, -\frac{1}{e}\right)$

$\lim_{x\to\infty} f(x) = \infty$

41. 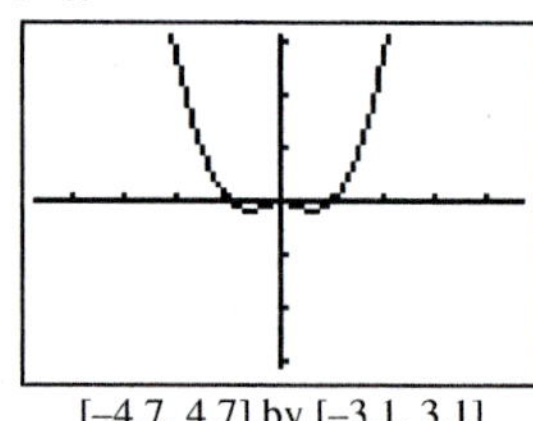

[–4.7, 4.7] by [–3.1, 3.1]

Domain: $(-\infty, 0) \cup (0, \infty)$

Range: $[-0.18, \infty)$

Discontinuous at $x = 0$

Decreasing on $(-\infty, -0.61]$, $(0, 0.61]$;

Increasing on $[-0.61, 0)$, $[0.61, \infty)$

Symmetric across y-axis

Bounded below

Local minima at $(-0.61, -0.18)$ and $(0.61, -0.18)$

No asymptotes

$\lim_{x\to\infty} f(x) = \infty$, $\lim_{x\to-\infty} f(x) = \infty$

43. $x = \log 4 \approx 0.6021$

45. $x = \dfrac{\ln 3}{\ln 1.05} \approx 22.5171$

47. $x = 10^{-7} = 0.0000001$

49. $\log_2 x = 2$, so $x = 2^2 = 4$

51. Multiply both sides by $2 \cdot 3^x$, leaving $(3^x)^2 - 1 = 10 \cdot 3^x$, or $(3^x)^2 - 10 \cdot 3^x - 1 = 0$. This is quadratic in 3^x, leading to $3^x = \dfrac{10 \pm \sqrt{100 + 4}}{2} = 5 \pm \sqrt{26}$. Only $5 + \sqrt{26}$ is positive, so the only answer is $x = \log_3(5 + \sqrt{26}) \approx 2.1049$.

53. $\log[(x + 2)(x - 1)] = 4$, so $(x + 2)(x - 1) = 10^4$.

The solutions to this quadratic equation are

$x = \frac{1}{2}(-1 \pm \sqrt{40{,}009})$, but of these two numbers, only the positive one, $x = \frac{1}{2}(\sqrt{40{,}009} - 1) \approx 99.5112$, works in the original equation.

55. $\log_2 x = \dfrac{\ln x}{\ln 2}$

57. $\log_5 x = \dfrac{\log x}{\log 5}$

59. Increasing, intercept at $(1, 0)$. The answer is (c).

61. Intercept at $(-1, 0)$. The answer is (b).

63. $A = 450(1 + 0.046)^3 \approx \$515.00$

65. $A = Pe^{rt}$

67. $PV = \dfrac{550\left(1 - \left(1 + \frac{0.055}{12}\right)^{(-12)(5)}\right)}{\left(\frac{0.055}{12}\right)} \approx \$28{,}794.06$

69. $20e^{-3k} = 50$, so $k = -\frac{1}{3}\ln\frac{5}{2} \approx -0.3054$.

71. $P(t) = 2.0956 \cdot 1.01218^t$, where x is the number of years since 1900. In 2005, $P(105) = 2.0956 \cdot 1.01218^{105} \approx 7.5$ million.

73. **(a)** $f(0) = 90$ units.

(b) $f(2) \approx 32.8722$ units.

(c) 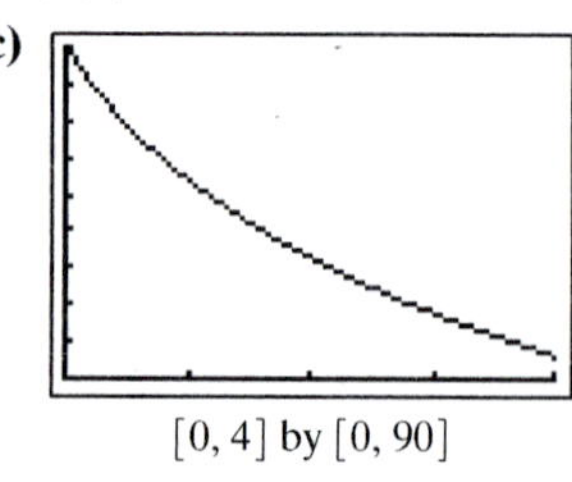

[0, 4] by [0, 90]

75. **(a)** $P(t) = 89{,}000(1 - 0.018)^t = 89{,}000(0.982)^t$.

(b) $P(t) = 50{,}000$ when $t = \dfrac{\ln(50/89)}{\ln 0.982} \approx 31.74$ years.

77. **(a)** $P(t) = 20 \cdot 2^t$, where t is time in months. (Other possible answers: $20 \cdot 2^{12t}$ if t is in years, or $20 \cdot 2^{t/30}$ if t is in days).

(b) $P(12) = 81{,}920$ rabbits after 1 year.
$P(60) \approx 2.3058 \times 10^{19}$ rabbits after 5 years.

(c) Solve $20 \cdot 2^t = 10{,}000$ to find $t = \log_2 500 \approx 8.9658$ months — 8 months and about 29 days.

79. **(a)** $S(t) = S_0 \cdot \left(\frac{1}{2}\right)^{t/1.5}$, where t is time in seconds.

(b) $S(1.5) = S_0/2$. $S(3) = S_0/4$.

(c) If $1\text{ g} = S(60) = S_0 \cdot \left(\frac{1}{2}\right)^{60/1.5} = S_0 \cdot \left(\frac{1}{2}\right)^{40}$, then $S_0 = 2^{40} \approx 1.0995 \times 10^{12}\text{ g} = 1.0995 \times 10^9\text{ kg} = 1{,}099{,}500$ metric tons.

81. Let a_1 = the amplitude of the ground motion of the Feb 4 quake, and let a_2 = the amplitude of the ground motion of the May 30 quake. Then:

$$6.1 = \log\frac{a_1}{T} + B \text{ and } 6.9 = \log\frac{a_2}{T} + B$$
$$\left(\log\frac{a_2}{T} + B\right) - \left(\log\frac{a_1}{T} + B\right) = 6.9 - 6.1$$
$$\log\frac{a_2}{T} - \log\frac{a_1}{T} = 0.8$$
$$\log\frac{a_2}{a_1} = 0.8$$
$$\frac{a_2}{a_1} = 10^{0.8}$$
$$a_2 \approx 6.31\, a_1$$

The ground amplitude of the deadlier quake was approximately 6.31 times stronger.

83. Solve $1500\left(1 + \frac{0.08}{4}\right)^{4t} = 3750$: $(1.02)^{4t} = 2.5$, so $t = \frac{1}{4}\frac{\ln 2.5}{\ln 1.02} \approx 11.5678$ years — round to 11 years 9 months (the next full compounding period).

85. $t = 133.83 \ln\frac{700}{250} \approx 137.7940$ — about 11 years 6 months.

87. $r = \left(1 + \frac{0.0825}{12}\right)^{12} - 1 \approx 8.57\%$

89. $I = 12 \cdot 10^{(-0.0125)(25)} = 5.84$ lumens

91. $\log_b x = \frac{\log x}{\log b}$. This is a vertical stretch if $\frac{1}{10} < b < 10$ (so that $|\log b| < 1$), and a shrink if $0 < b < \frac{1}{10}$ or $b > 10$. (There is also a reflection if $0 < b < 1$.)

93. (a) $P(0) = 16$ students.

(b) $P(t) = 800$ when $1 + 99e^{-0.4t} = 2$, or $e^{0.4t} = 99$, so $t = \frac{1}{0.4}\ln 99 \approx 11.4878$ — about $11\frac{1}{2}$ days.

(c) $P(t) = 400$ when $1 + 99e^{-0.4t} = 4$, or $e^{0.4t} = 33$, so $t = \frac{1}{0.4}\ln 33 \approx 8.7413$ — about 8 or 9 days.

95. The model is $T = 20 + 76e^{-kt}$, and $T(8) = 65 = 20 + 76e^{-8k}$. Then $e^{-8k} = \frac{45}{76}$, so $k = -\frac{1}{8}\ln\frac{45}{76} \approx 0.0655$. Finally, $T = 25$ when $25 = 20 + 76e^{-kt}$, so $t = -\frac{1}{k}\ln\frac{5}{76} \approx 41.54$ minutes.

97. (a) Matching up with the formula $S = R\frac{(1+i)^n - 1}{i}$, where $i = r/k$, with r being the rate and k being the number of payments per year, we find $r = 9\%$.

(b) $k = 4$ payments per year.

(c) Each payment is $R = \$100$.

99. (a) Grace's balance will always remain \$1000, since interest is not added to it. Every year she receives 5% of that \$1000 in interest; after t years, she has been paid $5t\%$ of the \$1000 investment, meaning that altogether she has $1000 + 1000 \cdot 0.05t = 1000(1 + 0.05t)$.

(b) The table is shown below; the second column gives values of $1000e^{0.05t}$. The effects of compounding continuously show up immediately.

Years	Not Compounded	Compounded
0	1000.00	1000.00
1	1050.00	1051.27
2	1100.00	1105.17
3	1150.00	1161.83
4	1200.00	1221.40
5	1250.00	1284.03
6	1300.00	1349.86
7	1350.00	1419.07
8	1400.00	1491.82
9	1450.00	1568.31
10	1500.00	1648.72

Chapter 3 Project

Answers are based on the sample data shown in the table.

2. Writing each maximum height as a (rounded) percentage of the previous maximum height produces the following table.

Bounce Number	Percentage Return
0	N/A
1	79%
2	77%
3	76%
4	78%
5	79%

The average is 77.8%.

3.

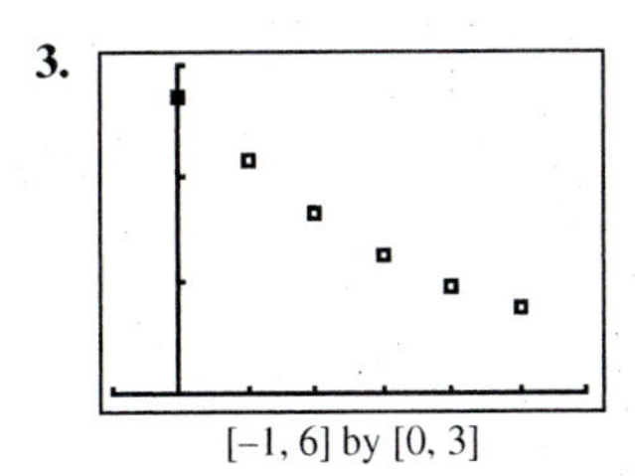

[–1, 6] by [0, 3]

4. Each successive height will be predicted by multiplying the previous height by the same percentage of rebound. The rebound height can therefore be predicted by the equation $y = HP^x$ where x is the bounce number. From the sample data, $H = 2.7188$ and $P \approx 0.778$.

5. $y = HP^x$ becomes $y \approx 2.7188 \cdot 0.778^x$.

6. The regression equation is $y \approx 2.733 \cdot 0.776^x$. Both H and P are close to, though not identical with, the values in the earlier equation.

7. A different ball could be dropped from the same original height, but subsequent maximum heights would in general change because the rebound percentage changed. So P would change in the equation.

8. H would be changed by varying the height from which the ball was dropped. P would be changed by using a different type of ball or a different bouncing surface.

9. $y = HP^x$
$= H(e^{\ln P})^x$
$= He^{(\ln P)x}$
$= 2.7188\, e^{-0.251x}$

10. $\ln y = \ln (HP^x)$
$= \ln H + x \ln P$
This is a linear equation.

11.

Bounce Number	ln (Height)
0	1.0002
1	0.76202
2	0.50471
3	0.23428
4	−0.01705
5	−0.25125

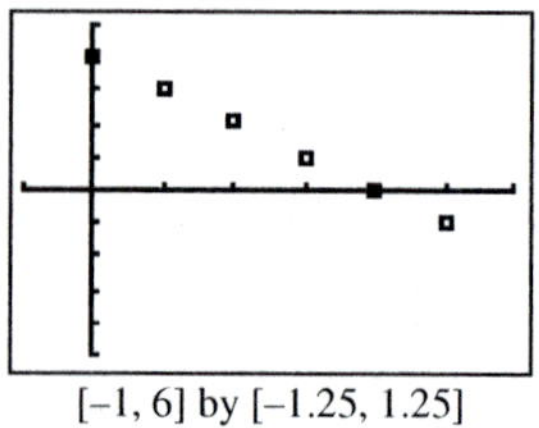

[−1, 6] by [−1.25, 1.25]

The linear regression produces $Y = \ln y \approx -0.253x + 1.005$. Since $\ln y \approx (\ln P)x + \ln H$, the slope of the line is $\ln P$ and the Y-intercept (that is, the $\ln y$-intercept) is $\ln H$.

Chapter 4
Trigonometric Functions

Section 4.1 Angles and Their Measures

Exploration 1

1. $2\pi r$
2. 2π radians (2π lengths of thread)
3. No, not quite, since the distance πr would require a piece of thread π times as long, and $\pi > 3$.
4. π radians

Quick Review 4.1

1. $C = 2\pi \cdot 2.5 = 5\pi$ in.

3. $r = \dfrac{1}{2\pi} \cdot 12 = \dfrac{6}{\pi}$ m

5. **(a)** $s = 47.52$ ft **(b)** $s = 39.77$ km

7. $60\dfrac{\text{mi}}{\text{hr}} \cdot 5280\dfrac{\text{ft}}{\text{mi}} \cdot \dfrac{1}{3600}\dfrac{\text{hr}}{\text{sec}} = 88$ ft/sec

9. $8.8\dfrac{\text{ft}}{\text{sec}} \cdot \dfrac{1}{5280}\dfrac{\text{mi}}{\text{ft}} \cdot 3600\dfrac{\text{sec}}{\text{hr}} = 6$ mph

Section 4.1 Exercises

1. $23°12' = \left(23 + \dfrac{12}{60}\right)^\circ = 23.2°$

3. $118°44'15'' = \left(118 + \dfrac{44}{60} + \dfrac{15}{3600}\right)^\circ = 118.7375°$

5. $21.2° = 21°(60 \cdot 0.2)' = 21°12'$

7. $118.32° = 118°(60 \cdot 0.32)' = 118°19.2'$
$= 118°19'(60 \cdot 0.2)'' = 118°19'12''$

For #9–15, use the formula $s = r\theta$, and the equivalent forms $r = s/\theta$ and $\theta = s/r$.

9. $60° \cdot \dfrac{\pi}{180°} = \dfrac{\pi}{3}$ rad

11. $120° \cdot \dfrac{\pi}{180°} = \dfrac{2\pi}{3}$ rad

13. $71.72° \cdot \dfrac{\pi}{180°} \approx 1.2518$ rad

15. $61°24' = \left(61 + \dfrac{24}{60}\right)^\circ = 61.4° \cdot \dfrac{\pi}{180°} \approx 1.0716$ rad

17. $\dfrac{\pi}{6} \cdot \dfrac{180°}{\pi} = 30°$

19. $\dfrac{\pi}{10} \cdot \dfrac{180°}{\pi} = 18°$

21. $\dfrac{7\pi}{9} \cdot \dfrac{180°}{\pi} = 140°$

23. $2 \cdot \dfrac{180}{\pi} \approx 114.59°$

25. $s = 50$ in.

27. $r = 6/\pi$ ft

29. $\theta = 3$ radians

31. $r = \dfrac{360}{\pi}$ cm

33. $\theta = s_1/r_1 = \dfrac{9}{11}$ rad and $s_2 = r_2\theta = 36$

35. The angle is $10° \cdot \dfrac{\pi}{180°} = \dfrac{\pi}{18}$ rad, so the curved side measures $\dfrac{11\pi}{18}$ in. The two straight sides measures 11 in. each, so the perimeter is $11 + 11 + \dfrac{11\pi}{18} \approx 24$ inches.

37. Five pieces of track form a semicircle, so each arc has a central angle of $\pi/5$ radians. The inside arc length is $r_i(\pi/5)$ and the outside arc length is $r_o(\pi/5)$. Since $r_o(\pi/5) - r_i(\pi/5) = 3.4$ inches, we conclude that $r_o - r_i = 3.4(5/\pi) \approx 5.4$ inches.

39. **(a)** NE is 45°. **(b)** NNE is 22.5°. **(c)** WSW is 247.5°.

41. ESE is closest at 112.5°.

43. The angle between them is $\theta = 9°42' = 9.7° \approx 0.1693$ radians, so the distance is about $s = r\theta = (25)(0.1693) \approx 4.23$ statute miles.

45. $v = 44$ ft/sec and $r = 13$ in., so
$\omega = v/r = \left(44\dfrac{\text{ft}}{\text{sec}} \cdot 60\dfrac{\text{sec}}{\text{min}}\right) \div \left(13 \text{ in.} \cdot \dfrac{1}{12}\dfrac{\text{ft}}{\text{in.}} \cdot 2\pi\dfrac{\text{rad}}{\text{rev}}\right)$
≈ 387.85 rpm.

47. $\omega = 2000$ rpm and $r = 5$ in., so
$v = r\omega = \left(5 \text{ in.} \cdot 12\dfrac{\text{teeth}}{\text{in.}}\right) \cdot \left(2000\dfrac{\text{rev}}{\text{min}} \cdot 2\pi\dfrac{\text{rad}}{\text{rev}} \cdot \dfrac{1}{60}\dfrac{\text{min}}{\text{sec}}\right) \approx 12{,}566.37$ teeth per second.

49.

47°
4 mi
38°
2 mi

51. $895 \text{ stat mi} \cdot \dfrac{10{,}800}{3956\pi}\dfrac{\text{naut mi}}{\text{stat mi}} \approx 778$ nautical miles

53. **(a)** $s = r\theta = (4)(4\pi) = 16\pi \approx 50.265$ in., or $\frac{4}{3}\pi \approx 4.189$ ft.

(b) $r\theta = 2\pi \approx 6.283$ ft.

55. **(a)** $\omega_1 = 120 \frac{\text{rev}}{\text{min}} \cdot 2\pi \frac{\text{rad}}{\text{rev}} \cdot \frac{1 \text{ min}}{60 \text{ sec}} = 4\pi \text{ rad/sec}$

(b) $v = R\omega_1 = (7 \text{ cm})\left(4\pi \frac{\text{rad}}{\text{sec}}\right) = 28\pi \text{ cm/sec}$

(c) $\omega_2 = v/r = \left(28\pi \frac{\text{cm}}{\text{sec}}\right) \div (4 \text{ cm}) = 7\pi \text{ rad/sec}$

57. True. In the amount of time it takes for the merry-go-round to complete one revolution, horse B travels a distance of $2\pi r$, where r is B's distance from the center. In the same time, horse A travels a distance of $2\pi(2r) = 2(2\pi r)$ — twice as far as B.

59. $x° = x°\left(\frac{\pi \text{ rad}}{180°}\right) = \frac{\pi x}{180}$. The answer is C.

61. Let n be the number of revolutions per minute.
$\left(\frac{26\pi \text{ in.}}{1 \text{ rev}}\right)\left(\frac{n \text{ rev}}{1 \text{ min}}\right)\left(\frac{60 \text{ min}}{1 \text{ hr}}\right)\left(\frac{1 \text{ mi}}{63{,}360 \text{ in.}}\right)$
$\approx 0.07735\, n$ mph.
Solving $0.07735\, n = 10$ yields $n \approx 129$.
The answer is B.

In #63–65, we need to "borrow" 1° and change it to 60′ in order to complete the subtraction.

63. $122°25' - 84°23' = 38°02'$

65. $93°16' - 87°39' = 92°76' - 87°39' = 5°37'$

In #67–69, find the difference in the latitude. Convert this difference to minutes; this is the distance in nautical miles. The Earth's diameter is not needed.

67. The difference in latitude is $34°03' - 32°43' = 1°20'$ $= 80$ minutes of arc, which is 80 naut mi.

69. The difference in latitude is $44°59' - 29°57' = 15°02'$ $= 902$ minutes of arc, which is 902 naut mi.

71. The whole circle's area is πr^2; the sector with central angle θ makes up $\theta/2\pi$ of that area, or $\frac{\theta}{2\pi} \cdot \pi r^2 = \frac{1}{2}\theta r^2$.

73.

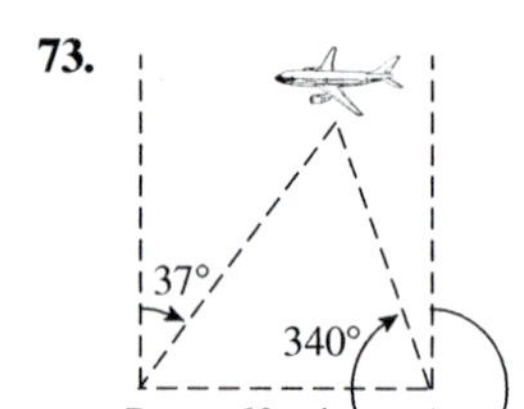

Section 4.2 Trigonometric Functions of Acute Angles

Exploration 1

1. sin and csc, cos and sec, and tan and cot.

2. $\tan\theta$

3. $\sec\theta$

4. 1

5. $\sin\theta$ and $\cos\theta$

Exploration 2

1. Let $\theta = 60°$. Then

$\sin\theta = \frac{\sqrt{3}}{2} \approx 0.866$ $\quad\csc\theta = \frac{2}{\sqrt{3}} \approx 1.155$

$\cos\theta = \frac{1}{2}$ $\quad\sec\theta = 2$

$\tan\theta = \sqrt{3} \approx 1.732$ $\quad\cot\theta = \frac{1}{\sqrt{3}} \approx 0.577$

2. The values are the same, but for different functions. For example, sin 30° is the same as cos 60°, cot 30° is the same as tan 60°, etc.

3. The value of a trig function at θ is the same as the value of its co-function at $90° - \theta$.

Quick Review 4.2

1. $x = \sqrt{5^2 + 5^2} = \sqrt{50} = 5\sqrt{2}$

3. $x = \sqrt{10^2 - 8^2} = 6$

5. $8.4 \text{ ft} \cdot 12 \frac{\text{in}}{\text{ft}} = 100.8$ in.

7. $a = (0.388)(20.4) = 7.9152$ km

9. $\alpha = 13.3 \cdot \frac{2.4}{31.6} \approx 1.0101$ (no units)

Section 4.2 Exercises

1. $\sin\theta = \frac{4}{5}, \cos\theta = \frac{3}{5}, \tan\theta = \frac{4}{3}, \csc\theta = \frac{5}{4}, \sec\theta = \frac{5}{3}, \cot\theta = \frac{3}{4}$.

3. $\sin\theta = \frac{12}{13}, \cos\theta = \frac{5}{13}, \tan\theta = \frac{12}{5}; \csc\theta = \frac{13}{12}, \sec\theta = \frac{13}{5}, \cot\theta = \frac{5}{12}$.

5. The hypotenuse length is $\sqrt{7^2 + 11^2} = \sqrt{170}$, so $\sin\theta = \frac{7}{\sqrt{170}}, \cos\theta = \frac{11}{\sqrt{170}}, \tan\theta = \frac{7}{11}; \csc\theta = \frac{\sqrt{170}}{7}, \sec\theta = \frac{\sqrt{170}}{11}, \cot\theta = \frac{11}{7}$.

7. The opposite side length is $\sqrt{11^2 - 8^2} = \sqrt{57}$, so $\sin\theta = \frac{\sqrt{57}}{11}, \cos\theta = \frac{8}{11}, \tan\theta = \frac{\sqrt{57}}{8}; \csc\theta = \frac{11}{\sqrt{57}}, \sec\theta = \frac{11}{8}, \cot\theta = \frac{8}{\sqrt{57}}$.

9. Using a right triangle with hypotenuse 7 and legs 3 (opposite) and $\sqrt{7^2 - 3^2} = \sqrt{40} = 2\sqrt{10}$ (adjacent), we have $\sin\theta = \frac{3}{7}, \cos\theta = \frac{2\sqrt{10}}{7}, \tan\theta = \frac{3}{2\sqrt{10}}; \csc\theta = \frac{7}{3}, \sec\theta = \frac{7}{2\sqrt{10}}, \cot\theta = \frac{2\sqrt{10}}{3}$.

11. Using a right triangle with hypotenuse 11 and legs 5 (adjacent) and $\sqrt{11^2 - 5^2} = \sqrt{96} = 4\sqrt{6}$ (opposite), we have $\sin\theta = \frac{4\sqrt{6}}{11}$, $\cos\theta = \frac{5}{11}$, $\tan\theta = \frac{4\sqrt{6}}{5}$; $\csc\theta = \frac{11}{4\sqrt{6}}$, $\sec\theta = \frac{11}{5}$, $\cot\theta = \frac{5}{4\sqrt{6}}$.

13. Using a right triangle with legs 5 (opposite) and 9 (adjacent) and hypotenuse $\sqrt{5^2 + 9^2} = \sqrt{106}$, we have $\sin\theta = \frac{5}{\sqrt{106}}$, $\cos\theta = \frac{9}{\sqrt{106}}$, $\tan\theta = \frac{5}{9}$; $\csc\theta = \frac{\sqrt{106}}{5}$, $\sec\theta = \frac{\sqrt{106}}{9}$, $\cot\theta = \frac{9}{5}$.

15. Using a right triangle with legs 3 (opposite) and 11 (adjacent) and hypotenuse $\sqrt{3^2 + 11^2} = \sqrt{130}$, we have $\sin\theta = \frac{3}{\sqrt{130}}$, $\cos\theta = \frac{11}{\sqrt{130}}$, $\tan\theta = \frac{3}{11}$; $\csc\theta = \frac{\sqrt{130}}{3}$, $\sec\theta = \frac{\sqrt{130}}{11}$, $\cot\theta = \frac{11}{3}$.

17. Using a right triangle with hypotenuse 23 and legs 9 (opposite) and $\sqrt{23^2 - 9^2} = \sqrt{448} = 8\sqrt{7}$ (adjacent), we have $\sin\theta = \frac{9}{23}$, $\cos\theta = \frac{8\sqrt{7}}{23}$, $\tan\theta = \frac{9}{8\sqrt{7}}$; $\csc\theta = \frac{23}{9}$, $\sec\theta = \frac{23}{8\sqrt{7}}$, $\cot\theta = \frac{8\sqrt{7}}{9}$.

19. $\frac{\sqrt{3}}{2}$

21. $\sqrt{3}$

23. $\frac{1}{\sqrt{2}} = \frac{\sqrt{2}}{2}$

25. $\sec 45° = 1/\cos 45° \approx 1.4142$. Squaring this result yields 2.0000, so $\sec 45° = \sqrt{2}$.

27. $\csc(\pi/3) = 1/\sin(\pi/3) \approx 1.1547$. Squaring this result yields 1.3333 or essentially 4/3, so $\csc(\pi/3) = \sqrt{4/3} = 2/\sqrt{3} = 2\sqrt{3}/3$.

For #29–39, the answers marked with an asterisk (*) should be found in DEGREE mode; the rest should be found in RADIAN mode. Since most calculators do not have the secant, cosecant, and cotangent functions built in, the reciprocal versions of these functions are shown.

29. ≈ 0.961*

31. ≈ 0.943*

33. ≈ 0.268

35. $\frac{1}{\cos 49°} \approx 1.524$*

37. $\frac{1}{\tan 0.89} \approx 0.810$

39. $\frac{1}{\tan(\pi/8)} \approx 2.414$

41. $\theta = 30° = \frac{\pi}{6}$

43. $\theta = 60° = \frac{\pi}{3}$

45. $\theta = 60° = \frac{\pi}{3}$

47. $\theta = 30° = \frac{\pi}{6}$

49. $x = \frac{15}{\sin 34°} \approx 26.82$

51. $y = \frac{32}{\tan 57°} \approx 20.78$

53. $y = 6/\sin 35° \approx 10.46$

For #55–57, choose whichever of the following formulas is appropriate:

$$a = \sqrt{c^2 - b^2} = c\sin\alpha = c\cos\beta = b\tan\alpha = \frac{b}{\tan\beta}$$

$$b = \sqrt{c^2 - a^2} = c\cos\alpha = c\sin\beta = a\tan\beta = \frac{a}{\tan\alpha}$$

$$c = \sqrt{a^2 + b^2} = \frac{a}{\cos\beta} = \frac{a}{\sin\alpha} = \frac{b}{\sin\beta} = \frac{b}{\cos\alpha}$$

If one angle is given, subtract from 90° to find the other angle.

55. $b = \frac{a}{\tan\alpha} = \frac{12.3}{\tan 20°} \approx 33.79$,

$c = \frac{a}{\sin\alpha} = \frac{12.3}{\sin 20°} \approx 35.96$, $\beta = 90° - \alpha = 70°$

57. $b = a\tan\beta = 15.58\tan 55° \approx 22.25$,

$c = \frac{a}{\cos\beta} = \frac{15.58}{\cos 55°} \approx 27.16$, $\alpha = 90° - \beta = 35°$

59. 0. As θ gets smaller and smaller, the side opposite θ gets smaller and smaller, so its ratio to the hypotenuse approaches 0 as a limit.

61. $h = 55\tan 75° \approx 205.26$ ft

63. $A = 12 \cdot \frac{5}{\sin 54°} \approx 74.16$ ft^2

65. $AC = 100\tan 75°12'42'' \approx 378.80$ ft

67. False. This is only true if θ is an acute angle in a right triangle. (Then it is true by definition.)

69. $\sec 90° = \frac{1}{\cos 90°} = \frac{1}{0}$ is undefined. The answer is E.

71. If the unknown slope is m, then $m\sin\theta = -1$, so $m = -\frac{1}{\sin\theta} = -\csc\theta$. The answer is D.

73. For angles in the first quadrant, sine values will be increasing, cosine values will be decreasing and only tangent values can be greater than 1. Therefore, the first column is tangent, the second column is sine, and the third column is cosine.

75. The distance d_A from A to the mirror is $5\cos 30°$; the distance from B to the mirror is $d_B = d_A - 2$. Then

$$PB = \frac{d_B}{\cos\beta} = \frac{d_A - 2}{\cos 30°} = 5 - \frac{2}{\cos 30°}$$
$$= 5 - \frac{4}{\sqrt{3}} \approx 2.69 \text{ m.}$$

77. One possible proof:

$$(\sin\theta)^2 + (\cos\theta)^2 = \left(\frac{a}{c}\right)^2 + \left(\frac{b}{c}\right)^2$$
$$= \frac{a^2}{c^2} + \frac{b^2}{c^2}$$
$$= \frac{a^2 + b^2}{c^2}$$
$$= \frac{c^2}{c^2} \quad \text{(Pythagorean theorem: } a^2 + b^2 = c^2.)$$
$$= 1$$

Section 4.3 Trigonometry Extended: The Circular Functions

Exploration 1

1. The side opposite θ in the triangle has length y and the hypotenuse has length r. Therefore

$\sin\theta = \frac{\text{opp}}{\text{hyp}} = \frac{y}{r}$.

2. $\cos\theta = \frac{\text{adj}}{\text{hyp}} = \frac{x}{r}$

3. $\tan\theta = \frac{\text{opp}}{\text{adj}} = \frac{y}{x}$

4. $\cot\theta = \frac{x}{y}$; $\sec\theta = \frac{r}{x}$; $\csc\theta = \frac{r}{y}$

Exploration 2

1. The x-coordinates on the unit circle lie between -1 and 1, and $\cos t$ is always an x-coordinate on the unit circle.
2. The y-coordinates on the unit circle lie between -1 and 1, and $\sin t$ is always a y-coordinate on the unit circle.
3. The points corresponding to t and $-t$ on the number line are wrapped to points above and below the x-axis with the same x-coordinates. Therefore $\cos t$ and $\cos(-t)$ are equal.
4. The points corresponding to t and $-t$ on the number line are wrapped to points above and below the x-axis with exactly opposite y-coordinates. Therefore $\sin t$ and $\sin(-t)$ are opposites.
5. Since 2π is the distance around the unit circle, both t and $t + 2\pi$ get wrapped to the same point.
6. The points corresponding to t and $t + \pi$ get wrapped to points on either end of a diameter on the unit circle. These points are symmetric with respect to the origin and therefore have coordinates (x, y) and $(-x, -y)$. Therefore $\sin t$ and $\sin(t + \pi)$ are opposites, as are $\cos t$ and $\cos(t + \pi)$.
7. By the observation in (6), $\tan t$ and $\tan(t + \pi)$ are ratios of the form $\frac{y}{x}$ and $\frac{-y}{-x}$, which are either equal to each other or both undefined.
8. The sum is always of the form $x^2 + y^2$ for some (x, y) on the unit circle. Since the equation of the unit circle is $x^2 + y^2 = 1$, the sum is always 1.
9. Answers will vary. For example, there are similar statements that can be made about the functions cot, sec, and csc.

Quick Review 4.3

1. $-30°$
3. $45°$
5. $\tan\frac{\pi}{6} = \frac{1}{\sqrt{3}} = \frac{\sqrt{3}}{3}$
7. $\csc\frac{\pi}{4} = \sqrt{2}$
9. Using a right triangle with hypotenuse 13 and legs 5 (opposite) and $\sqrt{13^2 - 5^2} = 12$ (adjacent), we have $\sin\theta = \frac{5}{13}$, $\cos\theta = \frac{12}{13}$, $\tan\theta = \frac{5}{12}$; $\csc\theta = \frac{13}{5}$, $\sec\theta = \frac{13}{12}$, $\cot\theta = \frac{12}{5}$.

Section 4.3 Exercises

1. The $450°$ angle lies on the positive–y axis $(450° - 360° = 90°)$, while the others are all coterminal in Quadrant II.

In #3–11, recall that the distance from the origin is $r = \sqrt{x^2 + y^2}$.

3. $\sin\theta = \frac{2}{\sqrt{5}}$, $\cos\theta = -\frac{1}{\sqrt{5}}$, $\tan\theta = -2$; $\csc\theta = \frac{\sqrt{5}}{2}$, $\sec\theta = -\sqrt{5}$, $\cot\theta = -\frac{1}{2}$.
5. $\sin\theta = -\frac{1}{\sqrt{2}}$, $\cos\theta = -\frac{1}{\sqrt{2}}$, $\tan\theta = 1$; $\csc\theta = -\sqrt{2}$, $\sec\theta = -\sqrt{2}$, $\cot\theta = 1$.
7. $\sin\theta = \frac{4}{5}$, $\cos\theta = \frac{3}{5}$, $\tan\theta = \frac{4}{3}$; $\csc\theta = \frac{5}{4}$, $\sec\theta = \frac{5}{3}$, $\cot\theta = \frac{3}{4}$.
9. $\sin\theta = 1$, $\cos\theta = 0$, $\tan\theta$ undefined; $\csc\theta = 1$, $\sec\theta$ undefined, $\cot\theta = 0$.
11. $\sin\theta = -\frac{2}{\sqrt{29}}$, $\cos\theta = \frac{5}{\sqrt{29}}$, $\tan\theta = -\frac{2}{5}$; $\csc\theta = -\frac{\sqrt{29}}{2}$, $\sec\theta = \frac{\sqrt{29}}{5}$, $\cot\theta = -\frac{5}{2}$.

For #13–15, determine the quadrant(s) of angles with the given measures, and then use the fact that $\sin t$ is positive when the terminal side of the angle is above the x-axis (in Quadrants I and II) and $\cos t$ is positive when the terminal side of the angle is to the right of the y-axis (in quadrants I and IV). Note that since $\tan t = \sin t/\cos t$, the sign of $\tan t$ can be determined from the signs of $\sin t$ and $\cos t$: if $\sin t$ and $\cos t$ have the same sign, the answer to (c) will be '+'; otherwise it will be '–'. Thus $\tan t$ is positive in Quadrants I and III.

13. These angles are in Quadrant I. **(a)** + (i.e., $\sin t > 0$). **(b)** + (i.e., $\cos t > 0$). **(c)** + (i.e., $\tan t > 0$).
15. These angles are in Quadrant III. **(a)** –. **(b)** –. **(c)** +.

For #17–19, use strategies similar to those for the previous problem set.

17. $143°$ is in Quadrant II, so $\cos 143°$ is negative.
19. $\frac{7\pi}{8}$ rad is in Quadrant II, so $\cos\frac{7\pi}{8}$ is negative.
21. A $(2, 2)$; $\tan 45° = \frac{y}{x} = 1 \Rightarrow y = x$.
23. C $(-\sqrt{3}, -1)$; $\frac{7\pi}{6}$ is in Quadrant III, so x and y are both negative. $\tan\frac{7\pi}{6} = \frac{1}{\sqrt{3}}$.

For #25–35, recall that the reference angle is the acute angle formed by the terminal side of the angle in standard position and the x-axis.

25. The reference angle is 60°. A right triangle with a 60° angle at the origin has the point $P(-1, \sqrt{3})$ as one vertex, with hypotenuse length $r = 2$, so $\cos 120° = \frac{x}{r} = -\frac{1}{2}$.

27. The reference angle is the given angle, $\frac{\pi}{3}$. A right triangle with a $\frac{\pi}{3}$ radian angle at the origin has the point $P(1, \sqrt{3})$ as one vertex, with hypotenuse length $r = 2$, so $\sec\frac{\pi}{3} = \frac{r}{x} = 2$.

29. The reference angle is $\frac{\pi}{6}$ (in fact, the given angle is coterminal with $\frac{\pi}{6}$). A right triangle with a $\frac{\pi}{6}$ radian angle at the origin has the point $P(\sqrt{3}, 1)$ as one vertex, with hypotenuse length $r = 2$, so $\sin\frac{13\pi}{6} = \frac{y}{r} = \frac{1}{2}$.

31. The reference angle is $\frac{\pi}{4}$ (in fact, the given angle is coterminal with $\frac{\pi}{4}$). A right triangle with a $\frac{\pi}{4}$ radian angle at the origin has the point $P(1, 1)$ as one vertex, so $\tan\frac{-15\pi}{4} = \frac{y}{x} = 1$.

33. $\cos\frac{23\pi}{6} = \cos\frac{11\pi}{6} = \frac{\sqrt{3}}{2}$

35. $\sin\frac{11\pi}{3} = \sin\frac{5\pi}{3} = -\frac{\sqrt{3}}{2}$

37. $-450°$ is coterminal with 270°, on the negative y-axis. **(a)** -1 **(b)** 0 **(c)** Undefined

39. 7π radians is coterminal with π radians, on the negative x-axis. **(a)** 0 **(b)** -1 **(c)** 0

41. $\frac{-7\pi}{2}$ radians is coterminal with $\frac{\pi}{2}$ radians, on the positive y-axis. **(a)** 1 **(b)** 0 **(c)** Undefined

43. Since $\cot\theta > 0$, $\sin\theta$ and $\cos\theta$ have the same sign, so $\sin\theta = +\sqrt{1 - \cos^2\theta} = \frac{\sqrt{5}}{3}$, and $\tan\theta = \frac{\sin\theta}{\cos\theta} = \frac{\sqrt{5}}{2}$.

45. $\cos\theta = +\sqrt{1 - \sin^2\theta} = \frac{\sqrt{21}}{5}$, so $\tan\theta = \frac{\sin\theta}{\cos\theta} = -\frac{2}{\sqrt{21}}$ and $\sec\theta = \frac{1}{\cos\theta} = \frac{5}{\sqrt{21}}$.

47. Since $\cos\theta < 0$ and $\cot\theta < 0$, $\sin\theta$ must be positive. With $x = -4$, $y = 3$, and $r = \sqrt{4^2 + 3^2} = 5$, we have $\sec\theta = -\frac{5}{4}$ and $\csc\theta = \frac{5}{3}$.

49. $\sin\left(\frac{\pi}{6} + 49{,}000\pi\right) = \sin\left(\frac{\pi}{6}\right) = \frac{1}{2}$

51. $\cos\left(\frac{5{,}555{,}555\pi}{2}\right) = \cos\left(\frac{\pi}{2}\right) = 0$

53. The calculator's value of the irrational number π is necessarily an approximation. When multiplied by a very large number, the slight error of the original approximation is magnified sufficiently to throw the trigonometric functions off.

55. $\mu = \frac{\sin 83°}{\sin 36°} \approx 1.69$

57. **(a)** When $t = 0$, $d = 0.4$ in.
(b) When $t = 3$, $d = 0.4e^{-0.6}\cos 12 \approx 0.1852$ in.

59. The difference in the elevations is 600 ft, so $d = 600/\sin\theta$. Then:
(a) $d = 600\sqrt{2} \approx 848.53$ ft.
(b) $d = 600$ ft.
(c) $d \approx 933.43$ ft.

61. True. Any angle in a triangle measures between 0° and 180°. Acute angles (<90°) determine reference triangles in Quadrant I, where the cosine is positive, while obtuse angles (>90°) determine reference triangles in Quadrant II, where the cosine is negative.

63. If $\sin\theta = 0.4$, then $\sin(-\theta) + \csc\theta = -\sin\theta + \frac{1}{\sin\theta} = -0.4 + \frac{1}{0.4} = 2.1$. The answer is E.

65. $(\sin t)^2 + (\cos t)^2 = 1$ for all t. The answer is A.

67. Since $\sin\theta > 0$ and $\tan\theta < 0$, the terminal side must be in Quadrant II, so $\theta = \frac{5\pi}{6}$.

69. Since $\tan\theta < 0$ and $\sin\theta < 0$, the terminal side must be in Quadrant IV, so $\theta = \frac{7\pi}{4}$.

71. The two triangles are congruent: both have hypotenuse 1, and the corresponding angles are congruent—the smaller acute angle has measure t in both triangles, and the two acute angles in a right triangle add up to $\pi/2$.

73. One possible answer: Starting from the point (a, b) on the unit circle—at an angle of t, so that $\cos t = a$—then measuring a quarter of the way around the circle (which corresponds to adding $\pi/2$ to the angle), we end at $(-b, a)$, so that $\sin(t + \pi/2) = a$. For (a, b) in Quadrant I, this is shown in the figure above; similar illustrations can be drawn for the other quadrants.

75. Starting from the point (a, b) on the unit circle—at an angle of t, so that $\cos t = a$—then measuring a quarter of the way around the circle (which corresponds to adding $\pi/2$ to the angle), we end at $(-b, a)$, so that $\sin(t + \pi/2) = a$. This holds true when (a, b) is in Quadrant II, just as it did for Quadrant I.

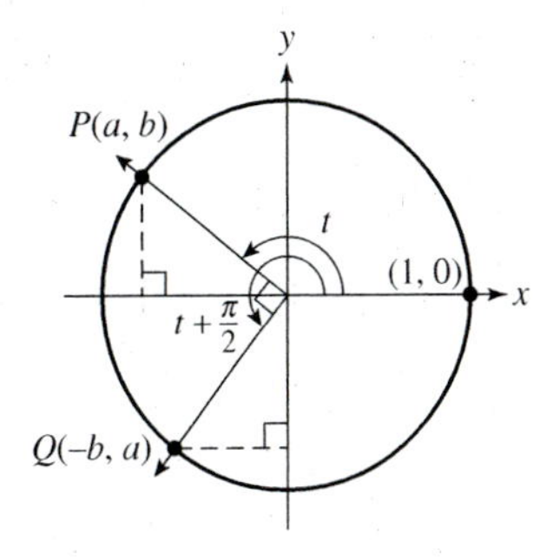

77. Seven decimal places are shown so that the slight differences can be seen. The magnitude of the relative error is less than 1% when $|\theta| < 0.2441$ (approximately). This can be seen by extending the table to larger values of θ, or by graphing $\left|\frac{\sin\theta - \theta}{\sin\theta}\right| - 0.01$.

77.

θ	$\sin\theta$	$\sin\theta - \theta$	$\left\lvert\frac{\sin\theta - \theta}{\sin\theta}\right\rvert$
−0.03	−0.0299955	0.0000045	0.0001500
−0.02	−0.0199987	0.0000013	0.0000667
−0.01	−0.0099998	0.0000002	0.0000167
0	0	0	—
0.01	0.0099998	−0.0000002	0.0000167
0.02	0.0199987	−0.0000013	0.0000667
0.03	0.0299955	−0.0000045	0.0001500

79. This Taylor polynomial is generally a very good approximation for $\sin\theta$—in fact, the relative error (see #77) is less than 1% for $|\theta| < 1$ (approx.). It is better for θ close to 0; it is slightly larger than $\sin\theta$ when $\theta < 0$ and slightly smaller when $\theta > 0$.

79.

θ	$\sin\theta$	$\theta - \frac{1}{6}\theta^3$	$\sin\theta - \left(\theta - \frac{1}{6}\theta^3\right)$
−0.3	−0.2955202	−0.2955000	−0.0000202
−0.2	−0.1986693	−0.1986667	−0.0000027
−0.1	−0.0998334	−0.0998333	−0.0000001
0	0	0	0
0.1	0.0998334	0.0998333	0.0000001
0.2	0.1986693	0.1986667	0.0000027
0.3	0.2955202	0.2955000	0.0000202

■ Section 4.4 Graphs of Sine and Cosine: Sinusoids

Exploration 1

1. $\pi/2$ (at the point $(0, 1)$)
2. $3\pi/2$ (at the point $(0, -1)$)
3. Both graphs cross the x-axis when the y-coordinate on the unit circle is 0.
4. (Calculator exploration)
5. The sine function tracks the y-coordinate of the point as it moves around the unit circle. After the point has gone completely around the unit circle (a distance of 2π), the same pattern of y-coordinates starts over again.
6. Leave all the settings as they are shown at the start of the Exploration, except change Y_{2T} to cos(T).

Quick Review 4.4

1. In order: $+, +, -, -$

3. In order: $+, -, +, -$

5. $-150° \cdot \frac{\pi}{180°} = -\frac{5\pi}{6}$

7. Starting with the graph of y_1, vertically stretch by 3 to obtain the graph of y_2.

9. Starting with the graph of y_1, vertically shrink by 0.5 to obtain the graph of y_2.

Section 4.4 Exercises

In #1–5, for $y = a \sin x$, the amplitude is $|a|$. If $|a| > 1$, there is a vertical stretch by a factor of $|a|$, and if $|a| < 1$, there is a vertical shrink by a factor of $|a|$. When $a < 0$, there is also a reflection across the x-axis.

1. Amplitude 2; vertical stretch by a factor of 2.

3. Amplitude 4; vertical stretch by a factor of 4, reflection across the x-axis.

5. Amplitude 0.73; vertical shrink by a factor of 0.73.

In #7–11, for $y = \cos bx$, the period is $2\pi/|b|$. If $|b| > 1$, there is a horizontal shrink by a factor of $1/|b|$, and if $|b| < 1$, there is a horizontal stretch by a factor of $1/|b|$. When $b < 0$, there is also a reflection across the y-axis. For $y = a \cos bx$, a has the same effects as in #1–5.

7. Period $2\pi/3$; horizontal shrink by a factor of $1/3$.

9. Period $2\pi/7$; horizontal shrink by a factor of $1/7$, reflection across the y-axis.

11. Period $2\pi/2 = \pi$; horizontal shrink by a factor of $1/2$. Also a vertical stretch by a factor of 3.

In #13–15, the amplitudes of the graphs for $y = a \sin bx$ and $y = a \cos bx$ are governed by a, while the period is governed by b, just as in #1–11. The frequency is 1/period.

13. For $y = 3 \sin (x/2)$, the amplitude is 3, the period is $2\pi/(1/2) = 4\pi$, and the frequency is $1/(4\pi)$.

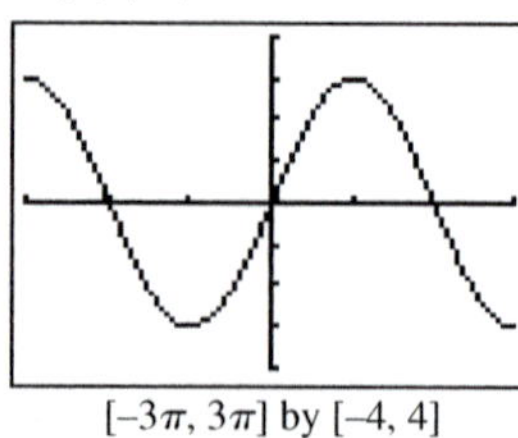

$[-3\pi, 3\pi]$ by $[-4, 4]$

15. For $y = -(3/2) \sin 2x$, the amplitude is $3/2$, the period is $2\pi/2 = \pi$, and the frequency is $1/\pi$.

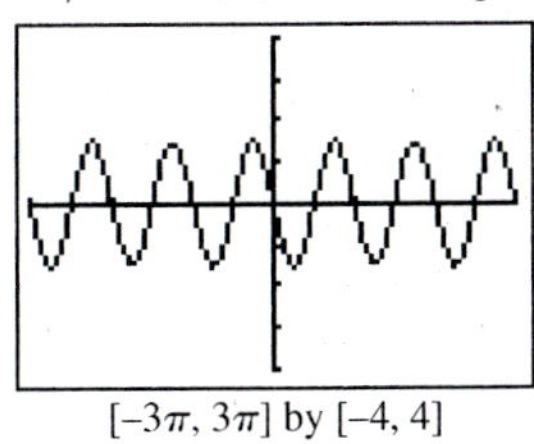

$[-3\pi, 3\pi]$ by $[-4, 4]$

Note: the frequency for each graph in #17–21 is $1/(2\pi)$.

17. Period 2π, amplitude $= 2$

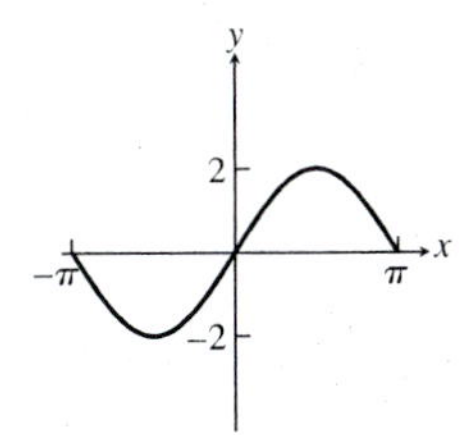

19. Period 2π, amplitude $= 3$

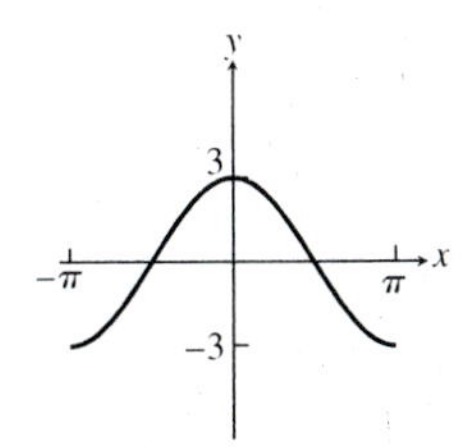

21. Period 2π, amplitude $= 0.5$

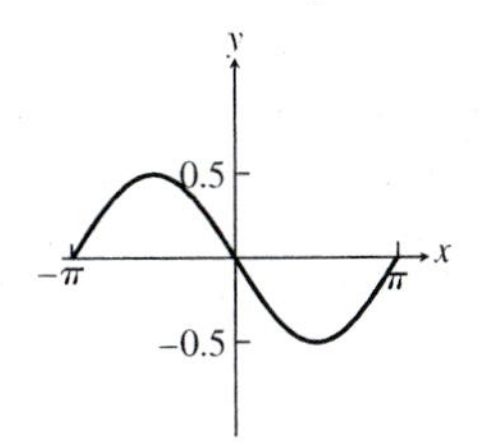

23. Period π, amplitude $= 5$, frequency $= 1/\pi$

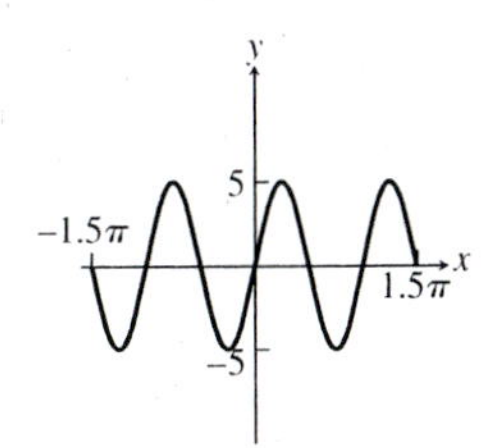

25. Period $2\pi/3$, amplitude $= 0.5$, frequency $= 3/2\pi$

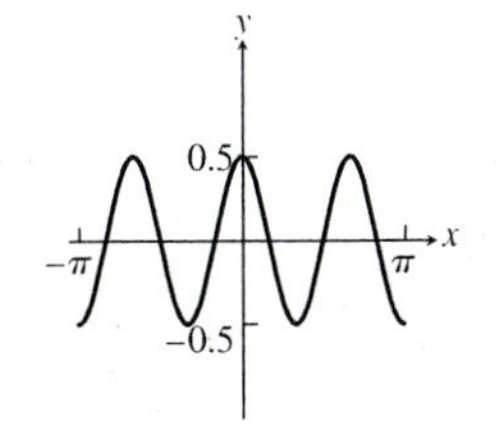

27. Period 8π, amplitude $= 4$, frequency $= 1/(8\pi)$

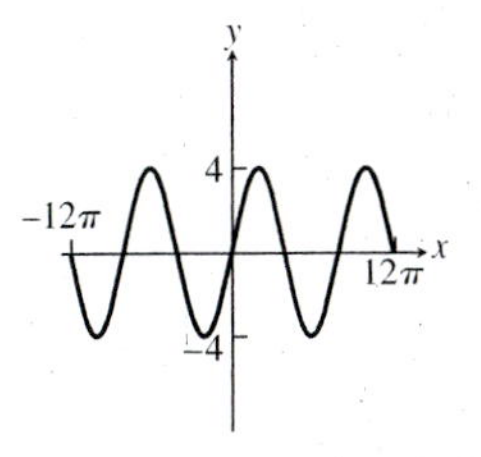

29. Period π; amplitude 1.5; $[-2\pi, 2\pi]$ by $[-2, 2]$

31. Period π; amplitude 3; $[-2\pi, 2\pi]$ by $[-4, 4]$

33. Period 6; amplitude 4; $[-3, 3]$ by $[-5, 5]$

35. Maximum: $2\left(\text{at } -\frac{3\pi}{2} \text{ and } \frac{\pi}{2}\right)$;

minimum: $-2\left(\text{at } -\frac{\pi}{2} \text{ and } \frac{3\pi}{2}\right)$.

Zeros: $0, \pm\pi, \pm 2\pi$.

37. Maximum: 1 (at $0, \pm\pi, \pm 2\pi$); minimum: $-1\left(\text{at } \pm\frac{\pi}{2} \text{ and } \pm\frac{3\pi}{2}\right)$. Zeros: $\pm\frac{\pi}{4}, \pm\frac{3\pi}{4}, \pm\frac{5\pi}{4}, \pm\frac{7\pi}{4}$.

39. Maximum: $1\left(\text{at } \pm\frac{\pi}{2}, \pm\frac{3\pi}{2}\right)$; minimum: -1 (at $0, \pm\pi, \pm 2\pi$).

Zeros: $\pm\frac{\pi}{4}, \pm\frac{3\pi}{4}, \pm\frac{5\pi}{4}, \pm\frac{7\pi}{4}$.

41. $y = \sin x$ has to be translated left or right by an odd multiple of π. One possibility is $y = \sin(x + \pi)$.

43. Starting from $y = \sin x$, horizontally shrink by $\frac{1}{3}$ and vertically shrink by 0.5. The period is $2\pi/3$. Possible window: $\left[-\frac{2\pi}{3}, \frac{2\pi}{3}\right]$ by $\left[-\frac{3}{4}, \frac{3}{4}\right]$.

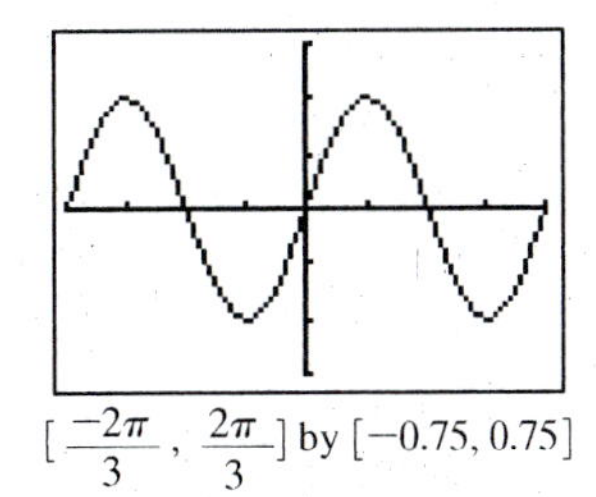

$\left[\frac{-2\pi}{3}, \frac{2\pi}{3}\right]$ by $[-0.75, 0.75]$

45. Starting from $y = \cos x$, horizontally stretch by 3, vertically shrink by $\frac{2}{3}$, reflect across x-axis. The period is 6π. Possible window: $[-6\pi, 6\pi]$ by $[-1, 1]$.

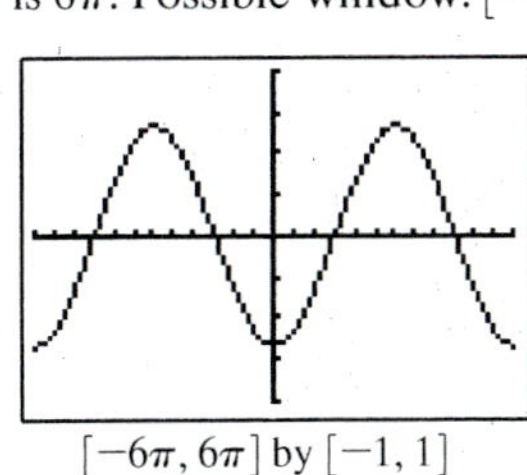

$[-6\pi, 6\pi]$ by $[-1, 1]$

47. Starting from $y = \cos x$, horizontally shrink by $\dfrac{3}{2\pi}$ and vertically stretch by 3. The period is 3. Possible window: $[-3, 3]$ by $[-3.5, 3.5]$.

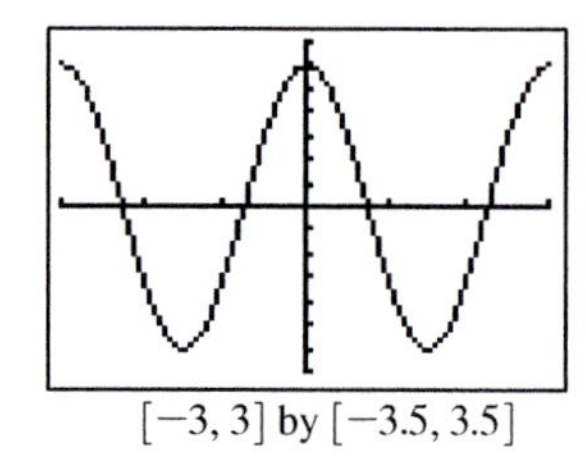

$[-3, 3]$ by $[-3.5, 3.5]$

49. Starting with y_1, vertically stretch by $\dfrac{5}{3}$.

51. Starting with y_1, horizontally shrink by $\dfrac{1}{2}$.

For #53–55, graph the functions or use facts about sine and cosine learned to this point.

53. (a) and (b)

55. (a) and (b) — both functions equal $\cos x$

In #57–59, for $y = a \sin(b(x - h))$, the amplitude is $|a|$, the period is $2\pi/|b|$, and the phase shift is h.

57. One possibility is $y = 3 \sin 2x$.

59. One possibility is $y = 1.5 \sin 12(x - 1)$.

61. Amplitude 2, period 2π, phase shift $\dfrac{\pi}{4}$, vertical translation 1 unit up.

63. Rewrite as $y = 5 \cos\left[3\left(x - \dfrac{\pi}{18}\right)\right] + 0.5$.

Amplitude 5, period $\dfrac{2\pi}{3}$, phase shift $\dfrac{\pi}{18}$, vertical translation $\dfrac{1}{2}$ units up.

65. Amplitude 2, period 1, phase shift 0, vertical translation 1 unit up.

67. Amplitude $\dfrac{7}{3}$, period 2π, phase shift $-\dfrac{5}{2}$, vertical translation 1 unit down.

69. $y = 2 \sin 2x$ $(a = 2, b = 2, h = 0, k = 0)$.

71. (a) There are two points of intersection in that interval.

(b) The coordinates are $(0, 1)$ and $(2\pi, 1.3^{-2\pi})$ $\approx (6.28, 0.19)$. In general, two functions intersect where $\cos x = 1$, i.e., $x = 2n\pi$, n an integer.

73. The height of the rider is modeled by $h = 30 - 25 \cos\left(\dfrac{2\pi}{40}t\right)$, where $t = 0$ corresponds to the time when the rider is at the low point. $h = 50$ when $\dfrac{-4}{5} = \cos\left(\dfrac{2\pi}{40}t\right)$. Then $\dfrac{2\pi}{40}t \approx 2.498$, so $t \approx 15.90$ sec.

75. (a) A model of the depth of the tide is $d = 2 \cos\left[\dfrac{\pi}{6.2}(t - 7.2)\right] + 9$, where t is hours since midnight. The first low tide is at 1:00 A.M. $(t = 1)$.

(b) At 4:00 A.M. $(t = 4)$: about 8.90 ft. At 9:00 P.M. $(t = 21)$: about 10.52 ft.

(c) 4:06 A.M. ($t = 4.1$ — halfway between 1:00 A.M. and 7:12 A.M.).

77. (a) The maximum d is approximately 21.4. The amplitude is $(21.4 - 7.2)/2 = 7.1$.

Scatterplot:

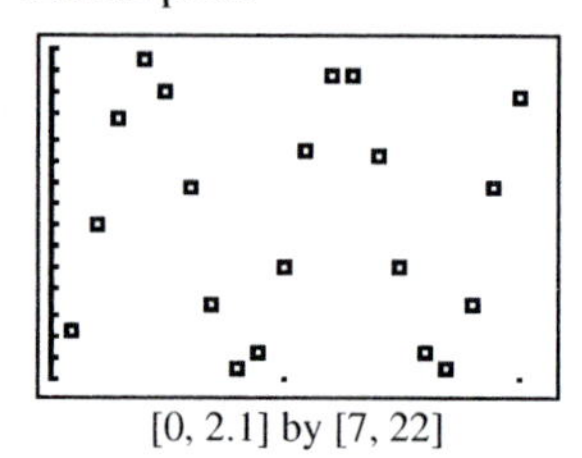

[0, 2.1] by [7, 22]

(b) The period appears to be slightly greater than 0.8, say 0.83.

(c) Since the function has a minimum at $t = 0$, we use an inverted cosine model: $d(t) = -7.1 \cos(2\pi t/0.83) + 14.3$.

(d)

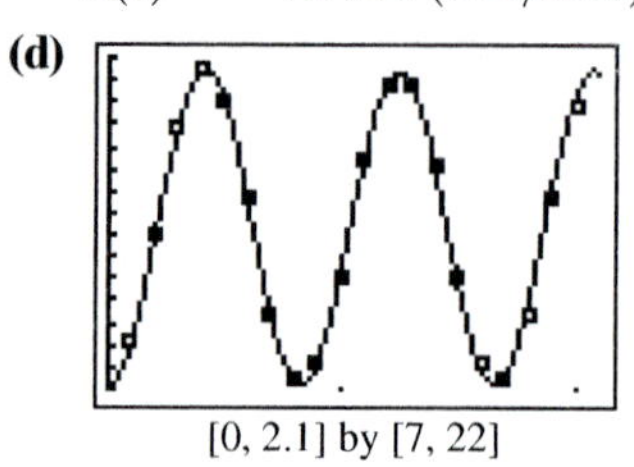

[0, 2.1] by [7, 22]

79. One possible answer is $T = 21.5 \cos\left(\dfrac{\pi}{6}(x - 7)\right) + 57.5$.

Start with the general form sinusoidal function $y = a \cos(b(x - h)) + k$, and find the variables a, b, h, and k as follows:

The amplitude is $|a| = \dfrac{79 - 36}{2} = 21.5$. We can arbitrarily choose to use the positive value, so $a = 21.5$.

The period is 12 months. $12 = \dfrac{2\pi}{|b|} \Rightarrow |b| = \dfrac{2\pi}{12} = \dfrac{\pi}{6}$.

Again, we can arbitrarily choose to use the positive value, so $b = \dfrac{\pi}{6}$.

The maximum is at month 7, so the phase shift $h = 7$.

The vertical shift $k = \dfrac{79 + 36}{2} = 57.5$.

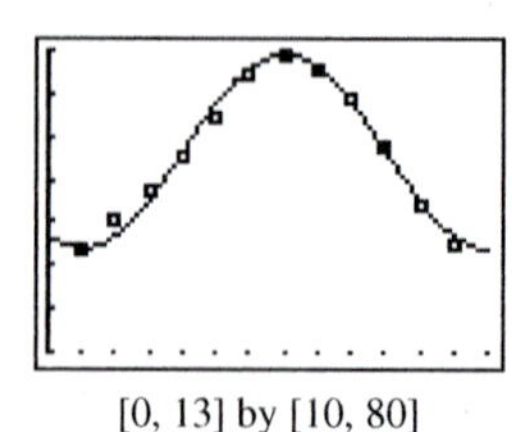

[0, 13] by [10, 80]

81. False. Since $y = \sin 2x$ is a horizontal *stretch* of $y = \sin 4x$ by a factor of 2, $y = \sin 2x$ has twice the period, not half. Remember, the period of $y = \sin bx$ is $2\pi/|b|$.

83. The minimum and maximum values differ by twice the amplitude. The answer is D.

85. For $f(x) = a \sin (bx + c)$, the period is $2\pi/|b|$, which here equals $2\pi/420 = \pi/210$. The answer is C.

87. (a)

$[-\pi, \pi]$ by $[-1.1, 1.1]$

(b) $\cos x \approx 0.0246x^4 + 0x^3 - 0.4410x^2 + 0x + 0.9703$. The coefficients given as "0" here may show up as very small numbers (e.g., 1.44×10^{-14}) on some calculators. Note that $\cos x$ is an even function, and only the even powers of x have nonzero (or a least "non-small") coefficients.

(c) The Taylor polynomial is
$1 - \frac{1}{2}x^2 + \frac{1}{24}x^4 = 1 - 0.5000x^2 + 0.04167x^4$; the coefficients are fairly similar.

89. (a) $p = \frac{2\pi}{524\pi} = \frac{1}{262}$ sec

(b) $f = 262 \frac{1}{\text{sec}}$ ("cycles per sec"), or 262 Hertz (Hz).

(c)

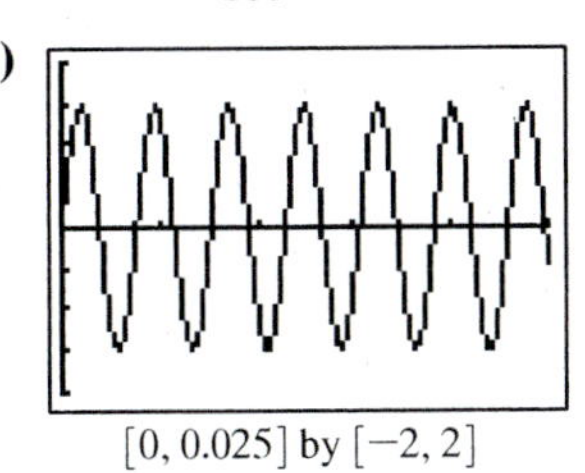

$[0, 0.025]$ by $[-2, 2]$

91. (a) $a - b$ must equal 1.

(b) $a - b$ must equal 2.

(c) $a - b$ must equal k.

For #93–95, note that A and C are one period apart. Meanwhile, B is located one-fourth of a period to the right of A, and the y-coordinate of B is the amplitude of the sinusoid.

93. The period of this function is π and the amplitude is 3. B and C are located (respectively) $\frac{\pi}{4}$ units and π units to the right of A. Therefore, $B = (0, 3)$ and $C = \left(\frac{3\pi}{4}, 0\right)$.

95. The period of this function is $\frac{2\pi}{3}$ and the amplitude is 2. B and C are located respectively $\frac{\pi}{6}$ units and $\frac{2\pi}{3}$ units to the right of A. Therefore, $B = \left(\frac{\pi}{4}, 2\right)$ and $C = \left(\frac{3\pi}{4}, 0\right)$

97. (a) Since $\sin(-\theta) = -\sin \theta$ (because sine is an odd function) $a \sin[-B(x - h)] + k = -a \sin[B(x - h)] + k$. Then any expression with a negative value of b can be rewritten as an expression of the same general form but with a positive coefficient in place of b.

(b) A sine graph can be translated a quarter of a period to the left to become a cosine graph of the same sinusoid. Thus $y = a \sin\left[b\left((x - h) + \frac{1}{4}\cdot\frac{2\pi}{b}\right)\right] + k$
$= a \sin\left[b\left(x - \left(h - \frac{\pi}{2b}\right)\right)\right] + k$ has the same graph as $y = a \cos[b(x - h)] + k$. We therefore choose $H = h - \frac{\pi}{2b}$.

(c) The angles $\theta + \pi$ and θ determine diametrically opposite points on the unit circle, so they have point symmetry with respect to the origin. The y-coordinates are therefore opposites, so $\sin(\theta + \pi) = -\sin \theta$.

(d) By the identity in (c). $y = a \sin[b(x - h) + \pi] + k$ $= -a \sin[b(x - h)] + k$. We therefore choose $H = h - \frac{\pi}{b}$.

(e) Part (b) shows how to convert $y = a \cos[b(x - h)] + k$ to $y = a \sin[b(x - H)] + k$, and parts (a) and (d) show how to ensure that a and b are positive.

■ Section 4.5 Graphs of Tangent, Cotangent, Secant, and Cosecant

Exploration 1

1. The graphs do not seem to intersect.

2. Set the expressions equal and solve for x:

$$-k \cos x = \sec x$$
$$-k \cos x = 1/\cos x$$
$$-k(\cos x)^2 = 1$$
$$(\cos x)^2 = -1/k$$

Since $k > 0$, this requires that the square of $\cos x$ be negative, which is impossible. This proves that there is no value of x for which the two functions are equal, so the graphs do not intersect.

Quick Review 4.5

1. Period π

3. Period 6π

For #5–7, recall that zeros of rational functions are zeros of the numerator, and vertical asymptotes are found at zeros of the denominator (provided the numerator and denominator have no common zeros).

5. Zero: 3. Asymptote: $x = -4$

7. Zero: -1. Asymptotes: $x = 2$ and $x = -2$

For #9, examine graphs to suggest the answer. Confirm by checking $f(-x) = f(x)$ for even functions and $f(-x) = -f(x)$ for odd functions.

9. Even: $(-x)^2 + 4 = x^2 + 4$

Section 4.5 Exercises

1. The graph of $y = 2 \csc x$ must be vertically stretched by 2 compared to $y = \csc x$, so $y_1 = 2 \csc x$ and $y_2 = \csc x$.

3. The graph of $y = 3 \csc 2x$ must be vertically stretched by 3 and horizontally shrunk by $\frac{1}{2}$ compared to $y = \csc x$, so $y_1 = 3 \csc 2x$ and $y_2 = \csc x$.

5. The graph of $y = \tan 2x$ results from shrinking the graph of $y = \tan x$ horizontally by a factor of $\frac{1}{2}$. There are vertical asymptotes at $x = \ldots, -\frac{\pi}{4}, \frac{\pi}{4}, \frac{3\pi}{4}, \ldots$.

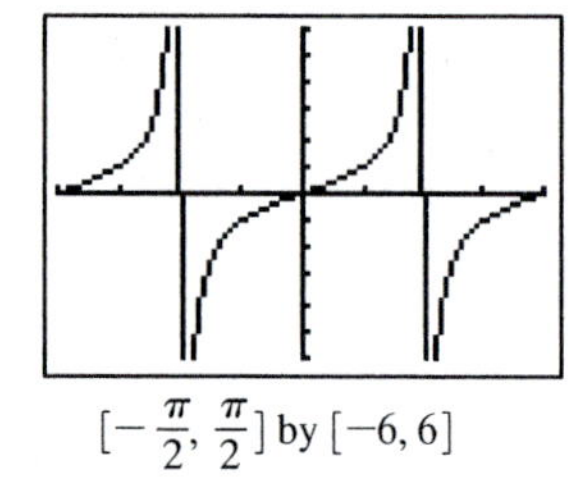

$[-\frac{\pi}{2}, \frac{\pi}{2}]$ by $[-6, 6]$

7. The graph of $y = \sec 3x$ results from shrinking the graph of $y = \sec x$ horizontally by a factor of $\frac{1}{3}$. There are vertical asymptotes at odd multiples of $\frac{\pi}{6}$.

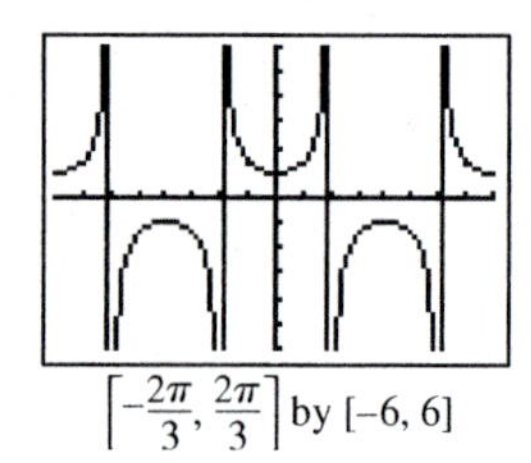

$\left[-\frac{2\pi}{3}, \frac{2\pi}{3}\right]$ by $[-6, 6]$

9. The graph of $y = 2 \cot 2x$ results from shrinking the graph of $y = \cot x$ horizontally by a factor of $\frac{1}{2}$ and stretching it vertically by a factor of 2. There are vertical asymptotes at $x = \ldots, -\pi, -\frac{\pi}{2}, 0, \frac{\pi}{2}, \ldots$.

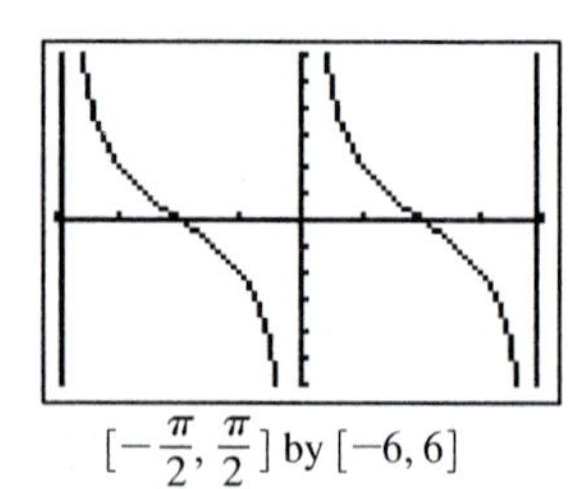

$[-\frac{\pi}{2}, \frac{\pi}{2}]$ by $[-6, 6]$

11. The graph of $y = \csc\left(\frac{x}{2}\right)$ results from horizontally stretching the graph of $y = \csc x$ by a factor of 2. There are vertical asymptotes at $x = \ldots, -4\pi, -2\pi, 0, 2\pi, \ldots$.

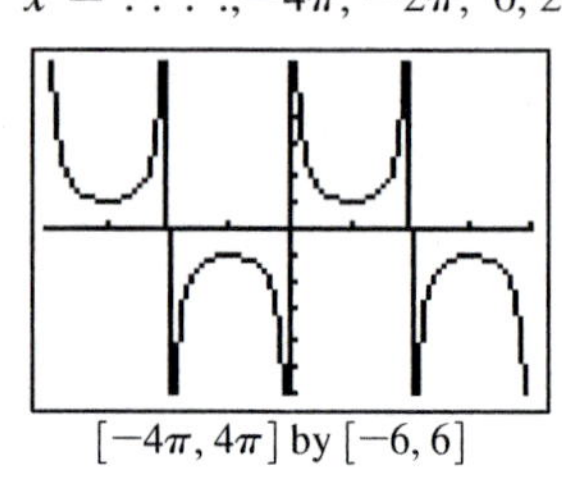

$[-4\pi, 4\pi]$ by $[-6, 6]$

13. Graph (a); Xmin $= -\pi$ and Xmax $= \pi$

15. Graph (c); Xmin $= -\pi$ and Xmax $= \pi$

17. Domain: All reals except integer multiples of π
Range: $(-\infty, \infty)$
Continuous on its domain
Decreasing on each interval in its domain
Symmetric with respect to the origin (odd)
Not bounded above or below
No local extrema
No horizontal asymptotes
Vertical asymptotes $x = k\pi$ for all integers k
End behavior: $\lim_{x\to\infty} \cot x$ and $\lim_{x\to-\infty} \cot x$ do not exist.

19. Domain: All reals except integer multiples of π
Range: $(-\infty, -1] \cup [1, \infty)$
Continuous on its domain
On each interval centered at $x = \frac{\pi}{2} + 2k\pi$ (k an integer):
decreasing on the left half of the interval and increasing on the right half
On each interval centered at $\frac{3\pi}{2} + 2k\pi$: increasing on the left half of the interval and decreasing on the right half
Symmetric with respect to the origin (odd)
Not bounded above or below
Local minimum 1 at each $x = \frac{\pi}{2} + 2k\pi$, local maximum -1 at each $x = \frac{3\pi}{2} + 2k\pi$, where k is an even integer in both cases
No horizontal asymptotes
Vertical asymptotes: $x = k\pi$ for all integers k
End behavior: $\lim_{x\to\infty} \csc x$ and $\lim_{x\to-\infty} \csc x$ do not exist.

21. Starting with $y = \tan x$, vertically stretch by 3.

23. Starting with $y = \csc x$, vertically stretch by 3.

25. Starting with $y = \cot x$, horizontally stretch by 2, vertically stretch by 3, and reflect across x-axis.

27. Starting with $y = \tan x$, horizontally shrink by $\frac{2}{\pi}$ and reflect across x-axis and shift up by 2 units.

29. $\sec x = 2$

$\cos x = \frac{1}{2}$

$x = \frac{\pi}{3}$

31. $\cot x = -\sqrt{3}$

$\tan x = -\frac{\sqrt{3}}{3}$

$x = \frac{5\pi}{6}$

33. $\csc x = 1$

$\sin x = 1$

$x = \frac{5\pi}{2}$

35. $\tan x = 1.3$

$x \approx 0.92$

37. $\cot x = -0.6$

$\tan x = -\frac{1}{0.6}$

$x \approx -1.03 + 2\pi$

≈ 5.25

39. $\csc x = 2$

$\sin x = \frac{1}{2}$

$x \approx 0.52$ or

$x \approx \pi - 0.52$

≈ 2.62

41. (a) One explanation: If O is the origin, the right triangles with hypotenuses $\overline{OP_1}$ and $\overline{OP_2}$, and one leg (each) on the x-axis, are congruent, so the legs have the same lengths. These lengths give the magnitudes of the coordinates of P_1 and P_2; therefore, these coordinates differ only in sign. Another explanation: The reflection of point (a, b) across the origin is $(-a, -b)$.

(b) $\tan t = \frac{\sin t}{\cos t} = \frac{b}{a}$.

(c) $\tan(t - \pi) = \frac{\sin(t - \pi)}{\cos(t - \pi)} = \frac{-b}{-a} = \frac{b}{a} = \tan t$.

(d) Since points on opposite sides of the unit circle determine the same tangent ratio, $\tan(t \pm \pi) = \tan t$ for all numbers t in the domain. Other points on the unit circle yield triangles with different tangent ratios, so no smaller period is possible.

(e) The tangent function repeats every π units; therefore, so does its reciprocal, the cotangent (see also #43).

43. For any x, $\left(\frac{1}{f}\right)(x + p) = \frac{1}{f(x + p)} = \frac{1}{f(x)} = \left(\frac{1}{f}\right)(x)$. This is not true for any smaller value of p, since this is the smallest value that works for f.

45. (a) $d = 350 \sec x = \frac{350}{\cos x}$ ft

(b) $d \approx 16{,}831$ ft

For #47–49, the equations can be rewritten (as shown), but generally are easiest to solve graphically.

47. $\sin^2 x = \cos x$; $x \approx \pm 0.905$

49. $\cos^2 x = \frac{1}{5}$; $x \approx \pm 1.107$ or $x \approx \pm 2.034$

51. False. $f(x) = \tan x$ is increasing only over intervals on which it is defined, that is, intervals bounded by consecutive asymptotes.

53. The cotangent curves are shaped like the tangent curves, but they are mirror images. The reflection of $\tan x$ in the x-axis is $-\tan x$. The answer is A.

55. $y = k/\sin x$ and the range of $\sin x$ is $[-1, 1]$. The answer is D.

57. On the interval $[-\pi, \pi]$, $f > g$ on about $(-0.44, 0) \cup (0.44, \pi)$.

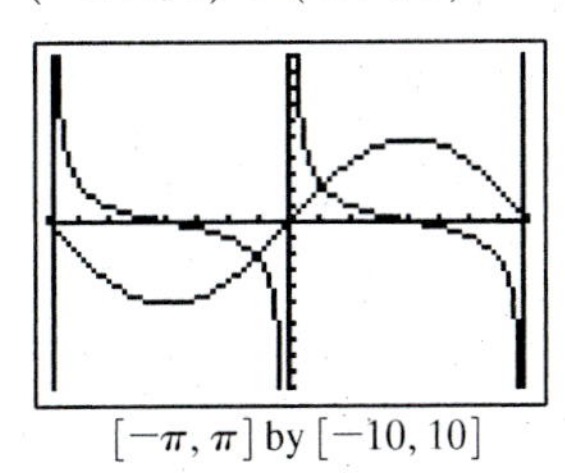

$[-\pi, \pi]$ by $[-10, 10]$

59. $\cot x$ is not defined at 0; the definition of "increasing on (a, b)" requires that the function be defined everywhere in (a, b). Also, choosing $a = -\pi/4$ and $b = \pi/4$, we have $a < b$ but $f(a) = 1 > f(b) = -1$.

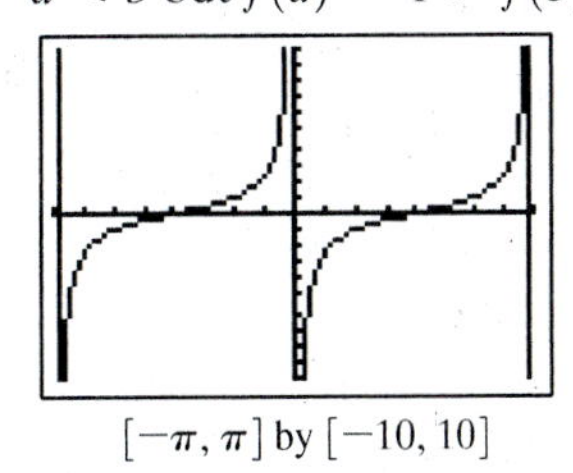

$[-\pi, \pi]$ by $[-10, 10]$

61. $\csc x = \sec\left(x - \frac{\pi}{2}\right)$ (or $\csc x = \sec\left(x - \left(\frac{\pi}{2} + n\pi\right)\right)$ for any integer n) This is a translation to the right of $\frac{\pi}{2}\left(\text{or } \frac{\pi}{2} + n\pi\right)$ units.

63. $d = 30 \sec x = \frac{30}{\cos x}$

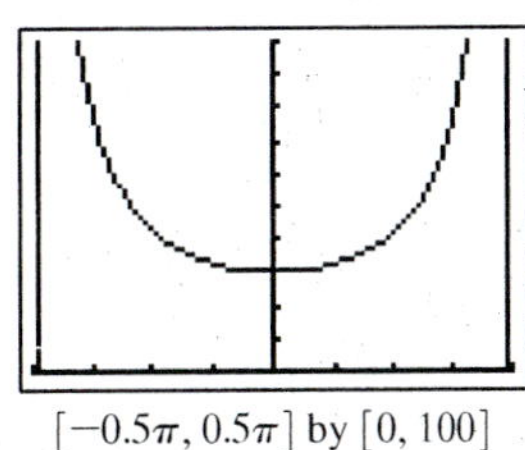

$[-0.5\pi, 0.5\pi]$ by $[0, 100]$

65. $0.058 \frac{\text{N}}{\text{m}} = \frac{1}{2}(1.5 \text{ m})\left(1050 \frac{\text{kg}}{\text{m}^3}\right)\left(9.8 \frac{\text{m}}{\text{sec}^2}\right)$

$(4.7 \times 10^{-6} \text{ m})\sec\phi \approx 0.03627 \sec\phi \frac{\text{kg}}{\text{sec}^2}$, so

$\sec\phi \approx 1.5990$, and $\phi \approx 0.8952$ radians $\approx 51.29°$.

Section 4.6 Graphs of Composite Trigonometric Functions

Exploration 1

$y = 3 \sin x + 2 \cos x$

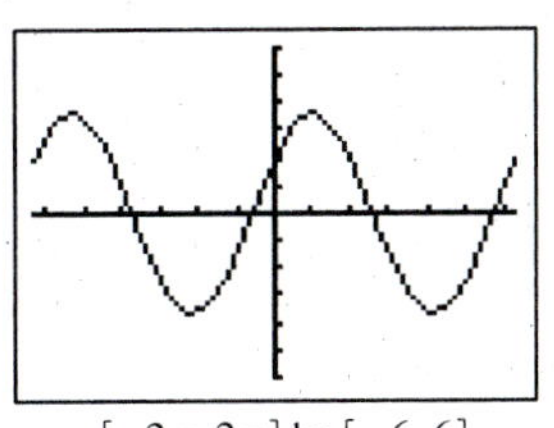

$[-2\pi, 2\pi]$ by $[-6, 6]$

Sinusoid

$y = 2 \sin x - 3 \cos x$

$[-2\pi, 2\pi]$ by $[-6, 6]$

Sinusoid

$y = 2\sin 3x - 4\cos 2x$

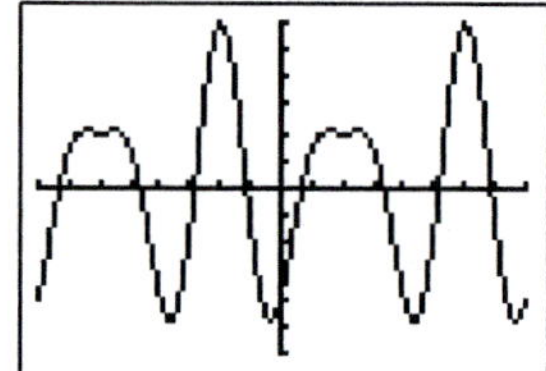

$[-2\pi, 2\pi]$ by $[-6, 6]$
Not a Sinusoid

$y = 2\sin(5x + 1) - 5\cos 5x$

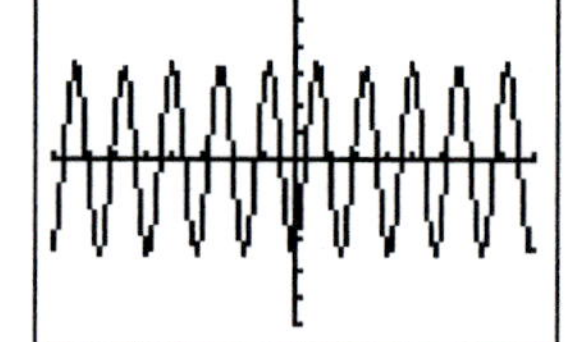

$[-2\pi, 2\pi]$ by $[-6, 6]$
Sinusoid

$y = \cos\left(\frac{7x - 2}{5}\right) + \sin\left(\frac{7x}{5}\right)$

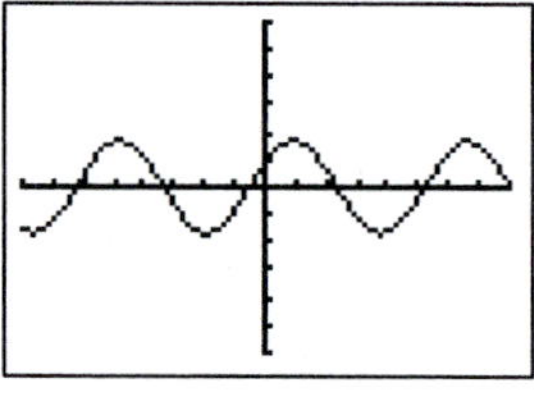

$[-2\pi, 2\pi]$ by $[-6, 6]$
Sinusoid

$y = 3\cos 2x + 2\sin 7x$

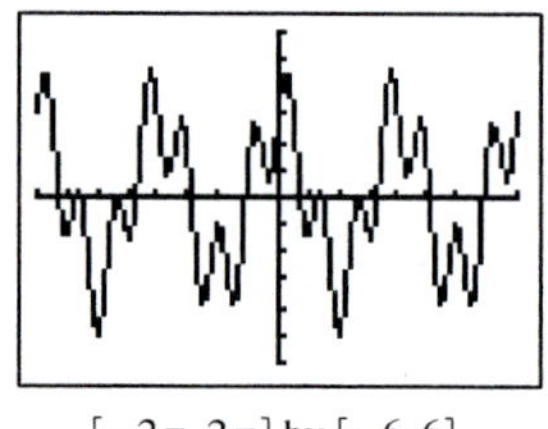

$[-2\pi, 2\pi]$ by $[-6, 6]$
Not a Sinusoid

Quick Review 4.6

1. Domain: $(-\infty, \infty)$; range: $[-3, 3]$
3. Domain: $[1, \infty)$; range: $[0, \infty)$
5. Domain: $(-\infty, \infty)$; range: $[-2, \infty)$
7. As $x \to -\infty$, $f(x) \to \infty$; as $x \to \infty$, $f(x) \to 0$.
9. $f \circ g(x) = (\sqrt{x})^2 - 4 = x - 4$, domain: $[0, \infty)$.
   $g \circ f(x) = \sqrt{x^2 - 4}$, domain: $(-\infty, -2] \cup [2, \infty)$.

Section 4.6 Exercises

1. Periodic.

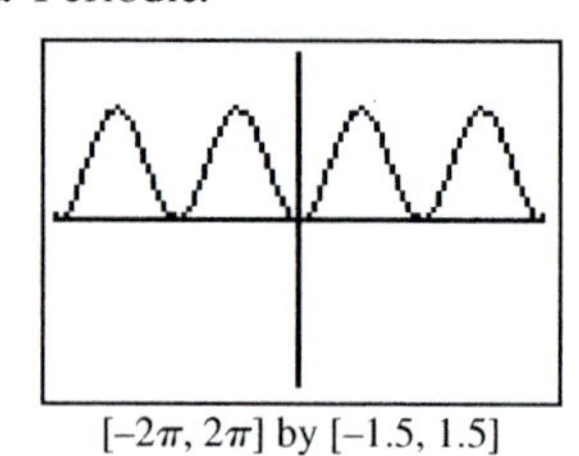

$[-2\pi, 2\pi]$ by $[-1.5, 1.5]$

3. Not periodic.

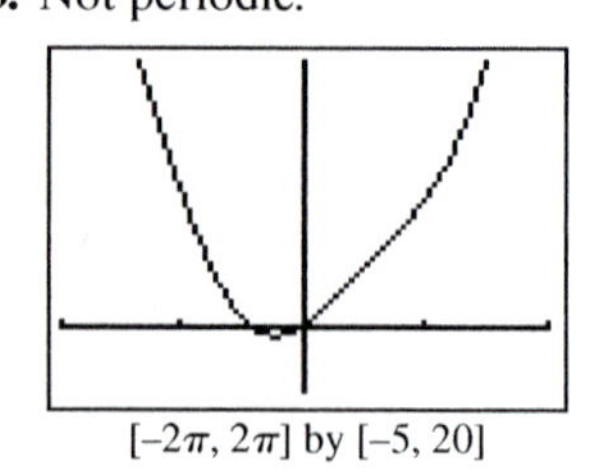

$[-2\pi, 2\pi]$ by $[-5, 20]$

5. Not periodic.

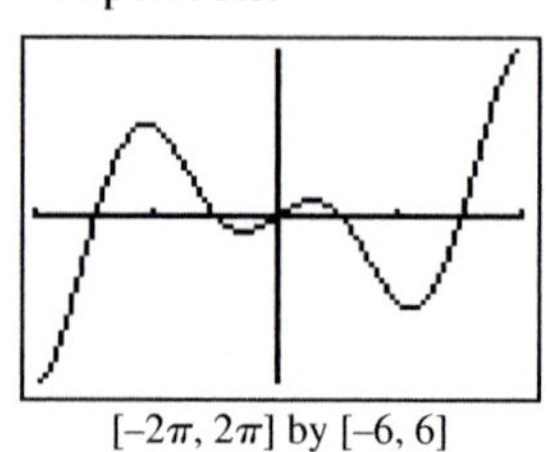

$[-2\pi, 2\pi]$ by $[-6, 6]$

7. Periodic.

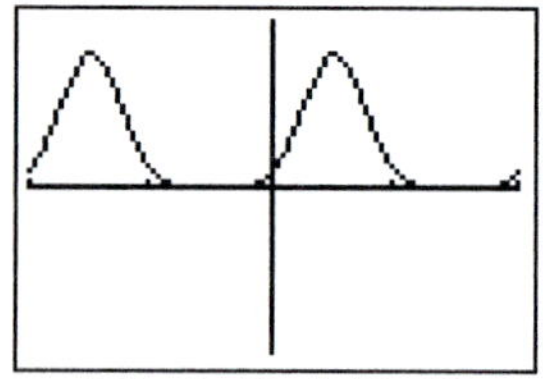

$[-2\pi, 2\pi]$ by $[-10, 10]$

9. Since the period of $\cos x$ is 2π, we have $\cos^2(x + 2\pi) = (\cos(x + 2\pi))^2 = (\cos x)^2 = \cos^2 x$. The period is therefore an exact divisor of 2π, and we see graphically that it is π. A graph for $-\pi \le x \le \pi$ is shown:

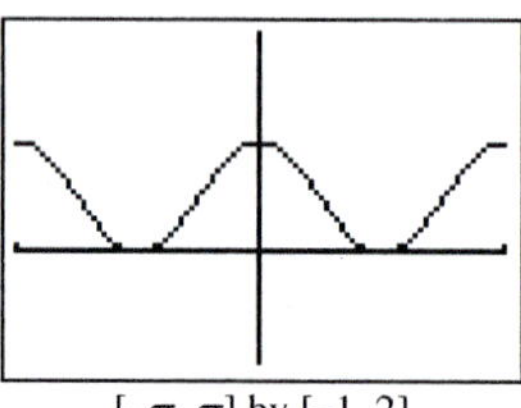

$[-\pi, \pi]$ by $[-1, 2]$

11. Since the period of $\cos x$ is 2π, we have $\sqrt{\cos^2(x + 2\pi)} = \sqrt{(\cos(x + 2\pi))^2} = \sqrt{(\cos x)^2} = \sqrt{\cos^2 x}$. The period is therefore an exact divisor of 2π, and we see graphically that it is π. A graph for $-\pi \le x \le \pi$ is shown:

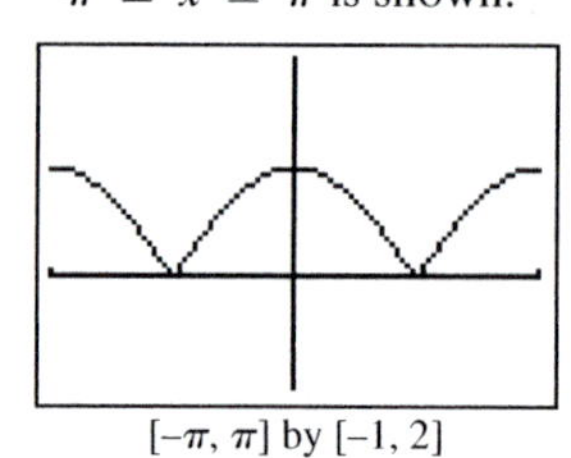

$[-\pi, \pi]$ by $[-1, 2]$

13. Domain: $(-\infty, \infty)$. Range: $[0, 1]$.

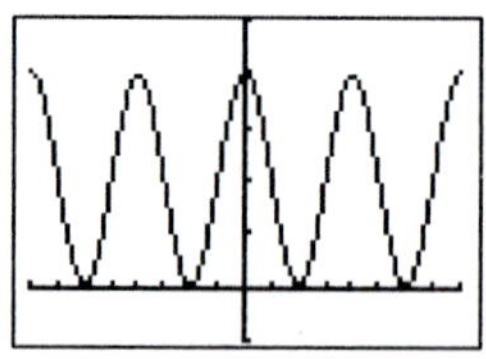

$[-2\pi, 2\pi]$ by $[-0.25, 1.25]$

15. Domain: all $x \neq n\pi$, n an integer. Range: $[0, \infty)$.

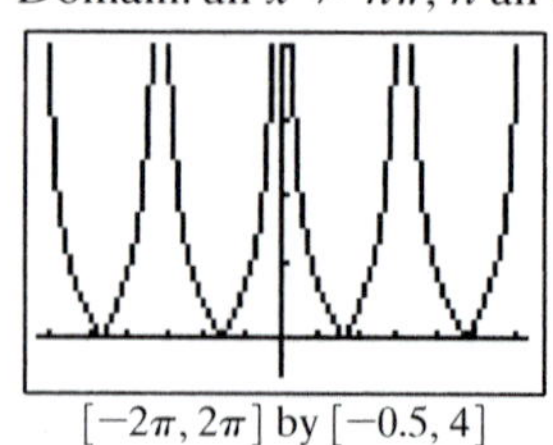

$[-2\pi, 2\pi]$ by $[-0.5, 4]$

17. Domain: all $x \neq \frac{\pi}{2} + n\pi$, n an integer. Range: $(-\infty, 0]$.

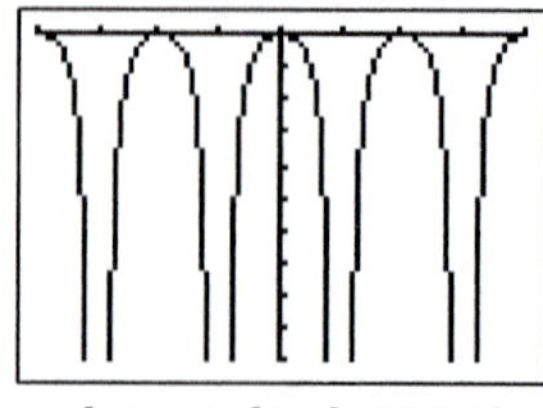

$[-2\pi, 2\pi]$ by $[-10, 0.2]$

In #19–21, the linear equations are found by setting the cosine term equal to ± 1.

19. $2x - 1 \leq y \leq 2x + 1$

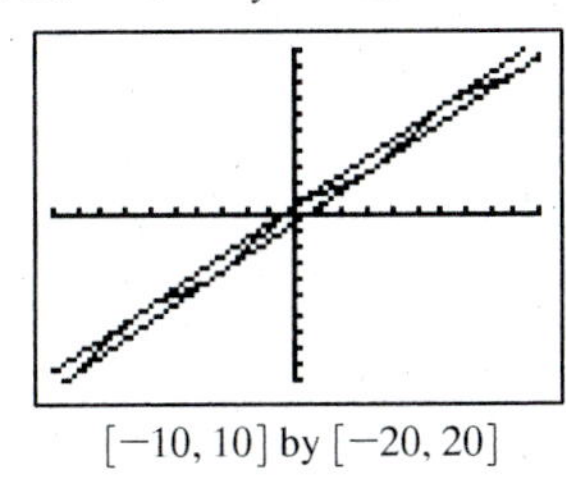

$[-10, 10]$ by $[-20, 20]$

21. $1 - 0.3x \leq y \leq 3 - 0.3x$

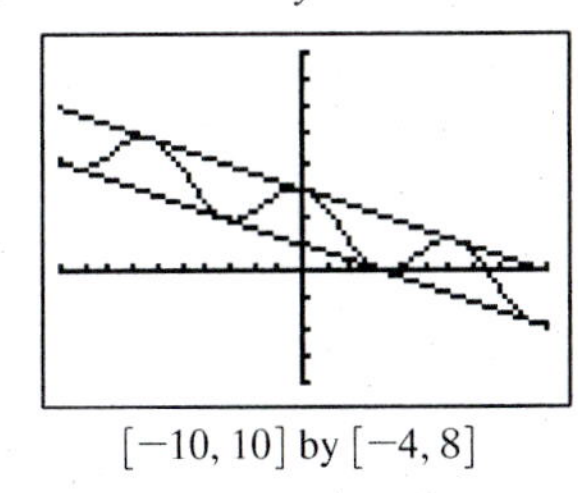

$[-10, 10]$ by $[-4, 8]$

For #23–27, the function $y_1 + y_2$ is a sinusoid if both y_1 and y_2 are sine or cosine functions with the same period.

23. Yes (period 2π)

25. Yes (period 2)

27. No

For #29–33, graph the function. Estimate a as the amplitude of the graph (i.e., the height of the maximum). Notice that the value of b is always the coefficient of x in the original functions. Finally, note that $a \sin[b(x - h)] = 0$ when $x = h$, so estimate h using a zero of $f(x)$ where $f(x)$changes from negative to positive.

29. $A \approx 3.61$, $b = 2$, and $h \approx 0.49$, so $f(x) \approx 3.61 \sin[2(x - 0.49)]$.

31. $A \approx 2.24$, $b = \pi$, and $h \approx 0.35$, so $f(x) \approx 2.24 \sin[\pi(x - 0.35)]$.

33. $A \approx 2.24$, $b = 1$, and $h \approx -1.11$, so $f(x) \approx 2.24 \sin(x + 1.11)$.

35. The period is 2π.

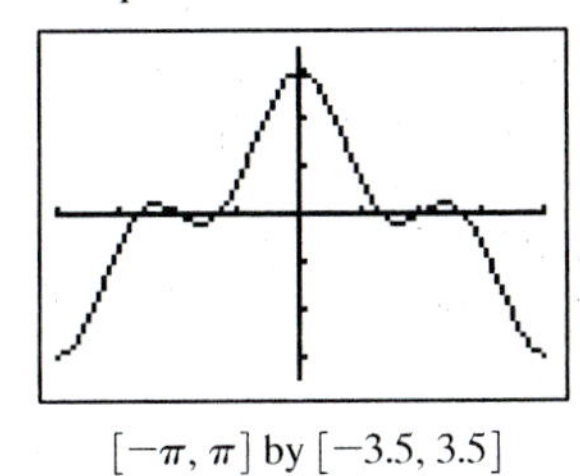

$[-\pi, \pi]$ by $[-3.5, 3.5]$

37. The period is 2π.

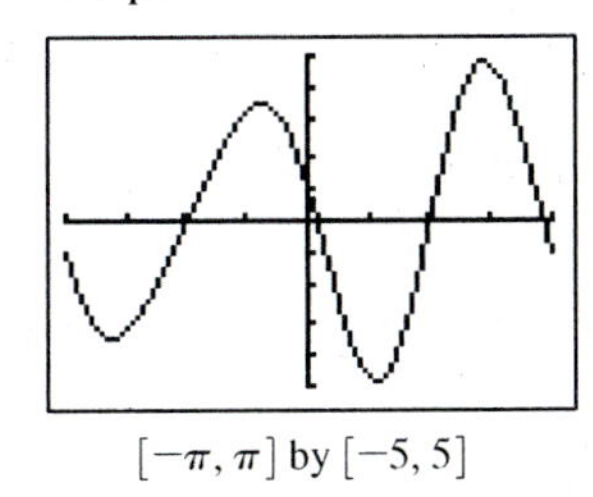

$[-\pi, \pi]$ by $[-5, 5]$

39. (a)

41. (c)

43. The damping factor is e^{-x}, which goes to zero as x gets large. So damping occurs as $x \to \infty$.

45. The amplitude, $\sqrt{5}$, is constant. So there is no damping.

47. The damping factor is x^3, which goes to zero as x goes to zero. So damping occurs as $x \to 0$.

49. f oscillates up and down between 1.2^{-x} and -1.2^{-x}. As $x \to \infty$, $f(x) \to 0$.

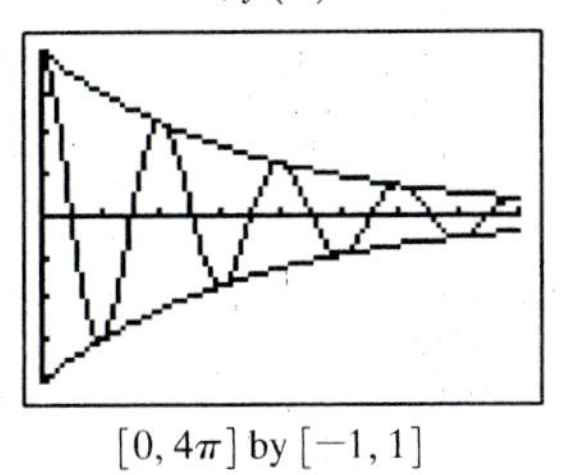

$[0, 4\pi]$ by $[-1, 1]$

51. f oscillates up and down between $\frac{1}{x}$ and $-\frac{1}{x}$. As $x \to \infty$, $f(x) \to 0$.

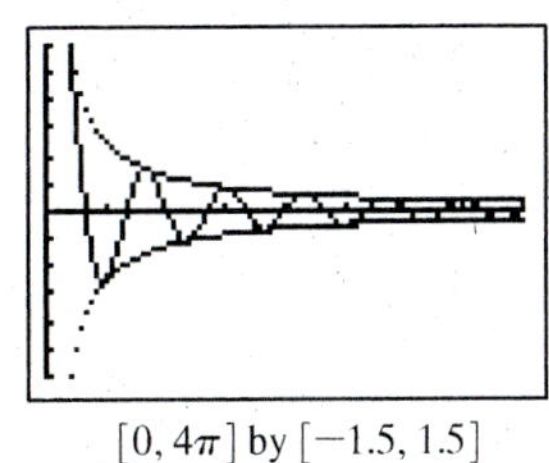

$[0, 4\pi]$ by $[-1.5, 1.5]$

53. Period 2π: $\sin[3(x + 2\pi)] + 2\cos[2(x + 2\pi)] = \sin(3x + 6\pi) + 2\cos(2x + 4\pi) = \sin 3x + 2\cos 2x$. The graph, shows that no $p < 2\pi$ could be the period.

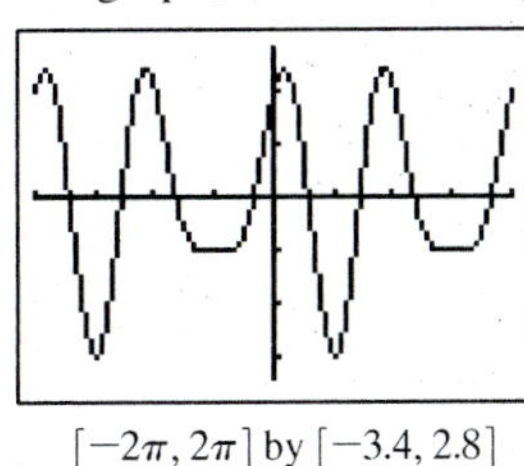

$[-2\pi, 2\pi]$ by $[-3.4, 2.8]$

55. Period 2π:
$2\sin[3(x + 2\pi) + 1] - \cos[5(x + 2\pi) - 1]$
$= 2\sin(3x + 1 + 6\pi) - \cos(5x - 1 + 10\pi)$
$= 2\sin(3x + 1) - \cos(5x - 1)$. The graph, shows that no $p < 2\pi$ could be the period.

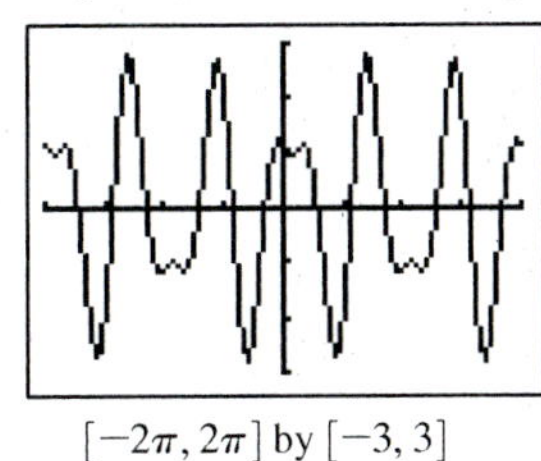

$[-2\pi, 2\pi]$ by $[-3, 3]$

57. Period 2π: $\left|\sin\left[\frac{1}{2}(x + 2\pi)\right]\right| + 2 =$

$\left|\sin\left(\frac{1}{2}x + \pi\right)\right| + 2 = \left|-\sin\frac{1}{2}x\right| + 2 = \left|\sin\frac{1}{2}x\right| + 2.$

The graph, shows that no $p < 2\pi$ could be the period.

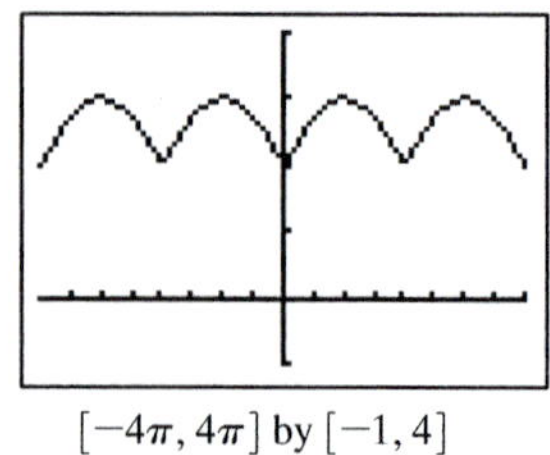

$[-4\pi, 4\pi]$ by $[-1, 4]$

59. Not periodic

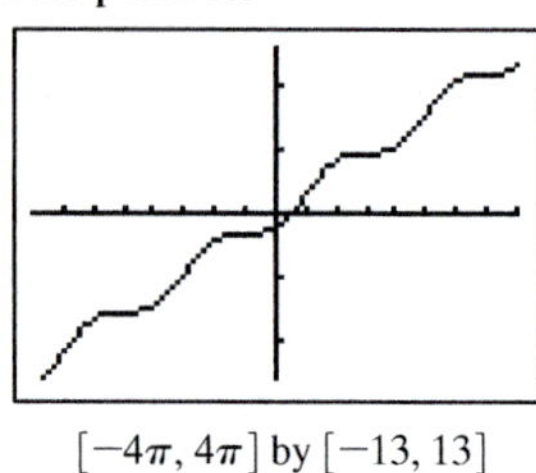

$[-4\pi, 4\pi]$ by $[-13, 13]$

61. Not periodic

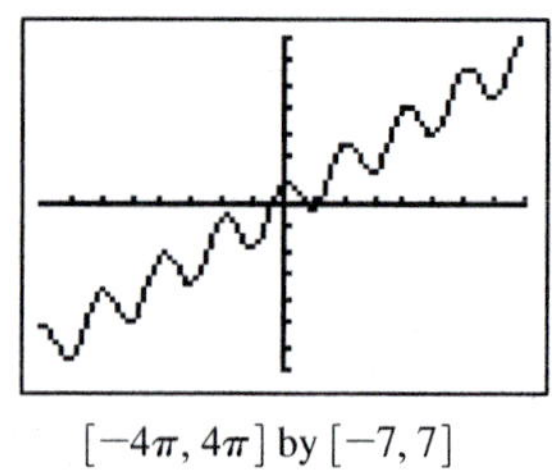

$[-4\pi, 4\pi]$ by $[-7, 7]$

For #63–69, graphs may be useful to suggest the domain and range.

63. There are no restrictions on the value of x, so the domain is $(-\infty, \infty)$. Range: $(-\infty, \infty)$.

65. There are no restrictions on the value of x, so the domain is $(-\infty, \infty)$. Range: $[1, \infty)$.

67. $\sin x$ must be nonnegative, so the domain is $\cdots \cup [-2\pi, -\pi] \cup [0, \pi] \cup [2\pi, 3\pi] \cup \cdots$; that is, all x with $2n\pi \le x \le (2n + 1)\pi$, n an integer. Range: $[0, 1]$.

69. There are no restrictions on the value of x, since $|\sin x| \ge 0$, so the domain is $(-\infty, \infty)$. Range: $[0, 1]$.

71. (a)

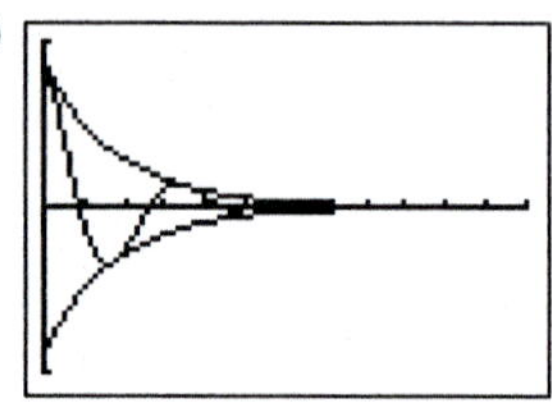

$[0, 12]$ by $[-0.5, 0.5]$

(b) For $t > 0.51$ (approximately).

73. No. This is suggested by a graph of $y = \sin x^3$; there is no other section of the graph that looks like the section between -1 and 1. In particular, there is only one zero of the function in that interval (at $x = 0$); nowhere else can we find an interval this long with only one zero.

75. (a) — this is obtained by adding x to all parts of the inequality $-1 \le \sin x \le 1$. In the second, after subtracting x from both sides, we are left with $-\sin x \le \sin x$, which is false when $\sin x$ is negative.

77. Graph (d), shown on $[-2\pi, 2\pi]$ by $[-4, 4]$

79. Graph (b), shown on $[-2\pi, 2\pi]$ by $[-4, 4]$

81. False. The behavior near zero, with a relative minimum of 0 at $x = 0$, is not repeated anywhere else.

83. The negative portions of the graph of $y = \sin x$ are reflected in the x-axis for $y = |\sin x|$. This halves the period. The answer is B.

85. $f(-x) = -x + \sin(-x) = -x - \sin x = -f(x)$. The answer is D.

87. (a) Answers will vary — for example,

on a TI-81: $\frac{\pi}{47.5} = 0.0661\ldots \approx 0.07$;

on a TI-82: $\frac{\pi}{47} = 0.0668\ldots \approx 0.07$;

on a TI-85: $\frac{\pi}{63} = 0.0498\ldots \approx 0.05$;

on a TI-92: $\frac{\pi}{119} = 0.0263\ldots \approx 0.03$.

(b) Period: $p = \pi/125 = 0.0251\ldots$. For any of the TI graphers, there are from 1 to 3 cycles between each pair of pixels; the graphs produced are therefore inaccurate, since so much detail is lost.

89. Domain: $(-\infty, \infty)$. Range: $[-1, 1]$. Horizontal asymptote: $y = 1$. Zeros at $\ln\left(\frac{\pi}{2} + n\pi\right)$, n a non-negative integer.

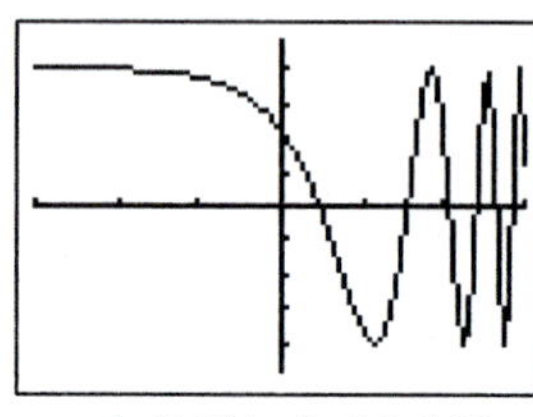

$[-3, 3]$ by $[-1.2, 1.2]$

91. Domain: $[0, \infty)$. Range: $(-\infty, \infty)$. Zeros at $n\pi$, n a nonnegative integer.

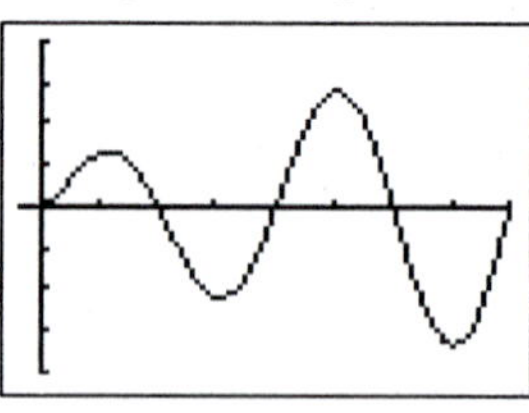

$[-0.5, 4\pi]$ by $[-4, 4]$

93. Domain: $(-\infty, 0) \cup (0, \infty)$. Range: approximately $[-0.22, 1)$. Horizontal asymptote: $y = 0$. Zeros at $n\pi$, n a non-zero integer.

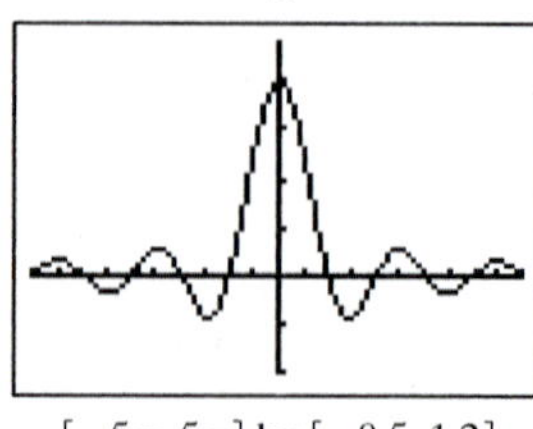

$[-5\pi, 5\pi]$ by $[-0.5, 1.2]$

95. Domain: $(-\infty, 0) \cup (0, \infty)$. Range: approximately $[-0.22, 1)$. Horizontal asymptote: $y = 1$. Zeros at $\frac{1}{n\pi}$, n a non-zero integer.

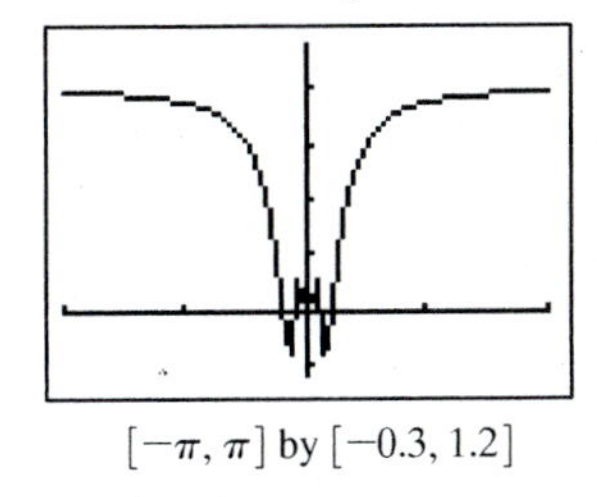

$[-\pi, \pi]$ by $[-0.3, 1.2]$

Section 4.7 Inverse Trigonometric Functions

Exploration 1

1. $\tan\theta = \frac{x}{1} = x$
2. $\tan^{-1} x = \tan^{-1}\left(\frac{x}{1}\right) = \theta$
3. $\sqrt{1 + x^2}$ (by the Pythagorean theorem)
4. $\sin(\tan^{-1}(x)) = \sin(\theta) = \frac{x}{\sqrt{1 + x^2}}$
5. $\sec(\tan^{-1}(x)) = \sec(\theta) = \sqrt{1 + x^2}$
6. The hypotenuse is positive in either quadrant. The ratios in the six basic trig functions are the same in every quadrant, so the functions are still valid regardless of the sign of x. (Also, the sign of the answer in (4) is negative, as it should be, and the sign of the answer in (5) is negative, as it should be.)

Quick Review 4.7

1. $\sin x$: positive; $\cos x$: positive; $\tan x$: positive

3. $\sin x$: negative; $\cos x$: negative; $\tan x$: positive

5. $\sin\frac{\pi}{6} = \frac{1}{2}$ **7.** $\cos\frac{2\pi}{3} = -\frac{1}{2}$

9. $\sin\frac{-\pi}{6} = -\frac{1}{2}$

Section 4.7 Exercises

For #1–11, keep in mind that the inverse sine and inverse tangent functions return values in $\left[-\frac{\pi}{2}, \frac{\pi}{2}\right]$, and the inverse cosine function gives values in $[0, \pi]$. A calculator may also be useful to suggest the exact answer. (A useful trick is to compute, e.g., $\sin^{-1}(\sqrt{3}/2)\pi$ and observe that this is ≈ 0.333, suggesting the answer $\pi/3$.)

1. $\sin^{-1}\left(\frac{\sqrt{3}}{2}\right) = \frac{\pi}{3}$ **3.** $\tan^{-1}(0) = 0$

5. $\cos^{-1}\left(\frac{1}{2}\right) = \frac{\pi}{3}$ **7.** $\tan^{-1}(-1) = -\frac{\pi}{4}$

9. $\sin^{-1}\left(-\frac{1}{\sqrt{2}}\right) = -\frac{\pi}{4}$ **11.** $\cos^{-1}(0) = \frac{\pi}{2}$

13. approx. 21.22° **15.** approx. −85.43°

17. approx. 1.172 **19.** approx. −0.478

21. $y = \tan^{-1}(x^2)$ is equivalent to $\tan y = x^2$, $-\pi/2 < y < \pi/2$. For x^2 to get very large, y has to approach $\pi/2$. So $\lim_{x\to\infty} \tan^{-1}(x^2) = \pi/2$ and $\lim_{x\to-\infty} \tan^{-1}(x^2) = \pi/2$.

23. $\cos\left(\sin^{-1}\frac{1}{2}\right) = \cos\frac{\pi}{6} = \frac{\sqrt{3}}{2}$

25. $\sin^{-1}\left(\cos\frac{\pi}{4}\right) = \sin^{-1}\left(\frac{\sqrt{2}}{2}\right) = \frac{\pi}{4}$

27. $\cos\left(2\sin^{-1}\frac{1}{2}\right) = \cos\left(2\cdot\frac{\pi}{6}\right) = \frac{1}{2}$

29. $\arcsin\left(\cos\frac{\pi}{3}\right) = \arcsin\frac{1}{2} = \frac{\pi}{6}$

31. $\cos(\tan^{-1}\sqrt{3}) = \cos\left(\frac{\pi}{3}\right) = \frac{1}{2}$

33. Domain: $[-1, 1]$
Range: $[-\pi/2, \pi/2]$
Continuous
Increasing
Symmetric with respect to the origin (odd)
Bounded
Absolute maximum of $\pi/2$, absolute minimum of $-\pi/2$
No asymptotes
No end behavior (bounded domain)

35. Domain: $(-\infty, \infty)$
Range: $(-\pi/2, \pi/2)$
Continuous
Increasing
Symmetric with respect to the origin (odd)
Bounded
No local extrema
Horizontal asymptotes: $y = \pi/2$ and $y = -\pi/2$
End behavior: $\lim_{x\to\infty} \tan^{-1} x = \pi/2$ and $\lim_{x\to-\infty} \tan^{-1} x = -\pi/2$

37. Domain: $\left[-\frac{1}{2}, \frac{1}{2}\right]$. Range: $\left[-\frac{\pi}{2}, \frac{\pi}{2}\right]$. Starting from $y = \sin^{-1} x$, horizontally shrink by $\frac{1}{2}$.

39. Domain: $(-\infty, \infty)$. Range: $\left(-\frac{5\pi}{2}, \frac{5\pi}{2}\right)$. Starting from $y = \tan^{-1} x$, horizontally stretch by 2 and vertically stretch by 5 (either order).

41. First set $\theta = \sin^{-1} x$ and solve $\sin\theta = 1$, yielding $\theta = \frac{\pi}{2} + 2n\pi$ for integers n. Since $\theta = \sin^{-1} x$ must be in $\left[-\frac{\pi}{2}, \frac{\pi}{2}\right]$, we have $\sin^{-1} x = \frac{\pi}{2}$, so $x = 1$.

43. Divide both sides of the equation by 2, leaving $\sin^{-1} x = \frac{1}{2}$, so $x = \sin\frac{1}{2} \approx 0.479$.

45. For any x in $[0, \pi]$, $\cos(\cos^{-1}x)$, $= x$. Hence, $x = \frac{1}{3}$.

47. Draw a right triangle with horizontal leg 1, vertical leg x (if $x > 0$, draw the vertical leg "up"; if $x < 0$, draw it down), and hypotenuse $\sqrt{1+x^2}$. The acute angle adjacent to the leg of length 1 has measure $\theta = \tan^{-1} x$ (take $\theta < 0$ if $x < 0$), so $\sin\theta = \sin(\tan^{-1} x) = \dfrac{x}{\sqrt{1+x^2}}$.

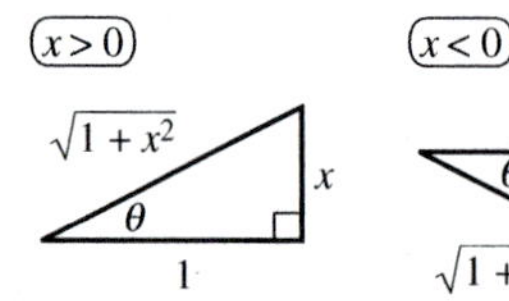

49. Draw a right triangle with horizontal leg $\sqrt{1-x^2}$, vertical leg x (if $x > 0$, draw the vertical leg "up"; if $x < 0$, draw it down), and hypotenuse 1. The acute angle adjacent to the horizontal leg has measure $\theta = \arcsin x$ (take $\theta < 0$ if $x < 0$), so

$\tan\theta = \tan(\arcsin x) = \dfrac{x}{\sqrt{1-x^2}}$.

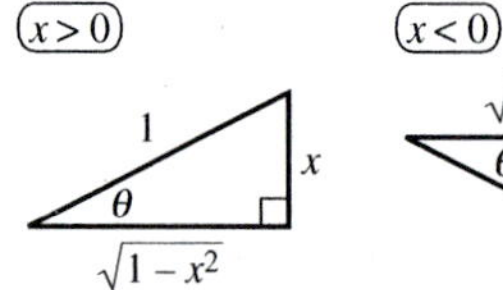

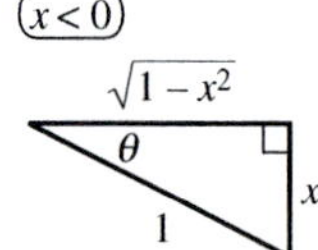

51. Draw a right triangle with horizontal leg 1, vertical leg $2x$ (up or down as $x > 0$ or $x < 0$), and hypotenuse $\sqrt{1+4x^2}$. The acute angle adjacent to the leg of length 1 has measure $\theta = \arctan 2x$ (take $\theta < 0$ if $x < 0$), so

$\cos\theta = \cos(\arctan 2x) = \dfrac{1}{\sqrt{1+4x^2}}$.

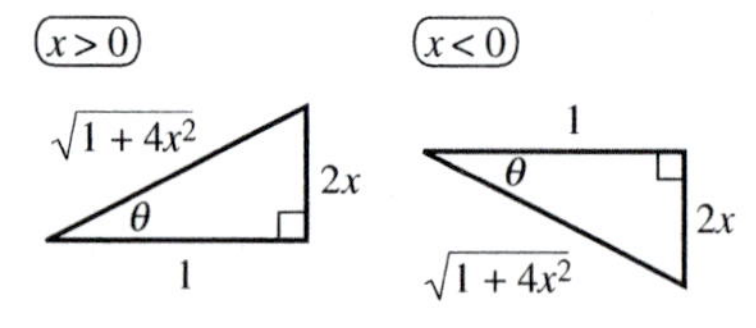

53. (a) Call the smaller (unlabeled) angle in the lower left α; then $\tan\alpha = \dfrac{2}{x}$, or $\alpha = \tan^{-1}\dfrac{2}{x}$ (since α is acute). Also, $\theta + \alpha$ is the measure of one acute angle in the right triangle formed by a line parallel to the floor and the wall; for this triangle $\tan(\theta+\alpha) = \dfrac{14}{x}$. Then

$\theta + \alpha = \tan^{-1}\dfrac{14}{x}$ (since $\theta + \alpha$ is acute), so

$\theta = \tan^{-1}\dfrac{14}{x} - \alpha = \tan^{-1}\dfrac{14}{x} - \tan^{-1}\dfrac{2}{x}$.

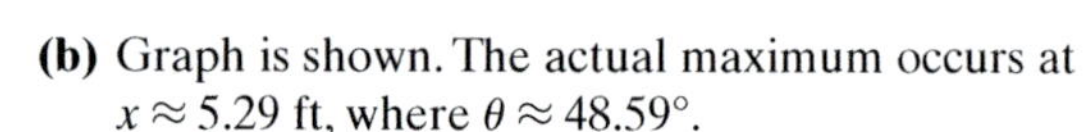
(b) Graph is shown. The actual maximum occurs at $x \approx 5.29$ ft, where $\theta \approx 48.59°$.

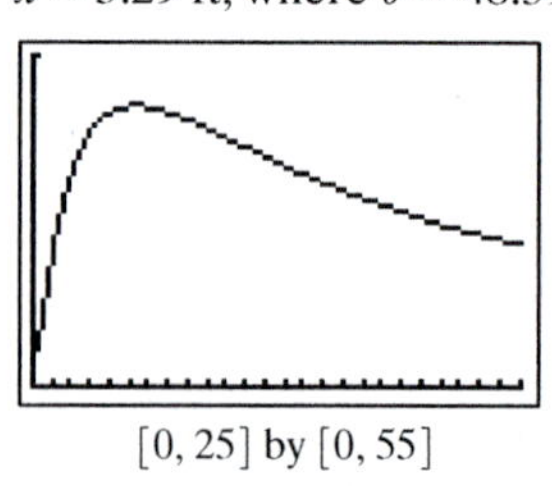

[0, 25] by [0, 55]

(c) Either $x \approx 1.83$ or $x \approx 15.31$—these round to 2 ft or 15 ft.

55. (a) $\theta = \tan^{-1}\dfrac{s}{500}$.

(b) As s changes from 10 to 20 ft, θ changes from about 1.1458° to 2.2906°—it almost exactly doubles (a 99.92% increase). As s changes from 200 to 210 ft, θ changes from about 21.80° to 22.78°—an increase of less than 1°, and a very small relative change (only about 4.25%).

(c) The x-axis represents the height and the y-axis represents the angle: the angle cannot grow past 90° (in fact, it *approaches* but never exactly equals 90°).

57. False. This is only true for $-1 \le x \le 1$, the domain of the $\sin^{-1}$ function. For $x < -1$ and for $x > 1$, $\sin(\sin^{-1} x)$ is undefined.

59. $\cos(5\pi/6) = -\sqrt{3}/2$, so $\cos^{-1}(-\sqrt{3}/2) = 5\pi/6$. The answer is E.

61. $\sec(\tan^{-1} x) = \sqrt{1 + \tan^2(\tan^{-1} x)} = \sqrt{1+x^2}$. The answer is C.

63. The cotangent function restricted to the interval $(0, \pi)$ is one-to-one and has an inverse. The unique angle y between 0 and π (non-inclusive) such that $\cot y = x$ is called the inverse cotangent (or arccotangent) of x, denoted $\cot^{-1} x$ or $\operatorname{arccot} x$. The domain of $y = \cot^{-1} x$ is $(-\infty, \infty)$ and the range is $(0, \pi)$.

65. (a) Domain all reals, range $[-\pi/2, \pi/2]$, period 2π.

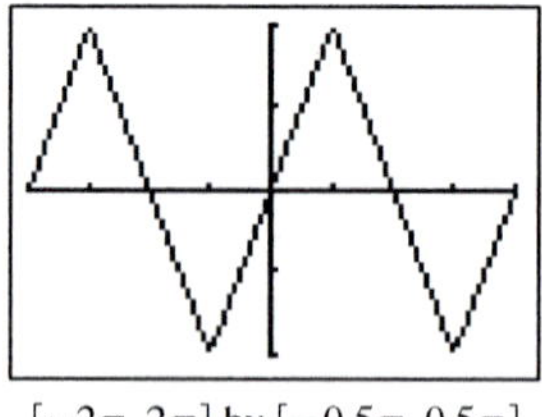

$[-2\pi, 2\pi]$ by $[-0.5\pi, 0.5\pi]$

(b) Domain all reals, range $[0, \pi]$, period 2π.

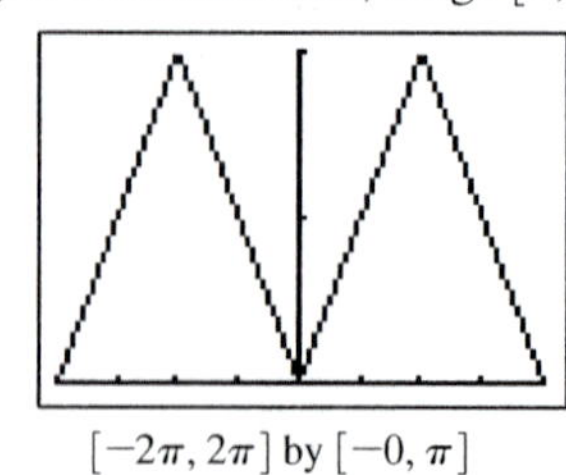

$[-2\pi, 2\pi]$ by $[-0, \pi]$

(c) Domain all reals except $\pi/2 + n\pi$ (n an integer), range $(-\pi/2, \pi/2)$, period π. Discontinuity is not removable.

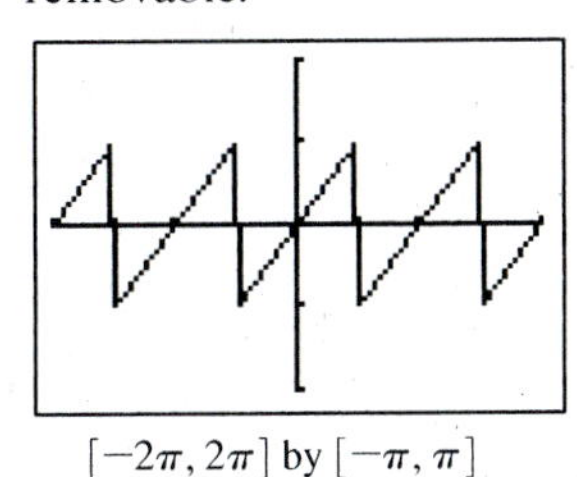

$[-2\pi, 2\pi]$ by $[-\pi, \pi]$

67. $y = \frac{\pi}{2} - \tan^{-1} x.$

(Note that $y = \tan^{-1}\left(\frac{1}{x}\right)$ does not have the correct range for negative values of x.)

69. In order to transform the arctangent function to a function that has horizontal asymptotes at $y = 24$ and $y = 42$, we need to find a and d that will satisfy the equation $y = a \tan^{-1} x + d$. In other words, we are shifting the horizontal asymptotes of $y = \tan^{-1} x$ from $y = -\frac{\pi}{2}$ and $y = \frac{\pi}{2}$ to the new asymptotes $y = 24$ and $y = 42$.

Solving $y = a \tan^{-1} x + d$ and $y = 24$ for $\tan^{-1} x$ in terms of a and d yields $24 = a \tan^{-1} x + d$; so, $\frac{24 - d}{a} = \tan^{-1}x$. We know that $y = 24$ is the lower horizontal asymptote and thus it corresponds to $y = -\frac{\pi}{2}$.

So, $\frac{24 - d}{a} = \tan^{-1}x = -\frac{\pi}{2} \Rightarrow \frac{24 - d}{a} = -\frac{\pi}{2}$. Solving this for d in terms of a yields $d = 24 + \left(\frac{\pi}{2}\right)a$.

Solving $y = a \tan^{-1} x + d$ and $y = 42$ for $\tan^{-1} x$ in terms of a and d yields $42 = a \tan^{-1} x + d$; so, $\frac{42 - d}{a} = \tan^{-1}x$. We know that $y = 42$ is the upper horizontal asymptote and thus it corresponds to $y = \frac{\pi}{2}$.

So, $\frac{42 - d}{a} = \tan^{-1}x = \frac{\pi}{2} \Rightarrow \frac{42 - d}{a} = \frac{\pi}{2}$. Solving this for d in terms of a yields $d = 42 - \left(\frac{\pi}{2}\right)a$.

If $d = 24 + \left(\frac{\pi}{2}\right)a$ and $d = 42 - \left(\frac{\pi}{2}\right)a$, then $24 + \left(\frac{\pi}{2}\right)a = 42 - \left(\frac{\pi}{2}\right)a$. So, $18 = \pi a$, and $a = \frac{18}{\pi}$.

Substitute this value for a into either of the two equations for d to get: $d = 24 + \left(\frac{\pi}{2}\right)\left(\frac{18}{\pi}\right) = 24 + 9 = 33$ or $d = 42 - \left(\frac{\pi}{2}\right)\left(\frac{18}{\pi}\right) = 42 - 9 = 33$.

The arctangent function with horizontal asymptotes at $y = 24$ and $y = 42$ will be $y = \frac{18}{\pi}\tan^{-1}x + 33$.

71. (a) The horizontal asymptote of the graph on the left is $y = \frac{\pi}{2}$.

(b) The two horizontal asymptotes of the graph on the right are $y = \frac{\pi}{2}$ and $y = \frac{3\pi}{2}$.

(c) The graph of $y = \cos^{-1}\left(\frac{1}{x}\right)$ will look like the graph on the left.

(d) The graph on the left is increasing on both connected intervals.

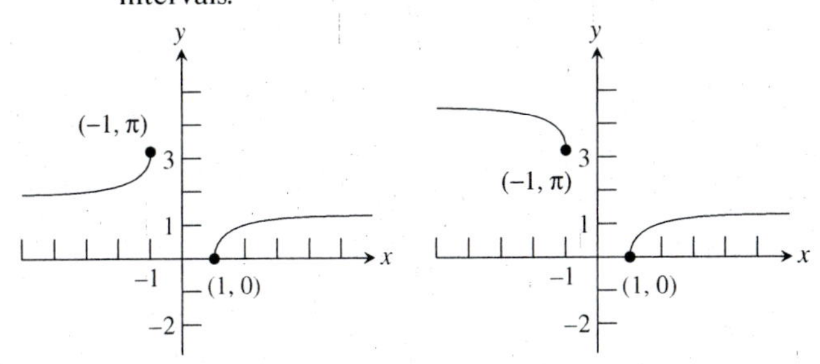

Section 4.8 Solving Problems with Trigonometry

Exploration 1

1. The parametrization should produce the unit circle.
2. The grapher is actually graphing the unit circle, but the y-window is so large that the point never seems to get above or below the x-axis. It is flattened vertically.
3. Since the grapher is plotting points along the unit circle, it covers the circle at a constant speed. Toward the extremes its motion is mostly vertical, so not much horizontal progress (which is all that we see) occurs. Toward the middle, the motion is mostly horizontal, so it moves faster.
4. The directed distance of the point from the origin at any T is exactly $\cos T$, and $d = \cos t$ models simple harmonic motion.

Quick Review 4.8

1. $b = 15 \cot 31° \approx 24.964$, $c = 15 \csc 31° \approx 29.124$
3. $b = 28 \cot 28° - 28 \cot 44° \approx 23.665$, $c = 28 \csc 28° \approx 59.642$, $a = 28 \csc 44° \approx 40.308$
5. complement: 58°, supplement: 148°
7. 45°
9. Amplitude: 3; period: π

Section 4.8 Exercises

All triangles in the supplied figures are right triangles.

1. $\tan 60° = \frac{h}{300 \text{ ft}}$, so $h = 300 \tan 60° = 300\sqrt{3} \approx 519.62$ ft.

3. Let d be the length of the horizontal leg. Then $\tan 10° = \frac{120 \text{ ft}}{d}$, so $d = \frac{120}{\tan 10°} = 120 \cot 10° \approx 680.55$ ft.

5. Let ℓ be the wire length (the hypotenuse); then $\cos 80° = \frac{5 \text{ ft}}{\ell}$, so $\ell = \frac{5}{\cos 80°} = 5 \sec 80° \approx 28.79$ ft.

Let h be the tower height (the vertical leg); then $\tan 80° = \frac{h}{5 \text{ ft}}$, so $h = 5 \tan 80° \approx 28.36$ ft.

7. $\tan 80°1'12'' = \frac{h}{185 \text{ ft}}$, so $h = 185 \tan 80°1'12'' \approx$ 1051 ft.

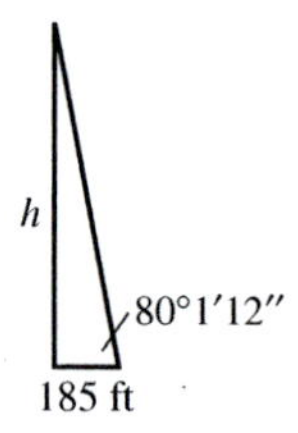

9. $\tan 83°12' = \frac{h}{100 \text{ ft}}$, so $h = 100 \tan 83°12' \approx 839$ ft.

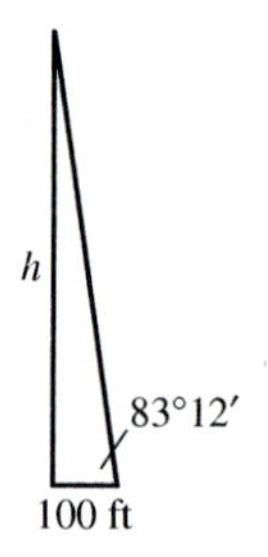

11. $\tan 55° = \frac{h}{10 \text{ m}}$, so $h = 10 \tan 55° \approx 14.3$ m

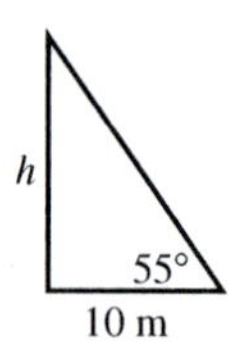

13. $\tan 35° = \frac{LP}{4.25 \text{ mi}}$, so $LP = 4.25 \tan 35° \approx 2.98$ mi.

15. Let x be the elevation of the bottom of the deck, and h be the height of the deck. Then $\tan 30° = \frac{x}{200 \text{ ft}}$ and $\tan 40° = \frac{x + h}{200 \text{ ft}}$, so $x = 200 \tan 30°$ ft and $x + h = 200 \tan 40°$ ft. Therefore $h = 200(\tan 40° - \tan 30°) \approx 52.35$ ft.

17. The two legs of the right triangle are the same length (30 knots · 2 hr = 60 naut mi), so both acute angles are 45°. The length of the hypotenuse is the distance: $60\sqrt{2} \approx 84.85$ naut mi. The bearing is $95° + 45° = 140°$.

19. The difference in elevations is 1097 ft. If the width of the canyon is w, then $\tan 19° = \frac{1097 \text{ ft}}{w}$, so $w = 1097 \cot 19° \approx 3186$ ft.

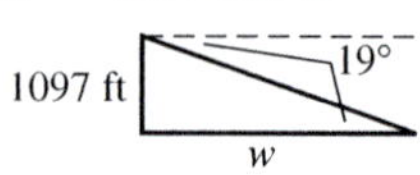

21. The acute angle in the triangle has measure $180° - 117° = 63°$, so $\tan 63° = \frac{\ell}{325 \text{ ft}}$. Then $\ell = 325 \tan 63° \approx 638$ ft.

23. If h is the height of the vertical span, $\tan 15° = \frac{h}{36.5 \text{ ft}}$, so $h = 36.5 \tan 15° \approx 9.8$ ft.

25. Let d be the distance from the boat to the shore, and let x be the short leg of the smaller triangle. For the two triangles, the larger acute angles are 70° and 80°. Then $\tan 80° = \frac{d}{x}$ and $\tan 70° = \frac{d}{x + 550}$, or $x = d \cot 80°$ and $x + 550 = d \cot 70°$. Therefore $d = \frac{550}{\cot 70° - \cot 80°} \approx 2931$ ft.

27. (a) Frequency: $\frac{\omega}{2\pi} = \frac{16\pi}{2\pi} = 8$ cycles/sec.

(b) $d = 6 \cos 16\pi t$ inches.

(c) When $t = 2.85$, $d \approx 1.854$; this is about 4.1 in. left of the starting position (when $t = 0$, $d = 6$).

29. The frequency is 2 cycles/sec, so $\omega = 2 \cdot 2\pi = 4\pi$ radians/sec. Assuming the initial position is $d = 3$ cm: $d = 3 \cos 4\pi t$.

31. (a) The amplitude is $a = 25$ ft, the radius of the wheel.

(b) $k = 33$ ft, the height of the center of the wheel.

(c) $\frac{\omega}{2\pi} = \frac{1}{20}$ rotations/sec, so $\omega = \pi/10$ radians/sec.

33. (a) Given a period of 12, we have $12 = \frac{2\pi}{|b|}$. $12|b| = 2\pi$ so $|b| = \frac{2\pi}{12} = \frac{\pi}{6}$. We select the positive value so $b = \frac{\pi}{6}$.

(b) Using the high temperature of 82 and a low temperature of 48, we find $|a| = \frac{82 - 48}{2} = \frac{34}{2}$ so $|a| = 17$ and we will select the positive value.

$k = \frac{82 + 48}{2} = 65$

(c) h is halfway between the times of the minimum and maximum. Using the maximum at time $t = 7$ and the minimum at time $t = 1$, we have $\frac{7 - 1}{2} = 3$. So, $h = 1 + 3 = 4$.

(d) The fit is very good for $y = 17 \sin\left(\frac{\pi}{6}(t - 4)\right) + 65$.

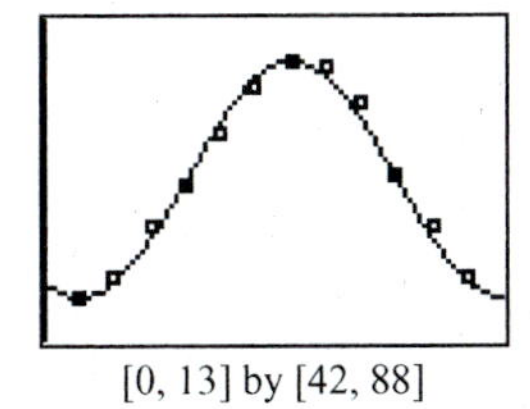

[0, 13] by [42, 88]

(e) There are several ways to find when the mean temperature will be 70°. *Graphical solution*: Graph the line $t = 70$ with the curve shown above, and find the intersection of the two curves. The two intersections are at $t \approx 4.57$ and $t \approx 9.43$.

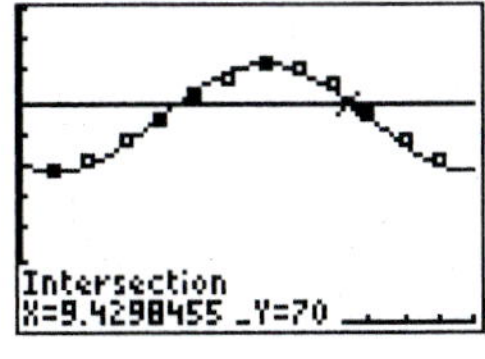

[0, 13] by [42, 88]

Algebraic solution: Solve $17 \sin\left(\frac{\pi}{6}(t - 4)\right) + 65 = 70$ for t.

$$17 \sin\left(\frac{\pi}{6}(t - 4)\right) + 65 = 70$$

$$\sin\left(\frac{\pi}{6}(t - 4)\right) = \frac{5}{17}$$

$$\frac{\pi}{6}(t - 4) = \sin^{-1}\left(\frac{5}{17}\right)$$

$$\frac{\pi}{6}(t - 4) \approx 0.299, 2.842$$

Note: $\sin \theta = \sin(\pi - \theta)$

$$t \approx 4.57, 9.43$$

Using either method to find t, find the day of the year as follows: $\frac{4.57}{12} \cdot 365 \approx 139$ and $\frac{9.43}{12} \cdot 365 \approx 287$. These represent May 19 and October 14.

35. (a) Solve this graphically by finding the zero of the function $P = 2t - 7 \sin\left(\frac{\pi t}{3}\right)$. The zero occurs at approximately 2.3. The function is positive to the right of the zero. So, the shop began to make a profit in March.

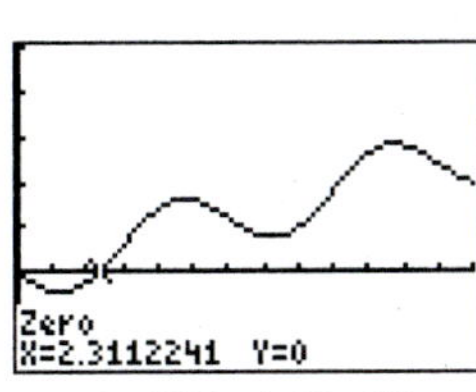

[0, 13] by [−20, 50]

(b) Solve this graphically by finding the maximum of the function $P = 2t - 7 \sin\left(\frac{\pi t}{3}\right)$. The maximum occurs at approximately 10.76, so the shop enjoyed its greatest profit in November.

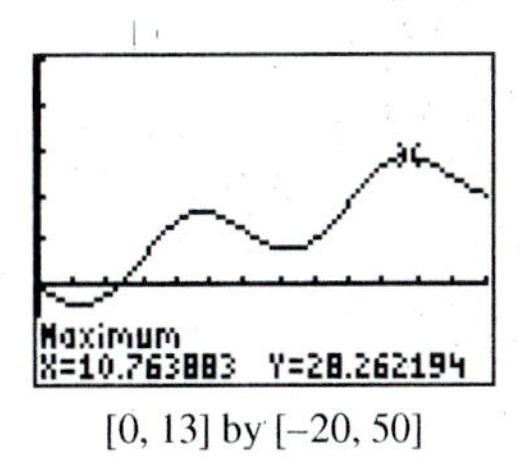

[0, 13] by [−20, 50]

37. True. The frequency and the period are reciprocals: $f = 1/T$. So the higher the frequency, the shorter the period.

39. If the building height in feet is x, then $\tan 58° = x/50$. So $x = 50 \tan 58° \approx 80$. The answer is D.

41. Model the tide level as a sinusoidal function of time, t. 6 hr, 12 min = 372 min is a half-period, and the amplitude is half of $13 - 9 = 4$. So use the model $f(t) = 2 \cos(\pi t/372) - 11$ with $t = 0$ at 8:15 PM. This takes on a value of -10 at $t = 124$. The answer is D.

43. (a)

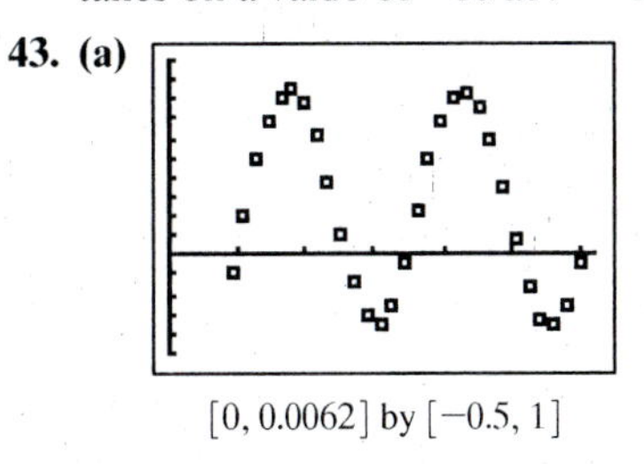

$[0, 0.0062]$ by $[-0.5, 1]$

(b) The first is the best. This can be confirmed by graphing all three equations.

(c) About $\frac{2464}{2\pi} = \frac{1232}{\pi} \approx 392$ oscillations/sec.

45. The 7-gon can be split into 14 congruent right triangles with a common vertex at the center. The legs of these triangles measure a and 2.5. The angle at the center is $\frac{2\pi}{14} = \frac{\pi}{7}$, so $a = 2.5 \cot \frac{\pi}{7} \approx 5.2$ cm

47. Choosing point E in the center of the rhombus, we have ΔAEB with right angle at E, and $m\angle EAB = 21°$. Then $AE = 18 \cos 21°$ in., $BE = 18 \sin 21°$ in., so that $AC = 2AE \approx 33.6$ in. and $BD = 2\,BE \approx 12.9$ in.

49. $\theta = \tan^{-1} 0.06 \approx 3.4$

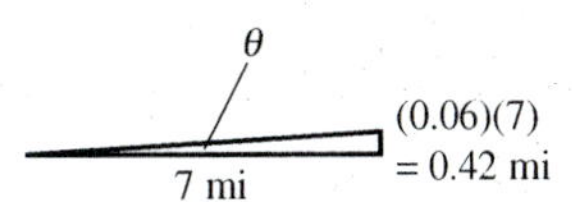

51. (a)

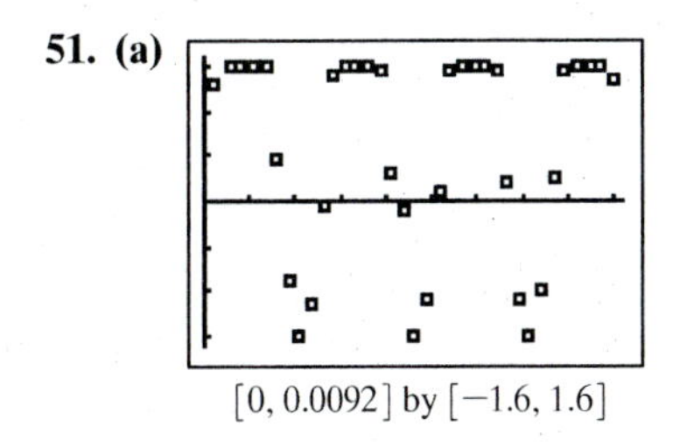

$[0, 0.0092]$ by $[-1.6, 1.6]$

(b) One pretty good match is $y = 1.51971 \sin[2467(t - 0.0002)]$ (that is, $a = 1.51971, b = 2467, h = 0.0002$). Answers will vary but should be close to these values. A good estimate of a can be found by noting the highest and lowest values of "Pressure" from the data. For the value of b, note the time between maxima (approx. $0.0033728 - 0.0008256 = 0.0025472$ sec); this is the period, so $b \approx \frac{2\pi}{0.0025472} \approx 2467$. Finally, since 0.0008256 is the location of the first peak after $t = 0$, choose h so that $2467(0.0008256 - h) \approx \frac{\pi}{2}$. This gives $h \approx 0.0002$.

(c) Frequency: about $\frac{2467}{2\pi} \approx \frac{1}{0.0025472} \approx 393$ Hz. It appears to be a G.

(d) Exercise 41 had $b \approx 2464$, so the frequency is again about 392 Hz; it also appears to be a G.

Chapter 4 Review

1. On the positive y-axis (between quadrants I and II); $\frac{5\pi}{2} \cdot \frac{180°}{\pi} = 450°$.

3. Quadrant III; $-135° \cdot \frac{\pi}{180°} = -\frac{3\pi}{4}$.

5. Quadrant I; $78° \cdot \frac{\pi}{180°} = \frac{13\pi}{30}$.

7. Quadrant I; $\frac{\pi}{12} \cdot \frac{180°}{\pi} = 15°$.

9. $270°$ or $\frac{3\pi}{2}$ radians

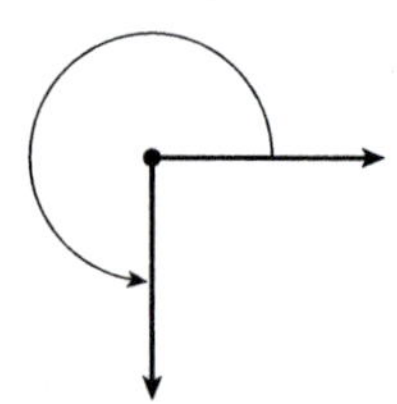

For #11–15, it may be useful to plot the given points and draw the terminal side to determine the angle. Be sure to make your sketch on a "square viewing window."

11. $\theta = \tan^{-1}\left(\frac{1}{\sqrt{3}}\right) = 30° = \frac{\pi}{6}$ radians

13. $\theta = 120° = \frac{2\pi}{3}$ radians

15. $\theta = 360° + \tan^{-1}(-2) \approx 296.565 \approx 5.176$ radians

17. $\sin 30° = \frac{1}{2}$

19. $\tan(-135°) = 1$

21. $\sin \frac{5\pi}{6} = \frac{1}{2}$

23. $\sec\left(-\frac{\pi}{3}\right) = 2$

25. $\csc 270° = -1$

27. $\cot(-90°) = 0$

29. Reference angle: $\frac{\pi}{6} = 30°$; use a 30–60 right triangle with side lengths $\sqrt{3}$, $(-)1$, and 2 (hypotenuse).
$\sin\left(-\frac{\pi}{6}\right) = -\frac{1}{2}, \cos\left(-\frac{\pi}{6}\right) = \frac{\sqrt{3}}{2}, \tan\left(-\frac{\pi}{6}\right) = -\frac{1}{\sqrt{3}};$
$\csc\left(-\frac{\pi}{6}\right) = -2, \sec\left(-\frac{\pi}{6}\right) = \frac{2}{\sqrt{3}}; \cot\left(-\frac{\pi}{6}\right) = -\sqrt{3}.$

31. Reference angle: 45°; use a 45–45 right triangle with side lengths $(-)1$, $(-)1$, and $\sqrt{2}$ (hypotenuse).
$\sin(-135°) = -\frac{1}{\sqrt{2}}, \cos(-135°) = -\frac{1}{\sqrt{2}},$
$\tan(-135°) = 1; \csc(-135°) = -\sqrt{2},$
$\sec(-135°) = -\sqrt{2}, \cot(-135°) = 1.$

33. The hypotenuse length is 13 cm, so $\sin \alpha = \frac{5}{13}$,
$\cos \alpha = \frac{12}{13}, \tan \alpha = \frac{5}{12}, \csc \alpha = \frac{13}{5}, \sec \alpha = \frac{13}{12},$
$\cot \alpha = \frac{12}{5}.$

For #35, since we are using a right triangle, we assume that θ is acute.

35. Draw a right triangle with legs 8 (adjacent) and 15, and hypotenuse $\sqrt{8^2 + 15^2} = \sqrt{289} = 17$.
$\sin \theta = \frac{15}{17}, \cos \theta = \frac{8}{17}, \tan \theta = \frac{15}{8}; \csc \theta = \frac{17}{15},$
$\sec \theta = \frac{17}{8}, \cot \theta = \frac{8}{15}.$

37. $x \approx 4.075$ radians

For #39–43, choose whichever of the following formulas is appropriate:

$$a = \sqrt{c^2 - b^2} = c \sin \alpha = c \cos \beta = b \tan \alpha = \frac{b}{\tan \beta}$$

$$b = \sqrt{c^2 - a^2} = c \cos \alpha = c \sin \beta = a \tan \beta = \frac{a}{\tan \alpha}$$

$$c = \sqrt{a^2 + b^2} = \frac{a}{\cos \beta} = \frac{a}{\sin \alpha} = \frac{b}{\sin \beta} = \frac{b}{\cos \alpha}$$

If one angle is given, subtract from 90° to find the other angle. If neither α nor β is given, find the value of one of the trigonometric functions, then use a calculator to approximate the value of one angle, then subtract from 90° to find the other.

39. $a = c \sin \alpha = 15 \sin 35° \approx 8.604, b = c \cos \alpha = 15 \cos 35° \approx 12.287, \beta = 90° - \alpha = 55°$

41. $b = a \tan \beta = 7 \tan 48° \approx 7.774, c = \frac{a}{\cos \beta} = \frac{7}{\cos 48°} \approx 10.461, \alpha = 90° - \beta = 42°$

43. $a = \sqrt{c^2 - b^2} = \sqrt{7^2 - 5^2} = \sqrt{24} = 2\sqrt{6} \approx 4.90$. For the angles, we know $\cos \alpha = \frac{5}{7}$; using a calculator, we find $\alpha \approx 44.42°$, so that $\beta = 90° - \alpha \approx 45.58°$.

45. $\sin x < 0$ and $\cos x < 0$: Quadrant III

47. $\sin x > 0$ and $\cos x < 0$: Quadrant II

49. The distance $OP = 3\sqrt{5}$, so $\sin\theta = \frac{2}{\sqrt{5}}$, $\cos\theta = -\frac{1}{\sqrt{5}}$, $\tan\theta = -2$; $\csc\theta = \frac{\sqrt{5}}{2}$, $\sec\theta = -\sqrt{5}$, $\cot\theta = -\frac{1}{2}$.

51. $OP = \sqrt{34}$, so $\sin\theta = -\frac{3}{\sqrt{34}}$, $\cos\theta = -\frac{5}{\sqrt{34}}$, $\tan\theta = \frac{3}{5}$; $\csc\theta = -\frac{\sqrt{34}}{3}$, $\sec\theta = -\frac{\sqrt{34}}{5}$, $\cot\theta = \frac{5}{3}$.

53. Starting from $y = \sin x$, translate left π units.

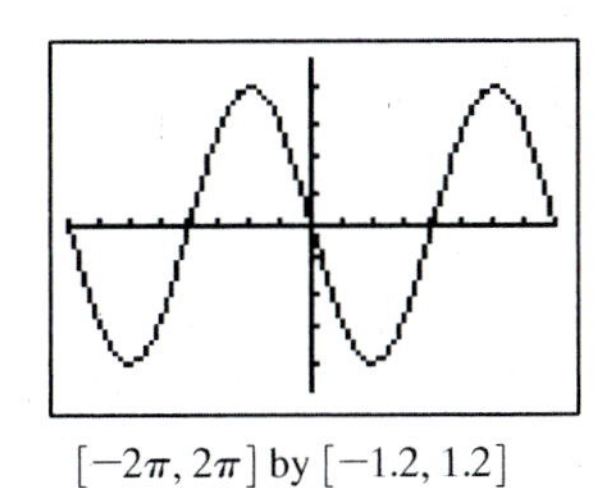

$[-2\pi, 2\pi]$ by $[-1.2, 1.2]$

55. Starting from $y = \cos x$, translate left $\frac{\pi}{2}$ units, reflect across x-axis, and translate up 4 units.

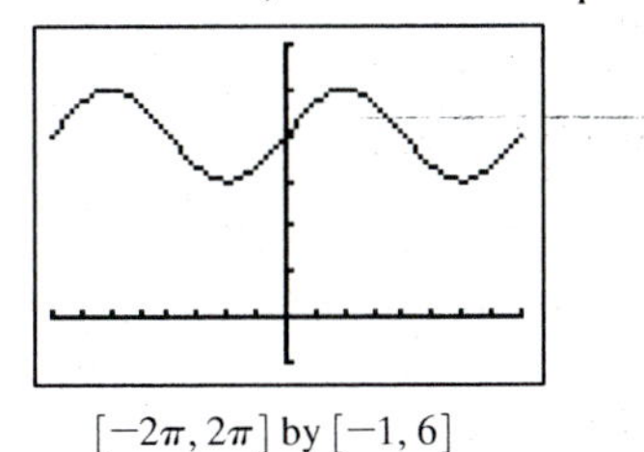

$[-2\pi, 2\pi]$ by $[-1, 6]$

57. Starting from $y = \tan x$, horizontally shrink by $\frac{1}{2}$.

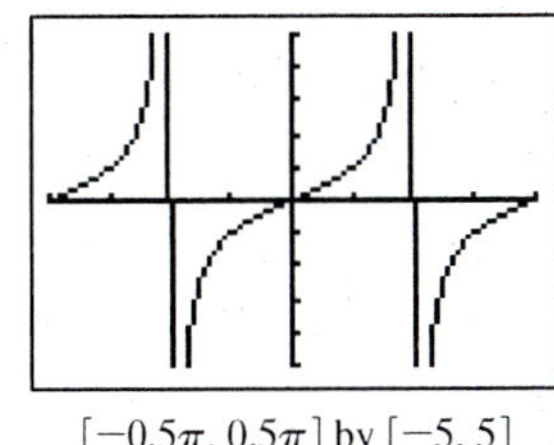

$[-0.5\pi, 0.5\pi]$ by $[-5, 5]$

59. Starting from $y = \sec x$, horizontally stretch by 2, vertically stretch by 2, and reflect across x-axis (in any order).

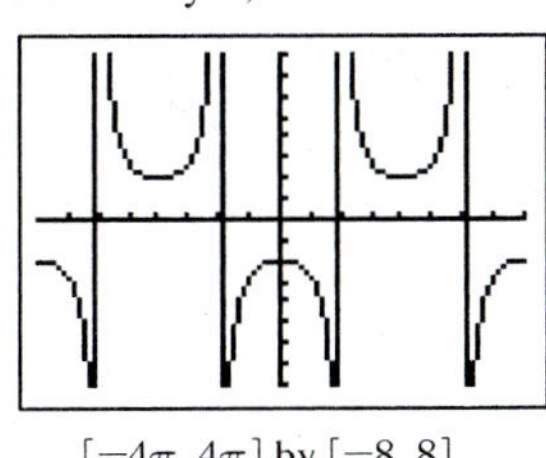

$[-4\pi, 4\pi]$ by $[-8, 8]$

For #61–65, recall that for $y = a\sin[b(x - h)]$ or $y = a\cos[b(x - h)]$, the amplitude is $|a|$, the period is $\frac{2\pi}{|b|}$, and the phase shift is h. The domain is always $(-\infty, \infty)$, and the range is $[-|a|, |a|]$.

61. $f(x) = 2\sin 3x$. Amplitude: 2; period: $\frac{2\pi}{3}$; phase shift: 0; domain: $(-\infty, \infty)$; range: $[-2, 2]$.

63. $f(x) = 1.5\sin\left[2\left(x - \frac{\pi}{8}\right)\right]$. Amplitude: 1.5; period: π; phase shift: $\frac{\pi}{8}$; domain: $(-\infty, \infty)$; range: $[-1.5, 1.5]$.

65. $y = 4\cos\left[2\left(x - \frac{1}{2}\right)\right]$. Amplitude: 4; period: π; phase shift: $\frac{1}{2}$; domain: $(-\infty, \infty)$; range: $[-4, 4]$.

For #67, graph the function. Estimate a as the amplitude of the graph (i.e., the height of the maximum). Notice that the value of b is always the coefficient of x in the original functions. Finally, note that $a\sin[b(x - h)] = 0$ when $x = h$, so estimate h using a zero of $f(x)$ where $f(x)$ changes from negative to positive.

67. $a \approx 4.47$, $b = 1$, and $h \approx 1.11$, so $f(x) \approx 4.47\sin(x - 1.11)$.

69. $\approx 49.996° \approx 0.873$ radians

71. $45° = \frac{\pi}{4}$ radians

73. Starting from $y = \sin^{-1}x$, horizontally shrink by $\frac{1}{3}$. Domain: $\left[-\frac{1}{3}, \frac{1}{3}\right]$. Range: $\left[-\frac{\pi}{2}, \frac{\pi}{2}\right]$.

75. Starting from $y = \sin^{-1}x$, translate right 1 unit, horizontally shrink by $\frac{1}{3}$, translate up 2 units. Domain: $\left[0, \frac{2}{3}\right]$. Range: $\left[2 - \frac{\pi}{2}, 2 + \frac{\pi}{2}\right]$.

77. $x = \frac{5\pi}{6}$

79. $x = \frac{3\pi}{4}$

81. $\frac{3\pi}{2}$

83. As $|x| \to \infty$, $\frac{\sin x}{x^2} \to 0$.

85. $\tan(\tan^{-1} 1) = \tan\frac{\pi}{4} = 1$

87. $\tan(\sin^{-1}\frac{3}{5}) = \frac{\sin\theta}{\cos\theta}$, where θ is an angle in $\left[-\frac{\pi}{2}, \frac{\pi}{2}\right]$ with $\sin\theta = \frac{3}{5}$. Then $\cos\theta = \sqrt{1 - \sin^2\theta} = \sqrt{0.64} = 0.80$ and $\tan\theta = 0.75$

89. Periodic; period π. Domain $x \neq \frac{\pi}{2} + n\pi$, n an integer. Range: $[1, \infty)$.

91. Not periodic. Domain: $x \neq \frac{\pi}{2} + n\pi$, n an integer. Range: $[-\infty, \infty)$.

93. $s = r\theta = (2)\left(\frac{2\pi}{3}\right) = \frac{4\pi}{3}$

95. $\tan 78° = \dfrac{h}{100 \text{ m}}$, so $h = 100 \tan 78° \approx 470$ m.

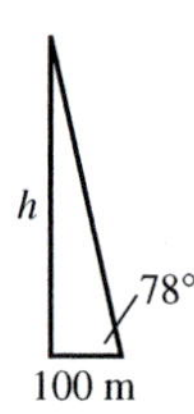

97. $\tan 42° = \dfrac{150 \text{ ft}}{x}$ and $\tan 18° = \dfrac{150 \text{ ft}}{d + x}$, so $x = 150 \cot 42°$ and $d + x = 150 \cot 18°$. Then $d = 150 \cot 18° - 150 \cot 42° \approx 295$ ft.

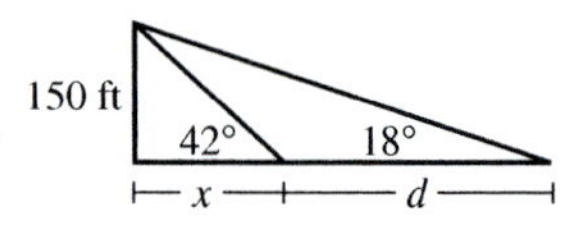

99. See figure below.

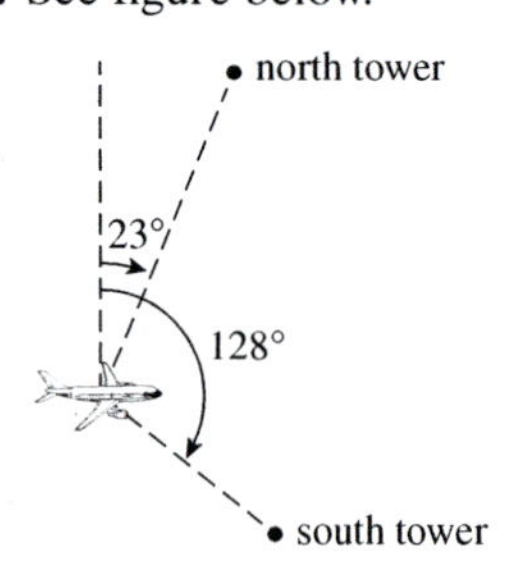

101. $\tan 72°24' = \dfrac{h}{62 \text{ ft}}$, so $h = 62 \tan 72°24' \approx 195.4$ ft

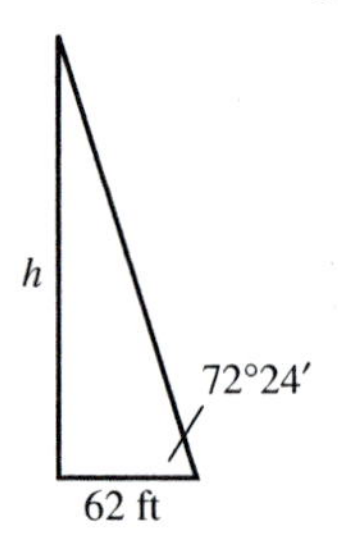

103. $s = r\theta = (44 \text{ in.})\left(6° \cdot \dfrac{\pi}{180°}\right) = \dfrac{22\pi}{15} \approx 4.6$ in.

105. *Solve algebraically*: Set $T(x) = 32$ and solve for x.

$$37.3 \sin\left[\frac{2\pi}{365}(x - 114)\right] + 26 = 32$$

$$\sin\left[\frac{2\pi}{365}(x - 114)\right] = \frac{6}{37.3}$$

$$\frac{2\pi}{365}(x - 114) = \sin^{-1}\left(\frac{6}{37.3}\right)$$

$$\frac{2\pi}{365}(x - 114) \approx 0.162, 2.98$$

Note: $\sin \theta = \sin(\pi - \theta)$

$$x \approx 123, 287$$

Solve graphically: Graph $T(x) = 32$ and $T(x) = 37.3 \sin\left[\dfrac{2\pi}{365}(x - 114)\right] + 25$ on the same set of axes, and then determine the intersections.

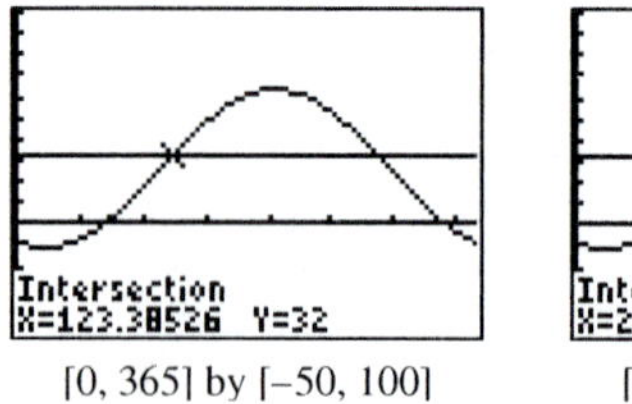

[0, 365] by [−50, 100]

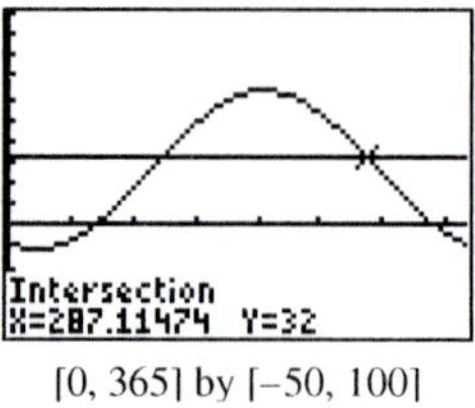

[0, 365] by [−50, 100]

Using either method, we would expect the average temperature to be 32°F on day 123 (May 3) and day 287 (October 14).

Chapter 4 Project

Solutions are based on the sample data shown in the table.

1.

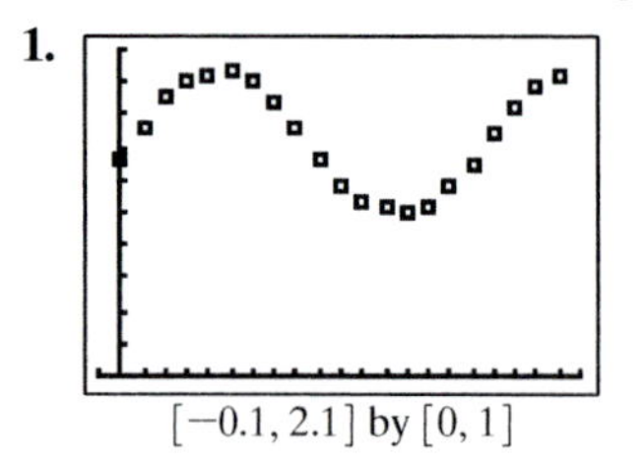
[−0.1, 2.1] by [0, 1]

2. The peak value seems to occur between $x = 0.4$ and $x = 0.5$, so let $h = 0.45$. The difference of the two extreme values is $0.931 - 0.495 = 0.436$, so let $a \approx 0.436/2 \approx 0.22$. The average of the two extreme values is $(0.931 + 0.495)/2 = 0.713$, so let $k = 0.71$. The time interval from $x = 0.5$ to $x = 1.3$, which equals 0.8, is right around a half-period, so let $b = \pi/0.8 \approx 3.93$. Then the equation is $y \approx 0.22 \cos(3.93(x - 0.45)) + 0.71$.
3. The constant a represents half the distance the pendulum bob swings as it moves from its highest point to its lowest point. And k represents the distance from the detector to the pendulum bob when it is in mid-swing.
4. Since the sine and cosine functions differ only by a phase shift, only h would change.
5. The regression yields $y \approx 0.22 \sin(3.87x - 0.16) + 0.71$. Most calculator/computer regression models are expressed in the form $y = a \sin(bx + f) + k$, where $-f/b = h$ in the equation $y = a \sin(b(x - h)) + k$. Here, the regression equation can be rewritten as $y \approx 0.22 \sin(3.87(x - 0.04)) + 0.71$. The difference in the two values of h for the cosine and sine models is 0.41, which is right around a quarter-period, as it should be.

Chapter 5
Analytic Trigonometry

Section 5.1 Fundamental Identities

Exploration 1

1. $\cos\theta = 1/\sec\theta$, $\sec\theta = 1/\cos\theta$, and $\tan\theta = \sin\theta/\cos\theta$
2. $\sin\theta = 1/\csc\theta$ and $\tan\theta = 1/\cot\theta$
3. $\csc\theta = 1/\sin\theta$, $\cot\theta = 1/\tan\theta$, and $\cot\theta = \cos\theta/\sin\theta$

Quick Review 5.1

For #1–3, use a calculator.

1. 1.1760 rad = 67.380°
3. 2.4981 rad = 143.130°
5. $a^2 - 2ab + b^2 = (a-b)^2$
7. $2x^2 - 3xy - 2y^2 = (2x+y)(x-2y)$
9. $\frac{1}{x}\cdot\frac{y}{y} - \frac{2}{y}\cdot\frac{x}{x} = \frac{y-2x}{xy}$
11. $\frac{x+y}{\frac{1}{x}+\frac{1}{y}} = (x+y)\cdot\left(\frac{xy}{x+y}\right) = xy$

Section 5.1 Exercises

1. $\sec^2\theta = 1 + \tan^2\theta = 1 + (3/4)^2 = 25/16$, so $\sec\theta = \pm 5/4$. Then $\cos\theta = 1/\sec\theta = \pm 4/5$. But $\sin\theta, \tan\theta > 0$ implies $\cos\theta > 0$. So $\cos\theta = 4/5$. Finally,

$$\tan\theta = \frac{3}{4}$$
$$\frac{\sin\theta}{\cos\theta} = \frac{3}{4}$$
$$\sin\theta = \frac{3}{4}\cos\theta = \frac{3}{4}\left(\frac{4}{5}\right) = \frac{3}{5}.$$

3. $\tan^2\theta = \sec^2\theta - 1 = 4^2 - 1 = 15$, so $\tan\theta = \pm\sqrt{15}$. But $\sec\theta > 0$, $\sin\theta < 0$ implies $\tan\theta < 0$, so $\tan\theta = -\sqrt{15}$. And $\cot\theta = 1/\tan\theta = -1/\sqrt{15} = -\sqrt{15}/15$.
5. $\cos(\pi/2 - \theta) = \sin\theta = 0.45$
7. $\cos(-\theta) = \cos\theta = \sin(\pi/2 - \theta) = -\sin(\theta - \pi/2) = -0.73$
9. $\tan x\cos x = \frac{\sin x}{\cos x}\cdot\cos x = \sin x$
11. $\sec y\sin\left(\frac{\pi}{2} - y\right) = \frac{1}{\cos y}\cdot\cos y = 1$
13. $\frac{1+\tan^2 x}{\csc^2 x} = \frac{\sec^2 x}{\csc^2 x} = \frac{1/\cos^2 x}{1/\sin^2 x} = \frac{\sin^2 x}{\cos^2 x} = \tan^2 x$
15. $\cos x - \cos^3 x = \cos x(1 - \cos^2 x) = \cos x\sin^2 x$
17. $\sin x\csc(-x) = \sin x\cdot\frac{1}{\sin(-x)} = -1$
19. $\cot(-x)\cot\left(\frac{\pi}{2} - x\right) = \frac{\cos(-x)}{\sin(-x)}\cdot\frac{\cos\left(\frac{\pi}{2}-x\right)}{\sin\left(\frac{\pi}{2}-x\right)}$
$= \frac{\cos(-x)}{\sin(-x)}\cdot\frac{\sin(x)}{\cos(x)} = -1$
21. $\sin^2(-x) + \cos^2(-x) = 1$
23. $\frac{\tan\left(\frac{\pi}{2} - x\right)\csc x}{\csc^2 x} = \frac{\cot x}{\csc x} = \frac{\cos x}{\sin x}\cdot\frac{\sin x}{1} = \cos x$
25. $(\sec^2 x + \csc^2 x) - (\tan^2 x + \cot^2 x)$
$= (\sec^2 x - \tan^2 x) + (\csc^2 x - \cot^2 x) = 1 + 1 = 2$
27. $(\sin x)(\tan x + \cot x) = (\sin x)\left(\frac{\sin x}{\cos x} + \frac{\cos x}{\sin x}\right)$
$= \sin x\left(\frac{\sin^2 x + \cos^2 x}{(\cos x)(\sin x)}\right) = \frac{1}{\cos x} = \sec x$
29. $(\sin x)(\cos x)(\tan x)(\sec x)(\csc x)$
$= (\sin x)(\cos x)\left(\frac{\sin x}{\cos x}\right)\left(\frac{1}{\cos x}\right)\left(\frac{1}{\sin x}\right) = \frac{\sin x}{\cos x}$
$= \tan x$
31. $\frac{\tan x}{\csc^2 x} + \frac{\tan x}{\sec^2 x}$
$= \left(\frac{\sin x}{\cos x}\right)(\sin^2 x) + \left(\frac{\sin x}{\cos x}\right)\cdot\cos^2 x$
$= \left(\frac{\sin x}{\cos x}\right)(\sin^2 x + \cos^2 x) = \frac{\sin x}{\cos x} = \tan x.$
33. $\frac{1}{\sin^2 x} + \frac{\sec^2 x}{\tan^2 x} = \csc^2 x + \frac{1}{\cos^2 x\left(\frac{\sin^2 x}{\cos^2 x}\right)}$
$= \csc^2 x + \frac{1}{\sin^2 x} = 2\csc^2 x$
35. $\frac{\sin x}{\cot^2 x} - \frac{\sin x}{\cos^2 x} = (\sin x)(\tan^2 x) - (\sin x)(\sec^2 x)$
$= (\sin x)(\tan^2 x - \sec^2 x) = (\sin x)(-1) = -\sin x$
37. $\frac{\sec x}{\sin x} - \frac{\sin x}{\cos x} = \frac{\sec x\cos x - \sin^2 x}{\sin x\cos x} = \frac{1 - \sin^2 x}{\sin x\cos x}$
$= \frac{\cos^2 x}{\sin x\cos x} = \frac{\cos x}{\sin x} = \cot x$
39. $\cos^2 x + 2\cos x + 1 = (\cos x + 1)^2$
41. $1 - 2\sin x + (1 - \cos^2 x) = 1 - 2\sin x + \sin^2 x$
$= (1 - \sin x)^2$
43. $\cos x - 2\sin^2 x + 1 = \cos x - 2 + 2\cos^2 x + 1$
$= 2\cos^2 x + \cos x - 1 = (2\cos x - 1)(\cos x + 1)$

45. $4\tan^2 x - \dfrac{4}{\cot x} + \sin x \csc x$

$= 4\tan^2 x - 4\tan x + \sin x \cdot \dfrac{1}{\sin x}$

$= 4\tan^2 x - 4\tan x + 1 = (2\tan x - 1)^2$

47. $\dfrac{1 - \sin^2 x}{1 + \sin x} = \dfrac{(1 - \sin x)(1 + \sin x)}{1 + \sin x} = 1 - \sin x$

49. $\dfrac{\sin^2 x}{1 + \cos x} = \dfrac{1 - \cos^2 x}{1 + \cos x} = \dfrac{(1 - \cos x)(1 + \cos x)}{1 + \cos x}$
$= 1 - \cos x$

51. $(\cos x)(2\sin x - 1) = 0$, so either $\cos x = 0$ or $\sin x = \dfrac{1}{2}$. Then $x = \dfrac{\pi}{2} + n\pi$ or $x = \dfrac{\pi}{6} + 2n\pi$ or $x = \dfrac{5\pi}{6} + 2n\pi$, n an integer. On the interval: $x = \left\{\dfrac{\pi}{6}, \dfrac{\pi}{2}, \dfrac{5\pi}{6}, \dfrac{3\pi}{2}\right\}$

53. $(\tan x)(\sin^2 x - 1) = 0$, so either $\tan x = 0$ or $\sin^2 x = 1$. Then $x = n\pi$ or $x = \dfrac{\pi}{2} + n\pi$, n an interger. However, $\tan x$ excludes $x = \dfrac{\pi}{2} + n\pi$, so we have only $x = n\pi$, n an integer. On the interval: $x = \{0, \pi\}$

55. $\tan x = \pm\sqrt{3}$, so $x = \pm\dfrac{\pi}{3} + n\pi$, n an integer.

On the interval: $x = \left\{\dfrac{\pi}{3}, \dfrac{2\pi}{3}, \dfrac{4\pi}{3}, \dfrac{5\pi}{3}\right\}$

57. $(2\cos x - 1)^2 = 0$, so $\cos x = \dfrac{1}{2}$; therefore $x = \pm\dfrac{\pi}{3} + 2n\pi$, n an integer.

59. $(\sin\theta)(\sin\theta - 2) = 0$, so $\sin\theta = 0$ or $\sin\theta = 2$. Then $\theta = n\pi$, n an integer.

61. $\cos(\sin x) = 1$ if $\sin x = n\pi$. Only $n = 0$ gives a value between -1 and $+1$, so $\sin x = 0$, or $x = n\pi$, n an integer.

63. $\cos^{-1} 0.37 \approx 1.1918$, so the solution set is $\{\pm 1.1918 + 2n\pi \mid n = 0, \pm 1, \pm 2, \ldots\}$.

65. $\sin^{-1} 0.30 \approx 0.3047$ and $\pi - 0.3047 \approx 2.8369$, so the solution set is $\{0.3047 + 2n\pi$ or $2.8369 + 2n\pi \mid n = 0, \pm 1, \pm 2, \ldots\}$.

67. $\sqrt{0.4} \approx 0.63246$, and $\cos^{-1} 0.63246 \approx 0.8861$, so the solution set is $\{\pm 0.8861 + n\pi \mid n = 0, \pm 1, \pm 2, \ldots\}$.

69. $\sqrt{1 - \cos^2\theta} = |\sin\theta|$

71. $\sqrt{9\sec^2\theta - 9} = 3|\tan\theta|$

73. $\sqrt{81\tan^2\theta + 81} = 9|\sec\theta|$

75. True. Since cosine is an even function, so is secant, and thus $\sec(x - \pi/2) = \sec(\pi/2 - x)$, which equals $\csc x$ by one of the cofunction identities.

77. $\tan x \sec x = \tan x/\cos x = \sin x/\cos^2 x \neq \sin x$. The answer is D.

79. $(\sec\theta + 1)(\sec\theta - 1) = \sec^2\theta - 1 = \tan^2\theta$. The answer is C.

81. $\sin x$, $\cos x = \pm\sqrt{1 - \sin^2 x}$, $\tan x = \pm\dfrac{\sin x}{\sqrt{1 - \sin^2 x}}$,

$\csc x = \dfrac{1}{\sin x}$, $\sec x = \pm\dfrac{1}{\sqrt{1 - \sin^2 x}}$

$\cot x = \pm\dfrac{\sqrt{1 - \sin^2 x}}{\sin x}$

83. The two functions are parallel to each other, separated by 1 unit for every x. At any x, the distance between the two graphs is $\sin^2 x - (-\cos^2 x) = \sin^2 x + \cos^2 x = 1$.

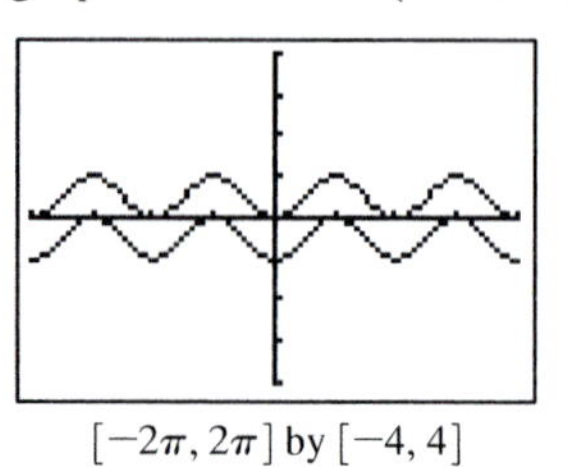

$[-2\pi, 2\pi]$ by $[-4, 4]$

85. (a)

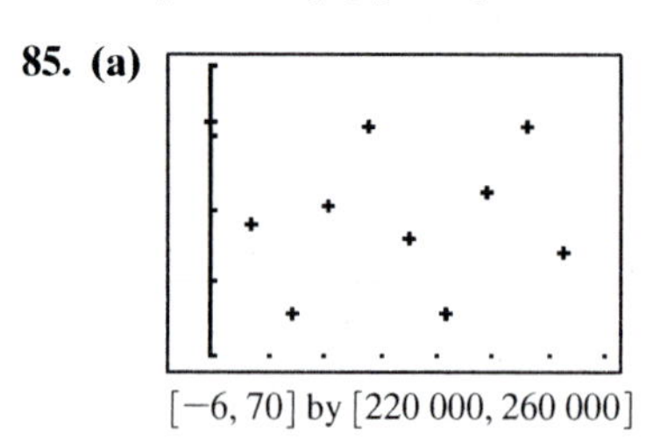

$[-6, 70]$ by $[220\,000, 260\,000]$

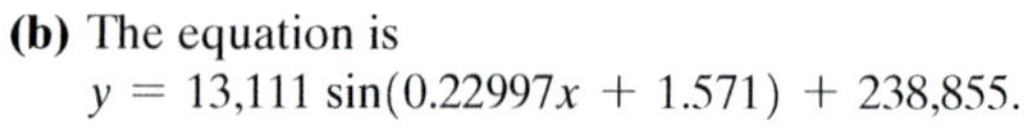

(b) The equation is
$y = 13{,}111\sin(0.22997x + 1.571) + 238{,}855$.

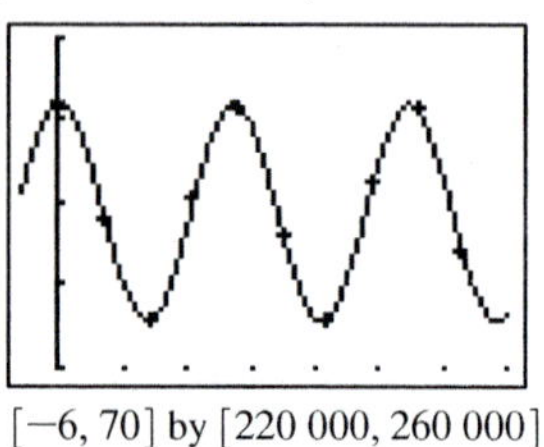

$[-6, 70]$ by $[220\,000, 260\,000]$

(c) $(2\pi)/0.22998 \approx 27.32$ days. This is the number of days that it takes the Moon to make one complete orbit of the Earth (known as the Moon's sidereal period).

(d) 225,744 miles

(e) $y = 13{,}111\cos(-0.22997x) + 238{,}855$, or
$y = 13{,}111\cos(0.22997x) + 238{,}855$.

87. Factor the left-hand side:

$\sin^4\theta - \cos^4\theta = (\sin^2\theta - \cos^2\theta)(\sin^2\theta + \cos^2\theta)$
$= (\sin^2\theta - \cos^2\theta) \cdot 1$
$= \sin^2\theta - \cos^2\theta$

89. Use the hint:

$\sin(\pi - x) = \sin(\pi/2 - (x - \pi/2))$
$= \cos(x - \pi/2)$ Cofunction identity
$= \cos(\pi/2 - x)$ Since cos is even
$= \sin x$ Cofunction identity

91. Since A, B, and C are angles of a triangle, $A + B = \pi - C$. So: $\sin(A + B) = \sin(\pi - C)$
$= \sin C$

■ Section 5.2 Proving Trigonometric Identities

Exploration 1

1. The graphs lead us to conclude that this is not an identity.

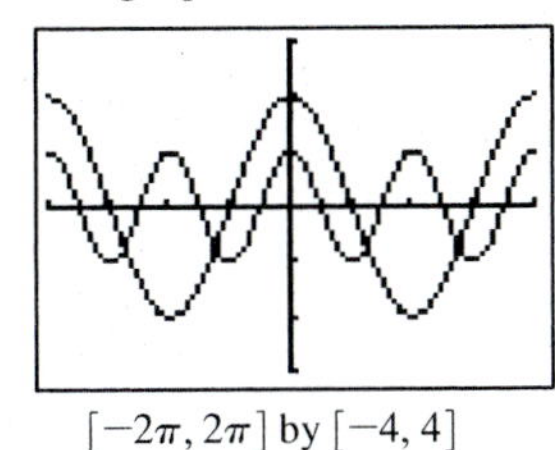

$[-2\pi, 2\pi]$ by $[-4, 4]$

2. For example, $\cos(2 \cdot 0) = 1$, whereas $2\cos(0) = 2$.
3. Yes.
4. The graphs lead us to conclude that this is an identity.

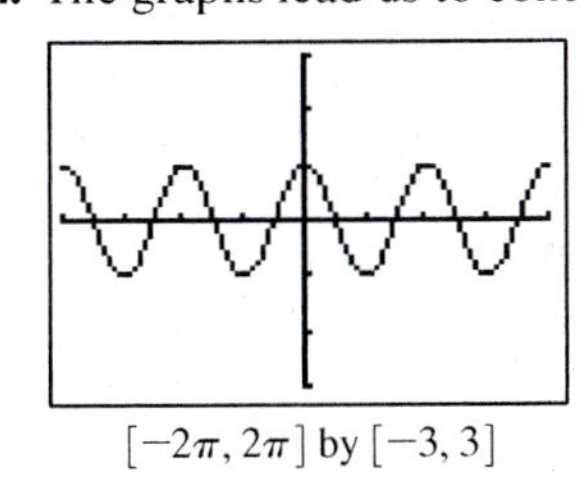

$[-2\pi, 2\pi]$ by $[-3, 3]$

5. No. The graph window can not show the full graphs, so they could differ outside the viewing window. Also, the function values could be so close that the graphs *appear* to coincide.

Quick Review 5.2

1. $\csc x + \sec x = \frac{1}{\sin x} + \frac{1}{\cos x} = \frac{\sin x + \cos x}{\sin x \cos x}$

3. $\cos x \cdot \frac{1}{\sin x} + \sin x \cdot \frac{1}{\cos x} = \frac{\cos^2 x + \sin^2 x}{\sin x \cos x}$
$= \frac{1}{\sin x \cos x}$

5. $\frac{\sin x}{1/\sin x} + \frac{\cos x}{1/\cos x} = \sin^2 x + \cos^2 x = 1$

7. No. (Any negative x.)

9. No. (Any x for which $\sin x < 0$, e.g. $x = -\pi/2$.)

11. Yes.

Section 5.2 Exercises

1. One possible proof:
$$\frac{x^3 - x^2}{x} - (x - 1)(x + 1) = \frac{x(x^2 - x)}{x} - (x^2 - 1)$$
$$= x^2 - x - (x^2 - 1)$$
$$= -x + 1$$
$$= 1 - x$$

3. One possible proof:
$$\frac{x^2 - 4}{x - 2} - \frac{x^2 - 9}{x + 3}$$
$$= \frac{(x + 2)(x - 2)}{x - 2} - \frac{(x + 3)(x - 3)}{x + 3}$$
$$= x + 2 - (x - 3)$$
$$= 5$$

5. $\frac{\sin^2 x + \cos^2 x}{\csc x} = \frac{1}{\csc x} = \sin x$. Yes.

7. $\cos x \cdot \cot x = \frac{\cos x}{1} \cdot \frac{\cos x}{\sin x} = \frac{\cos^2 x}{\sin x}$. No.

9. $(\sin^3 x)(1 + \cot^2 x) = (\sin^3 x)(\csc^2 x) = \frac{\sin^3 x}{\sin^2 x} = \sin x$. Yes.

11. $(\cos x)(\tan x + \sin x \cot x)$
$= \cos x \cdot \frac{\sin x}{\cos x} + \cos x \sin x \cdot \frac{\cos x}{\sin x} = \sin x + \cos^2 x$

13. $(1 - \tan x)^2 = 1 - 2\tan x + \tan^2 x$
$= (1 + \tan^2 x) - 2\tan x = \sec^2 x - 2\tan x$

15. One possible proof:
$$\frac{(1 - \cos u)(1 + \cos u)}{\cos^2 u} = \frac{1 - \cos^2 u}{\cos^2 u}$$
$$= \frac{\sin^2 u}{\cos^2 u}$$
$$= \tan^2 u$$

17. $\frac{\cos^2 x - 1}{\cos x} = \frac{-\sin^2 x}{\cos x} = \frac{\sin x}{\cos x} \cdot \sin x = -\tan x \sin x$

19. Multiply out the expression on the left side.

21. $(\cos t - \sin t)^2 + (\cos t + \sin t)^2$
$= \cos^2 t - 2\cos t \sin t + \sin^2 t + \cos^2 t$
$+ 2\cos t \sin t + \sin^2 t = 2\cos^2 t + 2\sin^2 t = 2$

23. $\frac{1 + \tan^2 x}{\sin^2 x + \cos^2 x} = \frac{\sec^2 x}{1} = \sec^2 x$

25. $\frac{\cos\beta}{1 + \sin\beta} = \frac{\cos^2\beta}{\cos\beta(1 + \sin\beta)} = \frac{1 - \sin^2\beta}{\cos\beta(1 + \sin\beta)}$
$= \frac{(1 - \sin\beta)(1 + \sin\beta)}{\cos\beta(1 + \sin\beta)} = \frac{1 - \sin\beta}{\cos\beta}$

27. $\frac{\tan^2 x}{\sec x + 1} = \frac{\sec^2 x - 1}{\sec x + 1} = \sec x - 1 = \frac{1}{\cos x} - 1$
$= \frac{1 - \cos x}{\cos x}$

29. $\cot^2 x - \cos^2 x = \left(\frac{\cos x}{\sin x}\right)^2 - \cos^2 x$
$= \frac{\cos^2 x(1 - \sin^2 x)}{\sin^2 x} = \cos^2 x \cdot \frac{\cos^2 x}{\sin^2 x}$
$= \cos^2 x \cot^2 x$

31. $\cos^4 x - \sin^4 x = (\cos^2 x + \sin^2 x)(\cos^2 x - \sin^2 x)$
$= 1(\cos^2 x - \sin^2 x) = \cos^2 x - \sin^2 x$

33. $(x \sin\alpha + y\cos\alpha)^2 + (x\cos\alpha - y\sin\alpha)^2$
$= (x^2\sin^2\alpha + 2xy\sin\alpha\cos\alpha + y^2\cos^2\alpha)$
$+ (x^2\cos^2\alpha - 2xy\cos\alpha\sin\alpha + y^2\sin^2\alpha)$
$= x^2\sin^2\alpha + y^2\cos^2\alpha + x^2\cos^2\alpha + y^2\sin^2\alpha$
$= (x^2 + y^2)(\sin^2\alpha + \cos^2\alpha) = x^2 + y^2$

35. $\frac{\tan x}{\sec x - 1} = \frac{\tan x(\sec x + 1)}{\sec^2 x - 1} = \frac{\tan x(\sec x + 1)}{\tan^2 x}$
$= \frac{\sec x + 1}{\tan x}$. See also #26.

37. $\frac{\sin x - \cos x}{\sin x + \cos x} = \frac{(\sin x - \cos x)(\sin x + \cos x)}{(\sin x + \cos x)^2}$
$= \frac{\sin^2 x - \cos^2 x}{\sin^2 x + 2\sin x\cos x + \cos^2 x} = \frac{\sin^2 x - (1 - \sin^2 x)}{1 + 2\sin x\cos x}$
$= \frac{2\sin^2 x - 1}{1 + 2\sin x\cos x}$

39. $\frac{\sin t}{1 - \cos t} + \frac{1 + \cos t}{\sin t} = \frac{\sin^2 t + (1 + \cos t)(1 - \cos t)}{(\sin t)(1 - \cos t)}$
$= \frac{\sin^2 t + 1 - \cos^2 t}{(\sin t)(1 - \cos t)} = \frac{1 - \cos^2 t + 1 - \cos^2 t}{(\sin t)(1 - \cos t)}$
$= \frac{2(1 - \cos^2 t)}{(\sin t)(1 - \cos t)} = \frac{2(1 + \cos t)}{\sin t}$

41. $\sin^2 x\cos^3 x = \sin^2 x\cos^2 x\cos x$
$= \sin^2 x(1 - \sin^2 x)\cos x = (\sin^2 x - \sin^4 x)\cos x$

43. $\cos^5 x = \cos^4 x\cos x = (\cos^2 x)^2\cos x$
$= (1 - \sin^2 x)^2\cos x = (1 - 2\sin^2 x + \sin^4 x)\cos x$

45. $\frac{\tan x}{1 - \cot x} + \frac{\cot x}{1 - \tan x}$
$= \frac{\tan x}{1 - \cot x}\cdot\frac{\sin x}{\sin x} + \frac{\cot x}{1 - \tan x}\cdot\frac{\cos x}{\cos x}$
$= \left(\frac{\sin^2 x/\cos x}{\sin x - \cos x} + \frac{\cos^2 x/\sin x}{\cos x - \sin x}\right)\frac{\sin x\cos x}{\sin x\cos x}$
$= \frac{\sin^3 x - \cos^3 x}{\sin x\cos x(\sin x - \cos x)}$
$= \frac{\sin^2 x + \sin x\cos x + \cos^2 x}{\sin x\cos x}$
$= \frac{1 + \sin x\cos x}{\sin x\cos x} = \frac{1}{\sin x\cos x} + 1 = \csc x\sec x + 1.$
This involves rewriting $a^3 - b^3$ as $(a - b)(a^2 + ab + b^2)$, where $a = \sin x$ and $b = \cos x$.

47. $\frac{2\tan x}{1 - \tan^2 x} + \frac{1}{2\cos^2 x - 1}$
$= \frac{2\tan x}{1 - \tan^2 x}\cdot\frac{\cos^2 x}{\cos^2 x} + \frac{1}{\cos^2 x - \sin^2 x}$
$= \frac{2\sin x\cos x}{\cos^2 x - \sin^2 x} + \frac{\cos^2 x + \sin^2 x}{\cos^2 x - \sin^2 x}$
$= \frac{2\sin x\cos x + \cos^2 x + \sin^2 x}{(\cos x - \sin x)(\cos x + \sin x)}$
$= \frac{(\cos x + \sin x)^2}{(\cos x - \sin x)(\cos x + \sin x)} = \frac{\cos x + \sin x}{\cos x - \sin x}$

49. $\cos^3 x = (\cos^2 x)(\cos x) = (1 - \sin^2 x)(\cos x)$

51. $\sin^5 x = (\sin^4 x)(\sin x) = (\sin^2 x)^2(\sin x)$
$= (1 - \cos^2 x)^2(\sin x)$
$= (1 - 2\cos^2 x + \cos^4 x)(\sin x)$

53. (d) — multiply out: $(1 + \sec x)(1 - \cos x)$
$= 1 - \cos x + \sec x - \sec x\cos x$
$= 1 - \cos x + \frac{1}{\cos x} - \frac{1}{\cos x}\cdot\cos x$
$= 1 - \cos x + \frac{1}{\cos x} - 1 = \frac{1 - \cos^2 x}{\cos x} = \frac{\sin^2 x}{\cos x}$
$= \frac{\sin x}{\cos x}\cdot\sin x = \tan x\sin x.$

55. (c) — put over a common denominator:
$\frac{1}{1 + \sin x} + \frac{1}{1 - \sin x} = \frac{1 - \sin x + 1 + \sin x}{1 - \sin^2 x}$
$= \frac{2}{\cos^2 x} = 2\sec^2 x.$

57. (b) — multiply and divide by $\sec x + \tan x$:
$\frac{1}{\sec x - \tan x}\cdot\frac{\sec x + \tan x}{\sec x + \tan x} = \frac{\sec x + \tan x}{\sec^2 x - \tan^2 x}$
$= \frac{\sec x + \tan x}{1}.$

59. True. If x is in the domain of both sides of the equation, then $x \ge 0$. The equation $(\sqrt{x})^2 = x$ holds for all $x \ge 0$, so it is an identity.

61. A proof is
$\frac{\sin x}{1 - \cos x} = \frac{\sin x}{1 - \cos x}\cdot\frac{1 + \cos x}{1 + \cos x}$
$= \frac{\sin x(1 + \cos x)}{1 - \cos^2 x}$
$= \frac{\sin x(1 + \cos x)}{\sin^2 x}$
$= \frac{1 + \cos x}{\sin x}$
The answer is E.

63. k must equal 1, so $f(x) \neq 0$. The answer is B.

65. $\sin x$; $\cos x\tan x = \cos x\cdot\frac{\sin x}{\cos x} = \sin x$

67. 1; $\frac{\csc x}{\sin x} - \frac{\cot x\csc x}{\sec x} = \frac{1/\sin x}{\sin x} - \frac{\cos x/\sin^2 x}{1/\cos x}$
$= \frac{1}{\sin^2 x} - \frac{\cos^2 x}{\sin^2 x} = \frac{1 - \cos^2 x}{\sin^2 x} = \frac{\sin^2 x}{\sin^2 x} = 1$

69. 1; $(\sec^2 x)(1 - \sin^2 x) = \left(\frac{1}{\cos x}\right)^2(\cos^2 x) = 1$

71. If A and B are complementary angles, then
$\sin^2 A + \sin^2 B = \sin^2 A + \sin^2(\pi/2 - A)$
$= \sin^2 A + \cos^2 A$
$= 1$

73. Multiply and divide by $1 - \sin t$ under the radical:
$\sqrt{\frac{1 - \sin t}{1 + \sin t}\cdot\frac{1 - \sin t}{1 - \sin t}} = \sqrt{\frac{(1 - \sin t)^2}{1 - \sin^2 t}}$
$= \sqrt{\frac{(1 - \sin t)^2}{\cos^2 t}} = \frac{|1 - \sin t|}{|\cos t|}$ since $\sqrt{a^2} = |a|$.
Now, since $1 - \sin t \ge 0$, we can dispense with the absolute value in the numerator, but it must stay in the denominator.

75. $\sin^6 x + \cos^6 x = (\sin^2 x)^3 + \cos^6 x$
$= (1 - \cos^2 x)^3 + \cos^6 x$
$= (1 - 3\cos^2 x + 3\cos^4 x - \cos^6 x) + \cos^6 x$
$= 1 - 3\cos^2 x(1 - \cos^2 x) = 1 - 3\cos^2 x\sin^2 x.$

77. One possible proof: $\ln|\tan x| = \ln\frac{|\sin x|}{|\cos x|}$
$= \ln|\sin x| - \ln|\cos x|.$

79. (a) They are not equal. Shown is the window $[-2\pi, 2\pi,]$ by $[-2, 2]$; graphing on nearly any viewing window does not show any apparent difference — but using TRACE, one finds that the y coordinates are not identical. Likewise, a table of values will show slight differences; for example, when $x = 1$, $y_1 = 0.53988$ while $y_2 = 0.54030$

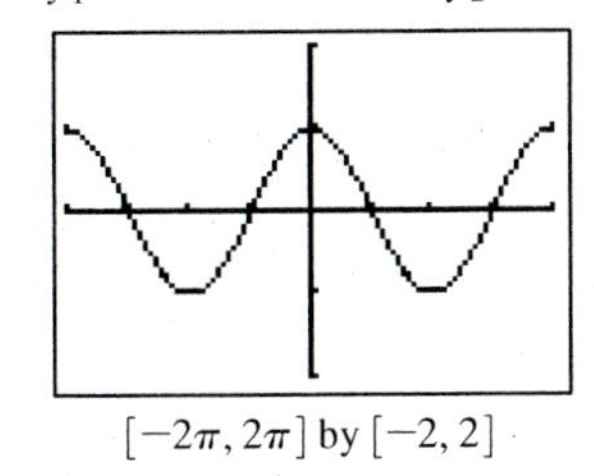

$[-2\pi, 2\pi]$ by $[-2, 2]$

(b) One choice for h is 0.001 (shown). The function y_3 is a combination of three sinusoidal functions ($1000 \sin(x + 0.001)$, $1000 \sin x$, and $\cos x$), all with period 2π.

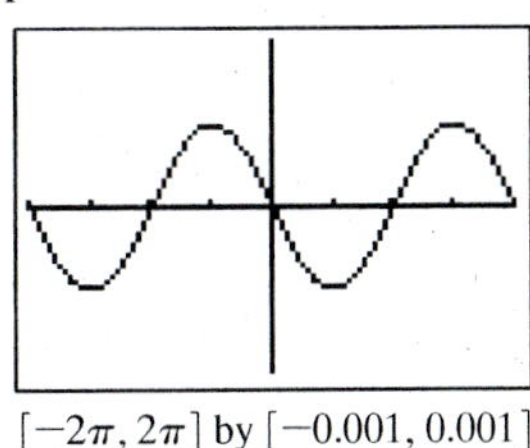

$[-2\pi, 2\pi]$ by $[-0.001, 0.001]$

81. In the decimal window, the x coordinates used to plot the graph on the calculator are (e.g.) 0, 0.1, 0.2, 0.3, etc. — that is, $x = n/10$, where n is an integer. Then $10\pi x = \pi n$, and the sine of integer multiples of π is 0; therefore, $\cos x + \sin 10\pi x = \cos x + \sin \pi n = \cos x + 0$ $= \cos x$. However, for other choices of x, such as $x = \frac{1}{\pi}$, we have $\cos x + \sin 10\pi x = \cos x + \sin 10 \neq \cos x$.

■ Section 5.3 Sum and Difference Identities

Exploration 1

1. $\sin(u + v) = -1$, $\sin u + \sin v = 1$. No.

2. $\cos(u + v) = 1$, $\cos u + \cos v = 2$. No.

3. $\tan(\pi/3 + \pi/3) = -\sqrt{3}$, $\tan \pi/3 + \tan \pi/3 = 2\sqrt{3}$. (Many other answers are possible.)

Quick Review 5.3

1. $15° = 45° - 30°$

3. $165° = 180° - 15° = 180° + 30° - 45° = 210° - 45°$

5. $\frac{5\pi}{12} = 4 \cdot \frac{\pi}{6} - \frac{\pi}{4} = \frac{2\pi}{3} - \frac{\pi}{4}$

7. No. $(f(x) + f(y) = \ln x + \ln y = \ln(xy)$ $= f(xy) \neq f(x + y))$

9. Yes. $(f(x + y) = 32(x + y) = 32x + 32y$ $= f(x) + f(y))$

Section 5.3 Exercises

1. $\sin 15° = \sin(45° - 30°)$
$= \sin 45° \cos 30° - \cos 45° \sin 30°$
$= \frac{\sqrt{2}}{2} \cdot \frac{\sqrt{3}}{2} - \frac{\sqrt{2}}{2} \cdot \frac{1}{2} = \frac{\sqrt{6} - \sqrt{2}}{4}$

3. $\sin 75° = \sin(45° + 30°)$
$= \sin 45° \cos 30° + \cos 45° \sin 30°$
$= \frac{\sqrt{2}}{2} \cdot \frac{\sqrt{3}}{2} + \frac{\sqrt{2}}{2} \cdot \frac{1}{2} = \frac{\sqrt{6} + \sqrt{2}}{4}$

5. $\cos \frac{\pi}{12} = \cos\left(\frac{\pi}{3} - \frac{\pi}{4}\right) = \cos \frac{\pi}{3} \cos \frac{\pi}{4} + \sin \frac{\pi}{3} \sin \frac{\pi}{4}$
$= \frac{1}{2} \cdot \frac{\sqrt{2}}{2} + \frac{\sqrt{3}}{2} \cdot \frac{\sqrt{2}}{2} = \frac{\sqrt{2} + \sqrt{6}}{4}$

7. $\tan \frac{5\pi}{12} = \tan\left(\frac{2\pi}{3} - \frac{\pi}{4}\right) = \frac{\tan(2\pi/3) - \tan(\pi/4)}{1 + \tan(2\pi/3)\tan(\pi/4)}$
$= \frac{-\sqrt{3} - 1}{1 - \sqrt{3}} = \frac{\sqrt{3} + 1}{\sqrt{3} - 1} = \frac{(\sqrt{3} + 1)^2}{3 - 1} = 2 + \sqrt{3}$

9. $\cos \frac{7\pi}{12} = \cos\left(\frac{5\pi}{6} - \frac{\pi}{4}\right)$
$= \cos \frac{5\pi}{6} \cos \frac{\pi}{4} + \sin \frac{5\pi}{6} \sin \frac{\pi}{4} = -\frac{\sqrt{3}}{2} \cdot \frac{\sqrt{2}}{2} + \frac{1}{2} \cdot \frac{\sqrt{2}}{2}$
$= \frac{\sqrt{2} - \sqrt{6}}{4}$

In #11–21, match the given expression with the sum and difference identities.

11. $\sin(42° - 17°) = \sin 25°$

13. $\sin\left(\frac{\pi}{5} + \frac{\pi}{2}\right) = \sin \frac{7\pi}{10}$

15. $\tan(19° + 47°) = \tan 66°$

17. $\cos\left(\frac{\pi}{7} - x\right) = \cos\left(x - \frac{\pi}{7}\right)$

19. $\sin(3x - x) = \sin 2x$

21. $\tan(2y + 3x)$

23. $\sin\left(x - \frac{\pi}{2}\right) = \sin x \cos \frac{\pi}{2} - \cos x \sin \frac{\pi}{2}$
$= \sin x \cdot 0 - \cos x \cdot 1 = -\cos x$

25. $\cos\left(x - \frac{\pi}{2}\right) = \cos x \cos \frac{\pi}{2} + \sin x \sin \frac{\pi}{2}$
$= \cos x \cdot 0 + \sin x \cdot 1 = \sin x$

27. $\sin\left(x + \frac{\pi}{6}\right) = \sin x \cos \frac{\pi}{6} + \cos x \sin \frac{\pi}{6}$
$= \sin x \cdot \frac{\sqrt{3}}{2} + \cos x \cdot \frac{1}{2}$

29. $\tan\left(\theta + \frac{\pi}{4}\right) = \frac{\tan \theta + \tan(\pi/4)}{1 - \tan \theta \tan(\pi/4)} = \frac{\tan \theta + 1}{1 - \tan \theta \cdot 1}$
$= \frac{1 + \tan \theta}{1 - \tan \theta}$

31. Equations B and F.

33. Equations D and H.

35. Rewrite as $\sin 2x \cos x - \cos 2x \sin x = 0$; the left side equals $\sin(2x - x) = \sin x$, so $x = n\pi$, n an integer.

37. $\sin\left(\frac{\pi}{2} - u\right) = \sin \frac{\pi}{2} \cos u - \cos \frac{\pi}{2} \sin u$
$= 1 \cdot \cos u - 0 \cdot \sin u = \cos u$.

39. $\cot\left(\frac{\pi}{2} - u\right) = \frac{\cos(\pi/2 - u)}{\sin(\pi/2 - u)} = \frac{\sin u}{\cos u} = \tan u$ using the first two cofunction identities.

41. $\csc\left(\frac{\pi}{2} - u\right) = \frac{1}{\sin(\pi/2 - u)} = \frac{1}{\cos u} = \sec u$ using the second cofunction identity.

43. To write $y = 3 \sin x + 4 \cos x$ in the form $y = a \sin(bx + c)$, rewrite the formula using the formula for the sine of a sum:

$$\begin{aligned} y &= a((\sin bx \cos c) + (\cos bx \sin c)) \\ &= a \sin bx \cos c + a \cos bx \sin c \\ &= (a \cos c)\sin bx + (a \sin c)\cos bx. \end{aligned}$$

Then compare the coefficients: $a \cos c = 3, b = 1, a \sin c = 4$.

Solve for a as follows:

$$\begin{aligned} (a \cos c)^2 + (a \sin c)^2 &= 3^2 + 4^2 \\ a^2 \cos^2 c + a^2 \sin^2 c &= 25 \\ a^2(\cos^2 c + \sin^2 c) &= 25 \\ a^2 &= 25 \\ a &= \pm 5 \end{aligned}$$

If we choose a to be positive, then $\cos c = 3/5$ and $\sin c = 4/5. c = \cos^{-1}(3/5) = \sin^{-1}(4/5)$. So the sinusoid is $y = 5 \sin(x + \cos^{-1}(3/5)) \approx 5 \sin(x + 0.9273)$.

45. Follow the steps shown in Exercise 43 to compare the coefficients in $y = (a \cos c)\sin bx + (a \sin c)\cos bx$ to the coefficients in $y = \cos 3x + 2 \sin 3x$: $a \cos c = 2$, $b = 3, a \sin c = 1$.

Solve for a as follows:

$$\begin{aligned} (a \cos c)^2 + (a \sin c)^2 &= 1^2 + 2^2 \\ a^2 (\cos^2 c + \sin^2 c) &= 5 \\ a &= \pm\sqrt{5} \end{aligned}$$

If we choose a to be positive, then $\cos c = 2/\sqrt{5}$ and $\sin c = 1/\sqrt{5}$. So the sinusoid is

$y = \sqrt{5}\sin(3x - \cos^{-1}(2/\sqrt{5})) \approx 2.236 \sin(3x - 0.4636)$.

47. $\sin(x - y) + \sin(x + y)$

$$\begin{aligned} &= (\sin x \cos y - \cos x \sin y) + (\sin x \cos y + \cos x \sin y) \\ &= 2 \sin x \cos y \end{aligned}$$

49.

$$\begin{aligned} \cos 3x &= \cos[(x + x) + x] \\ &= \cos(x + x) \cos x - \sin(x + x) \sin x \\ &= (\cos x \cos x - \sin x \sin x) \cos x \\ &\quad - (\sin x \cos x + \cos x \sin x) \sin x \\ &= \cos^3 x - \sin^2 x \cos x - 2 \cos x \sin^2 x \\ &= \cos^3 x - 3 \sin^2 x \cos x \end{aligned}$$

51. $\cos 3x + \cos x = \cos(2x + x) + \cos(2x - x)$; use #48 with x replaced with $2x$ and y replaced with x.

53. $\tan(x + y) \tan(x - y)$

$$= \left(\frac{\tan x + \tan y}{1 - \tan x \tan y}\right) \cdot \left(\frac{\tan x - \tan y}{1 + \tan x \tan y}\right)$$

$= \frac{\tan^2 x - \tan^2 y}{1 - \tan^2 x \tan^2 y}$ since both the numerator and denominator are factored forms for differences of squares.

55. $\frac{\sin (x + y)}{\sin (x - y)}$

$$\begin{aligned} &= \frac{\sin x \cos y + \cos x \sin y}{\sin x \cos y - \cos x \sin y} \\ &= \frac{\sin x \cos y + \cos x \sin y}{\sin x \cos y - \cos x \sin y} \cdot \frac{1/(\cos x \cos y)}{1/(\cos x \cos y)} \\ &= \frac{(\sin x \cos y)/(\cos x \cos y) + (\cos x \sin y)/(\cos x \cos y)}{(\sin x \cos y)/(\cos x \cos y) - (\cos x \sin y)/(\cos x \cos y)} \\ &= \frac{(\sin x/\cos x) + (\sin y/\cos y)}{(\sin x/\cos x) - (\sin y/\cos y)} \\ &= \frac{\tan x + \tan y}{\tan x - \tan y} \end{aligned}$$

57. False. For example, $\cos 3\pi + \cos 4\pi = 0$, but 3π and 4π are not supplementary. And even though $\cos (3\pi/2) + \cos (3\pi/2) = 0$, $3\pi/2$ is not supplementary with itself.

59. $y = \sin x \cos 2x + \cos x \sin 2x = \sin (x + 2x) = \sin 3x$. The answer is A.

61. For all u, v, $\tan(u + v) = \frac{\tan u + \tan v}{1 - \tan u \tan v}$. The answer is B.

63.

$$\begin{aligned} \tan(u - v) &= \frac{\sin(u - v)}{\cos(u - v)} \\ &= \frac{\sin u \cos v - \cos u \sin v}{\cos u \cos v + \sin u \sin v} \\ &= \frac{\dfrac{\sin u \cos v}{\cos u \cos v} - \dfrac{\cos u \sin v}{\cos u \cos v}}{\dfrac{\cos u \cos v}{\cos u \cos v} + \dfrac{\sin u \sin v}{\cos u \cos v}} \\ &= \frac{\dfrac{\sin u}{\cos u} - \dfrac{\sin v}{\cos v}}{1 + \dfrac{\sin u \sin v}{\cos u \cos v}} \\ &= \frac{\tan u - \tan v}{1 + \tan u \tan v} \end{aligned}$$

65. The identity would involve $\tan\left(\frac{3\pi}{2}\right)$, which does not exit.

$$\begin{aligned} \tan\left(x - \frac{3\pi}{2}\right) &= \frac{\sin\left(x - \frac{3\pi}{2}\right)}{\cos\left(x - \frac{3\pi}{2}\right)} \\ &= \frac{\sin x \cos \frac{3\pi}{2} - \cos x \sin \frac{3\pi}{2}}{\cos x \cos \frac{3\pi}{2} + \sin x \sin \frac{3\pi}{2}} \\ &= \frac{\sin x \cdot 0 - \cos x \cdot (-1)}{\cos x \cdot 0 + \sin x \cdot (-1)} \\ &= -\cot x \end{aligned}$$

67.

$$\begin{aligned} \frac{\cos(x + h) - \cos x}{h} &= \frac{\cos x \cos h - \sin x \sin h - \cos x}{h} \\ &= \frac{\cos x(\cos h - 1) - \sin x \sin h}{h} \\ &= \cos x\left(\frac{\cos h - 1}{h}\right) - \sin x \frac{\sin h}{h} \end{aligned}$$

69.

$$\begin{aligned} \sin(A + B) &= \sin(\pi - C) \\ &= \sin \pi \cos C - \cos \pi \sin C \\ &= 0 \cdot \cos C - (-1) \sin C \\ &= \sin C \end{aligned}$$

71. $\tan A + \tan B + \tan C = \frac{\sin A}{\cos A} + \frac{\sin B}{\cos B} + \frac{\sin C}{\cos C}$

$= \frac{\sin A(\cos B \cos C) + \sin B(\cos A \cos C)}{\cos A \cos B \cos C}$

$+ \frac{\sin C(\cos A \cos B)}{\cos A \cos B \cos C}$

$= \frac{\cos C(\sin A \cos B + \cos A \sin B) + \sin C(\cos A \cos B)}{\cos A \cos B \cos C}$

$= \frac{\cos C \sin(A + B) + \sin C(\cos(A + B) + \sin A \sin B)}{\cos A \cos B \cos C}$

$= \frac{\cos C \sin(\pi - C) + \sin C(\cos(\pi - C) + \sin A \sin B)}{\cos A \cos B \cos C}$

$= \frac{\cos C \sin C + \sin C(-\cos C) + \sin C \sin A \sin B}{\cos A \cos B \cos C}$

$= \frac{\sin A \sin B \sin C}{\cos A \cos B \cos C}$

$= \tan A \tan B \tan C$

73. This equation is easier to deal with after rewriting it as $\cos 5x \cos 4x + \sin 5x \sin 4x = 0$. The left side of this equation is the expanded form of $\cos(5x - 4x)$, which of course equals $\cos x$; the graph shown is simply $y = \cos x$. The equation $\cos x = 0$ is easily solved on the interval $[-2\pi, 2\pi]$: $x = \pm\frac{\pi}{2}$ or $x = \pm\frac{3\pi}{2}$. The original graph is so crowded that one cannot see where crossings occur.

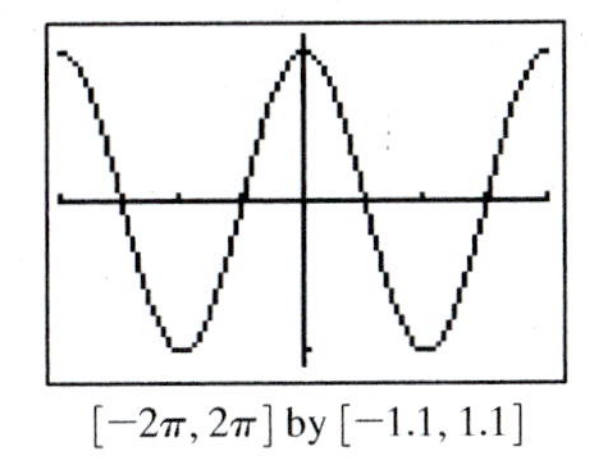

$[-2\pi, 2\pi]$ by $[-1.1, 1.1]$

75. $B = B_{\text{in}} + B_{\text{ref}}$

$= \frac{E_0}{c}\cos\left(\omega t - \frac{\omega x}{c}\right) + \frac{E_0}{c}\cos\left(\omega t + \frac{\omega x}{c}\right)$

$= \frac{E_0}{c}\left(\cos \omega t \cos \frac{\omega x}{c} + \sin \omega t \sin \frac{\omega x}{c}\right.$

$\left. + \cos \omega t \cos \frac{\omega x}{c} - \sin \omega t \sin \frac{\omega x}{c}\right)$

$= \frac{E_0}{c}\left(2 \cos \omega t \cos \frac{\omega x}{c}\right) = 2\frac{E_0}{c}\cos \omega t \cos \frac{\omega x}{c}$

■ Section 5.4 Multiple-Angle Identities

Exploration 1

1. $\sin^2 \frac{\pi}{8} = \frac{1 - \cos(\pi/4)}{2}$

$= \frac{1 - (\sqrt{2}/2)}{2} \cdot \frac{2}{2}$

$= \frac{2 - \sqrt{2}}{4}$

2. $\sin \frac{\pi}{8} = \pm\sqrt{\frac{2 - \sqrt{2}}{4}} = \frac{\sqrt{2 - \sqrt{2}}}{2}$.

We take the positive square root because $\frac{\pi}{8}$ is a first-quadrant angle.

3. $\sin^2 \frac{9\pi}{8} = \frac{1 - \cos(9\pi/4)}{2}$

$= \frac{1 - (\sqrt{2}/2)}{2} \cdot \frac{2}{2}$

$= \frac{2 - \sqrt{2}}{4}$

4. $\sin \frac{9\pi}{8} = \pm\sqrt{\frac{2 - \sqrt{2}}{4}} = \frac{-\sqrt{2 - \sqrt{2}}}{2}$.

We take the negative square root because $\frac{\pi}{8}$ is a third-quadrant angle.

Quick Review 5.4

1. $\tan x = 1$ when $x = \frac{\pi}{4} + n\pi$, n an integer

3. Either $\cos x = 0$ or $\sin x = 1$. The latter implies the former, so $x = \frac{\pi}{2} + n\pi$, n an integer.

5. $\sin x = -\cos x$ when $x = -\frac{\pi}{4} + n\pi$, n an integer

7. Either $\sin x = \frac{1}{2}$ or $\cos x = -\frac{1}{2}$. Then $x = \frac{\pi}{6} + 2n\pi$ or $x = \frac{5\pi}{6} + 2n\pi$ or $x = \pm\frac{2\pi}{3} + 2n\pi$, n an integer.

9. The trapezoid can be viewed as a rectangle and two triangles; the area is then

$A = (2)(3) + \frac{1}{2}(1)(3) + \frac{1}{2}(2)(3) = 10.5$ square units.

Section 5.4 Exercises

1. $\cos 2u = \cos(u + u) = \cos u \cos u - \sin u \sin u$
$= \cos^2 u - \sin^2 u$

3. Starting with the result of #1: $\cos 2u = \cos^2 u - \sin^2 u$
$= (1 - \sin^2 u) - \sin^2 u = 1 - 2\sin^2 u$

5. $2 \sin x \cos x - 2 \sin x = 0$, so $2 \sin x(\cos x - 1) = 0$; $\sin x = 0$ or $\cos x = 1$ when $x = 0$ or $x = \pi$.

7. $2\sin^2 x + \sin x - 1 = 0$, so $(2 \sin x - 1)(\sin x + 1)$
$= 0$; $\sin x = \frac{1}{2}$ or $\sin x = -1$ when $x = \frac{\pi}{6}$, $x = \frac{5\pi}{6}$ or $x = \frac{3\pi}{2}$.

9. $2 \sin x \cos x - \frac{\sin x}{\cos x} = 0$, so $\frac{\sin x}{\cos x}(2\cos^2 x - 1) = 0$, or $\frac{\sin x \cos 2x}{\cos x} = 0$. Then $\sin x = 0$ or $\cos 2x = 0$ (but $\cos x \neq 0$), so $x = 0$, $x = \frac{\pi}{4}$, $x = \frac{3\pi}{4}$, $x = \pi$, $x = \frac{5\pi}{4}$ or $x = \frac{7\pi}{4}$.

For #11–13, any one of the last several expressions given is an answer to the question. In some cases, other answers are possible, as well.

11. $\sin 2\theta + \cos \theta = 2 \sin \theta \cos \theta + \cos \theta$
$= (\cos \theta)(2 \sin \theta + 1)$

13. $\sin 2\theta + \cos 3\theta$
$= 2\sin\theta\cos\theta + \cos 2\theta\cos\theta - \sin 2\theta\sin\theta$
$= 2\sin\theta\cos\theta + (\cos^2\theta - \sin^2\theta)\cos\theta - 2\sin^2\theta\cos\theta$
$= 2\sin\theta\cos\theta + \cos^3\theta - 3\sin^2\theta\cos\theta$
$= 2\sin\theta\cos\theta + 4\cos^3\theta - 3\cos\theta$

15. $\sin 4x = \sin 2(2x) = 2\sin 2x\cos 2x$

17. $2\csc 2x = \dfrac{2}{\sin 2x} = \dfrac{2}{2\sin x\cos x}$
$= \dfrac{1}{\sin^2 x}\cdot\dfrac{\sin x}{\cos x} = \csc^2 x\tan x$

19. $\sin 3x = \sin 2x\cos x + \cos 2x\sin x = 2\sin x\cos^2 x$
$+ (2\cos^2 x - 1)\sin x = (\sin x)(4\cos^2 x - 1)$

21. $\cos 4x = \cos 2(2x) = 1 - 2\sin^2 2x$
$= 1 - 2(2\sin x\cos x)^2 = 1 - 8\sin^2 x\cos^2 x$

23. $2\cos^2 x + \cos x - 1 = 0$, so $\cos x = -1$ or $\cos x = \dfrac{1}{2}$,
$x = \dfrac{\pi}{3}, x = \pi$ or $x = \dfrac{5\pi}{3}$

25. $\cos 3x = \cos 2x\cos x - \sin 2x\sin x$
$= (1 - 2\sin^2 x)\cos x$
$- (2\sin x\cos x)\sin x$
$= \cos x - 2\sin^2 x\cos x$
$- 2\sin^2 x\cos x$
$= \cos x - 4\sin^2 x\cos x$
Thus the left side can be written as $2(\cos x)(1 - 2\sin^2 x)$ $= 2\cos x\cos 2x$. This equals 0 in $[0, 2\pi)$ when
$x = \dfrac{\pi}{4}, x = \dfrac{\pi}{2}, x = \dfrac{3\pi}{4}, x = \dfrac{5\pi}{4}, x = \dfrac{3\pi}{2}$, or $x = \dfrac{7\pi}{4}$.

27. $\sin 2x + \sin 4x = \sin 2x + 2\sin 2x\cos 2x$
$= (\sin 2x)(1 + 2\cos 2x) = 0$. Then $\sin 2x = 0$ or
$\cos 2x = -\dfrac{1}{2}$; the solutions in $[0, 2\pi)$ are
$x = 0, x = \dfrac{\pi}{3}, x = \dfrac{\pi}{2}$,
$x = \dfrac{2\pi}{3}, x = \pi, x = \dfrac{4\pi}{3}, x = \dfrac{3\pi}{2}$, or $x = \dfrac{5\pi}{3}$.

29. Using results from #25, $\sin 2x - \cos 3x$
$= (2\sin x\cos x) - (\cos x - 4\sin^2 x\cos x)$
$= (\cos x)(4\sin^2 x + 2\sin x - 1) = 0$.
$\cos x = 0$ when $x = \dfrac{\pi}{2}$ or $x = \dfrac{3\pi}{2}$, while the second factor equals zero when $\sin x = \dfrac{-1 \pm \sqrt{5}}{4}$. It turns out — as can be observed by noting, e.g., that $\sin^{-1}\left(\dfrac{-1+\sqrt{5}}{4}\right) \approx 0.31415926$ — that this means $x = 0.1\pi, x = 0.9\pi, x = 1.3\pi$, or $x = 1.7\pi$.

31. $\sin 15° = \pm\sqrt{\dfrac{1 - \cos 30°}{2}} = \pm\sqrt{\dfrac{1}{2}\left(1 - \dfrac{\sqrt{3}}{2}\right)}$
$= \pm\dfrac{1}{2}\sqrt{2 - \sqrt{3}}$. Since $\sin 15° > 0$, take the positive square root.

33. $\cos 75° = \pm\sqrt{\dfrac{1 + \cos 150°}{2}} = \pm\sqrt{\dfrac{1}{2}\left(1 - \dfrac{\sqrt{3}}{2}\right)}$
$= \pm\dfrac{1}{2}\sqrt{2 - \sqrt{3}}$. Since $\cos 75° > 0$, take the positive square root.

35. $\tan\dfrac{7\pi}{12} = \dfrac{1 - \cos(7\pi/6)}{\sin(7\pi/6)} = \dfrac{1 + \sqrt{3}/2}{-1/2} = -2 - \sqrt{3}$.

37. **(a)** Starting from the right side: $\dfrac{1}{2}(1 - \cos 2u)$
$= \dfrac{1}{2}[1 - (1 - 2\sin^2 u)] = \dfrac{1}{2}(2\sin^2 u) = \sin^2 u$.

(b) Starting from the right side: $\dfrac{1}{2}(1 + \cos 2u)$
$= \dfrac{1}{2}[1 + (2\cos^2 u - 1)] = \dfrac{1}{2}(2\cos^2 u) = \cos^2 u$.

39. $\sin^4 x = (\sin^2 x)^2 = \left[\dfrac{1}{2}(1 - \cos 2x)\right]^2$
$= \dfrac{1}{4}(1 - 2\cos 2x + \cos^2 2x)$
$= \dfrac{1}{4}\left[1 - 2\cos 2x + \dfrac{1}{2}(1 + \cos 4x)\right]$
$= \dfrac{1}{8}(2 - 4\cos 2x + 1 + \cos 4x)$
$= \dfrac{1}{8}(3 - 4\cos 2x + \cos 4x)$

41. $\sin^3 2x = \sin 2x\sin^2 2x = \sin 2x\cdot\dfrac{1}{2}(1 - \cos 4x)$
$= \dfrac{1}{2}(\sin 2x)(1 - \cos 4x)$

43. $\cos^2 x = \dfrac{1 - \cos x}{2}$, so $2\cos^2 x + \cos x - 1 = 0$. Then $\cos x = -1$ or $\cos x = \dfrac{1}{2}$. In the interval $[0, 2\pi)$, $x = \dfrac{\pi}{3}$, $x = \pi$, or $x = \dfrac{5\pi}{3}$. General solution: $= \pm\dfrac{\pi}{3} + 2n\pi$ or $x = \pi + 2n\pi$, n an integer.

45. The right side equals $\tan^2(x/2)$; the only way that $\tan(x/2) = \tan^2(x/2)$ is if either $\tan(x/2) = 0$ or $\tan(x/2) = 1$. In $[0, 2\pi)$, this happens when $x = 0$ or $x = \dfrac{\pi}{2}$. The general solution is $x = 2n\pi$ or $x = \dfrac{\pi}{2} + 2n\pi$, n an integer.

47. False. For example, $f(x) = 2\sin x$ has period 2π and $g(x) = \cos x$ has period 2π, but the product $f(x)g(x) = 2\sin x\cos x = \sin 2x$ has period π.

49. $f(2x) = \sin 2x = 2\sin x\cos x = 2f(x)g(x)$. The answer is D.

51. $\sin 2x = \cos x$
$2\sin x\cos x = \cos x$
$2\sin x = 1$ or $\cos x = 0$
$\sin x = \dfrac{1}{2}$ or $\cos x = 0$
$x = \dfrac{\pi}{6}$ or $\dfrac{5\pi}{6}$ $\quad x = \dfrac{\pi}{2}$ or $\dfrac{3\pi}{2}$
The answer is E.

53. **(a)** In the figure, the triangle with side lengths $x/2$ and R is a right triangle, since R is given as the perpendicular distance. Then the tangent of the angle $\theta/2$ is the ratio "opposite over adjacent": $\tan\frac{\theta}{2} = \frac{x/2}{R}$ Solving for x gives the desired equation. The central angle θ is $2\pi/n$ since one full revolution of 2π radians is divided evenly into n sections.

(b) $5.87 \approx 2R\tan\frac{\theta}{2}$, where $\theta = 2\pi/11$, so

$R \approx 5.87/(2\tan\frac{\pi}{11}) \approx 9.9957$. $R = 10$.

55. **(a)**

1 ft θ θ 1 ft
1 ft

The volume is 10 ft times the area of the end. The end is made up of two identical triangles, with area $\frac{1}{2}(\sin\theta)(\cos\theta)$ each, and a rectangle with area $(1)(\cos\theta)$. The total volume is then $10\cdot(\sin\theta\cos\theta + \cos\theta) = 10(\cos\theta)(1 + \sin\theta)$.

Considering only $-\frac{\pi}{2} \le \theta \le \frac{\pi}{2}$, the maximum value occurs when $\theta \approx 0.52$ (in fact, it happens exactly at $\theta = \frac{\pi}{6}$). The maximum value is about 12.99 ft^3.

57. $$\csc 2u = \frac{1}{\sin 2u} = \frac{1}{2\sin u\cos u} = \frac{1}{2}\cdot\frac{1}{\sin u}\cdot\frac{1}{\cos u} = \frac{1}{2}\csc u\sec u$$

59. $$\sec 2u = \frac{1}{\cos 2u} = \frac{1}{1 - 2\sin^2 u} = \left(\frac{1}{1 - 2\sin^2 u}\right)\left(\frac{\csc^2 u}{\csc^2 u}\right) = \frac{\csc^2 u}{\csc^2 u - 2}$$

61. $$\sec 2u = \frac{1}{\cos 2u} = \frac{1}{\cos^2 u - \sin^2 u} = \left(\frac{1}{\cos^2 u - \sin^2 u}\right)\left(\frac{\sec^2 u\csc^2 u}{\sec^2 u\csc^2 u}\right) = \frac{\sec^2 u\csc^2 u}{\csc^2 u - \sec^2 u}$$

63. **(a)** The following is a scatter plot of the days past January 1 as x-coordinates (L1) and the time (in 24 hour mode) as y-coordinates (L2) for the time of day that astronomical twilight began in northeastern Mali in 2005.

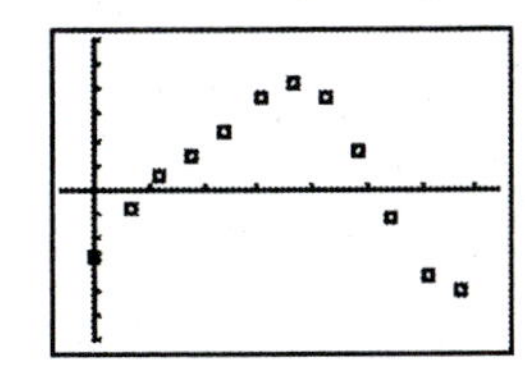

[–30, 370] by [–60, 60]

(b) The sine regression curve through the points defined by L1 and L2 is $y = 41.656\sin(0.015x - 0.825) - 1.473$. This is a fairly good fit, but not really as good as one might expect from data generated by a sinusoidal physical model.

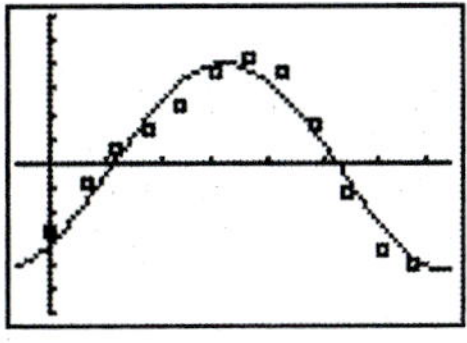

[–30, 370] by [–60, 60]

(c) Using the formula L2 $-$ Y1(L1) (where Y1 is the sine regression curve), the residual list is: $\{3.64, 7.56, 3.35, -5.94, -9.35, -3.90, 5.12, 9.43, 3.90, -4.57, -9.72, -3.22\}$.

(d) The following is a scatter plot of the days past January 1 as x-coordinates (L1) and the residuals (the difference between the actual number of minutes (L2) and the number of minutes predicted by the regression curve (Y1)) as y-coordinates (L3) for the time of day that astronomical twilight began in northeastern Mali in 2005.
The sine regression curve through the points defined by L1 and L3 is $y = 8.856\sin(0.0346x + 0.576) - 0.331$. (Note: Round L3 to 2 decimal places to obtain this answer.) This is another fairly good fit, which indicates that the residuals are not due to chance. There is a periodic variation that is most probably due to physical causes.

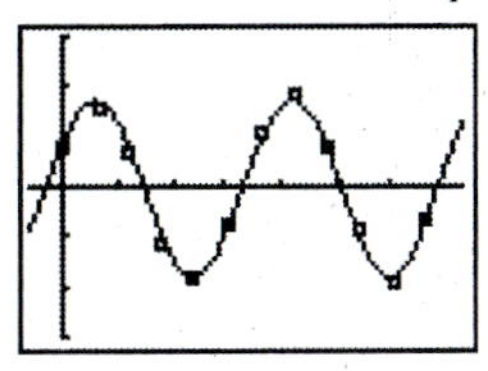

[–30, 370] by [–15, 15]

(e) The first regression indicates that the data are periodic and nearly sinusoidal. The second regression indicates that the *variation* of the data around the predicted values is also periodic and nearly sinusoidal. Periodic variation around periodic models is a predictable consequence of bodies orbiting bodies, but ancient astronomers had a difficult time reconciling the data with their simpler models of the universe.

Section 5.5 The Law of Sines

Exploration 1

1. If $BC \le AB$, the segment will not reach from point B to the dotted line. On the other hand, if $BC > AB$, then a circle of radius BC will intersect the dotted line in a unique point. (Note that the line only extends to the left of point A.)
2. A circle of radius BC will be tangent to the dotted line at C if $BC = h$, thus determining a unique triangle. It will miss the dotted line entirely if $BC < h$, thus determining zero triangles.
3. The second point (C_2) is the reflection of the first point (C_1) on the other side of the altitude.
4. $\sin C_2 = \sin(\pi - C_1) = \sin\pi\cos C_1 - \cos\pi\sin C_1 = \sin C_1$.
5. If $BC \ge AB$, then BC can only extend to the right of the altitude, thus determining a unique triangle.

Quick Review 5.5

1. $a = bc/d$

3. $c = ad/b$

5. $\frac{7 \sin 48°}{\sin 23°} \approx 13.314°$

7. $x = \sin^{-1} 0.3 \approx 17.458°$

9. $x = 180° - \sin^{-1}(-0.7) \approx 224.427°$

Section 5.5 Exercises

1. Given: $b = 3.7, B = 45°, A = 60°$ — an AAS case.
$C = 180° - (A + B) = 75°;$
$$\frac{a}{\sin A} = \frac{b}{\sin B} \Rightarrow a = \frac{b \sin A}{\sin B} = \frac{3.7 \sin 60°}{\sin 45°} \approx 4.5;$$
$$\frac{b}{\sin B} = \frac{c}{\sin C} \Rightarrow c = \frac{b \sin C}{\sin B} = \frac{3.7 \sin 75°}{\sin 45°} \approx 5.1$$

3. Given: $A = 100°, C = 35°, a = 22$ — an AAS case.
$B = 180° - (A + C) = 45°;$
$$b = \frac{a \sin B}{\sin A} = \frac{22 \sin 45°}{\sin 100°} \approx 15.8;$$
$$c = \frac{a \sin C}{\sin A} = \frac{22 \sin 35°}{\sin 100°} \approx 12.8$$

5. Given: $A = 40°, B = 30°, b = 10$ — an AAS case.
$C = 180° - (A + B) = 110°;$
$$a = \frac{b \sin A}{\sin B} = \frac{10 \sin 40°}{\sin 30°} \approx 12.9;$$
$$c = \frac{b \sin C}{\sin B} = \frac{10 \sin 110°}{\sin 30°} \approx 18.8$$

7. Given: $A = 33°, B = 70°, b = 7$ — an AAS case.
$C = 180° - (A + B) = 77°;$
$$a = \frac{b \sin A}{\sin B} = \frac{7 \sin 33°}{\sin 70°} \approx 4.1;$$
$$c = \frac{b \sin C}{\sin B} = \frac{7 \sin 77°}{\sin 70°} \approx 7.3$$

9. Given: $A = 32°, a = 17, b = 11$ — an SSA case.
$h = b \sin A \approx 5.8; h < b < a$, so there is one triangle.
$$B = \sin^{-1}\left(\frac{b \sin A}{a}\right) = \sin^{-1}(0.342\ldots) \approx 20.1°$$
$C = 180° - (A + B) \approx 127.9°;$
$$c = \frac{a \sin C}{\sin A} = \frac{17 \sin 127.9°}{\sin 32°} \approx 25.3$$

11. Given: $B = 70°, b = 14, c = 9$ — an SSA case.
$h = c \sin B \approx 8.5; h < c < b$, so there is one triangle.
$$C = \sin^{-1}\left(\frac{c \sin B}{b}\right) = \sin^{-1}(0.604\ldots) \approx 37.2°$$
$A = 180° - (B + C) \approx 72.8°;$
$$a = \frac{b \sin A}{\sin B} = \frac{14 \sin 72.8°}{\sin 70°} \approx 14.2$$

13. Given: $A = 36°, a = 2, b = 7. h = b \sin A \approx 4.1;$
$a < h$, so no triangle is formed.

15. Given: $C = 36°, a = 17, c = 16. h = a \sin C \approx 10.0;$
$h < c < a$, so there are two triangles.

17. Given: $C = 30°, a = 18, c = 9. h = a \sin C = 9;$
$h = c$, so there is one triangle.

19. Given: $A = 64°, a = 16, b = 17. h = b \sin A \approx 15.3;$
$h < a < b$, so there are two triangles.
$$B_1 = \sin^{-1}\left(\frac{b \sin A}{a}\right) = \sin^{-1}(0.954\ldots) \approx 72.7°$$
$C_1 = 180° - (A + B_1) \approx 43.3°;$
$$c_1 = \frac{a \sin C_1}{\sin A} = \frac{16 \sin 43.3°}{\sin 64°} \approx 12.2$$
Or (with B obtuse):
$B_2 = 180° - B_1 \approx 107.3°;$
$C_2 = 180° - (A + B_2) \approx 8.7°;$
$$c_2 = \frac{a \sin C_2}{\sin A} \approx 2.7$$

21. Given: $C = 68°, a = 19, c = 18. h = a \sin C \approx 17.6;$
$h < c < a$, so there are two triangles.
$$A_1 = \sin^{-1}\left(\frac{a \sin C}{c}\right) = \sin^{-1}(0.978\ldots) \approx 78.2°$$
$B_1 = 180° - (A + C) \approx 33.8°;$
$$b_1 = \frac{c \sin B_1}{\sin C} = \frac{18 \sin 33.8°}{\sin 68°} \approx 10.8$$
Or (with A obtuse):
$A_2 = 180° - A_1 \approx 101.8°;$
$B_2 = 180° - (A_2 + C) \approx 10.2°;$
$$b_2 = \frac{c \sin B_2}{\sin C} \approx 3.4$$

23. $h = 10 \sin 42° \approx 6.691$, so:

(a) $6.691 < b < 10.$

(b) $b \approx 6.691$ or $b \geq 10.$

(c) $b < 6.691$

25. **(a)** No: this is an SAS case

(b) No: only two pieces of information given.

27. Given: $A = 61°, a = 8, b = 21$ — an SSA case.
$h = b \sin A = 18.4; a < h$, so no triangle is formed.

29. Given: $A = 136°, a = 15, b = 28$ — an SSA case.
$h = b \sin A \approx 19.5; a < h$, so no triangle is formed.

31. Given: $B = 42°, c = 18, C = 39°$ — an AAS case.
$A = 180° - (B + C) = 99°;$
$$a = \frac{c \sin A}{\sin C} = \frac{18 \sin 99°}{\sin 39°} \approx 28.3;$$
$$b = \frac{c \sin B}{\sin C} = \frac{18 \sin 42°}{\sin 39°} \approx 19.1$$

33. Given: $C = 75°, b = 49, c = 48.$ — an SSA case.
$h = b \sin C \approx 47.3; h < c < b$, so there are two triangles.
$$B_1 = \sin^{-1}\left(\frac{b \sin C}{c}\right) = \sin^{-1}(0.986\ldots) \approx 80.4°$$
$A_1 = 180° - (B + C) \approx 24.6°;$
$$a_1 = \frac{c \sin A_1}{\sin C} = \frac{48 \sin 24.6°}{\sin 75°} \approx 20.7$$
Or (with B obtuse):
$B_2 = 180° - B_1 \approx 99.6°;$
$A_2 = 180° - (B_2 + C) \approx 5.4°;$
$$a_2 = \frac{c \sin A_2}{\sin C} \approx 4.7$$

35. Cannot be solved by law of sines (an SAS case).

37. Given: $c = AB = 56, A = 72°, B = 53°$ — an ASA case, so $C = 180° - (A + B) = 55°$

(a) $AC = b = \dfrac{c \sin B}{\sin C} = \dfrac{56 \sin 53°}{\sin 55°} \approx 54.6$ ft.

(b) $h = b \sin A(= a \sin B) \approx 51.9$ ft.

39. Given: $c = 16, C = 90° - 62 = 28°$, $B = 90° + 15° = 105°$ — an AAS case. $A = 180° - (B + C) = 47°$, so
$a = \dfrac{c \sin A}{\sin C} = \dfrac{16 \sin 47°}{\sin 28°} \approx 24.9$ ft.

41.

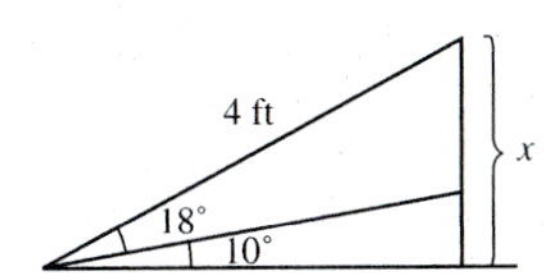

The length of the brace is the leg of the larger triangle. $\sin 28° = \dfrac{x}{4}$, so $x = 1.9$ ft.

43. Consider the triangle with vertices at the top of the flagpole (A) and the two observers (B and C). Then $a = 600$, $B = 19°$, and $C = 21°$ (an ASA case), so
$A = 180° - (B + C) = 140°$;
$b = \dfrac{a \sin B}{\sin A} = \dfrac{600 \sin 19°}{\sin 140°} \approx 303.9$;
$c = \dfrac{a \sin C}{\sin A} = \dfrac{600 \sin 21°}{\sin 140°} \approx 334.5$
and finally $h = b \sin C = c \sin B \approx 108.9$ ft.

45. Given: $c = 20, B = 52°, C = 33°$ — an AAS case.
$A = 180° - (B + C) = 95°$, so
$a = \dfrac{c \sin A}{\sin C} = \dfrac{20 \sin 95°}{\sin 33°} \approx 36.6$ mi, and
$b = \dfrac{c \sin B}{\sin C} = \dfrac{20 \sin 52°}{\sin 33°} \approx 28.9$ mi.

47. True. By the law of sines, $\dfrac{\sin A}{a} = \dfrac{\sin B}{b}$,
which is equivalent to $\dfrac{\sin A}{\sin B} = \dfrac{a}{b}$ (since $\sin A, \sin B \neq 0$).

49. The third angle is 32°. By the Law of Sines, $\dfrac{\sin 32°}{12.0} = \dfrac{\sin 53°}{x}$, which can be solved for x.
The answer is C.

51. The longest side is opposite the largest angle, while the shortest side is opposite the smallest angle. By the Law of Sines, $\dfrac{\sin 50°}{9.0} = \dfrac{\sin 70°}{x}$, which can be solved for x.
The answer is A.

53. (a) Given any triangle with side lengths a, b, and c, the law of sines says that $\dfrac{\sin A}{a} = \dfrac{\sin B}{b} = \dfrac{\sin C}{c}$.
But we can also find another triangle(using ASA) with two angles the same as the first (in which case the third angle is also the same) and a different side length — say, a'. Suppose that $a' = ka$ for some constant k. Then for this new triangle, we have $\dfrac{\sin A}{a'} = \dfrac{\sin B}{b'} = \dfrac{\sin C}{c'}$. Since $\dfrac{\sin A}{a'} = \dfrac{\sin A}{ka} = \dfrac{1}{k} \cdot \dfrac{\sin A}{a}$, we can see that $\dfrac{\sin B}{b'} = \dfrac{1}{k} \cdot \dfrac{\sin B}{b}$, so that $b' = kb$ and similarly, $c' = kc$. So for any choice of a positive constant k, we can create a triangle with angles A, B, and C.

(b) Possible answers: $a = 1, b = \sqrt{3}, c = 2$ (or any set of three numbers proportional to these).

(c) Any set of three identical numbers.

55. (a) $h = AB \sin A$

(b) $BC < AB \sin A$

(c) $BC \geq AB$ or $BC = AB \sin A$

(d) $AB \sin A < BC < AB$

57. Given: $c = 4.1, B = 25°, C = 36.5° - 25° = 11.5°$. An AAS case: $A = 180° - (B + C) = 143.5°$, so
$AC = b = \dfrac{c \sin B}{\sin C} = \dfrac{4.1 \sin 25°}{\sin 11.5°} \approx 8.7$ mi, and
$BC = a = \dfrac{c \sin A}{\sin C} = \dfrac{4.1 \sin 143.5°}{\sin 11.5°} \approx 12.2$ mi.
The height is $h = a \sin 25° = b \sin 36.5° \approx 5.2$ mi.

Section 5.6 The Law of Cosines

Exploration 1

1. The semiperimeters are 154 and 150.
$A = \sqrt{154(154 - 115)(154 - 81)(154 - 112)}$
$+ \sqrt{150(150 - 112)(150 - 102)(150 - 86)}$
$= 8475.742818 \text{ paces}^2$

2. 41,022.59524 square feet

3. 0.0014714831 square miles

4. 0.94175 acres

5. The estimate of "a little over an acre" seems questionable, but the roughness of their measurement system does not provide firm evidence that it is incorrect. If Jim and Barbara wish to make an issue of it with the owner, they would be well-advised to get some more reliable data.

6. Yes. In fact, any polygonal region can be subdivided into triangles.

Quick Review 5.6

1. $A = \cos^{-1}\left(\dfrac{3}{5}\right) \approx 53.130°$

3. $A = \cos^{-1}(-0.68) \approx 132.844°$

5. (a) $\cos A = \dfrac{81 - x^2 - y^2}{-2xy} = \dfrac{x^2 + y^2 - 81}{2xy}$.

(b) $A = \cos^{-1}\left(\dfrac{x^2 + y^2 - 81}{2xy}\right)$

7. One answer: $(x - 1)(x - 2) = x^2 - 3x + 2$.
Generally: $(x - a)(x - b) = x^2 - (a + b)x + ab$ for any two positive numbers a and b.

9. One answer: $(x - i)(x + i) = x^2 + 1$

Section 5.6 Exercises

1. Given: $B = 131°, c = 8, a = 13$ — an SAS case.
$b = \sqrt{a^2 + c^2 - 2ac \cos B} \approx \sqrt{369.460} \approx 19.2$;
$C = \cos^{-1}\left(\dfrac{a^2 + b^2 - c^2}{2ab}\right) \approx \cos^{-1}(0.949) \approx 18.3°$;
$A = 180° - (B + C) \approx 30.7°$.

3. Given: $a = 27, b = 19, c = 24$ — an SSS case.
$A = \cos^{-1}\left(\frac{b^2 + c^2 - a^2}{2bc}\right) \approx \cos^{-1}(0.228) \approx 76.8°;$
$B = \cos^{-1}\left(\frac{a^2 + c^2 - b^2}{2ac}\right) \approx \cos^{-1}(0.728) \approx 43.2°;$
$C = 180° - (A + B) \approx 60°.$

5. Given: $A = 55°, b = 12, c = 7$ — an SAS case.
$a = \sqrt{b^2 + c^2 - 2bc \cos A} \approx \sqrt{96.639} \approx 9.8;$
$B = \cos^{-1}\left(\frac{a^2 + c^2 - b^2}{2ac}\right) \approx \cos^{-1}(0.011) \approx 89.3°;$
$C = 180° - (A + B) \approx 35.7°.$

7. Given: $a = 12, b = 21, C = 95°$ — an SAS case.
$c = \sqrt{a^2 + b^2 - 2ab \cos C} \approx \sqrt{628.926} \approx 25.1;$
$A = \cos^{-1}\left(\frac{b^2 + c^2 - a^2}{2bc}\right) \approx \cos^{-1}(0.879) \approx 28.5°;$
$B = 180° - (A + C) \approx 56.5°.$

9. No triangles possible $(a + c = b)$

11. Given: $a = 3.2, b = 7.6, c = 6.4$ — an SSS case.
$A = \cos^{-1}\left(\frac{b^2 + c^2 - a^2}{2bc}\right) \approx \cos^{-1}(0.909) \approx 24.6°;$
$B = \cos^{-1}\left(\frac{a^2 + c^2 - b^2}{2ac}\right) \approx \cos^{-1}(^{-}0.160) \approx 99.2°;$
$C = 180° - (A + B) \approx 56.2°.$

13. Given: $A = 42°, a = 7, b = 10$ — an SSA case. Solve the quadratic equation $7^2 = 10^2 + c^2 - 2(10)c \cos 42°$, or $c^2 - (14.862\ldots)c + 51 = 0$; there are two positive solutions: ≈ 9.487 *or* 5.376. Since $\cos B = \frac{a^2 + c^2 - b^2}{2ac}$:
$c_1 \approx 9.487, B_1 \approx \cos^{-1}(0.294) \approx 72.9°$, and
$C_1 = 180° - (A + B_1) \approx 65.1°$,
or
$c_2 \approx 5.376, B_2 \approx \cos^{-1}(-0.294) \approx 107.1°$, and
$C_2 = 180° - (A + B_2) \approx 30.9°.$

15. Given: $A = 63°, a = 8.6, b = 11.1$ — an SSA case. Solve the quadratic equation
$8.6^2 = 11.1^2 + c^2 - 2(11.1)c \cos 63°$, or
$c^2 - (10.079)c + 49.25 = 0$; there are no real solutions, so there is no triangle.

17. Given: $A = 47°, b = 32, c = 19$ — an SAS case.
$a = \sqrt{b^2 + c^2 - 2bc \cos A} \approx \sqrt{555.689} \approx 23.573,$
so Area $\approx \sqrt{49431.307} \approx 222.33$ ft^2 (using Heron's formula). Or, use $A = \frac{1}{2}bc \sin A$.

19. Given: $B = 101°, a = 10, c = 22$ — an SAS case.
$b = \sqrt{a^2 + c^2 - 2ac \cos B} \approx \sqrt{667.955} \approx 25.845,$
so Area $\approx \sqrt{11659.462} \approx 107.98$ cm^2 (using Heron's formula). Or, use $A = \frac{1}{2}ac \sin B$.

For #21–27, a triangle can be formed if $a + b < c, a + c < b$, and $b + c < a$.

21. $s = \frac{17}{2}$; Area $= \sqrt{66.9375} \approx 8.18$

23. No triangle is formed $(a + b = c)$.

25. $a = 36.4$; Area $= \sqrt{46{,}720.3464} \approx 216.15$

27. $s = 42.1$; Area $= \sqrt{98{,}629.1856} \approx 314.05$

29. Let $a = 4, b = 5$, and $c = 6$. The largest angle is opposite the largest side, so we call it C. Since
$\cos C = \frac{a^2 + b^2 - c^2}{2ab}, C = \cos^{-1}\left(\frac{1}{8}\right) \approx 82.819°$
≈ 1.445 radians.

31. Following the method of Example 3, divide the hexagon into 6 triangles. Each has two 12-inch sides that form a 60° angle.
$6 \times \frac{1}{2}(12)(12)\sin 60° = 216\sqrt{3} \approx 374.1$ square inches

33.

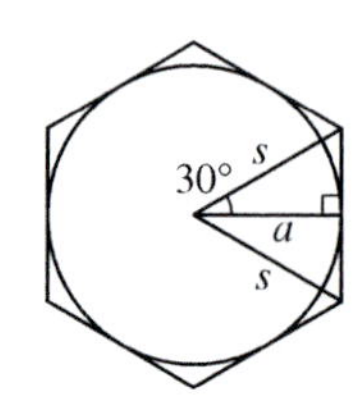

In the figure, $a = 12$ and so $s = 12 \sec 30° = 8\sqrt{3}$. The area of the hexagon is
$6 \times \frac{1}{2}(8\sqrt{3})(8\sqrt{3})\sin 60° = 288\sqrt{3}$
≈ 498.8 square inches.

35. Given: $C = 54°, BC = a = 160, AC = b = 110$ — an SAS case. $AB = c = \sqrt{a^2 + b^2 - 2ab \cos C}$
$\approx \sqrt{17{,}009.959} \approx 130.42$ ft.

37. **(a)** $c = \sqrt{40^2 + 60^2 - 2(40)(60) \cos 45°}$
$\approx \sqrt{1805.887} \approx 42.5$ ft.

(b) The home-to-second segment is the hypotenuse of a right triangle, so the distance from the pitcher's rubber to second base is $60\sqrt{2} - 40 \approx 44.9$ ft.

(c) $B = \cos^{-1}\left(\frac{a^2 + c^2 - b^2}{2ac}\right) \approx \cos^{-1}(-0.057)$
$\approx 93.3°.$

39. **(a)** Using right ΔACE, $m\angle CAE = \tan^{-1}\left(\frac{6}{18}\right)$
$= \tan^{-1}\left(\frac{1}{3}\right) \approx 18.435°.$

(b) Using $A \approx 18.435°$, we have an SAS case, so
$DF = \sqrt{9^2 + 12^2 - 2(9)(12) \cos A} \approx \sqrt{20.084}$
≈ 4.5 ft.

(c) $EF = \sqrt{18^2 + 12^2 - 2(18)(12) \cos A} \approx \sqrt{58.168}$
≈ 7.6 ft.

41. $AB = \sqrt{73^2 + 65^2 - 2(73)(65) \cos 8°}$
$\approx \sqrt{156.356} \approx 12.5$ yd.

43. $AB = c = \sqrt{2^2 + 3^2} = \sqrt{13}, AC = b = \sqrt{1^2 + 3^2}$
$= \sqrt{10}$, and $BC = a = \sqrt{1^2 + 2^2} = \sqrt{5}$, so
$m\angle CAB = A = \cos^{-1}\left(\frac{b^2 + c^2 - a^2}{2bc}\right)$
$= \cos^{-1}\left(\frac{9}{\sqrt{130}}\right) \approx 37.9°.$

45. True. By the Law of Cosines, $b^2 + c^2 - 2bc \cos A = a^2$, which is a positive number. Since $b^2 + c^2 - 2bc \cos A > 0$, it follows that $b^2 + c^2 > 2bc \cos A$.

47. Following the method of Example 3, divide the dodecagon into 12 triangles. Each has two 12-inch sides that form a 30° angle.

$12 \times \frac{1}{2}(12)(12)\sin 30° = 432$

The answer is B.

49. After 30 minutes, the first boat has traveled 12 miles and the second has traveled 16 miles. By the Law of Cosines, the two boats are $\sqrt{12^2 + 16^2 - 2(12)(16)\cos 110°} \approx 23.05$ miles apart. The answer is C.

51. Consider that a n-sided regular polygon inscribed within a circle can divide into n equilateral triangles, each with equal area of $\frac{r^2}{2}\sin\frac{360°}{n}$. (The two equal sides of the equilateral triangle are of length r, the radius of the circle.) Then, the area of the polygon is exactly $\frac{nr^2}{2}\sin\frac{360°}{n}$.

53. **(a)** Ship A: $\frac{30.2 - 15.1}{1 \text{ hr}} = 15.1$ knots;

Ship B: $\frac{37.2 - 12.4}{2 \text{ hrs}} = 12.4$ knots

(b) $\cos A = \frac{b^2 + c^2 - a^2}{2bc}$

$= \frac{(15.1)^2 + (12.4)^2 - (8.7)^2}{2(15.1)(12.4)}$

$A = 35.18°$

(c) $c^2 = a^2 + b^2 - 2ab\cos C$

$= (49.6)^2 + (60.4)^2 - 2(49.6)(60.4)\cos(35.18°)$

≈ 1211.04, so the boats are 34.8 nautical miles apart at noon.

55. Let P be the center of the circle. Then, $\cos P = \frac{5^2 + 5^2 - 7^2}{2(5)(5)} = 0.02$, so $P \approx 88.9°$. The area of the segment is $\pi r^2 \cdot \frac{88.9°}{360°} \approx 25\pi \cdot (0.247) \approx 19.39 \text{ in}^2$.

The area of the triangle, however, is $\frac{1}{2}(5)(5)\sin(88.9)$ $\approx 12.50 \text{ in}^2$, so the area of the shaded region is approx. 6.9 in^2.

Chapter 5 Review

1. $2\sin 100° \cos 100° = \sin 200°$

3. 1; the expression simplifies to $(\cos 2\theta)^2 + (2\sin\theta\cos\theta)^2 = (\cos 2\theta)^2 + (\sin 2\theta)^2 = 1$.

5. $\cos 3x = \cos(2x + x) = \cos 2x \cos x - \sin 2x \sin x$

$= (\cos^2 x - \sin^2 x)\cos x - (2\sin x\cos x)\sin x$

$= \cos^3 x - 3\sin^2 x\cos x$

$= \cos^3 x - 3(1 - \cos^2 x)\cos x$

$= \cos^3 x - 3\cos x + 3\cos^3 x$

$= 4\cos^3 x - 3\cos x$

7. $\tan^2 x - \sin^2 x = \sin^2 x\left(\frac{1 - \cos^2 x}{\cos^2 x}\right)$

$= \sin^2 x \cdot \frac{\sin^2 x}{\cos^2 x} = \sin^2 x \tan^2 x$

9. $\csc x - \cos x \cot x = \frac{1}{\sin x} - \cos x \cdot \frac{\cos x}{\sin x}$

$= \frac{1 - \cos^2 x}{\sin x} = \frac{\sin^2 x}{\sin x} = \sin x$

11. Recall that $\tan\theta\cot\theta = 1$. $\frac{1 + \tan\theta}{1 - \tan\theta} + \frac{1 + \cot\theta}{1 - \cot\theta}$

$= \frac{(1 + \tan\theta)(1 - \cot\theta) + (1 + \cot\theta)(1 - \tan\theta)}{(1 - \tan\theta)(1 - \cot\theta)}$

$= \frac{(1 + \tan\theta - \cot\theta - 1) + (1 + \cot\theta - \tan\theta - 1)}{(1 - \tan\theta)(1 - \cot\theta)}$

$= \frac{0}{(1 - \tan\theta)(1 - \cot\theta)} = 0$

13. $\cos^2\frac{t}{2} = \left[\pm\sqrt{\frac{1}{2}(1 + \cos t)}\right]^2 = \frac{1}{2}(1 + \cos t)$

$= \left(\frac{1 + \cos t}{2}\right)\left(\frac{\sec t}{\sec t}\right) = \frac{1 + \sec t}{2\sec t}$

15. $\frac{\cos\phi}{1 - \tan\phi} + \frac{\sin\phi}{1 - \cot\phi}$

$= \left(\frac{\cos\phi}{1 - \tan\phi}\right)\left(\frac{\cos\phi}{\cos\phi}\right) + \left(\frac{\sin\phi}{1 - \cot\phi}\right)\left(\frac{\sin\phi}{\sin\phi}\right)$

$= \frac{\cos^2\phi}{\cos\phi - \sin\phi} + \frac{\sin^2\phi}{\sin\phi - \cos\phi} = \frac{\cos^2\phi - \sin^2\phi}{\cos\phi - \sin\phi}$

$= \cos\phi + \sin\phi$

17. $\sqrt{\frac{1 - \cos y}{1 + \cos y}} = \sqrt{\frac{(1 - \cos y)^2}{(1 + \cos y)(1 - \cos y)}}$

$= \sqrt{\frac{(1 - \cos y)^2}{1 - \cos^2 y}} = \sqrt{\frac{(1 - \cos y)^2}{\sin^2 y}}$

$= \frac{|1 - \cos y|}{|\sin y|} = \frac{1 - \cos y}{|\sin y|}$ — since $1 - \cos y \geq 0$, we can drop that absolute value.

19. $\tan\left(u + \frac{3\pi}{4}\right) = \frac{\tan u + \tan(3\pi/4)}{1 - \tan u \tan(3\pi/4)}$

$= \frac{\tan u + (-1)}{1 - \tan u(-1)} = \frac{\tan u - 1}{1 + \tan u}$

21. $\tan\frac{1}{2}\beta = \frac{1 - \cos\beta}{\sin\beta} = \frac{1}{\sin\beta} - \frac{\cos\beta}{\sin\beta} = \csc\beta - \cot\beta$

23. Yes: $\sec x - \sin x\tan x = \frac{1}{\cos x} - \frac{\sin^2 x}{\cos x}$

$= \frac{1 - \sin^2 x}{\cos x} = \frac{\cos^2 x}{\cos x} = \cos x$.

25. Many answers are possible, for example,

$\sin 3x + \cos 3x$

$= (3\sin x - 4\sin^3 x) + (4\cos^3 x - 3\cos x)$

$= 3(\sin x - \cos x) - 4(\sin^3 x - \cos^3 x)$

$= (\sin x - \cos x)[3 - 4(\sin^2 x + \sin x\cos x + \cos^2 x)]$

$= (\sin x - \cos x)(3 - 4 - 4\sin x\cos x)$

$= (\cos x - \sin x)(1 + 4\sin x\cos x)$. Check other answers with a grapher.

27. Many answers are possible, for example,

$\cos^2 2x - \sin 2x = 1 - \sin^2 2x - \sin 2x$

$= 1 - 4\sin^2 x\cos^2 x - 2\sin x\cos x$. Check other answers with a grapher.

In #29–33, n represents any integer.

29. $\sin 2x = 0.5$ when $2x = \frac{\pi}{6} + 2n\pi$ or $2x = \frac{5\pi}{6} + 2n\pi$, so $x = \frac{\pi}{12} + n\pi$ or $x = \frac{5\pi}{12} + n\pi$.

31. $\tan x = -1$ when $x = -\frac{\pi}{4} + n\pi$

33. If $\tan^{-1} x = 1$, then $x = \tan 1$.

35. $x \approx 1.12$ or $x \approx 5.16$

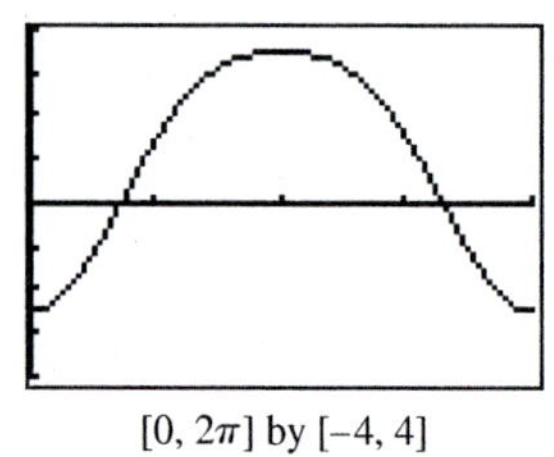

[0, 2π] by [−4, 4]

37. $x \approx 1.15$

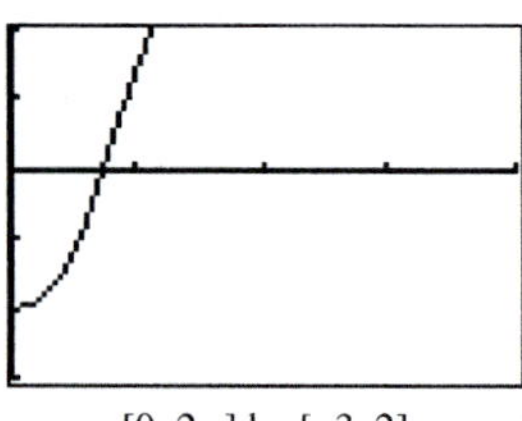

[0, 2π] by [−3, 2]

39. $\cos x = \frac{1}{2}$, so $x = \frac{\pi}{3}$ or $x = \frac{5\pi}{3}$.

41. The left side factors to $(\sin x - 3)(\sin x + 1) = 0$; only $\sin x = -1$ is possible, so $x = \frac{3\pi}{2}$.

43. $\sin(\cos x) = 1$ only if $\cos x = \frac{\pi}{2} + 2n\pi$. No choice of n gives a value in $[-1, 1]$, so there are no solutions.

For #45–47, use graphs to suggest the intervals. To find the endpoints of the intervals, treat the inequalities as equations and solve.

45. $\cos 2x = \frac{1}{2}$ has solutions $x = \frac{\pi}{6}, x = \frac{5\pi}{6}, x = \frac{7\pi}{6}$, and $x = \frac{11\pi}{6}$ in interval $[0, 2\pi)$. The solution set for the inequality is, $0 \le x < \frac{\pi}{6}$ or $\frac{5\pi}{6} < x < \frac{7\pi}{6}$ or $\frac{11\pi}{6} < x < 2\pi$; that is, $\left[0, \frac{\pi}{6}\right) \cup \left(\frac{5\pi}{6}, \frac{7\pi}{6}\right) \cup \left(\frac{11\pi}{6}, 2\pi\right)$.

47. $\cos x = \frac{1}{2}$ has solutions $x = \frac{\pi}{3}$ and $x = \frac{5\pi}{3}$ in the interval $[0, 2\pi)$. The solution set for the inequality is $\frac{\pi}{3} < x < \frac{5\pi}{3}$; that is, $\left(\frac{\pi}{3}, \frac{5\pi}{3}\right)$.

49. $y = 5 \sin(3x + \cos^{-1}(3/5)) \approx 5 \sin(3x + 0.9273)$

51. Given: $A = 79°, B = 33°, a = 7$ — an AAS case.

$C = 180° - (A + B) = 68°$

$b = \frac{a \sin B}{\sin A} = \frac{7 \sin 33°}{\sin 79°} \approx 3.9;$

$c = \frac{a \sin C}{\sin A} = \frac{7 \sin 68°}{\sin 79°} \approx 6.6.$

53. Given: $a = 8, b = 3, B = 30°$ — an SSA case. Using the Law of Sines: $h = a \sin B = 4$; $b < h$, so no triangle is formed. Using the Law of Cosines: Solve the quadratic equation $3^2 = 8^2 + c^2 - 2(8)c \cos 30°$, or $c^2 - (8\sqrt{3})c + 55 = 0$; there are no real solutions.

55. Given: $A = 34°, B = 74°, c = 5$ — an ASA case.

$C = 180° - (A + B) = 72°;$

$a = \frac{c \sin A}{\sin C} = \frac{5 \sin 34°}{\sin 72°} \approx 2.9;$

$b = \frac{c \sin B}{\sin C} = \frac{5 \sin 74°}{\sin 72°} \approx 5.1.$

57. Given: $a = 5, b = 7, c = 6$ — an SSS case.

$A = \cos^{-1}\left(\frac{b^2 + c^2 - a^2}{2bc}\right) \approx \cos^{-1}(0.714) \approx 44.4°;$

$B = \cos^{-1}\left(\frac{a^2 + c^2 - b^2}{2ac}\right) \approx \cos^{-1}(0.2) \approx 78.5°;$

$C = 180° - (A + B) \approx 57.1°.$

59. $s = \frac{1}{2}(3 + 5 + 6) = 7;$

Area $= \sqrt{s(s-a)(s-b)(s-c)}$
$= \sqrt{7(7-3)(7-5)(7-6)}$
$= \sqrt{56} \approx 7.5$

61. $h = 12 \sin 28° \approx 5.6$, so:

(a) $\approx 5.6 < b < 12$.

(b) $b \approx 5.6$ or $b \ge 12$.

(c) $b < 5.6$.

63. Given: $c = 1.75, A = 33°, B = 37°$ — an ASA case, so

$C = 180° - (A + B) = 110°;$

$a = \frac{c \sin A}{\sin C} = \frac{1.75 \sin 33°}{\sin 110°} \approx 1.0;$

$b = \frac{c \sin B}{\sin C} = \frac{1.75 \sin 37°}{\sin 110°} \approx 1.1,$

and finally, the height is $h = b \sin A = a \sin b \approx 0.6$ mi.

65. Let $a = 8, b = 9$, and $c = 10$. The largest angle is opposite the largest side, so we call it C. Since $\cos C = \frac{a^2 + b^2 - c^2}{2ab}$, $C = \cos^{-1}\left(\frac{5}{16}\right)$ $\approx 71.790°$, 1.253 rad.

67. (a) The point (x, y) has coordinates $(\cos\theta, \sin\theta)$, so the bottom is $b_1 = 2$ units wide, the top is $b_2 = 2x = 2\cos\theta$ units wide, and the height is $h = y = \sin\theta$ units. Either use the formula for the area of a trapezoid, $A = \frac{1}{2}(b_1 + b_2)h$, or notice that the trapezoid can be split into two triangles and a rectangle. Either way:

$A(\theta) = \sin\theta + \sin\theta\cos\theta = \sin\theta(1 + \cos\theta)$
$= \sin\theta + \frac{1}{2}\sin 2\theta.$

(b) The maximizing angle is $\theta = \frac{\pi}{3} = 60°$; the maximum area is $\frac{3}{4}\sqrt{3} \approx 1.30$ square units.

69. **(a)** Split the quadrilateral in half to leave two (identical) right triangles, with one leg 4000 mi, hypotenuse $4000 + h$ mi, and one acute angle $\theta/2$.

Then $\cos\frac{\theta}{2} = \frac{4000}{4000 + h}$; solve for h to leave

$$h = \frac{4000}{\cos(\theta/2)} - 4000 = 4000\sec\frac{\theta}{2} - 4000 \text{ miles.}$$

(b) $\cos\frac{\theta}{2} = \frac{4000}{4200}$, so $\theta = 2\cos^{-1}\left(\frac{20}{21}\right) \approx 0.62 \approx 35.51°$.

71. The hexagon is made up of 6 equilateral triangles; using Heron's formula (or some other method), we find that each triangle has area $\sqrt{24(24 - 16)^3} = \sqrt{12{,}288} = 64\sqrt{3}$. The hexagon's area is therefore, $384\sqrt{3}$ cm^2, and the radius of the circle is 16 cm, so the area of the circle is 256π cm^2, and the area outside the hexagon is $256\pi - 384\sqrt{3} \approx 139.140$ cm^2.

73. The volume of a cylinder with radius r and height h is $V = \pi r^2 h$, so the wheel of cheese has volume $\pi(9^2)(5) = 405\pi$ cm^3; a 15° wedge would have fraction $\frac{15}{360} = \frac{1}{24}$ of that volume, or $\frac{405\pi}{24} \approx 53.01$ cm^3.

75. **(a)** By the product-to-sum formula in 74 (c),

$$2\sin\frac{u+v}{2}\cos\frac{u-v}{2}$$
$$= 2\cdot\frac{1}{2}\left(\sin\frac{u+v+u-v}{2} + \sin\frac{u+v-(u-v)}{2}\right)$$
$$= \sin u + \sin v$$

(b) By the product-to-sum formula in 74 (c),

$$2\sin\frac{u-v}{2}\cos\frac{u+v}{2}$$
$$= 2\cdot\frac{1}{2}\left(\sin\frac{u-v+u+v}{2} + \sin\frac{u-v-(u+v)}{2}\right)$$
$$= \sin u + \sin(-v)$$
$$= \sin u - \sin v$$

(c) By the product-to-sum formula in 74 (b),

$$2\cos\frac{u+v}{2}\cos\frac{u-v}{2}$$
$$= 2\cdot\frac{1}{2}\left(\cos\frac{u+v-(u-v)}{2} + \cos\frac{u+v+u-v}{2}\right)$$
$$= \cos v + \cos u$$
$$= \cos u + \cos v$$

(d) By the product-to-sum formula in 74 (a),

$$-2\sin\frac{u+v}{2}\sin\frac{u-v}{2}$$
$$= -2\cdot\frac{1}{2}\left(\cos\frac{u+v-(u-v)}{2} - \cos\frac{u+v+u-v}{2}\right)$$
$$= -(\cos v - \cos u)$$
$$= \cos u - \cos v$$

77. **(a)** Any inscribed angle that intercepts an arc of 180° is a right angle.

(b) Two inscribed angles that intercept the same arc are congruent.

(c) In right $\Delta A'BC$, $\sin A' = \frac{opp}{hyp} = \frac{a}{d}$.

(d) Because $\angle A'$ and $\angle A$ are congruent, $\frac{\sin A}{a} = \frac{\sin A'}{a} = \frac{a/d}{a} = \frac{1}{d}$.

(e) Of course. They both equal $\frac{\sin A}{a}$ by the Law of Sines.

Chapter 5 Project

1.

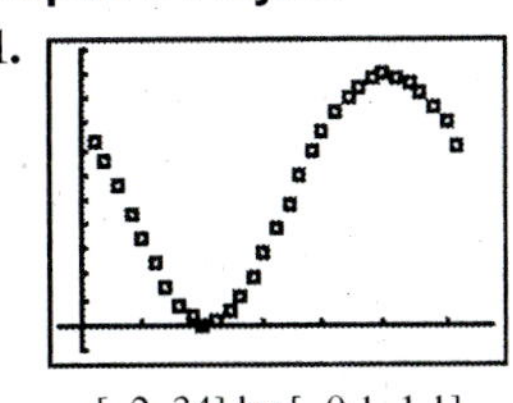

[–2, 34] by [–0.1, 1.1]

2. We set the amplitude as half the difference between the maximum value, 1.00, and the minimum value, 0.00, so $a = 0.5$. And we set the average value as the average of the maximum and minimum values, so $k = 0.5$. Since $\cos(b(x - h))$ takes on its maximum value at h, we set $h = 29$. Experimenting with the graph suggests that b should be about $2\pi/30.5$. So the equation is

$$y \approx 0.5\cos\left(\frac{2\pi}{30.5}(x - 29)\right) + 0.5.$$

5. Using the identities from Questions 3 and 4 together,

$$y \approx 0.5\cos\left(\frac{2\pi}{30.5}(x - 29)\right) + 0.5$$
$$= 0.5\cos\left(\frac{2\pi}{30.5}(29 - x)\right) + 0.5$$
$$= 0.5\sin\left(\frac{\pi}{2} - \frac{2\pi}{30.5}(29 - x)\right) + 0.5$$
$$= 0.5\sin\left(\frac{2\pi}{30.5}(x - 21.375)\right) + 0.5,$$

which is equivalent to
$y \approx 0.5\sin(0.21x - 4.4) + 0.5$.
Sinusoidal regression yields
$y \approx 0.5\sin(0.21x + 1.87) + 0.49$.

Chapter 6
Applications of Trigonometry

Section 6.1 Vectors in the Plane

Exploration 1

1. Use the HMT rule, which states that if an arrow has initial point (x_1, y_1) and terminal point (x_2, y_2), it represents the vector $\langle x_2 - x_1, y_2 - y_1\rangle$. If the initial point is $(2, 3)$ and the terminal point is $(7, 5)$, the vector is $\langle 7 - 2, 5 - 3\rangle = \langle 5, 2\rangle$.
2. Use the HMT rule, which states that if an arrow has initial point (x_1, y_1) and terminal point (x_2, y_2), it represents the vector $\langle x_2 - x_1, y_2 - y_1\rangle$. If the initial point is $(3, 5)$ and the terminal point is (x_2, y_2), the vector is $\langle x_2 - 3, y_2 - 5\rangle$. Using the given vector $\langle -3, 6\rangle$, we have $x_2 - 3 = -3$ and $y_2 - 5 = 6$.
$x_2 - 3 = -3 \Rightarrow x_2 = 0$; $y_2 - 5 = 6 \Rightarrow y_2 = 11$.
The terminal point is $(0, 11)$.
3. Use the HMT rule, which states that if an arrow has initial point (x_1, y_1) and terminal point (x_2, y_2), it represents the vector $\langle x_2 - x_1, y_2 - y_1\rangle$. If the initial point P is $(4, -3)$ and the terminal point Q is (x_2, y_2), the vector $\overrightarrow{PQ}$ is $\langle x_2 - 4, y_2 - (-3)\rangle$. Using the given vector $\overrightarrow{PQ}\langle 2, -4\rangle$, we have $x_2 - 4 = 2$ and $y_2 + 3 = -4$.
$x_2 - 4 = 2 \Rightarrow x_2 = 6$; $y_2 + 3 = -4 \Rightarrow y_2 = -7$.
The point Q is $(6, -7)$.
4. Use the HMT rule, which states that if an arrow has initial point (x_1, y_1) and terminal point (x_2, y_2), it represents the vector $\langle x_2 - x_1, y_2 - y_1\rangle$. If the initial point P is (x_1, y_1) and the terminal point Q is $(4, -3)$, the vector $\overrightarrow{PQ}$ is $\langle 4 - x_1, -3 - y_1\rangle$. Using the given vector $\overrightarrow{PQ}\langle 2, -4\rangle$, we have $4 - x_1 = 2$ and $-3 - y_1 = -4$.
$4 - x_1 = 2 \Rightarrow x_1 = 2$; $-3 - y_1 = -4 \Rightarrow y_1 = 1$.
The point P is $(2, 1)$.

Quick Review 6.1

1. $x = 9\cos 30° = \dfrac{9\sqrt{3}}{2}$, $y = 9\sin 30° = 4.5$

3. $x = 7\cos 220° \approx -5.36$, $y = 7\sin 220° \approx -4.5$

For #5, use a calculator.

5. $\theta \approx 33.85°$

For #7–9, the angle determined by $P(x, y)$ involves $\tan^{-1}(y/x)$. Since this will always be between $-180°$ and $+180°$, you may need to add 180° or 360° to put the angle in the correct quadrant.

7. $\theta = \tan^{-1}\left(\dfrac{9}{5}\right) \approx 60.95°$

9. $\theta = 180° + \tan^{-1}\left(\dfrac{5}{2}\right) \approx 248.20°$

Section 6.1 Exercises

For #1–3, recall that two vectors are equivalent if they have the same magnitude and direction. If R has coordinates (a, b) and S has coordinates (c, d), then the magnitude of $\overrightarrow{RS}$ is $|\overrightarrow{RS}| = \sqrt{(c-a)^2 + (d-b)^2} = RS$, the distance from R to S. The direction of $\overrightarrow{RS}$ is determined by the coordinates $(c - a, d - b)$.

1. If $R = (-4, 7)$ and $S = (-1, 5)$, then, using the HMT rule, $\overrightarrow{RS} = \langle -1 - (-4), 5 - 7\rangle = \langle 3, -2\rangle$.
If $P = (0, 0)$ and $Q = (3, -2)$, then, using the HMT rule, $\overrightarrow{PQ} = \langle 3 - 0, -2 - 0\rangle = \langle 3, -2\rangle$.
Both vectors represent $\langle 3, -2\rangle$ by the HMT rule.
3. If $R = (2, 1)$ and $S = (0, -1)$, then, using the HMT rule, $\overrightarrow{RS} = \langle 0 - 2, -1 - 1\rangle = \langle -2, -2\rangle$.
If $P = (1, 4)$ and $Q = (-1, 2)$, then, using the HMT rule, $\overrightarrow{PQ} = \langle -1 - 1, 2 - 4\rangle = \langle -2, -2\rangle$.
Both vectors represent $\langle -2, -2\rangle$ by the HMT rule.

5. $\overrightarrow{PQ} = \langle 3 - (-2), 4 - 2\rangle = \langle 5, 2\rangle$,
$|\overrightarrow{PQ}| = \sqrt{5^2 + 2^2} = \sqrt{29}$

7. $\overrightarrow{QR} = \langle -2 - 3, 5 - 4\rangle = \langle -5, 1\rangle$,
$|\overrightarrow{QR}| = \sqrt{(-5)^2 + 1^2} = \sqrt{26}$

9. $2\overrightarrow{QS} = 2\langle 2 - 3, -8 - 4\rangle = \langle -2, -24\rangle$,
$|2\overrightarrow{QS}| = \sqrt{(-2)^2 + (-24)^2} = \sqrt{580} = 2\sqrt{145}$

11. $3\overrightarrow{QR} + \overrightarrow{PS} = 3\langle -5, 1\rangle + \langle 4, -10\rangle = \langle -11, -7\rangle$,
$|3\overrightarrow{QR} + \overrightarrow{PS}| = \sqrt{(-11)^2 + (-7)^2} = \sqrt{170}$

13. $\langle -1, 3\rangle + \langle 2, 4\rangle = \langle 1, 7\rangle$

15. $\langle -1, 3\rangle - \langle 2, -5\rangle = \langle -3, 8\rangle$

17. $2\langle -1, 3\rangle + 3\langle 2, -5\rangle = \langle 4, -9\rangle$

19. $-2\langle -1, 3\rangle - 3\langle 2, 4\rangle = \langle -4, -18\rangle$

21. $\dfrac{\mathbf{u}}{|\mathbf{u}|} = \left\langle \dfrac{-2}{\sqrt{(-2)^2 + 4^2}}, \dfrac{4}{\sqrt{(-2)^2 + 4^2}}\right\rangle$
$\approx -0.45\mathbf{i} + 0.89\mathbf{j}$

23. $\dfrac{\mathbf{w}}{|\mathbf{w}|} = \left\langle \dfrac{-1}{\sqrt{(-1)^2 + (-2)^2}}, \dfrac{-2}{\sqrt{(-1)^2 + (-2)^2}}\right\rangle$
$\approx -0.45\mathbf{i} - 0.89\mathbf{j}$

For #25–27, the unit vector in the direction of $\mathbf{v} = \langle a, b\rangle$ is

$$\frac{1}{|\mathbf{v}|}\mathbf{v} = \left\langle \frac{a}{\sqrt{a^2 + b^2}}, \frac{b}{\sqrt{a^2 + b^2}}\right\rangle = \frac{a}{\sqrt{a^2 + b^2}}\mathbf{i} + \frac{b}{\sqrt{a^2 + b^2}}\mathbf{j}.$$

25. **(a)** $\left\langle \dfrac{2}{\sqrt{5}}, \dfrac{1}{\sqrt{5}}\right\rangle$

(b) $\frac{2}{\sqrt{5}}\mathbf{i} + \frac{1}{\sqrt{5}}\mathbf{j}$

27. (a) $\left\langle -\frac{4}{\sqrt{41}}, -\frac{5}{\sqrt{41}} \right\rangle$

(b) $-\frac{4}{\sqrt{41}}\mathbf{i} + \left(-\frac{5}{\sqrt{41}}\right)\mathbf{j} = -\frac{4}{\sqrt{41}}\mathbf{i} - \frac{5}{\sqrt{41}}\mathbf{j}$

29. $\mathbf{v} = \langle 18 \cos 25°, 18 \sin 25° \rangle \approx \langle 16.31, 7.61 \rangle$

31. $\mathbf{v} = \langle 47 \cos 108°, 47 \sin 108° \rangle \approx \langle -14.52, 44.70 \rangle$

33. $|\mathbf{u}| = \sqrt{3^2 + 4^2} = 5, \alpha = \cos^{-1}\left(\frac{3}{5}\right) \approx 53.13°$

35. $|\mathbf{u}| = \sqrt{3^2 + (-4)^2} = 5, \alpha = 360° - \cos^{-1}\left(\frac{3}{5}\right)$

$\approx 306.87°$

37. Since $(7 \cos 135°)\mathbf{i} + (7 \sin 135°)\mathbf{j} = (|\mathbf{u}| \cos \alpha)\mathbf{i} + (|\mathbf{u}| \sin \alpha)\mathbf{j}$, $|\mathbf{u}| = 7$ and $\alpha = 135°$.

For #39, first find the unit vector in the direction of **u**. Then multiply by the magnitude of **v**, $|\mathbf{v}|$.

39. $\mathbf{v} = |\mathbf{v}| \cdot \frac{\mathbf{u}}{|\mathbf{u}|} = 2\left\langle \frac{3}{\sqrt{3^2 + (-3)^2}}, \frac{-3}{\sqrt{3^2 + (-3)^2}} \right\rangle$

$= 2\left\langle \frac{\sqrt{2}}{2}, \frac{-\sqrt{2}}{2} \right\rangle = \langle \sqrt{2}, -\sqrt{2} \rangle$

41. A bearing of 335° corresponds to a direction angle of 115°. $\mathbf{v} = 530 \langle \cos 115°, \sin 115° \rangle \approx \langle -223.99, 480.34 \rangle$

43. (a) A bearing of 340° corresponds to a direction angle of 110°. $\mathbf{v} = 325 \langle \cos 110°, \sin 110° \rangle \approx \langle -111.16, 305.40 \rangle$

(b) The wind bearing of 320° corresponds to a direction angle of 130°. The wind vector is $\mathbf{w} = 40 \langle \cos 130°, \sin 130° \rangle \approx \langle -25.71, 30.64 \rangle$.
Actual velocity vector: $\mathbf{v} + \mathbf{w} \approx \langle -136.87, 336.04 \rangle$.
Actual speed: $\|\mathbf{v} + \mathbf{w}\| \approx \sqrt{136.87^2 + 336.04^2}$
≈ 362.84 mph.
Actual direction: $\theta = 180° + \tan^{-1}\left(\frac{336.04}{-136.87}\right)$
$\approx 112.16°$, so the bearing is about 337.84°.

45. (a) $\mathbf{v} = 10 \langle \cos 70°, \sin 70° \rangle \approx \langle 3.42, 9.40 \rangle$

(b) The horizontal component is the (constant) horizontal speed of the basketball as it travels toward the basket. The vertical component is the vertical velocity of the basketball, affected by both the initial speed and the downward pull of gravity.

47. We need to choose $\mathbf{w} = \langle a, b \rangle = k \langle \cos 33°, \sin 33° \rangle$, so that $k \cos(33° - 15°) = k \cos 18° = 2.5$. (Redefine "horizontal" to mean the parallel to the inclined plane; then the towing vector makes an angle of 18° with the "horizontal.") Then $k = \frac{2.5}{\cos 18°} \approx 2.63$ lb, so that $\mathbf{w} \approx \langle 2.20, 1.43 \rangle$.

49. $\mathbf{F} = \langle 50 \cos 45°, 50 \sin 45° \rangle + \langle 75 \cos(-30°), 75 \sin(-30°) \rangle \approx \langle 100.31, -2.14 \rangle$, so $|\mathbf{F}| \approx 100.33$ lb and $\theta \approx -1.22°$.

51. Ship heading: $\langle 12 \cos 90°, 12 \sin 90° \rangle = \langle 0, 12 \rangle$
Current heading: $\langle 4 \cos 225°, 4 \sin 225° \rangle \approx \langle -2.83, -2.83 \rangle$
The ship's actual velocity vector is $\langle -2.83, 9.17 \rangle$, so its speed is $\approx \sqrt{(-2.83)^2 + 9.17^2} \approx 9.6$ mph and the direction angle is $\cos^{-1}\left(\frac{-2.83}{9.6}\right) \approx 107.14°$, so the bearing is about 342.86°.

53. Let w be the speed of the ship. The ship's velocity (in still water) is $\langle w \cos 270°, w \sin 270° \rangle = \langle 0, -w \rangle$. Let z be the speed of the current. Then, the current velocity is $\langle z \cos 135°, z \sin 135° \rangle \approx \langle -0.71z, 0.71z \rangle$. The position of the ship after two hours is $\langle 20 \cos 240°, 20 \sin 240° \rangle \approx \langle -10, -17.32 \rangle$. Putting all this together we have:
$2\langle 0, -w \rangle + 2\langle -0.71z, 0.71z \rangle = \langle -10, -17.32 \rangle$,
$\langle -1.42z, -2w + 1.42z \rangle = \langle -10, -17.32 \rangle$, so $z \approx 7.07$ and $w \approx 13.66$. The speed of the ship is about 13.66 mph, and the speed of the current is about 7.07 mph.

55. True. Vectors **u** and **−u** have the same length but opposite directions. Thus, the length of **−u** is also 1.

57. $|\langle 2, -1 \rangle| = \sqrt{2^2 + (-1)^2} = \sqrt{5}$
The answer is D.

59. The x-component is $3 \cos 30°$, and the y-component is $3 \sin 30°$. The answer is A.

61. (a) Let A be the point (a_1, a_2) and B be the point (b_1, b_2). Then $\overrightarrow{OA}$ is the vector $\langle a_1, a_2 \rangle$ and $\overrightarrow{OB}$ is the vector $\langle b_1, b_2 \rangle$.
So, $\overrightarrow{BA} = \langle a_1 - b_1, a_2 - b_2 \rangle$
$= \langle a_1, a_2 \rangle - \langle b_1, b_2 \rangle$
$= \overrightarrow{OA} - \overrightarrow{OB}$

(b) $x\,\overrightarrow{OA} + y\,\overrightarrow{OB} = x(\overrightarrow{OC} + \overrightarrow{CA}) + y(\overrightarrow{OC} + \overrightarrow{CB})$ (from part (a))
$= x\,\overrightarrow{OC} + x\,\overrightarrow{CA} + y\,\overrightarrow{OC} + y\,\overrightarrow{CB}$
$= (x + y)\,\overrightarrow{OC} + x\,\overrightarrow{CA} + y\,\overrightarrow{CB}$
$= \overrightarrow{OC} + x\,\overrightarrow{CA} + y\,\overrightarrow{CB}$ (since $x + y = 1$)
$= \overrightarrow{OC} + y\frac{|\overrightarrow{BC}|}{|\overrightarrow{CA}|} \cdot \overrightarrow{CA} + y\,\overrightarrow{CB}$ $\left(\text{since } x = y\frac{|\overrightarrow{BC}|}{|\overrightarrow{CA}|}\right)$
$= \overrightarrow{OC} + y\left(|\overrightarrow{BC}| \cdot \frac{\overrightarrow{CA}}{|\overrightarrow{CA}|} + \overrightarrow{CB}\right)$

$\frac{\overrightarrow{CA}}{|\overrightarrow{CA}|}$ is a unit vector, and $\overrightarrow{BC}$ points in the same direction.

$= \overrightarrow{OC} + y\left(|\overrightarrow{BC}| \cdot \frac{\overrightarrow{BC}}{|\overrightarrow{BC}|} + \overrightarrow{CB}\right)$
$= \overrightarrow{OC} + y(\overrightarrow{BC} + \overrightarrow{CB})$
$= \overrightarrow{OC}$

63. Use the result of Exercise 61. First we show that if C is on the line segment AB, then there is a real number t so that $\frac{|\overrightarrow{BC}|}{|\overrightarrow{CA}|} = \frac{t}{1 - t}$. (Convince yourself that $t = \frac{BC}{BC + CA}$ works.) Then $\overrightarrow{OC} = t\,\overrightarrow{OA} + (1 - t)\,\overrightarrow{OB}$.
A similar argument can be used in the cases where B is on the line segment AC or A is on the line segment BC.

Suppose there is a real number t so that $\overrightarrow{OC} = t\,\overrightarrow{OA} + (1 - t)\,\overrightarrow{OB}$. We also know $\overrightarrow{OC} = \overrightarrow{OB} + \overrightarrow{BC}$ and $\overrightarrow{OC} = \overrightarrow{OA} + \overrightarrow{AC}$. So we have $t\,\overrightarrow{OA} + (1 - t)\overrightarrow{OB} = \overrightarrow{OB} + \overrightarrow{BC}$ and $t\,\overrightarrow{OA} + (1 - t)\overrightarrow{OB} = \overrightarrow{OA} + \overrightarrow{AC}$. Therefore, $t\,(\overrightarrow{OA} - \overrightarrow{OB}) = \overrightarrow{BC}$ and $(t - 1)(\overrightarrow{OA} - \overrightarrow{OB}) = \overrightarrow{AC}$. Hence $\overrightarrow{BC}$ and $\overrightarrow{AC}$ have the same or opposite directions, so C must lie on the line L through the two points A and B.

Section 6.2 Dot Product of Vectors

Exploration 1

1. $\mathbf{u} = \langle -2 - x, 0 - y\rangle = \langle -2 - x, -y\rangle$
   $\mathbf{v} = \langle 2 - x, 0 - y\rangle = \langle 2 - x, -y\rangle$
2. $\mathbf{u}\cdot\mathbf{v} = (-2 - x)(2 - x) + (-y)(-y) = -4 + x^2 + y^2$
   $= -4 + 4 = 0$
   Therefore, $\theta = 90°$.
3. Answers will vary.

Quick Review 6.2

1. $|\mathbf{u}| = \sqrt{2^2 + (-3)^2} = \sqrt{13}$
3. $|\mathbf{u}| = \sqrt{\cos^2 35° + \sin^2 35°} = 1$
5. $\overrightarrow{AB} = \langle 1 - (-2), \sqrt{3} - 0\rangle = \langle 3, \sqrt{3}\rangle$
7. $\overrightarrow{AB} = \langle 1 - 2, -\sqrt{3} - 0\rangle = \langle -1, -\sqrt{3}\rangle$
9. $\mathbf{u} = |\mathbf{u}|\cdot\dfrac{\mathbf{v}}{|\mathbf{v}|} = \dfrac{2\cdot\langle 2, 3\rangle}{\sqrt{2^2 + 3^2}} = \dfrac{\langle 4, 6\rangle}{\sqrt{13}}$
   $= \left\langle \dfrac{4}{\sqrt{13}}, \dfrac{6}{\sqrt{13}}\right\rangle$

Section 6.2 Exercises

1. $60 + 12 = 72$
3. $-12 - 35 = -47$
5. $12 + 18 = 30$
7. $-14 + 0 = -14$
9. $|\mathbf{u}| = \sqrt{\mathbf{u}\cdot\mathbf{u}} = \sqrt{25 + 144} = 13$
11. $|\mathbf{u}| = \sqrt{\mathbf{u}\cdot\mathbf{u}} = \sqrt{16} = 4$
13. $\mathbf{u}\cdot\mathbf{v} = 4 - 15 = -11$, $|\mathbf{u}| = \sqrt{16 + 9} = 5$,
    $|\mathbf{v}| = \sqrt{1 + 25} = \sqrt{26}$, $\theta = \cos^{-1}\left(\dfrac{-11}{5\sqrt{26}}\right) \approx 115.6°$
15. $\mathbf{u}\cdot\mathbf{v} = -6 + 15 = 9$,
    $|\mathbf{u}| = \sqrt{4 + 9} = \sqrt{13}$,
    $|\mathbf{u}| = \sqrt{9 + 25} = \sqrt{34}$,
    $\theta = \cos^{-1}\left(\dfrac{9}{\sqrt{13}\cdot\sqrt{34}}\right) \approx 64.65°$
17. $\mathbf{u}\cdot\mathbf{v} = -6 - 6\sqrt{3}$, $|\mathbf{u}| = \sqrt{9 + 9} = \sqrt{18}$,
    $|\mathbf{v}| = \sqrt{4 + 12} = \sqrt{16} = 4$,
    $\theta = \cos^{-1}\left(\dfrac{-6 - 6\sqrt{3}}{\sqrt{18}\cdot 4}\right) = 165°$
19. $\mathbf{u}$ has direction angle $\dfrac{\pi}{4}$ and $\mathbf{v}$ has direction angle $\dfrac{3\pi}{2}$ (which is equivalent to $-\dfrac{\pi}{2}$), so the angle between the vectors is $\dfrac{\pi}{4} - \left(-\dfrac{\pi}{2}\right) = \dfrac{3\pi}{4}$ or 135°.
21. $\mathbf{u}\cdot\mathbf{v} = -24 + 20 = -4$, $|\mathbf{u}| = \sqrt{64 + 25} = \sqrt{89}$,
    $|\mathbf{v}| = \sqrt{9 + 16} = 5$,
    $\theta = \cos^{-1}\left(\dfrac{-4}{5\sqrt{89}}\right) \approx 94.86°$
23. $\mathbf{u}\cdot\mathbf{v} = \langle 2, 3\rangle\cdot\left\langle \dfrac{3}{2}, -1\right\rangle = 2\left(\dfrac{3}{2}\right) + 3(-1)$
    $= 3 - 3 = 0$
    Since $\mathbf{u}\cdot\mathbf{v} = 0$, $\mathbf{u}$ and $\mathbf{v}$ are orthogonal.

For #25–27, first find $\text{proj}_{\mathbf{v}}\mathbf{u}$. Then use the fact that $\mathbf{u}\cdot\mathbf{v} = 0$ when $\mathbf{u}$ and $\mathbf{v}$ are orthogonal.

25. $\text{proj}_{\mathbf{v}}\mathbf{u} = \left(\dfrac{\langle -8, 3\rangle\cdot\langle -6, -2\rangle}{36 + 4}\right)\langle -6, -2\rangle$
    $= \left(\dfrac{42}{40}\right)\langle -6, -2\rangle = \dfrac{21}{20}\langle -6, -2\rangle$
    $= -\dfrac{21}{10}\langle 3, 1\rangle$
    $\mathbf{u} = -\dfrac{21}{10}\langle 3, 1\rangle + \dfrac{17}{10}\langle -1, 3\rangle$
27. $\text{proj}_{\mathbf{v}}\mathbf{u} = \left(\dfrac{\langle 8, 5\rangle\cdot\langle -9, -2\rangle}{81 + 4}\right)\langle -9, -2\rangle$
    $= \left(\dfrac{-82}{85}\right)\langle -9, -2\rangle = \dfrac{82}{85}\langle 9, 2\rangle$
    $\mathbf{u} = \dfrac{82}{85}\langle 9, 2\rangle + \dfrac{29}{85}\langle -2, 9\rangle$
29.

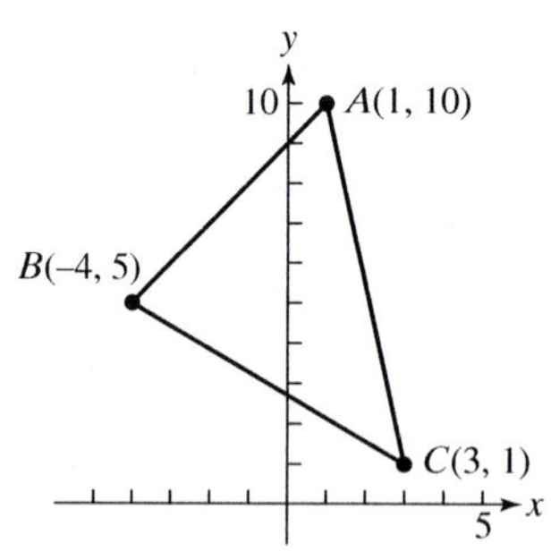

$\overrightarrow{CA}\cdot\overrightarrow{CB} = \langle 1 - 3, 10 - 1\rangle\cdot\langle -4 - 3, 5 - 1\rangle$
$= \langle -2, 9\rangle\cdot\langle -7, 4\rangle = 14 + 36 = 50$,

$|\overrightarrow{CA}| = \sqrt{4 + 81} = \sqrt{85}$, $|\overrightarrow{CB}| = \sqrt{49 + 16} = \sqrt{65}$,
$C = \cos^{-1}\left(\dfrac{40}{\sqrt{85}\cdot\sqrt{65}}\right) \approx 47.73°$
$\overrightarrow{BC}\cdot\overrightarrow{BA} = \langle 7, -4\rangle\cdot\langle 1 - (-4), 10 - 5\rangle$
$= \langle 7, -4\rangle\cdot\langle 5, 5\rangle = 35 - 20 = 15$,
$|\overrightarrow{BC}| = \sqrt{65}$, $|\overrightarrow{BA}| = \sqrt{50}$, $B = \cos^{-1}\left(\dfrac{15}{\sqrt{65}\cdot\sqrt{50}}\right)$
$\approx 74.74°$
$A = 180° - B - C \approx 180° - 74.74° - 47.73° = 57.53°$

For #31, use the relationship $\mathbf{u}\cdot\mathbf{v} = |\mathbf{u}|\,|\mathbf{v}|\cos\theta$.

31. $\mathbf{u}\cdot\mathbf{v} = 3\cdot 8\cos 150° \approx -20.78$

For #33–37, vectors are orthogonal if $\mathbf{u}\cdot\mathbf{v} = 0$ and are parallel if $\mathbf{u} = k\mathbf{v}$ for some constant k.

33. Parallel: $-2\left\langle -\frac{10}{4}, -\frac{3}{2}\right\rangle = \left\langle \frac{10}{2}, 3\right\rangle = \langle 5, 3\rangle$

35. Neither: $\mathbf{u}\cdot\mathbf{v} = -120 \neq 0$ and
$$\frac{-15}{4}\mathbf{v} = \frac{-15}{4}\langle -4, 5\rangle = \left\langle 15, \frac{-75}{4}\right\rangle \neq \mathbf{u}$$

37. Orthogonal: $\mathbf{u}\cdot\mathbf{v} = -60 + 60 = 0$

For #39–41 part (b), first find the direction(s) of $\overrightarrow{AP}$ and then find the unit vectors. Then find P by adding the coordinates of A to the components of a unit vector.

39. (a) A is $(4, 0)$ and B is $(0, -3)$.

(b) The line is parallel to $\overrightarrow{AB} = \langle 0 - 4, -3 - 0\rangle = \langle -4, -3\rangle$, so the direction of $\overrightarrow{AP}$ is $\mathbf{u} = \langle 3, -4\rangle$ or $\mathbf{v} = \langle -3, 4\rangle$.

$$\overrightarrow{AP} = \frac{\mathbf{u}}{|\mathbf{u}|} = \frac{\langle 3,-4\rangle}{\sqrt{9+16}} = \left\langle \frac{3}{5}, -\frac{4}{5}\right\rangle \text{ or}$$

$$\overrightarrow{AP} = \frac{\mathbf{v}}{|\mathbf{v}|} = \frac{\langle -3, 4\rangle}{\sqrt{9+16}} = \left\langle -\frac{3}{5}, \frac{4}{5}\right\rangle.$$

So, P is $(4.6, -0.8)$ or $(3.4, 0.8)$.

41. (a) A is $(7, 0)$ and B is $(0, -3)$.

(b) The line is parallel to $\overrightarrow{AB} = \langle 0 - 7, -3 - 0\rangle = \langle -7, -3\rangle$, so the direction of $\overrightarrow{AP}$ is $\mathbf{u} = \langle 3, -7\rangle$ or $\mathbf{v} = \langle -3, 7\rangle$.

$$\overrightarrow{AP} = \frac{\mathbf{u}}{|\mathbf{u}|} = \frac{\langle 3, -7\rangle}{\sqrt{9+49}} = \left\langle \frac{3}{\sqrt{58}}, -\frac{7}{\sqrt{58}}\right\rangle \text{ or}$$

$$\overrightarrow{AP} = \frac{\mathbf{v}}{|\mathbf{v}|} = \frac{\langle -3, 7\rangle}{\sqrt{58}} = \left\langle -\frac{3}{\sqrt{58}}, \frac{7}{\sqrt{58}}\right\rangle.$$

So P is $\left(7 + \frac{3}{\sqrt{58}}, -\frac{7}{\sqrt{58}}\right) \approx (-7.39, 0.92)$ or

$\left(7 - \frac{3}{\sqrt{58}}, \frac{7}{\sqrt{58}}\right) \approx (6.61, 0.92)$.

43. $2v_1 + 3v_2 = 10, v_1^2 + v_2^2 = 17$. Since $v_1 = 5 - \frac{3}{2}v_2$,

$\left(5 - \frac{3}{2}v_2\right)^2 + v_2^2 = 17, 25 - 15v_2 + \frac{9}{4}v_2^2 + v_2^2 = 17,$

$\frac{13}{4}v_2^2 - 15v_2 + 8 = 0, 13v_2^2 - 60v_2 + 32 = 0,$

$(v_2 - 4)(13v_2 - 8) = 0$, so $v_2 = 4$ or $v_2 = \frac{8}{13}$.

Therefore, $\mathbf{v} \approx \langle -1, 4\rangle$ or $\mathbf{v} = \left\langle \frac{53}{13}, \frac{8}{13}\right\rangle \approx \langle 4.07, 0.62\rangle$.

45. $\mathbf{v} = (\cos 60°)\mathbf{i} + (\sin 60°)\mathbf{j} = \frac{1}{2}\mathbf{i} + \frac{\sqrt{3}}{2}\mathbf{j}$

$$\mathbf{F}_1 = \text{proj}_\mathbf{v}\mathbf{F} = (\mathbf{F}\cdot\mathbf{v})\mathbf{v} = \left(-160\,\mathbf{j}\cdot\left(\frac{1}{2}\mathbf{i} + \frac{\sqrt{3}}{2}\mathbf{j}\right)\right)\mathbf{v}$$

$$= -80\sqrt{3}\left(\frac{1}{2}\mathbf{i} + \frac{\sqrt{3}}{2}\mathbf{j}\right) = -40\sqrt{3}\,\mathbf{i} - 120\,\mathbf{j}.$$

The magnitude of the force is

$|\mathbf{F}_1| = \sqrt{(-40\sqrt{3})^2 + (-120)^2} = \sqrt{19{,}200}$
≈ 138.56 pounds.

47. (a) $\mathbf{v} = (\cos 12°)\mathbf{i} + (\sin 12°)\mathbf{j}$
$\mathbf{F} = -2000\mathbf{j}$
$\mathbf{F}_1 = \text{proj}_\mathbf{v}\mathbf{F} = (\mathbf{F}\cdot\mathbf{v})\mathbf{v}$
$= (\langle 0, -2000\rangle \cdot \langle \cos 12°, \sin 12°\rangle)\,\langle \cos 12°, \sin 12°\rangle$
$= (-2000 \sin 12°)\langle \cos 12°, \sin 12°\rangle$.
Since $\langle \cos 12°, \sin 12°\rangle$ is a unit vector,the magnitude of the force being extended is $|\mathbf{F}_1| = 2000 \sin 12° \approx 415.82$ pounds.

(b) We are looking for the gravitational force exerted perpendicular to the street. A unit vector perpendicular to the street is $\mathbf{w} = \langle \cos(-78°), \sin(-78°)\rangle$, so $\mathbf{F}_2 = \text{proj}_\mathbf{w}\mathbf{F} = (\mathbf{F}\cdot\mathbf{w})\mathbf{w}$
$= (-2000 \sin(-78°))\,\langle \cos(-78°), \sin(-78°)\rangle$
Since $\langle \cos(-78°), \sin(-78°)\rangle$ is a unit vector, the magnitude of the force perpendicular to the street is $-2000 \sin(-78°) \approx 1956.30$ pounds.

49. Since the car weighs 2600 pounds, the force needed to lift the car is $\langle 0, 2600\rangle$.
$W = \mathbf{F}\cdot\overrightarrow{AB} = \langle 0, 2600\rangle \cdot \langle 0, 5.5\rangle = 14{,}300$ foot-pounds

51. $\mathbf{F} = 12\cdot\frac{\langle 1, 2\rangle}{|\langle 1, 2\rangle|} = \frac{12}{\sqrt{5}}\langle 1, 2\rangle$

$W = \mathbf{F}\cdot\overrightarrow{AB} = \frac{12}{\sqrt{5}}\langle 1, 2\rangle \cdot \langle 4, 0\rangle = \frac{48}{\sqrt{5}}$
≈ 21.47 foot-pounds

53. $\mathbf{F} = 30\cdot\frac{\langle 2, 2\rangle}{|\langle 2, 2\rangle|} = \frac{30}{\sqrt{2^2 + 2^2}}\langle 2, 2\rangle = 15\sqrt{2}\,\langle 1, 1\rangle$

Since we want to move 3 feet along the line $y = \frac{1}{2}x$, we solve for x and y by using the Pythagorean theorem:
$x^2 + y^2 = 3^2, x^2 + \left(\frac{1}{2}x\right)^2 = 9, \frac{5}{4}x^2 = 9,$

$x = \frac{6}{\sqrt{5}}, y = \frac{3}{\sqrt{5}}$

$\overrightarrow{AB} = \left\langle \frac{6}{\sqrt{5}}, \frac{3}{\sqrt{5}}\right\rangle$

$W = \mathbf{F}\cdot\overrightarrow{AB} = \langle 15\sqrt{2}, 15\sqrt{2}\rangle \cdot \left\langle \frac{6}{\sqrt{5}}, \frac{3}{\sqrt{5}}\right\rangle$

$= 135\sqrt{\frac{2}{5}} = 27\sqrt{10} \approx 85.38$ foot-pounds

55. $W = \mathbf{F}\cdot\overrightarrow{AB} = |\mathbf{F}|\,|\overrightarrow{AB}| \cos\theta = 200\sqrt{13}\cos 30°$
$= 200\sqrt{13}\cdot\frac{\sqrt{3}}{2} = 100\sqrt{39} \approx 624.5$ foot-pounds

57. (a) Let $\mathbf{u} = \langle u_1, u_2\rangle$, $\mathbf{v} = \langle v_1, v_2\rangle$, and $\mathbf{w} = \langle w_1, w_2\rangle$.
$\mathbf{0}\cdot u = \langle 0, 0\rangle \cdot \langle u_1, u_2\rangle = 0\cdot u_1 + 0\cdot u_2 = 0$

(b) $\mathbf{u}\cdot(\mathbf{v} + \mathbf{w}) = \langle u_1, u_2\rangle \cdot (\langle v_1, v_2\rangle + \langle w_1, w_2\rangle)$
$= \langle u_1, u_2\rangle \cdot \langle v_1 + w_1, v_2 + w_2\rangle$
$= u_1(v_1 + w_1) + u_2(v_2 + w_2)$
$= u_1v_1 + u_2v_2 + u_1w_1 + u_2w_2$
$= \langle u_1, u_2\rangle \cdot \langle v_1, v_2\rangle + \langle u_1, u_2\rangle \cdot \langle w_1, w_2\rangle$
$= \mathbf{u}\cdot\mathbf{v} + \mathbf{u}\cdot\mathbf{w}$

(c) $(\mathbf{u} + \mathbf{v})\cdot\mathbf{w} = (\langle u_1, u_2\rangle + \langle v_1, v_2\rangle)\cdot\langle w_1, w_2\rangle$
$= \langle u_1 + v_1, u_2 + v_2\rangle \cdot \langle w_1, w_2\rangle$
$= (u_1 + v_1)\,w_1 + (u_2 + v_2)w_2$
$= u_1w_1 + u_2w_2 + v_1w_1 + v_2w_2$
$= \langle u_1, u_2\rangle \cdot \langle w_1, w_2\rangle + \langle v_1, v_2\rangle \cdot \langle w_1, w_2\rangle$
$= \mathbf{u}\cdot\mathbf{w} + \mathbf{v}\cdot\mathbf{w}$

(d) $(c\mathbf{u})\cdot\mathbf{v} = (c\langle u_1, u_2\rangle)\cdot\langle v_1, v_2\rangle = \langle cu_1, cu_2\rangle \cdot \langle v_1, v_2\rangle$
$= cu_1v_1 + cu_2v_2 = \langle u_1, u_2\rangle \cdot \langle cv_1, cv_2\rangle = \mathbf{u}\cdot(c\mathbf{v})$
$= c(u_1v_1 + u_2v_2) = c(\langle u_1, u_2\rangle \cdot \langle v_1, v_2\rangle) = c(\mathbf{u}\cdot\mathbf{v})$

59.

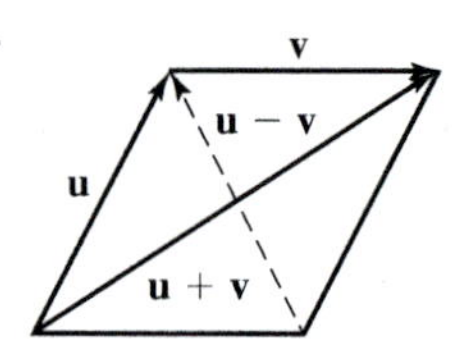

As the diagram indicates, the long diagonal of the parallelogram can be expressed as the vector $\mathbf{u} + \mathbf{v}$ while the short diagonal can be expressed as the vector $\mathbf{u} - \mathbf{v}$. The sum of the squares of the diagonals is
$|\mathbf{u} + \mathbf{v}|^2 + |\mathbf{u} - \mathbf{v}|^2 = (\mathbf{u} + \mathbf{v}) \cdot (\mathbf{u} + \mathbf{v})$
$+ (\mathbf{u} - \mathbf{v}) \cdot (\mathbf{u} - \mathbf{v}) = \mathbf{u} \cdot \mathbf{u} + \mathbf{u} \cdot \mathbf{v} + \mathbf{v} \cdot \mathbf{u} + \mathbf{v} \cdot \mathbf{v}$
$+ \mathbf{u} \cdot \mathbf{u} - \mathbf{u} \cdot \mathbf{v} - \mathbf{u} \cdot \mathbf{v} + \mathbf{v} \cdot \mathbf{v} = 2|\mathbf{u}|^2 + 2|\mathbf{v}|^2$
$+ 2\mathbf{u} \cdot \mathbf{v} - 2\mathbf{u} \cdot \mathbf{v} = 2|\mathbf{u}|^2 + 2|\mathbf{v}|^2$, which is the sum of the squares of the sides.

61. False. If either $\mathbf{u}$ or $\mathbf{v}$ is the zero vector, then $\mathbf{u} \cdot \mathbf{v} = 0$ and so $\mathbf{u}$ and $\mathbf{v}$ are orthogonal, but they do not count as perpendicular.

63. $\mathbf{u} \cdot \mathbf{v} = 0$, so the vectors are perpendicular. The answer is D.

65. $$\text{proj}_{\mathbf{v}}\mathbf{u} = \left(\frac{\mathbf{u} \cdot \mathbf{v}}{|\mathbf{v}|^2}\right)\mathbf{v} = \left(\frac{3 + 0}{2^2}\right)\langle 2, 0\rangle = \left(\frac{3}{4}\right)\langle 2, 0\rangle = \left\langle \frac{3}{2}, 0\right\rangle$$

The answer is A.

67. (a) $2 \cdot 0 + 5 \cdot 2 = 10$ and $2 \cdot 5 + 5 \cdot 0 = 10$

(b) $\overrightarrow{AP} = \langle 3 - 0, 7 - 2\rangle = \langle 3, 5\rangle$
$\overrightarrow{AB} = \langle 5 - 0, 0 - 2\rangle = \langle 5, -2\rangle$

$$\mathbf{w}_1 = \text{proj}_{\overrightarrow{AB}}\overrightarrow{AP} = \left(\frac{\langle 3, 5\rangle \cdot \langle 5, -27\rangle}{5^2 + (-2)^2}\right)\langle 5, -2\rangle = \left(\frac{15 - 10}{29}\right)\langle 5, -2\rangle = \frac{5}{29}\langle 5, -2\rangle$$

$$\mathbf{w}_2 = \overrightarrow{AP} - \text{proj}_{\overrightarrow{AB}}\overrightarrow{AP} = \langle 3, 5\rangle - \frac{5}{29}\langle 5, -2\rangle = \left\langle 3 - \frac{25}{29}, 5 + \frac{10}{29}\right\rangle = \frac{1}{29}\langle 62, 155\rangle$$

(c) $\mathbf{w}_2$ is a vector from a point on $\overrightarrow{AB}$ to point P. Since $\mathbf{w}_2$ is perpendicular to $\overrightarrow{AB}$, $|\mathbf{w}_2|$ is the shortest distance from $\overrightarrow{AB}$ to P.

$$|\mathbf{w}_2| = \sqrt{\left(\frac{62}{29}\right)^2 + \left(\frac{155}{29}\right)^2} = \sqrt{\frac{27{,}869}{29^2}} = \frac{31\sqrt{29}}{29}$$

(d) Consider Figure 6.19. To find the distance from a point P to a line L, we must first find $\mathbf{u}_1 = \text{proj}_{\mathbf{v}}\mathbf{u}$. In this case,

$$\text{proj}_{\overrightarrow{AB}}\overrightarrow{AP} = \left(\frac{\langle x_0, y_0 - 2\rangle \cdot \langle 5, -2\rangle}{(\sqrt{5^2 + (-2)^2})^2}\right)\langle 5, -2\rangle = \left(\frac{5x_0 - 2y_0 + 4}{29}\right)\langle 5, -2\rangle = \left\langle \frac{25x_0 - 10y_0 + 20}{29}, \frac{-10x_0 + 4y_0 - 8}{29}\right\rangle$$ and

$$\overrightarrow{AP} - \text{proj}_{\overrightarrow{AB}}\overrightarrow{AP} = \langle x_0, y_0 - 2\rangle - \left\langle \frac{25x_0 - 10y_0 + 20}{29}, \frac{-10x_0 + 4y_0 - 8}{29}\right\rangle$$
$$= \frac{1}{29}(\langle 29x_0, 29(y_0 - 2)\rangle - \langle 25x_0 - 10y_0 + 20, -10x_0 + 4y_0 - 8\rangle)$$
$$= \frac{1}{29}\langle 4x_0 + 10y_0 - 20, 10x_0 + 25y_0 - 50\rangle$$

So, the distance is the magnitude of this vector.

$$d = \frac{1}{29}\sqrt{(4x_0 + 10y_0 - 20)^2 + (10x_0 + 25y_0 - 50)^2}$$
$$= \frac{\sqrt{2^2(2x_0 + 5y_0 - 10)^2 + 5^2(2x_0 + 5y_0 - 10)^2}}{29}$$
$$= \frac{\sqrt{29(2x_0 + 5y_0 - 10)^2}}{29}$$
$$= \frac{|(2x_0 + 5y_0 - 10)|}{\sqrt{29}}$$

(e) In the general case, $\overrightarrow{AB} = \left\langle \frac{c}{a}, \frac{-c}{b}\right\rangle$ and $\overrightarrow{AP} = \left\langle x_0, y_0 - \frac{c}{b}\right\rangle$, so $\text{proj}_{\overrightarrow{AB}}\overrightarrow{AP}$

$$= \frac{\left(\frac{x_0 c}{a} - \frac{(bcy_0 - c^2)}{b^2}\right)\left\langle \frac{c}{a}, -\frac{c}{b}\right\rangle}{\left|\frac{c}{a}\right|^2 + \left|-\frac{c}{b}\right|^2}$$

$$= \left\langle \frac{b^2x_0 - aby_0 + ac}{a^2 + b^2}, \frac{-abx_0 + a^2y_0 - \frac{a^2c}{b}}{a^2 + b^2}\right\rangle$$

$$\overrightarrow{AP} - \text{proj}_{\overrightarrow{AB}}\overrightarrow{AP} = \left\langle x_0, y_0 - \frac{c}{b}\right\rangle - \left\langle \frac{b^2x_0 - aby_0 + ac}{a^2 + b^2}, \frac{-abx_0 + a^2y_0 - \frac{a^2c}{b}}{a^2 + b^2}\right\rangle$$
$$= \left\langle \frac{a^2x_0 + aby_0 - ac}{a^2 + b^2}, \frac{abx_0 + b^2y_0 - bc}{a^2 + b^2}\right\rangle$$

The magnitude of this vector, $|\overrightarrow{AP} - \text{proj}_{\overrightarrow{AB}}\overrightarrow{AP}|$, is the distance from point P to L: $\frac{ax_0 + by_0 - c}{\sqrt{a^2 + b^2}}$.

69. One possible answer:
If $a\mathbf{u} + b\mathbf{v} = c\mathbf{u} + d\mathbf{v}$
$a\mathbf{u} - c\mathbf{u} + b\mathbf{v} - d\mathbf{v} = 0$
$(a - c)\mathbf{u} + (b - d)\mathbf{v} = 0$
Since $\mathbf{u}$ and $\mathbf{v}$ are not parallel, the only way for this equality to hold true for all vectors $\mathbf{u}$ and $\mathbf{v}$ is if $(a - c) = 0$ and $(b - d) = 0$, which indicates that $a = c$ and $b = d$.

Section 6.3 Parametric Equations and Motion

Exploration 1

1.

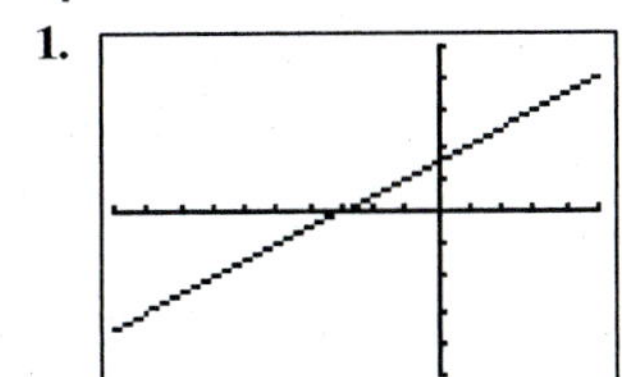

$[-10, 5]$ by $[-5, 5]$

2. $0.5(17) + 1.5 = 10$, so the point $(17, 10)$ is on the graph, $t = -8$

3. $0.5(-23) + 1.5 = -10$, so the point $(-23, -10)$ is on the graph, $t = 12$.

4. $x = a = 1 - 2t, 2t = 1 - a, t = \frac{1}{2} - \frac{a}{2}$.

   Alternatively, $b = 2 - t$, so $t = 2 - b$.

5. Choose Tmin and Tmax so that Tmin ≤ -2 and Tmax ≥ 5.5.

Exploration 2

1. It looks like the line in Figure 6.32.

2. The graph is a vertical line segment that extends from $(400, 0)$ to $(400, 20)$.

3. For 19° and 20°, the ball does not clear the fence, as shown below.

   19°:

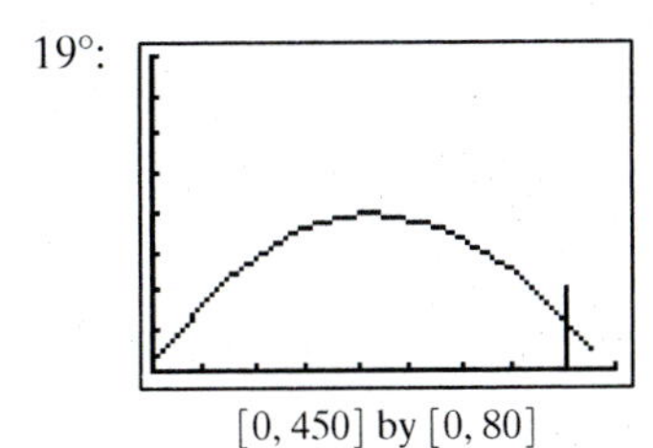

$[0, 450]$ by $[0, 80]$

20°:

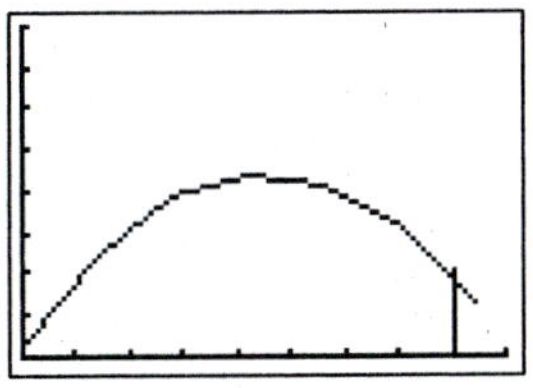

$[0, 450]$ by $[0, 80]$

For 21° and 22°, the ball clears the fence, as shown below.

21°

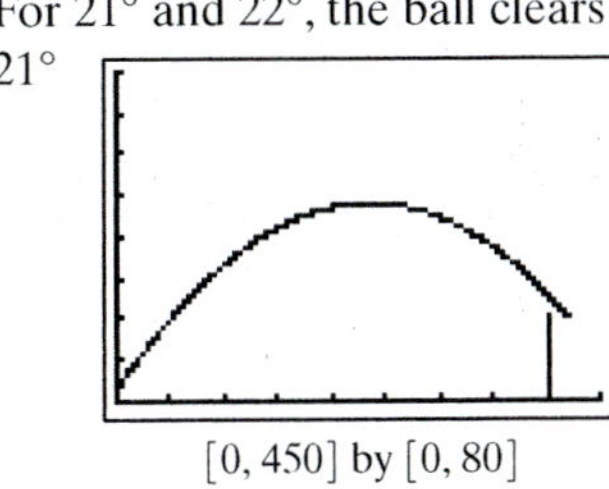

$[0, 450]$ by $[0, 80]$

22°:

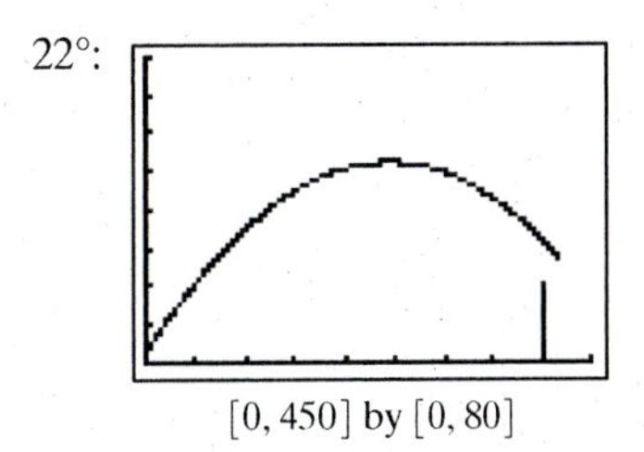

$[0, 450]$ by $[0, 80]$

Quick Review 6.3

1. (a) $\overrightarrow{OA} = \langle -3, -2 \rangle$

   (b) $\overrightarrow{OB} = \langle 4, 6 \rangle$

   (c) $\overrightarrow{AB} = \langle 4 - (-3), 6 - (-2) \rangle = \langle 7, 8 \rangle$

3. $m = \dfrac{6 - (-2)}{4 - (-3)} = \dfrac{8}{7}$

   $y + 2 = \frac{8}{7}(x + 3)$ or $y - 6 = \frac{8}{7}(x - 4)$

5. Graph $y = \pm\sqrt{8x}$.

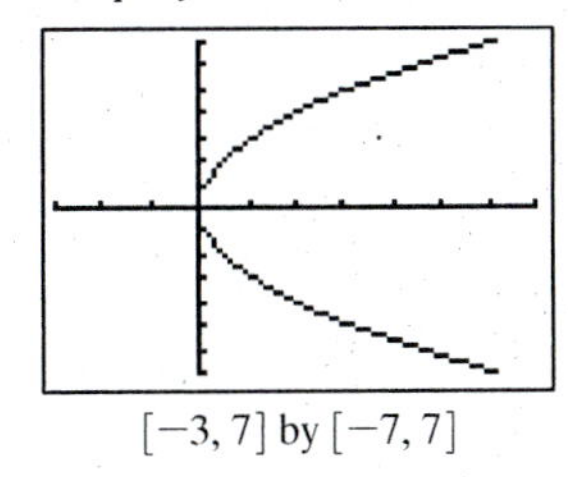

$[-3, 7]$ by $[-7, 7]$

7. $x^2 + y^2 = 4$

9. $\dfrac{600 \text{ rotations}}{1 \text{ min}} \cdot \dfrac{1 \text{ min}}{60 \text{ sec}} \cdot \dfrac{2\pi \text{ rad}}{1 \text{ rotation}} = 20\pi \text{ rad/sec}$

Section 6.3 Exercises

1. (b) $[-5, 5]$ by $[-5, 5]$

3. (a) $[-5, 5]$ by $[-5, 5]$

5. (a)

t	-2	-1	0	1	2
x	0	1	2	3	4
y	$-\frac{1}{2}$	-2	undef.	4	$\frac{5}{2}$

(b)

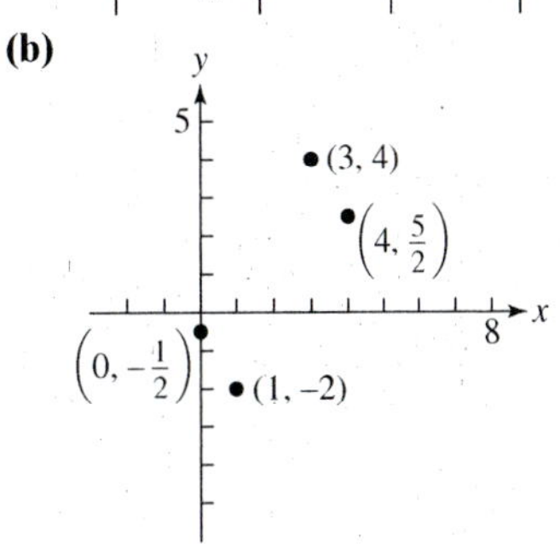

7.

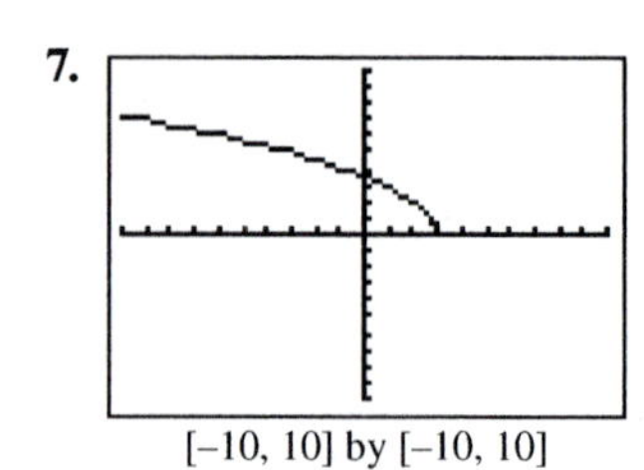

[−10, 10] by [−10, 10]

9.

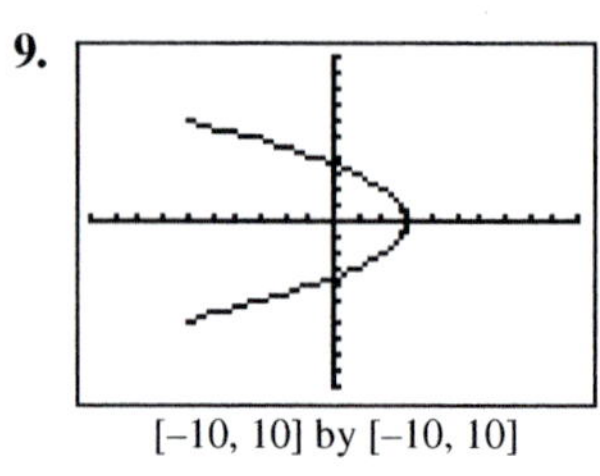

[−10, 10] by [−10, 10]

11. $x = 1 + y$, so $y = x - 1$: line through $(0, -1)$ and $(1, 0)$

13. $t = \frac{1}{2}x + \frac{3}{2}$, so $y = 9 - 4\left(\frac{1}{2}x + \frac{3}{2}\right)$;
$y = -2x + 3$, $3 \le x \le 7$: line segment with endpoints $(3, -3)$ and $(7, -11)$

15. $x = (y - 1)^2$: parabola that opens to right with vertex at $(0, 1)$

17. $y = x^3 - 2x + 3$: cubic polynomial

19. $x = 4 - y^2$; parabola that opens to left with vertex at $(4, 0)$

21. $t = x + 3$, so
$y = \frac{2}{x + 3}$, on domain: $[-8, -3) \cup (-3, 2]$

23. $x^2 + y^2 = 25$, circle of radius 5 centered at $(0, 0)$

25. $x^2 + y^2 = 4$, three-fourths of a circle of radius 2 centered at $(0, 0)$ (not in Quadrant II)

27. $\overrightarrow{OA} = \langle -2, 5 \rangle$, $\overrightarrow{OB} = \langle 4, 2 \rangle$, $\overrightarrow{OP} = \langle x, y \rangle$
$\overrightarrow{OP} - \overrightarrow{OA} = t(\overrightarrow{OB} - \overrightarrow{OA})$
$\langle x + 2, y - 5 \rangle = t\langle 6, -3 \rangle$
$x + 2 = 6t \Rightarrow x = 6t - 2$
$y - 5 = 3t \Rightarrow y = -3t + 5$

For #29–31, many answers are possible; one or two of the simplest are given.

29. Two possibilities are $x = t + 3$, $y = 4 - \frac{7}{3}t$, $0 \le t \le 3$, or $x = 3t + 3$, $y = 4 - 7t$, $0 \le t \le 1$.

31. One possibility is $x = 5 + 3 \cos t$, $y = 2 + 3 \sin t$, $0 \le t \le 2\pi$.

33. In Quadrant I, we need $x > 0$ and $y > 0$, so $2 - |t| > 0$ and $t - 0.5 > 0$. Then $-2 < t < 2$ and $t > 0.5$, so $0.5 < t < 2$. This is not changed by the additional requirement that $-3 \le t \le 3$.

35. In Quadrant III, we need $x < 0$ and $y < 0$, so $2 - |t| < 0$ and $t - 0.5 < 0$. Then ($t < -2$ or $t > 2$) and $t < 0.5$, so $t < -2$. With the additional requirement that $-3 \le t \le 3$, this becomes $-3 \le t < -2$.

37. (a) One good window is $[-20, 300]$ by $[-1, 10]$. If your grapher allows, use "Simultaneous" rather than "Sequential" plotting. Note that 100 yd is 300 ft. To show the whole race, use $0 \le t \le 13$ (upper limit may vary), since Ben finishes in $12.91\overline{6}$ sec. Note that it is the *process* of graphing (during which one observes Ben passing Jerry and crossing "the finish line" first), not the final product (which is two horizontal lines) which is needed; for that reason, no graph is shown here.

(b) After 3 seconds, Jerry is at $20(3) = 60$ ft and Ben is at $24(3) - 10 = 62$ ft. Ben is ahead by 2 ft.

39. (a) $y = -16t^2 + v_0t + s_0 = -16t^2 + 0t + 1000$
$= -16t^2 + 1000$

(b) Graph and trace: $x = 1$ and $y = -16t^2 + 1000$ with $0 \le t \le 6$, on the window $[0, 2]$ by $[0, 1200]$. Use something like 0.2 or less for Tstep. This graph will appear as a vertical line from $(1, 424)$ to $(1, 1000)$; it is not shown here because the simulation is accomplished by the *tracing*, not by the *picture*.

(c) When $t = 4$, $y = -16(4)^2 + 1000 = 744$ ft; the food containers are 744 ft above the ground after 4 sec.

41. Possible answers:

(a) $0 < t < \frac{\pi}{2}$ (t in radians)

(b) $0 < t < \pi$

(c) $\frac{\pi}{2} < t < \frac{3\pi}{2}$

43. (a) $x = 400$ when $t \approx 2.80$ — about 2.80 sec.

(b) When $t \approx 2.80$ sec, $y \approx 7.18$ ft.

(c) Reaching up, the outfielder's glove should be at or near the height of the ball as it approaches the wall. If hit at an angle of 20°, the ball would strike the wall about 19.74 ft up (after 2.84 sec) — the outfielder could not catch this.

45. (a) Yes: $x = (5 + 120 \cos 30°)t$; this equals 350 when $t \approx 3.21$. At this time, the ball is at a height of $y = -16t^2 + (120 \sin 30°)t + 4 \approx 31.59$ ft.

(b) The ball clears the wall with about 1.59 ft to spare (when $t \approx 3.21$).

47. No: $x = (30 \cos 70°)t$ and $y = -16t^2 + (30 \sin 70°)t + 3$. The dart lands when $y = 0$, which happens when $t \approx 1.86$ sec. At this point, the dart is about 19.11 ft from Tony, just over 10 in. short of the target.

49. The parametric equations for this motion are $x = (v + 160 \cos 20°)\, t$ and $y = -16t^2 + (160 \sin 20°)t + 4$, where v is the velocity of the wind (in ft/sec) — it should be positive if the wind is in the direction of the hit, and negative if the wind is against the ball.

To solve this algebraically, eliminate the parameter t as follows:

$t = \frac{x}{v + 160 \cos 20°}$. So $y = -16\left(\frac{x}{v + 160 \cos 20°}\right)^2$
$+ 160 \sin 20°\left(\frac{x}{v + 160 \cos 20°}\right) + 4.$

Substitute $x = 400$ and $y = 30$:

$$30 = -16\left(\frac{400}{v + 160 \cos 20°}\right)^2 + 160 \sin 20°\left(\frac{400}{v + 160 \cos 20°}\right) + 4.$$

Let $u = \frac{400}{v + 160 \cos 20°}$, so the equation becomes $-16u^2 + 54.72u - 26 = 0$. Using the quadratic formula, we find that $u = \frac{-54.72 \pm \sqrt{54.72^2 - 4(-16)(-26)}}{-32} \approx 0.57, 2.85$. Solving $0.57 = \frac{400}{v + 160 \cos 20°}$ and $2.85 = \frac{400}{v + 160 \cos 20°}$, $v \approx 551, v \approx -10$. A wind speed of 551 ft/sec (375.7 mph) is unrealistic, so we eliminate that solution. So the wind will be blowing against the ball in order for the ball to hit within a few inches of the top of the wall.

To verify this graphically, graph the equation

$$30 = -16\left(\frac{400}{v + 160 \cos 20°}\right)^2 + 160 \sin 20°\left(\frac{400}{v + 160 \cos 20°}\right) + 4,$$ and find the zero.

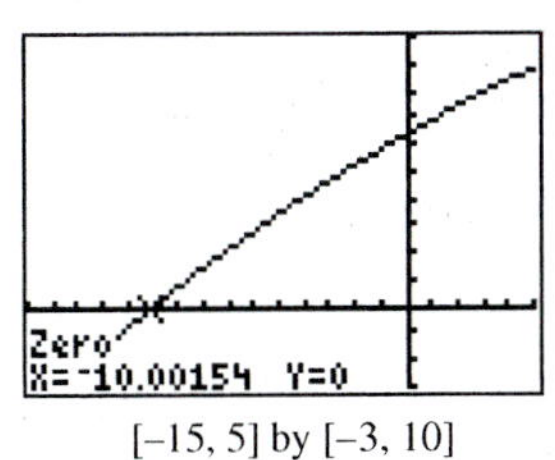

[–15, 5] by [–3, 10]

51. $x = 35 \cos\left(\frac{\pi}{6}t\right)$ and $y = 50 + 35 \sin\left(\frac{\pi}{6}t\right)$

53. (a) When $t = \pi$(or 3π, or 5π, etc.), $y = 2$. This corresponds to the highest points on the graph.

(b) The x-intercepts occur where $y = 0$, which happens when $t = 0, 2\pi, 4\pi$, etc. The x coordinates at those times are (respectively) $0, 2\pi, 4\pi$, etc., so these are 2π units apart.

55. The particle begins at -10, moves right to $+2.25$ (at $t = 1.5$), then changes direction and ends at -4.

57. The particle begins at -5, moves right to about $+0.07$ (at time $t \approx 0.15$), changes direction and moves left to about -20.81 (at time $t \approx 4.5$), then changes direction and ends at $+7$.

59. True. Eliminate t from the first set:

$t = x_1 + 1$
$y_1 = 3(x_1 + 1) + 1$
$y_1 = 3x_1 + 4$

Eliminate t from the second set:

$t = \frac{3}{2}x_2 + 2$
$y_2 = 2\left(\frac{3}{2}x_2 + 2\right)$
$y_2 = 3x_2 + 4$

Both sets correspond to the rectangular equation $y = 3x + 4$.

61. $x = (-1)^2 - 4 = -3, y = -1 + \frac{1}{-1} = -2$

The answer is A.

63. Set $-16t^2 + 80t + 7$ equal to 91 and solve either graphically or using the quadratic formula. The answer is D.

65. (a)

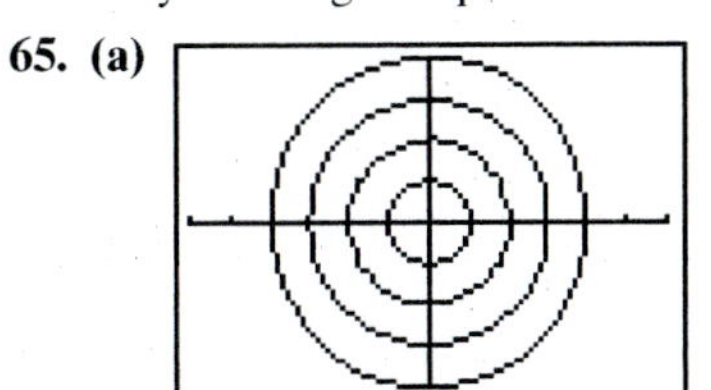
[–6, 6] by [–4, 4]

(b) $x^2 + y^2 = (a \cos t)^2 + (a \sin t)^2$
$= a^2 \cos^2 t + a^2 \sin^2 t$
$= a^2$

The radius of the circles are $a = \{1, 2, 3, 4\}$, centered at $(0, 0)$.

(c)

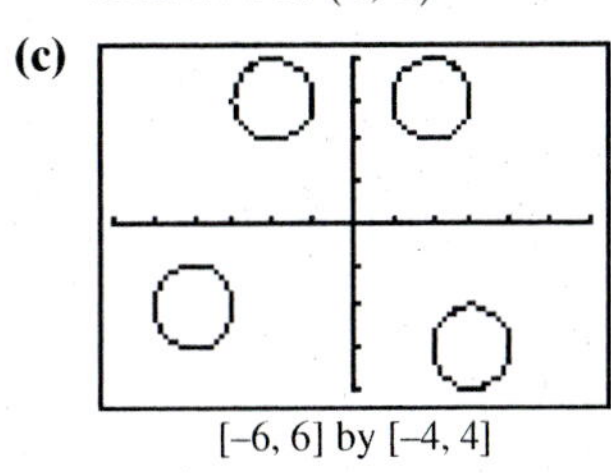
[–6, 6] by [–4, 4]

(d) $x - h = a \cos t$ and $y - k = a \sin t$, so
$(x - h)^2 + (y - k)^2$
$= (a \cos t)^2 + (a \sin t)^2$
$= a^2 \cos^2 t + a^2 \sin^2 t$
$= a^2$

The graph is the circle of radius a centered at (h, k).

(e) If $(x + 1)^2 + (y - 4)^2 = 9$, then $a = 3, h = -1$, and $k = 4$.
As a result, $x = 3 \cos t - 1$ and $y = 3 \sin t + 4$.

67. (a) Jane is traveling in a circle of radius 20 feet and center $(0, 20)$, which yields $x_1 = 20 \cos(nt)$ and $y_1 = 20 + 20 \sin(nt)$. Since the Ferris wheel is making one revolution (2π) every 12 seconds,

$2\pi = 12n$, so $n = \frac{2\pi}{12} = \frac{\pi}{6}$.

Thus,

$x_1 = 20 \cos\left(\frac{\pi}{6}t\right)$ and $y_1 = 20 + 20 \sin\left(\frac{\pi}{6}t\right)$, in radian mode.

(b) Since the ball was released at 75 ft in the positive x-direction and gravity acts in the negative y-direction at 16 ft/s^2, we have $x_2 = at + 75$ and $y_2 = -16t^2 + bt$, where a is the initial speed of the ball in the x-direction and b is the initial speed of the ball in the y-direction. The initial velocity vector of the ball is $60\langle\cos 120°, \sin 120°\rangle = \langle -30, 30\sqrt{3}\rangle$, so $a = -30$ and $b = 30\sqrt{3}$. As a result $x_2 = -30t + 75$ and $y_2 = -16t^2 + 30\sqrt{3}t$ are the parametric equations for the ball.

(c)

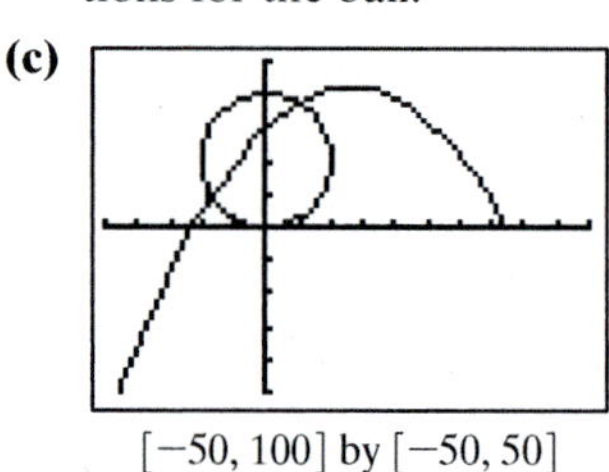

$[-50, 100]$ by $[-50, 50]$

Our graph shows that Jane and the ball will be close to each other but not at the exact same point at $t = 2.1$ seconds.

(d) $d(t) = \sqrt{(x_1 - x_2)^2 + (y_1 - y_2)^2}$

$$= \sqrt{\left(20\cos\left(\frac{\pi}{6}t\right) + 30t - 75\right)^2 + \left(20 + 20\sin\left(\frac{\pi}{6}t\right) + 16t^2 - (30\sqrt{3})t\right)^2}$$

(e)

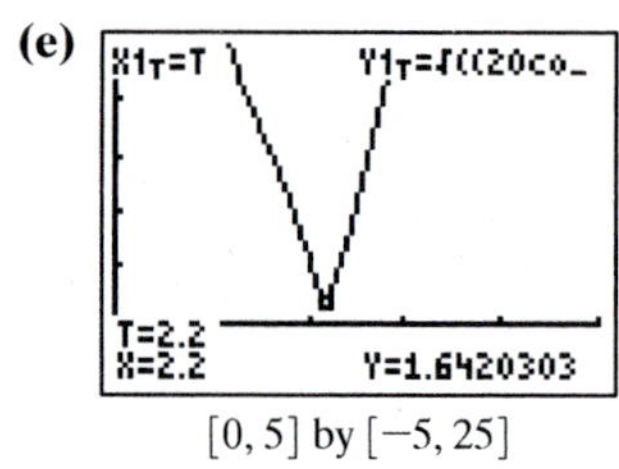

$[0, 5]$ by $[-5, 25]$

The minimum distance occurs at $t = 2.2$, when $d(t) = 1.64$ feet.

69. Chang's position: $x_1 = 20\cos\left(\frac{\pi}{6}t\right)$ and $y_1 = 20 + 20\sin\left(\frac{\pi}{6}t\right)$. Kuan's position: $x_2 = 15 + 15\cos\left(\frac{\pi}{4}t\right)$ and $y_2 = 15 + 15\sin\left(\frac{\pi}{4}t\right)$.

Find (graphically) the minimum of $d(t) = \sqrt{(x_1 - x_2)^2 + (y_1 - y_2)^2}$. It occurs when $t \approx 21.50$ sec; the minimum distance is about 4.11 ft.

71. (a) $x(0) = 0c + (1 - 0)a = a$ and
$y(0) = 0d + (1 - 0)b = b$

(b) $x(1) = 1c + (1 - 1)a = c$ and
$y(1) = 1d + (1 - 1)b = d$

73. Since the relationship between x and y is linear and one unit of time ($t = 1$) separates the two points, $t = \frac{1}{3} \approx 0.33$ will divide the segment into three equal pieces. Similarly, $t = \frac{1}{4} = 0.25$ will divide the segment into four equal pieces.

Section 6.4 Polar Coordinates

Exploration 1

2. $\left(2, \frac{\pi}{3}\right) = (1, \sqrt{3})$

$\left(-1, \frac{\pi}{2}\right) = (0, -1)$

$(2, \pi) = (-2, 0)$

$\left(-5, \frac{3\pi}{2}\right) = (0, 5)$

$(3, 2\pi) = (3, 0)$

3. $(-1, -\sqrt{3}) = \left(-2, \frac{\pi}{3}\right)$

$(0, 2) = \left(2, \frac{\pi}{2}\right)$

$(3, 0) = (3, 0)$

$(-1, 0) = (1, \pi)$

$(0, -4) = \left(4, \frac{3\pi}{2}\right)$

Quick Review 6.4

1. (a) Quadrant II

(b) Quadrant III

3. Possible answers: $7\pi/4, -9\pi/4$

5. Possible answers: $520°, -200°$

7. $(x - 3)^2 + y^2 = 4$

9. $a^2 = 12^2 + 10^2 - 2(12)(10)\cos 60°$
$a \approx 11.14$

Section 6.4 Exercises

1. $\left(-\frac{3}{2}, \frac{3\sqrt{3}}{2}\right)$

3. $(-1, -\sqrt{3})$

5. (a)

θ	$\frac{\pi}{4}$	$\frac{\pi}{2}$	$\frac{5\pi}{6}$	π	$\frac{4\pi}{3}$	2π
r	$\frac{3\sqrt{2}}{2}$	3	1.5	0	$\frac{-3\sqrt{3}}{2}$	0

(b)

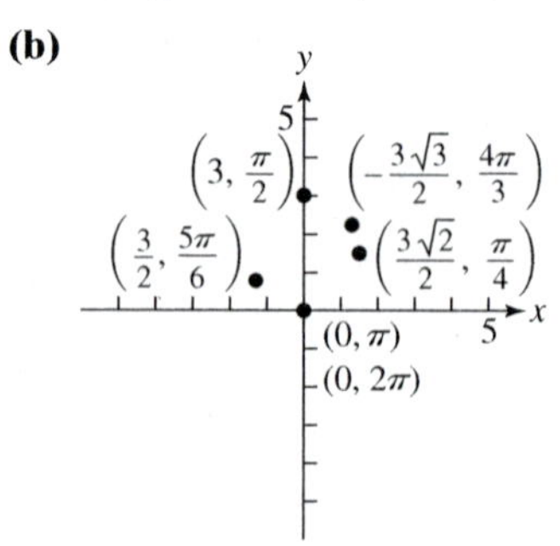

7.

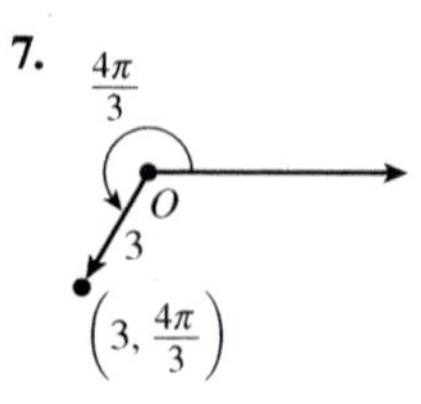

9.

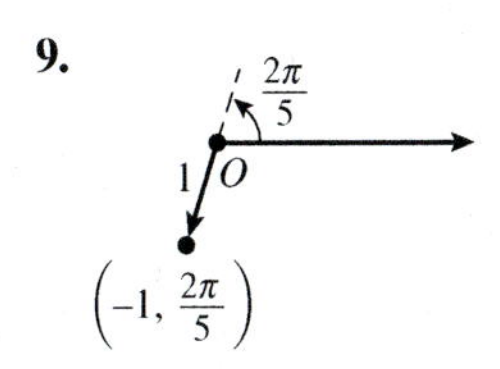

11. (2, 30°) 2 30° O

13. 120° O 2 (−2, 120°)

15. $\left(\frac{3}{4}, \frac{3}{4}\sqrt{3}\right)$

17. $(-2.70, 1.30)$

19. $(2, 0)$

21. $(0, -2)$

23. $\left(2, \frac{\pi}{6} + 2n\pi\right)$ and $\left(-2, \frac{\pi}{6} + (2n + 1)\pi\right)$, n an integer

25. $(1.5, -20° + 360n°)$ and $(-1.5, 160° + 360n°)$, n an integer

27. (a) $\left(\sqrt{2}, \frac{\pi}{4}\right)$ or $\left(-\sqrt{2}, \frac{5\pi}{4}\right)$

(b) $\left(\sqrt{2}, \frac{\pi}{4}\right)$ or $\left(-\sqrt{2}, -\frac{3\pi}{4}\right)$

(c) The answers from (a), and also $\left(\sqrt{2}, \frac{9\pi}{4}\right)$ or $\left(-\sqrt{2}, \frac{13\pi}{4}\right)$

29. (a) $(\sqrt{29}, \tan^{-1}(-2.5) + \pi) \approx (\sqrt{29}, 1.95)$ or $(-\sqrt{29}, \tan^{-1}(-2.5) + 2\pi) \approx (-\sqrt{29}, 5.09)$

(b) $(-\sqrt{29}, \tan^{-1}(-2.5)) \approx (-\sqrt{29}, -1.19)$ or $(\sqrt{29}, \tan^{-1}(-2.5) + \pi) \approx (\sqrt{29}, 1.95)$

(c) The answers from (a), plus $(\sqrt{29}, \tan^{-1}(-2.5) + 3\pi) \approx (\sqrt{29}, 8.23)$ or $(-\sqrt{29}, \tan^{-1}(-2.5) + 4\pi) \approx (-\sqrt{29}, 11.38)$

31. (b)

33. (c)

35. $x = 3$ — a vertical line

37. $r^2 + 3r \sin \theta = 0$, or $x^2 + y^2 + 3y = 0$. Completing the square gives $x^2 + \left(y + \frac{3}{2}\right)^2 = \frac{9}{4}$ — a circle centered at $\left(0, -\frac{3}{2}\right)$ with radius $\frac{3}{2}$.

39. $r^2 - r \sin \theta = 0$, or $x^2 + y^2 - y = 0$. Completing the square gives $x^2 + \left(y - \frac{1}{2}\right)^2 = \frac{1}{4}$ — a circle centered at $\left(0, \frac{1}{2}\right)$ with radius $\frac{1}{2}$.

41. $r^2 - 2r \sin \theta + 4r \cos \theta = 0$, or $x^2 + y^2 - 2y + 4x = 0$. Completing the square gives $(x + 2)^2 + (y - 1)^2 = 5$ — a circle centered at $(-2, 1)$ with radius $\sqrt{5}$.

43. $r = 2/\cos \theta = 2 \sec \theta$ — a vertical line

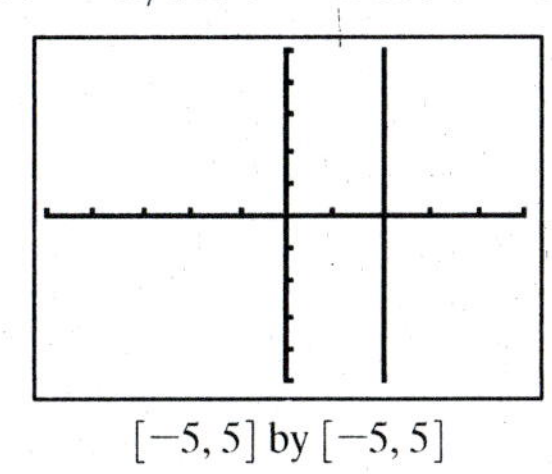

$[-5, 5]$ by $[-5, 5]$

45. $r = \dfrac{5}{2 \cos \theta - 3 \sin \theta}$

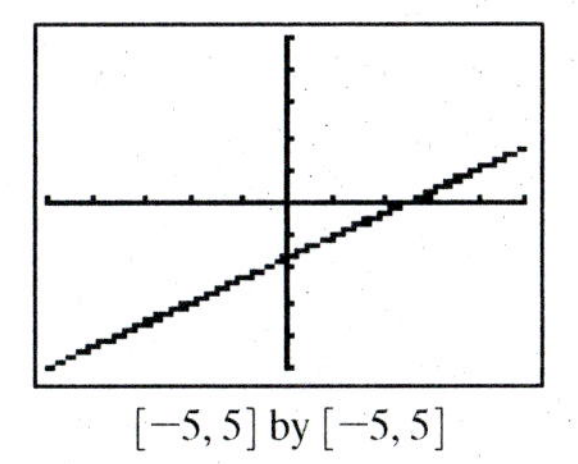

$[-5, 5]$ by $[-5, 5]$

47. $r^2 - 6r \cos \theta = 0$, so $r = 6 \cos \theta$

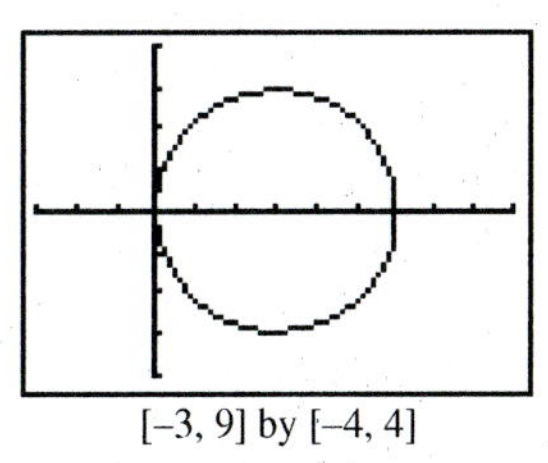

$[-3, 9]$ by $[-4, 4]$

49. $r^2 + 6r \cos \theta + 6r \sin \theta = 0$, so $r = -6 \cos \theta - 6 \sin \theta$

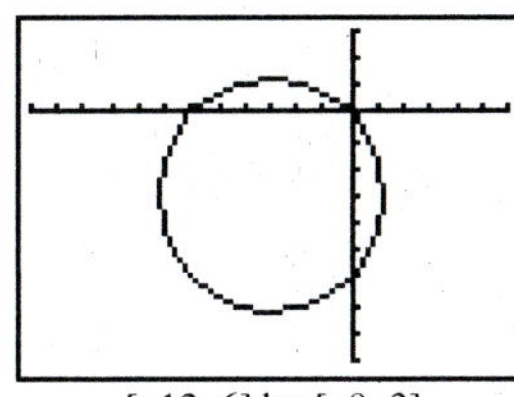

$[-12, 6]$ by $[-9, 3]$

51. $d = \sqrt{4^2 + 2^2 - 2 \cdot 4 \cdot 2 \cos (12° - 72°)} = \sqrt{20 - 16 \cos 60°} = \sqrt{12} = 2\sqrt{3}$ mi

53. Using the Pythagorean theorem, the center-to-vertex distance is $\frac{a}{\sqrt{2}}$. The four vertices are then $\left(\frac{a}{\sqrt{2}}, \frac{\pi}{4}\right)$, $\left(\frac{a}{\sqrt{2}}, \frac{3\pi}{4}\right)$, $\left(\frac{a}{\sqrt{2}}, \frac{5\pi}{4}\right)$, and $\left(\frac{a}{\sqrt{2}}, \frac{7\pi}{4}\right)$. Other polar coordinates for these points are possible, of course.

55. False. Point (r, θ) is the same as point $(r, \theta + 2n\pi)$ for any integer n. So each point has an infinite number of distinct polar coordinates.

57. For point (r, θ), changing the sign on r and adding 3π to θ constitutes a twofold reflection across the orgin. The answer is C.

59. For point (r, θ), changing the sign on r and subtracting 180° from θ constitutes a twofold reflection across the origin. The answer is A.

61. (a) If $\theta_1 - \theta_2$ is an odd integer multiple of π, then the distance is $|r_1 + r_2|$. If $\theta_1 - \theta_2$ is an even integer multiple of π, then the distance is $|r_1 - r_2|$.

(b) Consider the triangle formed by O_1, P_1, and P_2 (ensuring that the angle at the origin is less than 180°), then by the Law of Cosines,
$\overline{P_1P_2}^2 = \overline{OP_1}^2 + \overline{OP_2}^2 - 2 \cdot \overline{OP_1} \cdot \overline{OP_2} \cos\theta$,
where θ is the angle between $\overline{OP_1}$ and $\overline{OP_2}$. In polar coordinates, this formula translates very nicely into $d^2 = r_1^2 + r_2^2 - 2r_1r_2\cos(\theta_2 - \theta_1)$ (or $\cos(\theta_1 - \theta_2)$ since $\cos(\theta_2 - \theta_1) = \cos(\theta_1 - \theta_2)$), so

$d = \sqrt{r_1^2 + r_2^2 - 2r_1 r_2 \cos(\theta_2 - \theta_1)}$.

(c) Yes. If $\theta_1 - \theta_2$ is an odd integer multiple of π, then $\cos(\theta_1 - \theta_2) = -1 \Rightarrow d = \sqrt{r_1^2 + r_2^2 + 2r_1r_2}$ $= |r_1 + r_2|$. If $\theta_1 - \theta_2$ is an even integer multiple of π, then $\cos(\theta_1 - \theta_2) = 1 \Rightarrow$

$d = \sqrt{r_1^2 + r_2^2 - 2r_1r_2} = |r_1 - r_2|$.

63. $d = \sqrt{2^2 + 5^2 - 2(2)(5)\cos 120°} \approx 6.24$

65. $d = \sqrt{(-3)^2 + (-5)^2 - 2(-3)(-5)\cos 135°} \approx 7.43$

67. Since $x = r\cos\theta$ and $y = r\sin\theta$, the parametric equation would be $x = f(\theta)\cos(\theta)$ and $y = f(\theta)\sin(\theta)$.

69. $x = 5(\cos\theta)(\sin\theta)$
$y = 5\sin^2\theta$

71. $x = 4(\cos\theta)(\csc\theta) = 4\cot\theta$
$y = 4(\sin\theta)(\csc\theta) = 4$

Section 6.5 Graphs of Polar Equations

Exploration 1

Answers will vary.

Exploration 2

1. If $r^2 = 4\cos(2\theta)$, then r does not exist when $\cos(2\theta) < 0$. Since $\cos(2\theta) < 0$ whenever θ is in the interval $\left(\frac{\pi}{4} + n\pi, \frac{3\pi}{4} + n\pi\right)$, n is any integer, the domain of r does not include these intervals.
2. $-r\sqrt{\cos(2\theta)}$ draws the same graph, but in the opposite direction.
3. $(r)^2 - 4\cos(-2\theta) = r^2 - 4\cos(2\theta)$ (since $\cos(\theta) = \cos(-\theta)$)
4. $(-r)^2 - 4\cos(-2\theta) = r^2 - 4\cos(2\theta)$
5. $(-r)^2 - 4\cos(2\theta) = r^2 - 4\cos(2\theta)$

Quick Review 6.5

For #1–3, use your grapher's TRACE function to solve.

1. Minimum: −3 at $x = \left\{\frac{\pi}{2}, \frac{3\pi}{2}\right\}$; Maximum: 3 at $x = \{0, \pi, 2\pi\}$

3. Minimum: 0 at $x = \left\{\frac{\pi}{4}, \frac{3\pi}{4}, \frac{5\pi}{4}, \frac{7\pi}{4}\right\}$; Maximum: 2 at $x = \{0, \pi, 2\pi\}$

5. (a) No **(b)** No **(c)** Yes

7. $\sin(\pi - \theta) = \sin\theta$

9. $\cos 2(\pi + \theta) = \cos(2\pi + 2\theta) = \cos 2\theta$
$= \cos^2\theta - \sin^2\theta$

Section 6.5 Exercises

1. (a)

θ	0	$\pi/4$	$\pi/2$	$3\pi/4$	π	$5\pi/4$	$3\pi/2$	$7\pi/4$
r	3	0	−3	0	3	0	−3	0

(b)

$\left(0, \left\{\frac{\pi}{4}, \frac{3\pi}{4}, \frac{5\pi}{4}, \frac{7\pi}{4}\right\}\right)$ $\left(-3, \frac{3\pi}{2}\right)$ $(3, 0)$ $(3, \pi)$ $\left(-3, \frac{\pi}{2}\right)$

3. $k = \pi$

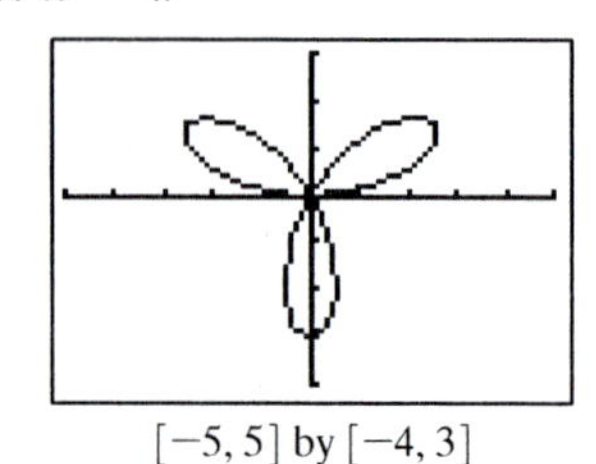

[−5, 5] by [−4, 3]

5. $k = 2\pi$

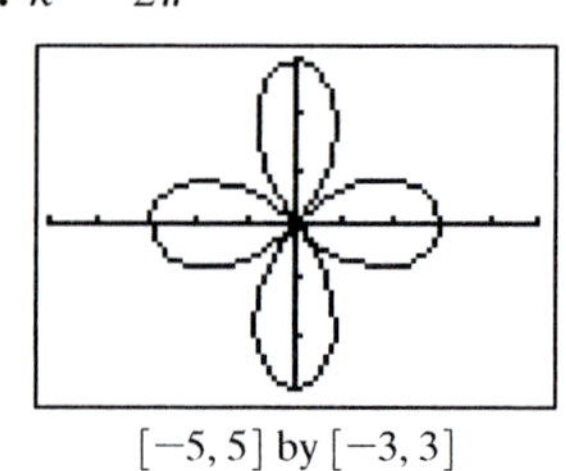

[−5, 5] by [−3, 3]

7. r_1 is not shown (this is a 12-petal rose). r_2 is not shown (this is a 6-petal rose), r_3 is graph (b).

9. Graph (b) is $r = 2 - 2\cos\theta$: Taking $\theta = 0$ and $\theta = \frac{\pi}{2}$, we get $r = 2$ and $r = 4$ from the first equation, and $r = 0$ and $r = 2$ from the second. No graph matches the first of these (r, θ) pairs, but (b) matches the latter (and any others one might choose).

11. Graph (a) is $r = 2 - 2\sin\theta$ — where $\theta = \frac{\pi}{2}$, $2 + 2\cos\theta = 2$, but $\left(2, \frac{\pi}{2}\right)$ is clearly not on graph (a); meanwhile $2 - 2\sin\frac{\pi}{2} = 0$, and $\left(0, \frac{\pi}{2}\right)$ (the origin) is part of graph (a).

13. Symmetric about the y-axis: replacing (r, θ) with $(r, \pi - \theta)$ gives the same equation, since $\sin(\pi - \theta) = \sin\theta$.

15. Symmetric about the x-axis: replacing (r, θ) with $(r, -\theta)$ gives the same equation, since $\cos(-\theta) = \cos\theta$.

17. All three symmetries. Polar axis: replacing (r, θ) with $(r, -\theta)$ gives the same equation, since $\cos(-2\theta) = \cos 2\theta$. y-axis: replacing (r, θ) with $(r, \pi - \theta)$ gives the same equation, since $\cos[2(\pi - \theta)] = \cos(2\pi - 2\theta)$ $= \cos(-2\theta) = \cos 2\theta$. Pole: replacing (r, θ) with $(r, \theta + \pi)$ gives the same equation, since $\cos[2(\theta + \pi)]$ $= \cos(2\theta + 2\pi) = \cos 2\theta$.

19. Symmetric about the y-axis: replacing (r, θ) with $(r, \pi - \theta)$ gives the same equation, since $\sin(\pi - \theta) = \sin \theta$.

21. Maximum r is 5 — when $\theta = 2n\pi$ for any integer n.

23. Maximum r is 3 (along with -3) — when $\theta = 2n\pi/3$ for any integer n.

25. Domain: All reals
Range: $[-3, 3]$
Symmetric about the x-axis, y-axis, and origin
Continuous
Bounded
Maximum r-value: 3
No asymptotes

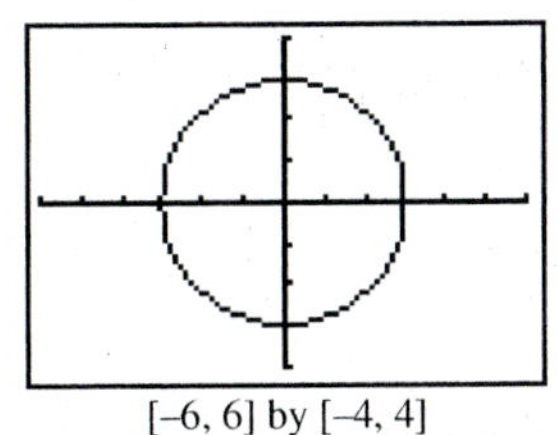

[–6, 6] by [–4, 4]

27. Domain: $\theta = \pi/3 + n\pi$, n is an integer
Range: $(-\infty, \infty)$
Symmetric about the origin
Continuous
Unbounded
Maximum r-value: none
No asymptotes

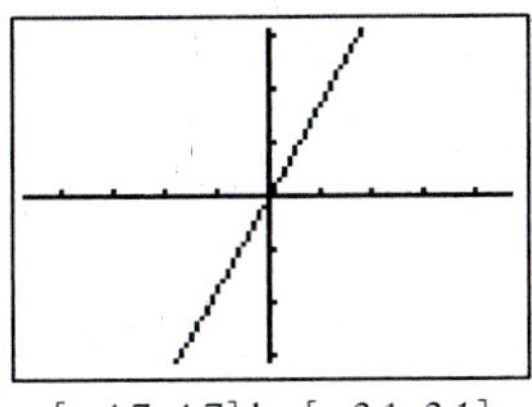

$[-4.7, 4.7]$ by $[-3.1, 3.1]$

29. Domain: All reals
Range: $[-2, 2]$
Symmetric about the y-axis
Continuous
Bounded
Maximum r-value: 2
No asymptotes

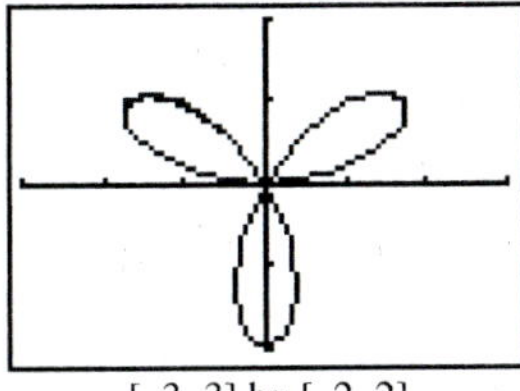

[–3, 3] by [–2, 2]

31. Domain: All reals
Range: $[1, 9]$
Symmetric about the y-axis
Continuous
Bounded
Maximum r-value: 9
No asymptotes

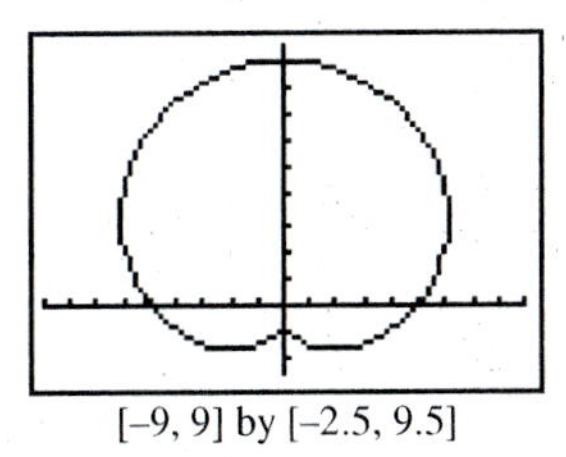

[–9, 9] by [–2.5, 9.5]

33. Domain: All reals
Range: $[0, 8]$
Symmetric about the x-axis
Continuous
Bounded
Maximum r-value: 8
No asymptotes

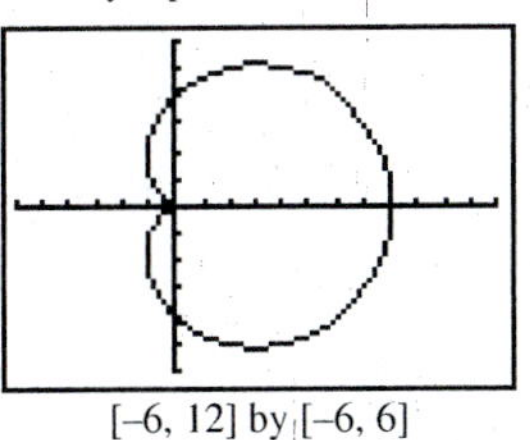

[–6, 12] by [–6, 6]

35. Domain: All reals
Range: $[3, 7]$
Symmetric about the x-axis
Continuous
Bounded
Maximum r-value: 7
No asymptotes

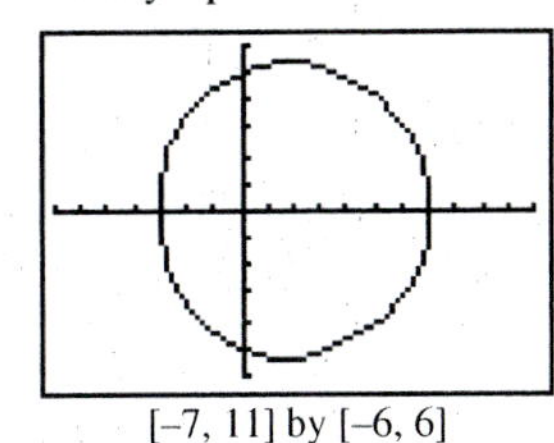

[–7, 11] by [–6, 6]

37. Domain: All reals
Range: $[-3, 7]$
Symmetric about the x-axis
Continuous
Bounded
Maximum r-value: 7
No asymptotes

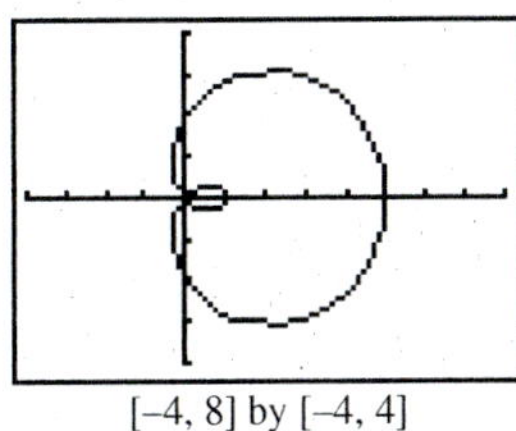

[–4, 8] by [–4, 4]

39. Domain: All reals
Range: $[0, 2]$
Symmetric about the x-axis
Continuous
Bounded
Maximum r-value: 2
No asymptotes

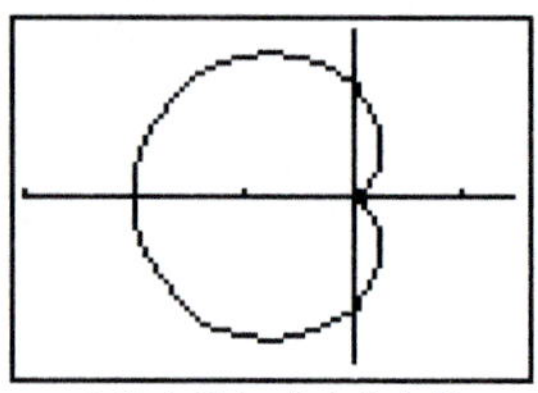

$[-3, 1.5]$ by $[-1.5, 1.5]$

41. Domain: All reals
Range: $[0, \infty)$
Continuous
No symmetry
Unbounded
Maximum r-value: none
No asymptotes
Graph for $\theta \geq 0$:

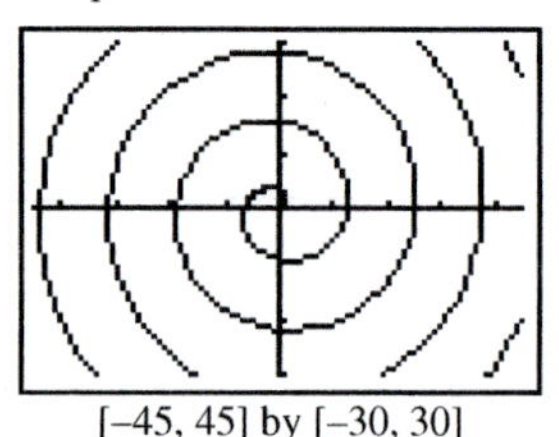

$[-45, 45]$ by $[-30, 30]$

43. Domain: $\left[0, \frac{\pi}{2}\right] \cup \left[\frac{3\pi}{2}, 2\pi\right]$
Range: $[0, 1]$
Symmetric about the origin
Continuous on each interval in domain
Bounded
Maximum r-value: 1
No asymptotes

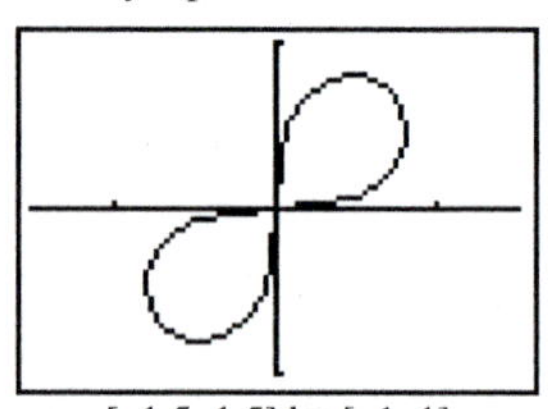

$[-1.5, 1.5]$ by $[-1, 1]$

For #45–47, recall that the petal length is the maximum r-value over the interval that creates the petal.

45. $r = -2$ when $\theta = \left\{\frac{3\pi}{4}, \frac{7\pi}{4}\right\}$ and $r = 6$ when $\theta = \left\{\frac{\pi}{4}, \frac{5\pi}{4}\right\}$. There are four petals with lengths $\{6, 2, 6, 2\}$.

47. $r = -3$ when $\theta = \left\{0, \frac{2\pi}{5}, \frac{4\pi}{5}, \frac{6\pi}{5}, \frac{8\pi}{5}\right\}$ and $r = 5$ when $\theta = \left\{\frac{\pi}{5}, \frac{3\pi}{5}, \pi, \frac{7\pi}{5}, \frac{9\pi}{5}\right\}$.
There are ten petals with lengths $\{3, 5, 3, 5, 3, 5, 3, 5, 3, 5\}$.

49. r_1 and r_2 produce identical graphs — r_1 begins at $(1, 0)$ and r_2 begins at $(-1, 0)$.

51. r_2 and r_3 produce identical graphs — r_1 begins at $(3, 0)$ and r_3 begins at $(-3, 0)$.

53. **(a)** A 4-petal rose curve with 2 short petals of length 1 and 2 long petals of length 3.

(b) Symmetric about the origin.

(c) Maximum r-value: 3.

55. **(a)** A 6-petal rose curve with 3 short petals of length 2 and 3 long petals of length 4.

(b) Symmetric about the x-axis.

(c) Maximum r-value: 4.

57. Answers will vary but generally students should find that a controls the length of the rose petals and n controls both the number of rose petals and symmetry. If n is odd, n rose petals are formed, with the cosine curve symmetric about the polar x-axis and sine curve symmetric about the y-axis. If n is even, $2n$ rose petals are formed, with both the cosine and sine functions having symmetry about the polar x-axis, y-axis, and origin.

59.

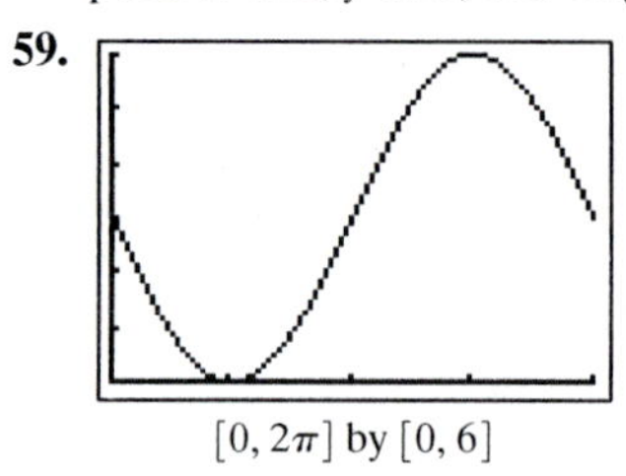

$[0, 2\pi]$ by $[0, 6]$

$y = 3 - 3 \sin x$ has minimum and maximum values of 0 and 6 on $[0, 2\pi]$. So the range of the polar function $r = 3 - 3 \sin \theta$ is also $[0, 6]$.

61. False. The spiral $r = \theta$ is unbounded, since a point on the curve can be found at any arbitrarily large distance from the origin by setting θ numerically equal to that distance.

63. With $r = a \cos n\theta$, if n is even there are $2n$ petals. The answer is D.

65. When $\cos \theta = -1$, $r = 5$. The answer is B.

67. **(a)** Symmetry about the polar x-axis: $r - a \cos (n\theta) = 0$ $\Rightarrow r - a \cos (-n\theta) = r - a \cos (n\theta)$ (since $\cos (\theta)$ is even, i.e., $\cos (\theta) = \cos (-\theta)$ for all θ) $= 0$.

(b) No symmetry about y-axis: $r - a \cos (n\theta) = 0 \Rightarrow$ $-r - a \cos (-n\theta) = -r - a \cos (n\theta)$ (since $\cos (\theta)$ is even) $\neq r - a \cos (n\theta)$ unless $r = 0$. As a result, the equation is not symmetric about the y-axis.

(c) No symmetry about origin: $r - a \cos (n\theta) = 0 \Rightarrow$ $-r - a \cos (n\theta) \neq -r - a \cos (n\theta)$ unless $r = 0$. As a result, the equation is not symmetric about the origin.

(d) Since $|\cos (n\theta)| \leq 1$ for all θ, the maximum r-value is $|a|$.

(e) Domain: All reals
Range: $[-|a|, |a|]$
Symmetric about the x-axis
Continuous
Bounded
Maximum r-value: 2
No asymptotes

69. (a) For r_1: $0 \le \theta \le 4\pi$ (or any interval that is 4π units long). For r_2: same answer.

(b) r_1: 10 (overlapping) petals. r_2: 14 (overlapping) petals.

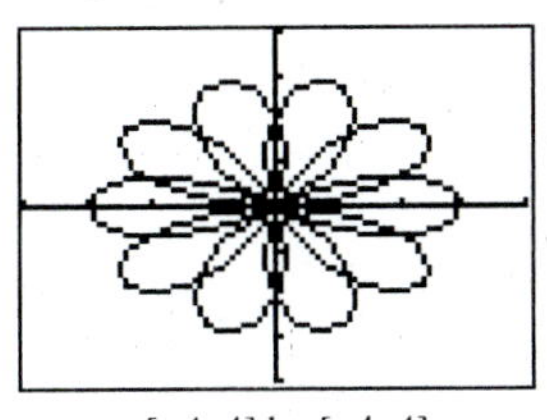

$[-4, 4]$ by $[-4, 4]$

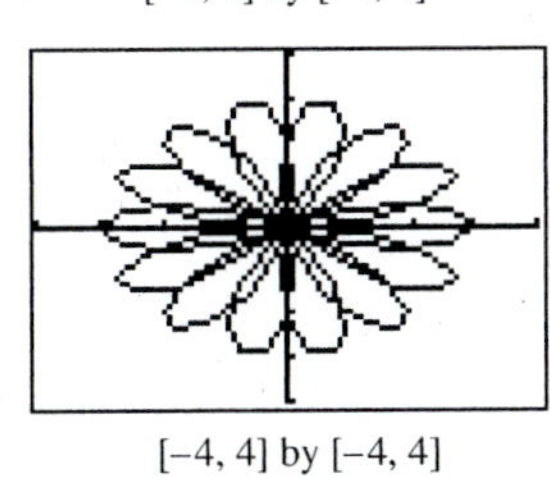

$[-4, 4]$ by $[-4, 4]$

71. Starting with the graph of r_1, if we rotate counterclockwise (centered at the origin) by $\pi/4$ radians (45°), we get the graph of r_2; rotating r_1 counterclockwise by $\pi/3$ radians (60°) gives the graph of r_3.

(a)

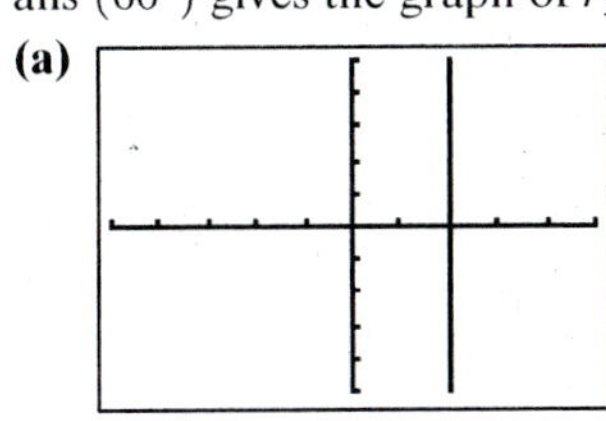

$[-5, 5]$ by $[-5, 5]$

(b)

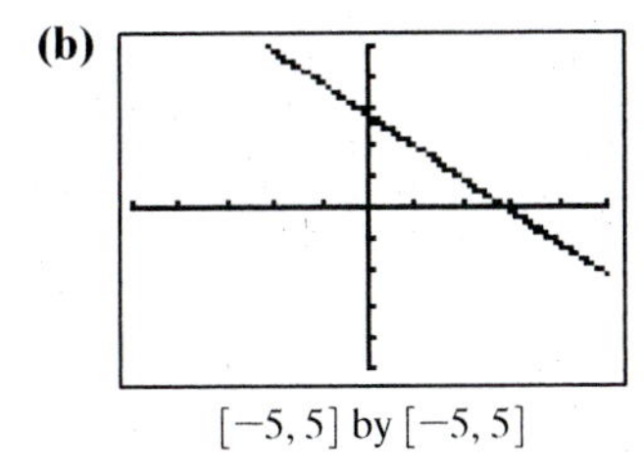

$[-5, 5]$ by $[-5, 5]$

(c)

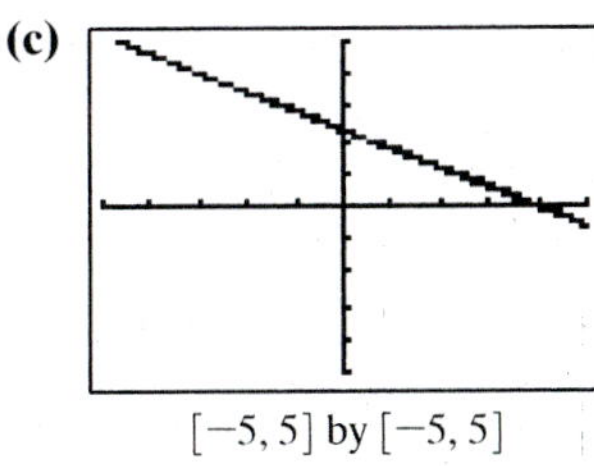

$[-5, 5]$ by $[-5, 5]$

73. The second graph is the result of rotating the first graph clockwise (centered at the origin) through an angle of α. The third graph results from rotating the first graph counterclockwise through the same angle. One possible explanation: the radius r achieved, for example, when $\theta = 0$ in the first equation is achieved instead when $\theta = -\alpha$ for the second equation, and when $\theta = \alpha$ for the third equation.

Section 6.6 DeMoivre's Theorem and *n*th Roots

Quick Review 6.6

1. Using the quadratic equation to find the roots of $x^2 + 13 = 4x$, we have $x^2 - 4x + 13 = 0$ with $a = 1$, $b = -4$, and $c = 13$.

$$x = \frac{-(-4) \pm \sqrt{(-4)^2 - 4(1)(13)}}{2(1)}$$

$$= \frac{4 \pm \sqrt{16 - 52}}{2} = \frac{4 \pm \sqrt{-36}}{2} = \frac{4 \pm 6i}{2}$$

$$x = \frac{4 + 6i}{2} = \frac{4}{2} + \frac{6i}{2} = 2 + 3i \text{ and}$$

$$x = \frac{4 - 6i}{2} = \frac{4}{2} - \frac{6i}{2} = 2 - 3i$$

The roots are $2 + 3i$ and $2 - 3i$.

3. $(1 + i)^5 = (1 + i) \cdot [(1 + i)^2]^2 = (1 + i) \cdot (2i)^2$
$= -4(1 + i) = -4 - 4i$

For #5–7, use the given information to find a point P on the terminal side of the angle, which in turn determines the quadrant of the terminal side.

5. $P(-\sqrt{3}, 1)$, in quadrant II: $\theta = \dfrac{5\pi}{6}$

7. $P(-1, -\sqrt{3},)$, in quadrant III: $\theta = \dfrac{4\pi}{3}$

9. $x^3 = 1$ when $x = 1$

Section 6.6 Exercises

1.

For #3–11, $a + bi = r(\cos\theta + i\sin\theta)$, where $r = |a + bi| = \sqrt{a^2 + b^2}$ and θ is chosen so that $\cos\theta = \dfrac{a}{\sqrt{a^2 + b^2}}$ and $\sin\theta = \dfrac{b}{\sqrt{a^2 + b^2}}$.

3. $r = |3i| = 3$; $\cos\theta = 0$ and $\sin\theta = 1$, so $\theta = \dfrac{\pi}{2}$:

$$3i = 3\left(\cos\frac{\pi}{2} + i\sin\frac{\pi}{2}\right)$$

5. $r = |2 + 2i| = 2\sqrt{2}$; $\cos\theta = \dfrac{\sqrt{2}}{2}$ and $\sin\theta = \dfrac{\sqrt{2}}{2}$,

so $\theta = \dfrac{\pi}{4}$: $2 + 2i = 2\sqrt{2}\left(\cos\dfrac{\pi}{4} + i\sin\dfrac{\pi}{4}\right)$

7. $r = |-2 + 2i\sqrt{3}| = 4$; $\cos\theta = -\dfrac{1}{2}$ and $\sin\theta = \dfrac{\sqrt{3}}{2}$,

so $\theta = \dfrac{2\pi}{3}$: $-2 + 2i\sqrt{3} = 4\left(\cos\dfrac{2\pi}{3} + i\sin\dfrac{2\pi}{3}\right)$

9. $r = |3 + 2i| = \sqrt{13}$; $\cos\theta = \dfrac{3}{\sqrt{13}}$ and $\sin\theta = \dfrac{2}{\sqrt{13}}$, so $\theta \approx 0.588$: $3 + 2i \approx \sqrt{13}(\cos 0.59 + i\sin 0.59)$

11. $r = 3$; $30° = \dfrac{\pi}{6}$; $3\left(\cos\dfrac{\pi}{6} + i\sin\dfrac{\pi}{6}\right)$

13. $3\left(\dfrac{\sqrt{3}}{2} - \dfrac{1}{2}i\right) = \dfrac{3}{2}\sqrt{3} - \dfrac{3}{2}i$

15. $5\left(\dfrac{1}{2} - \dfrac{\sqrt{3}}{2}i\right) = \dfrac{5}{2} - \dfrac{5}{2}\sqrt{3}\,i$

17. $\sqrt{2}\left(-\dfrac{\sqrt{3}}{2} - \dfrac{1}{2}i\right) = \dfrac{\sqrt{6}}{2} - \dfrac{\sqrt{2}}{2}i$

19. $(2\cdot 7)[\cos(25° + 130°) + i\sin(25° + 130°)] = 14(\cos 155° + i\sin 155°)$

21. $(5\cdot 3)\left[\cos\left(\dfrac{\pi}{4} + \dfrac{5\pi}{3}\right) + i\sin\left(\dfrac{\pi}{4} + \dfrac{5\pi}{3}\right)\right] = 15\left(\cos\dfrac{23\pi}{12} + i\sin\dfrac{23\pi}{12}\right)$

23. $\dfrac{2}{3}[\cos(30° - 60°) + i\sin(30° - 60°)] = \dfrac{2}{3}[\cos(-30°) + i\sin(-30°)] = \dfrac{2}{3}(\cos 30° - i\sin 30°)$

25. $\dfrac{6}{3}[\cos(5\pi - 2\pi) + i\sin(5\pi - 2\pi)] = 2(\cos 3\pi + i\sin 3\pi) = 2(\cos\pi + i\sin\pi)$

27. (a) $3 - 2i \approx \sqrt{13}\,[\cos(5.695) + i\sin(5.695)]$ and $1 + i = \sqrt{2}\left(\cos\dfrac{\pi}{4} + i\sin\dfrac{\pi}{4}\right)$, so

$$\sqrt{13}\left[\cos(5.695) + i\sin(5.695)\right]\cdot\sqrt{2}\left[\cos\frac{\pi}{4} + i\sin\frac{\pi}{4}\right]$$
$$= \sqrt{26}\left[\cos\left(5.695 + \frac{\pi}{4}\right) + i\sin\left(5.695 + \frac{\pi}{4}\right)\right] = 5 + i$$
$$\frac{\sqrt{13}[\cos(5.695) + i\sin(5.695)]}{\sqrt{2}[\cos(\pi/4) + i\sin(\pi/4)]} \approx \sqrt{6.5}\left[\cos\left(5.695 - \frac{\pi}{4}\right) + i\sin\left(5.695 - \frac{\pi}{4}\right)\right] = \frac{1}{2} - \frac{5}{2}i$$

(b) $(3 - 2i)(1 + i) = 3 + 3i - 2i - 2i^2 = 5 + i$

$$\frac{3 - 2i}{1 + i} = \frac{3 - 2i}{1 + i}\cdot\frac{1 - i}{1 - i} = \frac{1 - 5i}{2} = \frac{1}{2} - \frac{5}{2}i$$

29. (a) $3 + i \approx \sqrt{10}[\cos(0.321) + i\sin(0.321)]$ and $5 - 3i \approx \sqrt{34}[\cos(-0.540) + i\sin(-0.540)]$, so

$$\sqrt{10}[\cos(0.321) + i\sin(0.321)]\cdot\sqrt{34}[\cos(-0.540) + i\sin(-0.540)]$$
$$= 2\sqrt{85}[\cos(-0.219) + i\sin(-0.219)] = 18 - 4i$$
$$\frac{\sqrt{10}[\cos(0.321) + i\sin(0.321)]}{\sqrt{34}[\cos(-0.540) + i\sin(-0.540)]} \approx \sqrt{\frac{5}{17}}[\cos(0.862) + i\sin(0.862)] \approx 0.35 + 0.41i$$

(b) $(3 + i)(5 - 3i) = 15 - 9i + 5i - 3i^2 = 18 - 4i$

$$\frac{3 + i}{5 - 3i} = \frac{3 + i}{5 - 3i}\cdot\frac{5 + 3i}{5 + 3i} = \frac{(3 + i)(5 + 3i)}{34}$$
$$\frac{1}{17}(6 + 7i) \approx 0.35 + 0.41i$$

31. $\left(\cos\dfrac{\pi}{4} + i\sin\dfrac{\pi}{4}\right)^3 = \cos\dfrac{3\pi}{4} + i\sin\dfrac{3\pi}{4} = -\dfrac{\sqrt{2}}{2} + i\dfrac{\sqrt{2}}{2}$

33. $\left[2\left(\cos\dfrac{3\pi}{4} + i\sin\dfrac{3\pi}{4}\right)\right]^3 = 8\left(\cos\dfrac{9\pi}{4} + i\sin\dfrac{9\pi}{4}\right) = 4\sqrt{2} + 4\sqrt{2}i$

35. $(1 + i)^5 = \left[\sqrt{2}\left(\cos\dfrac{\pi}{4} + i\sin\dfrac{\pi}{4}\right)\right]^5$

$$= (\sqrt{2})^5\left(\cos\frac{5\pi}{4} + i\sin\frac{5\pi}{4}\right) = 4\sqrt{2}\left(\cos\frac{5\pi}{4} + i\sin\frac{5\pi}{4}\right) = -4 - 4i$$

37. $(1 - \sqrt{3}i)^3 = \left[2\left(\cos\dfrac{5\pi}{3} + i\sin\dfrac{5\pi}{3}\right)\right]^3 = 8(\cos 5\pi + i\sin 5\pi) = 8(\cos\pi + i\sin\pi) = -8$

For #39–43, the cube roots of $r(\cos\theta + i\sin\theta)$ are $= \sqrt[3]{r}\left(\cos\dfrac{\theta + 2k\pi}{3} + i\sin\dfrac{\theta + 2k\pi}{3}\right)$, $k = 0, 1, 2$.

39. $\sqrt[3]{2}\left(\cos\dfrac{2k\pi + 2\pi}{3} + i\sin\dfrac{2k\pi + 2\pi}{3}\right) = \sqrt[3]{2}\left(\cos\dfrac{2\pi(k + 1)}{3} + i\sin\dfrac{2\pi(k + 1)}{3}\right)$, $k = 0, 1, 2$:

$$\sqrt[3]{2}\left(\cos\frac{2\pi}{3} + i\sin\frac{2\pi}{3}\right) = \sqrt[3]{2}\left(-\frac{1}{2} + \frac{\sqrt{3}}{2}i\right) = \frac{1 + \sqrt{3}i}{\sqrt[3]{4}},$$
$$\sqrt[3]{2}\left(\cos\frac{4\pi}{3} + i\sin\frac{4\pi}{3}\right) = \sqrt[3]{2}\left(-\frac{1}{2} - \frac{\sqrt{3}}{2}i\right) = \frac{-1 - \sqrt{3}i}{\sqrt[3]{4}},$$
$$\sqrt[3]{2}(\cos 2\pi + i\sin 2\pi) = \sqrt[3]{2}$$

41. $\sqrt[3]{3}\left(\cos\dfrac{2k\pi + 4\pi/3}{3} + i\sin\dfrac{2k\pi + 4\pi/3}{3}\right) = \sqrt[3]{3}\left(\cos\dfrac{2\pi(3k + 2)}{9} + i\sin\dfrac{2\pi(3k + 2)}{9}\right)$, $k = 0, 1, 2$:

$$\sqrt[3]{3}\left(\cos\frac{4\pi}{9} + i\sin\frac{4\pi}{9}\right),\ \sqrt[3]{3}\left(\cos\frac{10\pi}{9} + i\sin\frac{10\pi}{9}\right),$$
$$\sqrt[3]{3}\left(\cos\frac{16\pi}{9} + i\sin\frac{16\pi}{9}\right)$$

43. $3 - 4i \approx 5(\cos 5.355 + i\sin 5.355)$

$\sqrt[3]{5}\left(\cos\dfrac{2k\pi + 5.355}{3} + i\sin\dfrac{2k\pi + 5.355}{3}\right)$, $k = 0, 1, 2$:

$\approx \sqrt[3]{5}(\cos 1.79 + i\sin 1.79)$,
$\approx \sqrt[3]{5}(\cos 3.88 + i\sin 3.88)$,
$\approx \sqrt[3]{5}(\cos 5.97 + i\sin 5.97)$

For #45–49, the fifth roots of $r(\cos\theta + i\sin\theta)$ are

$$\sqrt[5]{r}\left(\cos\frac{\theta + 2k\pi}{5} + i\sin\frac{\theta + 2k\pi}{5}\right), k = 0, 1, 2, 3, 4.$$

45. $\cos\frac{2k\pi + \pi}{5} + i\sin\frac{2k\pi + \pi}{5}$

$= \cos\frac{\pi(2k + 1)}{5} + i\sin\frac{\pi(2k + 1)}{5}, k = 0, 1, 2, 3, 4:$

$\cos\frac{\pi}{5} + i\sin\frac{\pi}{5}, \cos\frac{3\pi}{5} + i\sin\frac{3\pi}{5}, -1,$

$\cos\frac{7\pi}{5} + i\sin\frac{7\pi}{5}, \cos\frac{9\pi}{5} + i\sin\frac{9\pi}{5}$

47. $\sqrt[5]{2}\left(\cos\frac{2k\pi + \pi/6}{5} + i\sin\frac{2k\pi + \pi/6}{5}\right)$

$= \sqrt[5]{2}\left(\cos\frac{\pi(12k + 1)}{30} + \sin\frac{\pi(12k + 1)}{30}\right),$

$k = 0, 1, 2, 3, 4:$

$\sqrt[5]{2}\left(\cos\frac{\pi}{30} + i\sin\frac{\pi}{30}\right), \sqrt[5]{2}\left(\cos\frac{13\pi}{30} + i\sin\frac{13\pi}{30}\right),$

$\sqrt[5]{2}\left(\cos\frac{5\pi}{6} + i\sin\frac{5\pi}{6}\right), \sqrt[5]{2}\left(\cos\frac{37\pi}{30} + i\sin\frac{37\pi}{30}\right),$

$\sqrt[5]{2}\left(\cos\frac{49\pi}{30} + i\sin\frac{49\pi}{30}\right)$

49. $\sqrt[5]{2}\left(\cos\frac{2k\pi + \pi/2}{5} + i\sin\frac{2k\pi + \pi/2}{5}\right)$

$= \sqrt[5]{2}\left(\cos\frac{\pi(4k + 1)}{10} + i\sin\frac{\pi(4k + 1)}{10}\right),$

$k = 0, 1, 2, 3, 4:$

$\sqrt[5]{2}\left(\cos\frac{\pi}{10} + i\sin\frac{\pi}{10}\right), \sqrt[5]{2}\left(\cos\frac{\pi}{2} + i\sin\frac{\pi}{2}\right) = \sqrt[5]{2}i,$

$\sqrt[5]{2}\left(\cos\frac{9\pi}{10} + i\sin\frac{9\pi}{10}\right), \sqrt[5]{2}\left(\cos\frac{13\pi}{10} + i\sin\frac{13\pi}{10}\right),$

$\sqrt[5]{2}\left(\cos\frac{17\pi}{10} + i\sin\frac{17\pi}{10}\right)$

For #51–55, the nth roots of $r(\cos\theta + i\sin\theta)$are

$$\sqrt[n]{r}\left(\cos\frac{\theta + 2k\pi}{n} + i\sin\frac{\theta + 2k\pi}{n}\right), k = 0, 1, 2, \ldots, n - 1.$$

51. $1 + i = \sqrt{2}\left(\cos\frac{\pi}{4} + i\sin\frac{\pi}{4}\right)$, so the roots are

$\sqrt[4]{\sqrt{2}}\left(\cos\frac{2k\pi + \pi/4}{4} + i\sin\frac{2k\pi + \pi/4}{4}\right)$

$= \sqrt[8]{2}\left(\cos\frac{\pi(8k + 1)}{16} + i\sin\frac{\pi(8k + 1)}{16}\right),$

$k = 0, 1, 2, 3:$

$\sqrt[8]{2}\left(\cos\frac{\pi}{16} + i\sin\frac{\pi}{16}\right), \sqrt[8]{2}\left(\cos\frac{9\pi}{16} + i\sin\frac{9\pi}{16}\right),$

$\sqrt[8]{2}\left(\cos\frac{17\pi}{16} + i\sin\frac{17\pi}{16}\right), \sqrt[8]{2}\left(\cos\frac{25\pi}{16} + i\sin\frac{25\pi}{16}\right)$

53. $2 + 2i = 2\sqrt{2}\left(\cos\frac{\pi}{4} + i\sin\frac{\pi}{4}\right)$, so the roots are

$\sqrt[3]{2\sqrt{2}}\left(\cos\frac{2k\pi + \pi/4}{3} + i\sin\frac{2k\pi + \pi/4}{3}\right)$

$= \sqrt{2}\left(\cos\frac{\pi(8k + 1)}{12} + i\sin\frac{\pi(8k + 1)}{12}\right), k = 0, 1, 2:$

$\sqrt{2}\left(\cos\frac{\pi}{12} + i\sin\frac{\pi}{12}\right), -1 + i,$

$\sqrt{2}\left(\cos\frac{17\pi}{12} + i\sin\frac{17\pi}{12}\right)$

55. $-2i = 2\left(\cos\frac{3\pi}{2} + i\sin\frac{3\pi}{2}\right)$, so the roots are

$\sqrt[6]{2}\left(\cos\frac{2k\pi + 3\pi/2}{6} + i\sin\frac{2k\pi + 3\pi/2}{6}\right)$

$= \sqrt[6]{2}\left(\cos\frac{\pi(4k + 3)}{12} + i\sin\frac{\pi(4k + 3)}{12}\right),$

$k = 0, 1, 2, 3, 4, 5:$

$\frac{1 + i}{\sqrt[6]{4}}, \sqrt[6]{2}\left(\cos\frac{7\pi}{12} + i\sin\frac{7\pi}{12}\right),$

$\sqrt[6]{2}\left(\cos\frac{11\pi}{12} + i\sin\frac{11\pi}{12}\right), \sqrt[6]{2}\left(\cos\frac{5\pi}{4} + i\sin\frac{5\pi}{4}\right),$

$\sqrt[6]{2}\left(\cos\frac{19\pi}{12} + i\sin\frac{19\pi}{12}\right), \sqrt[6]{2}\left(\cos\frac{23\pi}{12} + i\sin\frac{23\pi}{12}\right)$

For #57–59, the nth roots of unity are

$$\cos\frac{2k\pi}{n} + i\sin\frac{2k\pi}{n}, k = 0, 1, 2, \ldots, n - 1.$$

57. $1, -\frac{1}{2} \pm \frac{\sqrt{3}}{2}i$

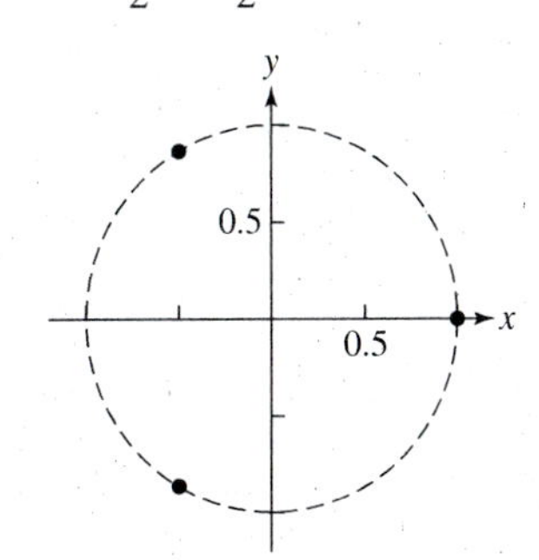

59. $\pm 1, \frac{1}{2} \pm \frac{\sqrt{3}}{2}i, -\frac{1}{2} \pm \frac{\sqrt{3}}{2}i$

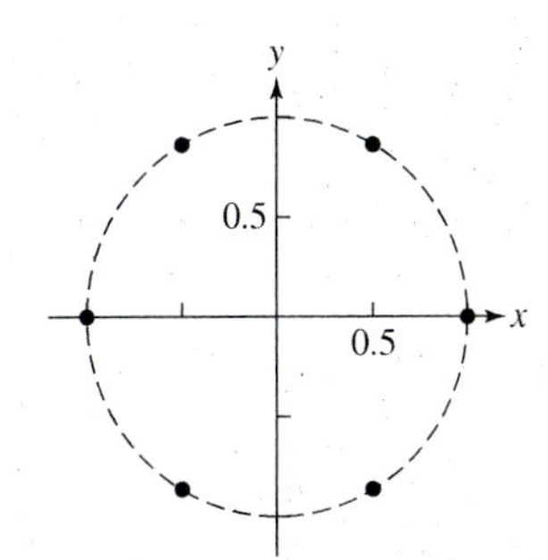

61. $z = (1 + \sqrt{3}i)^3 = \left[2\left(\cos\frac{\pi}{3} + i\sin\frac{\pi}{3}\right)\right]^3$

$= 8(\cos\pi + i\sin\pi) = -8$; the cube roots are -2 and $1 \pm \sqrt{3}i$.

63. $\frac{z_1}{z_2} = \frac{r_1(\cos\theta_1 + i\sin\theta_1)}{r_2(\cos\theta_2 + i\sin\theta_2)} = \frac{r_1}{r_2}\cdot\frac{\cos\theta_1 + i\sin\theta_1}{\cos\theta_2 + i\sin\theta_2}\cdot\frac{\cos\theta_2 - i\sin\theta_2}{\cos\theta_2 - i\sin\theta_2}$

$= \frac{r_1}{r_2}\cdot\frac{\cos\theta_1\cos\theta_2 + \sin\theta_1\sin\theta_2 + i(\sin\theta_1\cos\theta_2 - \cos\theta_1\sin\theta_2)}{(\cos\theta_2)^2 + (\sin\theta_2)^2}$

$= \frac{r_1}{r_2}\cdot[\cos\theta_1\cos\theta_2 + \sin\theta_1\sin\theta_2 + i(\sin\theta_1\cos\theta_2 - \cos\theta_1\sin\theta_2)].$

Now use the angle difference formulas:

$\cos(\theta_1 - \theta_2) = \cos\theta_1\cos\theta_2 + \sin\theta_1\sin\theta_2$ and $\sin(\theta_1 - \theta_2) = \sin\theta_1\cos\theta_2 - \cos\theta_1\sin\theta_2$.

So $\frac{z_1}{z_2} = \frac{r_1}{r_2}[\cos(\theta_1 - \theta_2) + i\sin(\theta_1 - \theta_2)]$.

65. False. If $z = r(\cos\theta + i\sin\theta)$, then it is also true that $z = r[\cos(\theta + 2n\pi) + i\sin(\theta + 2n\pi)]$ for any integer n. For example, here are two trigonometric forms for $1 + i$: $\sqrt{2}\left(\cos\frac{\pi}{4} + i\sin\frac{\pi}{4}\right)$, $\sqrt{2}\left(\cos\frac{9\pi}{4} + i\sin\frac{9\pi}{4}\right)$.

67. $2\left(\cos\frac{2\pi}{3} + i\sin\frac{2\pi}{3}\right) = 2\left(-\frac{1}{2} + i\frac{\sqrt{3}}{2}\right) = -1 + \sqrt{3}i$

The answer is B.

69. $\sqrt{2}\left(\cos\frac{\pi}{4} + i\sin\frac{\pi}{4}\right)\cdot\sqrt{2}\left(\cos\frac{7\pi}{4} + i\sin\frac{7\pi}{4}\right)$

$= (\sqrt{2}\cdot\sqrt{2})\left[\cos\left(\frac{\pi}{4} + \frac{7\pi}{4}\right) + i\sin\left(\frac{\pi}{4} + \frac{7\pi}{4}\right)\right]$

$= 2(\cos 2\pi + i\sin 2\pi)$

$= 2$

The answer is A.

71. (a) $a + bi = r(\cos\theta + i\sin\theta)$, where $r = \sqrt{a^2 + b^2}$ and $\theta = \tan^{-1}\left(\frac{b}{a}\right)$. Then $a + (-bi) = r'(\cos\theta' + i\sin\theta')$, where $r' = \sqrt{a^2 + (-b)^2}$ and $\theta' = \tan^{-1}\left(\frac{-b}{a}\right)$. Since $r' = \sqrt{a^2 + b^2} = r$ and $\theta' = \tan^{-1}\left(\frac{-b}{a}\right) = -\tan^{-1}\left(\frac{b}{a}\right) = -\theta$, we have $a - bi = r(\cos(-\theta) + i\sin(-\theta))$

(b) $z\cdot\bar{z} = r[\cos\theta + i\sin\theta]\cdot r[\cos(-\theta) + i\sin(-\theta)]$

$= r^2[\cos\theta\cos(-\theta) + i(\sin(-\theta))(\cos\theta) + i(\sin\theta)(\cos(-\theta)) - (\sin\theta)(\sin(-\theta))]$

Since $\sin\theta$ is an odd function (i.e., $\sin(-\theta) = -\sin(\theta)$) and $\cos\theta$ is an even function (i.e., $\cos(-\theta) = \cos\theta$), we have

$z\cdot\bar{z} = r^2[\cos^2\theta - i(\sin\theta)(\cos\theta) + i(\sin\theta)(\cos\theta) + \sin^2\theta]$

$= r^2[\cos^2\theta + \sin^2\theta]$

$= r^2$

(c) $\frac{z}{\bar{z}} = \frac{r[\cos\theta + i\sin\theta]}{r[\cos(-\theta) + i\sin(-\theta)]} = \cos[\theta - (-\theta)] + i\sin[\theta - (-\theta)]$

$= \cos(2\theta) + i\sin(2\theta)$

(d) $-z = -(a + bi) = -a + (-bi) = r(\cos\theta + i\sin\theta)$, where $r = \sqrt{(-a^2) + (-b)^2}$ and $\theta = \tan^{-1}\left(\frac{-b}{-a}\right) = \tan^{-1}\left(\frac{b}{a}\right)$

Recall, however, that $(-a, -b)$ is in the quadrant directly opposite the quadrant that holds (a, b) (i.e., if (a, b) is in Quadrant I, $(-a, -b)$ is in Quadrant III, and if (a, b) is in Quadrant II, $(-a, -b)$ is in Quadrant IV). Thus,

$-z = \sqrt{a^2 + b^2}(\cos(\theta + \pi) + i\sin(\theta + \pi)) = r(\cos(\theta + \pi) + i\sin(\theta + \pi))$.

73. (a)

```
25√(2)*(cos(-π/4
)+isin(-π/4))*14
*(cos(π/3)+isin(
π/3))
    478.11+128.11i
```

(b)

```
2√(2)*(cos(135)+
isin(135))/(6*(c
os(300)+isin(300
)))
          -.46-.12i
```

(c)

```
(cos(3π/4)+isin(
3π/4))^8
    1.00+2.00E-13i
```

75. Using $\tan^{-1}\left(\frac{1}{\sqrt{2}}\right) \approx 0.62$, we have $\sqrt{2} + i \approx \sqrt{3}\,(\cos 0.62 + i \sin 0.62)$, so graph $x(t) = (\sqrt{3})^t \cos(0.62t)$ and $y(t) = (\sqrt{3})^t \sin(0.62t)$.
Use Tmin = 0, Tmax = 4, Tstep = 1.
Shown is [−7, 2] by [−0, 6].

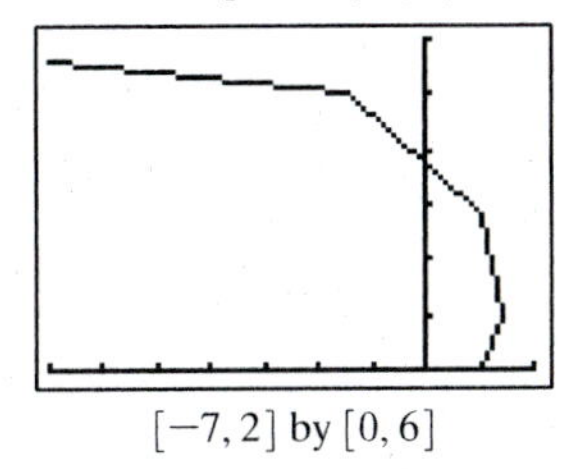

[−7, 2] by [0, 6]

77. Suppose that $z_1 = r_1(\cos\theta_1 + i\sin\theta_1)$ and $z_2 = r_2(\cos\theta_2 + i\sin\theta_2)$. Each triangle's angle at the origin has the same measure: For the smaller triangle, this angle has measure θ_1; for the larger triangle, the side from the origin out to z_2 makes an angle of θ_2 with the x-axis, while the side from the origin to z_1z_2 makes an angle of $\theta_1 + \theta_2$, so that the angle between is θ_1 as well.
The corresponding side lengths for the sides adjacent to these angles have the same ratio: two longest side have lengths $|z_1| = r_1$ (for the smaller) and $|z_1z_2| = r_1r_2$, for a ratio of r_2. For the other two sides, the lengths are 1 and r_2, again giving a ratio of r_2. Finally, the Law of Sines can be used to show that the remaining side have the same ratio.

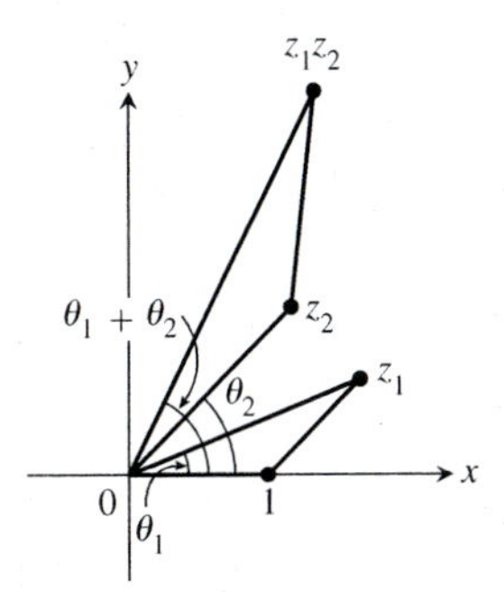

79. The solutions are the cube roots of 1:
$\cos\left(\frac{2\pi k}{3}\right) + i\sin\left(\frac{2\pi k}{3}\right)$, $k = 0, 1, 2$ or
$1, -\frac{1}{2} + \frac{\sqrt{3}}{2}i, -\frac{1}{2} - \frac{\sqrt{3}}{2}i$

81. The solutions are the cube roots of −1:
$\cos\left(\frac{\pi + 2\pi k}{3}\right) + i\sin\left(\frac{\pi + 2\pi k}{3}\right)$, $k = 0, 1, 2$ or
$-1, \frac{1}{2} + \frac{\sqrt{3}}{2}i, \frac{1}{2} - \frac{\sqrt{3}}{2}i$

83. The solutions are the fifth roots of −1:
$\cos\left(\frac{\pi + 2\pi k}{5}\right) + i\sin\left(\frac{\pi + 2\pi k}{5}\right)$, $k = 0, 1, 2, 3, 4$, or
$-1, \approx 0.81 + 0.59i, 0.81 - 0.59i, -0.31 + 0.95i, -0.31 - 0.95i$

Chapter 6 Review

1. $\mathbf{u} - \mathbf{v} = \langle 2 - 4, -1 - 2\rangle = \langle -2, -3\rangle$

3. $|\mathbf{u} + \mathbf{v}| = \sqrt{(2+4)^2 + (-1+2)^2} = \sqrt{37}$

5. $\mathbf{u}\cdot\mathbf{v} = 8 - 2 = 6$

7. $3\overrightarrow{AB} = 3\langle 3 - 2, 1 - (-1)\rangle = \langle 3, 6\rangle$;
$|3\overrightarrow{AB}| = \sqrt{3^2 + 6^2} = \sqrt{45} = 3\sqrt{5}$

9. $\overrightarrow{AC} + \overrightarrow{BD} = \langle -4 - 2, 2 - (-1)\rangle + \langle 1 - 3, -5 - 1\rangle$
$= \langle -8, -3\rangle$; $|\overrightarrow{AC} + \overrightarrow{BD}| = \sqrt{8^2 + 3^2} = \sqrt{73}$

11. (a) $\frac{\overrightarrow{AB}}{|\overrightarrow{AB}|} = \frac{\langle -2, 1\rangle}{\sqrt{(-2)^2 + 1^2}} = \frac{\langle -2, 1\rangle}{\sqrt{5}} = \left\langle -\frac{2}{\sqrt{5}}, \frac{1}{\sqrt{5}}\right\rangle$

(b) $-3\cdot\frac{\overrightarrow{AB}}{|\overrightarrow{BA}|} = -3\left\langle\frac{-2}{\sqrt{5}}, \frac{1}{\sqrt{5}}\right\rangle = \left\langle\frac{6}{\sqrt{5}}, -\frac{3}{\sqrt{5}}\right\rangle$

For #13, the direction angle θ of $\langle a, b\rangle$ has $\tan\theta = \frac{b}{a}$; start with $\tan^{-1}\left(\frac{b}{a}\right)$, and add (or subtract) 180° if the angle is not in the correct quadrant. The angle between two vectors is the absolute value of the difference between their angles; if this difference is greater than 180°, subtract it from 360°.

13. (a) $\theta_u = \tan^{-1}\left(\frac{3}{4}\right) \approx 0.64$, $\theta_v = \tan^{-1}\left(\frac{5}{2}\right) \approx 1.19$

(b) $\theta_v - \theta_u \approx 0.55$.

15. $(-2.5\cos 25°, -2.5\sin 25°) \approx (-2.27, -1.06)$

17. $(2\cos(-\pi/4), 2\sin(-\pi/4)) = (\sqrt{2}, -\sqrt{2})$

19. $\left(1, -\frac{2\pi}{3} + (2n+1)\pi\right)$ and $\left(-1, -\frac{2\pi}{3} + 2n\pi\right)$, n an integer

21. (a) $\left(-\sqrt{13}, \pi + \tan^{-1}\left(-\frac{3}{2}\right)\right) \approx (-\sqrt{13}, 2.16)$ or
$\left(\sqrt{13}, 2\pi + \tan^{-1}\left(-\frac{3}{2}\right)\right) \approx (\sqrt{13}, 5.30)$

(b) $\left(\sqrt{13}, \tan^{-1}\left(-\frac{3}{2}\right)\right) \approx (\sqrt{13}, -0.98)$ or
$\left(-\sqrt{13}, \pi + \tan^{-1}\left(-\frac{3}{2}\right)\right) \approx (-\sqrt{13}, 2.16)$

(c) The answers from (a), and also
$\left(-\sqrt{13}, 3\pi + \tan^{-1}\left(-\frac{3}{2}\right)\right) \approx (-\sqrt{13}, 8.44)$ or
$\left(\sqrt{13}, 4\pi + \tan^{-1}\left(-\frac{3}{2}\right)\right) \approx (\sqrt{13}, 11.58)$

23. (a) $(5, 0)$ or $(-5, \pi)$ or $(5, 2\pi)$

(b) $(-5, -\pi)$ or $(5, 0)$ or $(-5, \pi)$

(c) The answers from (a), and also $(-5, 3\pi)$ or $(5, 4\pi)$

25. $t = -\frac{1}{5}x + \frac{3}{5}$, so $y = 4 + 3\left(-\frac{1}{5}x + \frac{3}{5}\right)$
$= -\frac{3}{5}x + \frac{29}{5}$:
Line through $\left(0, \frac{29}{5}\right)$ with slope $m = -\frac{3}{5}$

27. $t = y + 1$, so $x = 2(y+1)^2 + 3$: Parabola that opens to the right with vertex at $(3, -1)$.

29. $x + 1 = e^{2t}, t = \dfrac{\ln(x+1)}{2}$, so $y = e^{\frac{1}{2}\ln(x+1)} = e^{\ln\sqrt{x+1}}$
$= \sqrt{x+1}$: square root function starting at $(-1, 0)$

31. $m = \dfrac{4-(-2)}{3-(-1)} = \dfrac{6}{4} = \dfrac{3}{2}$, so $\Delta x = 2$ when $\Delta y = 3$.
One possibility for the parametrization of the line is: $x = 2t + 3,\ y = 3t + 4$.

33. $a = -3, b = 4, |z_1| = \sqrt{3^2 + 4^2} = 5$

35. $6(\cos 30° + i \sin 30°) = 6\left(\dfrac{\sqrt{3}}{2} + \dfrac{1}{2}i\right) = 3\sqrt{3} + 3i$

37. $2.5\left(\cos\dfrac{4\pi}{3} + i\sin\dfrac{4\pi}{3}\right) = 2.5\left(-\dfrac{1}{2} - \dfrac{\sqrt{3}}{2}i\right)$
$= -1.25 - 1.25\sqrt{3}\,i$

39. $3 - 3i = 3\sqrt{2}\left(\cos\dfrac{7\pi}{4} + i\sin\dfrac{7\pi}{4}\right)$. Other representations would use angles $\dfrac{7\pi}{4} + 2n\pi$, n an integer.

41. $3 - 5i = \sqrt{34}\left\{\cos\left[\tan^{-1}\left(-\dfrac{5}{3}\right)\right] + i\sin\left[\tan^{-1}\left(-\dfrac{5}{3}\right)\right]\right\}$
$\approx \sqrt{34}\,[\cos(-1.03) + i\sin(-1.03)]$
$\approx \sqrt{34}[\cos(5.25 + i\sin(5.25)]$
Other representations would use angles $\approx 5.25 + 2n\pi$, n an integer.

43. $z_1 z_2 = (3)(4)[\cos(30° + 60°) + i\sin(30° + 60°)]$
$= 12(\cos 90° + i\sin 90°)$
$z_1/z_2 = \dfrac{3}{4}[\cos(30° - 60°) + i\sin(30° - 60°)]$
$= \dfrac{3}{4}[\cos(-30°) + i\sin(-30°)]$
$= \dfrac{3}{4}(\cos 330° + i\sin 330°)$

45. **(a)** $\left[3\left(\cos\dfrac{\pi}{4} + i\sin\dfrac{\pi}{4}\right)\right]^5 = 3^5\left(\cos\dfrac{5\pi}{4} + i\sin\dfrac{5\pi}{4}\right)$
$= 243\left(\cos\dfrac{5\pi}{4} + i\sin\dfrac{5\pi}{4}\right)$
(b) $-\dfrac{243\sqrt{2}}{2} - \dfrac{243\sqrt{2}}{2}i$

47. **(a)** $\left[5\left(\cos\dfrac{5\pi}{3} + i\sin\dfrac{5\pi}{3}\right)\right]^3 = 5^3(\cos 5\pi + i\sin 5\pi)$
$= 125(\cos\pi + i\sin\pi)$
(b) $-125 + 0i = -125$

For #49–51, the nth roots of $r(\cos\theta + i\sin\theta)$ are
$\sqrt[n]{r}\left(\cos\dfrac{2k\pi + \theta}{n} + i\sin\dfrac{2k\pi + \theta}{n}\right), k = 0, 1, 2, \ldots, n - 1.$

49. $3 + 3i = 3\sqrt{2}\left(\cos\dfrac{\pi}{4} + i\sin\dfrac{\pi}{4}\right)$, so the roots are
$\sqrt[4]{3\sqrt{2}}\left(\cos\dfrac{2k\pi + \pi/4}{4} + i\sin\dfrac{2k\pi + \pi/4}{4}\right)$
$= \sqrt[8]{18}\left(\cos\dfrac{\pi(8k+1)}{16} + i\sin\dfrac{\pi(8k+1)}{16}\right)$,
$k = 0, 1, 2, 3$:
$\sqrt[8]{18}\left(\cos\dfrac{\pi}{16} + i\sin\dfrac{\pi}{16}\right)$,
$\sqrt[8]{18}\left(\cos\dfrac{9\pi}{16} + i\sin\dfrac{9\pi}{16}\right)$,
$\sqrt[8]{18}\left(\cos\dfrac{17\pi}{16} + i\sin\dfrac{17\pi}{16}\right)$,
$\sqrt[8]{18}\left(\cos\dfrac{25\pi}{16} + i\sin\dfrac{25\pi}{16}\right)$

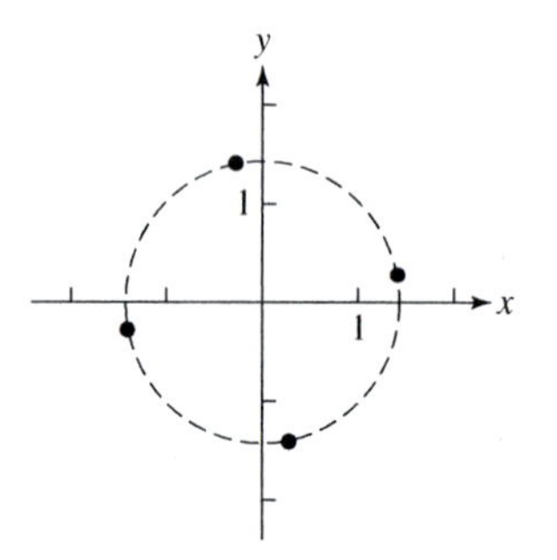

51. $1 = \cos 0 + i\sin 0$, so the roots are
$\cos\dfrac{2k\pi + 0}{5} + i\sin\dfrac{2k\pi + 0}{5} = \cos\dfrac{2k\pi}{5} + i\sin\dfrac{2k\pi}{5}$,
$k = 0, 1, 2, 3, 4$:
$\cos 0 + i\sin 0 = 1$, $\cos\dfrac{2\pi}{5} + i\sin\dfrac{2\pi}{5}$,
$\cos\dfrac{4\pi}{5} + i\sin\dfrac{4\pi}{5}$, $\cos\dfrac{6\pi}{5} + i\sin\dfrac{6\pi}{5}$
$\cos\dfrac{8\pi}{5} + i\sin\dfrac{8\pi}{5}$

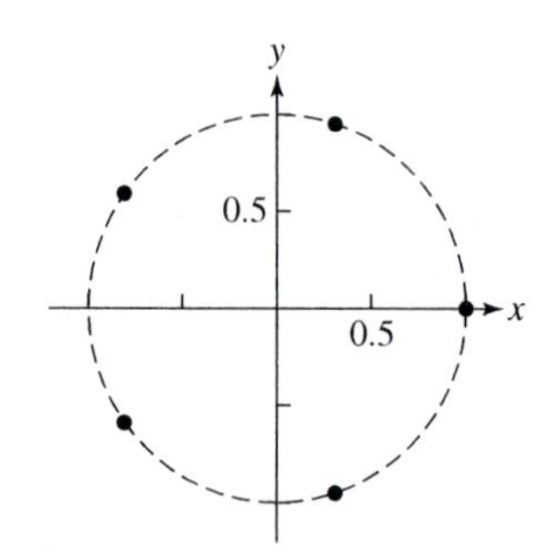

53. Graph (b)

55. Graph (a)

57. Not shown

59. Graph (c)

61. $x^2 + y^2 = 4$ — a circle with center $(0, 0)$ and radius 2

63. $r^2 + 3r\cos\theta + 2r\sin\theta = x^2 + y^2 + 3x + 2y = 0$.
Completing the square: $\left(x + \dfrac{3}{2}\right)^2 + (y+1)^2 = \dfrac{13}{4}$
— a circle of radius $\dfrac{\sqrt{13}}{2}$ with center $\left(-\dfrac{3}{2}, -1\right)$

65. $r = \dfrac{-4}{\sin\theta} = -4\csc\theta$

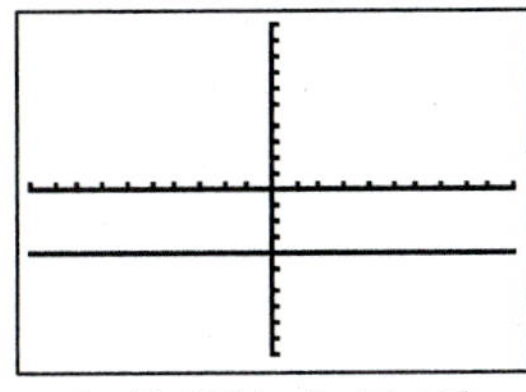

$[-10, 10]$ by $[-10, 10]$

67. $(r\cos\theta - 3)^2 + (r\sin\theta + 1)^2 = 10$, so $r^2(\cos^2\theta + \sin^2\theta) + r(-6\cos\theta + 2\sin\theta) + 10 = 10$, or $r = 6\cos\theta - 2\sin\theta$

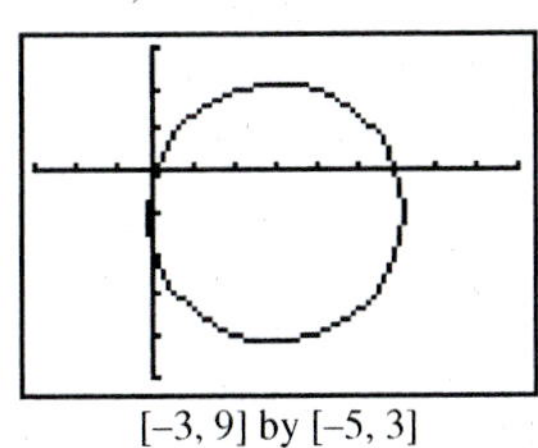

$[-3, 9]$ by $[-5, 3]$

69.

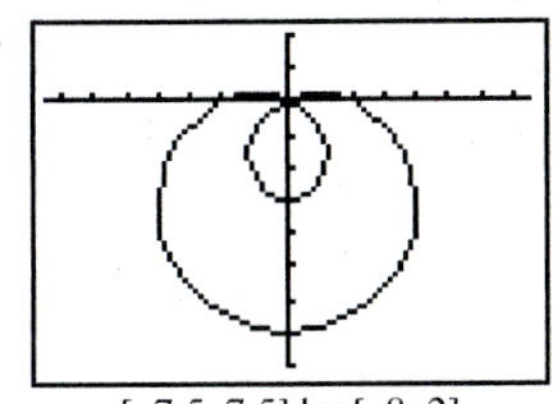

$[-7.5, 7.5]$ by $[-8, 2]$

Domain: All reals
Range: $[-3, 7]$
Symmetric about the y-axis
Continuous
Bounded
Maximum r-value: 7
No asymptotes

71.

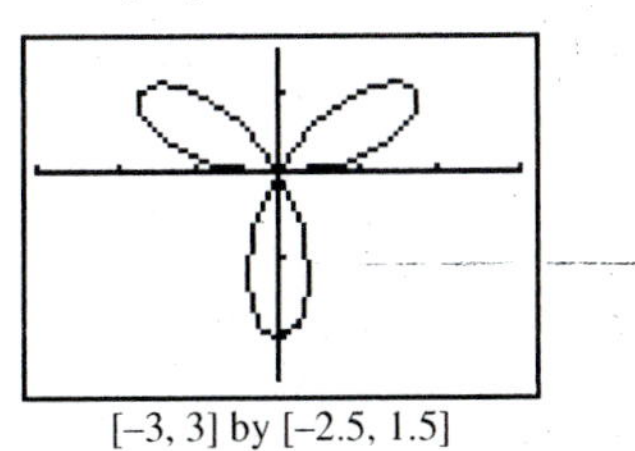

$[-3, 3]$ by $[-2.5, 1.5]$

Domain: All reals
Range: $[-2, 2]$
Symmetric about the y-axis
Continuous
Bounded
Maximum r-value: 2
No asymptotes

73. (a) $r = a\sec\theta \Rightarrow \dfrac{r}{\sec\theta} = a \Rightarrow r\cos\theta = a \Rightarrow x = a.$

(b) $r = b\csc\theta \Rightarrow \dfrac{r}{\csc\theta} = b \Rightarrow r\sin\theta = b \Rightarrow y = b.$

(c) $y = mx + b \Rightarrow r\sin\theta = mr\cos\theta + b \Rightarrow$
$r(\sin\theta - m\cos\theta) = b \Rightarrow r = \dfrac{b}{\sin\theta - m\cos\theta}.$

The domain of r is any value of θ for which $\sin\theta \neq m\cos\theta \Rightarrow \tan\theta \neq m \Rightarrow \theta \neq \arctan(m)$.

(d)

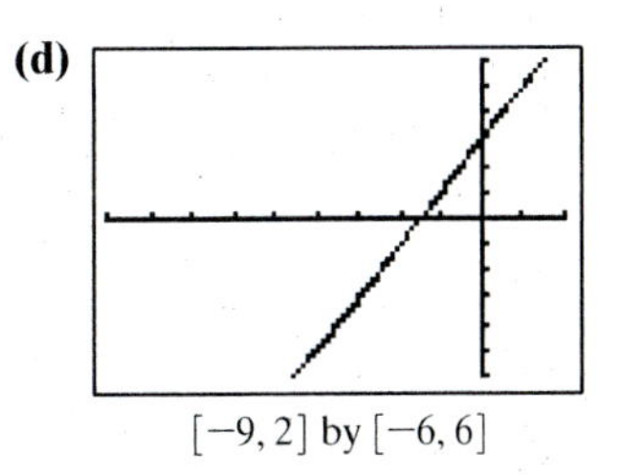

$[-9, 2]$ by $[-6, 6]$

75. (a) $\mathbf{v} = 480\langle\sin 285°, \cos 285°\rangle \approx \langle -463.64, 124.23\rangle$

(b) The wind vector is $\mathbf{w} = 30\langle\sin 265°, \cos 265°\rangle \approx \langle -29.89, -2.61\rangle$. Actual velocity vector: $\mathbf{v} + \mathbf{w} \approx \langle -493.53, 121.62\rangle$. Actual speed: $\|\mathbf{v} + \mathbf{w}\| \approx \sqrt{493.53^2 + 121.62^2} \approx 508.29$ mph. Actual bearing: $360° + \tan^{-1}\left(\dfrac{-493.53}{121.62}\right) \approx 283.84°$

77.

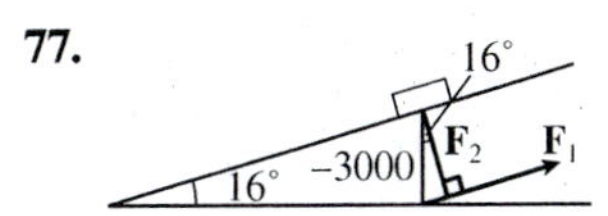

$\mathbf{F}_1$ Force to keep car from going downhill

$\mathbf{F}_2$ Force perpendicular to the street

(a) $\mathbf{F}_1 = -3000\sin 16° \approx -826.91$, so the force required to keep the car from rolling down the hill is approximately 826.91 pounds.

(b) $\mathbf{F}_2 = -3000\cos 16° \approx 2883.79$, so the force perpendicular to the street is approximately 2883.79 pounds.

79. (a) $h = -16t^2 + v_0t + s_0 = -16t^2 + 245t + 200$

(b) Graph and trace: $x = 17$ and $y = -16t^2 + 245t + 200$ with $0 \le t \le 16.1$ (upper limit may vary) on $[0, 18]$ by $[0, 1200]$. This graph will appear as a vertical line from about $(17, 0)$ to about $(17, 1138)$. Tracing shows how the arrow begins at a height of 200 ft, rises to over 1000 ft, then falls back to the ground.

(c) Graph $x = t$ and $y = -16t^2 + 245t + 200$ with $0 \le t \le 16.1$ (upper limit may vary).

$[0, 18]$ by $[0, 1200]$

(d) When $t = 4$, $h = 924$ ft.

(e) When $t \approx 7.66$, the arrow is at its peak: about 1138 ft.

(f) The arrow hits the ground ($h = 0$) after about 16.09 sec.

81. $x = 40\sin\left(\dfrac{2\pi}{15}t\right)$, $y = 50 - 40\cos\left(\dfrac{2\pi}{15}t\right)$, assuming the wheel turns counterclockwise.

83. (a)

$[-7.5, 7.5]$ by $[-5, 5]$

(b) All 4's should be changed to 5's.

85. $x = (66\cos 12°)t$ and $y = -16t^2 + (66\sin 12°)t + 3.5$. $y = 0$ when $t \approx 1.06$ sec (and also when $t \approx -0.206$, but that is not appropriate in this problem). When $t \approx 1.06$ sec, $x \approx 68.65$ ft.

87. If we assume that the initial height is 0 ft, then $x = (85 \cos 56°)t$ and $y = -16t^2 + (85 \sin 56°)t$. [If the assumed initial height is something other than 0 ft, add that amount to y.]

(a) Find graphically: the maximum y value is about 77.59 ft (after about 2.20 seconds).

(b) $y = 0$ when $t \approx 4.404$ sec

89. Kathy's position: $x_1 = 60 \cos\left(\frac{\pi}{6}t\right)$ and

$$y_1 = 60 + 60 \sin\left(\frac{\pi}{6}t\right)$$

Ball's position:
$x_2 = -80 + (100 \cos 70°)t$ and
$y_2 = -16t^2 + (100 \sin 70°)t$
Find (graphically) the minimum of
$d(t) = \sqrt{(x_1 - x_2)^2 + (y_1 - y_2)^2}$. It occurs when $t \approx 2.64$ sec; the minimum distance is about 17.65 ft.

Chapter 6 Project

Answers are based on the sample data shown in the table.

1.

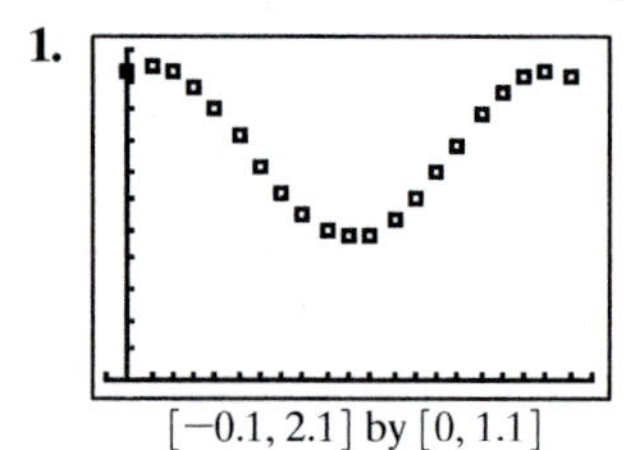

$[-0.1, 2.1]$ by $[0, 1.1]$

2. Sinusoidal regression produces $y \approx 0.28 \sin(3.46x + 1.20) + 0.75$ or, with a phase shift of 2π, $y \approx 0.28 \sin(3.46x - 5.09) + 0.75$ $\approx 0.28 \sin(3.46(x - 1.47)) + 0.75$.

3.

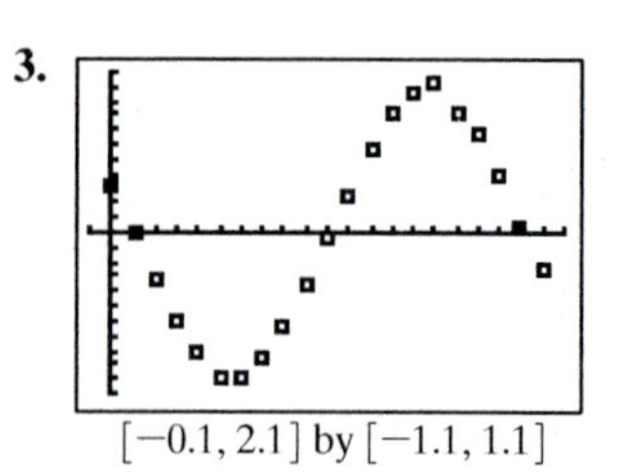

$[-0.1, 2.1]$ by $[-1.1, 1.1]$

The curve $y \approx 0.9688 \cos(3.46(x - 1.47))$ closely fits the data.

4. The distance and velocity both vary sinusoidally, with the same period but a phase shift of 90° — like the x- and y-coordinates of a point moving around a circle. A scatter plot of distance versus time should have the shape of a circle (or ellipse).

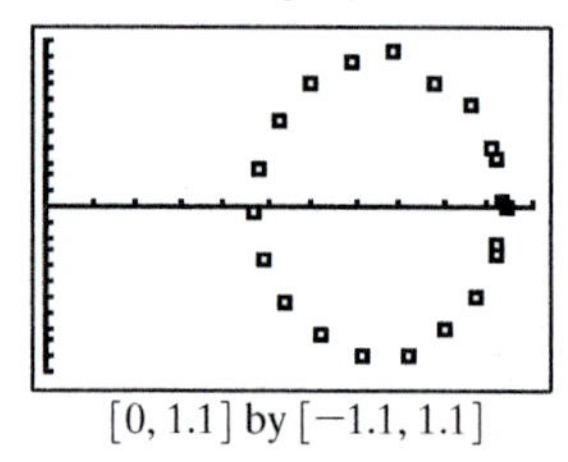

$[0, 1.1]$ by $[-1.1, 1.1]$

5.

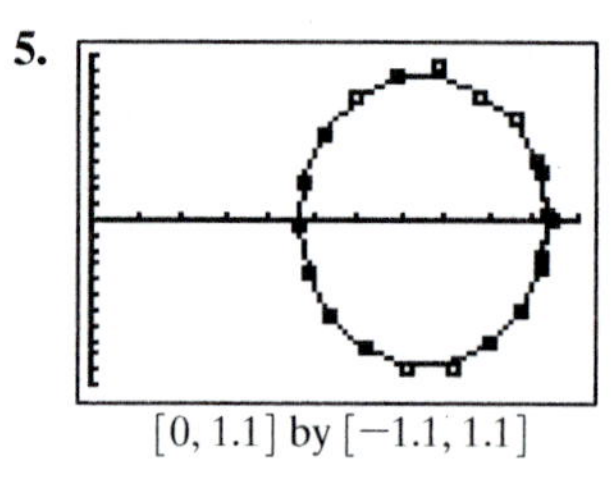

$[0, 1.1]$ by $[-1.1, 1.1]$

Chapter 7 Systems and Matrices

■ Section 7.1 Solving Systems of Two Equations

Exploration 1

1.

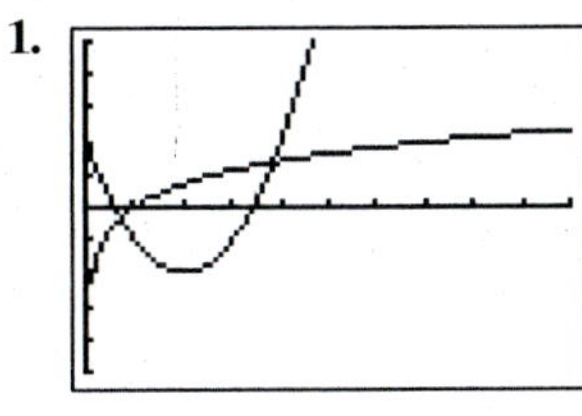

[0, 10] by [−5, 5]

2.

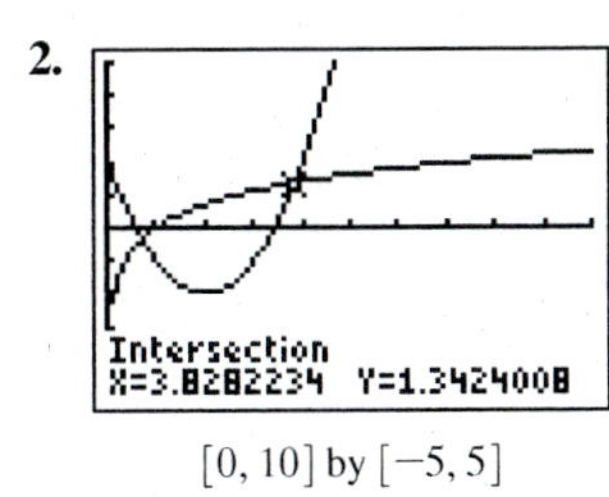

[0, 10] by [−5, 5]

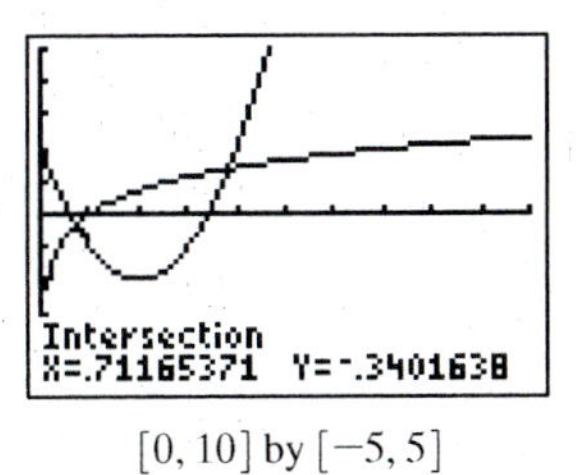

[0, 10] by [−5, 5]

3. The function $\ln x$ is only defined for $x > 0$, so all solutions must be positive. As x approaches infinity, $x^2 - 4x + 2$ is going to infinity much more quickly than $\ln x$ is going to infinity, hence will always be larger than $\ln x$ for x-values greater than 4.

Quick Review 7.1

1. $3y = 5 - 2x$

$y = \frac{5}{3} - \frac{2}{3}x$

3. $(3x + 2)(x - 1) = 0$

$3x + 2 = 0 \quad \text{or} \quad x - 1 = 0$

$3x = -2 \qquad x = 1$

$x = -\frac{2}{3}$

5. $x^3 - 4x = 0$

$x(x^2 - 4) = 0$

$x(x - 2)(x + 2) = 0$

$x = 0, x = 2, x = -2$

7. $m = -\frac{4}{5}, y - 2 = -\frac{4}{5}(x + 1)$

$y = -\frac{4}{5}x - \frac{4}{5} + 2$

$y = \frac{-4x + 6}{5}$

9. $-2(2x + 3y) = -2(5)$

$-4x - 6y = -10$

Section 7.1 Exercises

1. (a) No: $5(0) - 2(4) \neq 8$.

(b) Yes: $5(2) - 2(1) = 8$ and $2(2) - 3(1) = 1$.

(c) No: $2(-2) - 3(-9) \neq 1$.

In #3–11, there may be more than one good way to choose the variable for which the substitution will be made. One approach is given. In most cases, the solution is only shown up to the point where the value of the first variable is found.

3. $(x, y) = (9, -2)$: Since $y = -2$, we have $x - 4 = 5$, so $x = 9$.

5. $(x, y) = \left(\frac{50}{7}, -\frac{10}{7}\right)$: $y = 20 - 3x$, so $x - 2(20 - 3x) = 10$, or $7x = 50$, so $x = \frac{50}{7}$.

7. $(x, y) = \left(-\frac{1}{2}, 2\right)$: $x = (3y - 7)/2$, so $2(3y - 7) + 5y = 8$, or $11y = 22$, so $y = 2$.

9. No solution: $x = 3y + 6$, so $-2(3y + 6) + 6y = 4$, or $-12 = 4$ — not true.

11. $(x, y) = (\pm 3, 9)$; The second equation gives $y = 9$, so, $x^2 = 9$, or $x = \pm 3$.

13. $(x, y) = \left(-\frac{3}{2}, \frac{27}{2}\right)$ or $(x, y) = \left(\frac{1}{3}, \frac{2}{3}\right)$: $6x^2 + 7x - 3 = 0$, so $x = -\frac{3}{2}$ or $x = \frac{1}{3}$. Substitute these values into $y = 6x^2$.

15. $(x, y) = (0, 0)$ or $(x, y) = (3, 18)$: $3x^2 = x^3$, so $x = 0$ or $x = 3$. Substitute these values into $y = 2x^2$.

17. $(x, y) = \left(\frac{-1 + 3\sqrt{89}}{10}, \frac{3 + \sqrt{89}}{10}\right)$ and $\left(\frac{-1 - 3\sqrt{89}}{10}, \frac{3 - \sqrt{89}}{10}\right)$: $x - 3y = -1$, so $x = 3y + 1$. Substitute $x = 3y + 1$ into $x^2 + y^2 = 9$: $(3y - 1)^2 + y^2 = 9 \Rightarrow 10y^2 - 6y - 8 = 0$. Using the quadratic formula, we find that $y = \frac{3 \pm \sqrt{89}}{10}$.

In the following, $\mathbf{E}_1$ and $\mathbf{E}_2$ refer to the first and second equations, respectively.

19. $(x, y) = (8, -2)$: $\mathbf{E}_1 + \mathbf{E}_2$ leaves $2x = 16$, so $x = 8$.

21. $(x, y) = (4, 2)$: $2\mathbf{E}_1 + \mathbf{E}_2$ leaves $11x = 44$, so $x = 44$.

23. No solution: $3\mathbf{E}_1 + 2\mathbf{E}_2$ leaves $0 = -72$, which is false.

25. There are infinitely many solutions, any pair $\left(x, \frac{2}{3}x - \frac{5}{3}\right)$: $3\mathbf{E}_1 + \mathbf{E}_2$ leaves $0 = 0$, which is always true. As long as (x, y) satisfies one equation, it will also satisfy the other.

27. $(x, y) = (0, 1)$ or $(x, y) = (3, -2)$

29. No solution

31. One solution

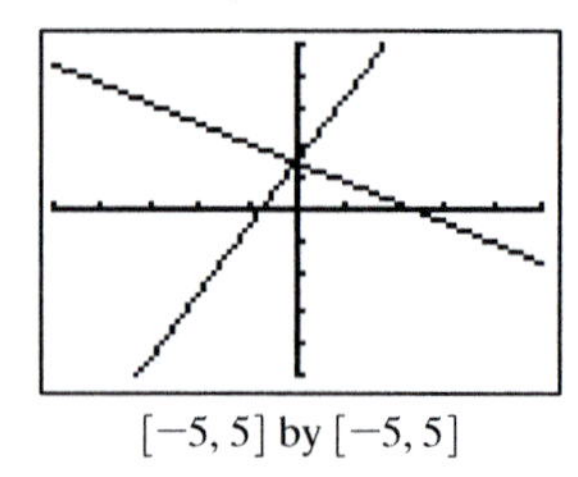

$[-5, 5]$ by $[-5, 5]$

33. Infinitely many solutions

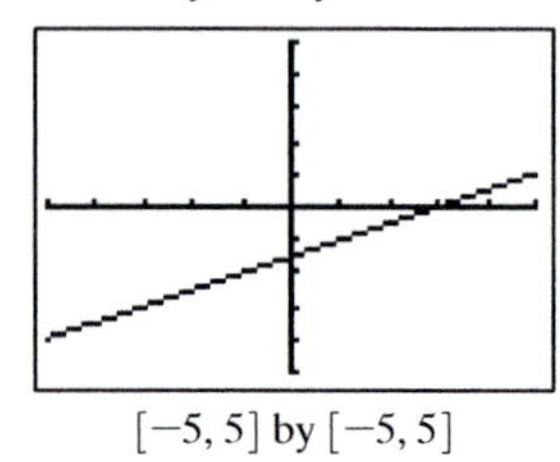

$[-5, 5]$ by $[-5, 5]$

35. $(x, y) \approx (0.69, -0.37)$

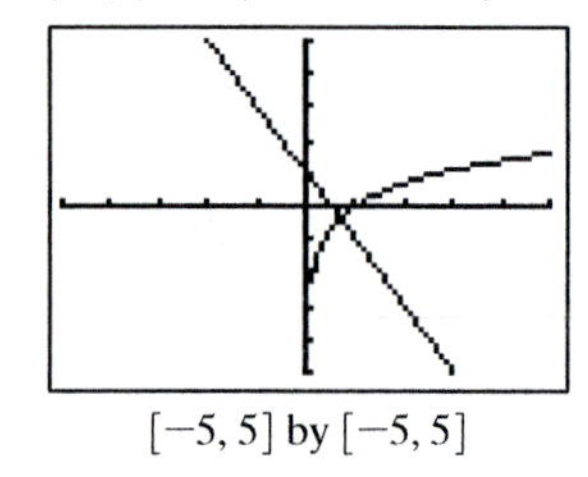

$[-5, 5]$ by $[-5, 5]$

37. $(x, y) \approx (-2.32, -3.16)$ or $(0.47, -1.77)$ or $(1.85, -1.08)$

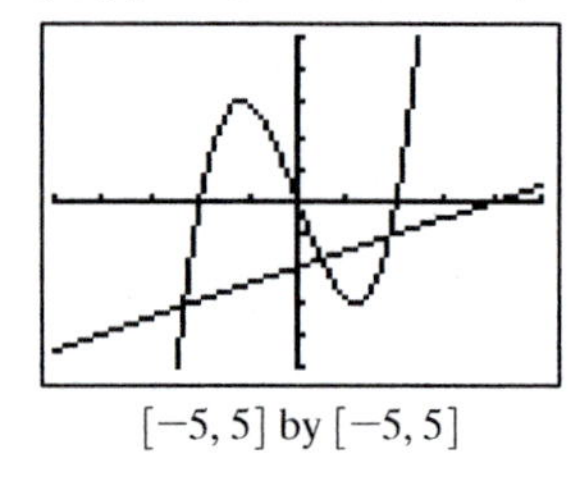

$[-5, 5]$ by $[-5, 5]$

39. $(x, y) = (-1.2, 1.6)$ or $(2, 0)$

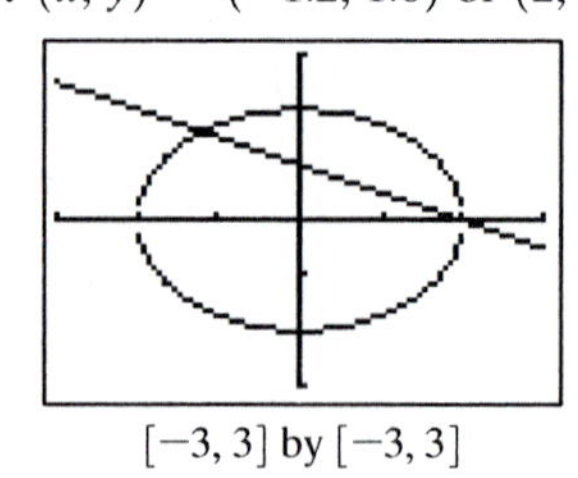

$[-3, 3]$ by $[-3, 3]$

41. $(x, y) \approx (2.05, 2.19)$ or $(-2.05, 2.19)$

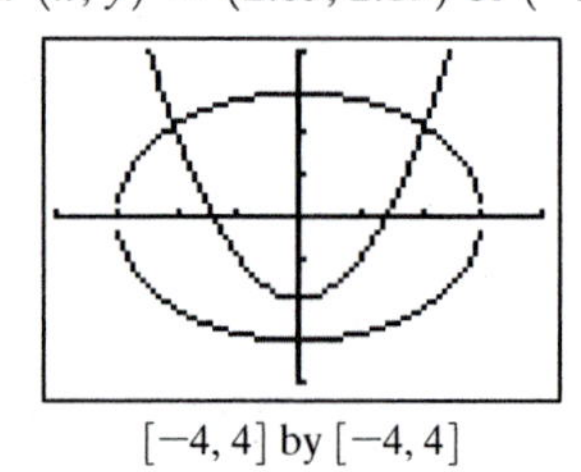

$[-4, 4]$ by $[-4, 4]$

43. $(x, p) = (3.75, 143.75)$: $200 - 15x = 50 + 25x$, so $40x = 150$.

45. In this problem, the graphs are representative of the expenditures (in billions of dollars) for benefits and administrative costs from federal hospital and medical insurance trust funds for several years, where x is the number of years past 1980.

(a) The following is a scatter plot of the data with the quadratic regression equation $y = -0.0938x^2 + 15.0510x - 28.2375$ superimposed on it.

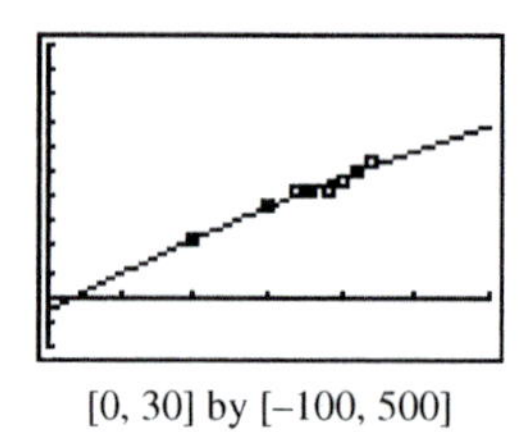

[0, 30] by [−100, 500]

(b) The following is a scatter plot of the data with the logistic regression equation $y = \dfrac{353.6473}{(1 + 8.6873e^{-0.1427x})}$ superimposed on it.

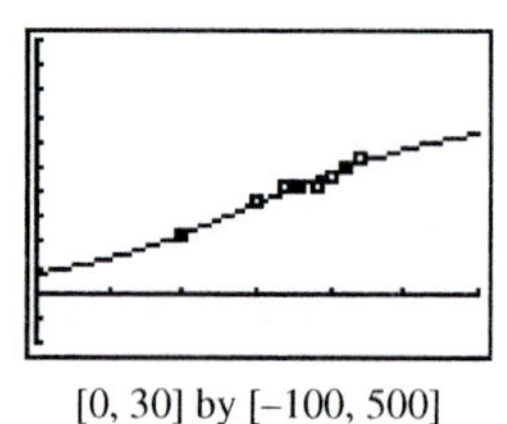

[0, 30] by [−100, 500]

(c) Quadratic regression model

Graphical solution: Graph the line $y = 300$ with the quadratic regression curve $y = -0.0938x^2 + 15.0510x - 28.2375$ and find the intersection of the two curves. The two intersect at $x \approx 26.03$. The expenditures will be 300 billion dollars sometime in the year 2006.

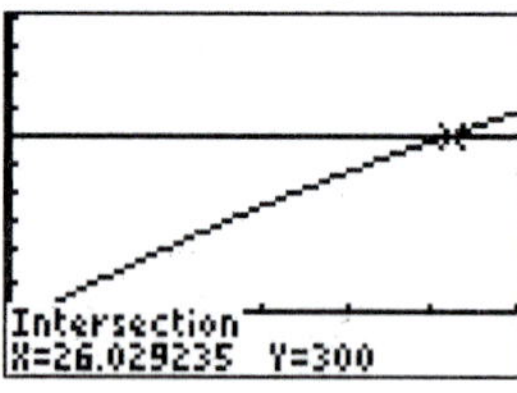

[0, 30] by [−100, 500]

Another graphical solution would be to find where the graph of the difference of the two curves is equal to 0.

Note: The quadratic and the line intersect in two points, but the second point is an unrealistic answer. This would be sometime in the year 2114.

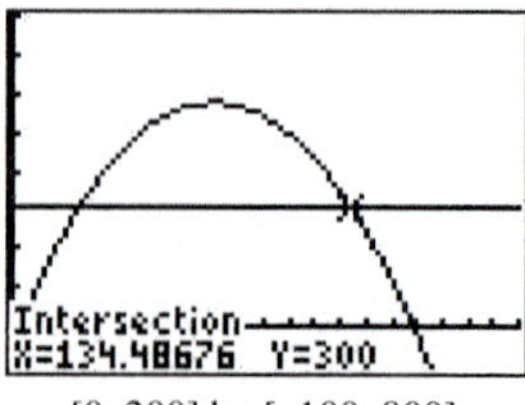

[0, 200] by [−100, 800]

Algebraic solution: Solve $300 = -0.0938x^2 + 15.0510x - 28.2375$ for x.
Use the quadratic formula to solve the equation $-0.0938x^2 + 15.0510x - 328.2375 = 0$.
$a = -0.0938 \quad b = 15.0510 \quad c = -328.2375$

$$x = \frac{-(15.0510) \pm \sqrt{(15.0510)^2 - 4(-0.0938)(-328.2375)}}{2(-0.0938)}$$
$$= \frac{-(15.0510) \pm \sqrt{226.5326 - 123.1547}}{-0.1876}$$
$$= \frac{-(15.0510) \pm 10.1675}{-0.1876}$$

$x \approx 26.03$ and $x \approx 134.43$

We select $x = 26.03$, which indicates that the expenditures will be 300 billion dollars sometime in the year 2006.

Logistic regression model

Graphical solution: Graph the line $y = 300$ with the logistic regression curve $y = \frac{353.6473}{(1 + 8.6873e^{-0.1427x})}$ and find the intersection of the two curves. The two intersect at $x \approx 27.21$.
The expenditures will be 300 billion dollars sometime in the year 2007.

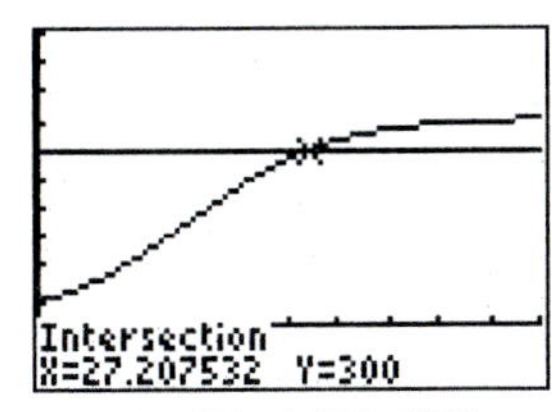

[0, 50] by [−100, 500]

Another graphical solution would be to find where the graph of the difference of the two curves is equal to 0.

Algebraic solution: Solve

$$300 = \frac{353.6473}{(1 + 8.6873e^{-0.1427x})} \text{ for } x.$$
$$300(1 + 8.86873e^{-0.1427x}) = 353.6473$$
$$8.8687e^{-0.1427x} = \frac{353.6473}{300} - 1$$
$$8.8687e^{-0.1427x} = 1.1788 - 1 = 0.1788$$
$$e^{-0.1427x} = \frac{0.1788}{8.8687} = 0.0202$$
$$-0.1427x = \ln 0.0202$$
$$x = \frac{\ln 0.0202}{-0.1427} \approx 27.34$$

The expenditures will be 300 billion dollars sometime in the year 2007.

(d) The long-range implication of using the quadratic regression equation is that the expenditures will eventually fall to zero.
(The graph of the function is a parabola with vertex at about (80, 576) and it opens downward. So, eventually the curve will cross the x-axis and the expenditures will be 0. This will happen when $x \approx 158$.)

(e) The long-range implication of using the logistic regression equation is that the expenditures will eventually level off at about 354 billion dollars.
(We notice that as x gets larger, $e^{-0.1427x}$ approaches 0. Therefore, the denominator of the function approaches 1 and the function itself approaches 353.65, which is about 354.)

47. In this problem, the graphs are representative of the population (in thousands) of the states of Arizona and Massachusetts for several years, where x is the number of years past 1980.

(a) The following is a scatter plot of the Arizona data with the linear regression equation $y = 127.6351x + 2587.0010$ superimposed on it.

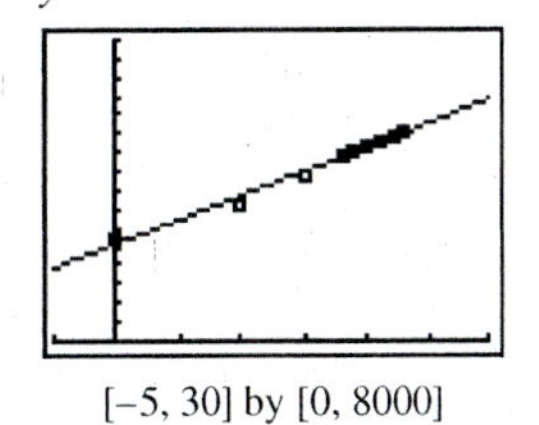
[−5, 30] by [0, 8000]

(b) The following is a scatter plot of the Massachusetts data with the linear regression equation $y = 31.3732x + 5715.9742$ superimposed on it.

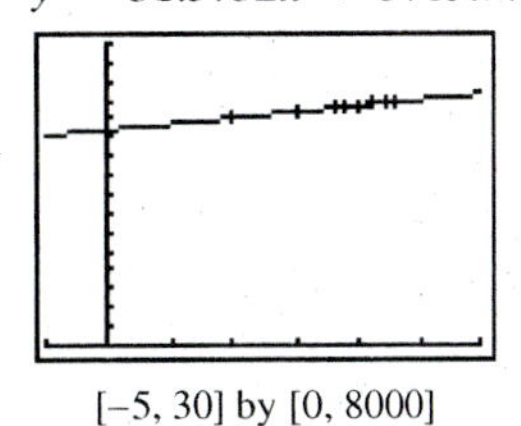
[−5, 30] by [0, 8000]

(c) *Graphical solution:* Graph the two linear equations $y = 127.6351x + 2587.0010$ and $y = 31.3732x + 5715.9742$ on the same axes and find the point of intersection. The two curves intersect at $x \approx 32.5$.
The population of the two states will be the same sometime in the year 2012.

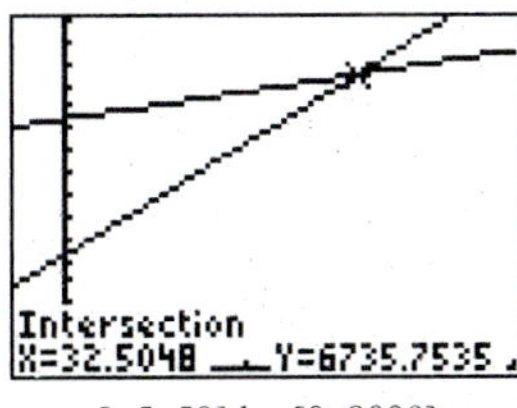

[−5, 50] by [0, 8000]

Another graphical solution would be to find where the graph of the difference of the two curves is equal to 0.

Algebraic solution: Solve
$127.6351x + 2587.0010 = 31.3732x + 5715.9742$ for x.

$$127.6351x + 2587.0010 = 31.3732x + 5715.9742$$
$$96.2619x = 3128.9732$$
$$x = \frac{3128.9732}{96.2619} \approx 32.5$$

The population of the two states will be the same sometime in the year 2012.

49. $200 = 2(x + y)$ and $500 = xy$. Then $y = 100 - x$, so $500 = x(100 - x)$, and therefore $x = 50 \pm 20\sqrt{5}$, and $y = 50 \mp 20\sqrt{5}$. Both answers correspond to a rectangle with approximate dimensions 5.28 m $\times$ 94.72 m.

51. If r is Hank's rowing speed (in miles per hour) and c is the speed of the current, $\frac{24}{60}(r - c) = 1$ and $\frac{13}{60}(r + c) = 1$. Therefore $r = c + \frac{5}{2}$ (from the first equation); substituting gives $\frac{13}{60}\left(2c + \frac{5}{2}\right) = 1$, so $2c = \frac{60}{13} - \frac{5}{2} = \frac{55}{26}$, and $c = \frac{55}{52} \approx 1.06$ mph. Finally, $r = c + \frac{5}{2} = \frac{185}{52} \approx 3.56$ mph.

53. $m + \ell = 1.74$ and $\ell = m + 0.16$, so $2m + 0.16 = 1.74$. Then $m = \$0.79$ (79 cents) and $\ell = \$0.95$ (95 cents).

55. $4 = -a + b$ and $6 = 2a + b$, so $b = a + 4$ and $6 = 3a + 4$. Then $a = \frac{2}{3}$ and $b = \frac{14}{3}$.

57. (a) Let $C(x) =$ the amount charged by each rental company, and let $x =$ the number of miles driven by Pedro.
Company A: $C(x) = 40 + 0.10x$
Company B: $C(x) = 25 + 0.15x$
Solving these two equations for x,

$$40 + 0.10x = 25 + 0.15x$$
$$15 = 0.05x$$
$$300 = x$$

Pedro can drive 300 miles to be charged the same amount by the two companies.

(b) One possible answer: If Pedro is making only a short trip, Company B is better because the flat fee is less. However, if Pedro drives the rental van over 300 miles, Company A's plan is more economical for his needs.

59. False. A system of two linear equations in two variables has either 0, 1, or infinitely many solutions.

61. Using $(x, y) = (3, -2)$,
$2(3) - 3(-2) = 12$
$3 + 2(-2) = -1$
The answer is C.

63. Two parabolas can intersect in 0, 1, 2, 3, or 4 places, or infinitely many places if the parabolas completely coincide. The answer is D.

65. (a)

$$\frac{x^2}{4} + \frac{y^2}{9} = 1$$
$$9x^2 + 4y^2 = 36$$
$$4y^2 = 36 - 9x^2$$
$$y^2 = \frac{36 - 9x^2}{4}$$
$$y = \frac{3}{2}\sqrt{4 - x^2},\ y = -\frac{3}{2}\sqrt{4 - x^2}$$

(b)

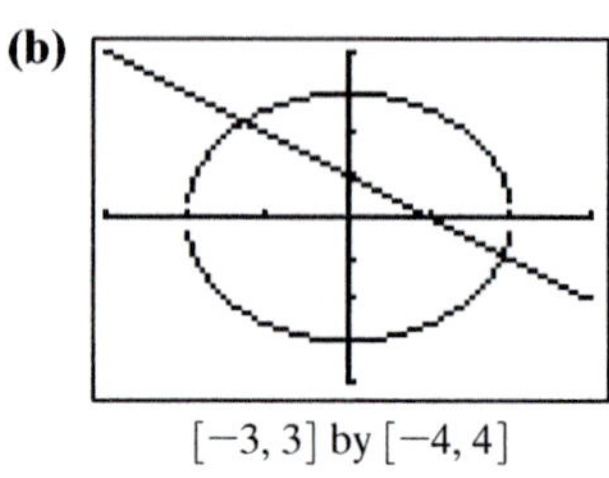

$[-3, 3]$ by $[-4, 4]$

$(x, y) \approx (-1.29, 2.29)$ or $(1.91, -0.91)$

(c) $\frac{(-1.29)^2}{4} + \frac{(2.29)^2}{9} \approx 0.9987 \approx 1$ and $(-1.29) + (2.29) = 1$, so the first solution checks.
$\frac{(1.91)^2}{4} + \frac{(-0.91)^2}{9} \approx 1.004 \approx 1$ and $(1.91) + (-0.91) = 1$, so the second solution checks.

67. Subtract the second equation from the first, leaving $-3y = -10$, or $y = \frac{10}{3}$. Then $x^2 = 4 - \frac{10}{3} = \frac{2}{3}$, so $x = \pm\sqrt{\frac{2}{3}}$.

69. The vertex of the parabola $R = (100 - 4x)x = 4x(25 - x)$ has first coordinate $x = 12.5$ units.

Section 7.2 Matrix Algebra

Exploration 1

1. $a_{11} = 3(1) - (1) = 2$ Set $i = j = 1$.
$a_{12} = 3(1) - (2) = 1$ Set $i = 1, j = 2$.
$a_{21} = 3(2) - (1) = 5$ Set $i = 2, j = 1$.
$a_{22} = 3(2) - (2) = 4$ Set $i = j = 2$.

So, $A = \begin{bmatrix} 2 & 1 \\ 5 & 4 \end{bmatrix}$. Similar computations show that $B = \begin{bmatrix} -1 & 2 \\ 2 & 5 \end{bmatrix}$.

2. The additive inverse of A is $-A$ and $-A = \begin{bmatrix} -2 & -1 \\ -5 & -4 \end{bmatrix}$.

$$A + (-A) = \begin{bmatrix} 2 & 1 \\ 5 & 4 \end{bmatrix} + \begin{bmatrix} -2 & -1 \\ -5 & -4 \end{bmatrix} = \begin{bmatrix} 0 & 0 \\ 0 & 0 \end{bmatrix} = [0]$$

The order of $[0]$ is 2×2.

3.
$$3A - 2B = 3\begin{bmatrix} 2 & 1 \\ 5 & 4 \end{bmatrix} - 2\begin{bmatrix} -1 & 2 \\ 2 & 5 \end{bmatrix} = \begin{bmatrix} 6 & 3 \\ 15 & 12 \end{bmatrix} - \begin{bmatrix} -2 & 4 \\ 4 & 10 \end{bmatrix} = \begin{bmatrix} 8 & -1 \\ 11 & 2 \end{bmatrix}$$

Exploration 2

1. $\det(A) = -a_{12}a_{21}a_{33} + a_{13}a_{21}a_{32} + a_{11}a_{22}a_{33} - a_{13}a_{22}a_{31} - a_{11}a_{23}a_{32} + a_{12}a_{23}a_{31}$
Each element contains an element from each row and each column due to a definition of a determinant. Regardless of the row or column "picked" to apply the definition, all other elements of the matrix are eventually factored into the multiplication.

2.
$$\begin{vmatrix} a_{11} & a_{12} & a_{13} \\ a_{21} & a_{22} & a_{23} \\ a_{31} & a_{32} & a_{33} \end{vmatrix} = a_{11}(-1)^2\begin{vmatrix} a_{22} & a_{23} \\ a_{32} & a_{33} \end{vmatrix} + a_{12}(-1)^3\begin{vmatrix} a_{21} & a_{23} \\ a_{31} & a_{33} \end{vmatrix} + a_{13}(-1)^4\begin{vmatrix} a_{21} & a_{22} \\ a_{31} & a_{32} \end{vmatrix}$$
$$= a_{11}(a_{22}a_{33} - a_{23}a_{32}) - a_{12}(a_{21}a_{33} - a_{23}a_{31}) + a_{13}(a_{21}a_{32} - a_{22}a_{31})$$
$$= a_{11}a_{22}a_{33} - a_{11}a_{23}a_{32} - a_{12}a_{21}a_{33} + a_{12}a_{23}a_{31} + a_{13}a_{21}a_{32} - a_{13}a_{22}a_{31}$$

The two expressions are exactly equal.

3. Recall that A_{ij} is $(-1)^{i+j}M_{ij}$ where M_{ij} is the determinant of the matrix obtained by deleting the row and column containing a_{ij}. Let $A = k \times k$ square matrix with zeros in the ith row. Then: $\det(A) =$

$$\text{ith row} \rightarrow \begin{vmatrix} a_{11} & a_{12} & \cdots & a_{1k} \\ a_{21} & a_{22} & \cdots & a_{2k} \\ \vdots & \vdots & & \vdots \\ 0 & 0 & \cdots & 0 \\ \vdots & \vdots & & \vdots \\ a_{k1} & a_{k2} & \cdots & a_{kk} \end{vmatrix}$$

$= 0 \cdot A_{i1} + 0 \cdot A_{i2} + \cdots + 0 \cdot A_{ik} = 0 + 0 + \cdots + 0$
$= 0$

Quick Review 7.2

1. (a) $(3, 2)$

(b) $(x, -y)$

3. (a) $(-2, 3)$

(b) (y, x)

5. $(3 \cos \theta, 3 \sin \theta)$

7. $\sin(\alpha + \beta) = \sin \alpha \cos \beta + \cos \alpha \sin \beta$

9. $\cos(\alpha + \beta) = \cos \alpha \cos \beta - \sin \alpha \sin \beta$

Section 7.2 Exercises

1. 2×3; not square

3. 3×2; not square

5. 3×1; not square

7. $a_{13} = 3$

9. $a_{32} = 4$

11. (a) $\begin{bmatrix} 3 & 0 \\ -3 & 1 \end{bmatrix}$ **(b)** $\begin{bmatrix} 1 & 6 \\ 1 & 9 \end{bmatrix}$ **(c)** $\begin{bmatrix} 6 & 9 \\ -3 & 15 \end{bmatrix}$

(d) $2A - 3B = 2\begin{bmatrix} 2 & 3 \\ -1 & 5 \end{bmatrix} - 3\begin{bmatrix} 1 & -3 \\ -2 & -4 \end{bmatrix}$

$= \begin{bmatrix} 4 & 6 \\ -2 & 10 \end{bmatrix} - \begin{bmatrix} 3 & -9 \\ -6 & -12 \end{bmatrix} = \begin{bmatrix} 1 & 15 \\ 4 & 22 \end{bmatrix}$

13. (a) $\begin{bmatrix} 1 & 1 \\ -2 & 0 \\ -1 & 0 \end{bmatrix}$ **(b)** $\begin{bmatrix} -7 & 1 \\ 2 & -2 \\ 5 & 2 \end{bmatrix}$ **(c)** $\begin{bmatrix} -9 & 3 \\ 0 & -3 \\ 6 & 3 \end{bmatrix}$

(d) $2A - 3B = 2\begin{bmatrix} -3 & 1 \\ 0 & -1 \\ 2 & 1 \end{bmatrix} - 3\begin{bmatrix} 4 & 0 \\ -2 & 1 \\ -3 & -1 \end{bmatrix}$

$= \begin{bmatrix} -6 & 2 \\ 0 & -2 \\ 4 & 2 \end{bmatrix} - \begin{bmatrix} 12 & 0 \\ -6 & 3 \\ -9 & -3 \end{bmatrix} = \begin{bmatrix} -18 & 2 \\ 6 & -5 \\ 13 & 5 \end{bmatrix}$

15. (a) $\begin{bmatrix} -3 \\ 1 \\ 4 \end{bmatrix}$ **(b)** $\begin{bmatrix} -1 \\ 1 \\ -4 \end{bmatrix}$ **(c)** $\begin{bmatrix} -6 \\ 3 \\ 0 \end{bmatrix}$ **(d)** $2A - 3B = 2\begin{bmatrix} -2 \\ 1 \\ 0 \end{bmatrix} - 3\begin{bmatrix} -1 \\ 0 \\ 4 \end{bmatrix} = \begin{bmatrix} -4 \\ 2 \\ 0 \end{bmatrix} - \begin{bmatrix} -3 \\ 0 \\ 12 \end{bmatrix} = \begin{bmatrix} -1 \\ 2 \\ -12 \end{bmatrix}$

17. (a) $AB = \begin{bmatrix} (2)(1) + (3)(-2) & (2)(-3) + (3)(-4) \\ (-1)(1) + (5)(-2) & (-1)(-3) + (5)(-4) \end{bmatrix} = \begin{bmatrix} -4 & -18 \\ -11 & -17 \end{bmatrix}$

(b) $BA = \begin{bmatrix} (1)(2) + (-3)(-1) & (1)(3) + (-3)(5) \\ (-2)(2) + (-4)(-1) & (-2)(3) + (-4)(5) \end{bmatrix} = \begin{bmatrix} 5 & -12 \\ 0 & -26 \end{bmatrix}$

19. (a) $AB = \begin{bmatrix} (2)(1) + (0)(-3) + (1)(0) & (2)(2) + (0)(1) + (1)(-2) \\ (1)(1) + (4)(-3) + (-3)(0) & (1)(2) + (4)(1) + (-3)(-2) \end{bmatrix} = \begin{bmatrix} 2 & 2 \\ -11 & 12 \end{bmatrix}$

(b) $BA = \begin{bmatrix} (1)(2) + (2)(1) & (1)(0) + (2)(4) & (1)(1) + (2)(-3) \\ (-3)(2) + (1)(1) & (-3)(0) + (1)(4) & (-3)(1) + (1)(-3) \\ (0)(2) + (-2)(1) & (0)(0) + (-2)(4) & (0)(1) + (-2)(-3) \end{bmatrix} = \begin{bmatrix} 4 & 8 & -5 \\ -5 & 4 & -6 \\ -2 & -8 & 6 \end{bmatrix}$

21. (a) $AB = \begin{bmatrix} (-1)(2) + (0)(-1) + (2)(4) & (-1)(1) + (0)(0) + (2)(-3) & (-1)(0) + (0)(2) + (2)(-1) \\ (4)(2) + (1)(-1) + (-1)(4) & (4)(1) + (1)(0) + (-1)(-3) & (4)(0) + (1)(2) + (-1)(-1) \\ (2)(2) + (0)(-1) + (1)(4) & (2)(1) + (0)(0) + (1)(-3) & (2)(0) + (0)(2) + (1)(-1) \end{bmatrix}$

$= \begin{bmatrix} 6 & -7 & -2 \\ 3 & 7 & 3 \\ 8 & -1 & -1 \end{bmatrix}$

(b) $BA = \begin{bmatrix} (2)(-1) + (1)(4) + (0)(2) & (2)(0) + (1)(1) + (0)(0) & (2)(2) + (1)(-1) + (0)(1) \\ (-1)(-1) + (0)(4) + (2)(2) & (-1)(0) + (0)(1) + (2)(0) & (-1)(2) + (0)(-1) + (2)(1) \\ (4)(-1) + (-3)(4) + (-1)(2) & (4)(0) + (-3)(1) + (-1)(0) & (4)(2) + (-3)(-1) + (-1)(1) \end{bmatrix}$

$= \begin{bmatrix} 2 & 1 & 3 \\ 5 & 0 & 0 \\ -18 & -3 & 10 \end{bmatrix}$

23. (a) $AB = [(2)(-5) + (-1)(4) + (3)(2)] = [-8]$

(b) $BA = \begin{bmatrix} (-5)(2) & (-5)(-1) & (-5)(3) \\ (4)(2) & (4)(-1) & (4)(3) \\ (2)(2) & (2)(-1) & (2)(3) \end{bmatrix} = \begin{bmatrix} -10 & 5 & -15 \\ 8 & -4 & 12 \\ 4 & -2 & 6 \end{bmatrix}$

25. (a) AB is not possible.

(b) $BA = [(-3)(-1) + (5)(3) \quad (-3)(2) + (5)(4)] = [18 \ \ 14]$

27. **(a)** $AB = \begin{bmatrix} (0)(1)+(0)(2)+(1)(-1) & (0)(2)+(0)(0)+(1)(3) & (0)(1)+(0)(1)+(1)(4) \\ (0)(1)+(1)(2)+(0)(-1) & (0)(2)+(1)(0)+(0)(3) & (0)(1)+(1)(1)+(0)(4) \\ (1)(1)+(0)(2)+(0)(-1) & (1)(2)+(0)(0)+(0)(3) & (1)(1)+(0)(1)+(0)(4) \end{bmatrix} = \begin{bmatrix} -1 & 3 & 4 \\ 2 & 0 & 1 \\ 1 & 2 & 1 \end{bmatrix}$

(b) $BA = \begin{bmatrix} (1)(0)+(2)(0)+(1)(1) & (1)(0)+(2)(1)+(1)(0) & (1)(1)+(2)(0)+(1)(0) \\ (2)(0)+(0)(0)+(1)(1) & (2)(0)+(0)(1)+(1)(0) & (2)(1)+(0)(0)+(1)(0) \\ (-1)(0)+(3)(0)+(4)(1) & (-1)(0)+(3)(1)+(4)(0) & (-1)(1)+(3)(0)+(4)(0) \end{bmatrix} = \begin{bmatrix} 1 & 2 & 1 \\ 1 & 0 & 2 \\ 4 & 3 & -1 \end{bmatrix}$

In #29–31, use the fact that two matrices are equal only if all entries are equal.

29. $a = 5, b = 2$

31. $a = -2, b = 0$

33. $AB = \begin{bmatrix} (2)(0.8)+(1)(-0.6) & (2)(-0.2)+(1)(0.4) \\ (3)(0.8)+(4)(-0.6) & (3)(-0.2)+(4)(0.4) \end{bmatrix}$

$= \begin{bmatrix} 1 & 0 \\ 0 & 1 \end{bmatrix},$

$BA = \begin{bmatrix} (0.8)(2)+(-0.2)(3) & (0.8)(1)+(-0.2)(4) \\ (-0.2)(2)+(0.4)(3) & (0.6)(1)+(-0.4)(4) \end{bmatrix}$

$= \begin{bmatrix} 1 & 0 \\ 0 & 1 \end{bmatrix}$, so A and B are inverses.

35. $\begin{bmatrix} 2 & 3 \\ 2 & 2 \end{bmatrix}^{-1} = \dfrac{1}{(2)(2)-(2)(3)}\begin{bmatrix} 2 & -3 \\ -2 & 2 \end{bmatrix}$

$= -\dfrac{1}{2}\begin{bmatrix} 2 & -3 \\ -2 & 2 \end{bmatrix} = \begin{bmatrix} -1 & 1.5 \\ 1 & -1 \end{bmatrix}$

37. No inverse: The determinant (found with a calculator) is 0.

39. $A = \begin{bmatrix} 1 & -1 & 1 & -1 \\ -1 & 1 & -1 & 1 \\ 1 & -1 & 1 & -1 \\ -1 & 1 & -1 & 1 \end{bmatrix};$

No inverse, $\det(A) = 0$ (found using a calculator)

41. Use row 2 or column 2 since they have the greatest number of zeros. Using column 2:

$\begin{vmatrix} 2 & 1 & 1 \\ -1 & 0 & 2 \\ 1 & 3 & -1 \end{vmatrix} = (1)(-1)^3\begin{vmatrix} -1 & 2 \\ 1 & -1 \end{vmatrix}$

$+ (0)(-1)^4\begin{vmatrix} 2 & 1 \\ 1 & -1 \end{vmatrix} + (3)(-1)^5\begin{vmatrix} 2 & 1 \\ -1 & 2 \end{vmatrix}$

$= (-1)(1-2) + 0 + (-3)(4+1)$

$= 1 + 0 - 15$

$= -14$

43. $3X = B - A$

$X = \dfrac{B-A}{3} = \dfrac{1}{3}\left(\begin{bmatrix} 4 \\ 2 \end{bmatrix} - \begin{bmatrix} 1 \\ 3 \end{bmatrix}\right) = \dfrac{1}{3}\begin{bmatrix} 3 \\ -1 \end{bmatrix} = \begin{bmatrix} 1 \\ -\frac{1}{3} \end{bmatrix}$

45. **(a)** The entries a_{ij} and a_{ji} are the same because each gives the distance between the same two cities.

(b) The entries a_{ii} are all 0 because the distance between a city and itself is 0.

47. **(a)** $B^T A = [\$0.80 \quad \$0.85 \quad \$1.00]\begin{bmatrix} 100 & 60 \\ 120 & 70 \\ 200 & 120 \end{bmatrix}$

$\begin{bmatrix} 0.80(100) & 0.80(60) \\ +\,0.85(120) & +\,0.85(70) \\ 1(200) & +\,1(120) \end{bmatrix}$

$= [382 \quad 227.50]$

(b) b_{1j} in matrix $B^T A$ represents the income Happy Valley Farms makes at grocery store j, selling all three types of eggs.

49. **(a)** Total revenue = sum of (price charged)(number sold)
$= AB^T$ or BA^T

(b) Profit = Total revenue − Total Cost
$= AB^T - CB^T$
$= (A - C)B^T$

51. **(a)** $[x' \quad y'] = [x \quad y]\begin{bmatrix} \cos\alpha & -\sin\alpha \\ \sin\alpha & \cos\alpha \end{bmatrix},$

$x, y = 1, \alpha = 30°$

$= [1 \quad 1]\begin{bmatrix} \frac{\sqrt{3}}{2} & -\frac{1}{2} \\ \frac{1}{2} & \frac{\sqrt{3}}{2} \end{bmatrix}$

$= \begin{bmatrix} \frac{\sqrt{3}+1}{2} & \frac{\sqrt{3}-1}{2} \end{bmatrix} \approx [1.37 \quad 0.37]$

(b) $[x \quad y] = [x' \quad y']\begin{bmatrix} \cos\alpha & \sin\alpha \\ -\sin\alpha & \cos\alpha \end{bmatrix},$

$x', y' = 1, \alpha = 30°$

$= [1 \quad 1]\begin{bmatrix} \frac{\sqrt{3}}{2} & \frac{1}{2} \\ -\frac{1}{2} & \frac{\sqrt{3}}{2} \end{bmatrix}$

$= \begin{bmatrix} \frac{\sqrt{3}-1}{2} & \frac{\sqrt{3}+1}{2} \end{bmatrix} \approx [0.37 \quad 1.37]$

53. Answers will vary. One possible answer is provided for each.

(a) $c(A + B) = c[a_{ij} + b_{ij}] = [ca_{ij} + cb_{ij}] = cA + cB$

(b) $(c + d)A = (c + d)[a_{ij}] = c[a_{ij}] + d[a_{ij}]$
$= cA + dA$

(c) $c(dA) = c[da_{ij}] = [cda_{ij}] = cd[a_{ij}] = cdA$

(d) $1 \cdot A = 1 \cdot [a_{ij}] = [a_{ij}] = A$

55. $A \cdot A^{-1} = \begin{bmatrix} a & b \\ c & d \end{bmatrix}\left(\dfrac{1}{ad - bc}\right)\begin{bmatrix} d & -b \\ -c & a \end{bmatrix}$

$= \left(\dfrac{1}{ad - bc}\right)\begin{bmatrix} a & b \\ c & d \end{bmatrix}\begin{bmatrix} d & -b \\ -c & a \end{bmatrix}$

(since $\dfrac{1}{ad - bc}$ is a scalar)

$= \left(\dfrac{1}{ad - bc}\right)\begin{bmatrix} ad - bc & -ab + ba \\ cd - cd & -bc + ad \end{bmatrix} = \left(\dfrac{1}{ad - bc}\right)$

$\begin{bmatrix} ad - bc & 0 \\ 0 & ad - bc \end{bmatrix}$

$= \begin{bmatrix} \dfrac{ad - bc}{ad - bc} & 0 \\ 0 & \dfrac{ad - bc}{ad - bc} \end{bmatrix}$

$= \begin{bmatrix} 1 & 0 \\ 0 & 1 \end{bmatrix} = I_2$

57. If (x, y) is reflected across the y-axis, then $(x, y) \Rightarrow (-x, y)$.

$[x' \quad y'] = [x \quad y]\begin{bmatrix} -1 & 0 \\ 0 & 1 \end{bmatrix}$

59. If (x, y) is reflected across the line $y = -x$, then $(x, y) \Rightarrow (-y, -x)$.

$[x' \quad y'] = [x \quad y]\begin{bmatrix} 0 & -1 \\ -1 & 0 \end{bmatrix}$

61. If (x, y) is horizontally stretched (or shrunk) by a factor of c, then $(x, y) \Rightarrow (cx, y)$.

$[x' \quad y'] = [x \quad y]\begin{bmatrix} c & 0 \\ 0 & 1 \end{bmatrix}$

63. False. The determinant can be negative. For example, the determinant of $A = \begin{bmatrix} 1 & 0 \\ 2 & -1 \end{bmatrix}$ is $1(-1) - 2(0) = -1$.

65. The matrix AB has the same number of rows as A and the same number of columns as B. The answer is B.

67. The value in row 1, column 3 is 3. The answer is D.

69. (a) Let $A = [a_{ij}]$ be an $n \times n$ matrix and let B be the same as matrix A, except that the ith row of B is the ith row of A multiplied by the scalar c. Then:

$$\det(B) = \begin{vmatrix} a_{11} & a_{12} & \dots & a_{1n} \\ a_{21} & a_{22} & \dots & a_{2n} \\ \vdots & & & \\ ca_{i1} & ca_{i2} & \dots & ca_{in} \\ \vdots & & & \\ a_{n1} & a_{n2} & \dots & a_{nn} \end{vmatrix} \quad (i\text{th row} \rightarrow ca_{i1} \dots)$$

$= ca_{i1}(-1)^{i+1}|A_{i1}| + ca_{i2}(-1)^{i+2}|A_{i2}| + \dots + ca_{in}(-1)^{in}|A_{in}|$

$= c(a_{i1}(-1)^{i+1}|A_{i1}| + a_{i2}(-1)^{i+2}|A_{i2}| + \dots + a_{in}(-1)^{i+n}|A_{in}|)$

$= c\det(A)$ (by definition of determinant)

(b) Use the 2×2 case as an example

$\det(A) = \begin{vmatrix} a_{11} & 0 \\ a_{21} & a_{22} \end{vmatrix} = a_{11}a_{22} - 0 = a_{11}a_{22}$

which is the product of the diagonal elements. Now consider the general case where A is an $n \times n$ matrix. Then:

$$\det(A) = \begin{vmatrix} a_{11} & 0 & 0 & \dots & 0 \\ a_{21} & a_{22} & 0 & \dots & 0 \\ \vdots & \vdots & & & \vdots \\ a_{n1} & a_{n2} & & \dots & a_{nn} \end{vmatrix}$$

$$= a_{11}(-1)^2\begin{vmatrix} a_{22} & 0 & 0 & \dots & 0 \\ a_{32} & a_{33} & 0 & \dots & 0 \\ \vdots & \vdots & \vdots & & \vdots \\ a_{n2} & a_{n3} & a_{n4} & \dots & a_{nn} \end{vmatrix}$$

$$= a_{11}(a_{22})(-1)^2\begin{vmatrix} a_{33} & 0 & \dots & 0 \\ a_{43} & a_{44} & \dots & 0 \\ \vdots & \vdots & & \vdots \\ a_{n3} & a_{n4} & \dots & a_{nn} \end{vmatrix}$$

$$= a_{11}a_{22}\dots a_{n-2\ n-2}(-1)^2\begin{vmatrix} a_{n-1\ n-1} & 0 \\ a_{n\ n-1} & a_{nn} \end{vmatrix}$$

$= a_{11}a_{22}\dots a_{n-2\ n-2}\,a_{n-1\ n-1}\,a_{nn}$, which is exactly the product of the diagonal elements (by induction).

71. (a) $A \cdot A^{-1} = \begin{bmatrix} \cos\alpha & -\sin\alpha \\ \sin\alpha & \cos\alpha \end{bmatrix}\begin{bmatrix} \cos\alpha & \sin\alpha \\ -\sin\alpha & \cos\alpha \end{bmatrix}$

$= \begin{bmatrix} \cos^2\alpha + \sin^2\alpha & \cos\alpha\sin\alpha - \sin\alpha\cos\alpha \\ \sin\alpha\cos\alpha - \cos\alpha\sin\alpha & \sin^2\alpha + \cos^2\alpha \end{bmatrix} = \begin{bmatrix} 1 & 0 \\ 0 & 1 \end{bmatrix} = I_2$

(b) From the diagram, we know that:

$x = r\cos\theta \qquad y = r\sin\theta$

$x' = r\cos(\theta - \alpha) \qquad y' = r\sin(\theta - \alpha)$

or $\cos(\theta - \alpha) = \dfrac{x'}{r} \qquad \sin(\theta - \alpha) = \dfrac{y'}{r}$

From algebra, we know that:

$x = r\cos(\theta + \alpha - \alpha) = r\cos(\alpha + (\theta - \alpha))$ and

$y = r\sin(\theta + \alpha - \alpha) = r\sin(\alpha + (\theta - \alpha))$

Using the trigonometric properties and substitution, we have:

$x = r(\cos\alpha\cos(\theta - \alpha) - \sin\alpha\sin(\theta - \alpha))$

$= r\cos\alpha\cos(\theta - \alpha) - r\sin\alpha\sin(\theta - \alpha)$

$= (r\cos\alpha)\left(\dfrac{x'}{r}\right) - (r\sin\alpha)\left(\dfrac{y'}{r}\right)$

$= x'\cos\alpha - y'\sin\alpha$

$y = r(\sin\alpha\cos(\theta - \alpha) + \cos\alpha\sin(\theta - \alpha))$

$= r\sin\alpha\cos(\theta - \alpha) + r\cos\alpha\sin(\theta - \alpha)$

$= (r\sin\alpha)\left(\dfrac{x'}{r}\right) + (r\cos\alpha)\left(\dfrac{y'}{r}\right)$

$= x'\sin\alpha + y'\cos\alpha.$

(c) $[x \quad y] = [x' \quad y']\begin{bmatrix} \cos\alpha & \sin\alpha \\ -\sin\alpha & \cos\alpha \end{bmatrix}$

which is $[x' \quad y']A^{-1}$, the inverse of A.

73. (a) $\det(xI_3 - A) = \begin{vmatrix} x - a_{11} & -a_{12} & -a_{13} \\ -a_{21} & x - a_{22} & -a_{23} \\ -a_{31} & -a_{32} & x - a_{33} \end{vmatrix}$

$= (x - a_{11})(-1)^2 \begin{vmatrix} x - a_{22} & -a_{23} \\ -a_{32} & x - a_{33} \end{vmatrix} + (-a_{12})(-1)^3 \begin{vmatrix} -a_{21} & -a_{23} \\ -a_{31} & x - a_{33} \end{vmatrix} + (-a_{13})(-1)^4 \begin{vmatrix} -a_{21} & x - a_{22} \\ -a_{31} & -a_{32} \end{vmatrix}$

$= (x - a_{11})((x - a_{22})(x - a_{33}) - a_{23}a_{32}) + a_{12}((-a_{21})(x - a_{33}) - a_{23}a_{31}) - a_{13}(a_{21}a_{32} + (a_{31})(x - a_{22}))$

$= (x - a_{11})(x^2 - a_{33}x - a_{22}x + a_{22}a_{33} - a_{23}a_{32}) + a_{12}(-a_{21}x + a_{21}a_{33} - a_{23}a_{31}) - a_{13}(a_{21}a_{32} + a_{31}x - a_{22}a_{31})$

$= x^3 - a_{33}x^2 - a_{22}x^2 + a_{22}a_{33}x - a_{23}a_{32}x - a_{11}x^2 + a_{11}a_{33}x + a_{11}a_{22}x - a_{11}a_{22}a_{33} + a_{11}a_{23}a_{32} - a_{12}a_{21}x + a_{12}a_{21}a_{33}$
$- a_{12}a_{23}a_{31} - a_{13}a_{21}a_{32} - a_{13}a_{31}x + a_{13}a_{22}a_{31}$

$= x^3 + (-a_{33} - a_{22} - a_{11})x^2 + (a_{22}a_{33} - a_{23}a_{32} + a_{11}a_{33} + a_{11}a_{22} - a_{12}a_{21} - a_{13}a_{31})x + (-a_{11}a_{22}a_{33} + a_{11}a_{23}a_{32}$
$+ a_{12}a_{21}a_{33} - a_{12}a_{23}a_{31} - a_{13}a_{21}a_{32} + a_{13}a_{22}a_{31})$

(b) The constant term equals $-\det(A)$.

(c) The coefficient of x^2 is the opposite of the sum of the elements of the main diagonal in A.

(d) $f(A) = \det(AI - A) = \det(A - A)$
$= \det([0]) = 0$

Section 7.3 Multivariate Linear Systems and Row Operations

Exploration 1

1. $x + y + z$ must equal the total number of liters in the mixture, namely 60 L.
2. $0.15x + 0.35y + 0.55z$ must equal total amount of acid in the mixture; since the mixture must be 40% acid and have 60 L of solution, the total amount of acid must be $0.40(60) = 24$ L.
3. The number of liters of 35% solution, y, must equal twice the number of liters of 55% solution, z. Hence $y = 2z$.
4. $\begin{bmatrix} 1 & 1 & 1 \\ 0.15 & 0.35 & 0.55 \\ 0 & 1 & -2 \end{bmatrix}\begin{bmatrix} x \\ y \\ z \end{bmatrix} = \begin{bmatrix} 60 \\ 24 \\ 0 \end{bmatrix}$

   $A = \begin{bmatrix} 1 & 1 & 1 \\ 0.15 & 0.35 & 0.55 \\ 0 & 1 & -2 \end{bmatrix}, X = \begin{bmatrix} x \\ y \\ z \end{bmatrix}, B = \begin{bmatrix} 60 \\ 24 \\ 0 \end{bmatrix}$
5. $X = A^{-1}B = \begin{bmatrix} 3.75 \\ 37.5 \\ 18.75 \end{bmatrix}$
6. 3.75 L of 15% acid, 37.5 L of 35% acid, and 18.75 L of 55% acid are required to make 60 L of a 40% acid solution.

Quick Review 7.3

1. $(40)(0.32) = 12.8$ liters
3. $(50)(1 - 0.24) = 38$ liters
5. $(-1, 6)$
7. $y = w - z + 1$
9. $\begin{bmatrix} 1 & 3 \\ -2 & -2 \end{bmatrix}^{-1} = \begin{bmatrix} -0.5 & -0.75 \\ 0.5 & 0.25 \end{bmatrix}$

Section 7.3 Exercises

1. $x - 3y + z = 0$ (1)
$2y + 3z = 1$ (2)
$z = -2$ (3)

Use $z = -2$ in equation (2).

$2y + 3(-2) = 1$
$2y = 7$
$y = \frac{7}{2}$

Use $z = -2$, $y = 7/2$ in equation (1).

$x - 3\left(\frac{7}{2}\right) + (-2) = 0$

$x = \frac{25}{2}$

So the solution is $(25/2, 7/2, -2)$.

3.
$x - y + z = 0$
$2x - 3z = -1$
$-x - y + 2z = -1$

$x - y + z = 0$
$-2y + z = -3$ ← $2x - 3z = -1$; $2(-x - y + 2z = -1)$
$-x - y + 2z = -1$

$x - y + z = 0$
$-2y + z = -3$
$-2y + 3z = -1$ ← $x - y + z = 0$; $-x - y + 2z = -1$

$x - y + z = 0$
$-2y + z = -3$
$2z = 2$ ← $-1(-2y + z = -3)$; $-2y + 3z = -1$

$x - y + z = 0$
$y - \frac{1}{2}z = \frac{3}{2}$ —— $-\frac{1}{2}(-2y + z = -3)$
$z = 1$ —— $\frac{1}{2}(2z = 2)$

$y - \frac{1}{2}(1) = \frac{3}{2}$; $y = 2$

$x - 2 + 1 = 0$; $x = 1$

The solution is $(1, 2, 1)$.

5.
$$\begin{aligned} x + y + z &= -3 \\ 4x - y &= -5 \\ -3x + 2y + z &= 4 \end{aligned}$$

$$\begin{aligned} x + y + z &= -3 \\ 4x - y &= -5 \\ -4x + y &= 7 \end{aligned} \qquad \begin{aligned} &-1(x + y + z = -3) \\ &-3x + 2y + z = 4 \end{aligned}$$

$$\begin{aligned} x + y + z &= -3 \\ 4x - y &= -5 \\ 0 &= 2 \end{aligned} \qquad \begin{aligned} &4x - y = -5 \\ &-4x + y = 7 \end{aligned}$$

The system has no solution.

7.
$$\begin{aligned} x + y - z &= 4 \\ y + w &= -4 \\ x - y &= 1 \\ x + z + w &= 1 \end{aligned}$$

$$\begin{aligned} 2y - z &= 3 \\ y + w &= -4 \\ x - y &= 1 \\ x + z + w &= 1 \end{aligned} \qquad \begin{aligned} &x + y - z = 4 \\ &-1(x - y = 1) \end{aligned}$$

$$\begin{aligned} 2y - z &= 3 \\ y + w &= -4 \\ x - y &= 1 \\ y + z + w &= 0 \end{aligned} \qquad \begin{aligned} &-1(x - y = 1) \\ &x + z + w = 1 \end{aligned}$$

$$\begin{aligned} -z - 2w &= 11 \\ y + w &= -4 \\ x - y &= 1 \\ y + z + w &= 0 \end{aligned} \qquad \begin{aligned} &2y - z = 3 \\ &-2(y + w = -4) \end{aligned}$$

$$\begin{aligned} -z - 2w &= 11 \\ y + w &= -4 \\ x - y &= 1 \\ z &= 4 \end{aligned} \qquad \begin{aligned} &-1(y + w = -4) \\ &y + z + w = 0 \end{aligned}$$

$$\begin{aligned} x - y &= 1 \\ y + w &= -4 \\ w + \frac{1}{2}z &= -\frac{11}{2} \\ z &= 4 \end{aligned} \qquad -\frac{1}{2}(-z - 2w = 11)$$

$$w + \frac{1}{2}(4) = -\frac{11}{2};\ w = -\frac{15}{2}$$

$$y + \left(-\frac{15}{2}\right) = -4;\ y = \frac{7}{2}$$

$$x - \frac{7}{2} = 1;\ x = \frac{9}{2}$$

So the solution is $\left(\frac{9}{2}, \frac{7}{2}, 4, -\frac{15}{2}\right)$.

9. $\begin{bmatrix} 2 & -6 & 4 \\ 1 & 2 & -3 \\ 0 & -8 & 4 \end{bmatrix}$

11. $\begin{bmatrix} 0 & -10 & 10 \\ 1 & 2 & -3 \\ -3 & 1 & -2 \end{bmatrix}$

13. R_{12}

15. $(-3)R_2 + R_3$

For #17–19, answers will vary depending on the exact sequence of row operations used. One possible sequence of row operations (not necessarily the shortest) is given. The answers shown are not necessarily the ones that might be produced by a grapher or other technology. In some cases, they are not the ones given in the text answers.

17. $\begin{bmatrix} 1 & 3 & -1 \\ 2 & 1 & 4 \\ -3 & 0 & 1 \end{bmatrix} \xrightarrow[(3)R_1 + R_3]{(-2)R_1 + R_2} \begin{bmatrix} 1 & 3 & -1 \\ 0 & -5 & 6 \\ 0 & 9 & -2 \end{bmatrix} \xrightarrow[(-1/5)R_2]{(9/5)R_2 + R_3} \begin{bmatrix} 1 & 3 & -1 \\ 0 & 1 & -1.2 \\ 0 & 0 & 8.8 \end{bmatrix} \xrightarrow{(5/44)R_3} \begin{bmatrix} 1 & 3 & -1 \\ 0 & 1 & -1.2 \\ 0 & 0 & 1 \end{bmatrix}$

19. $\begin{bmatrix} 1 & 2 & 3 & -4 \\ -2 & 6 & -6 & 2 \\ 3 & 12 & 6 & 12 \end{bmatrix} \xrightarrow[(-3)R_1 + R_3]{(2)R_1 + R_2} \begin{bmatrix} 1 & 2 & 3 & -4 \\ 0 & 10 & 0 & -6 \\ 0 & 6 & -3 & 24 \end{bmatrix} \xrightarrow[(-1/3)R_3]{(1/10)R_2} \begin{bmatrix} 1 & 2 & 3 & -4 \\ 0 & 1 & 0 & -0.6 \\ 0 & -2 & 1 & -8 \end{bmatrix} \xrightarrow{(2)R_2 + R_3}$
$\begin{bmatrix} 1 & 2 & 3 & -4 \\ 0 & 1 & 0 & -0.6 \\ 0 & 0 & 1 & -9.2 \end{bmatrix}$

In #21–23, reduced row echelon format is essentially unique, though the sequence of steps may vary from those shown.

21. $\begin{bmatrix} 1 & 0 & 2 & 1 \\ 3 & 2 & 4 & 7 \\ 2 & 1 & 3 & 4 \end{bmatrix} \xrightarrow[(-2)R_1 + R_3]{(-3)R_1 + R_2} \begin{bmatrix} 1 & 0 & 2 & 1 \\ 0 & 2 & -2 & 4 \\ 0 & 1 & -1 & 2 \end{bmatrix} \xrightarrow{(1/2)R_2} \begin{bmatrix} 1 & 0 & 2 & 1 \\ 0 & 1 & -1 & 2 \\ 0 & 1 & -1 & 2 \end{bmatrix} \xrightarrow{(-1)R_2 + R_3} \begin{bmatrix} 1 & 0 & 2 & 1 \\ 0 & 1 & -1 & 2 \\ 0 & 0 & 0 & 0 \end{bmatrix}$

23. $\begin{bmatrix} 1 & 2 & 3 & 1 \\ -3 & -5 & -7 & -4 \end{bmatrix} \xrightarrow{(3)R_1 + R_2} \begin{bmatrix} 1 & 2 & 3 & 1 \\ 0 & 1 & 2 & -1 \end{bmatrix} \xrightarrow{(-2)R_2 + R_1} \begin{bmatrix} 1 & 0 & -1 & 3 \\ 0 & 1 & 2 & -1 \end{bmatrix}$

25. $\begin{bmatrix} 2 & -3 & 1 & 1 \\ -1 & 1 & -4 & -3 \\ 3 & 0 & -1 & 2 \end{bmatrix}$

27. $\begin{bmatrix} 2 & -5 & 1 & -1 & -3 \\ 1 & 0 & -2 & 1 & 4 \\ 0 & 2 & -3 & -1 & 5 \end{bmatrix}$

In #29–31, the variable names (x, y, etc.) are arbitrary.

29. $3x + 2y = -1$
$-4x + 5y = 2$

31. $2x + z = 3$
$-x + y = 2$
$2y - 3z = -1$

33. $$\begin{bmatrix} 1 & -2 & 1 & 8 \\ 2 & 1 & -3 & -9 \\ -3 & 1 & 3 & 5 \end{bmatrix} \xrightarrow[(3)R_1 + R_3]{(-2)R_1 + R_2} \begin{bmatrix} 1 & -2 & 1 & 8 \\ 0 & 5 & -5 & -25 \\ 0 & -5 & 6 & 29 \end{bmatrix} \xrightarrow[(1/5)R_2]{(1)R_2 + R_3} \begin{bmatrix} 1 & -2 & 1 & 8 \\ 0 & 1 & -1 & -5 \\ 0 & 0 & 1 & 4 \end{bmatrix}$$

$x - 2y + z = 8$
$y - z = -5$
$z = 4$
$y - 4 = -5;\ y = -1$
$x - 2(-1) + 4 = 8;\ x = 2$
So the solution is $(2, -1, 4)$.

35. $(x, y, z) = (-2, 3, 1)$: $$\begin{bmatrix} 1 & 2 & -1 & 3 \\ 3 & 7 & -3 & 12 \\ -2 & -4 & 3 & -5 \end{bmatrix} \xrightarrow[(2)R_1 + R_3]{(-3)R_1 + R_2} \begin{bmatrix} 1 & 2 & -1 & 3 \\ 0 & 1 & 0 & 3 \\ 0 & 0 & 1 & 1 \end{bmatrix} \xrightarrow[(1)R_3 + R_1]{(-2)R_2 + R_1} \begin{bmatrix} 1 & 0 & 0 & -2 \\ 0 & 1 & 0 & 3 \\ 0 & 0 & 1 & 1 \end{bmatrix}$$

37. No solution: $$\begin{bmatrix} 1 & 1 & 3 & 2 \\ 3 & 4 & 10 & 5 \\ 1 & 2 & 4 & 3 \end{bmatrix} \xrightarrow[(-1)R_1 + R_3]{(-3)R_1 + R_2} \begin{bmatrix} 1 & 1 & 3 & 2 \\ 0 & 1 & 1 & -1 \\ 0 & 1 & 1 & 1 \end{bmatrix} \xrightarrow{(-1)R_2 + R_3} \begin{bmatrix} 1 & 1 & 3 & 2 \\ 0 & 1 & 1 & -1 \\ 0 & 0 & 0 & 2 \end{bmatrix}$$

39. $(x, y, z) = (2 - z, 1 + z, z)$ — the final matrix translates to $x + z = 2$ and $y - z = 1$.
$$\begin{bmatrix} 1 & 0 & 1 & 2 \\ 2 & 1 & 1 & 5 \end{bmatrix} \xrightarrow{(-2)R_1 + R_2} \begin{bmatrix} 1 & 0 & 1 & 2 \\ 0 & 1 & -1 & 1 \end{bmatrix}$$

41. No solution: $$\begin{bmatrix} 1 & 2 & 4 \\ 3 & 4 & 5 \\ 2 & 3 & 4 \end{bmatrix} \xrightarrow[(-2)R_1 + R_3]{(-3)R_1 + R_2} \begin{bmatrix} 1 & 2 & 4 \\ 0 & -2 & -7 \\ 0 & -1 & -4 \end{bmatrix} \xrightarrow{(1/2)R_2} \begin{bmatrix} 1 & 2 & 4 \\ 0 & -1 & -7/2 \\ 0 & -1 & -4 \end{bmatrix}$$

43. $(x, y, z) = (z + w + 2, 2z - w - 1, z, w)$ — the final matrix translates to $x - z - w = 2$ and $y - 2z + w = -1$.
$$\begin{bmatrix} 1 & 1 & -3 & 0 & 1 \\ 1 & 0 & -1 & -1 & 2 \\ 2 & 1 & -4 & -1 & 3 \end{bmatrix} \xrightarrow[(-2)R_2 + R_3]{(-1)R_2 + R_1} \begin{bmatrix} 0 & 1 & -2 & 1 & -1 \\ 1 & 0 & -1 & -1 & 2 \\ 0 & 1 & -2 & 1 & -1 \end{bmatrix} \xrightarrow[R_{12}]{(-1)R_1 + R_3} \begin{bmatrix} 1 & 0 & -1 & -1 & 2 \\ 0 & 1 & -2 & 1 & -1 \\ 0 & 0 & 0 & 0 & 0 \end{bmatrix}$$

45. $$\begin{bmatrix} 2 & 5 \\ 1 & -2 \end{bmatrix} \begin{bmatrix} x \\ y \end{bmatrix} = \begin{bmatrix} -3 \\ 1 \end{bmatrix}$$

47. $3x - y = -1$
$2x + 4y = 3$

49. $(x, y) = (-2, 3)$:
$$\begin{bmatrix} x \\ y \end{bmatrix} = \begin{bmatrix} 2 & -3 \\ 4 & 1 \end{bmatrix}^{-1} \begin{bmatrix} -13 \\ -5 \end{bmatrix} = \frac{1}{14} \begin{bmatrix} 1 & 3 \\ -4 & 2 \end{bmatrix} \begin{bmatrix} -13 \\ -5 \end{bmatrix} = \frac{1}{14} \begin{bmatrix} -28 \\ 42 \end{bmatrix} = \begin{bmatrix} -2 \\ 3 \end{bmatrix}.$$

51. $(x, y, z) = (-2, -5, -7)$; $$\begin{bmatrix} x \\ y \\ z \end{bmatrix} = \begin{bmatrix} 2 & -1 & 1 \\ 1 & 2 & -3 \\ 3 & -2 & 1 \end{bmatrix}^{-1} \begin{bmatrix} -6 \\ 9 \\ -3 \end{bmatrix} = \begin{bmatrix} -2 \\ -5 \\ -7 \end{bmatrix}$$

53. $(x, y, z, w) = (-1, 2, -2, 3)$; $$\begin{bmatrix} x \\ y \\ z \\ w \end{bmatrix} = \begin{bmatrix} 2 & -1 & 1 & 1 \\ 1 & 2 & -3 & 1 \\ 3 & -1 & -1 & 2 \\ -2 & 3 & 1 & -3 \end{bmatrix}^{-1} \begin{bmatrix} -3 \\ 12 \\ 3 \\ -3 \end{bmatrix} = \begin{bmatrix} -1 \\ 2 \\ -2 \\ 3 \end{bmatrix}$$

55. $(x, y, z) = (0, -10, 1)$: Solving up from the bottom gives $z = 1$; then $y - 2 = -12$, so $y = -10$; then $2x + 10 = 10$, so $x = 0$.
$$\begin{array}{l} 2x - y = 10 \\ x - z = -1 \\ y + z = -9 \end{array} \Rightarrow 2\mathbf{E}_2 - \mathbf{E}_1 \Rightarrow \begin{array}{l} 2x - y = 10 \\ y - 2z = -12 \\ y + z = -9 \end{array} \Rightarrow \mathbf{E}_3 - \mathbf{E}_2 \Rightarrow \begin{array}{l} 2x - y = 10 \\ y - 2z = -12 \\ 3z = 3 \end{array}$$

57. $(x, y, z, w) = (3, 3, -2, 0)$: Solving up from the bottom gives $w = 0$; then $-z + 0 = 2$, so $z = -2$; then $-3y + 4 = -5$, so $y = 3$; then $x + 6 - 4 = 5$, so $x = 3$.

$$\begin{array}{rl} x + 2y + 2z + w = 5 & \\ 2x + y + 2z = 5 & \Rightarrow E_2 - 2E_1 \Rightarrow \\ 3x + 3y + 3z + 2w = 12 & \Rightarrow E_3 - 3E_1 \Rightarrow \\ x + z + w = 1 & \Rightarrow E_4 - E_1 \Rightarrow \end{array} \quad \begin{array}{rl} x + 2y + 2z + w = 5 & \\ -3y - 2z - 2w = -5 & \\ -3y - 3z - w = -3 & \Rightarrow E_3 - E_2 \Rightarrow \\ -2y - z = -4 & \Rightarrow 3E_4 - 2E_2 \Rightarrow \end{array}$$

$$\begin{array}{rl} x + 2y + 2z + w = 5 & \\ -3y - 2z - 2w = -5 & \\ -z + w = 2 & \\ z + 4w = -2 & \Rightarrow E_4 + E_3 \Rightarrow \end{array} \quad \begin{array}{r} x + 2y + 2z + w = 5 \\ -3y - 2z - 2w = -5 \\ -z + w = 2 \\ 5w = 0 \end{array}$$

59. $(x, y, z) = \left(2 - \frac{3}{2}z, -\frac{1}{2}z - 4, z\right)$: z can be anything; once z is chosen, we have $2y + z = -8$, so $y = -\frac{1}{2}z - 4$; then $x - \left(-\frac{1}{2}z - 4\right) + z = 6$, so $x = 2 - \frac{3}{2}z$

$$\begin{array}{rl} x - y + z = 6 & \\ x + y + 2z = -2 & \Rightarrow E_2 - E_1 \Rightarrow \end{array} \quad \begin{array}{r} x - y + z = 6 \\ 2y + z = -8 \end{array}$$

61. $(x, y, z, w) = (-1 - 2w, w + 1, -w, w)$: w can be anything; once w is chosen, we have $-z - w = 0$, so $z = -w$; then $y - w = 1$, so $y = w + 1$; then $x + (w + 1) + (-w) + 2w = 0$, so $x = -1 - 2w$.

$$\begin{array}{rl} 2x + y + z + 4w = -1 & \Rightarrow E_1 - 2E_3 \Rightarrow \\ x + 2y + z + w = 1 & \Rightarrow E_2 - E_3 \Rightarrow \\ x + y + z + 2w = 0 & \end{array} \quad \begin{array}{rl} -y - z = -1 & \Rightarrow E_1 + E_2 \Rightarrow \\ y - w = 1 & \\ x + y + z + 2w = 0 & \end{array} \quad \begin{array}{r} -z - w = 0 \\ y - w = 1 \\ x + y + z + 2w = 0 \end{array}$$

63. $(x, y, z, w) = (-w - 2, 0.5 - z, z, w)$: z and w can be anything; once they are chosen, we have $-y - z = -0.5$, so $y = 0.5 - z$; then since $y + z = 0.5$ we have $x + 0.5 + w = -1.5$, so $x = -w - 2$.

$$\begin{array}{rl} 2x + y + z + 2w = -3.5 & \Rightarrow E_1 - 2E_2 \Rightarrow \\ x + y + z + w = -1.5 & \end{array} \quad \begin{array}{r} -y - z = -0.5 \\ x + y + z + w = -1.5 \end{array}$$

65. No solution: $E_1 + E_3$ gives $2x + 2y - z + 5w = 3$, which contradicts E_4.

67. $f(x) = 2x^2 - 3x - 2$: We have $f(-1) = a(-1)^2 + b(-1) + c = a - b + c = 3$, $f(1) = a + b + c = -3$, and $f(2) = 4a + 2b + c = 0$. Solving this system gives $(a, b, c) = (2, -3, -2)$.

$$\begin{array}{rl} a - b + c = 3 & \\ a + b + c = -3 & \Rightarrow E_2 - E_1 \Rightarrow \\ 4a + 2b + c = 0 & \Rightarrow E_3 - 4E_1 \Rightarrow \end{array} \quad \begin{array}{rl} a - b + c = 3 & \\ 2b = -6 & \\ 6b - 3c = -12 & \Rightarrow E_3 - 3E_2 \Rightarrow \end{array} \quad \begin{array}{r} a - b + c = 3 \\ 2b \quad = -6 \\ -3c = 6 \end{array}$$

69. $f(x) = (-c - 3)x^2 + x + c$, for any c — or $f(x) = ax^2 + x + (-a - 3)$, for any a: We have $f(-1) = a - b + c = -4$ and $f(1) = a + b + c = -2$. Solving this system gives $(a, b, c) = (-c - 3, 1, c) = (a, 1, -a - 3)$. Note that when $c = -3$ (or $a = 0$), this is simply the line through $(-1, -4)$ and $(1, -2)$.

$$\begin{array}{rl} a - b + c = -4 & \\ a + b + c = -2 & \Rightarrow E_2 - E_1 \Rightarrow \end{array} \quad \begin{array}{r} a - b + c = -4 \\ 2b = 2 \end{array}$$

71. In this problem, the graphs are representative of the population (in thousands) of the cities of Corpus Christi, TX, and Garland, TX, for several years, where x is the number of years past 1980.

(a) The following is a scatter plot of the Corpus Christi data with the linear regression equation $y = 2.0735x + 234.0268$ superimposed on it.

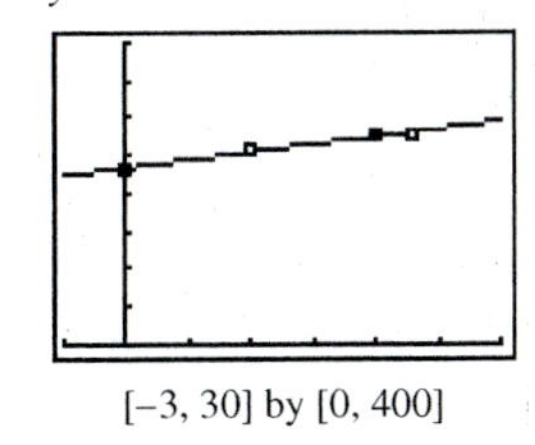

[–3, 30] by [0, 400]

(b) The following is a scatter plot of the Garland data with the linear regression equation $y = 3.5302x + 141.7246$ superimposed on it.

[–3, 30] by [0, 400]

(c) *Graphical solution:* Graph the two linear equations $y = 2.0735x + 234.0268$ and $y = 3.5302x + 141.7246$ on the same axis and find their point of intersection. The two curves intersect at $x \approx 63.4$. So, the population of the two cities will be the same sometime in the year 2043.

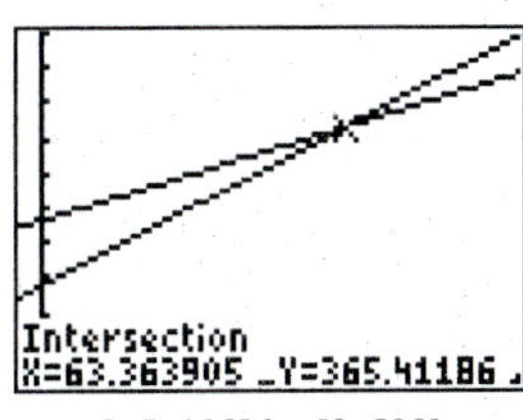

[–5, 100] by [0, 500]

Another graphical solution would be to find where the graph of the difference of the two curves is equal to 0.

Algebraic solution:

Solve $2.0735x + 234.0268 = 3.5302x + 141.7246$ for x.

$$\begin{aligned} 2.0735x + 234.0268 &= 3.5302x + 141.7246 \\ 1.4567x &= 92.3022 \\ x &= \frac{92.3022}{1.4567} \approx 63.4 \end{aligned}$$

The population of the two cities will be the same sometime in the year 2043.

73. $(x, y, z) = (825, 410, 165)$, where x is the number of children, y is the number of adults, and z is the number of senior citizens.

$$\begin{aligned} x + y + z &= 1400 & & & x + y + z &= 1400 \\ 25x + 100y + 75z &= 74{,}000 \Rightarrow \mathbf{E}_2 - 75\mathbf{E}_1 &\Rightarrow & & -50x + 25y &= -31{,}000 \\ x - y - z &= 250 \quad \Rightarrow \mathbf{E}_3 + \mathbf{E}_1 &\Rightarrow & & 2x &= 1650 \end{aligned}$$

75. $(x, y, z) = (14{,}500, 5500, 60{,}000)$ (all amounts in dollars), where x is the amount invested in CDs, y is the amount in bonds, and z is the amount in the growth fund.

$$\begin{aligned} x + y + z &= 80{,}000 & & & x + y + z &= 80{,}000 \\ 0.067x + 0.093y + 0.156z &= 10{,}843 \Rightarrow 1000\mathbf{E}_2 - 67\mathbf{E}_1 &\Rightarrow & & 26y + 89z &= 5{,}483{,}000 \\ 3x + 3y - z &= 0 \quad \Rightarrow \quad \mathbf{E}_3 - 3\mathbf{E}_1 &\Rightarrow & & -4z &= -240{,}000 \end{aligned}$$

77. $(x, y, z) \approx (0, 38{,}983.05, 11{,}016.95)$: If z dollars are invested in the growth fund, then $y = \frac{1}{295}(21{,}250{,}000 - 885z) \approx 72{,}033.898 - 3z$ dollars must be invested in bonds, and $x \approx 2z - 22{,}033.898$ dollars are invested in CDs. Since $x \geq 0$, we see that $z \geq 11016.95$ (approximately); the minimum value of z requires that $x = 0$ (this is logical, since if we wish to minimize z, we should put the rest of our money in bonds, since bonds have a better return than CDs). Then $y \approx 72{,}033.898 - 3z = 38{,}983.05$.

$$\begin{aligned} x + y + z &= 50{,}000 & & x + y + z &= 50{,}000 \\ 0.0575x + 0.087y + 0.146z &= 5000 \Rightarrow 10{,}000\mathbf{E}_2 - 575\mathbf{E}_1 \Rightarrow & & 295y + 885z &= 21{,}250{,}000 \end{aligned}$$

79. 22 nickels, 35 dimes, and 17 quarters:

$$\begin{bmatrix} 1 & 1 & 1 & 74 \\ 5 & 10 & 25 & 885 \\ 1 & -1 & 1 & 4 \end{bmatrix} \xrightarrow[(-1)R_1 + R_3]{(-5)R_1 + R_2} \begin{bmatrix} 1 & 1 & 1 & 74 \\ 0 & 5 & 20 & 515 \\ 0 & -2 & 0 & -70 \end{bmatrix} \xrightarrow[(-1/2)R_2]{R_{23}} \begin{bmatrix} 1 & 1 & 1 & 74 \\ 0 & 1 & 0 & 35 \\ 0 & 5 & 20 & 515 \end{bmatrix} \xrightarrow[(-5)R_2 + R_3]{(-1)R_2 + R_1}$$

$$\begin{bmatrix} 1 & 0 & 1 & 39 \\ 0 & 1 & 0 & 35 \\ 0 & 0 & 20 & 340 \end{bmatrix} \xrightarrow{(1/20)R_3} \begin{bmatrix} 1 & 0 & 1 & 39 \\ 0 & 1 & 0 & 35 \\ 0 & 0 & 1 & 17 \end{bmatrix} \xrightarrow{(-1)R_3 + R_1} \begin{bmatrix} 1 & 0 & 0 & 22 \\ 0 & 1 & 0 & 35 \\ 0 & 0 & 1 & 17 \end{bmatrix}$$

81. $(x, p) = \left(\frac{16}{3}, \frac{220}{3}\right)$: $\begin{bmatrix} x \\ p \end{bmatrix} = \begin{bmatrix} 5 & 1 \\ -10 & 1 \end{bmatrix}^{-1} \begin{bmatrix} 100 \\ 20 \end{bmatrix} = \frac{1}{15}\begin{bmatrix} 1 & -1 \\ 10 & 5 \end{bmatrix}\begin{bmatrix} 100 \\ 20 \end{bmatrix} = \frac{1}{15}\begin{bmatrix} 80 \\ 1100 \end{bmatrix} = \frac{1}{3}\begin{bmatrix} 16 \\ 220 \end{bmatrix}$

83. Adding one row to another is the same as multiplying that first row by 1 and then adding it to the other, so that it falls into the category of the second type of elementary row operations. Also, it corresponds to adding one equation to another in the original system.

85. False. For a nonzero square matrix to have an inverse, the determinant of the matrix must not be equal to zero.

87. $2(3) - (-1)(2) = 8$. The answer is D.

89. Twice the first row was added to the second row. The answer is D.

91. **(a)** The planes can intersect at exactly one point

(b) At least two planes are parallel, or else the line of each pair of intersecting planes is parallel to the third plane.

(c) Two or more planes can coincide, or else all three planes can intersect along a single line.

93. **(a)** $C(x) = (x - 3)(x - 5) - (-1)(-2)$
$= x^2 - 8x + 13.$

(b)

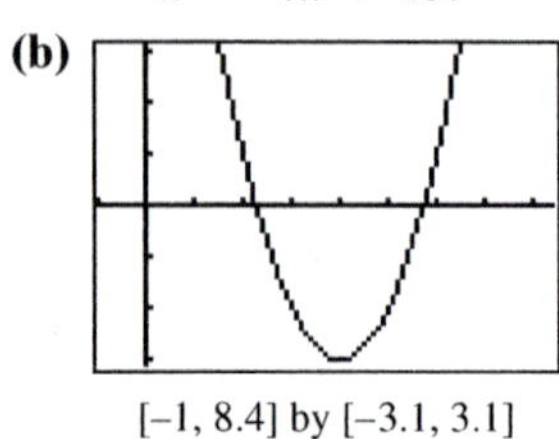

[-1, 8.4] by [-3.1, 3.1]

(c) $C(x) = 0$ when $x = 4 \pm \sqrt{13}$ — approx. 2.27 and 5.73.

(d) $\det A = 13$, and the y-intercept is $(0, 13)$. This is the case because $C(0) = (3)(5) - (1)(2) = \det A$.

(e) $a_{11} + a_{22} = 3 + 5 = 8$. The eigenvalues add to $(4 - \sqrt{13}) + (4 + \sqrt{13}) = 8$, also.

Section 7.4 Partial Fractions

Exploration 1

1. **(a)** $25 - 1 = A_1(5 - 5) + A_2(5 + 3)$
$24 = 8A_2$
$3 = A_2$

(b) $-15 - 1 = A_1(-3 - 5) + A_2(-3 + 3)$
$-16 = -8A_1$
$2 = A_1$

2. **(a)** $-4 + 4 + 4 = A_1(2 - 2)^2 + 2A_2(2 - 2) + 2A_3$
$4 = 2A_3$
$2 = A_3$

(b) $0 + 0 + 4 = A_1(0 - 2)^2 + 0 \cdot A_2(0 - 2) + 0 \cdot A_3$
$4 = 4A_1$
$1 = A_1$

(c) Suppose $x = 3$
$-9 + 6 + 4 = 1 \cdot (3 - 2)^2 + 2 \cdot 3(3 - 2) + 3A_2$
$1 = 1 + 6 + 3A_2$
$-6 = 3A_2$
$-2 = A_2$

Quick Review 7.4

1. $\dfrac{(x - 3) + 2(x - 1)}{(x - 1)(x - 3)} = \dfrac{3x - 5}{(x - 1)(x - 3)} = \dfrac{3x - 5}{x^2 - 4x + 3}$

3. $\dfrac{(x + 1)^2 + 3x(x + 1) + x}{x(x + 1)^2} = \dfrac{4x^2 + 6x + 1}{x(x + 1)^2} = \dfrac{4x^2 + 6x + 1}{x^3 + 2x^2 + x}$

5. Synthetic division by -2:

$$\begin{array}{r|rrrr} -2 & 3 & -6 & -2 & 7 \\ & & -6 & 0 & 4 \\ \hline & 3 & 0 & -2 & 3 \end{array}$$

$$\frac{f(x)}{d(x)} = 3x^2 - 2 + \frac{3}{x-2}$$

7. Possible real rational zeros:
$$\frac{\pm 1,\ \pm 2,\ \pm 3,\ \pm 4,\ \pm 6,\ \pm 12}{\pm 1}$$
From a graph, $x = -1$ and $x = 3$ seem reasonable:

$$\begin{array}{r|rrrrr} 1 & 1 & -2 & 1 & -8 & -12 \\ & & 1 & -3 & 4 & -12 \\ \hline & 1 & -3 & 4 & -12 & 0 \end{array}$$

$$\begin{array}{r|rrrr} -3 & 1 & -3 & 4 & -12 \\ & & -3 & 0 & -12 \\ \hline & 1 & 0 & 4 & 0 \end{array}$$

$$x^4 - 2x^2 + x^2 - 8x + 12 = (x+1)(x-3)(x^2+4)$$

In #9, equate coefficients.

9. $A = 3, B = -1, C = 1$

Section 7.4 Exercises

1. $\dfrac{x^2-7}{x(x^2-4)} = \dfrac{A_1}{x} + \dfrac{A_2}{x-2} + \dfrac{A_3}{x+2}$

3. $\dfrac{x^5 - 2x^4 + x - 1}{x^3(x-1)^2(x^2+9)}$
$$= \frac{A_1}{x} + \frac{A_2}{x^2} + \frac{A_3}{x^3} + \frac{A_4}{x-1} + \frac{A_5}{(x-1)^2} + \frac{B_1x + C_1}{x^2+9}$$

5. $\dfrac{-3}{x+4} + \dfrac{4}{x-2}$: $x + 22 = A(x-2) + B(x+4)$
$= (A+B)x + (-2A+4B)$

$$\begin{aligned} A + B &= 1 \\ -2A + 4B &= 22 \end{aligned} \Rightarrow \begin{bmatrix} A \\ B \end{bmatrix} = \begin{bmatrix} 1 & 1 \\ -2 & 4 \end{bmatrix}^{-1} \begin{bmatrix} 1 \\ 22 \end{bmatrix} = \begin{bmatrix} -3 \\ 4 \end{bmatrix}$$

7. $\dfrac{3}{x^2+1} + \dfrac{2x-1}{(x^2+1)^2}$: $3x^2 + 2x + 2$
$= (Ax+B)(x^2+1) + (Cx+D)$
$= Ax^3 + Bx^2 + (A+C)x + (B+D)$

$$\begin{aligned} A \quad\quad\quad &= 0 \\ B \quad\quad &= 3 \\ A \quad + C \quad &= 2 \\ B \quad + D &= 2 \end{aligned} \Rightarrow \begin{bmatrix} A \\ B \\ C \\ D \end{bmatrix}$$

$$= \begin{bmatrix} 1 & 0 & 0 & 0 \\ 0 & 1 & 0 & 0 \\ 1 & 0 & 1 & 0 \\ 0 & 1 & 0 & 1 \end{bmatrix}^{-1} \begin{bmatrix} 0 \\ 3 \\ 2 \\ 2 \end{bmatrix} = \begin{bmatrix} 0 \\ 3 \\ 2 \\ -1 \end{bmatrix}$$

9. $\dfrac{1}{x-2} + \dfrac{2}{(x-2)^2} + \dfrac{1}{(x-2)^3}$: $x^2 - 2x + 1$
$= A(x-2)^2 + B(x-2) + C$
$= Ax^2 + (-4A+B)x + (4A - 2B + C)$

$$\begin{aligned} A \quad\quad\quad &= 1 \\ -4A + B \quad &= -2 \\ 4A - 2B + C &= 1 \end{aligned} \Rightarrow \begin{bmatrix} 1 & 0 & 0 & 1 \\ -4 & 1 & 0 & -2 \\ 4 & -2 & 1 & 1 \end{bmatrix}$$

Using a grapher, we find that the reduced row echelon form of the augmented matrix is:

$$\begin{bmatrix} 1 & 0 & 0 & 1 \\ 0 & 1 & 0 & 2 \\ 0 & 0 & 1 & 1 \end{bmatrix} \Rightarrow \begin{bmatrix} A \\ B \\ C \end{bmatrix} = \begin{bmatrix} 1 \\ 2 \\ 1 \end{bmatrix}$$

11. $\dfrac{2}{x+3} - \dfrac{1}{(x+3)^2} + \dfrac{3x-1}{x^2+2} + \dfrac{x+2}{(x^2+2)^2}$:
$5x^5 + 22x^4 + 36x^3 + 53x^2 + 71x + 20$
$= A(x+3)(x^2+2)^2 + B(x^2+2)^2$
$+ (Cx+D)(x+3)^2(x^2+2) + (Ex+F)(x+3)^2$
$= (A+C)x^5 + (3A + B + 6C + D)x^4$
$+ (4A + 11C + 6D + E)x^3 + (12A + 4B + 12C$
$+ 11D + 6E + F)x^2 + (4A + 18C + 12D + 9E$
$+ 6F)x + (12A + 4B + 18D + 9F)$

$$\begin{aligned} A \quad\quad + C \quad\quad\quad\quad\quad\quad &= 5 \\ 3A + B + 6C + D \quad\quad\quad &= 22 \\ 4A \quad + 11C + 6D + E \quad &= 36 \\ 12A + 4B + 12C + 11D + 6E + F &= 53 \\ 4A \quad + 18C + 12D + 9E + 6F &= 71 \\ 12A + 4B \quad + 18D \quad + 9F &= 20 \end{aligned} \Rightarrow$$

$$\begin{bmatrix} 1 & 0 & 1 & 0 & 0 & 0 & 5 \\ 3 & 1 & 6 & 1 & 0 & 0 & 22 \\ 4 & 0 & 11 & 6 & 1 & 0 & 36 \\ 12 & 4 & 12 & 11 & 6 & 1 & 53 \\ 4 & 0 & 18 & 12 & 9 & 6 & 71 \\ 12 & 4 & 0 & 18 & 0 & 9 & 20 \end{bmatrix}$$

Using a grapher, we find that the reduced row echelon form of the augmented matrix is:

$$\begin{bmatrix} 1 & 0 & 0 & 0 & 0 & 0 & 2 \\ 0 & 1 & 0 & 0 & 0 & 0 & -1 \\ 0 & 0 & 1 & 0 & 0 & 0 & 3 \\ 0 & 0 & 0 & 1 & 0 & 0 & -1 \\ 0 & 0 & 0 & 0 & 1 & 0 & 1 \\ 0 & 0 & 0 & 0 & 0 & 1 & 2 \end{bmatrix} \Rightarrow \begin{bmatrix} A \\ B \\ C \\ D \\ E \\ F \end{bmatrix} = \begin{bmatrix} 2 \\ -1 \\ 3 \\ -1 \\ 1 \\ 2 \end{bmatrix}$$

13. $\dfrac{2}{(x-5)(x-3)} = \dfrac{A_1}{x-5} + \dfrac{A_2}{x-3}$, so $2 = A_1(x-3) + A_2(x-5)$. With $x = 5$, we see that $2 = 2A_1$, so $A_1 = 1$; with $x = 3$ we have $2 = -2A_2$, so $A_2 = -1$: $\dfrac{1}{x-5} - \dfrac{1}{x-3}$.

15. $\dfrac{4}{x^2-1} = \dfrac{A_1}{x-1} + \dfrac{A_2}{x+1}$, so $4 = A_1(x+1) + A_2(x-1)$. With $x = 1$, we see that $4 = 2A_1$, so $A_1 = 2$; with $x = -1$ we have $4 = -2A_2$, so $A_2 = -2$: $\dfrac{2}{x-1} - \dfrac{2}{x+1}$.

17. $\dfrac{1}{x^2+2x} = \dfrac{A_1}{x} + \dfrac{A_2}{x+2}$, so $1 = A_1(x+2) + A_2x$.

With $x = 0$, we see that $1 = 2A_1$, so $A_1 = \dfrac{1}{2}$; with $x = -2$, we have $1 = -2A_2$, so $A_2 = -\dfrac{1}{2}$:

$$\frac{1/2}{x} - \frac{1/2}{x+2} = \frac{1}{2x} - \frac{1/2}{x+2}.$$

19. $\dfrac{-x+10}{x^2+x-12} = \dfrac{A_1}{x-3} + \dfrac{A_2}{x+4}$, so $-x + 10$
$= A_1(x+4) + A_2(x-3)$. With $x = 3$, we see that $7 = 7A_1$, so $A_1 = 1$; with $x = -4$ we have $14 = -7A_2$, so $A_2 = -2$: $\dfrac{1}{x-3} - \dfrac{2}{x+4}$.

21. $\dfrac{x+17}{2x^2+5x-3} = \dfrac{A_1}{x+3} + \dfrac{A_2}{2x-1}$, so $x + 17$ $= A_1(2x-1) + A_2(x+3)$. With $x = -3$, we see that $14 = -7A_1$, so $A_1 = -2$; with $x = \dfrac{1}{2}$ we have $\dfrac{35}{2} = \dfrac{7}{2}A_2$, so $A_2 = 5$: $\dfrac{-2}{x+3} + \dfrac{5}{2x-1}$.

23. $\dfrac{2x^2+5}{(x^2+1)^2} = \dfrac{B_1x+C_1}{x^2+1} + \dfrac{B_2x+C_2}{(x^2+1)^2}$, so $2x^2+5$ $= (B_1x+C_1)(x^2+1) + B_2x + C_2$. Expanding the right side leaves $2x^2+5 = B_1x^3 + C_1x^2$ $+ (B_1+B_2)x + C_1 + C_2$; equating coefficients reveals that $B_1 = 0, C_1 = 2, B_1 + B_2 = 0$, and $C_1 + C_2 = 5$. This means that $B_2 = 0$ and $C_2 = 3$: $\dfrac{2}{x^2+1} + \dfrac{3}{(x^2+1)^2}$.

25. The denominator factors into $x(x-1)^2$, so $\dfrac{x^2-x+2}{x^3-2x^2+x} = \dfrac{A_1}{x} + \dfrac{A_2}{x-1} + \dfrac{A_3}{(x-1)^2}$. Then $x^2-x+2 = A_1(x-1)^2 + A_2x(x-1) + A_3x$. With $x = 0$, we have $2 = A_1$; with $x = 1$, we have $2 = A_3$; with $x = 2$, we have $4 = A_1 + 2A_2 + 2A_3 = 2 + 2A_2 + 4$, so $A_2 = -1$: $\dfrac{2}{x} - \dfrac{1}{x-1} + \dfrac{2}{(x-1)^2}$.

27. $\dfrac{3x^2-4x+3}{x^3-3x^2} = \dfrac{A_1}{x} + \dfrac{A_2}{x^2} + \dfrac{A_3}{x-3}$. Then $3x^2-4x+3 = A_1x(x-3) + A_2(x-3) + A_3x^2$. With $x = 0$, we have $3 = -3A_2$, so $A_2 = -1$; with $x = 3$, we have $18 = 9A_3$, so $A_3 = 2$; with $x = 1$, we have $2 = -2A_1 - 2A_2 + A_3 = -2A_1 + 2 + 2$, so $A_1 = 1$: $\dfrac{1}{x} - \dfrac{1}{x^2} + \dfrac{2}{x-3}$.

29. $\dfrac{2x^3+4x-1}{(x^2+2)^2} = \dfrac{B_1x+C_1}{x^2+2} + \dfrac{B_2x+C_2}{(x^2+2)^2}$. Then $2x^3+4x-1 = (B_1x+C_1)(x^2+2) + B_2x + C_2$. Expanding the right side and equating coefficients reveals that $B_1 = 2, C_1 = 0, 2B_1 + B_2 = 4$, and $2C_1 + C_2 = -1$. This means than $B_2 = 0$ and $C_2 = -1$: $\dfrac{2x}{x^2+2} - \dfrac{1}{(x^2+2)^2}$.

31. The denominator factors into $(x-1)(x^2+x+1)$, so $\dfrac{x^2+3x+2}{x^3-1} = \dfrac{A}{x-1} + \dfrac{Bx+C}{x^2+x+1}$. Then $x^2+3x+2 = A(x^2+x+1) + (Bx+C)(x-1)$. With $x = 1$, we have $6 = 3A$, so $A = 2$; with $x = 0$, $2 = A - C$, so $C = 0$. Finally, with $x = 2$, we have $12 = 7A + 2B$, so $B = -1$: $\dfrac{2}{x-1} - \dfrac{x}{x^2+x+1}$.

In #33–35, find the quotient and remainder via long division or other methods (note in particular that if the degree of the numerator and denominator are the same, the quotient is the ratio of the leading coefficients). Use the usual methods to find the partial fraction decomposition.

33. $\dfrac{2x^2+x+3}{x^2-1} = 2 + \dfrac{x+5}{x^2-1}$; $\dfrac{r(x)}{h(x)} = \dfrac{x+5}{x^2-1}$ $= \dfrac{A_1}{x-1} + \dfrac{A_2}{x+1}$, so $x + 5 = A_1(x+1)$ $+ A_2(x-1)$. With $x = 1$ and $x = -1$ (respectively), we find that $A_1 = 3$ and $A_2 = -2$: $\dfrac{3}{x-1} - \dfrac{2}{x+1}$.

Graph of $\dfrac{2x^2+x+3}{x^2-1}$:

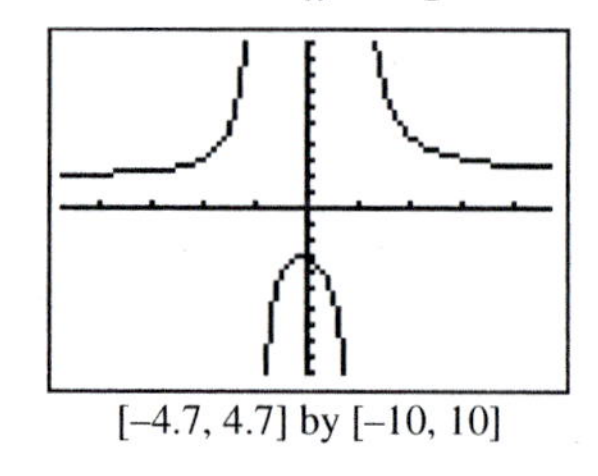

[–4.7, 4.7] by [–10, 10]

Graph of $y = 2$:

[–4.7, 4.7] by [–10, 10]

Graph of $\dfrac{3}{x-1}$:

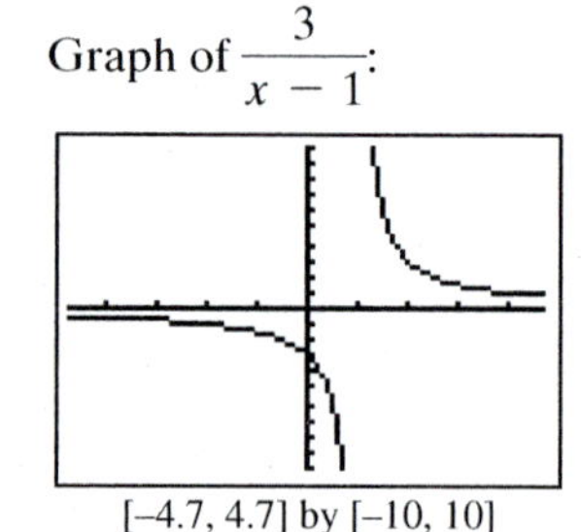

[–4.7, 4.7] by [–10, 10]

Graph of $-\dfrac{2}{x+1}$:

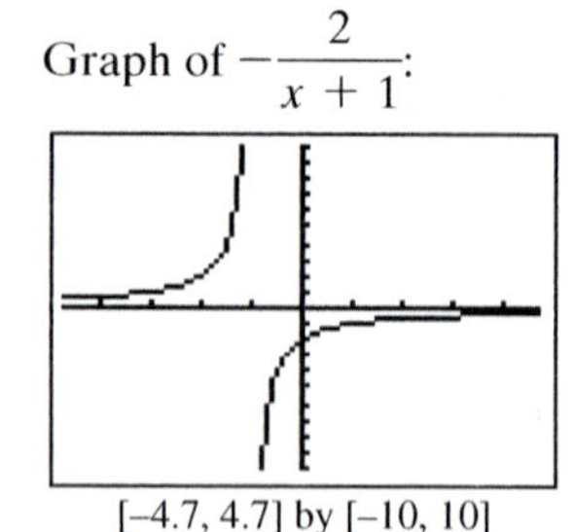

[–4.7, 4.7] by [–10, 10]

35. $\frac{x^3-2}{x^2+x} = x - 1 + \frac{x-2}{x^2+x}; \frac{r(x)}{h(x)} = \frac{x-2}{x^2+x}$

$= \frac{A_1}{x+1} + \frac{A_2}{x}$, so $x - 2 = A_1x + A_2(x+1)$. With $x = -1$ and $x = 0$ (respectively), we find that $A_1 = 3$ and $A_2 = -2$: $\frac{3}{x+1} - \frac{2}{x}$.

Graph of $y = \frac{x^3-2}{x^2+x}$:

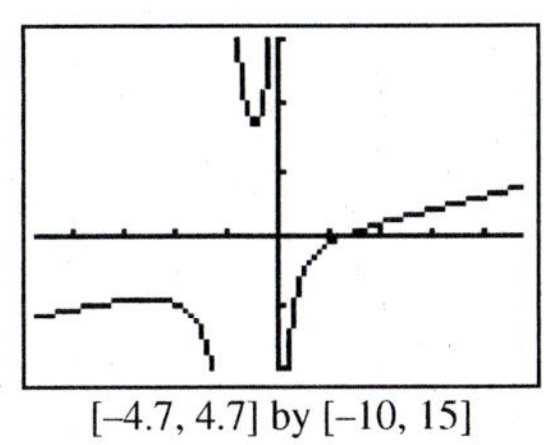

[–4.7, 4.7] by [–10, 15]

Graph of $y = x - 1$:

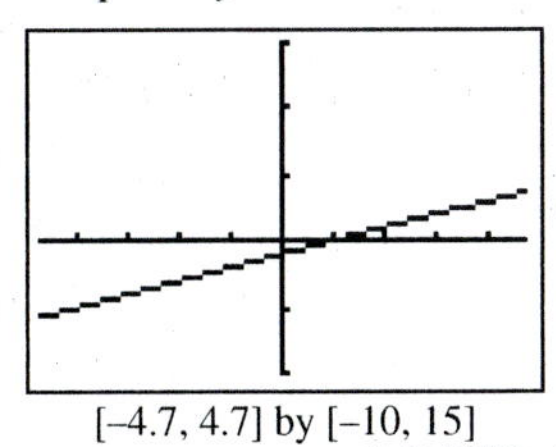

[–4.7, 4.7] by [–10, 15]

Graph of $y = \frac{3}{x+1}$:

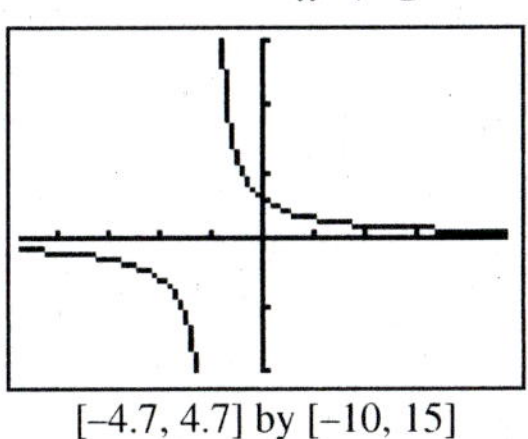

[–4.7, 4.7] by [–10, 15]

Graph of $y = -\frac{2}{x}$:

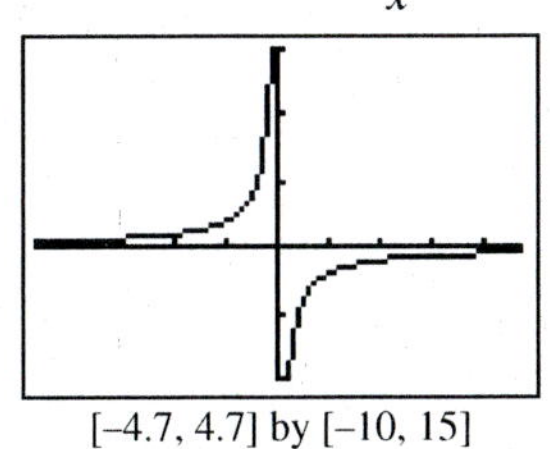

[–4.7, 4.7] by [–10, 15]

37. (c) **39.** (d) **41.** (a)

43. $\frac{-1}{ax} + \frac{1}{a(x-a)}$:

$1 = A(x-a) + Bx = (A+B)x - aA$

$A + B = 0$

$-aA = 1$

Since $A = -\frac{1}{a}, -\frac{1}{a} + B = 0, B = \frac{1}{a}$

$$\begin{bmatrix} A \\ B \end{bmatrix} = \begin{bmatrix} -\frac{1}{a} \\ \frac{1}{a} \end{bmatrix}$$

45. $\frac{-3}{(b-a)(x-a)} + \frac{3}{(b-a)(x-b)}$:

$3 = A(x-b) + B(x-a) = (A+B)x + (-bA - aB)$

$$\begin{matrix} A + B = 0 \\ -bA - aB = 3 \end{matrix} \Rightarrow \begin{bmatrix} 1 & 1 & 0 \\ -b & -a & 3 \end{bmatrix} \Rightarrow$$

$$\begin{bmatrix} 1 & 1 & 0 \\ 0 & b-a & 3 \end{bmatrix} \Rightarrow \begin{bmatrix} 1 & 1 & 0 \\ 0 & 1 & \frac{3}{b-a} \end{bmatrix} \Rightarrow$$

$$\begin{bmatrix} 1 & 0 & \frac{-3}{b-a} \\ 0 & 1 & \frac{3}{b-a} \end{bmatrix} \Rightarrow \begin{bmatrix} A \\ B \end{bmatrix} = \begin{bmatrix} \frac{-3}{b-a} \\ \frac{3}{b-a} \end{bmatrix}$$

47. True. The behavior of $f(x)$ near $x = 3$ is the same as the behavior of $y = \frac{1}{x-3}$, and $\lim_{x \to 3^-} \frac{1}{x-3} = -\infty$.

49. The denominator factor x^2 calls for the terms $\frac{A_1}{x}$ and $\frac{A_2}{x^2}$ in the partial fraction decomposition, and the denominator factor $x^2 + 2$ calls for the term $\frac{B_1x + C_1}{x^2+2}$.

The answer is E.

51. The y-intercept is -1, and because the denominators are both of degree 1, the expression changes sign at each asymptote. The answer is B.

53. (a) $x = 1$: $1 + 4 + 1 = A(1+1) + (B+C)(0)$

$6 = 2A$

$A = 3$

(b) $x = i$: $-1 + 4i + 1 = 3(-1+1) + (Bi + C)(i - 1)$

$4i = (Bi + C)(i - 1)$

$4i = -B - Bi + Ci - C$

$$\begin{matrix} -B - C = 0 \\ -Bi + Ci = 4i \end{matrix} \Rightarrow \begin{matrix} B + C = 0 \\ B - C = -4 \end{matrix} \Rightarrow$$

$$\begin{bmatrix} 1 & 1 & 0 \\ 1 & -1 & -4 \end{bmatrix} \Rightarrow \begin{bmatrix} 1 & 1 & 0 \\ 0 & -2 & -4 \end{bmatrix} \Rightarrow \begin{bmatrix} 1 & 1 & 0 \\ 0 & 1 & 2 \end{bmatrix} \Rightarrow$$

$$\begin{bmatrix} 1 & 0 & -2 \\ 0 & 1 & 2 \end{bmatrix} \Rightarrow \begin{bmatrix} B \\ C \end{bmatrix} = \begin{bmatrix} -2 \\ 2 \end{bmatrix}$$

$x = -i$

$-1 - 4i + 1 = A(-1 - 1) + (-Bi + C)(-i - 1)$

$-4i = -B + Bi - Ci - C$, which is the same as above.

$B = -2, C = 2$

55. $y = \frac{b}{(x-1)^2}$ has a greater effect on $f(x)$ at $x = 1$.

Section 7.5 Systems of Inequalities in Two Variables

Quick Review 7.5

1. x-intercept: $(3, 0)$; y-intercept: $(0, -2)$

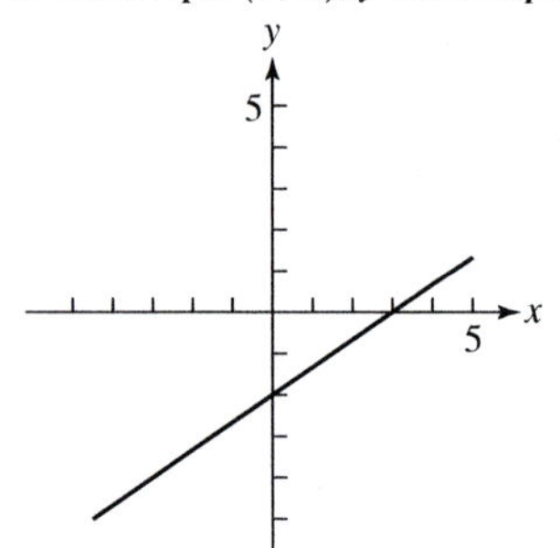

3. x-intercept: $(20, 0)$; y-intercept: $(0, 50)$

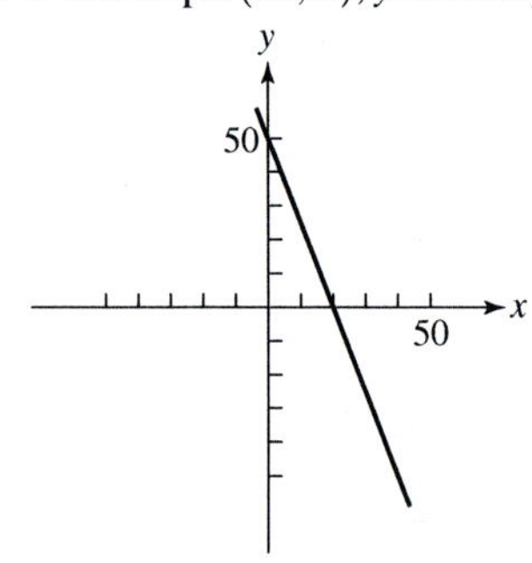

For #5–9, a variety of methods could be used. One is shown.

5. $\begin{bmatrix} 4 & 1 & 180 \\ 1 & 1 & 90 \end{bmatrix} \Rightarrow \begin{bmatrix} 1 & \frac{1}{4} & 45 \\ 0 & 1 & 60 \end{bmatrix} \Rightarrow \begin{bmatrix} 1 & 0 & 30 \\ 0 & 1 & 60 \end{bmatrix} \Rightarrow \begin{bmatrix} x \\ y \end{bmatrix} = \begin{bmatrix} 30 \\ 60 \end{bmatrix}$

7. $\begin{bmatrix} 4 & 1 & 180 \\ 10 & 5 & 800 \end{bmatrix} \Rightarrow \begin{bmatrix} 1 & \frac{1}{4} & 45 \\ 0 & 1 & 140 \end{bmatrix} \Rightarrow \begin{bmatrix} 1 & 0 & 10 \\ 0 & 1 & 140 \end{bmatrix} \Rightarrow \begin{bmatrix} x \\ y \end{bmatrix} = \begin{bmatrix} 10 \\ 140 \end{bmatrix}$

9. $\begin{bmatrix} 1 & 1 & 6 \\ 2 & 8 & 30 \end{bmatrix} \Rightarrow \begin{bmatrix} 1 & 1 & 6 \\ 0 & 1 & 3 \end{bmatrix} \Rightarrow \begin{bmatrix} 1 & 0 & 3 \\ 0 & 1 & 3 \end{bmatrix} \Rightarrow \begin{bmatrix} x \\ y \end{bmatrix} = \begin{bmatrix} 3 \\ 3 \end{bmatrix}$

Section 7.5 Exercises

1. Graph (c); boundary included

3. Graph (b); boundary included

5. Graph (e); boundary included

7.

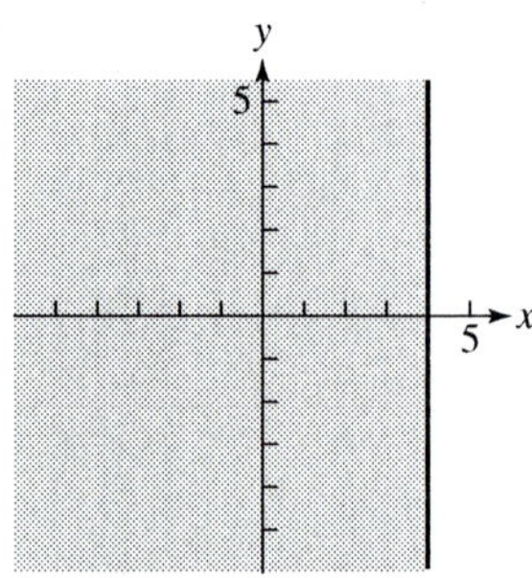

boundary line $x = 4$ included

9.

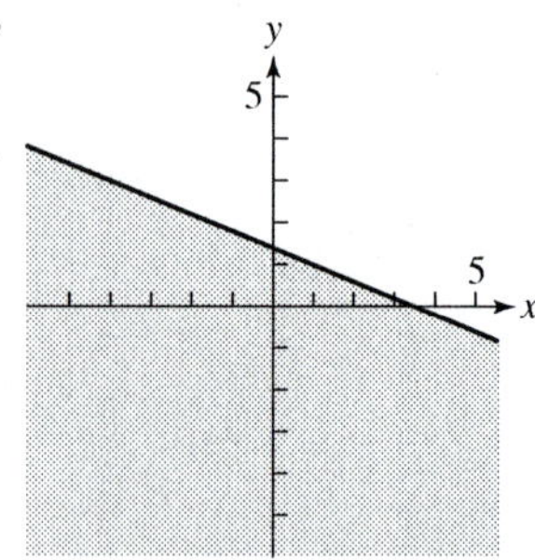

boundary line $2x + 5y = 7$ included

11.

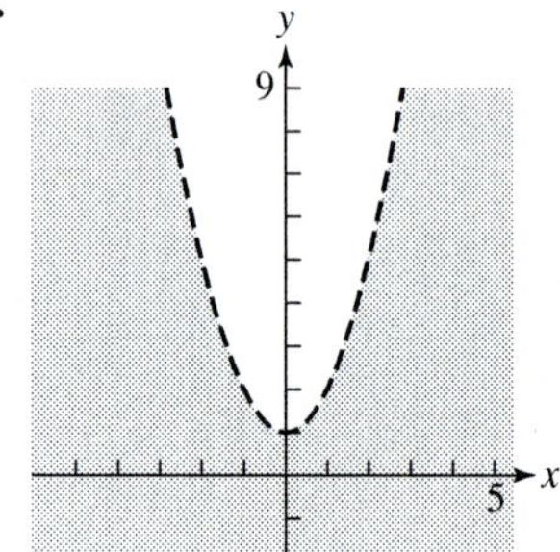

boundary curve $y = x^2 + 1$ excluded

13.

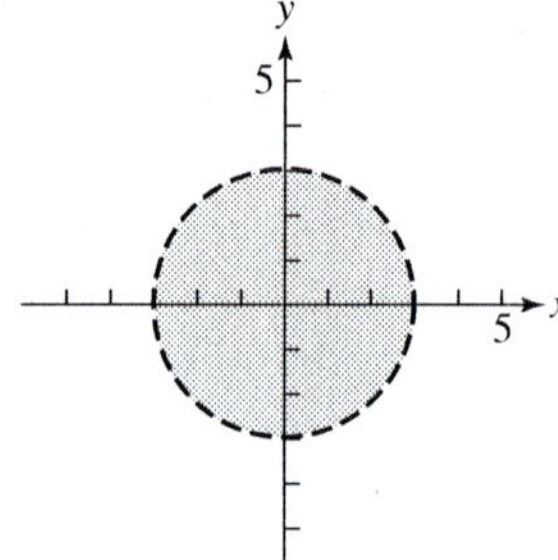

boundary circle $x^2 + y^2 = 9$ excluded

15.

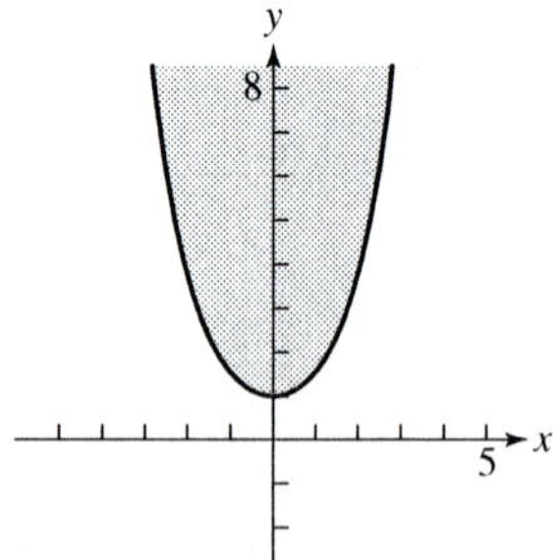

boundary curve $y = \dfrac{e^x + e^{-x}}{2}$ included

17.

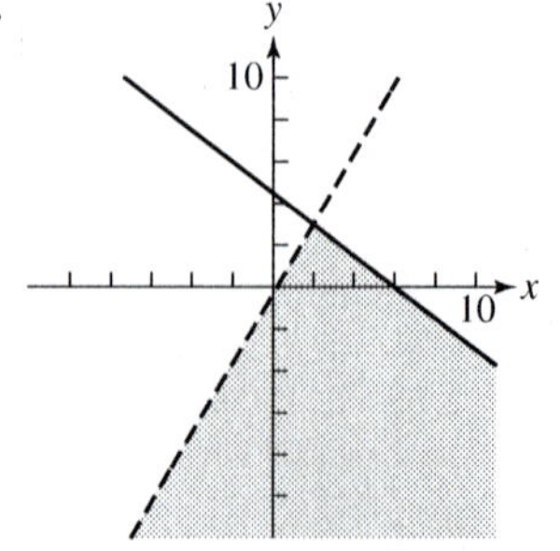

Corner at $(2, 3)$. Left boundary is excluded, the other is included.

19.

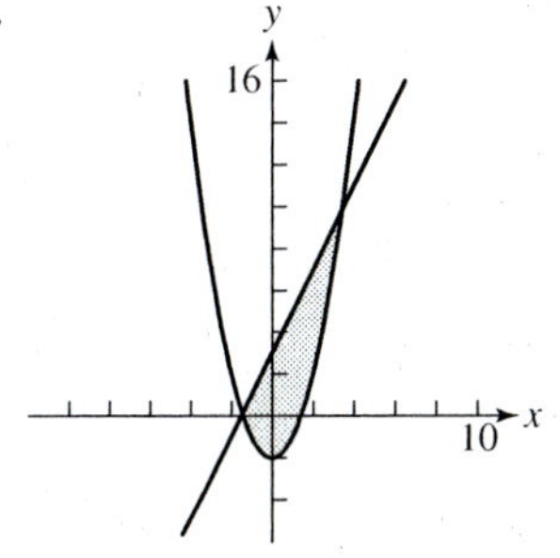

Corners at about $(-1.45, 0.10)$ and $(3.45, 9.90)$. Boundaries included.

21.

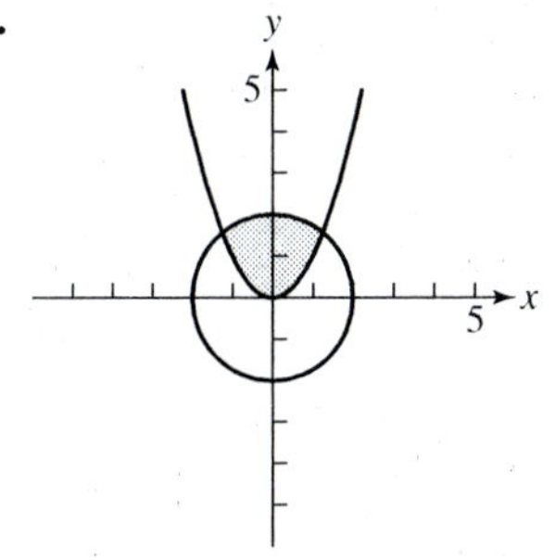

Corners at about $(\pm 1.25, 1.56)$ Boundaries included.

23.

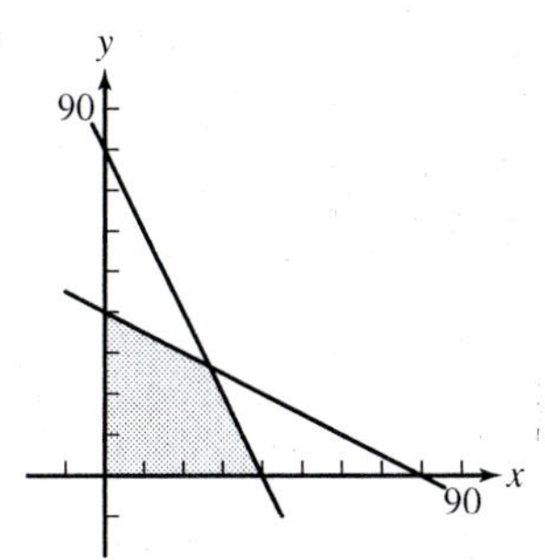

Corners at $(0, 40)$, $(26.7, 26.7)$, $(0, 0)$, and $(40, 0)$. Boundaries included.

25.

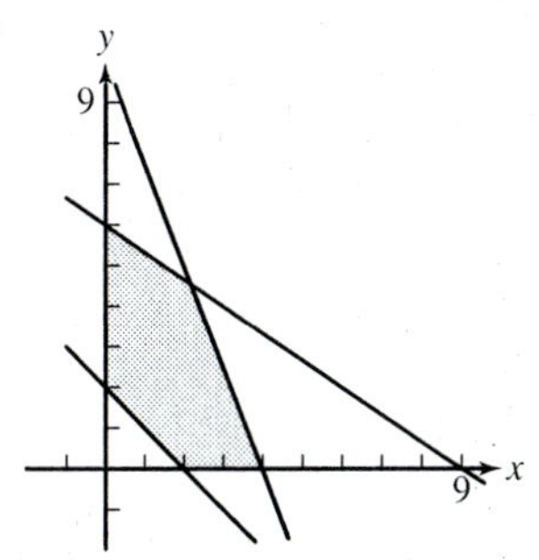

Corners at $(0, 2)$, $(0, 6)$, $(2.18, 4.55)$, $(4, 0)$, and $(2, 0)$. Boundaries included.

27. $x^2 + y^2 \le 4$

$y \ge -x^2 + 1$

For #29, first we must find the equations of the lines—then the inequalities.

29. line 1: $m = \dfrac{\Delta y}{\Delta x} = \dfrac{(5-3)}{(0-4)} = \dfrac{2}{-4} = \dfrac{-1}{2}$, $y = \dfrac{-1}{2}x + 5$

line 2: $m = \dfrac{\Delta y}{\Delta x} = \dfrac{(0-3)}{(6-4)} = \dfrac{-3}{2}$,

$(y - 0) = \dfrac{-3}{2}(x - 6)$, $y = \dfrac{-3}{2}x + 9$

line 3: $x = 0$

line 4: $y = 0$

$y \le \dfrac{-1}{2}x + 5$

$y \le \dfrac{-3}{2}x + 9$

$x \ge 0$

$y \ge 0$

For #31–35, use your grapher to determine the feasible area, and then solve for the corner points graphically or algebraically. Evaluate $f(x)$ at the corner points to determine maximum and minimum values.

31.

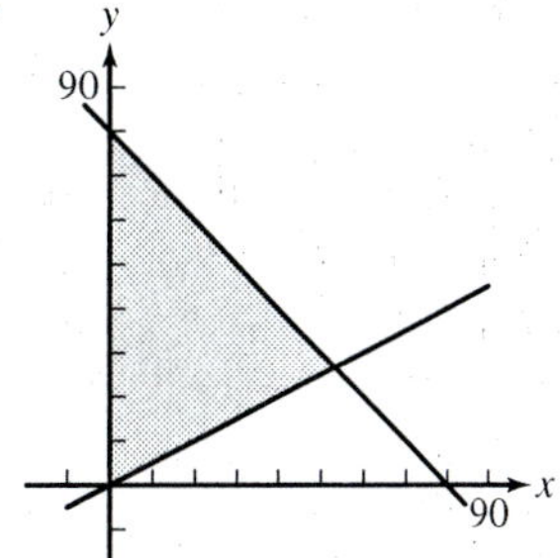

Corner points: $(0, 0)$

$(0, 80)$, the y-intercept of $x + y = 80$

$\left(\dfrac{160}{3}, \dfrac{80}{3}\right)$, the intersection of $x + y = 80$ and $x - 2y = 0$

(x, y)	$(0, 0)$	$(0, 80)$	$\left(\dfrac{160}{3}, \dfrac{80}{3}\right)$
f	0	240	$\dfrac{880}{3} \approx 293.33$

$f_{\text{min}} = 0$ [at $(0, 0)$]; $f_{\text{max}} \approx 293.33 \left[\text{at} \left(\dfrac{160}{3}, \dfrac{80}{3}\right)\right]$

33.

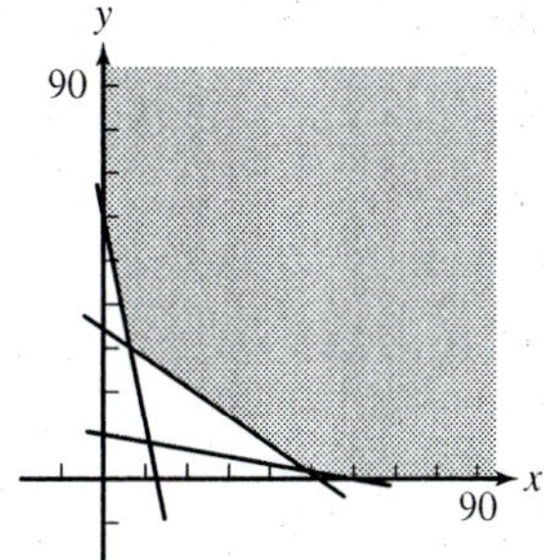

Corner points: $(0, 60)$ y-intercept of $5x + y = 60$

$(6, 30)$ intersection of $5x + y = 60$ and $4x + 6y = 204$

$(48, 2)$ intersection of $4x + 6y = 204$ and $x + 6y = 60$

$(60, 0)$ x-intercept of $x + 6y = 60$

(x, y)	$(0, 60)$	$(6, 30)$	$(48, 2)$	$(60, 0)$
f	240	162	344	420

$f_{\text{min}} = 162$ [at $(6, 30)$]; f_{max} = none (region is unbounded)

35.

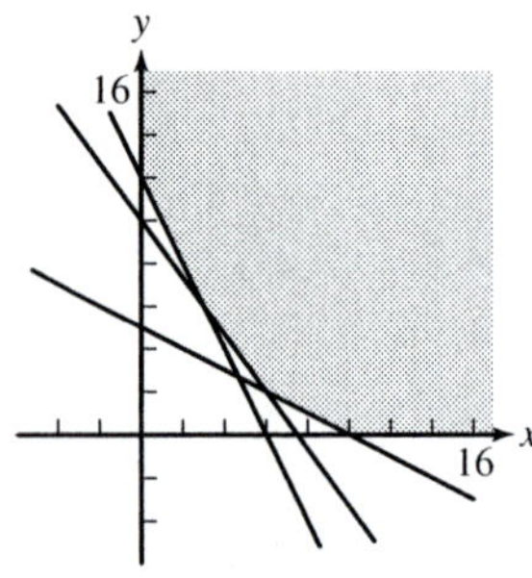

Corner points: (0, 12) y-intercept of $2x + y = 12$
(3, 6) intersection of $2x + y = 12$ and $4x + 3y = 30$
(6, 2) intersection of $4x + 3y = 30$ and $x + 2y = 10$
(10, 0) x-intercept of $x + 2y = 10$

(x, y)	(0, 12)	(3, 6)	(6, 2)	(10, 0)
f	24	27	34	50

$f_{\min} = 24$ [at (0, 12)]; $f_{\max} =$ none (region is unbounded)

For #37–39, first set up the equations; then solve.

37. Let $x =$ number of tons of ore R
$y =$ number of tons of ore S
$C =$ total cost $= 50x + 60y$, the objective function
$80x + 140y \geq 4000$ At least 4000 lb of mineral A
$160x + 50y \geq 3200$ At least 3200 lb of mineral B
$x \geq 0, y \geq 0$
The region of feasible points is the intersection of $80x + 140y \geq 4000$ and $160x + 50y \geq 3200$ in the first quadrant. The region has three corner points: (0, 64), (13.48, 20.87), and (50, 0). $C_{\min} = \$1{,}926.20$ when 13.48 tons of ore R and 20.87 tons of ore S are processed.

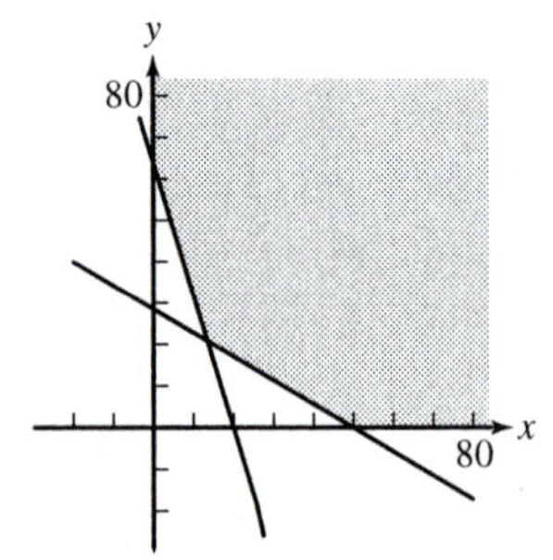

39. Let $x =$ number of operations performed by Refinery 1
$y =$ number of operations performed by Refinery 2
$C =$ total cost $= 300x + 600y$, the objective function
$x + y \geq 100$ At least 100 units of grade A
$2x + 4y \geq 320$ At least 320 units of grade B
$x + 4y \geq 200$ At least 200 units of grade C
$x \geq 0, y \geq 0$
The region of feasible points is the intersection of $x + y \geq 100, 2x + 4y \geq 320$, and $x + 4y \geq 200$ in the first quadrant. The corners are (0, 100), (40, 60), (120, 20), and (200, 0). $C_{\min} = \$48{,}000$, which can be obtained by using Refinery 1 to perform 40 operations and Refinery 2 to perform 60 operations, or using Refinery 1 to perform 120 operations and Refinery 2 to perform 20 operations, or any other combination of x and y such that $2x + 4y = 320$ with $40 \leq x \leq 120$.

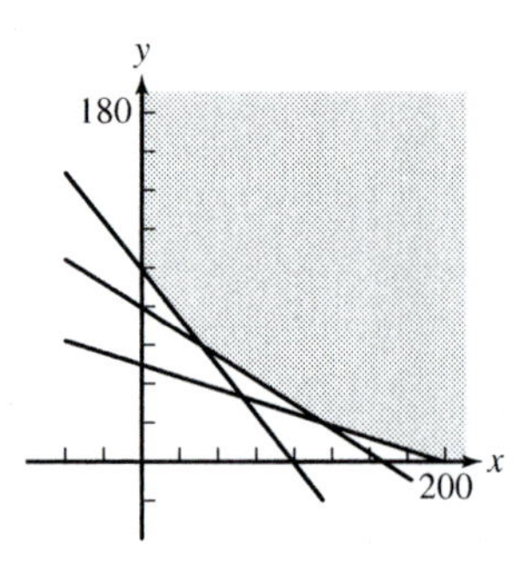

41. False. The graph is a half-plane.

43. The graph of $3x + 4y \geq 5$ is Regions I and II plus the boundary. The graph of $2x - 3y \leq 4$ is Regions I and IV plus the boundary. And the intersection of the regions is the graph of the system. The answer is A.

45. (3, 4) fails to satisfy $x + 3y \leq 12$. The answer is D.

47. (a) One possible answer: Two lines are parallel if they have exactly the same slope. Let l_1 be $5x + 8y = a$ and l_2 be $5x + 8y = b$. Then l_1 becomes $y = \frac{-5}{8} + \frac{a}{8}$ and l_2 becomes $y = \frac{-5}{8}x + \frac{b}{8}$. Since $M_{l1} = \frac{-5}{8} = M_{l2}$, the lines are parallel.

(b) One possible answer: If two lines are parallel, then a line l_2 going through the point (0, 10) will be further away from the origin then a line l_1 going through the point (0, 5). In this case f_1 could be expressed as $mx + 5$ and f_2 could be expressed as $mx + 10$. Thus, l is moving further away from the origin as f increases.

(c) One possible answer: The region is bounded and includes all its boundary points.

49. $4x^2 + 9y^2 = 36$
$9y^2 = 36 - 4x^2$
$y^2 = 4 - \frac{4}{9}x^2$
$y_1 = \sqrt{4 - \frac{4}{9}x^2} = 2\sqrt{1 - \frac{x^2}{9}}$
$y_2 = -\sqrt{4 - \frac{4}{9}x^2} = -2\sqrt{1 - \frac{x^2}{9}}$

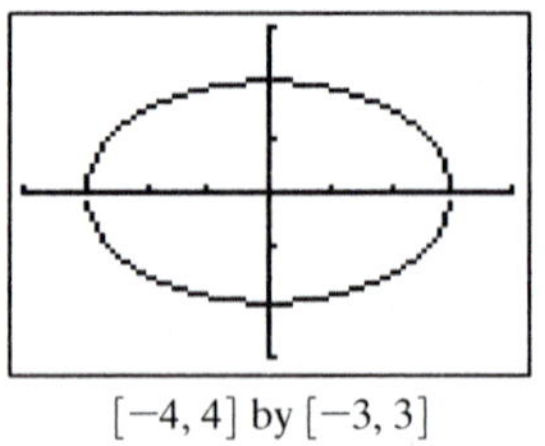

[−4, 4] by [−3, 3]

51. $4x^2 + 9y^2 \leq 36$
$9y^2 \leq 36 - 4x^2$
$y^2 \leq \frac{36 - 4x^2}{9}$
$y_1 \leq \sqrt{\frac{36 - 4x^2}{9}}$
$y_2 \geq -\sqrt{\frac{36 - 4x^2}{9}}$
$y_3 \geq x^2 - 1$

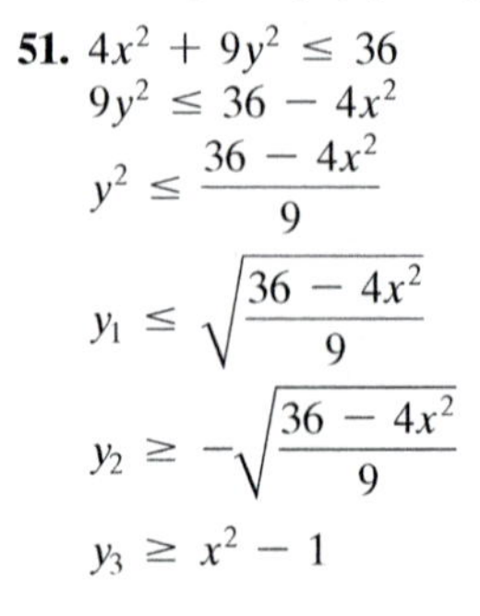

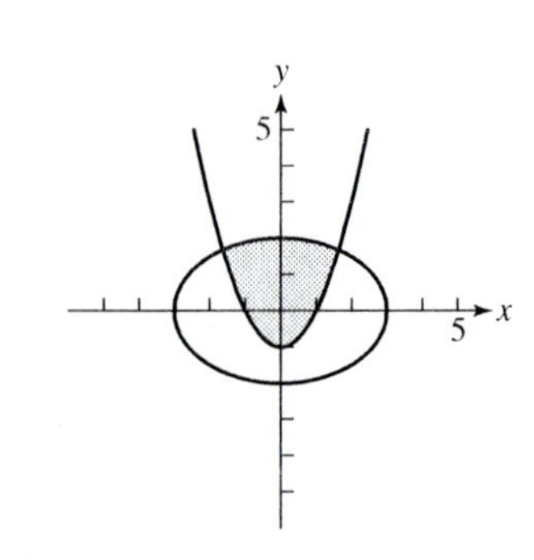

■ Chapter 7 Review

1. (a) $\begin{bmatrix} 1 & 2 \\ 8 & 3 \end{bmatrix}$ **(b)** $\begin{bmatrix} -3 & 4 \\ 0 & -3 \end{bmatrix}$ **(c)** $\begin{bmatrix} 2 & -6 \\ -8 & 0 \end{bmatrix}$ **(d)** $\begin{bmatrix} -7 & 11 \\ 4 & -6 \end{bmatrix}$

3. $AB = \begin{bmatrix} (-1)(3) + (4)(0) & (-1)(-1) + (4)(-2) & (-1)(5) + (4)(4) \\ (0)(3) + (6)(0) & (0)(-1) + (6)(-2) & (0)(5) + (6)(4) \end{bmatrix} = \begin{bmatrix} -3 & -7 & 11 \\ 0 & -12 & 24 \end{bmatrix}$; BA is not possible.

5. $AB = [(-1)(5) + (4)(2) \quad (-1)(-3) + (4)(1)] = [3 \ \ 7]$; BA is not possible.

7. $$AB = \begin{bmatrix} (0)(2) + (1)(1) + (0)(-2) & (0)(-3) + (1)(2) + (0)(1) & (0)(4) + (1)(-3) + (0)(-1) \\ (1)(2) + (0)(1) + (0)(-2) & (1)(-3) + (0)(2) + (0)(1) & (1)(4) + (0)(-3) + (0)(-1) \\ (0)(2) + (0)(1) + (1)(-2) & (0)(-3) + (0)(2) + (1)(1) & (0)(4) + (0)(-3) + (1)(-1) \end{bmatrix} = \begin{bmatrix} 1 & 2 & -3 \\ 2 & -3 & 4 \\ -2 & 1 & -1 \end{bmatrix}$$

$$BA = \begin{bmatrix} (2)(0) + (-3)(1) + (4)(0) & (2)(1) + (-3)(0) + (4)(0) & (2)(0) + (-3)(0) + (4)(1) \\ (1)(0) + (2)(1) + (-3)(0) & (1)(1) + (2)(0) + (-3)(0) & (1)(0) + (2)(0) + (-3)(1)] \\ (-2)(0) + (1)(1) + (-1)(0) & (-2)(1) + (1)(0) + (-1)(0) & (-2)(0) + (1)(0) + (-1)(1) \end{bmatrix} = \begin{bmatrix} -3 & 2 & 4 \\ 2 & 1 & -3 \\ 1 & -2 & -1 \end{bmatrix}$$

9. Carry out the multiplication of AB and BA and confirm that both products equal I_4.

11. Using a calculator:

$$\begin{bmatrix} 1 & 2 & 0 & -1 \\ 2 & -1 & 1 & 2 \\ 2 & 0 & 1 & 2 \\ -1 & 1 & 1 & 4 \end{bmatrix}^{-1} = \begin{bmatrix} -2 & -5 & 6 & -1 \\ 0 & -1 & 1 & 0 \\ 10 & 24 & -27 & 4 \\ -3 & -7 & 8 & -1 \end{bmatrix}$$

13. $$\begin{aligned} \det &= \begin{vmatrix} 1 & -3 & 2 \\ 2 & 4 & -1 \\ -2 & 0 & 1 \end{vmatrix} \\ &= (-2)(-1)^4 \begin{vmatrix} -3 & 2 \\ 4 & -1 \end{vmatrix} + 0 + (1)(-1)^6 \begin{vmatrix} 1 & -3 \\ 2 & 4 \end{vmatrix} \\ &= -2(3 - 8) + (4 - (-6)) \\ &= 10 + 10 \\ &= 20 \end{aligned}$$

For #15–17, one possible sequence of row operations is shown.

15. $$\begin{bmatrix} 1 & 0 & 2 \\ 3 & 1 & 5 \\ 1 & -1 & 3 \end{bmatrix} \xrightarrow[(-1)R_1 + R_3]{(-3)R_1 + R_2} \begin{bmatrix} 1 & 0 & 2 \\ 0 & 1 & -1 \\ 0 & -1 & 1 \end{bmatrix} \xrightarrow{(1)R_2 + R_3} \begin{bmatrix} 1 & 0 & 2 \\ 0 & 1 & -1 \\ 0 & 0 & 0 \end{bmatrix}$$

17. $$\begin{bmatrix} 1 & 2 & 3 & 1 \\ 2 & 3 & 3 & -2 \\ 1 & 2 & 4 & 6 \end{bmatrix} \xrightarrow[(-1)R_1 + R_3]{(-2)R_1 + R_2} \begin{bmatrix} 1 & 2 & 3 & 1 \\ 0 & -1 & -3 & -4 \\ 0 & 0 & 1 & 5 \end{bmatrix} \xrightarrow[(3)R_3 + R_2]{(2)R_2 + R_1} \begin{bmatrix} 1 & 0 & -3 & -7 \\ 0 & -1 & 0 & 11 \\ 0 & 0 & 1 & 5 \end{bmatrix} \xrightarrow[(3)R_3 + R_1]{(-1)R_2} \begin{bmatrix} 1 & 0 & 0 & 8 \\ 0 & 1 & 0 & -11 \\ 0 & 0 & 1 & 5 \end{bmatrix}$$

For #19–21, use any of the methods of this chapter. Solving for x (or y) and substituting is probably easiest for these systems.

19. $(x, y) = (1, 2)$: From $\mathbf{E}_1$, $y = 3x - 1$; substituting in $\mathbf{E}_2$ gives $x + 2(3x - 1) = 5$. Then $7x = 7$, so $x = 1$. Finally, $y = 2$.

21. No solution: From $\mathbf{E}_1$, $x = 1 - 2y$; substituting in $\mathbf{E}_2$ gives $4y - 4 = -2(1 - 2y)$, or $4y - 4 = 4y - 2$ — which is impossible.

23. $(x, y, z, w) = (2 - z - w, w + 1, z, w)$: Note that the last equation in the triangular system is not useful. z and w can be anything, then $y = w + 1$ and $x = 2 - z - w$.

$$\begin{aligned} x + z + w &= 2 \\ x + y + z &= 3 \Rightarrow \mathbf{E}_2 - \mathbf{E}_1 \Rightarrow \\ 3x + 2y + 3z + w &= 8 \Rightarrow \mathbf{E}_3 - 3\mathbf{E}_1 \Rightarrow \end{aligned} \qquad \begin{aligned} x + z + w &= 2 \\ y - w &= 1 \\ 2y - 2w &= 2 \Rightarrow \mathbf{E}_3 - 2\mathbf{E}_2 \Rightarrow \end{aligned} \qquad \begin{aligned} x + z + w &= 2 \\ y - w &= 1 \\ 0 &= 0 \end{aligned}$$

25. No solution: $\mathbf{E}_1$ and $\mathbf{E}_3$ are inconsistent.

$$\begin{aligned} x + y - 2z &= 2 \\ 3x - y + z &= 4 \\ -2x - 2y + 4z &= 6 \Rightarrow \mathbf{E}_3 + 2\mathbf{E}_1 \Rightarrow \end{aligned} \qquad \begin{aligned} x + y - 2z &= 2 \\ 3x - y + z &= 4 \\ 0 &= 10 \end{aligned}$$

27. $(x, y, z, w) = (1 - 2z + w, 2 + z - w, z, w)$: Note that the last two equations in the triangular system give no additional information. z and w can be anything, then $y = 2 + z - w$ and $x = 13 - 6(2 + z - w) + 4z - 5w = 1 - 2z + w$.

$$\begin{array}{llllll} -x - 6y + 4z - 5w = -13 & & -x - 6y + 4z - 5w = -13 & & & -x - 6y + 4z - 5w = -13 \\ 2x + y + 3z - w = 4 & \Rightarrow \mathbf{E}_2 + 2\mathbf{E}_1 \Rightarrow & -11y + 11z - 11w = -22 & \Rightarrow -\frac{1}{11}\mathbf{E}_2 \Rightarrow & & y - z + w = 2 \\ 2x + 2y + 2z = 6 & \Rightarrow \mathbf{E}_3 + 2\mathbf{E}_1 \Rightarrow & -10y + 10z - 10w = -20 & \Rightarrow -\frac{1}{10}\mathbf{E}_3 \Rightarrow & & y - z + w = 2 \\ -x - 3y + z - 2w = -7 & \Rightarrow \mathbf{E}_4 - \mathbf{E}_1 \Rightarrow & 3y - 3z + 3w = 6 & \Rightarrow \frac{1}{3}\mathbf{E}_4 \Rightarrow & & y - z + w = 2 \end{array}$$

29. $(x, y, z) = \left(\frac{9}{4}, -\frac{3}{4}, -\frac{7}{4}\right)$: $\begin{bmatrix} x \\ y \\ z \end{bmatrix} = \begin{bmatrix} 1 & 2 & 1 \\ 1 & -3 & 2 \\ 2 & -3 & 1 \end{bmatrix}^{-1} \begin{bmatrix} -1 \\ 1 \\ 5 \end{bmatrix} = \frac{1}{12}\begin{bmatrix} 3 & -5 & 7 \\ 3 & -1 & -1 \\ 3 & 7 & -5 \end{bmatrix} \begin{bmatrix} -1 \\ 1 \\ 5 \end{bmatrix}$.

31. There is no inverse, since the coefficient matrix, shown on the right, has determinant 0 (found with a calculator). Note that this does not necessarily mean there is no solution — there may be infinitely many solutions. However, by other means one can determine that there is no solution in this case.

$$\begin{bmatrix} 2 & 1 & 1 & -1 \\ 2 & -1 & 1 & -1 \\ -1 & 1 & -1 & 1 \\ 1 & -2 & 1 & -1 \end{bmatrix}$$

33. $(x, y, z, w) = (2 - w, z + 3, z, w)$ — z and w can be anything:

$$\begin{bmatrix} 1 & 2 & -2 & 1 & 8 \\ 2 & 3 & -3 & 2 & 13 \end{bmatrix} \xrightarrow{(-2)R_1 + R_2} \begin{bmatrix} 1 & 2 & -2 & 1 & 8 \\ 0 & -1 & 1 & 0 & -3 \end{bmatrix} \xrightarrow[(-1)R_2]{(2)R_2 + R_1} \begin{bmatrix} 1 & 0 & 0 & 1 & 2 \\ 0 & 1 & -1 & 0 & 3 \end{bmatrix}$$

35. $(x, y, z, w) = (-2, 1, 3, -1)$: $\begin{bmatrix} 1 & 2 & 4 & 6 & 6 \\ 3 & 4 & 8 & 11 & 11 \\ 2 & 4 & 7 & 11 & 10 \\ 3 & 5 & 10 & 14 & 15 \end{bmatrix} \xrightarrow[R_{24}]{(-2)R_1 + R_3} \begin{bmatrix} 1 & 2 & 4 & 6 & 6 \\ 3 & 5 & 10 & 14 & 15 \\ 0 & 0 & -1 & -1 & -2 \\ 3 & 4 & 8 & 11 & 11 \end{bmatrix} \xrightarrow[(-1)R_3]{(-1)R_4 + R_2}$

$$\begin{bmatrix} 1 & 2 & 4 & 6 & 6 \\ 0 & 1 & 2 & 3 & 4 \\ 0 & 0 & 1 & 1 & 2 \\ 3 & 4 & 8 & 11 & 11 \end{bmatrix} \xrightarrow[(-4)R_2 + R_4]{(-2)R_2 + R_1} \begin{bmatrix} 1 & 0 & 0 & 0 & -2 \\ 0 & 1 & 2 & 3 & 4 \\ 0 & 0 & 1 & 1 & 2 \\ 3 & 0 & 0 & -1 & -5 \end{bmatrix} \xrightarrow[(-2)R_3 + R_2]{(-3)R_1 + R_4} \begin{bmatrix} 1 & 0 & 0 & 0 & -2 \\ 0 & 1 & 0 & 1 & 0 \\ 0 & 0 & 1 & 1 & 2 \\ 0 & 0 & 0 & -1 & 1 \end{bmatrix}$$

$$\xrightarrow[(1)R_4 + R_3]{(1)R_4 + R_2} \begin{bmatrix} 1 & 0 & 0 & 0 & -2 \\ 0 & 1 & 0 & 0 & 1 \\ 0 & 0 & 1 & 0 & 3 \\ 0 & 0 & 0 & -1 & 1 \end{bmatrix} \xrightarrow{(-1)R_4} \begin{bmatrix} 1 & 0 & 0 & 0 & -2 \\ 0 & 1 & 0 & 0 & 1 \\ 0 & 0 & 1 & 0 & 3 \\ 0 & 0 & 0 & 1 & -1 \end{bmatrix}$$

37. $(x, p) \approx (7.57, 42.71)$: Solve $100 - x^2 = 20 + 3x$ to give $x \approx 7.57$ (the other solution, $x \approx -10.57$, makes no sense in this problem). Then $p = 20 + 3x \approx 42.71$.

39. $(x,y) \approx (0.14, -2.29)$

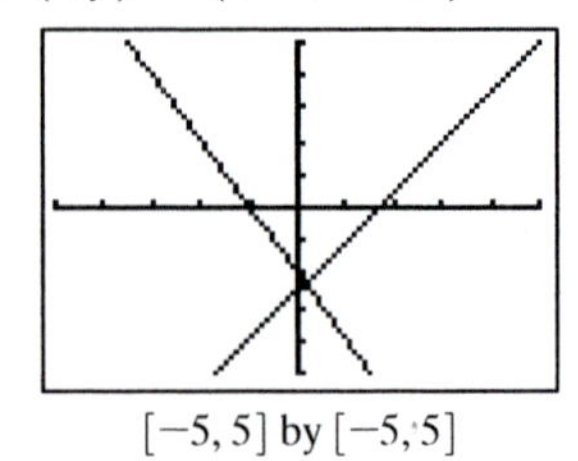

$[-5, 5]$ by $[-5, 5]$

41. $(x, y) = (-2, 1)$ or $(x, y) = (2, 1)$

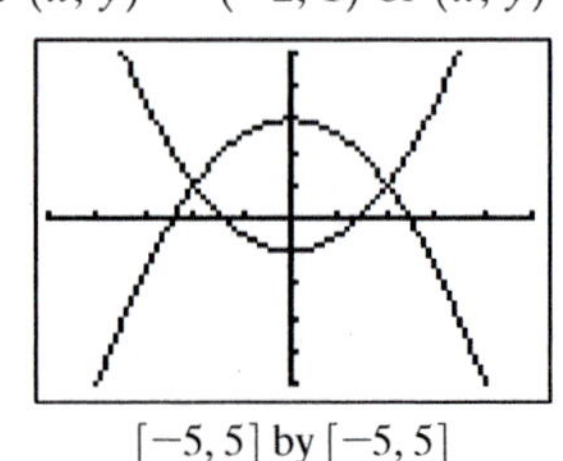

$[-5, 5]$ by $[-5, 5]$

43. $(x, y) \approx (2.27, 1.53)$

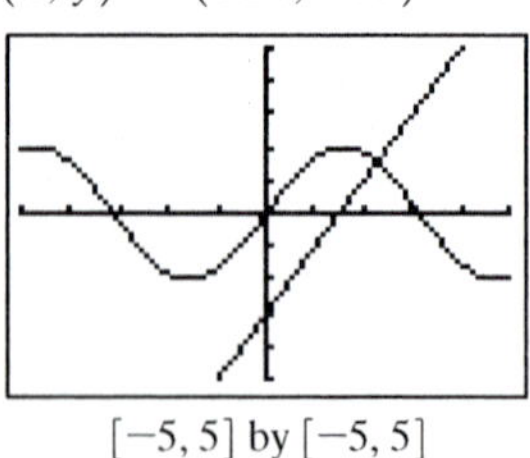

$[-5, 5]$ by $[-5, 5]$

45. $(a, b, c, d) = \left(\frac{17}{840}, -\frac{33}{280}, -\frac{571}{420}, \frac{386}{35}\right)$ $= (0.020\ldots, -0.117\ldots, -1.359\ldots, 11.028\ldots)$. In matrix form, the system is as shown below. Use a calculator to find the inverse matrix and multiply.

$$\begin{bmatrix} 8 & 4 & 2 & 1 \\ 64 & 16 & 4 & 1 \\ 216 & 36 & 6 & 1 \\ 729 & 81 & 9 & 1 \end{bmatrix} \begin{bmatrix} a \\ b \\ c \\ d \end{bmatrix} = \begin{bmatrix} 8 \\ 5 \\ 3 \\ 4 \end{bmatrix}$$

47. $\dfrac{3x-2}{x^2-3x-4} = \dfrac{A_1}{x+1} + \dfrac{A_2}{x-4}$, so $3x - 2$ $= A_1(x-4) + A_2(x+1)$. With $x = -1$, we see that $-5 = -5A_1$, so $A_1 = 1$; with $x = 4$ we have $10 = 5A_2$, so $A_2 = 2$: $\dfrac{1}{x+1} + \dfrac{2}{x-4}$.

49. The denominator factors into $(x+2)(x+1)^2$, so $\dfrac{3x+5}{x^3+4x^2+5x+2} = \dfrac{A_1}{x+2} + \dfrac{A_2}{x+1} + \dfrac{A_3}{(x+1)^2}$. Then $3x + 5 = A_1(x+1)^2 + A_2(x+2)(x+1)$ $+ A_3(x+2)$. With $x = -2$, we have $-1 = A_1$; with $x = -1$, we have $2 = A_3$; with $x = 0$, we have 5 $= A_1 + 2A_2 + 2A_3 = -1 + 2A_2 + 4$, so that $A_2 = 1$: $-\dfrac{1}{x+2} + \dfrac{1}{x+1} + \dfrac{2}{(x+1)^2}$.

51. The denominator factors into $(x+1)(x^2+1)$, so $\dfrac{5x^2-x-2}{x^3+x^2+x+1} = \dfrac{A}{x+1} + \dfrac{Bx+C}{x^2+1}$. Then $5x^2 - x - 2 = A(x^2+1) + (Bx+C)(x+1)$. With $x = -1$, we have $4 = 2A_1$, so $A = 2$; with $x = 0$, $-2 = A + C$, so $C = -4$. Finally, with $x = 1$ we have $2 = 2A + (B+C)(2) = 4 + 2B - 8$, so that $B = 3$: $\dfrac{2}{x+1} + \dfrac{3x-4}{x^2+1}$

53. (c)

55. (b)

57.

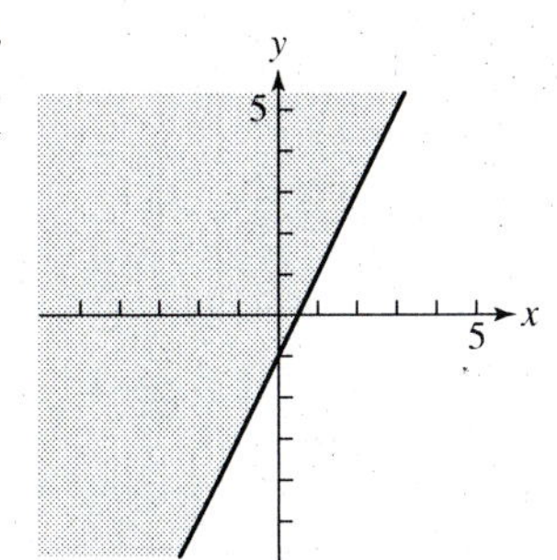

59. Corner points: $(0, 90)$, $(90, 0)$, $\left(\dfrac{360}{13}, \dfrac{360}{13}\right)$. Boundaries included.

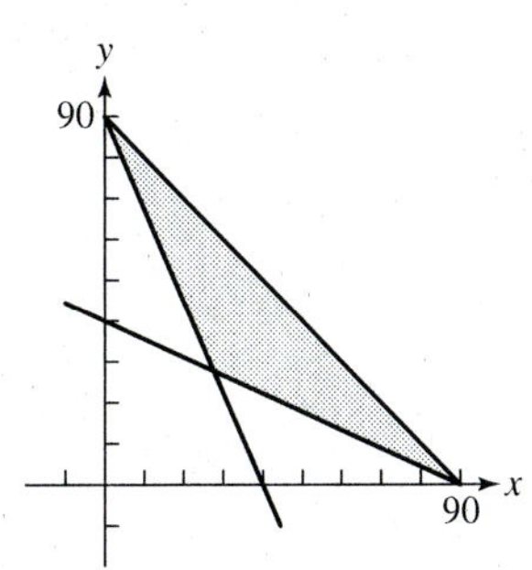

61. Corner points: approx. $(0.92, 2.31)$ and $(5.41, 3.80)$. Boundaries excluded.

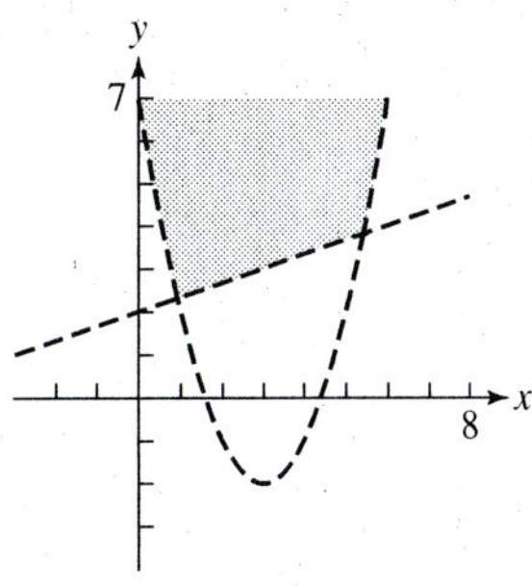

63. Corner points: approx. $(-1.25, 1.56)$ and $(1.25, 1.56)$. Boundaries included.

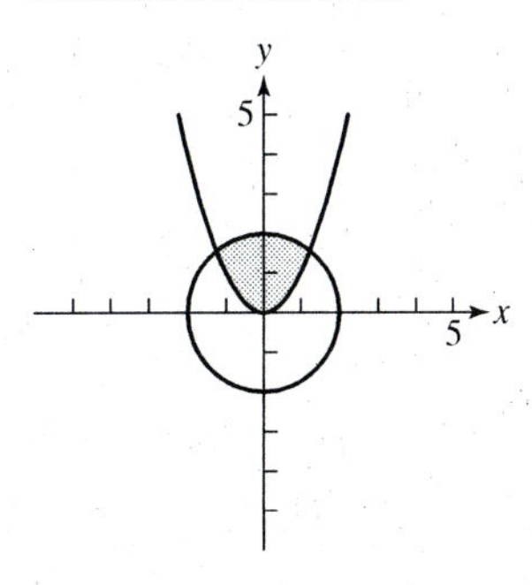

65. Corner points: $(0, 20)$, $(25, 0)$, and $(10, 6)$.

(x, y)	$(0, 20)$	$(10, 6)$	$(25, 0)$
f	120	106	175

$f_{\min} = 106$ [at $(10, 6)$]; $f_{\max} =$ none (unbounded)

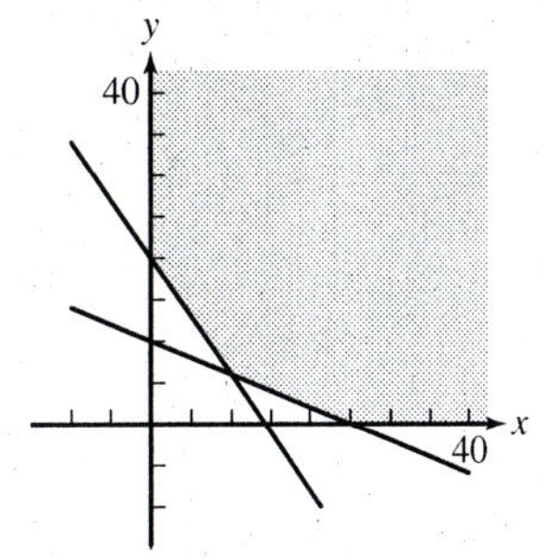

67. Corner points: $(4, 40)$, $(10, 25)$, and $(70, 10)$.

(x, y)	$(4, 40)$	$(10, 25)$	$(70, 10)$
f	292	205	280

$f_{\min} = 205$ [at $(10, 25)$]; $f_{\max} = 292$ [at $(4, 40)$]

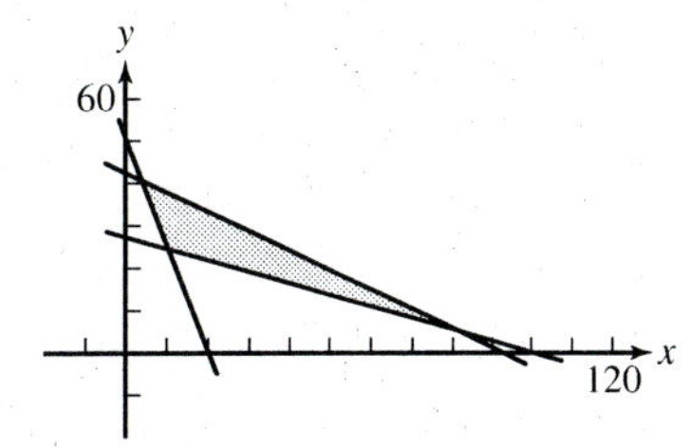

69. (a) $[1 \;\; 2]\begin{bmatrix} \cos 45° & -\sin 45° \\ \sin 45° & \cos 45° \end{bmatrix} \approx \begin{bmatrix} 2.12 \\ 0.71 \end{bmatrix}$

(b) $[1 \;\; 2]\begin{bmatrix} \cos 45° & \sin 45° \\ -\sin 45° & \cos 45° \end{bmatrix} \approx \begin{bmatrix} -0.71 \\ 2.12 \end{bmatrix}$

71. In this problem, the graphs are representative of the population (in thousands) of the states of Hawaii and Idaho for several years, where x is the number of years past 1980.

(a) The following is a scatter plot of the Hawaii data with the linear regression equation $y = 12.2614x + 979.5909$ superimposed on it.

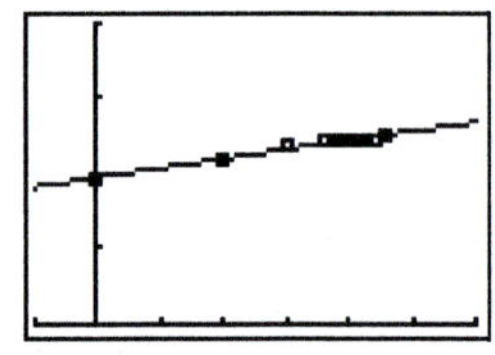

$[-5, 30]$ by $[0, 2000]$

(b) The following is a scatter plot of the Idaho data with the linear regression equation $y = 19.8270x + 893.9566$ superimposed on it.

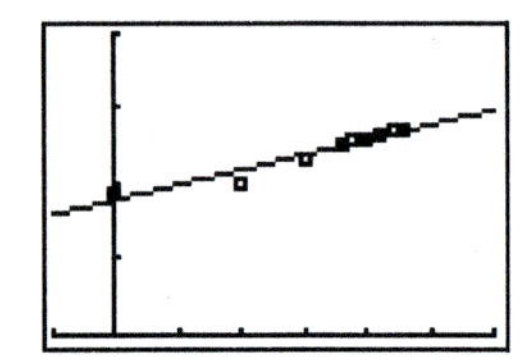

$[-5, 30]$ by $[0, 2000]$

(c) *Graphical solution:* Graph the two linear equations $y = 12.2614x + 979.5909$ and $y = 19.8270x + 893.9566$ on the same axis and find their point of intersection. The two curves intersect at $x \approx 11.3$.
The population of the two states will be the same sometime in the year 1991.

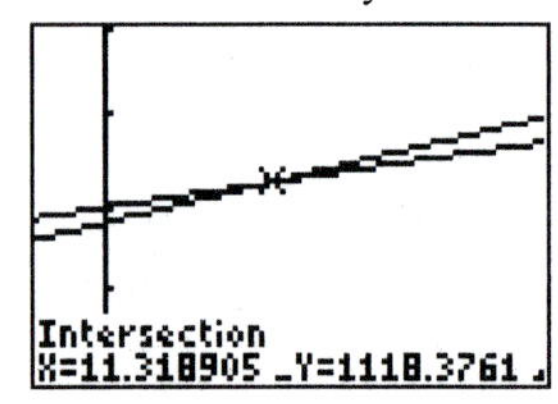

$[-5, 30]$ by $[0, 2000]$

Another graphical solution would be to find where the graph of the difference of the two curves is equal to 0.

Algebraic solution: Solve
$12.2614x + 979.5909 = 19.8270x + 893.9566$ for x.

$$12.2614x + 979.5909 = 19.8270x + 893.9566$$
$$7.5656x = 85.6343$$
$$x = \frac{85.6343}{7.5656} \approx 11.3$$

The population of the two states will be the same sometime in the year 1991.

73. (a) $N = [200 \quad 400 \quad 600 \quad 250]$

(b) $P = [\$80 \quad \$120 \quad \$200 \quad \$300]$

(c) $$NP^T = [200 \quad 400 \quad 600 \quad 250]\begin{bmatrix} \$\,80 \\ \$120 \\ \$200 \\ \$300 \end{bmatrix} = \$259{,}000$$

75. Let x be the number of vans, y be the number of small trucks, and z be the number of large trucks needed. The requirements of the problem are summarized above (along with the requirements that each of x, y, and z must be a non-negative integer).

$$8x + 15y + 22 \geq 115$$
$$3x + 10y + 20z \geq 85$$
$$2x + 6y + 5z \geq 35$$

The methods of this chapter do not allow complete solution of this problem. Solving this system of *inequalities* as if it were a system of *equations* gives $(x, y, z) = (1.77, 3.30, 2.34)$, which suggests the answer $(x, y, z) = (2, 4, 3)$; one can easily check that $(x, y, z) = (2, 4, 2)$ actually works, as does $(1, 3, 3)$. The first of these solutions requires 8 vehicles, while the second requires only 7. There are a number of other seven-vehicle answers (these can be found by trial and error): Use no vans, anywhere from 0 to 5 small trucks, and the rest should be large trucks — that is, (x, y, z) should be one of $(0, 0, 7)$, $(0, 1, 6)$, $(0, 2, 5)$, $(0, 3, 4)$, $(0, 4, 3)$, or $(0, 5, 2)$.

77. $(x, y, z) = (160000, 170000, 320000)$, where x is the amount borrowed at 4%, y is the amount borrowed at 6.5%, and z is the amount borrowed at 9%. Solve the system below.

$$\begin{aligned} x + y + z &= 650{,}000 \\ 0.04x + 0.065y + 0.09z &= 46{,}250 \\ 2x - z &= 0 \end{aligned}$$

One method to solve the system is to solve using Gaussian elimination:
Multiply equation 1 by -0.065 and add the result to equation 2, replacing equation 2:

$$\begin{aligned} x + y + z &= 650{,}000 \\ -0.025x + 0.025z &= 4000 \\ 2x - z &= 0 \end{aligned}$$

Divide equation 2 by 0.025 to simplify:

$$\begin{aligned} x + y + z &= 650{,}000 \\ -x + z &= 160{,}000 \\ 2x - z &= 0 \end{aligned}$$

Now add equation 2 to equation 3, replacing equation 3:

$$\begin{aligned} x + y + z &= 650{,}000 \\ -x + z &= 160{,}000 \\ x &= 160{,}000 \end{aligned}$$

Substitute $x = 160{,}000$ into equation 2 to solve for z: $z = 320{,}000$. Substitute these values into equation 1 to solve for y: $y = 170{,}000$.

79. Pipe A: 15 hours. Pipe B: $\frac{60}{11} \approx 5.45$ hours (about 5 hours 27.3 minutes). Pipe C: 12 hours. If x is the portion of the pool that A can fill in one hour, y is the portion that B fills in one hour, and z is the portion that C fills in one hour, then solving the system above gives

$$\begin{aligned} x + y + z &= 1/3 \\ x + y &= 1/4 \\ y + z &= 1/3.75 \end{aligned}$$

$$(x, y, z) = \left(\frac{1}{15}, \frac{11}{60}, \frac{1}{12}\right)$$

One method to solve the system is to use elimination: Subtract equation 2 from equation 1:

$$x + y + z = 1/3$$
$$z = 1/12$$
$$y + z = 4/15 \quad \text{(convert 1/3.75 to simpler form)}$$

Subtract equation 2 from equation 3:

$$x + y + z = 1/3$$
$$z = 1/12$$
$$y = 11/60$$

Substitute the values for y and z into equation 1 to solve for x: $x = 1/15$.

81. $n = p$ — the number of columns in A is the same as the number of rows in B.

Chapter 7 Project

1. The graphs are representative of the male and female population in the United States from 1990 to 2004, where x is the number of years after 1990.

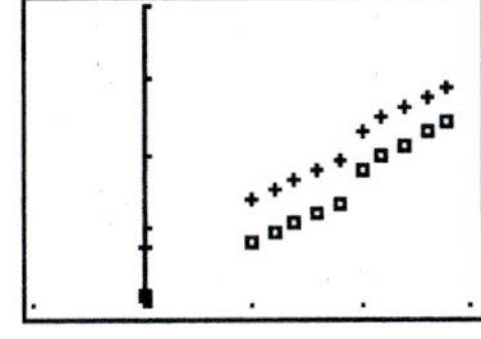

[−5, 15] by [120, 160]

The linear regression equation for the male population is $y \approx 1.7585x + 119.5765$.
The linear regression equation for the female population is $y \approx 1.6173x + 126.4138$.

2. The slope is the rate of change of people (in millions) per year. The y-intercept is the number of people (either males or females) in 1990.

3. Yes, the male population is predicted to eventually surpass the female population, because the males' regression line has a greater slope. But since 2000, the female population has always been greater. Since a span of only 15 years is represented, the data are most likely not enough to create a model for 100 or more years.

4.

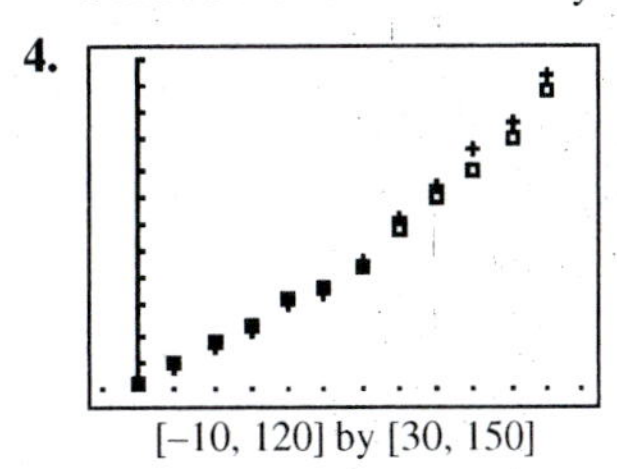

[−10, 120] by [30, 150]

5. Males: $y \approx \dfrac{412.574}{1 + 10.956e^{(-0.01539x)}}$

Females: $y \approx \dfrac{315.829}{1 + 9.031e^{(-0.01831x)}}$

The curves intersect at approximately (45, 64); this represents the time when the female population became greater than the male population.
The curves intersect at approximately (159, 212); this represents the time when the male population will again become greater the female population.

7. Approximately $\dfrac{138.1}{281.5} \approx 0.491 = 49.1\%$ male and 50.9% female

Chapter 8
Analytic Geometry in Two and Three Dimensions

■ Section 8.1 Conic Sections and Parabolas

Exploration 1

1. From Figure 8.4, we see that the axis of the parabola is $x = 0$. Thus, we want to find the point along $x = 0$ that is equidistant from both $(0, 1)$ and the line $y = -1$. Since the axis is perpendicular to the directrix, the point on the directrix closest to the parabola is $(0, 1)$ and $(0, -1)$, it must be located at $(0, 0)$.

2. Choose any point on the parabola (x, y). From figures 8.3 and 8.4, we see that the distance from (x, y) to the focus is $d_1 = \sqrt{(x-0)^2 + (y-1)^2} = \sqrt{x^2 + (y-1)^2}$ and the distance from (x, y) to the directrix is $d_2 = \sqrt{(x-x)^2 + (y-(-1))^2} = \sqrt{(y+1)^2}$.
Since d_1 must equal d_2, we have
$$d_1 = \sqrt{x^2 + (y-1)^2} = \sqrt{(y+1)^2} = d_2$$
$$x^2 + (y-1)^2 = (y+1)^2$$
$$x^2 + y^2 - 2y + 1 = y^2 + 2y + 1$$
$$x^2 = 4y$$
$$\frac{x^2}{4} = y \text{ or } x^2 = 4y$$

3. From the figure, we see that the first dashed line above $y = 0$ is $y = 1$, and we assume that each subsequent dashed line increases by $y = 1$. Using the equation above, we solve $\left\{1 = \frac{x^2}{4}, 2 = \frac{x^2}{4}, 3 = \frac{x^2}{4}, 4 = \frac{x^2}{4}, 5 = \frac{x^2}{4}, 6 = \frac{x^2}{4}\right\}$ to find:
$\{(-2\sqrt{6}, 6), (-2\sqrt{5}, 5), (-4, 4), (-2\sqrt{3}, 3), (-2\sqrt{2}, 2), (-2, 1), (0, 0), (2, 1), (2\sqrt{2}, 2), (2\sqrt{3}, 3), (4, 4), (2\sqrt{5}, 5), (2\sqrt{6}, 6)\}$

Exploration 2

1.

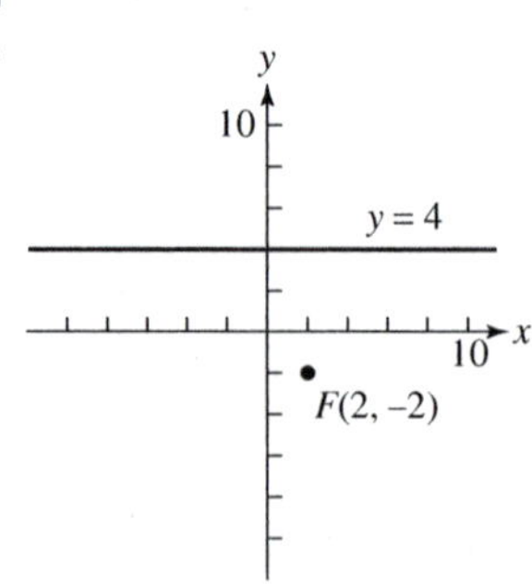

2.

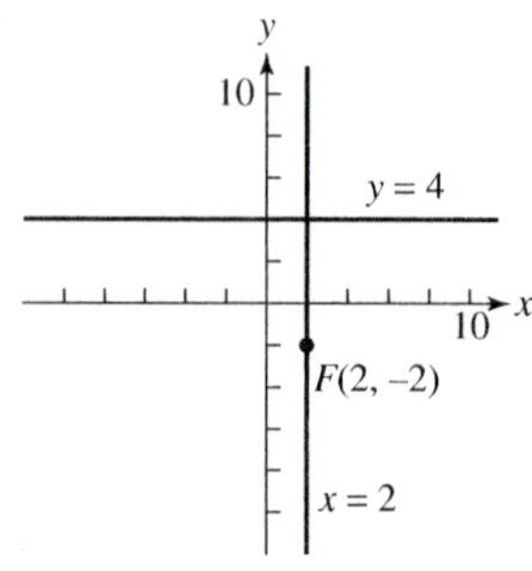

The equation of the axis is $x = 2$.

3.

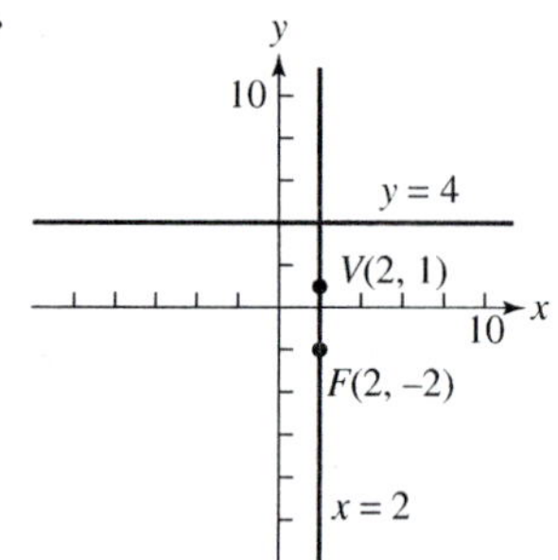

4. Since the focus $(h, k + p) = (2, -2)$ and the directrix $y = k - p = 4$, we have $k + p = -2$ and $k - p = 4$. Thus, $k = 1$, $p = -3$. As a result, the focal length p is -3 and the focal width $|4p|$ is 12.

5. Since the focal width is 12, each endpoint of the chord is 6 units away from the focus $(2, -2)$ along the line $y = -2$. The endpoints of the chord, then, are $(2 - 6, -2)$ and $(2 + 6, -2)$, or $(-4, -2)$ and $(8, -2)$.

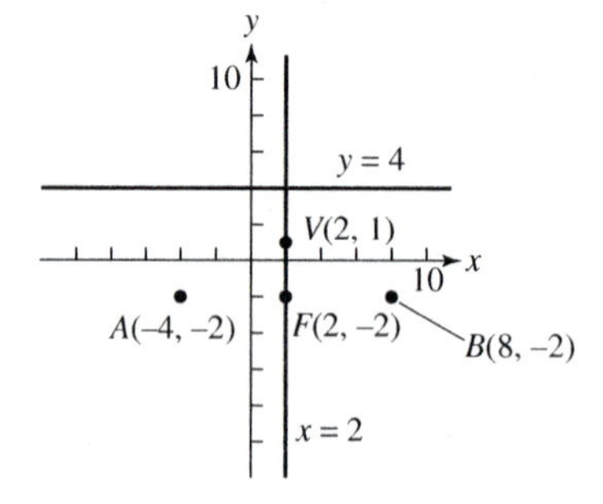

6.

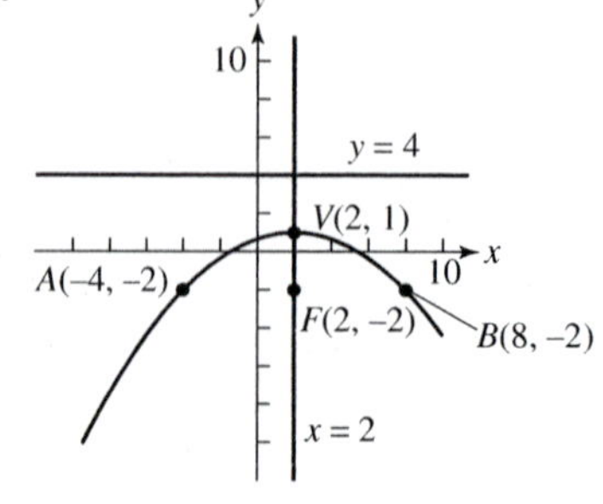

7. Downward
8. $h = 2, p = -3, k = 1$, so $(x - 2)^2 = -12(y - 1)$

Quick Review 8.1

1. $\sqrt{(2 - (-1))^2 + (5 - 3)^2} = \sqrt{9 + 4} = \sqrt{13}$
3. $y^2 = 4x, y = \pm 2\sqrt{x}$
5. $y + 7 = -(x^2 - 2x), y + 7 - 1 = -(x - 1)^2,$
   $y + 6 = -(x - 1)^2$
7. Vertex: $(1, 5)$. $f(x)$ can be obtained from $g(x)$ by stretching x^2 by 3, shifting up 5 units, and shifting right 1 unit.

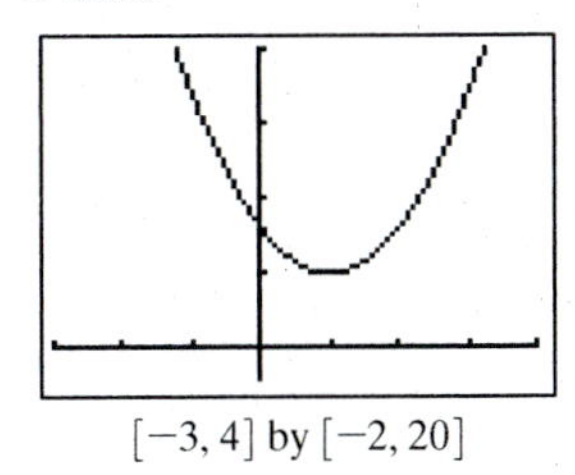

$[-3, 4]$ by $[-2, 20]$

9. $f(x) = a(x + 1)^2 + 3$, so $1 = a + 3, a = -2,$
   $f(x) = -2(x + 1)^2 + 3.$

Section 8.1 Exercises

1. $k = 0, h = 0, p = \frac{6}{4} = \frac{3}{2}$. Vertex: $(0, 0)$, Focus: $\left(0, \frac{3}{2}\right)$,

   Directrix: $y = -\frac{3}{2}$, Focal width: $\left|4p\right| = \left|4 \cdot \frac{3}{2}\right| = 6$

3. $k = 2, h = -3, p = \frac{4}{4} = 1$. Vertex: $(-3, 2)$,

   Focus: $(-2, 2)$, Directrix: $x = -3 - 1 = -4$,
   Focal width: $|4p| = |4(1)| = 4$.

5. $k = 0, h = 0, 4p = \frac{-4}{3}$, so $p = -\frac{1}{3}$. Vertex: $(0, 0)$,

   Focus: $\left(0, -\frac{1}{3}\right)$, Directrix: $y = \frac{1}{3}$, Focal width:

   $\left|4p\right| = \left|\left(\frac{-4}{3}\right)\right| = \frac{4}{3}$

7. (c)
9. (a)

For #11–29, recall that the standard form of the parabola is dependent on the vertex (h, k), the focal length p, the focal width $|4p|$, and the direction that the parabola opens.

11. $p = -3$ and the parabola opens to the left, so $y^2 = -12x$.
13. $-p = 4$ (so $p = -4$) and the parabola opens downward, so $x^2 = -16y$.
15. $p = 5$ and the parabola opens upward, so $x^2 = 20y$.
17. $h = 0, k = 0, |4p| = 8 \Rightarrow p = 2$ (since it opens to the right): $(y - 0)^2 = 8(x - 0)$; $y^2 = 8x$.
19. $h = 0, k = 0, |4p| = 6 \Rightarrow p = -\frac{3}{2}$ (since it opens downward): $(x - 0)^2 = -6(y - 0)$; $x^2 = -6y$
21. $h = -4, k = -4, -2 = -4 + p$, so $p = 2$ and the parabola opens to the right; $(y + 4)^2 = 8(x + 4)$
23. Parabola opens upward and vertex is halfway between focus and directrix on $x = h$ axis, so $h = 3$ and $k = \frac{4 + 1}{2} = \frac{5}{2}$; $1 = \frac{5}{2} - p$, so $p = \frac{3}{2}$.

    $(x - 3)^2 = 6\left(y - \frac{5}{2}\right)$
25. $h = 4, k = 3$; $6 = 4 - p$, so $p = -2$ and parabola opens to the left. $(y - 3)^2 = -8(x - 4)$
27. $h = 2, k = -1$; $|4p| = 16 \Rightarrow p = 4$ (since it opens upward): $(x - 2)^2 = 16(y + 1)$
29. $h = -1, k = -4$; $|4p| = 10 \Rightarrow p = -\frac{5}{2}$ (since it opens to the left): $(y + 4)^2 = -10(x + 1)$
31.

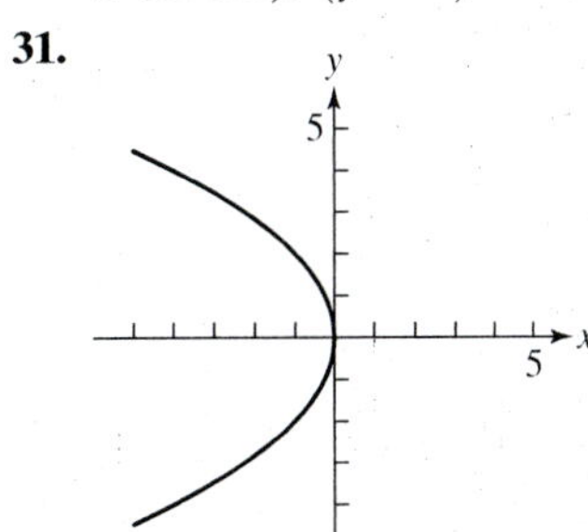

33.

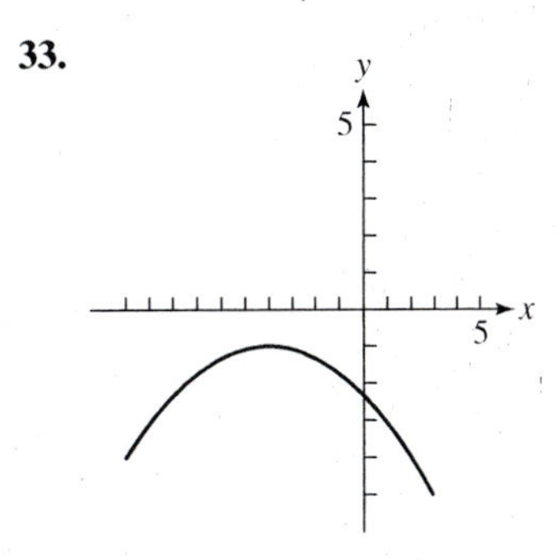

35.

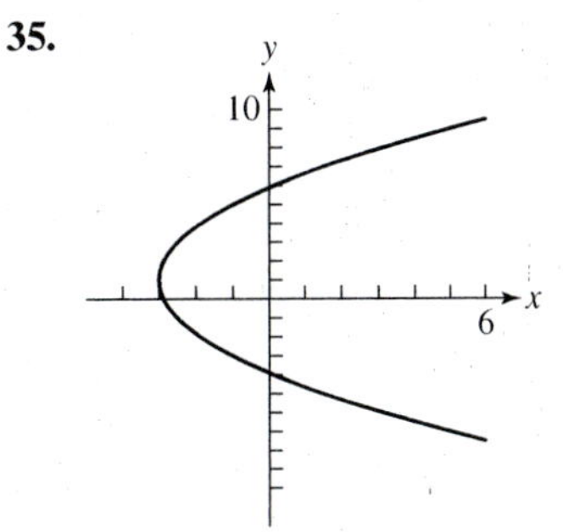

37.

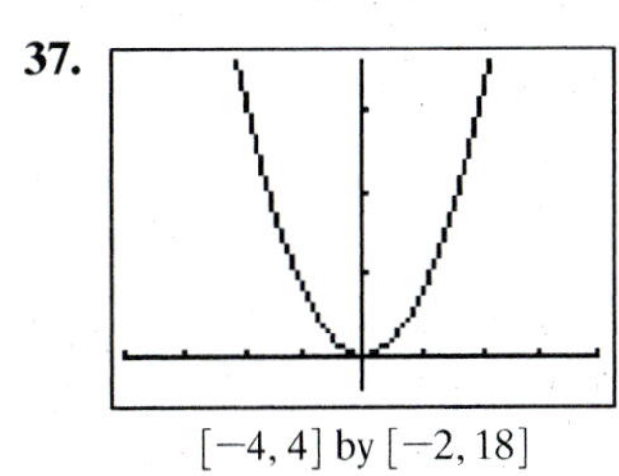

$[-4, 4]$ by $[-2, 18]$

39.

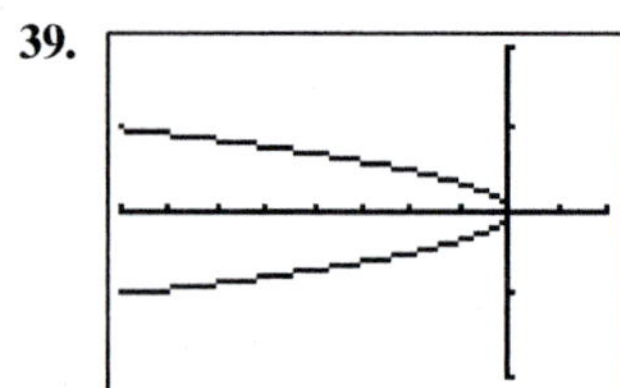

$[-8, 2]$ by $[-2, 2]$

41.

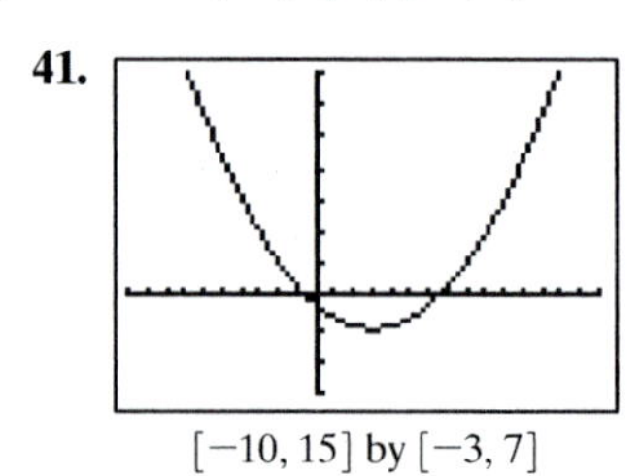

$[-10, 15]$ by $[-3, 7]$

43.

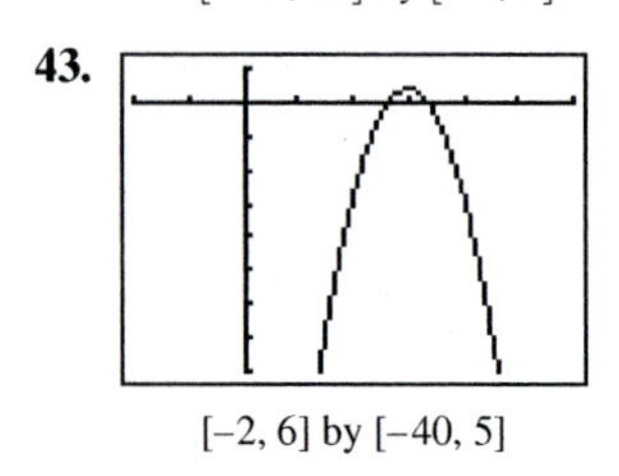

$[-2, 6]$ by $[-40, 5]$

45.

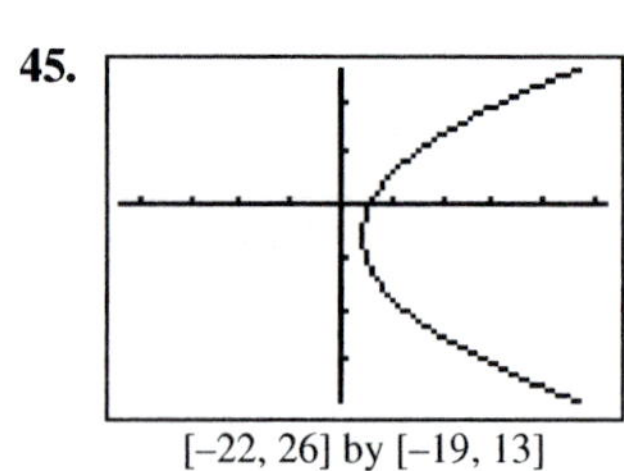

$[-22, 26]$ by $[-19, 13]$

47.

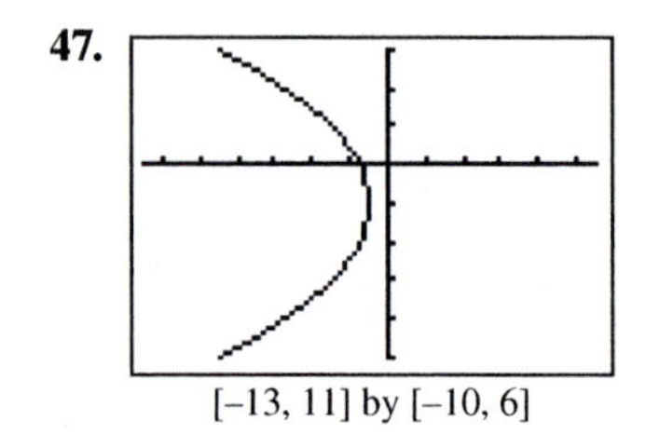

$[-13, 11]$ by $[-10, 6]$

49. Completing the square produces $y - 2 = (x + 1)^2$. The vertex is $(h, k) = (-1, 2)$, so the focus is $(h, k + p) = \left(-1, 2 + \frac{1}{4}\right) = \left(-1, \frac{9}{4}\right)$, and the directrix is $y = k - p = 2 - \frac{1}{4} = \frac{7}{4}$

51. Completing the square produces $8(x - 2) = (y - 2)^2$. The vertex is $(h, k) = (2, 2)$ so the focus is $(h + p, k) = (2 + 2, 2) = (4, 2)$, and the directrix is $x = h - p = 2 - 2 = 0$.

53. $h = 0$, $k = 2$, and the parabola opens to the left, so $(y - 2)^2 = 4p(x)$. Using $(-6, -4)$, we find $(-4 - 2)^2 = 4p(-6) \Rightarrow 4p = -\frac{36}{6} = -6$. The equation for the parabola is: $(y - 2)^2 = -6x$

55. $h = 2$, $k = -1$ and the parabola opens down so $(x - 2)^2 = 4p(y + 1)$. Using $(0, -2)$, we find that $(0 - 2)^2 = 4p(-2 + 1)$, so $4 = -4p$ and $p = -1$. The equation for the parabola is: $(x - 2)^2 = -4(y + 1)$.

57. One possible answer:
If p is replaced by $-p$ in the proof, then the result is $x^2 = -4py$, which is the correct result.

59. For the beam to run parallel to the axis of the mirror, the filament should be placed at the focus. As with Example 6, we must find p by using the fact that the points $(\pm 3, 2)$ must lie on the parabola. Then,

$$(\pm 3)^2 = 4p(2)$$
$$9 = 8p$$
$$p = \frac{9}{8} = 1.125 \text{ cm}$$

Because $p = 1.125$ cm, the filament should be placed 1.125 cm from the vertex along the axis of the mirror.

61. $4p = 10$, so $p = \frac{5}{2}$ and the focus is at $(0, p) = (0, 2.5)$. The electronic receiver is located 2.5 units from the vertex along the axis of the parabolic microphone.

63. Consider the roadway to be the axis. Then, the vertex of the parabola is $(300, 10)$ and the points $(0, 110)$ and $(600, 110)$ both lie on it. Using the standard formula, $(x - 300)^2 = 4p(y - 10)$. Solving for $4p$, we have $(600 - 300)^2 = 4p(110 - 10)$, or $4p = 900$, so the formula for the parabola is $(x - 300)^2 = 900(y - 10)$. The length of each cable is the distance from the parabola to the line $y = 0$. After solving the equation of the parabola for y ($y = \frac{1}{900}x^2 - \frac{2}{3}x + 110$), we determine that the length of each cable is

$$\sqrt{(x - x)^2 + \left(\frac{1}{900}x^2 - \frac{2}{3}x + 110 - 0\right)^2} = \frac{1}{900}x^2 - \frac{2}{3}x + 110.$$

Starting at the leftmost tower, the lengths of the cables are: $\approx \{79.44, 54.44, 35, 21.11, 12.78, 10, 12.78, 21.11, 35, 54.44, 79.44\}$.

65. False. Every point on a parabola is the same distance from its focus and its directrix.

67. The word "oval" does not denote a mathematically precise concept. The answer is D.

69. The focus of $y^2 = 4px$ is $(p, 0)$. Here $p = 3$, so the answer is B.

71. (a)–(c)

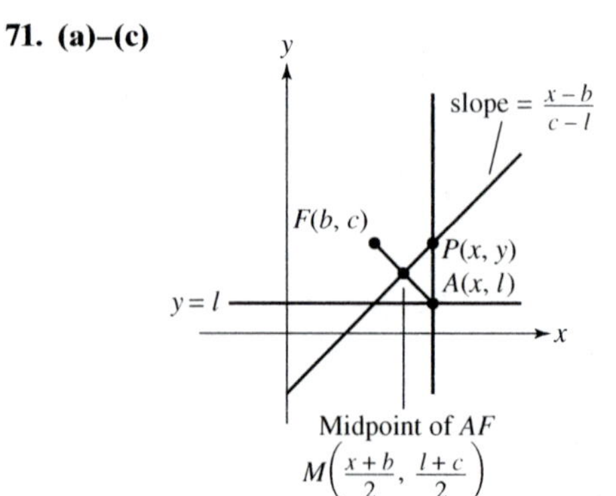

(d) As A moves, P traces out the curve of a parabola.

(e) With labels as shown, we can express the coordinates of P using the point-slope equation of the line PM:

$$y - \frac{\ell + c}{2} = \frac{x - b}{c - \ell}\left(x - \frac{x + b}{2}\right)$$

$$y - \frac{\ell + c}{2} = \frac{(x - b)^2}{2(c - \ell)}$$

$$2(c - \ell)\left(y - \frac{\ell + c}{2}\right) = (x - b)^2$$

This is the equation of a parabola with vertex at $\left(b, \frac{\ell + c}{2}\right)$ and focus at $\left(b, \frac{\ell + c}{2} + p\right)$ where $p = \frac{c - \ell}{2}$.

73. (a)

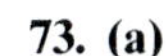

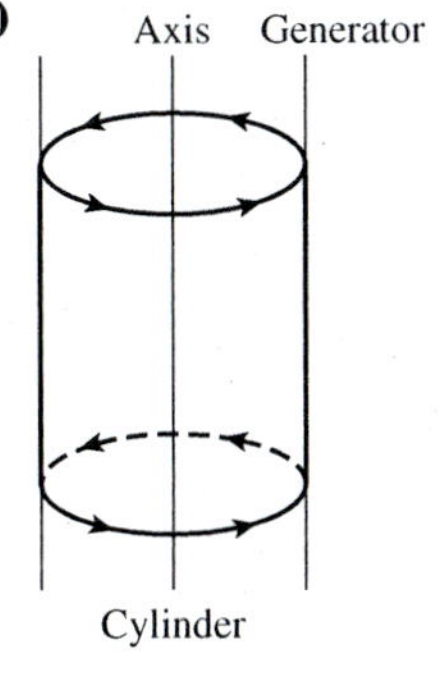

Cylinder

(b)

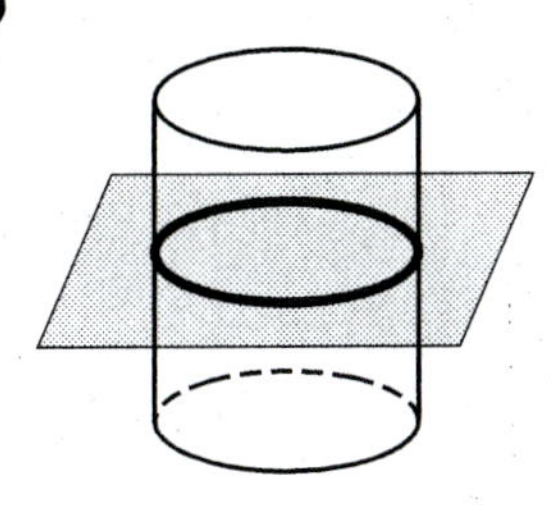

Circle

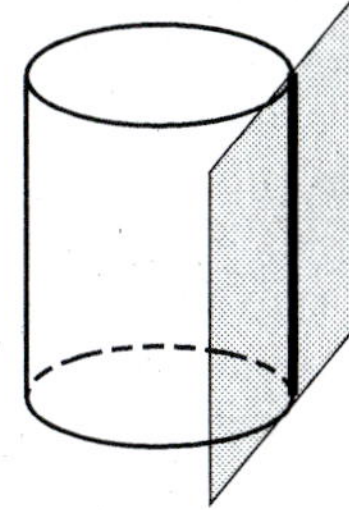

Single line

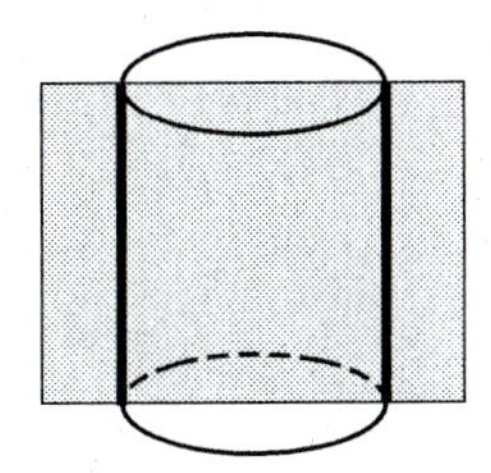

Two parallel lines

(c)

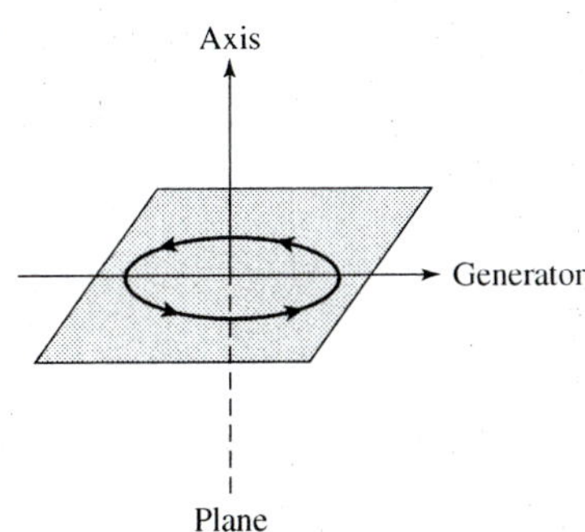

(d)

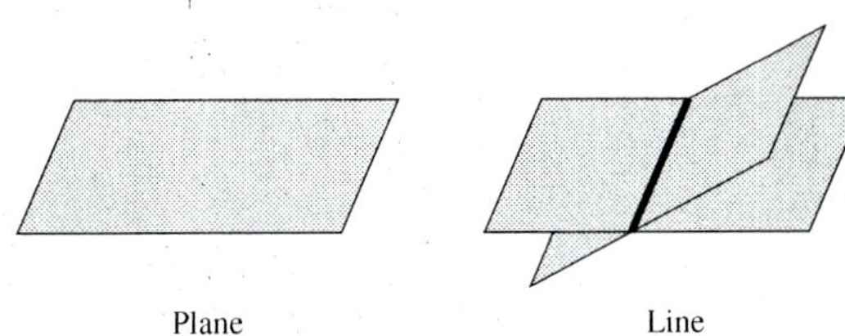

Plane Line

75. (a) The focus of the parabola $y = \frac{1}{4p}x^2$ is at $(0, p)$ so any line with slope m that passes through the focus must have equation $y = mx + p$.
The endpoints of a focal chord are the intersection points of the parabola $y = \frac{1}{4p}x^2$ and the line $y = mx + p$.
Solving the equation $\frac{1}{4p}x^2 - mx - p = 0$ using the quadratic formula, we have

$$x = \frac{m \pm \sqrt{m^2 - 4\left(\frac{1}{4p}\right)(-p)}}{2\left(\frac{1}{4p}\right)}$$

$$= \frac{m \pm \sqrt{m^2 + 1}}{\frac{1}{2p}} = 2p(m \pm \sqrt{m^2 + 1}).$$

(b) The y-coordinates of the endpoints of a focal chord are

$$y = \frac{1}{4p}(2p(m + \sqrt{m^2 + 1}))^2 \text{ and}$$

$$y = \frac{1}{4p}(2p(m - \sqrt{m^2 + 1}))^2$$

$$\frac{1}{4p}(4p^2)(m^2 + 2m\sqrt{m^2 + 1} + (m^2 + 1))$$

$$= \frac{1}{4p}(4p^2)(m^2 - 2m\sqrt{m^2 + 1} + (m^2 + 1))$$

$$= p(2m^2 + 2m\sqrt{m^2 + 1} + 1)$$

$$= p(2m^2 - 2m\sqrt{m^2 + 1} + 1)$$

Using the distance formula for $(2p(m - \sqrt{m^2 + 1}), p(2m^2 - 2m\sqrt{m^2 + 1} + 1))$ and $(2p(m + \sqrt{m^2 + 1}), p(2m^2 + 2m\sqrt{m^2 + 1} + 1))$, we know that the length of any focal chord is $\sqrt{(x_2 - x_1)^2 + (y_2 - y_1)^2}$

$$= \sqrt{(4p\sqrt{m^2 + 1})^2 + (4mp\sqrt{m^2 + 1})^2}$$

$$= \sqrt{(16m^2p^2 + 16p^2) + (16m^4p^2 + 16m^2p^2)}$$

$$= \sqrt{16m^4p^2 + 32m^2p^2 + 16p^2}$$

The quantity under the radical sign is smallest when $m = 0$. Thus the smallest focal chord has length $\sqrt{16p^2} = |4p|$.

■ Section 8.2 Ellipses

Exploration 1

1. The equations $x = -2 + 3\cos t$ and $y = 5 + 7\sin t$ can be rewritten as $\cos t = \frac{x+2}{3}$ and $\sin t = \frac{y-5}{7}$.
   Substituting these into the identity $\cos^2 t + \sin^2 t = 1$ yields the equation $\frac{(x+2)^2}{9} + \frac{(y-5)^2}{49} = 1$.

2. 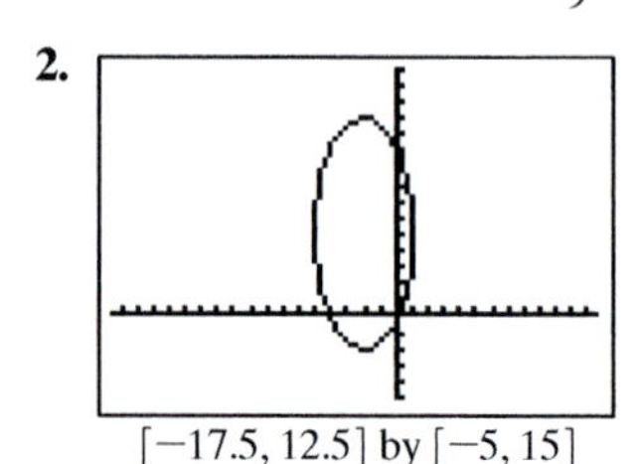

   [−17.5, 12.5] by [−5, 15]

3. Example 1: Since $\frac{x^2}{9} + \frac{y^2}{4} = 1$, a parametric solution is $x = 3\cos t$ and $y = 2\sin t$.
   Example 2: Since $\frac{y^2}{13} + \frac{x^2}{4} = 1$, a parametric solution is $y = \sqrt{13}\sin t$ and $x = 2\cos t$.
   Example 3: Since $\frac{(x-3)^2}{25} + \frac{(y+1)^2}{16} = 1$, a parametric solution is $x = 5\cos t + 3$ and $y = 4\sin t - 1$.

4. 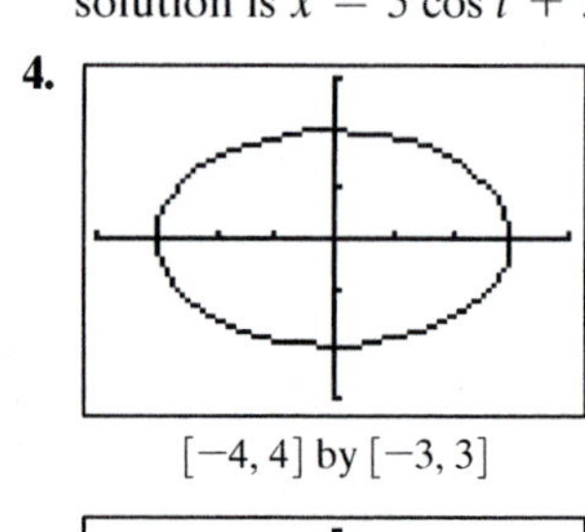

   [−4, 4] by [−3, 3]

   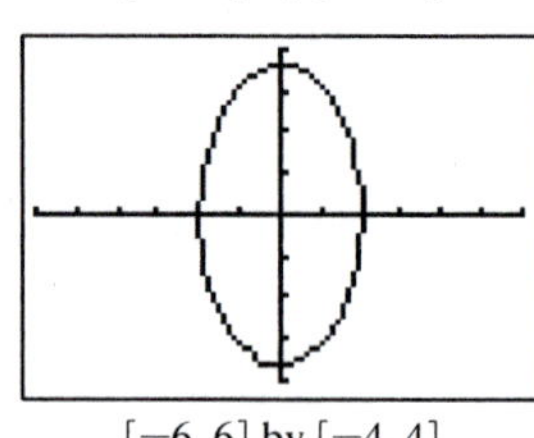

   [−6, 6] by [−4, 4]

   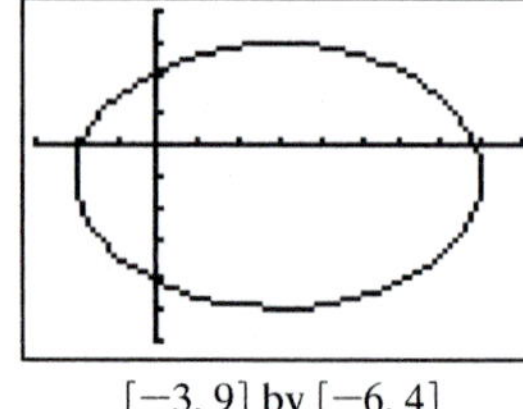

   [−3, 9] by [−6, 4]

   Answers may vary. In general, students should find that the eccentricity is equal to the ratio of the distance between foci over distance between vertices.

5. Example 1: The equations $x = 3\cos t$, $y = 2\sin t$ can be rewritten as $\cos t = \frac{x}{3}$, $\sin t = \frac{y}{2}$, which using $\cos^2 t + \sin^2 t = 1$ yield $\frac{x^2}{9} + \frac{y^2}{4} = 1$ or $4x^2 + 9y^2 = 36$.
   Example 2: The equations $x = 2\cos t$, $y = \sqrt{13}\sin t$ can be rewritten as $\cos t = \frac{x}{2}$, $\sin t = \frac{y}{\sqrt{13}}$, which using $\sin^2 t + \cos^2 t = 1$ yield $\frac{y^2}{13} + \frac{x^2}{4} = 1$.
   Example 3: By rewriting $x = 3 + 5\cos t$, $y = -1 + 4\sin t$ as $\cos t = \frac{x-3}{5}$, $\sin t = \frac{y+1}{4}$ and using $\cos^2 t + \sin^2 t = 1$, we obtain $\frac{(x-3)^2}{25} + \frac{(y+1)^2}{16} = 1$.

Exploration 2

Answers will vary due to experimental error. The theoretical answers are as follows.

2. $a = 9$ cm, $b = \sqrt{80} \approx 8.94$ cm, $c = 1$ cm, $e = 1/9 \approx 0.11$, $b/a \approx 0.99$.
3. $a = 8$ cm, $b = \sqrt{60} \approx 7.75$ cm, $c = 2$ cm, $e = 1/4 = 0.25$, $b/a \approx 0.97$;
   $a = 7$ cm, $b = \sqrt{40} \approx 6.32$ cm, $c = 3$ cm, $e = 3/7 \approx 0.43$, $b/a \approx 0.90$;
   $a = 6$ cm, $b = \sqrt{20} \approx 4.47$ cm, $c = 4$ cm, $e = 2/3 \approx 0.67$, $b/a \approx 0.75$.
4. The ratio b/a decreases slowly as $e = c/a$ increases rapidly. The ratio b/a is the height-to-width ratio, which measures the shape of the ellipse—when b/a is close to 1, the ellipse is nearly circular; when b/a is close to 0, the ellipse is elongated. The eccentricity ratio $e = c/a$ measures how off-center the foci are—when e is close to 0, the foci are near the center of the ellipse; when e is close to 1, the foci are far from the center and near the vertices of the ellipse. The foci must be extremely off-center for the ellipse to be significantly elongated.
5. 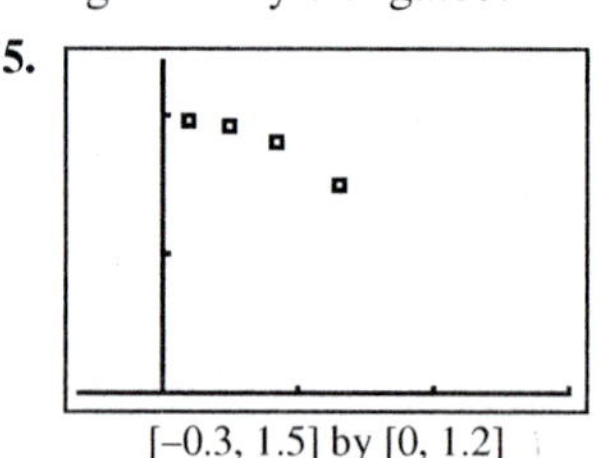

   [−0.3, 1.5] by [0, 1.2]

   $$\frac{b}{a} = \frac{\sqrt{a^2 - c^2}}{a} = \sqrt{1 - \frac{c^2}{a^2}} = \sqrt{1 - e^2}$$

   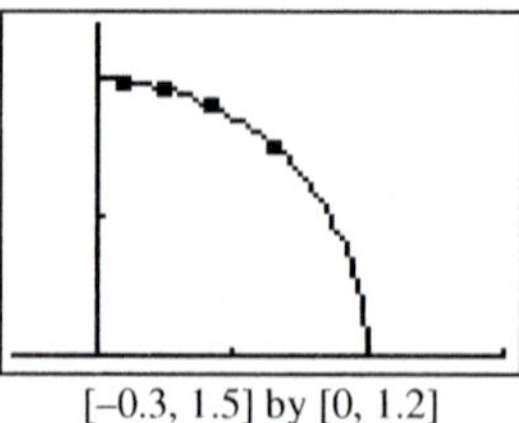

   [−0.3, 1.5] by [0, 1.2]

Quick Review 8.2

1. $\sqrt{(2-(-3))^2+(4-(-2))^2}=\sqrt{5^2+6^2}=\sqrt{61}$

3. $4y^2+9x^2=36,\ 4y^2=36-9x^2,$

$y=\pm\sqrt{\dfrac{36-9x^2}{4}}=\pm\dfrac{3}{2}\sqrt{4-x^2}$

5. $3x+12=(10-\sqrt{3x-8})^2$

$$\begin{aligned}3x+12&=100-20\sqrt{3x-8}+3x-8\\-80&=-20\sqrt{3x-8}\\4&=\sqrt{3x-8}\\16&=3x-8\\3x&=24\\x&=8\end{aligned}$$

7. $6x^2+12=(11-\sqrt{6x^2+1})^2$

$$\begin{aligned}6x^2+12&=121-22\sqrt{6x^2+1}+6x^2+1\\-110&=-22\sqrt{6x^2+1}\\6x^2+1&=25\\6x^2-24&=0\\x^2-4&=0\\x&=2,\ x=-2\end{aligned}$$

9. $2\left(x-\dfrac{3}{2}\right)^2-\dfrac{15}{2}=0$, so $x=\dfrac{3\pm\sqrt{15}}{2}$

Section 8.2 Exercises

1. $h=0, k=0, a=4, b=\sqrt{7}$, so $c=\sqrt{16-7}=3$
Vertices: $(4,0), (-4,0)$; Foci: $(3,0), (-3,0)$

3. $h=0, k=0, a=6, b=3\sqrt{3}$, so $c=\sqrt{36-27}=3$
Vertices: $(0,6), (0,-6)$; Foci: $(0,3), (0,-3)$

5. $\dfrac{x^2}{4}+\dfrac{y^2}{3}=1$. $h=0, k=0, a=2, b=\sqrt{3}$, so

$c=\sqrt{4-3}=1$
Vertices: $(2,0), (-2,0)$; Foci: $(1,0), (-1,0)$

7. (d)

9. (a)

11.

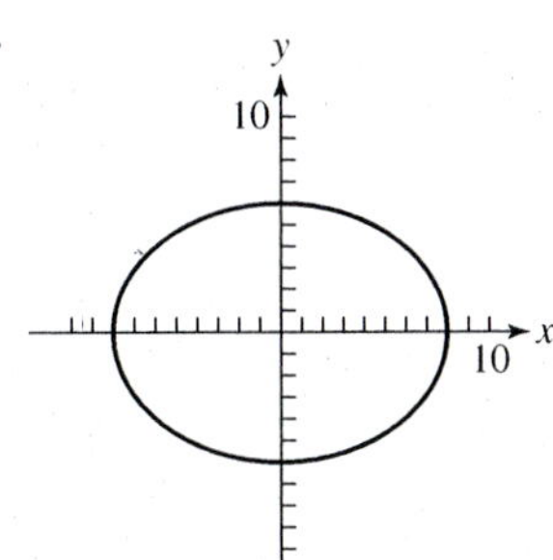

13.

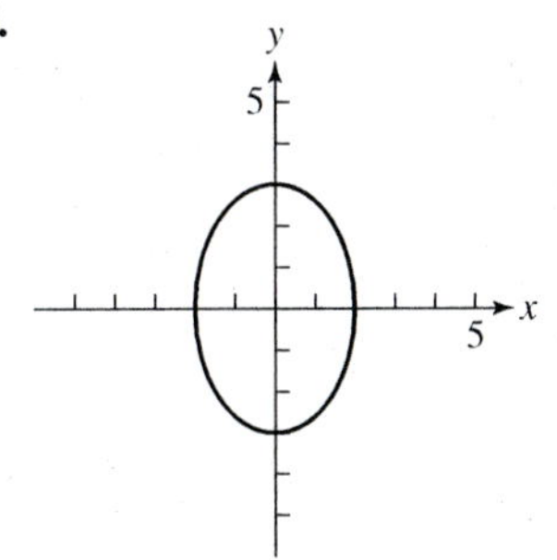

15.

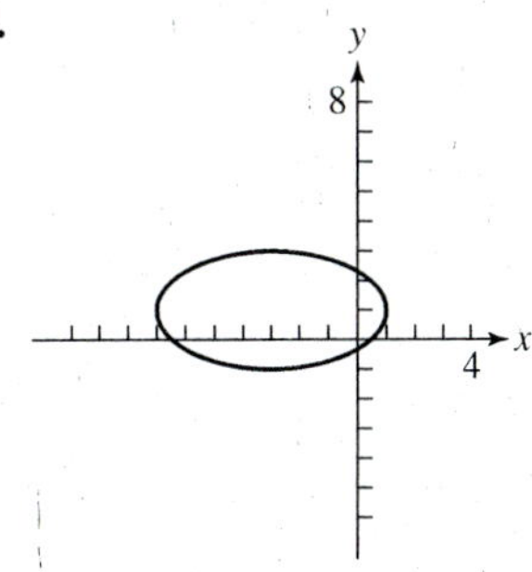

17.

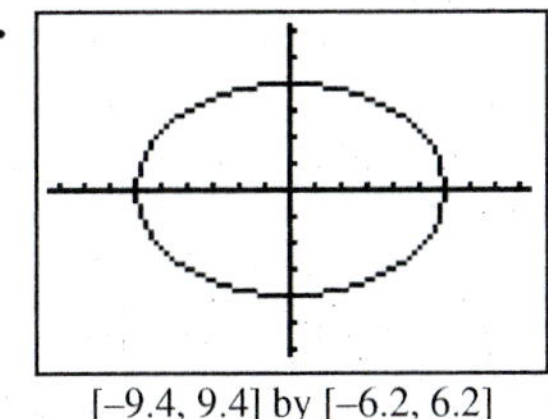

[−9.4, 9.4] by [−6.2, 6.2]

$y=\pm\dfrac{2}{3}\sqrt{-x^2+36}$

19.

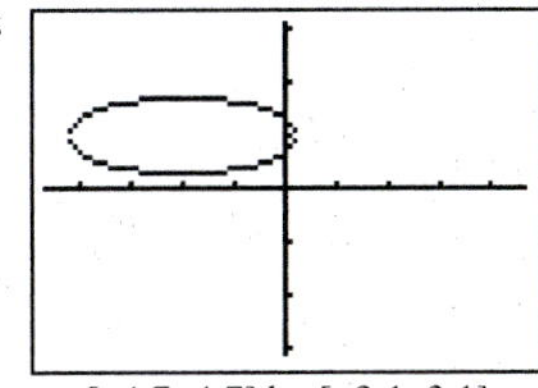

[−4.7, 4.7] by [−3.1, 3.1]

$y=1\pm\sqrt{-\dfrac{(x+2)^2}{10}+\dfrac{1}{2}}$

21. $\dfrac{x^2}{4}+\dfrac{y^2}{9}=1$

23. $c=2$ and $a=\dfrac{10}{2}=5$, so $b=\sqrt{a^2-c^2}=\sqrt{21}$:

$\dfrac{x^2}{25}+\dfrac{y^2}{21}=1$

25. $\dfrac{x^2}{16}+\dfrac{y^2}{25}=1$

27. $b=4;\ \dfrac{x^2}{16}+\dfrac{y^2}{36}=1$

29. $a=5;\ \dfrac{x^2}{25}+\dfrac{y^2}{16}=1$

31. The center (h,k) is $(1,2)$ (the midpoint of the axes); a and b are half the lengths of the axes (4 and 6, respectively): $\dfrac{(x-1)^2}{16}+\dfrac{(y-2)^2}{36}=1$

33. The center (h,k) is $(3,-4)$ (the midpoint of the major axis); $a=3$, half the lengths of the major axis. Since $c=2$ (half the distance between the foci),

$b=\sqrt{a^2-c^2}=\sqrt{5}$: $\dfrac{(x-3)^2}{9}+\dfrac{(y+4)^2}{5}=1$

35. The center (h, k) is $(3, -2)$ (the midpoint of the major axis); a and b are half the lengths of the axes (3 and 5, respectively):

$$\frac{(x-3)^2}{9} + \frac{(y+2)^2}{25} = 1$$

For #37–39, an ellipse with equation $\frac{(x-h)^2}{a^2} + \frac{(y-k)^2}{b^2} = 1$ has center (h, k), vertices $(h \pm a, k)$, and foci $(h \pm c, k)$ where $c = \sqrt{a^2 - b^2}$.

37. Center $(-1, 2)$; Vertices $(-1 \pm 5, 2) = (-6, 2), (4, 2)$; Foci $(-1 \pm 3, 2) = (-4, 2), (2, 2)$

39. Center $(7, -3)$; Vertices: $(7, -3 \pm 9) = (7, 6), (7, -12)$; Foci: $(7, -3 \pm \sqrt{17}) \approx (7, 1.12), (7, -7.12)$

41.

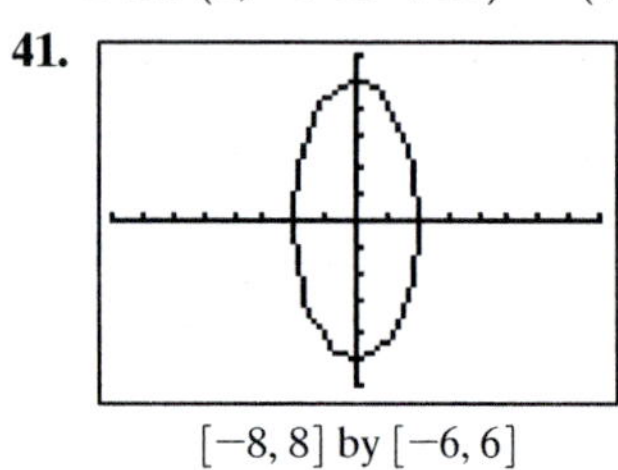

$[-8, 8]$ by $[-6, 6]$

$x = 2 \cos t,\ y = 5 \sin t$

43.

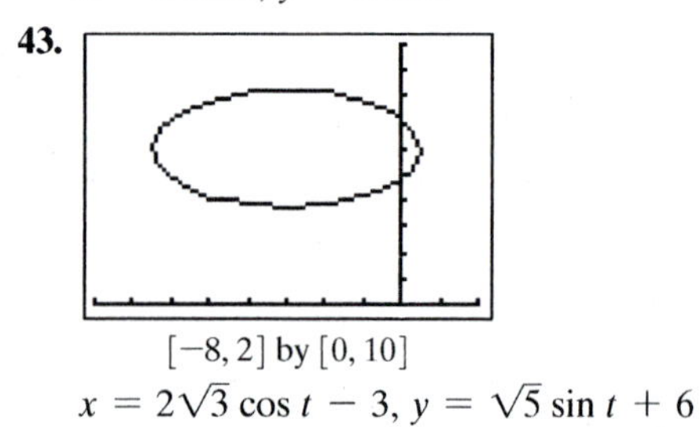

$[-8, 2]$ by $[0, 10]$

$x = 2\sqrt{3} \cos t - 3,\ y = \sqrt{5} \sin t + 6$

For #45–47, complete the squares in x and y, then put in standard form. (The first one is done in detail; the others just show the final form.)

45. $9x^2 + 4y^2 - 18x + 8y - 23 = 0$ can be rewritten as $9(x^2 - 2x) + 4(y^2 + 2y) = 23$. This is equivalent to $9(x^2 - 2x + 1) + 4(y^2 + 2y + 1) = 23 + 9 + 4$, or $9(x-1)^2 + 4(y+1)^2 = 36$. Divide both sides by 36 to obtain $\frac{(x-1)^2}{4} + \frac{(y+1)^2}{9} = 1$. Vertices: $(1, -4)$ and $(1, 2)$ Foci: $(1, -1 \pm \sqrt{5})$. Eccentricity: $\frac{\sqrt{5}}{3}$.

47. $\frac{(x+3)^2}{16} + \frac{(y-1)^2}{9} = 1$. Vertices: $(-7, 1)$ and $(1, 1)$.

Foci: $(-3 \pm \sqrt{7}, 1)$. Eccentricity: $\frac{\sqrt{7}}{4}$

49. The center (h, k) is $(2, 3)$ (given); a and b are half the lengths of the axes (4 and 3, respectively):

$$\frac{(x-2)^2}{16} + \frac{(y-3)^2}{9} = 1$$

51. Consider Figure 8.15(b); call the point $(0, c)$ F_1, and the point $(0, -c)$ F_2. By the definition of an ellipse, any point P (located at (x, y)) satisfies the equation $\overrightarrow{PF} + \overrightarrow{PF_2} = 2a$ thus, $\sqrt{(x-0)^2 + (y-c)^2} + \sqrt{(x-0)^2 + (y+c)^2} = \sqrt{x^2 + (y-c)^2} + \sqrt{x^2 + (y+c)^2} = 2a$

then $\sqrt{x^2 + (y-c)^2} = 2a - \sqrt{x^2 + (y+c)^2}$

$$\begin{aligned}
x^2 + (y-c)^2 &= 4a^2 - 4a\sqrt{x^2 + (y+c)^2} \\
&\quad + x^2 + (y+c)^2 \\
y^2 - 2cy + c^2 &= 4a^2 - 4a\sqrt{x^2 + (y+c)^2} \\
&\quad + y^2 + 2cy + c^2 \\
4a\sqrt{x^2 + (y+c)^2} &= 4a^2 + 4cy \\
a\sqrt{x^2 + (y+c)^2} &= a^2 + cy \\
a^2(x^2 + (y+c)^2) &= a^4 + 2a^2cy + c^2y^2 \\
a^2x^2 + (a^2 - c^2)y^2 &= a^2(a^2 - c^2) \\
a^2x^2 + b^2y^2 &= a^2b^2 \\
\frac{x^2}{b^2} + \frac{y^2}{a^2} &= 1
\end{aligned}$$

53. Since the Moon is furthest from the Earth at 252,710 miles and closest at 221,463, we know that $2a = 252{,}710 + 221.463$, or $a = 237{,}086.5$. Since $c + 221{,}463 = a$, we know $c = 15{,}623.5$ and $b = \sqrt{a^2 - c^2} = \sqrt{(237{,}086.5)^2 - (15{,}623.5)^2} \approx 236{,}571$.

From these, we calculate $e = \frac{c}{a} = \frac{15{,}623.5}{237{,}086.5} \approx 0.066$.

The orbit of the Moon is very close to a circle, but still takes the shape of an ellipse.

55. For Saturn, $c = ea = (0.0560)(1{,}427) \approx 79.9$ Gm. Saturn's perihelion is $a - c = 1427 - 79.9 \approx 1347$ Gm and its aphelion is $a + c = 1427 + 72.21 \approx 1507$ Gm.

57. For sungrazers, $a - c < 1.5(1.392) = 2.088$. The eccentricity of their ellipses is very close to 1.

59. $a = 8$ and $b = 3.5$, so $c = \sqrt{a^2 - b^2} = \sqrt{51.75}$. Foci at $(\pm\sqrt{51.75}, 0) \approx (\pm 7.19, 0)$.

61. Substitute $y^2 = 4 - x^2$ into the first equation:

$$\begin{aligned}
\frac{x^2}{4} + \frac{4 - x^2}{9} &= 1 \\
9x^2 + 4(4 - x^2) &= 36 \\
5x^2 &= 20 \\
x^2 &= 4 \\
x = \pm 2,\ y &= 0
\end{aligned}$$

Solution: $(-2, 0), (2, 0)$

63. (a)

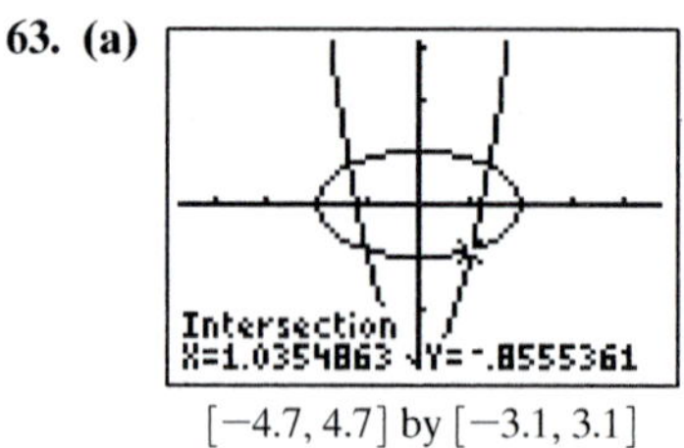

$[-4.7, 4.7]$ by $[-3.1, 3.1]$

Approximate solutions: $(\pm 1.04, -0.86), (\pm 1.37, 0.73)$

(b) $\left(\frac{\pm\sqrt{94 - 2\sqrt{161}}}{8}, -\frac{1 + \sqrt{161}}{16}\right)$, $\left(\frac{\pm\sqrt{94 + 2\sqrt{161}}}{8}, \frac{-1 + \sqrt{161}}{16}\right)$

65. False. The distance is $a - c = a(1 - c/a) = a(1 - e)$.

67. $\frac{x^2}{4} + \frac{y^2}{1} = 1$, so $c = \sqrt{a^2 - b^2} = \sqrt{2^2 - 1^2} = \sqrt{3}$. The answer is C.

69. Completing the square produces $\frac{(x-4)^2}{4} + \frac{(y-3)^2}{9} = 1$. The answer is B.

71. (a) When $a = b = r$, $A = \pi ab = \pi rr = \pi r^2$ and

$$P \approx \pi(2r)\left(3 - \frac{\sqrt{(3r+r)(r+3r)}}{r+r}\right)$$
$$= 2\pi r\left(3 - \frac{\sqrt{16r^2}}{2r}\right) = 2\pi r\left(3 - \frac{4r}{2r}\right)$$
$$= 2\pi r\,(3-2) = 2\pi r.$$

(b) One possibility: $\frac{x^2}{16} + \frac{y^2}{9} = 1$ with $A = 12\pi$ and $P \approx (21 - \sqrt{195})\pi \approx 22.10$, and $\frac{x^2}{100} + y^2 = 1$ with $A = 10\pi$ and $P \approx (33 - \sqrt{403})\pi \approx 40.61$.

73. (a) Graphing in parametric mode with Tstep $= \frac{\pi}{24}$.

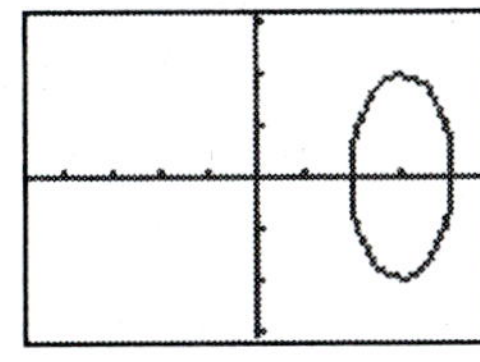

[–4.7, 4.7] by [–3.1, 3.1]

(b) The equations $x(t) = 3 + \cos(2t - 5)$ and $y(t) = -2\sin(2t - 5)$ can be rewritten as $\cos(2t - 5) = x - 3$ and $\sin(2t - 5) = -y/2$. Substituting these into the identity $\cos^2(2t - 5) + \sin^2(2t - 5) = 1$ yields the equation $y^2/4 + (x - 3)^2 = 1$. This is the equation of an ellipse with $x = 3$ as the focal axis. The center of the ellipse is $(3, 0)$ and the vertices are $(3, 2)$ and $(3, -2)$. The length of the major axis is 4 and the length of the minor axis is 2.

75. Write the equation in standard form by completing the squares and then dividing by the constant on the right-hand side.

$$Ax^2 + Dx + \frac{D^2}{4A} + Cy^2 + Ey + \frac{E^2}{4C} = \frac{D^2}{4A} + \frac{E^2}{4C} - F$$

$$\frac{x^2 + \frac{D}{A}x + \frac{D^2}{4A^2}}{C} + \frac{y^2 + \frac{E}{C}y + \frac{E^2}{4C^2}}{A}$$
$$= \frac{1}{AC}\left(\frac{D^2}{4A} + \frac{E^2}{4C} - F\right)$$

$$\frac{\left(x + \frac{D}{2A}\right)^2}{C} + \frac{\left(y + \frac{E}{2C}\right)^2}{A} = \frac{CD^2 + AE^2 - 4ACF}{4A^2C^2}$$

$$\left(\frac{4A^2C^2}{CD^2 + AE^2 - 4ACF}\right) \times \left[\frac{\left(x + \frac{D}{2A}\right)^2}{C} + \frac{\left(y + \frac{E}{2C}\right)^2}{A}\right] = 1$$

$$\frac{4A^2C\left(x + \frac{D}{2A}\right)^2}{CD^2 + AE^2 - 4ACF} + \frac{4AC^2\left(y + \frac{E}{2C}\right)^2}{CD^2 + AE^2 - 4ACF} = 1$$

Since $AC > 0$, $A \neq 0$ and $C \neq 0$ (we are not dividing by zero). Further, $AC > 0 \Rightarrow 4A^2C > 0$ and $4AC^2 > 0$ (either $A > 0$ and $C > 0$, or $A < 0$ and $C < 0$), so the equation represents an ellipse.

Section 8.3 Hyperbolas

Exploration 1

1. The equations $x = -1 + 3/\cos t = -1 + 3\sec t$ and $y = 1 + 2\tan t$ can be rewritten as $\sec t = \frac{x+1}{3}$ and $\tan t = \frac{y-1}{2}$. Substituting these into the identity $\sec^2 t - \tan^2 t = 1$ yields the equation $\frac{(x+1)^2}{9} - \frac{(y-1)^2}{4} = 1$.

2.

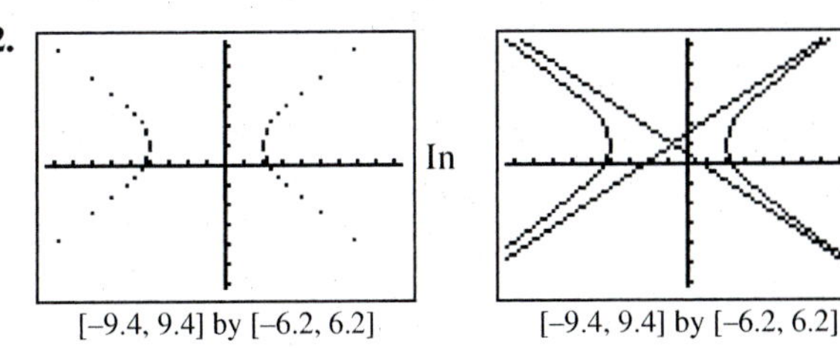

[–9.4, 9.4] by [–6.2, 6.2] In [–9.4, 9.4] by [–6.2, 6.2]

Connected graphing mode, pseudo-asymptotes appear because the grapher connects computed points by line segments regardless of whether this makes sense. Using Dot mode with a small Tstep will produce the best graphs.

3. Example 1: $x = 3/\cos(t)$, $y = 2\tan(t)$
Example 2: $x = 2\tan(t)$, $y = \sqrt{5}/\cos(t)$
Example 3: $x = 3 + 5/\cos(t)$, $y = -1 + 4\tan(t)$
Example 4: $x = -2 + 3/\cos(t)$, $y = 5 + 7\tan(t)$

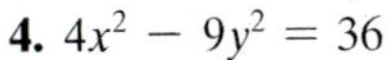

4. $4x^2 - 9y^2 = 36$

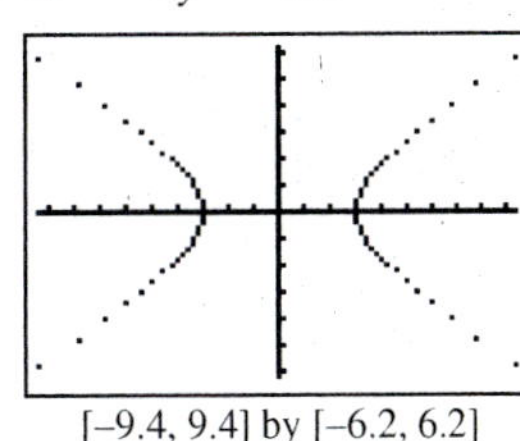

[–9.4, 9.4] by [–6.2, 6.2]

$\frac{y^2}{5} - \frac{x^2}{4} = 1$

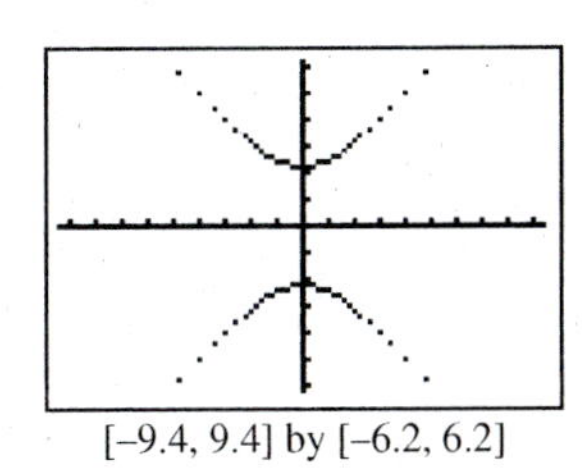

[–9.4, 9.4] by [–6.2, 6.2]

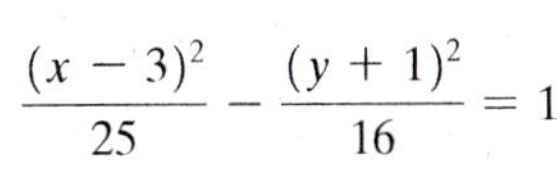

$\frac{(x-3)^2}{25} - \frac{(y+1)^2}{16} = 1$

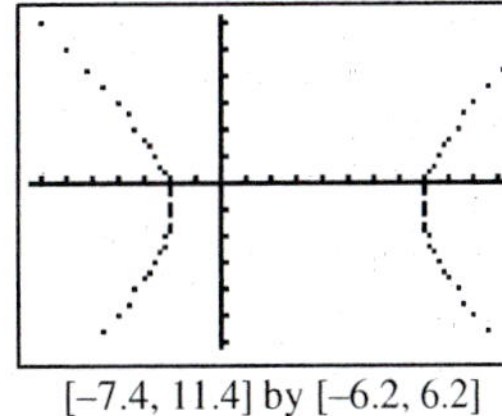

[–7.4, 11.4] by [–6.2, 6.2]

$\frac{(x+2)^2}{9} - \frac{(y-5)^2}{49} = 1$

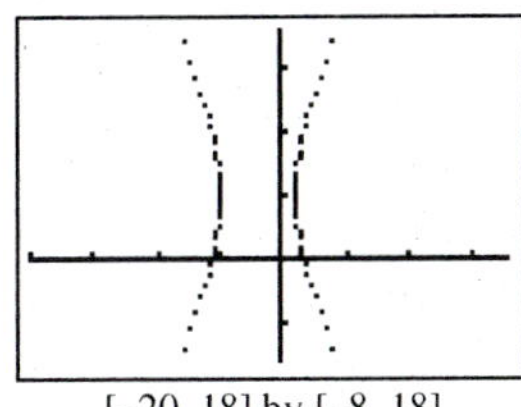

[–20, 18] by [–8, 18]

5. Example 1: The equations $x = 3/\cos t = 3 \sec t$, $y = 2 \tan t$ can be rewritten as $\sec t = \frac{x}{3}$, $\tan t = \frac{y}{2}$, which using the identity $\sec^2 t - \tan^2 t = 1$ yield $\frac{x^2}{9} - \frac{y^2}{4} = 1$.
Example 2: The equations $x = 2 \tan t$, $y = \sqrt{5}/\cos t = \sqrt{5} \sec t$ can be rewritten as $\tan t = \frac{x}{2}$, $\sec t = \frac{y}{\sqrt{5}}$, which using $\sec^2 t - \tan^2 t = 1$ yield $\frac{y^2}{5} - \frac{x^2}{4} = 1$.
Example 3: By rewriting $x = 3 + 5/\cos t$, $y = -1 + 4 \tan t$ as $\sec t = \frac{x-3}{5}$, $\tan t = \frac{y+1}{4}$ and using $\sec^2 t - \tan^2 t = 1$, we obtain $\frac{(x-3)^2}{25} - \frac{(y+1)^2}{16} = 1$.
Example 4: By rewriting $x = -2 + 3/\cos t$, $y = 5 + 7 \tan t$ as $\sec t = \frac{x+2}{3}$, $\tan t = \frac{y-5}{7}$ and using $\sec^2 t - \tan^2 t = 1$, we obtain $\frac{(x+2)^2}{9} - \frac{(y-5)^2}{49} = 1$.

Quick Review 8.3

1. $\sqrt{(-7-4)^2 + (-8-(-3))^2} = \sqrt{(-11)^2 + (-5)^2} = \sqrt{146}$

3. $9y^2 - 16x^2 = 144$
$9y^2 = 144 + 16x^2$
$y = \pm\frac{4}{3}\sqrt{9 + x^2}$

5. $\sqrt{3x + 12} = 10 + \sqrt{3x - 8}$
$3x + 12 = 100 + 20\sqrt{3x - 8} + 3x - 8$
$-80 = 20\sqrt{3x - 8}$
$-4 = \sqrt{3x - 8}$ no solution

7. $\sqrt{6x^2 + 12} = 1 + \sqrt{6x^2 + 1}$
$6x^2 + 12 = 1 + 2\sqrt{6x^2 + 1} + 6x^2 + 1$
$10 = 2\sqrt{6x^2 + 1}$
$25 = 6x^2 + 1$
$6x^2 - 24 = 0$
$x^2 - 4 = 0$
$x = 2, x = -2$

9. $c = a + 2$, $(a + 2)^2 - a^2 = \frac{16a}{3}$,
$a^2 + 4a + 4 - a^2 = \frac{16a}{3}$, $4a = 12$: $a = 3, c = 5$

Section 8.3 Exercises

For #1–5, recall the Pythagorean relation that $c^2 = a^2 + b^2$.

1. $a = 4, b = \sqrt{7}, c = \sqrt{16 + 7} = \sqrt{23}$;
Vertices: $(\pm 4, 0)$; Foci: $(\pm\sqrt{23}, 0)$

3. $a = 6, b = \sqrt{13}, c = \sqrt{36 + 13} = 7$;
Vertices: $(0, \pm 6)$; Foci: $(0, \pm 7)$

5. $\frac{x^2}{4} - \frac{y^2}{3} = 1$; $a = 2, b = \sqrt{3}, c = \sqrt{7}$;
Vertices: $(\pm 2, 0)$; Foci: $(\pm\sqrt{7}, 0)$

7. (c)

9. (a)

11. Transverse axis from $(-7, 0)$ to $(7, 0)$; asymptotes:
$y = \pm\frac{5}{7}x$,
$y = \pm\frac{5}{7}\sqrt{x^2 - 49}$

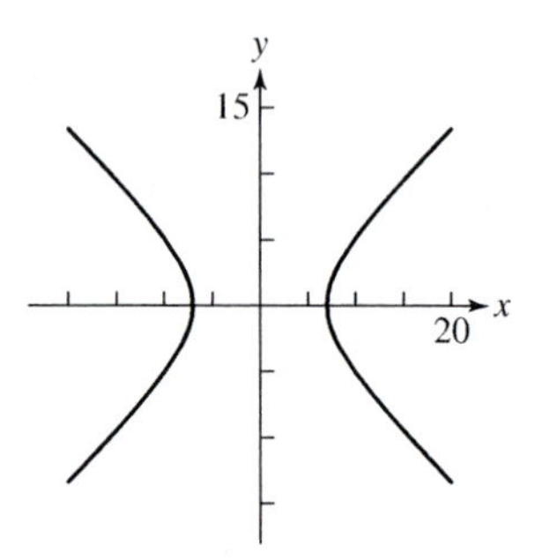

13. Transverse axis from $(0, -5)$ to $(0, 5)$; asymptotes:
$y = \pm\frac{5}{4}x$,
$y = \pm\frac{5}{4}\sqrt{x^2 + 16}$

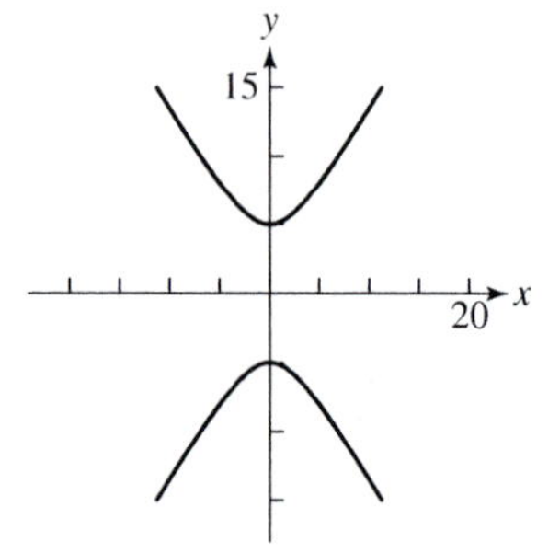

15. The center (h, k) is $(-3, 1)$. Since $a^2 = 16$ and $b^2 = 4$, we have $a = 4$ and $b = 2$. The vertices are at $(-3 \pm 4, 1)$ or $(-7, 1)$ and $(1, 1)$.

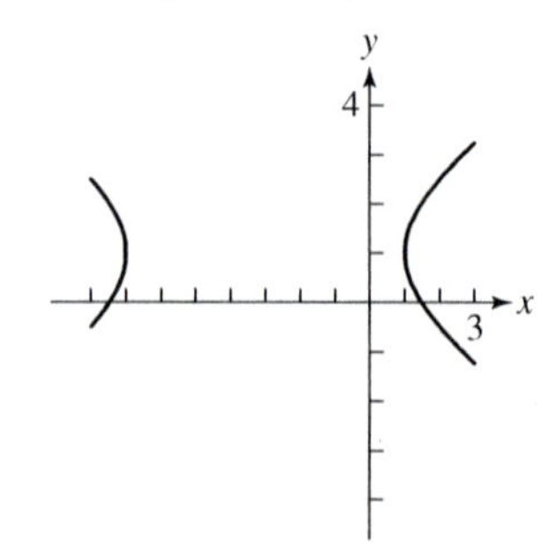

17.

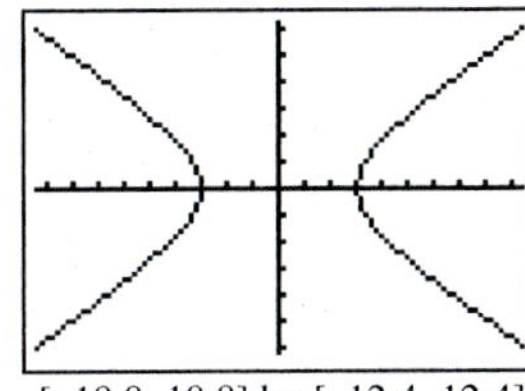

[–18.8, 18.8] by [–12.4, 12.4]

$y = \pm\frac{2}{3}\sqrt{x^2 - 36}$

19.

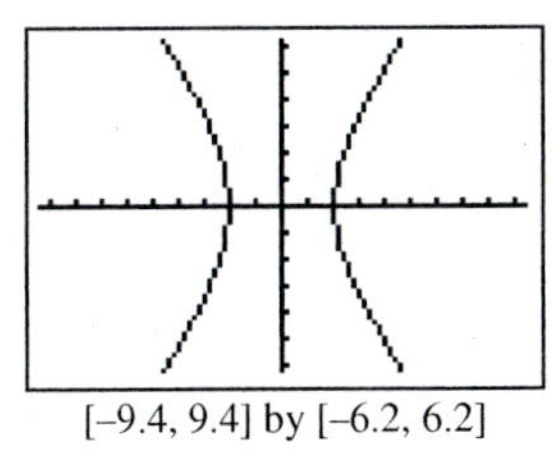

[–9.4, 9.4] by [–6.2, 6.2]

$y = \pm\frac{3}{2}\sqrt{x^2 - 4}$

21.

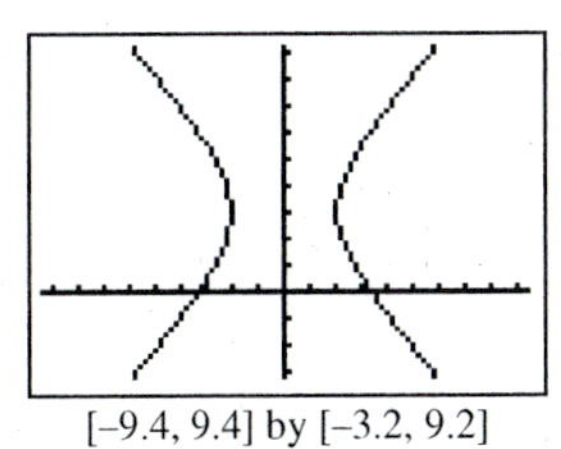

[–9.4, 9.4] by [–3.2, 9.2]

$y = 3 \pm \frac{1}{2}\sqrt{5x^2 - 20}$

23. $c = 3$ and $a = 2$, so $b = \sqrt{c^2 - a^2} = \sqrt{5}$: $\frac{x^2}{4} - \frac{y^2}{5} = 1$

25. $c = 15$ and $b = 4$, so $a = \sqrt{c^2 - b^2} = \sqrt{209}$:
$\frac{y^2}{16} - \frac{x^2}{209} = 1$

27. $a = 5$ and $c = ea = 10$, so $b = \sqrt{100 - 25} = 5\sqrt{5}$:
$\frac{x^2}{25} - \frac{y^2}{75} = 1$

29. $b = 5, a = \sqrt{c^2 - b^2} = \sqrt{169 - 25} = 12$:
$\frac{y^2}{144} - \frac{x^2}{25} = 1$

31. The center (h, k) is $(2, 1)$ (the midpoint of the transverse axis endpoints); $a = 2$, half the length of the transverse axis. And $b = 3$, half the length of the conjugate axis.
$\frac{(y-1)^2}{4} - \frac{(x-2)^2}{9} = 1$

33. The center (h, k) is $(2, 3)$ (the midpoint of the transverse axis); $a = 3$, half the length of the transverse axis.

Since $|b/a| = \frac{4}{3}$, $b = 4$: $\frac{(x-2)^2}{9} - \frac{(y-3)^2}{16} = 1$

35. The center (h, k) is $(-1, 2)$, the midpoint of the transverse axis. $a = 2$, half the length of the transverse axis. The center-to-focus distance is $c = 3$, so $b = \sqrt{c^2 - a^2}$
$= \sqrt{5}$: $\frac{(x+1)^2}{4} - \frac{(y-2)^2}{5} = 1$

37. The center (h, k) is $(-3, 6)$, the midpoint of the transverse axis. $a = 5$, half the length of the transverse axis. The center-to-focus distance $c = ea$
$= 2 \cdot 5 = 10$, so $b = \sqrt{c^2 - a^2} = \sqrt{100 - 25} = 5\sqrt{5}$
$\frac{(y-6)^2}{25} - \frac{(x+3)^2}{75} = 1$

For #39–41, a hyperbola with equation
$\frac{(x-h)^2}{a^2} - \frac{(y-k)^2}{b^2} = 1$ has center (h, k) vertices $(h \pm a, k)$, and foci $(h \pm c, k)$ where $c = \sqrt{a^2 + b^2}$.
A hyperbola with equation $\frac{(y-k)^2}{a^2} - \frac{(x-h)^2}{b^2} = 1$ has center (h, k), vertices $(h, k \pm a)$, and foci $(h, k \pm c)$ where again $c = \sqrt{a^2 + b^2}$.

39. Center $(-1, 2)$; Vertices: $(-1 \pm 12, 2) = (11, 2), (-13, 2)$; Foci: $(-1 \pm 13, 2) = (12, 2), (-14, 2)$

41. Center $(2, -3)$; Vertices: $(2, -3 \pm 8) = (2, 5), (2, -11)$; Foci: $(2, -3 \pm \sqrt{145})$

43.

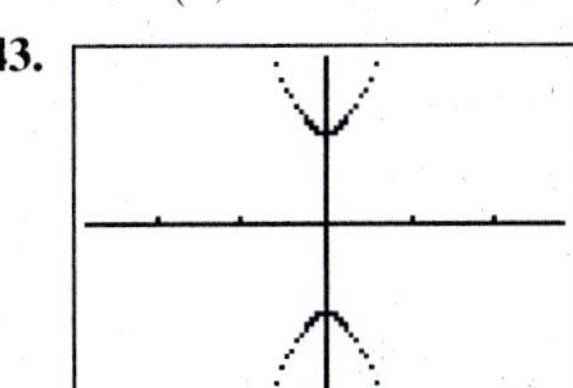

[–14.1, 14.1] by [–9.3, 9.3]

$y = 5/\cos t$, $x = 2 \tan t$

45.

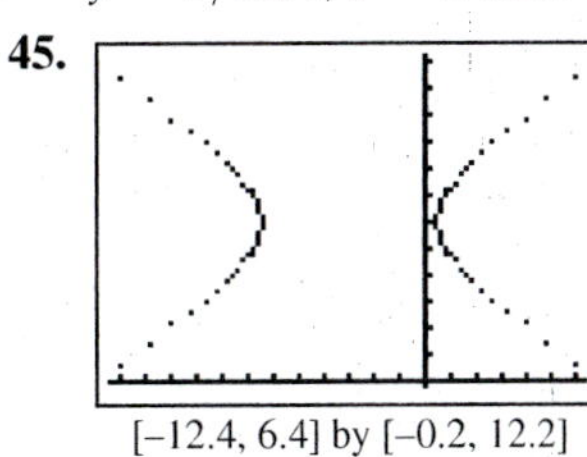

[–12.4, 6.4] by [–0.2, 12.2]

$x = -3 + 2\sqrt{3}/\cos t$, $y = 6 + \sqrt{5}\tan t$

47.

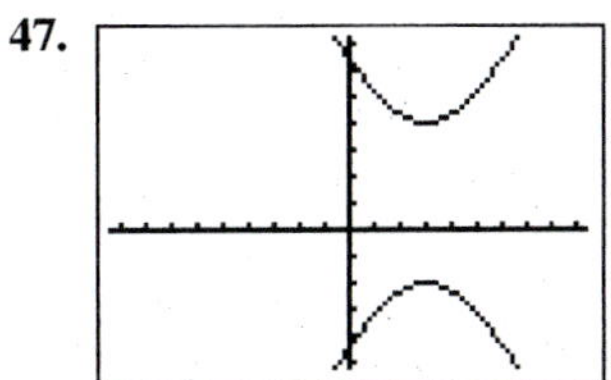

[–9.4, 9.4] by [–5.2, 7.2]

Divide the entire equation by 36. Vertices: $(3, -2)$ and $(3, 4)$, Foci: $(3, 1 \pm \sqrt{13})$, $e = \frac{\sqrt{13}}{3}$.

For #49, complete the squares in x and y, then write the equation in standard form. (The first one is done in detail; the other shows just the final form.) As in the previous problems, the values of h, k, a, and b can be "read" from the equation $\pm\frac{(x-h)^2}{a^2} \mp \frac{(y-k)^2}{b^2} = 1$. The asymptotes are $y - k = \pm\frac{b}{a}(x - h)$. If the x term is positive, the transverse axis endpoints are $(h \pm a, k)$; otherwise the endpoints are $(h, k \pm b)$.

49.

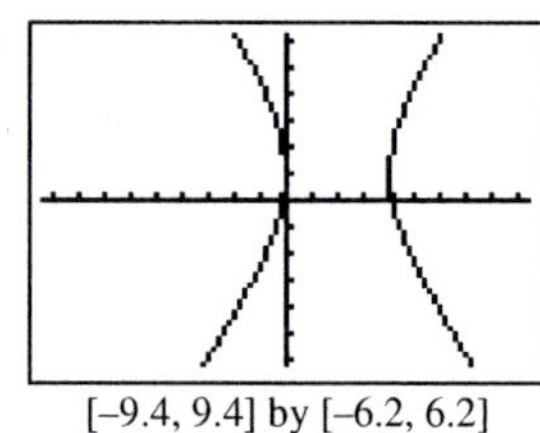

[−9.4, 9.4] by [−6.2, 6.2]

$9x^2 - 4y^2 - 36x + 8y - 4 = 0$ can be rewritten as $9(x^2 - 4x) - 4(y^2 - 2y) = 4$. This is equivalent to $9(x^2 - 4x + 4) - 4(y^2 - 2y + 1) = 4 + 36 - 4$, or $9(x - 2)^2 - 4(y - 1)^2 = 36$. Divide both sides by 36 to obtain $\frac{(x-2)^2}{4} - \frac{(y-1)^2}{9} = 1$. Vertices: $(0, 1)$ and $(4, 1)$. Foci: $(2 \pm \sqrt{13}, 1)$, $e = \frac{\sqrt{13}}{2}$

51. $a = 2$, $(h, k) = (0, 0)$ and the hyperbola opens to the left and right, so $\frac{x^2}{4} - \frac{y^2}{b^2} = 1$. Using $(3, 2)$: $\frac{9}{4} - \frac{4}{b^2} = 1$, $9b^2 - 16 = 4b^2$, $5b^2 = 16$, $b^2 = \frac{16}{5}$; $\frac{x^2}{4} - \frac{5y^2}{16} = 1$

53. Consider Figure 8.24(b). Label $(0, c)$ as point F_1, label $(0, -c)$ as point F_2 and consider any point $P(x, y)$ along the hyperbola. By definition, $PF_1 - PF_2 = \pm 2a$, with $c > a \geq 0$

$$\sqrt{(x-0)^2 + (y-(-c))^2} - \sqrt{(x-0)^2 + (y-c)^2} = \pm 2a$$

$$\sqrt{x^2 + (y+c)^2} = \pm 2a + \sqrt{x^2 + (y-c)^2}$$

$$x^2 + y^2 + 2cy + c^2 = 4a^2 \pm 4a\sqrt{x^2 + (y-c)^2} + x^2 + y^2 - 2cy + c^2$$

$$\pm a\sqrt{x^2 + (y-c)^2} = a^2 - cy$$

$$a^2(x^2 + y^2 - 2cy + c^2) = a^4 - 2a^2cy + c^2y^2$$

$$-a^2x^2 + (c^2 - a^2)y^2 = a^2(c^2 - a^2)$$

$$b^2y^2 - a^2x^2 = a^2b^2$$

$$\frac{y^2}{a^2} - \frac{x^2}{b^2} = 1$$

55.

$$c - a = 120,\ b^2 = 250a$$

$$c^2 - a^2 = b^2$$

$$(a + 120)^2 - a^2 = 250a$$

$$a^2 + 240a + 14{,}400 - a^2 = 250a$$

$$10a = 14{,}400$$

$$a = 1440 \text{ Gm}$$

$a = 1440$ Gm, $b = 600$ Gm, $c = 1560$, $e = \frac{1560}{1440} = \frac{13}{12}$.

The Sun is centered at focus $(c, 0) = (1560, 0)$.

57. The *Princess Ann* is located at the intersection of two hyperbolas: one with foci O and R, and the other with foci O and Q. For the first of these, the center is $(0, 40)$, so the center-to-focus distance is $c = 40$ mi. The transverse axis length is $2b = (323.27\ \mu\text{sec})(980\ \text{ft}/\mu\text{sec}) = 316{,}804.6$ ft ≈ 60 mi. Then $a \approx \sqrt{40^2 - 30^2} = \sqrt{700}$ mi. For the other hyperbola, $c = 100$ mi, $2a = (646.53\ \mu\text{sec})(980\ \text{ft}/\mu\text{sec}) = 633599.4$ ft ≈ 120 mi, and $b \approx \sqrt{100^2 - 60^2} = 80$ mi. The two equations are therefore

$$\frac{(y-40)^2}{900} - \frac{x^2}{700} = 1 \text{ and } \frac{(x-100)^2}{3600} - \frac{y^2}{6400} = 1.$$

The intersection of the upper branch of the first hyperbola and the right branch of the second hyperbola (found graphically) is approximately $(886.67, 1045.83)$. The ship is located about 887 miles east and 1046 miles north of point O – a bearing and distance of about 40.29° and 1371.11 miles, respectively.

59.

$$\frac{x^2}{4} - \frac{y^2}{9} = 1$$

$$x - \frac{2\sqrt{3}}{3}y = -2$$

Solve the second equation for x and substitute into the first equation.

$$x = \frac{2\sqrt{3}}{3}y - 2$$

$$\frac{1}{4}\left(\frac{2\sqrt{3}}{3}y - 2\right)^2 - \frac{y^2}{9} = 1$$

$$\frac{1}{4}\left(\frac{4}{3}y^2 - \frac{8\sqrt{3}}{3}y + 4\right) - \frac{y^2}{9} = 1$$

$$\frac{2}{9}y^2 - \frac{2\sqrt{3}}{3}y = 0$$

$$\frac{2}{9}y(y - 3\sqrt{3}) = 0$$

$y = 0$ or $y = 3\sqrt{3}$

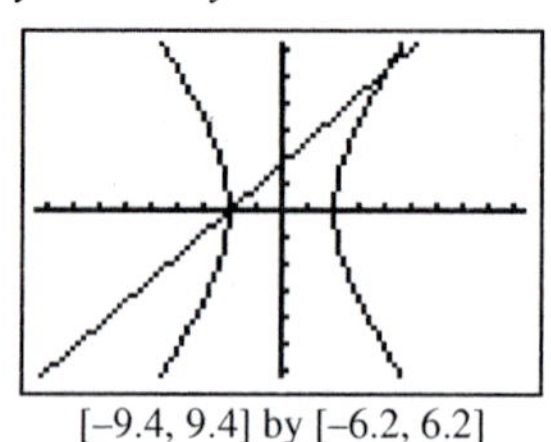

[−9.4, 9.4] by [−6.2, 6.2]

Solutions: $(-2, 0)$, $(4, 3\sqrt{3})$

61. (a)

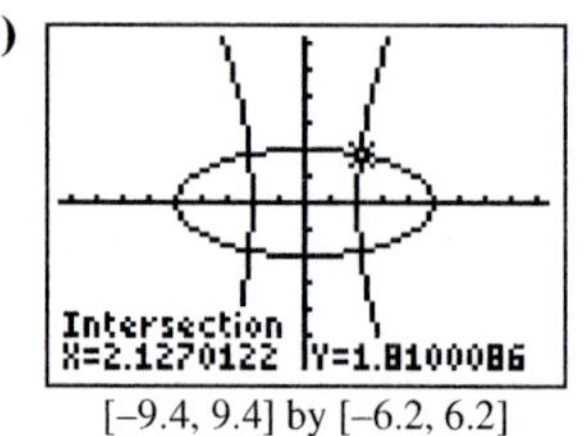

[−9.4, 9.4] by [−6.2, 6.2]

There are four solutions: $(\pm 2.13, \pm 1.81)$

(b) The exact solutions are $\left(\pm 10\sqrt{\frac{29}{641}}, \pm 10\sqrt{\frac{21}{641}}\right)$.

63. True. The distance is $c - a = a(c/a - 1) = a(e - 1)$.

65. $\frac{x^2}{4} - \frac{y^2}{1} = 1$, so $c = \sqrt{4 + 1}$ and the foci are each $\sqrt{5}$ units away horizontally from $(0, 0)$. The answer is B.

67. Completing the square twice, and dividing to obtain 1 on the right, turns the equation into $\frac{(y+3)^2}{4} - \frac{(x-2)^2}{12} = 1$. The answer is B.

69. (a–d)

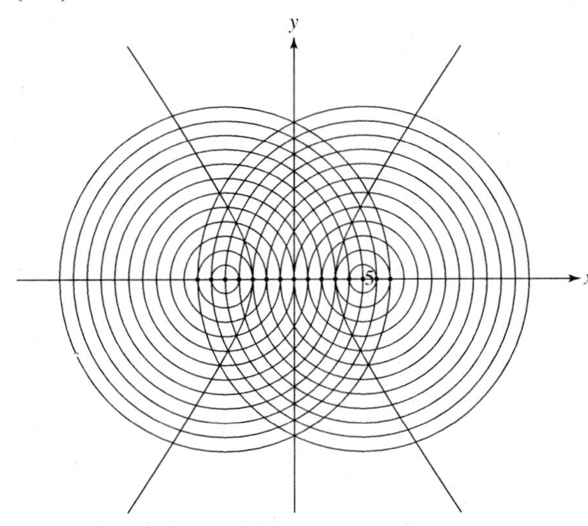

(e) $a = 3, c = 5, b = 4;$
$x^2/9 - y^2/16 = 1$

71. From Section 8.2, Question #75, we have $Ax^2 + Cy^2 + Dx + Ey + F = 0$ becomes

$$\frac{4A^2C\left(x + \frac{D}{2A}\right)^2}{CD^2 + AE^2 - 4ACF} + \frac{4AC^2\left(y + \frac{F}{2C}\right)^2}{CD^2 + AE^2 - 4ACF} = 1$$

Since $AC < 0$ means that either ($A < 0$ and $C > 0$) or ($A > 0$ and $C < 0$), either ($4A^2C < 0$ and $4AC^2 > 0$), or ($4A^2C > 0$ and $4AC^2 < 0$). In the equation above, that means that the $+$ sign will become a $(-)$ sign once all the values $A, B, C, D, E,$ and F are determined, which is exactly the equation of the hyperbola. Note that if $A > 0$ and $C < 0$, the equation becomes:

$$\frac{4AC^2\left(y + \frac{E}{2C}\right)^2}{CD^2 + AE^2 - 4ACF} - \frac{|4A^2C|\left(x + \frac{D}{2A}\right)^2}{CD^2 + AE^2 - 4ACF} = 1$$

If $A < 0$ and $C > 0$, the equation becomes:

$$\frac{4A^2C\left(x + \frac{D}{2A}\right)^2}{CD^2 + AE^2 - 4ACF} - \frac{|4AC^2|\left(y + \frac{E}{2C}\right)^2}{CD^2 + AE^2 - 4ACF} = 1$$

73. The asymptotes of the first hyperbola are $y = \pm\frac{b}{a}(x - h) + k$ and the asymptotes of the second hyperbola are $y = \pm\frac{b}{a}(x - h) + k$; they are the same.

[Note that in the second equation, the standard usage of $a + b$ has been revised.] The conjugate axis for hyperbola 1 is $2b$, which is the same as the transverse axis for hyperbola 2. The conjugate axis for hyperbola 2 is $2a$, which is the same as the transverse axis of hyperbola 1.

75. The standard forms involved multiples of $x, x^2, y,$ and y^2, as well as constants; therefore they can be rewritten in the general form $Ax^2 + Cy^2 + Dx + Ey + F = 0$ (none of the standard forms we have seen require a Bxy term). For example, rewrite $y = ax^2$ as $ax^2 - y = 0$; this is the general form with $A = a$ and $E = -1$, and all others 0. Similarly, the hyperbola $\frac{y^2}{b^2} - \frac{x^2}{a^2} = 1$ can be put in standard form with $A = -\frac{1}{a^2}, C = \frac{1}{b^2}, F = -1$, and $B = D = E = 0$.

■ Section 8.4 Translation and Rotation of Axes

Quick Review 8.4

1. $\cos 2\alpha = \frac{5}{13}$

3. $\cos 2\alpha = \frac{1}{2}$

5. $2\alpha = \frac{\pi}{2}$, so $\alpha = \frac{\pi}{4}$

7. $\cos 2\alpha = 2\cos^2\alpha - 1 = \frac{3}{5}, 2\cos^2\alpha = \frac{8}{5}, \cos^2\alpha = \frac{4}{5},$
$\cos\alpha = \frac{2}{\sqrt{5}}$

9. $\cos 2\alpha = 1 - 2\sin^2\alpha = \frac{5}{6}, -2\sin^2\alpha = -\frac{1}{6} \Rightarrow$
$\sin^2\alpha = \frac{1}{12} \Rightarrow \sin\alpha = \sqrt{\frac{1}{12}} \Rightarrow \sin\alpha = \frac{1}{\sqrt{12}}$

Section 8.4 Exercises

1. Use the quadratic formula with $a = 1, b = 10$, and $c = x^2 - 6x + 18$. Then $b^2 - 4ac = (10)^2 - 4(x^2 - 6x + 18) = -4x^2 + 24x + 28 = 4(-x^2 + 6x + 7)$, and

$$y = \frac{-10 \pm \sqrt{4(-x^2 + 6x + 7)}}{2} = -5 \pm \sqrt{-x^2 + 6x + 7}$$

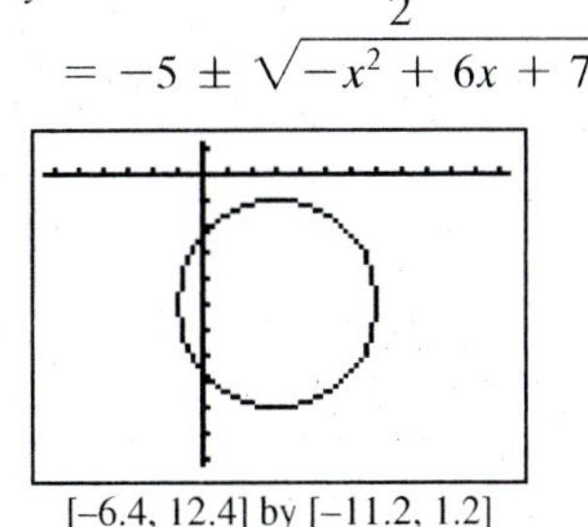
[–6.4, 12.4] by [–11.2, 1.2]

3. Use the quadratic formula with $a = 1, b = -8$, and $c = -8x + 8$. Then $b^2 - 4ac = (-8)^2 - 4(-8x + 8) = 32x + 32 = 32(x + 1)$, and

$$y = \frac{8 \pm \sqrt{32(x + 1)}}{2} = 4 \pm 2\sqrt{2x + 2}$$

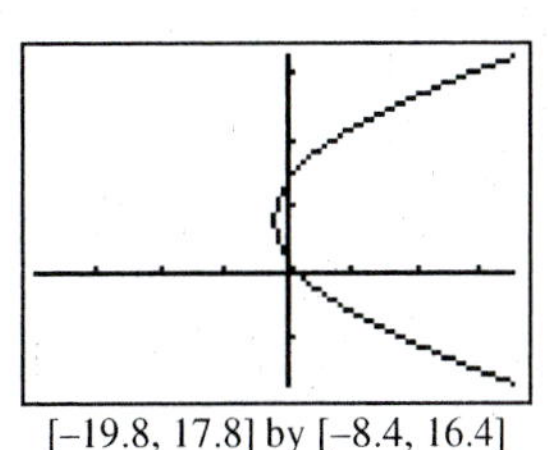
[–19.8, 17.8] by [–8.4, 16.4]

5. $-4xy + 16 = 0 \Rightarrow -4xy = -16 \Rightarrow y = 4/x$

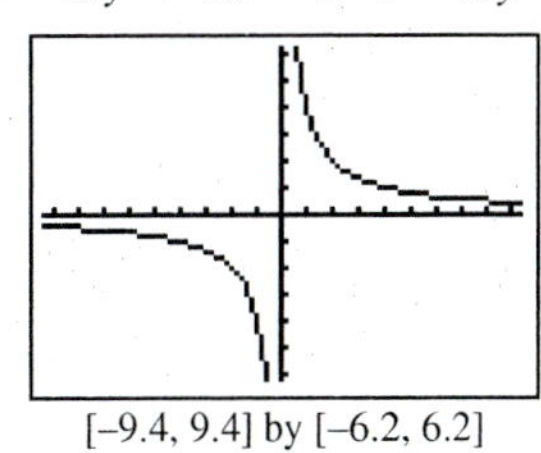
[–9.4, 9.4] by [–6.2, 6.2]

7. $xy - y - 8 = 0 \Rightarrow y(x - 1) = 8 \Rightarrow y = 8/(x - 1)$

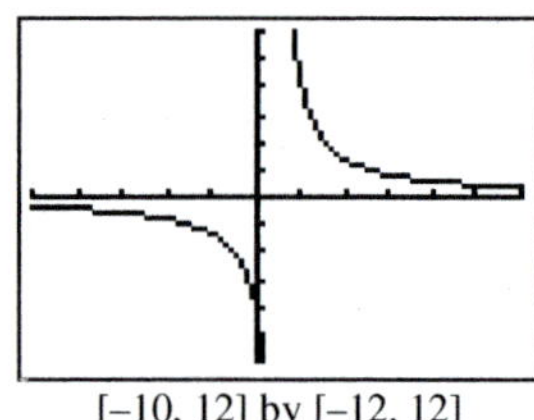

$[-10, 12]$ by $[-12, 12]$

9. Use the quadratic formula with $a = 3$, $b = 4 - x$, and $c = 2x^2 - 3x - 6$. Then $b^2 - 4ac$
$= (4 - x)^2 - 12(2x^2 - 3x - 6) = -23x^2 + 28x + 88$,
and $y = \dfrac{x - 4 \pm \sqrt{-23x^2 + 28x + 88}}{6}$

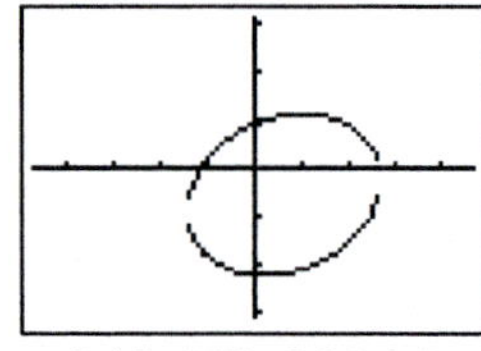

$[-4.7, 4.7]$ by $[-3.1, 3.1]$

11. Use the quadratic formula with $a = 8$, $b = 4 - 4x$, and $c = 2x^2 - 10x - 13$.
Then $b^2 - 4ac = (4 - 4x)^2 - 32(2x^2 - 10x - 13)$
$= -48x^2 + 288x + 432 = 48(-x^2 + 6x + 9)$, and
$y = \dfrac{4x - 4 \pm \sqrt{48(-x^2 + 6x + 9)}}{16}$

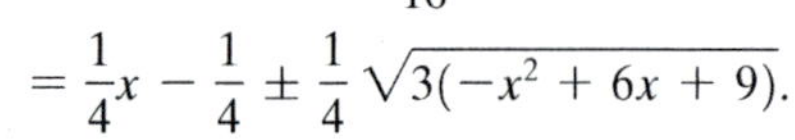

$= \dfrac{1}{4}x - \dfrac{1}{4} \pm \dfrac{1}{4}\sqrt{3(-x^2 + 6x + 9)}$.

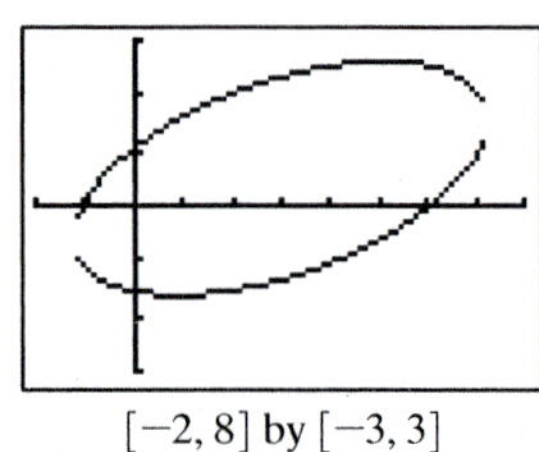

$[-2, 8]$ by $[-3, 3]$

13. $h = 0$, $k = 0$ and the parabola opens downward, so $4py = x^2$). Using $(2, -1)$: $-4p = 4$, $p = -1$. The standard form is $x^2 = -4y$.

15. $h = 0$, $k = 0$ and the hyperbola opens to the right and left, so $a = 3$, and $b = 4$. The standard form is $\dfrac{x^2}{9} - \dfrac{y^2}{16} = 1$.

For #17–19, recall that $x' = x - h$ and $y' = y - k$.

17. $(x', y') = (4, -1)$

19. $(x', y') = (5, -3 - \sqrt{5})$

21. $4(y^2 - 2y) - 9(x^2 + 2x) = 41$, so
$4(y - 1)^2 - 9(x + 1)^2 = 41 + 4 - 9 = 36$. Then
$\dfrac{(y - 1)^2}{9} - \dfrac{(x + 1)^2}{4} = 1$. This is a hyperbola, with $a = 3$, $b = 2$, and $c = \sqrt{13}$.
$\dfrac{(y')^2}{9} - \dfrac{(x')^2}{4} = 1$.

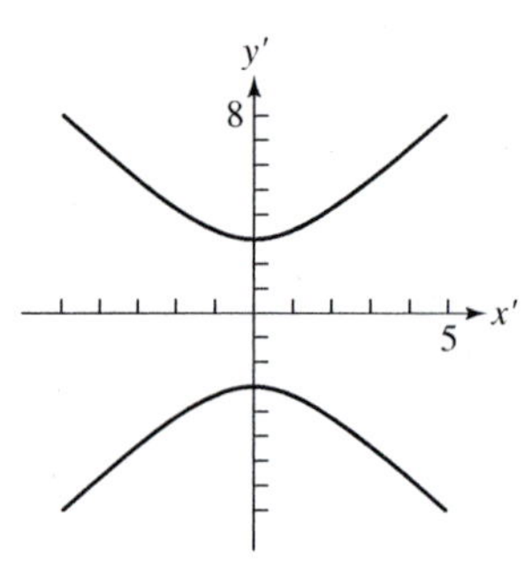

23. $y - 2 = (x + 1)^2$, a parabola. The vertex is $(h, k) = (-1, 2)$, so $y' = (x')^2$.

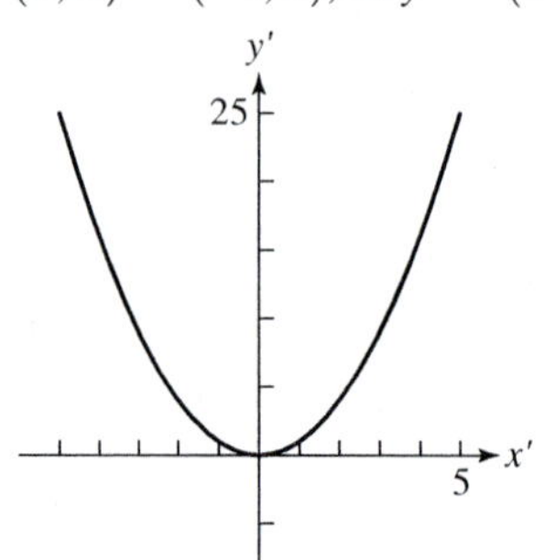

25. $9(x^2 - 2x) + 4(y^2 + 4y) = 11$, so
$9(x - 1)^2 + 4(y + 2)^2 = 11 + 9 + 16 = 36$. Then
$\dfrac{(x - 1)^2}{4} + \dfrac{(y + 2)^2}{9} = 1$.
This is an ellipse, with $a = 2$, $b = 3$, and $c = \sqrt{5}$.
Foci: $(1, -2 \pm \sqrt{5})$. Center $(1, -2)$, so
$\dfrac{(x')^2}{4} + \dfrac{(y')^2}{9} = 1$.

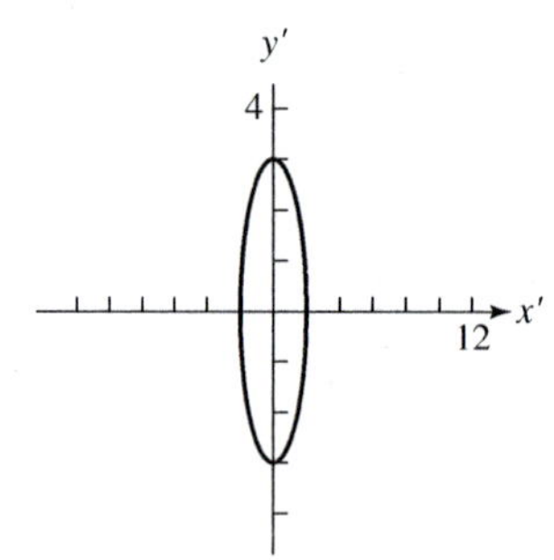

27. $8(x - 2) = (y - 2)^2$, a parabola. The vertex is $(h, k) = (2, 2)$, so $8x' = (y')^2$.

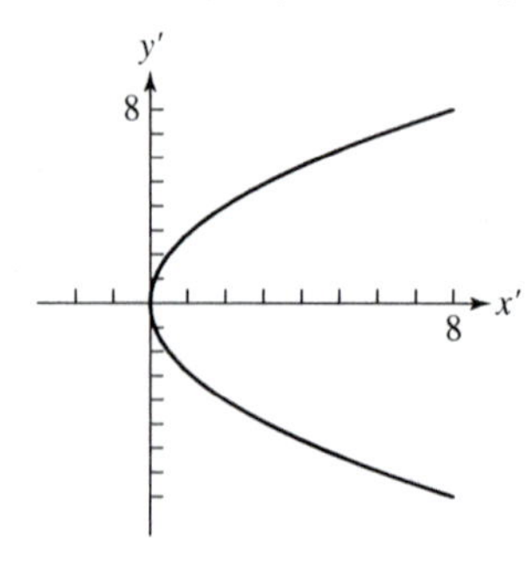

29. $2(x^2 + 2x) - y^2 = -6$, so $2(x + 1)^2 - y^2 = -6 + 2 = -4$, Then $\frac{y^2}{4} - \frac{(x + 1)^2}{2} = 1$. This is a hyperbola, with $a = \sqrt{2}$, $b = 2$, and $c = \sqrt{6}$. Foci: $(-1, \pm\sqrt{6})$. Center $(-1, 0)$, so $\frac{(y')^2}{4} - \frac{(x')^2}{2} = 1$.

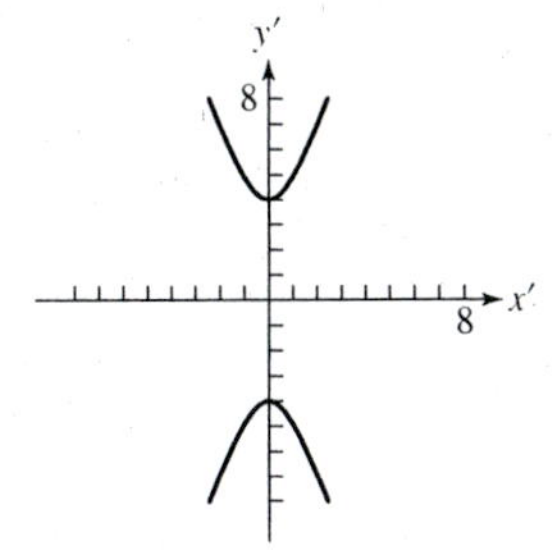

31. The horizontal distance from O to P is $x = h + x' = x' + h$, and the vertical distance from O to P is $y = k + y' = y' + k$.

For #33–35, recall that $x' = x \cos \alpha + y \sin \alpha$ and $y' = -x \sin \alpha + y \cos \alpha$.

33. $(x', y') = \left(-2 \cos \frac{\pi}{4} + 5 \sin \frac{\pi}{4}, 2 \sin \frac{\pi}{4} + 5 \cos \frac{\pi}{4}\right) = \left(\frac{3\sqrt{2}}{2}, \frac{7\sqrt{2}}{2}\right)$

35. $\alpha \approx 1.06$, $(x', y') = (-5 \cos (1.06) - 4 \sin (1.06), 5 \sin (1.06) - 4 \cos (1.06)) \approx (-5.94, 2.38)$

For #37–39, use the discriminant $B^2 - 4AC$ to determine the type of conic. Then use the relationship of $\cot 2\alpha = \frac{A - C}{B}$ to determine the angle of rotation.

37. $B^2 - 4AC = 1 > 0$, hyperbola; $\cot 2\alpha = 0$, so $\alpha = \frac{\pi}{4}$.

Translating, $\left(\frac{x' - y'}{\sqrt{2}}\right)\left(\frac{x' + y'}{\sqrt{2}}\right) = 8$,

$\frac{(x')^2}{16} - \frac{(y')^2}{16} = 1$, $y' = \pm\sqrt{(x')^2 - 16}$

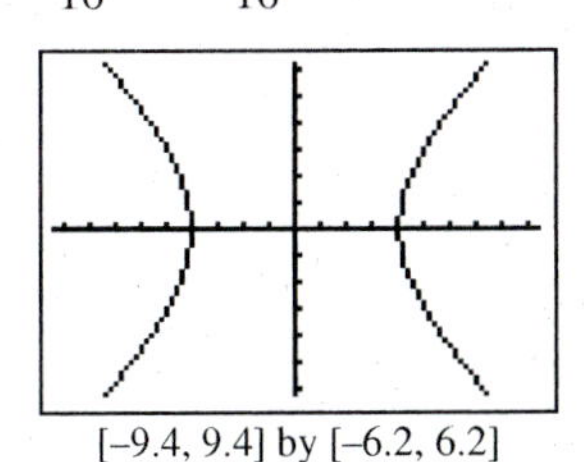

[–9.4, 9.4] by [–6.2, 6.2]

39. $B^2 - 4AC = 3 - 4(2)(1) = -5 < 0$, ellipse;

$\cot 2\alpha = \frac{1}{\sqrt{3}}$, $\alpha = \frac{\pi}{6}$. Translating,

$x = x' \cos \frac{\pi}{6} - y' \sin \frac{\pi}{6}$, $y = x' \sin \frac{\pi}{6} + y' \cos \frac{\pi}{6}$, so the equation becomes $\frac{5(x')^2}{2} + \frac{(y')^2}{2} = 10$,

$\frac{(x')^2}{4} + \frac{(y')^2}{20} = 1$

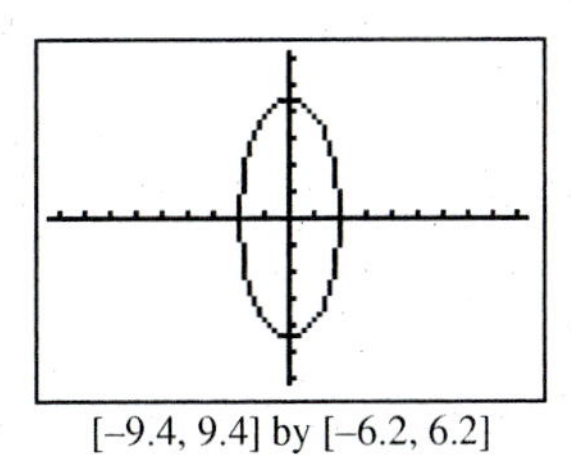

[–9.4, 9.4] by [–6.2, 6.2]

41. $B^2 - 4AC = -176 < 0$, ellipse. Use the quadratic formula with $a = 9$, $b = -20x$, and $c = 16x^2 - 40$. Then $b^2 - 4ac = (-20x)^2 - 4(9)(16x^2 - 40) = -176x^2 + 1440 = 16(-11x^2 + 90)$, and

$$y = \frac{20x \pm \sqrt{16(-11x^2 + 90)}}{18} = \frac{10x \pm 2\sqrt{90 - 11x^2}}{9}$$

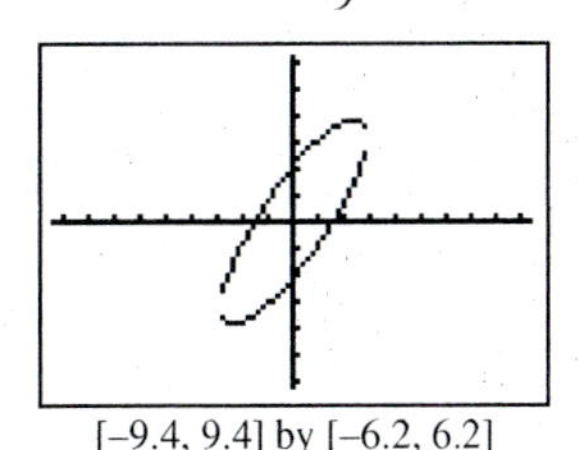

[–9.4, 9.4] by [–6.2, 6.2]

$\cot 2\alpha = -\frac{7}{20}$, $\alpha \approx 0.954 \approx 54.65°$

43. $B^2 - 4AC = 16 - 4(1)(10) = -24 < 0$; ellipse

45. $B^2 - 4AC = 36 - 4(9)(1) = 0$; parabola

47. $B^2 - 4AC = 16 - 4(8)(2) = -48 < 0$; ellipse

49. $B^2 - 4AC = 0 - 4(1)(-3) = 12 > 0$; hyperbola

51. $B^2 - 4AC = 4 - 4(4)(1) = -12 < 0$; ellipse

53. In the new coordinate system, the center $(x', y') = (0, 0)$, the vertices occur at $(\pm 3, 0)$ and the foci are located at $(\pm 3\sqrt{2}, 0)$. We use $x = x' \cos \frac{\pi}{4} - y' \sin \frac{\pi}{4}$, $y = x' \sin \frac{\pi}{4} + y' \cos \frac{\pi}{4}$ to translate "back." Under the "old" coordinate system, the center $(x, y) = (0, 0)$, the vertices occured at $\left(\frac{3\sqrt{2}}{2}, \frac{3\sqrt{2}}{2}\right)$ and $\left(-\frac{3\sqrt{2}}{2}, -\frac{3\sqrt{2}}{2}\right)$. and the foci are located at $(3, 3)$ and $(-3, -3)$.

55. Answers will vary. One possible answer: Using the geometric relationships illustrated, it is clear that $x = x' \cos \alpha - y' \cos\left(\frac{\pi}{2} - \alpha\right) = x' \cos \alpha - y' \sin \alpha$ and that $y = x' \cos\left(\frac{\pi}{2} - \alpha\right) + y' \cos \alpha = x' \sin \alpha + y' \cos \alpha$.

57. True. The Bxy term is missing and so the rotation angle α is zero.

59. Eliminating the cross-product term requires rotation, not translation. The answer is B.

61. Completing the square twice, and dividing to obtain 1 on the right, turns the equation into

$$\frac{(x - 1)^2}{16} + \frac{(y + 2)^2}{9} = 1$$

The vertices lie 4 units to the left and right of center $(1, -2)$. The answer is A.

63. (a) The rotated axes pass through the old origin with slopes of ± 1, so the equations are $y = \pm x$.

(b) The location of $(x'', y'') = (0, 0)$ in the xy system can be found by reversing the transformations. In the $x'y'$ system, $(x'', y'') = (0, 0)$ has coordinates $(h, k) = \left(\frac{21}{\sqrt{5}}, \frac{3\sqrt{5}}{10}\right)$. The coordinates of this point in the xy system are then given by the second set of rotation formulas; with $\cos\alpha = \frac{1}{\sqrt{5}}$, $\sin\alpha = \frac{2}{\sqrt{5}}$:

$$x = \frac{21}{\sqrt{5}}\left(\frac{1}{\sqrt{5}}\right) - \frac{3\sqrt{5}}{10}\left(\frac{2}{\sqrt{5}}\right) = \frac{18}{5}$$

$$y = \frac{21}{\sqrt{5}}\left(\frac{2}{\sqrt{5}}\right) + \frac{3\sqrt{5}}{10}\left(\frac{1}{\sqrt{5}}\right) = \frac{87}{10}$$

The $x''y''$ axes pass through the point $(x, y) = \left(\frac{18}{5}, \frac{87}{10}\right)$ with slopes of $\frac{2/\sqrt{5}}{1/\sqrt{5}} = 2$ and its negative reciprocal, $-\frac{1}{z}$. Using this information to write linear equations in point-slope form, and then converting to slope-intercept form, we obtain

$$y = 2x + \frac{3}{2}$$

$$y = -\frac{1}{2}x + \frac{21}{2}$$

65. First, consider the linear terms:

$$\begin{aligned} Dx + Ey &= D(x'\cos\alpha - y'\sin\alpha) \\ &\quad + E(x'\sin\alpha + y'\cos\alpha) \\ &= (D\cos\alpha + E\sin\alpha)x' \\ &\quad + (E\cos\alpha - D\sin\alpha)y' \end{aligned}$$

This shows that $Dx + Ey = D'x' + E'y'$, where $D' = D\cos\alpha + E\sin\alpha$ and $E' = E\cos\alpha - D\sin\alpha$.

Now, consider the quadratic terms:

$Ax^2 + Bxy + Cy^2 = A(x'\cos\alpha - y'\sin\alpha)^2 + B(x'\cos\alpha - y'\sin\alpha)(x'\sin\alpha + y'\cos\alpha) + C(x'\sin\alpha + y'\cos\alpha)^2$

$$\begin{aligned} &= A(x'^2\cos^2\alpha - 2x'y'\cos\alpha\sin\alpha + y'^2\sin^2\alpha) \\ &\quad + B(x'^2\cos\alpha\sin\alpha + x'y'\cos^2\alpha - x'y'\sin^2\alpha \\ &\quad - y'^2\sin\alpha\cos\alpha) + C(x'^2\sin^2\alpha + 2x'y'\sin\alpha\cos\alpha \\ &\quad + y'^2\cos^2\alpha) \\ &= (A\cos^2\alpha + B\cos\alpha\sin\alpha + C\sin^2\alpha)x'^2 \\ &\quad + [B(\cos^2\alpha - \sin^2\alpha) \\ &\quad + 2(C - A)(\sin\alpha\cos\alpha)]x'y' \\ &\quad + (C\cos^2\alpha - B\cos\alpha\sin\alpha + A\sin^2\alpha)y'^2 \\ &= (A\cos^2\alpha + B\cos\alpha\sin\alpha + C\sin^2\alpha)x'^2 \\ &\quad + [B\cos 2\alpha + (C - A)\sin 2\alpha]x'y' \\ &\quad + (C\cos^2\alpha - B\cos\alpha\sin\alpha + A\sin^2\alpha)y'^2 \end{aligned}$$

This shows that $Ax^2 + Bxy + Cy^2 = A'x'^2 + B'x'y' + C'y'^2$, where $A' = A\cos^2\alpha + B\cos\alpha\sin\alpha + C\sin^2\alpha$, $B' = B\cos 2\alpha + (C - A)\sin 2\alpha$, and $C' = C\cos^2\alpha - B\cos\alpha\sin\alpha + A\sin^2\alpha$.

The results above imply that if the formulas for A', B', C', D', and F' are applied, then $A'x'^2 + B'x'y' + C'y'^2 + D'x' + E'y' + F' = 0$ is equivalent to $Ax^2 + Bxy + Cy^2 + Dx + Ey + F = 0$. Therefore, the formulas are correct.

67. Making the substitutions $x = x'\cos\alpha - y'\sin\alpha$ and $y = x'\sin\alpha + y'\cos\alpha$, we find that:

$$B'x'y' = (B\cos^2\alpha - B\sin^2\alpha + 2C\sin\alpha\cos\alpha - 2A\sin\alpha\cos\alpha)x'y'$$

$$Ax'^2 = (A\cos^2\alpha + B\sin\alpha\cos\alpha + C\sin^2\alpha)(x')^2$$

$$Cy'^2 = (A\sin^2\alpha + C\cos^2\alpha - B\cos\alpha\sin\alpha)(y')^2$$

$$\begin{aligned} B'^2 - 4A'C' &= (B\cos(2\alpha) - (A - C)\sin(2\alpha))^2 \\ &\quad - 4(A\cos^2\alpha + B\sin\alpha\cos\alpha + C\sin^2\alpha) \\ &\quad (A\sin^2\alpha - B\sin\alpha\cos\alpha + C\cos^2\alpha) \\ &= \frac{1}{2}B^2\cos(4\alpha) + \frac{1}{2}B^2 + BC\sin(4\alpha) - BA\sin(4\alpha) \\ &\quad + \frac{1}{2}C^2 - \frac{1}{2}C^2\cos(4\alpha) - CA + CA\cos(4\alpha) \\ &\quad + \frac{1}{2}A^2 - \frac{1}{2}A^2\cos(4\alpha) - 4\left(\frac{1}{2}A\cos(2\alpha) + \frac{1}{2}A\right. \\ &\quad \left. + \frac{1}{2}B\sin(2\alpha) + \frac{1}{2}C - \frac{1}{2}C\cos(2\alpha)\right) \\ &\quad \left(\frac{1}{2}A - \frac{1}{2}A\cos(2\alpha) + \frac{1}{2}C\cos(2\alpha) + \frac{1}{2}C\right. \\ &\quad \left. - \frac{1}{2}B\sin(2\alpha)\right) \\ &= \frac{1}{2}B^2\cos(4\alpha) + \frac{1}{2}B^2 + BC\sin(4\alpha) - BA\sin(4\alpha) \\ &\quad + \frac{1}{2}C^2 - \frac{1}{2}C^2\cos(4\alpha) - CA + CA\cos(4\alpha) + \frac{1}{2}A^2 \\ &\quad - \frac{1}{2}A^2\cos(4\alpha) - BC\sin(4\alpha) + BA\sin(4\alpha) - 3AC \\ &\quad - \frac{1}{2}C^2 - \frac{1}{2}A^2 + \frac{1}{2}A^2\cos(4\alpha) + \frac{1}{2}B^2 - \frac{1}{2}B^2\cos(4\alpha) \\ &\quad + \frac{1}{2}C^2\cos(4\alpha) - AC\cos(4\alpha) \\ &= B^2 - 4AC. \end{aligned}$$

69. Intersecting lines: $x^2 + xy = 0$ can be rewritten as $x = 0$ (the y-axis) and $y = -x$

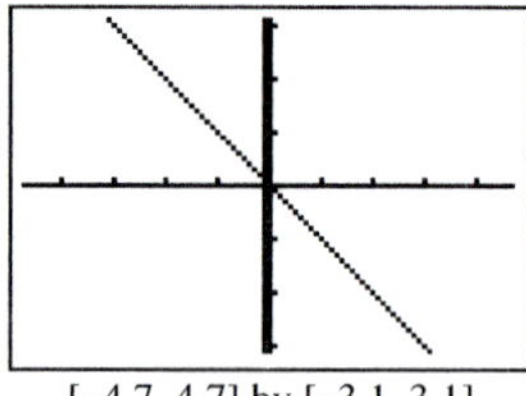

[–4.7, 4.7] by [–3.1, 3.1]

A plane containing the axis of a cone intersects the cone.

Parallel lines: $x^2 = 4$ can be rewritten as $x = \pm 2$ (a pair of vertical lines)

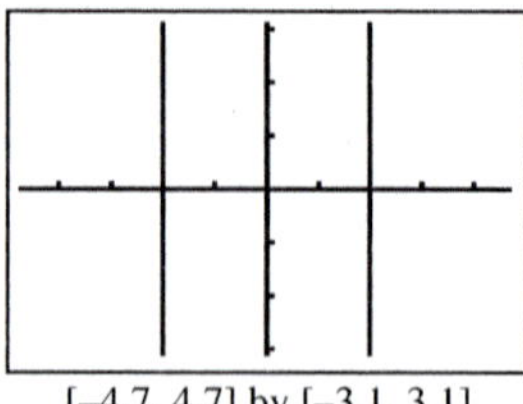

[–4.7, 4.7] by [–3.1, 3.1]

A degenerate cone is created by a generator that is parallel to the axis, producing a cylinder. A plane parallel to a generator of the cylinder intersects the cylinder and its interior.

One line: $y^2 = 0$ can be rewritten as $y = 0$ (the x-axis).

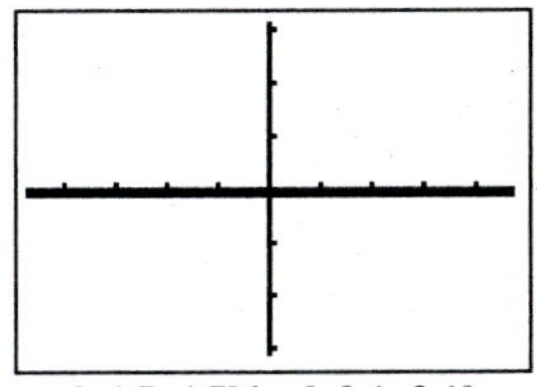

[–4.7, 4.7] by [–3.1, 3.1]

A plane containing a generator of a cone intersects the cone.

No graph: $x^2 = -1$

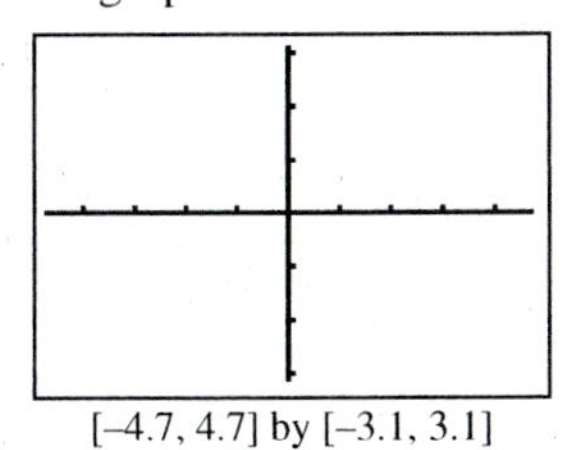

[–4.7, 4.7] by [–3.1, 3.1]

A plane parallel to a generator of a cylinder fails to intersect the cylinder.

Circle: $x^2 + y^2 = 9$

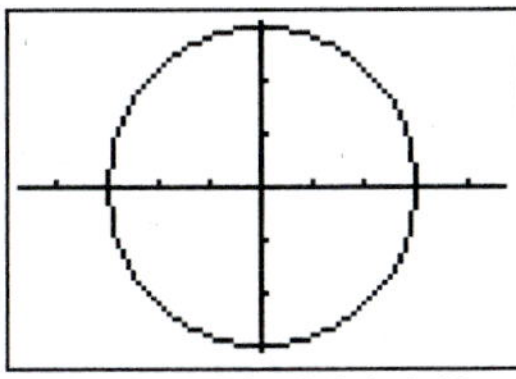

[–4.7, 4.7] by [–3.1, 3.1]

A plane perpendicular to the axis of a cone intersects the cone but not its vertex.

Point: $x^2 + y^2 = 0$, the point $(0, 0)$.

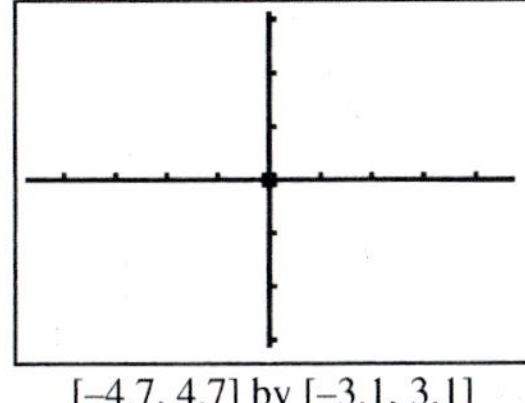

[–4.7, 4.7] by [–3.1, 3.1]

A plane perpendicular to the axis of a cone intersects the vertex of the cone.

No graph: $x^2 + y^2 = -1$

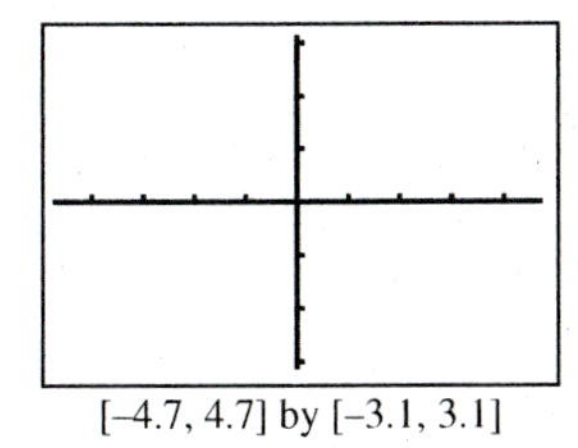

[–4.7, 4.7] by [–3.1, 3.1]

A degenerate cone is created by a generator that is perpendicular to the axis, producing a plane. A second plane perpendicular to the axis of this degenerate cone fails to intersect it.

■ Section 8.5 Polar Equations of Conics

Exploration 1

For $e = 0.7$ and $e = 0.8$, an ellipse; for $e = 1$, a parabola; for $e = 1.5$ and $e = 3$, a hyperbola.

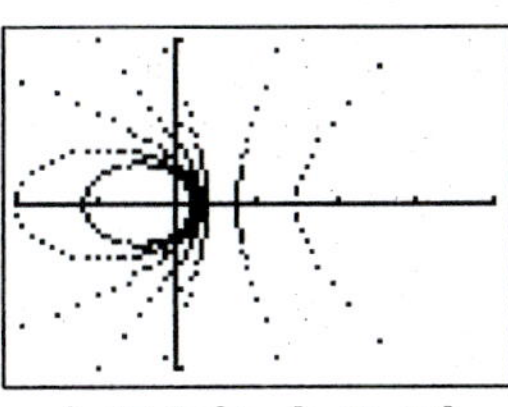

[−12, 24] by [−12, 12]

The five graphs all have a common focus, the pole $(0, 0)$, and a common directrix, the line $x = 3$. As the eccentricity e increases, the graphs move away from the focus and toward the directrix.

Quick Review 8.5

1. $r = -3$

3. $\theta = \dfrac{7\pi}{6}$ or $-\dfrac{5\pi}{6}$

5. $h = 0, k = 0, 4p = 16$, so $p = 4$
The focus is $(0, 4)$ and the directrix is $y = -4$.

7. $a = 3, b = 2, c = \sqrt{5}$; Foci: $(\pm\sqrt{5}, 0)$; Vertices: $(\pm 3, 0)$

9. $a = 4, b = 3, c = 5$; Foci: $(\pm 5, 0)$; Vertices: $(\pm 4, 0)$

Section 8.5 Exercises

1. $r = \dfrac{2}{1 - \cos\theta}$ — a parabola.

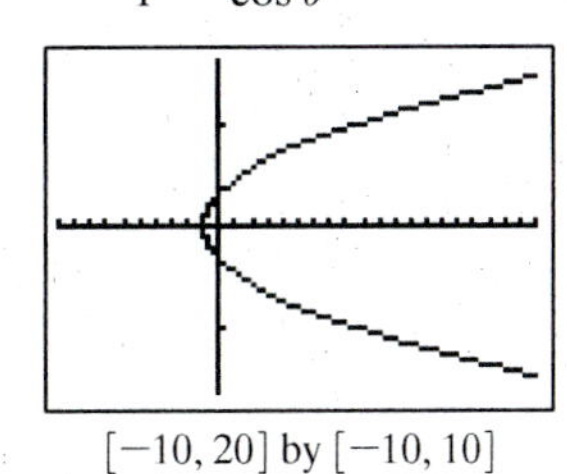

[−10, 20] by [−10, 10]

3. $r = \dfrac{\frac{12}{5}}{1 + \left(\frac{3}{5}\right)\sin\theta} = \dfrac{12}{5 + 3\sin\theta}$ — an ellipse.

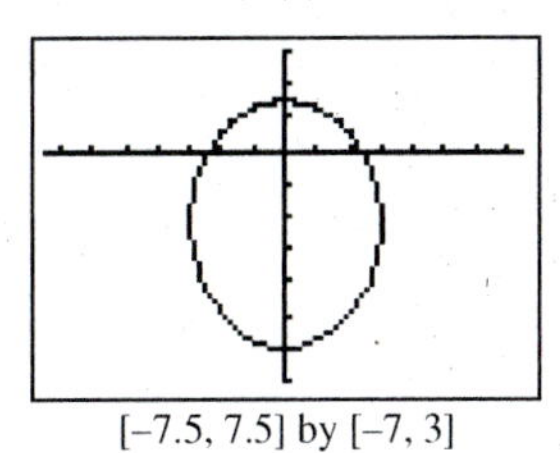

[–7.5, 7.5] by [–7, 3]

5. $r = \dfrac{\frac{7}{3}}{1 - \left(\frac{7}{3}\right)\sin\theta} = \dfrac{7}{3 - 7\sin\theta}$ — a hyperbola.

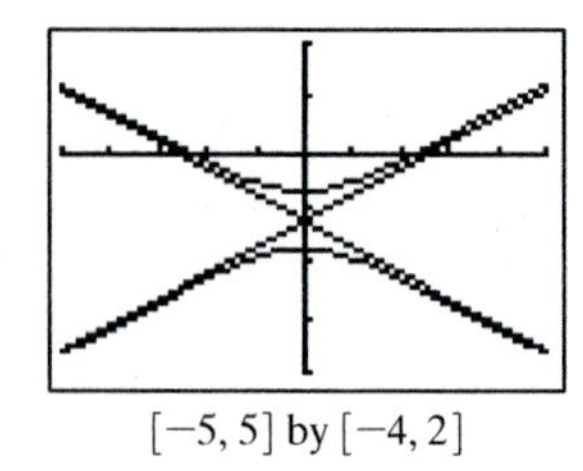

$[-5, 5]$ by $[-4, 2]$

7. Parabola with $e = 1$ and directrix $x = 2$.

9. Divide numerator and denominator by 2.

Parabola with $e = 1$ and directrix $y = -\dfrac{5}{2} = -2.5$.

11. Divide numerator and denominator by 6.

Ellipse with $e = \dfrac{5}{6}$ and directrix $y = 4$.

13. Divide numerator and denominator by 5.

Ellipse with $e = \dfrac{2}{5} = 0.4$ and directrix $x = 3$.

15. (b) $[-15, 5]$ by $[-10, 10]$

17. (f) $[-5, 5]$ by $[-3, 3]$

19. (c) $[-10, 10]$ by $[-5, 10]$

For #21–27, one must solve two equations $a = \dfrac{ep}{1+e}$ and $b = \dfrac{ep}{1-e}$ for e and p (given two constants a and b). The general solution to this is $e = \dfrac{b-a}{b+a}$ and $p = \dfrac{2ab}{b-a}$.

21. The directrix must be $x = p > 0$, since the right major-axis endpoint is closer to $(0, 0)$ than the left one, so the equation has the form $r = \dfrac{ep}{1 + e\cos\theta}$. Then $1.5 = \dfrac{ep}{1 + e\cos 0} = \dfrac{ep}{1+e}$ and $6 = \dfrac{ep}{1 + e\cos\pi} = \dfrac{ep}{1-e}$ (so $a = 1.5$ and $b = 6$). Therefore $e = \dfrac{3}{5} = 0.6$ and $p = 4$, so $r = \dfrac{2.4}{1 + (3/5)\cos\theta} = \dfrac{12}{5 + 3\cos\theta}$.

23. The directrix must be $y = p > 0$, since the upper major-axis endpoint is closer to $(0, 0)$ than the lower one, so the equation has the form $r = \dfrac{ep}{1 + e\cos\theta}$. Then $1 = \dfrac{ep}{1 + e\sin(\pi/2)} = \dfrac{ep}{1+e}$ and $3 = \dfrac{ep}{1 + e\sin(3\pi/2)} = \dfrac{ep}{1-e}$ (so $a = 1$ and $b = 3$). Therefore $e = \dfrac{1}{2} = 0.5$ and $p = 3$, so $r = \dfrac{1.5}{1 + (1/2)\sin\theta} = \dfrac{3}{2 + \sin\theta}$.

25. The directrix must be $x = p > 0$, since both transverse-axis endpoints have positive x coordinates, so the equation has the form $r = \dfrac{ep}{1 + e\cos\theta}$. Then $3 = \dfrac{ep}{1 + e\cos 0} = \dfrac{ep}{1+e}$ and $-15 = \dfrac{ep}{1 + e\cos\pi} = \dfrac{ep}{1-e}$ (so $a = 3$ and $b = -15$). Therefore $e = \dfrac{3}{2} = 1.5$ and $p = 5$, so $r = \dfrac{7.5}{1 + (3/2)\cos\theta} = \dfrac{15}{2 + 3\cos\theta}$.

27. The directrix must be $y = p > 0$, since both transverse-axis endpoints have positive y coordinates, so the equation has the form $r = \dfrac{ep}{1 + e\cos\theta}$. Then $2.4 = \dfrac{ep}{1 + e\sin(\pi/2)} = \dfrac{ep}{1+e}$ and $-12 = \dfrac{ep}{1 + e\sin(3\pi/2)} = \dfrac{ep}{1-e}$ (so $a = 2.4$ and $b = -12$). Therefore $e = \dfrac{3}{2} = 1.5$ and $p = 4$, so $r = \dfrac{6}{1 + (3/2)\sin\theta} = \dfrac{12}{2 + 3\sin\theta}$.

29. The directrix must be $x = p > 0$, so the equation has the form $r = \dfrac{ep}{1 + e\cos\theta}$. Then $0.75 = \dfrac{ep}{1 + e\cos 0} = \dfrac{ep}{1+e}$ and $3 = \dfrac{ep}{1 + e\cos\pi} = \dfrac{ep}{1-e}$ (so $a = 0.75$ and $b = 3$). Therefore $e = \dfrac{3}{5} = 0.6$ and $p = 2$, so $r = \dfrac{1.2}{1 + (3/5)\cos\theta} = \dfrac{6}{5 + 3\cos\theta}$.

31. $r = \dfrac{21}{5 - 2\cos\theta} = \dfrac{4.2}{1 - 0.4\cos\theta}$, so $e = 0.4$. The vertices are $(7, 0)$ and $(3, \pi)$, so $2a = 10$, $a = 5$, $c = ae = (0.4)(5) = 2$, so $b = \sqrt{a^2 - c^2} = \sqrt{25 - 4} = \sqrt{21}$.

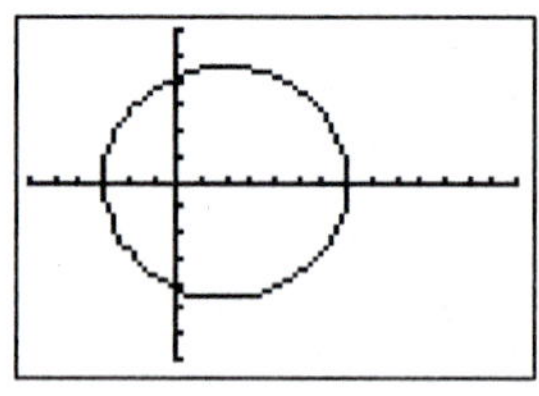

$[-6, 14]$ by $[-7, 6]$

$e = 0.4$, $a = 5$, $b = \sqrt{21}$, $c = 2$

33. $r = \dfrac{24}{4 + 2\sin\theta} = \dfrac{6}{1 + (1/2)\sin\theta}$, so $e = \dfrac{1}{2}$. The vertices are $(4, \frac{\pi}{2})$ and $(12, \frac{3\pi}{2})$, so $2a = 16$, $a = 8$. $c = ae = \dfrac{1}{2}\cdot 8 = 4$, so $b = \sqrt{a^2 - c^2} = \sqrt{64 - 16} = 4\sqrt{3}$.

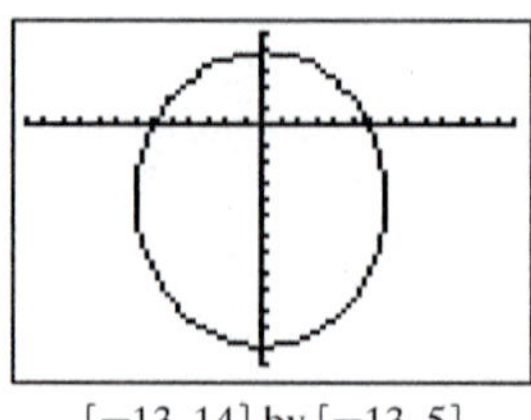

$[-13, 14]$ by $[-13, 5]$

$e = \dfrac{1}{2}$, $a = 8$, $b = 4\sqrt{3}$, $c = 4$

35. $r = \dfrac{16}{3 + 5\cos\theta} = \dfrac{16/3}{1 + (5/3)\cos\theta}$, so $e = \dfrac{5}{3}$. The vertices are $(2, 0)$ and $(-8, \pi)$, so $2a = 6$, $a = 3$, $c = ae$ $= \dfrac{5}{3}\cdot 3 = 5$ and $b = \sqrt{c^2 - a^2} = \sqrt{25 - 9} = 4$.

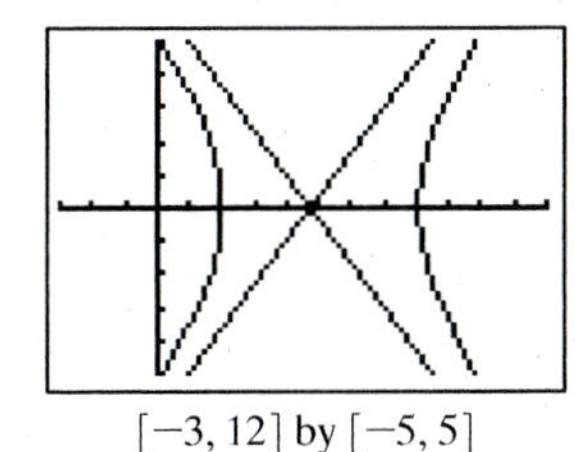

$[-3, 12]$ by $[-5, 5]$

$e = \dfrac{5}{3}$, $a = 3$, $b = 4$, $c = 5$

37. $r = \dfrac{4}{2 - \sin\theta} = \dfrac{2}{1 - (1/2)\sin\theta}$ so $e = \dfrac{1}{2}$ (an ellipse). The vertices are $\left(4, \dfrac{\pi}{2}\right)$ and $\left(\dfrac{4}{3}, \dfrac{3\pi}{2}\right)$ and the conic is symmetric around $x = 0$, so $x = 0$ is the semi-major axis and $2a = \dfrac{16}{3}$, so $a = \dfrac{8}{3}$. $c = ea = \dfrac{1}{2}\cdot\dfrac{8}{3} = \dfrac{4}{3}$ and $b = \sqrt{a^2 - c^2} = \sqrt{\left(\dfrac{8}{3}\right)^2 - \left(\dfrac{4}{3}\right)^2} = \dfrac{4\sqrt{3}}{3}$. The center $(h, k) = \left(0, \dfrac{12}{3} - \dfrac{8}{3}\right) = \left(0, \dfrac{4}{3}\right)$. The equation for the ellipse is

$$\frac{\left(y - \frac{4}{3}\right)^2}{\left(\frac{8}{3}\right)^2} + \frac{(x - 0)^2}{\left(\frac{4\sqrt{3}}{3}\right)^2} = \frac{9\left(y - \frac{4}{3}\right)^2}{64} + \frac{3x^2}{16} = 1$$

39. $r = \dfrac{4}{2 - 2\cos\theta} = \dfrac{2}{1 - \cos\theta}$, so $e = 1$ and $k = \dfrac{2}{e} = 2$. Since $k = 2p$, $p = 1$ and $4p = 4$, the vertex $(h, k) = (-1, 0)$ and the parabola opens to the right, so the equation is $y^2 = 4(x + 1)$.

41. Setting $e = 0.97$ and $a = 18.09$ AU,

$r = \dfrac{18.09\,(1 - 0.97^2)}{1 + 0.97\cos\theta}$

The perihelion of Halley's Comet is

$r = \dfrac{18.09\,(1 - 0.97^2)}{1 + 0.97} \approx 0.54$ AU

The aphelion of Halley's Comet is

$r = \dfrac{18.09\,(1 - 0.97^2)}{1 - 0.97} \approx 35.64$ AU

43. (a) The total radius of the orbit is $r = 250 + 1740 = 1990$ km $= 1{,}990{,}000$ m. Then $v \approx \sqrt{2{,}406{,}030} \approx 1551$ m/sec $= 1.551$ km/sec.

(b) The circumference of one orbit is $2\pi r \approx 12503.5$ km; one orbit therefore takes about 8061 seconds, or about 2 hr 14 min.

45. True. For a circle, $e = 0$. But when $e = 0$, the equation degenerates to $r = 0$, which yields a single point, the pole.

47. Conics are defined in terms of the ratio distance to focus : distance to directrix. The answer is D.

49. Conics written in polar form always have one focus at the pole. The answer is B.

51. (a) When $\theta = 0$, $\cos\theta = 1$, so $1 + e\cos\theta = 1 + e$. Then $\dfrac{a(1 - e^2)}{1 + e\cos\theta} = \dfrac{a(1 - e^2)}{1 + e} = a(1 - e)$

Similarly, when $\theta = \pi$, $\cos\theta = -1$, so $1 + e\cos\theta = 1 - e$. Then $\dfrac{a(1 - e^2)}{1 + e\cos\theta} = \dfrac{a(1 - e^2)}{1 - e} = a(1 + e)$

(b) $a(1 - e) = a\left(1 - \dfrac{c}{a}\right) = a - a\cdot\dfrac{c}{a} = a - c$

$a(1 + e) = a\left(1 + \dfrac{c}{a}\right) = a + a\cdot\dfrac{c}{a} = a + c$

(c)

Planet	Perihelion (in Au)	Aphelion
Mercury	0.307	0.467
Venus	0.718	0.728
Earth	0.983	1.017
Mars	1.382	1.665
Jupiter	4.953	5.452
Saturn	9.020	10.090

(d) The difference is greatest for Saturn.

53. If $r < 0$, then the point P can be expressed as the point $(r, \theta + \pi)$ then $PF = r$ and $PD = k - r\cos\theta$.

$$PF = ePD$$
$$r = e(k - r\cos\theta)$$
$$r = \frac{ke}{1 + e\cos\theta}$$

Recall that $P\ (r, \theta)$ can also be expressed as $(-r, \theta - \pi)$ then $PD = -r$ and $PF = -r\cos(\theta - \pi) - k$

$$PD = ePF$$
$$-r = e[-r\cos(\theta - \pi) - k]$$
$$-r = -er\cos(\theta - \pi) - ek$$
$$-r = er\cos\theta - ek$$
$$-r - er\cos\theta = -ek$$
$$r = \frac{ke}{1 + e\cos\theta}$$

55. Consider the polar equation $r = \dfrac{16}{5 - 3\cos\theta}$. To transform this to a Cartesian equation, rewrite the equation as $5r - 3r\cos\theta = 16$. Then use the substitutions $r = \sqrt{x^2 + y^2}$ and $x = r\cos\theta$ to obtain

$5\sqrt{x^2 + y^2} - 3x = 16.$

$5\sqrt{x^2 + y^2} = 3x + 16$; $25(x^2 + y^2) = 9x^2 + 96x + 256$

$25x^2 + 25y^2 = 9x^2 + 96x + 256$

$16x^2 - 96x + 25y^2 = 256$;

Completing the square on the x term gives

$16(x^2 - 6x + 9) + 25y^2 = 256 + 144$

$16(x - 3)^2 + 25y^2 = 400$;

The Cartesian equation is $\dfrac{(x - 3)^2}{25} + \dfrac{y^2}{16} = 1$.

57. Apply the formula $e \cdot PD = PF$ to a hyperbola with one focus at the pole and directrix $x = -k$, letting P be the vertex closest to the pole. Then $a + k = c + PD$ and $PF = c - a$. Using $e = \frac{c}{a}$, we have:

$$\begin{aligned} e \cdot PD &= PF \\ e(a + k - c) &= c - a \\ e(a + k - ae) &= ae - a \\ ae + ke - ae^2 &= ae - a \\ ke - ae^2 &= -a \\ ke &= ae^2 - a \\ ke &= a(e^2 - 1) \end{aligned}$$

Thus, the equation $r = \dfrac{ke}{1 - e\cos\theta}$ becomes $r = \dfrac{a(e^2 - 1)}{1 - e\cos\theta}$.

59. **(a)** Let $P(x, y)$ be a point on the hyperbola. The horizontal distance from P to the point $Q(a^2/c, y)$ on line L is $PQ = |a^2/c - x|$. The distance to the focus $(c, 0)$ is $PF = \sqrt{(x - c)^2 + y^2} = \sqrt{x^2 - 2cx + c^2 + y^2}$. To confirm that $PF/PQ = c/a$, cross-multiply to get $a\,PF = c\,PQ$; we need to confirm that $a\sqrt{x^2 - 2cx + c^2 + y^2} = |a^2 - cx|$. Square both sides: $a^2(x^2 - 2cx + c^2 + y^2) = a^4 - 2a^2cx + c^2x^2$. Substitute $a^2 + b^2$ for c^2, multiply out both sides, and cancel out terms, leaving $a^2y^2 + a^2b^2 = b^2x^2$. Since P is on the hyperbola, $x^2/a^2 - y^2/b^2 = 1$, or equivalent $b^2x^2 - a^2y^2 = a^2b^2$; this confirms the equality.

(b) According to the polar definition, the eccentricity is the ratio PF/PQ, which we found to be c/a in (a).

(c) Since $e = c/a$, $a/e = \dfrac{a}{c/a} = a^2/c$ and $ae = c$; the distance from F to L is $c - \dfrac{a^2}{c} = ea - \dfrac{a}{e}$ as desired.

Section 8.6 Three-Dimensional Cartesian Coordinates

Quick Review 8.6

1. $\sqrt{(x - 2)^2 + (y + 3)^2}$

3. P lies on the circle of radius 5 centered at $(2, -3)$

5. $\dfrac{\mathbf{v}}{|\mathbf{v}|} = \dfrac{\langle -4, 5\rangle}{\sqrt{41}} = \left\langle \dfrac{-4}{\sqrt{41}}, \dfrac{5}{\sqrt{41}} \right\rangle$

7. Circle of radius 5 centered at $(-1, 5)$

9. $(x + 1)^2 + (y - 3)^2 = 4$. Center: $(-1, 3)$, radius: 2

Section 8.6 Exercises

1.

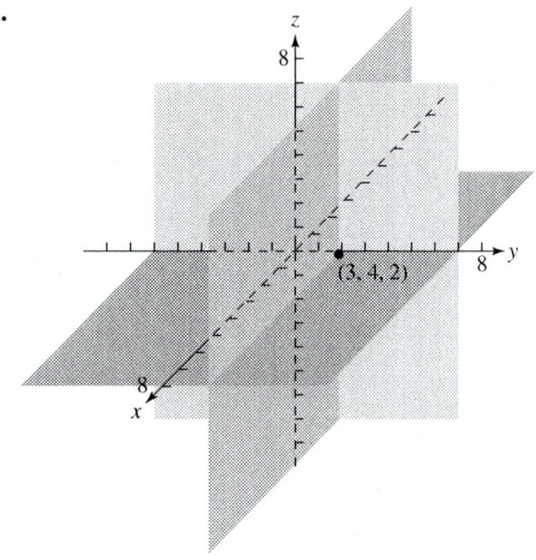

3.

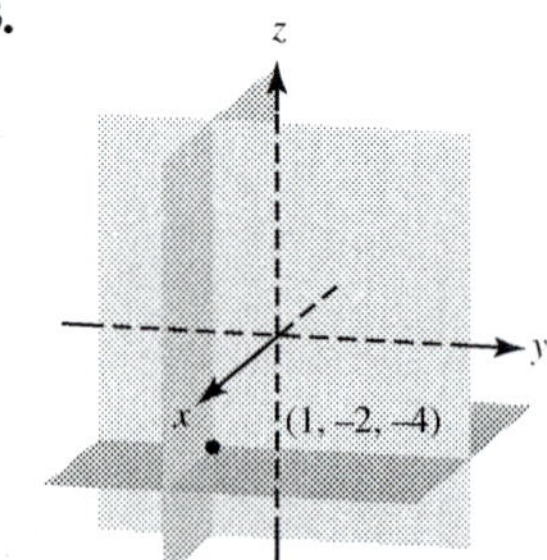

5. $\sqrt{(3 - (-1))^2 + (-4 - 2)^2 + (6 - 5)^2} = \sqrt{53}$

7. $\sqrt{(a - 1)^2 + (b - (-3))^2 + (c - 2)^2} = \sqrt{(a - 1)^2 + (b + 3)^2 + (c - 2)^2}$

9. $\left(\dfrac{3 - 1}{2}, \dfrac{-4 + 2}{2}, \dfrac{6 + 5}{2}\right) = \left(1, -1, \dfrac{11}{2}\right)$

11. $\left(\dfrac{2x - 2}{2}, \dfrac{2y + 8}{2}, \dfrac{2z + 6}{2}\right) = (x - 1, y + 4, z + 3)$

13. $(x - 5)^2 + (y + 1)^2 + (z + 2)^2 = 64$

15. $(x - 1)^2 + (y + 3)^2 + (z - 2)^2 = a$

17.

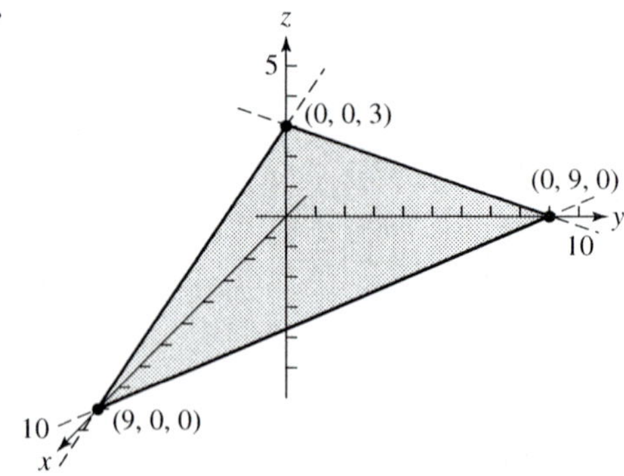

19.

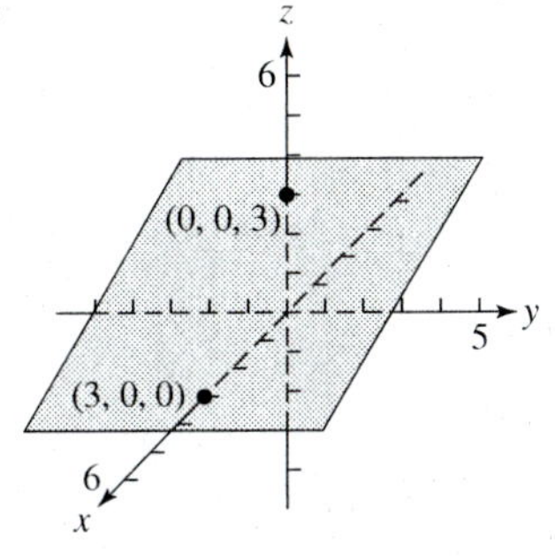

21.

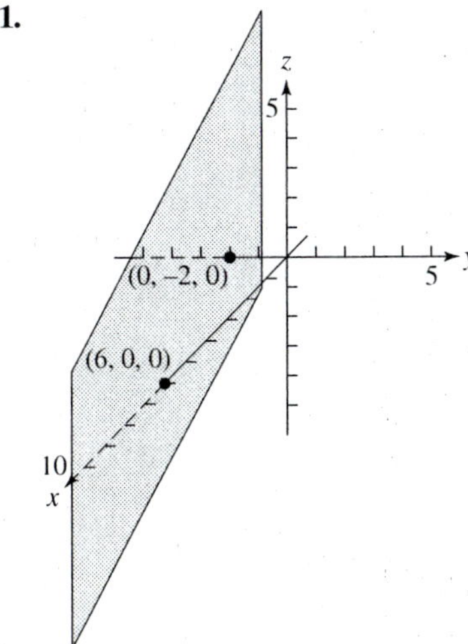

23. $\mathbf{r} + \mathbf{v} = \langle 1, 0, -3\rangle + \langle -3, 4, -5\rangle = \langle -2, 4, -8\rangle$

25. $\mathbf{v}\cdot\mathbf{w} = -12 - 12 - 60 = -84$

27. $\mathbf{r}\cdot(\mathbf{v} + \mathbf{w}) = \mathbf{r}\cdot(\langle -3, 4, -5\rangle + \langle 4, -3, 12\rangle)$
$= \langle 1, 0, -3\rangle\cdot\langle 1, 1, 7\rangle = 1 + 0 - 21 = -20$

29. $\dfrac{\mathbf{w}}{|\mathbf{w}|} = \dfrac{\langle 4, -3, 12\rangle}{\sqrt{4^2 + (-3)^2 + 12^2}} = \left\langle \dfrac{4}{13}, -\dfrac{3}{13}, \dfrac{12}{13}\right\rangle$

31. $\langle \mathbf{i}\cdot\mathbf{v}, \mathbf{j}\cdot\mathbf{v}, \mathbf{k}\cdot\mathbf{v}\rangle = \langle -3, 4, -5\rangle$

33. The plane's velocity relative to the air is
$\mathbf{v}_1 = -200\cos 20°\,\mathbf{i} + 200\sin 20°\,\mathbf{k}$
The air's velocity relative to the ground is
$\mathbf{v}_2 = -10\cos 45°\,\mathbf{i} - 10\sin 45°\,\mathbf{j}$
Adding these two vectors and converting to decimal values rounded to two places produces the plane's velocity relative to the ground:
$\mathbf{v} = -195.01\,\mathbf{i} - 7.07\,\mathbf{j} + 68.40\,\mathbf{k}$

For #35–37, the vector form is $\mathbf{r}_0 + t\mathbf{v}$ with $\mathbf{r}_0\langle x_0, y_0, z_0\rangle$, and the parametric form is $x = x_0 + ta$, $y = y_0 + tb$, $z = z_0 + tc$ where $\mathbf{v} = \langle a, b, c\rangle$.

35. Vector form: $\mathbf{r} = \langle 2, -1, 5\rangle + t\langle 3, 2, -7\rangle$; parametric form: $x = 2 + 3t$, $y = -1 + 2t$, $z = 5 - 7t$

37. Vector form: $\mathbf{r} = \langle 6, -9, 0\rangle + t\langle 1, 0, -4\rangle$; parametric form: $x = 6 + t$, $y = -9$, $z = -4t$

39. Midpoint of $\overline{BC}$: $\langle 1, 1, -1\rangle$. Distance from A to midpoint of $\overline{BC}$:
$\sqrt{(-1 - 1)^2 + (2 - 1)^2 + (4 - (-1))^2} = \sqrt{30}$

41. Direction vector: $\langle 0 - (-1), 6 - 2, -3 - 4\rangle$
$= \langle 1, 4, -7\rangle$, $\overrightarrow{OA} = \langle -1, 2, 4\rangle$,
$\mathbf{r} = \langle -1, 2, 4\rangle + t\langle 1, 4, -7\rangle$

43. Direction vector: $\langle 2 - (-1), -4 - 2, 1 - 4\rangle$
$= \langle 3, -6, -3\rangle$, $\overrightarrow{OA} = \langle -1, 2, 4\rangle$, so a vector equation of the line is $\mathbf{r} = \langle -1, 2, 4\rangle + t\langle 3, -6, -3\rangle$
$= \langle -1 + 3t, 2 - 6t, 4 - 3t\rangle$. This can be expressed in parametric form: $x = -1 + 3t$, $y = 2 - 6t$, $z = 4 - 3t$.

45. Midpoint of $\overline{AC}$: $\left(\dfrac{1}{2}, -1, \dfrac{5}{2}\right)$. Direction vector:
$\left\langle \dfrac{1}{2} - 0, -1 - 6, \dfrac{5}{2} - (-3)\right\rangle = \left\langle \dfrac{1}{2}, -7, \dfrac{11}{2}\right\rangle$,
$\overrightarrow{OB} = \langle 0, -6, -3\rangle$, so a vector equation of the line is
$\mathbf{r} = \langle 0, 6, -3\rangle + t\left\langle \dfrac{1}{2}, -7, \dfrac{11}{2}\right\rangle$
$= \left\langle \dfrac{1}{2}t, 6 - 7t, -3 + \dfrac{11}{2}t\right\rangle$. This can be expressed in parametric form: $x = \dfrac{1}{2}t$, $y = 6 - 7t$, $z = -3 + \dfrac{11}{2}t$.

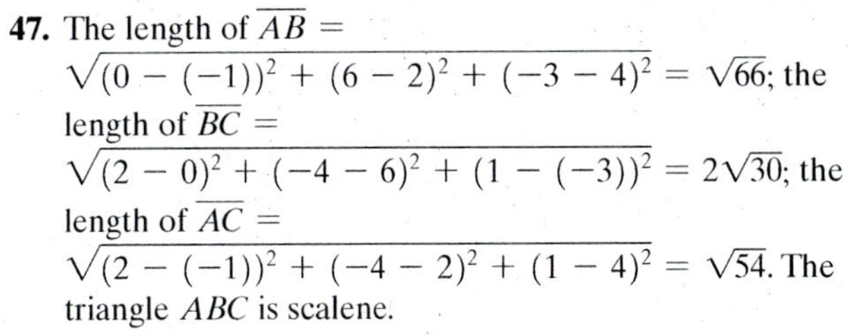

47. The length of $\overline{AB}$ =
$\sqrt{(0 - (-1))^2 + (6 - 2)^2 + (-3 - 4)^2} = \sqrt{66}$; the length of $\overline{BC}$ =
$\sqrt{(2 - 0)^2 + (-4 - 6)^2 + (1 - (-3))^2} = 2\sqrt{30}$; the length of $\overline{AC}$ =
$\sqrt{(2 - (-1))^2 + (-4 - 2)^2 + (1 - 4)^2} = \sqrt{54}$. The triangle ABC is scalene.

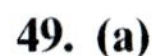

49. (a)

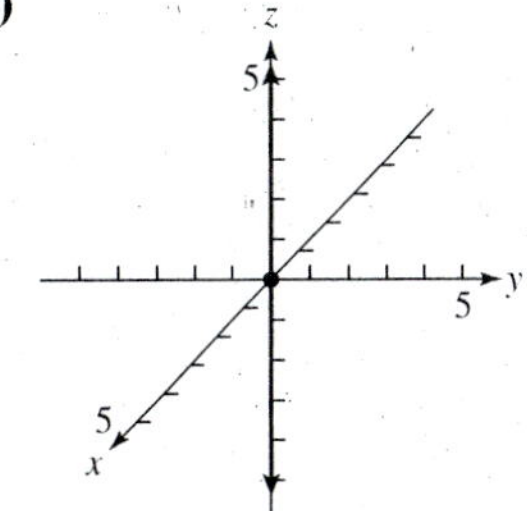

(b) the z-axis; a line through the origin in the direction $\mathbf{k}$.

51. (a)

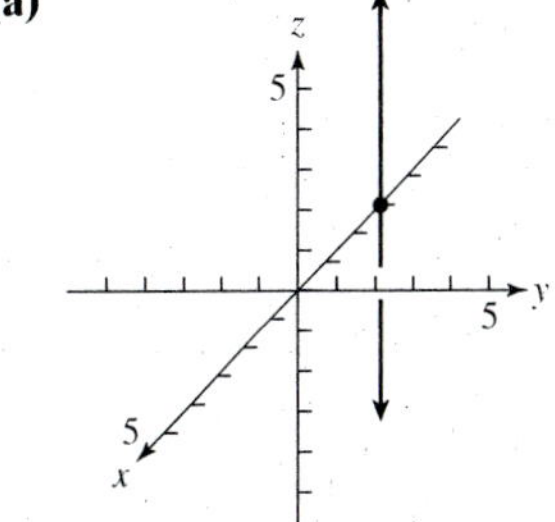

(b) the intersection of the xz plane (at $y = 0$) and yz plane (at $x = -3$); a line parallel to the z-axis through $(-3, 0, 0)$

53. Direction vector: $\langle x_2 - x_1, y_2 - y_1, z_2 - z_1\rangle$,
$\overrightarrow{OP} = \langle x_1, y_2, z_3\rangle$, so a vector equation of the line is
$\mathbf{r} = \langle x_1 + (x_2 - x_1)t, y_1 + (y_2 - y_1)t, z_1 + (z_2 - z_1)t\rangle$.

55.

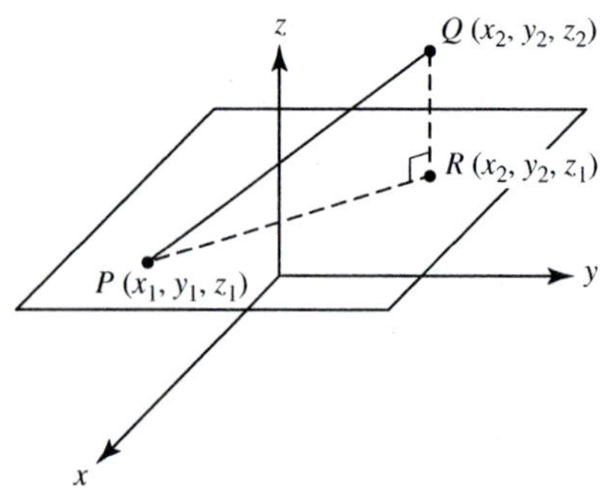

By the Pythagorean Theorem,
$d(P, Q) = \sqrt{(d(P, R))^2 + (d(R, Q)^2)}$

$= \sqrt{(\sqrt{(x_1 - x_2)^2 + (y_1 - y_2)^2})^2 + (|z_1 - z_2|)^2}$
$= \sqrt{(x_1 - x_2)^2 + (y_1 - y_2)^2 + (z_1 - z_2)^2}$

57. True. This is the equation of a vertical elliptic cylinder. The equation can be viewed as an equation in three variables, where the coefficient of z is zero.

59. The general form for a first-degree equation in three variables is $Ax + By + Cz + D = 0$. The answer is B.

61. The dot product of two vectors is a scalar. The answer is C.

63. (a) Each cross-section is its own ellipse.

$x = 0$: $\frac{y^2}{4} + \frac{z^2}{16} = 1$, an ellipse centered at $(0, 0)$ (in the yz plane) of "width" 4 and "height" 8.

$y = 0$: $\frac{x^2}{9} + \frac{z^2}{16} = 1$, an ellipse centered at $(0, 0)$ (in the xz plane) of "width" 6 and "height" 8.

$z = 0$: $\frac{x^2}{9} + \frac{y^2}{4} = 1$, an ellipse centered at $(0, 0)$ (in the xy plane) of "width" 6 and "height" 4.

(b) Algebraically, $z = \sqrt{1 - x^2 - y^2}$ has only positive values; $0 \leq z \leq 1$ and the "bottom" half of the sphere is never formed. The equation of the whole sphere is $x^2 + y^2 + z^2 = 1$.

(c)

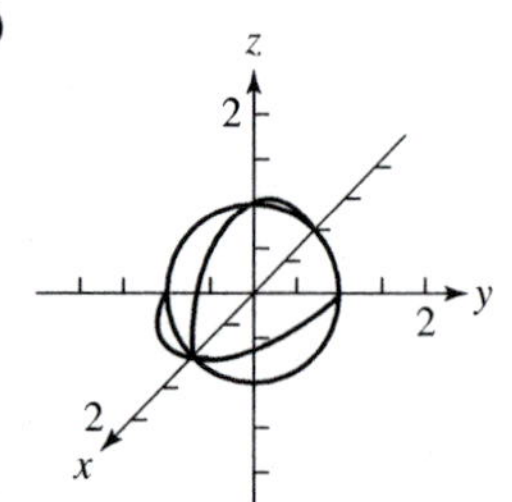

(d) A sphere is an ellipsoid in which all of the $x = 0$, $y = 0$, and $z = 0$ "slices" (i.e., the cross-sections of the coordinate planes) are circles. Since a circle is a degenerate ellipse, it follows that a sphere is a degenerate ellipsoid.

65. $\langle 2 - 3, -6 + 1, 1 - 4\rangle = \langle -1, -5, -3\rangle$

67. $\mathbf{i} \times \mathbf{j} = \langle 1, 0, 0\rangle \times \langle 0, 1, 0\rangle = \langle 0 - 0, 0 - 0, 1 - 0\rangle = \langle 0, 0, 1\rangle = \mathbf{k}$

Chapter 8 Review

1. $h = 0, k = 0, 4p = 12$, so $p = 3$.
Vertex: $(0, 0)$, focus: $(3, 0)$, directrix: $x = -3$, focal width: 12

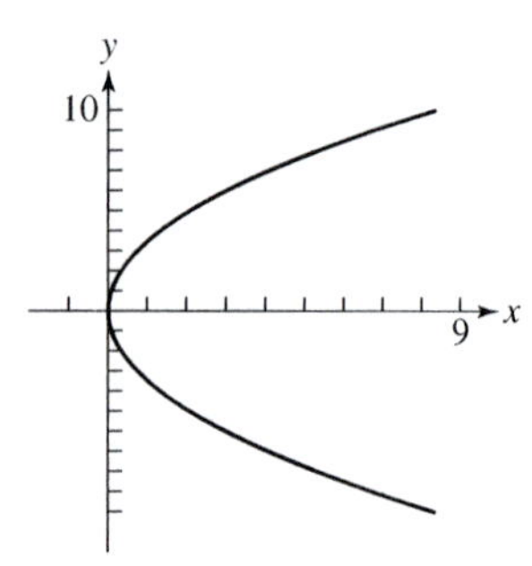

3. $h = -2, k = 1, 4p = -4$, so $p = -1$.
Vertex: $(-2, 1)$, focus: $(-2, 0)$, directrix: $y = 2$, focal width: 4

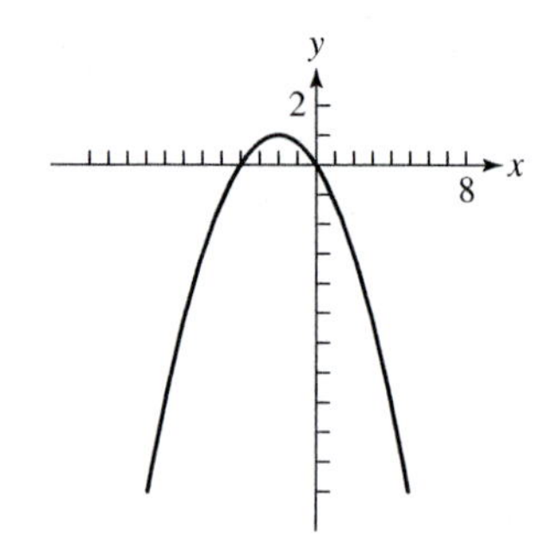

5. Ellipse. Center $(0, 0)$. Vertices: $(0, \pm 2\sqrt{2})$. Foci: $(0, \pm\sqrt{3})$ since $c = \sqrt{8 - 5} = \sqrt{3}$.

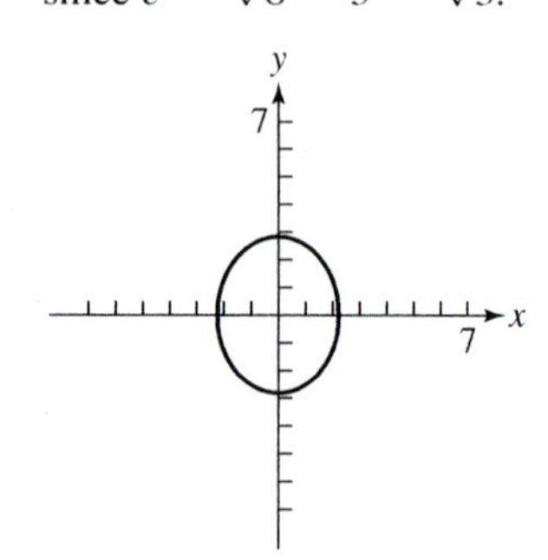

7. Hyperbola. Center: $(0, 0)$. Vertices: $(\pm 5, 0)$, $c = \sqrt{a^2 + b^2} = \sqrt{25 + 36} = \sqrt{61}$, so the foci are: $(\pm\sqrt{61}, 0)$

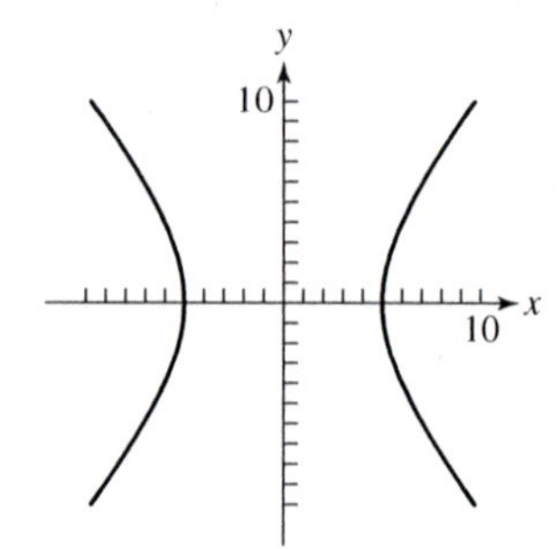

9. Hyperbola. Center: $(-3, 5)$. Vertices: $(-3 \pm 3\sqrt{2}, 5)$, $c = \sqrt{a^2 + b^2} = \sqrt{18 + 28} = \sqrt{46}$, so the foci are: $(-3 \pm \sqrt{46}, 5)$

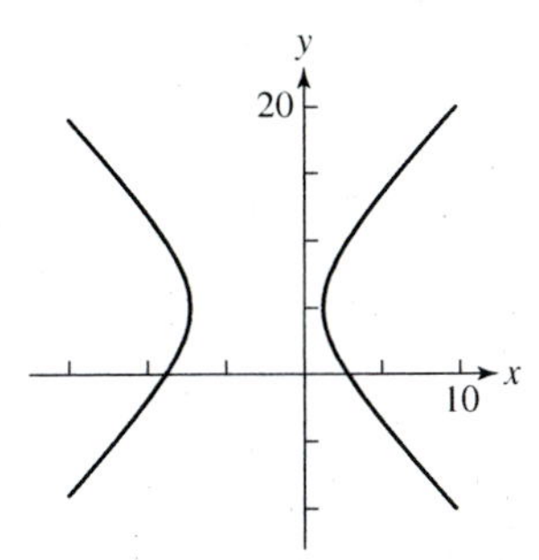

11. Ellipse. Center: $(2, -1)$. Vertices: $(2 \pm 4, -1) = (6, -1)$ and $(-2, -1)$, $c = \sqrt{a^2 - b^2} = \sqrt{16 - 7} = 3$, so the foci are: $(2 \pm 3, -1) = (5, -1)$ and $(-1, -1)$

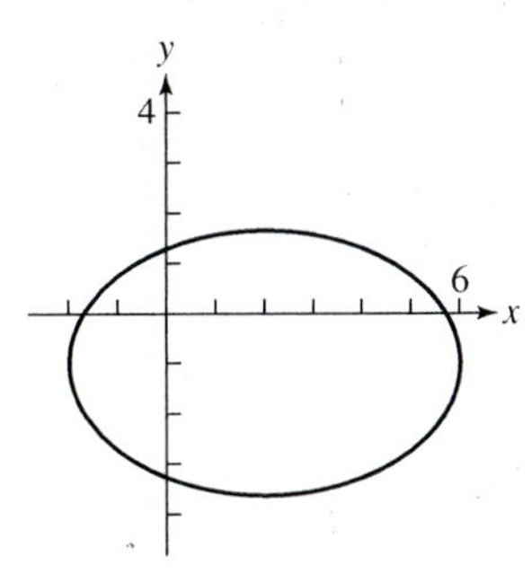

13. (b)

15. (h)

17. (f)

19. (c)

21. $B^2 - 4AC = 0 - 4(1)(0) = 0$, parabola $(x^2 - 6x + 9) = y + 3 + 9$, so $(x - 3)^2 = y + 12$

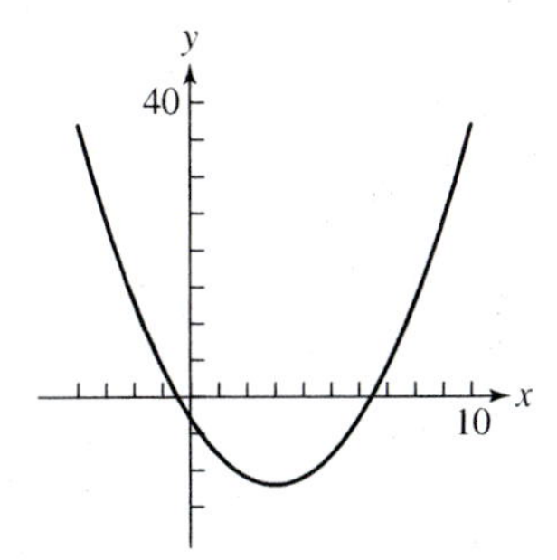

23. $B^2 - 4AC = 0 - 4(1)(-1) = 4 > 0$, hyperbola $(x^2 - 2x + 1) - (y^2 - 4y + 4)$ $= 1 - 4 + 6$

$$\frac{(x-1)^2}{3} - \frac{(y-2)^2}{3} = 1$$

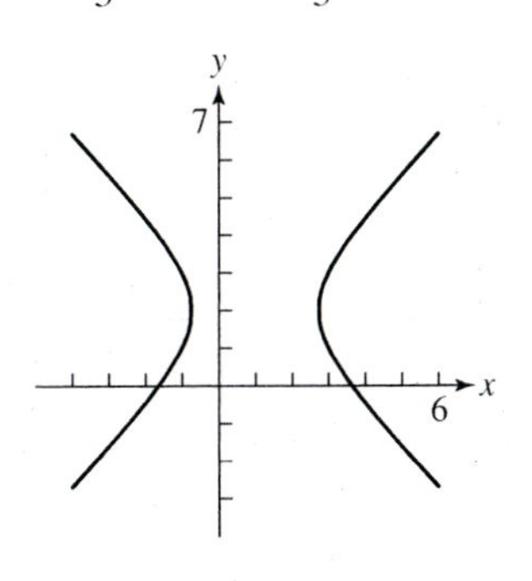

25. $B^2 - 4AC = 0 - 4(1)(0) = 0$, parabola $(y^2 - 4y + 4) = 6x + 13 + 4$, so $(y - 2)^2 = 6\left(x + \frac{17}{6}\right)$

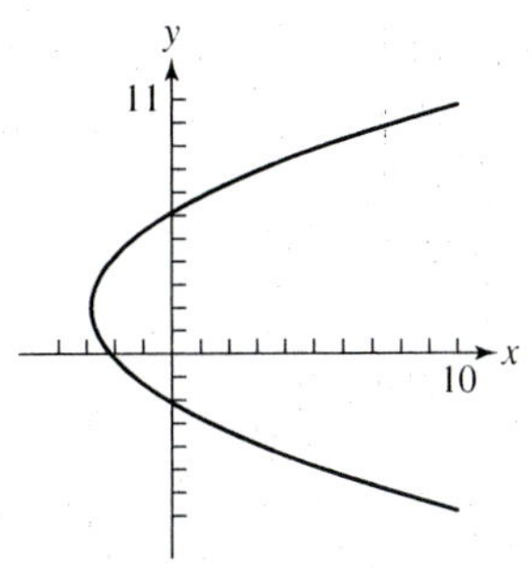

27. $B^2 - 4AC = 0 - 4(2)(-3) = 24 > 0$, hyperbola $2(x^2 - 6x + 9) - 3(y^2 + 8y + 16)$ $= 18 - 48 - 60$, so

$$\frac{(y+4)^2}{30} - \frac{(x-3)^2}{45} = 1$$

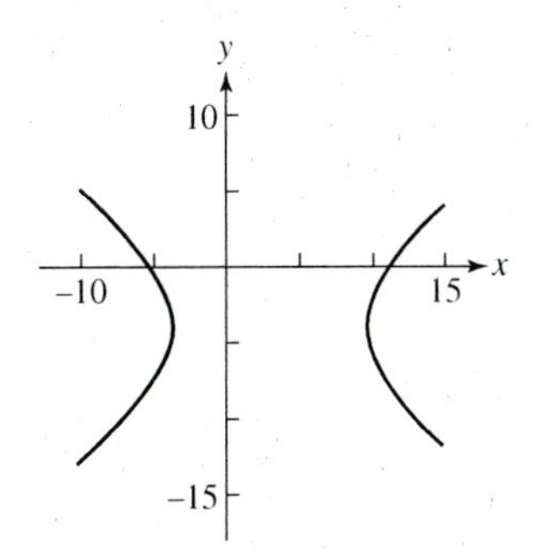

29. By definition, every point $P(x, y)$ that lies on the parabola is equidistant from the focus to the directrix. The distance between the focus and point P is: $\sqrt{(x-0)^2 + (y-p)^2} = \sqrt{x^2 + (y-p)^2}$, while the distance between the point P and the line $y = -p$ is: $\sqrt{(x-x)^2 + (y+p)^2} = \sqrt{(y+p)^2}$. Setting these equal:

$$\begin{aligned} \sqrt{x^2 + (y-p)^2} &= y + p \\ x^2 + (y-p)^2 &= (y+p)^2 \\ x^2 + y^2 - 2py + p^2 &= y^2 + 2py + p^2 \\ x^2 &= 4py \end{aligned}$$

31. Use the quadratic formula with $a = 6$, $b = -8x - 5$, and $c = 3x^2 - 5x + 20$. Then $b^2 - 4ac = (-8x - 5)^2$ $- 24(3x^2 - 5x + 20) = -8x^2 + 200x - 455$, and $y = \frac{1}{12}\left[8x + 5 \pm \sqrt{-8x^2 + 200x - 455}\right]$ – an ellipse

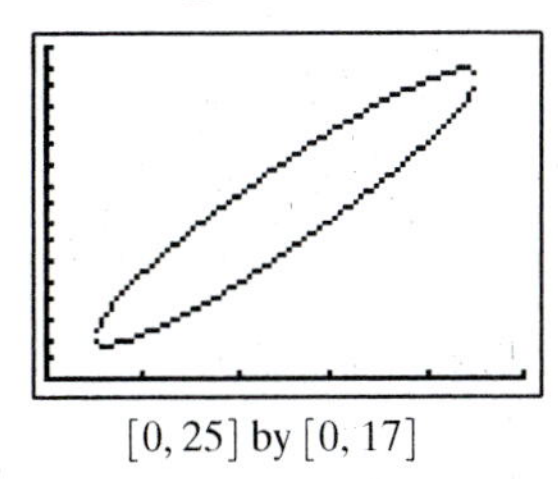

[0, 25] by [0, 17]

33. This is a *linear* equation in y: $(6 - 2x)y + (3x^2 - 5x - 10) = 0$. Subtract $3x^2 - 5x - 10$ and divide by $6 - 2x$, and we have $y = \dfrac{3x^2 - 5x - 10}{2x - 6}$ — a hyperbola.

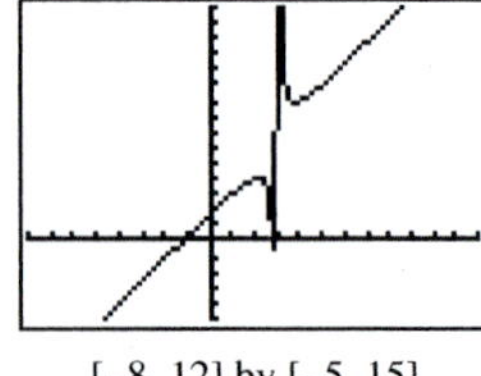

[−8, 12] by [−5, 15]

35. Use the quadratic formula with $a = -2$, $b = 7x + 20$, and $c = -3x^2 - x - 15$. Then $b^2 - 4ac = (7x + 20)^2 + 8(-3x^2 - x - 15) = 25x^2 + 272x + 280$, and $y = \dfrac{1}{4}\left[7x + 20 \pm \sqrt{25x^2 + 272x + 280}\right]$ a hyperbola.

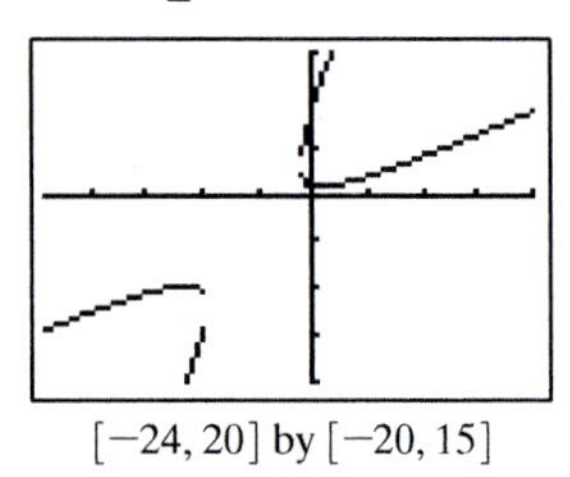

[−24, 20] by [−20, 15]

37. $h = 0$, $k = 0$, $p = 2$, and the parabola opens to the right as $y^2 = 8x$.

39. $h = -3$, $k = 3$, $p = k - y = 3 - 0 = 3$ (since $y = 0$ is the directrix) the parabola opens upward, so $(x + 3)^2 = 12(y - 3)$.

41. $h = 0$, $k = 0$, $c = 12$ and $a = 13$, so $b = \sqrt{a^2 - c^2} = \sqrt{169 - 144} = 5$. $\dfrac{x^2}{169} + \dfrac{y^2}{25} = 1$

43. $h = 0$, $k = 2$, $a = 3$, $c = 2 - h$ (so $c = 2$) and $b = \sqrt{a^2 - c^2} = \sqrt{9 - 4} = \sqrt{5}$

$\dfrac{x^2}{9} + \dfrac{(y - 2)^2}{5} = 1$

45. $h = 0$, $k = 0$, $c = 6$, $a = 5$, $b = \sqrt{c^2 - a^2} = \sqrt{36 - 25} = \sqrt{11}$, so

$\dfrac{y^2}{25} - \dfrac{x^2}{11} = 1$

47. $h = 2$, $k = 1$, $a = 3$, $\dfrac{b}{a} = \dfrac{4}{3}$ $\left(b = \dfrac{4}{3} \cdot 3 = 4\right)$, so

$\dfrac{(x - 2)^2}{9} - \dfrac{(y - 1)^2}{16} = 1$

49. $\dfrac{x}{5} = \cos t$ and $\dfrac{y}{2} = \sin t$, so $\dfrac{x^2}{25} + \dfrac{y^2}{4} = 1$ — an ellipse.

51. $x + 2 = \cos t$ and $y - 4 = \sin t$, so $(x + 2)^2 + (y - 4)^2 = 1$ — an ellipse (a circle).

53. $\dfrac{x}{3} = \sec t$ and $\dfrac{y}{5} = \tan t$, so $\dfrac{x^2}{9} - \dfrac{y^2}{25} = 1$ — a hyperbola.

55.

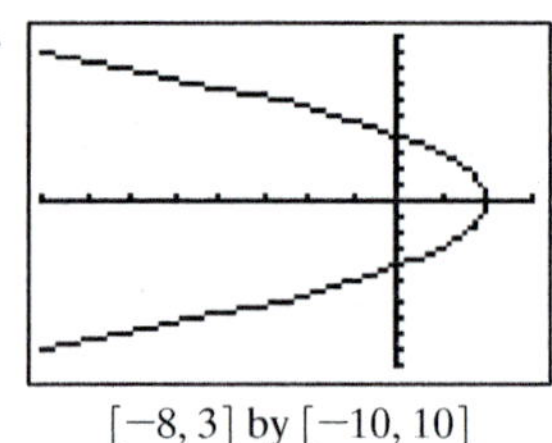

[−8, 3] by [−10, 10]

Parabola with vertex at $(2, 0)$, so $h = 2$, $k = 0$, $e = 1$. The graph crosses the y-axis, so $\left(4, \frac{\pi}{2}\right) = (0, 4)$ lies on the parabola. Substituting $(0, 4)$ into $y^2 = 4p(x - 2)$ we have $16 = 4p(-2)$, $p = -2$. $y^2 = -8(x - 2)$.

57.

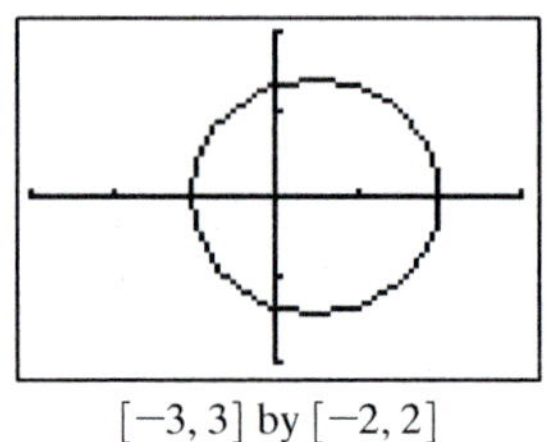

[−3, 3] by [−2, 2]

$e = \dfrac{1}{3}$, so an ellipse. In polar coordinates the vertices are $(2, 0)$ and $(1, \pi)$. Converting to Cartesian we have $(2, 0)$ and $(-1, 0)$, so $2a = 3$, $a = \dfrac{3}{2}$, $c = ea = \dfrac{1}{3} \cdot \dfrac{3}{2} = \dfrac{1}{2}$ and the center $(h, k) = \left(2 - \dfrac{3}{2}, 0\right) = \left(\dfrac{1}{2}, 0\right)$ (since it's symmetric about the polar x-axis). Solving for

$b = \sqrt{a^2 - c^2} = \sqrt{\left(\dfrac{3}{2}\right)^2 - \left(\dfrac{1}{2}\right)^2} = \sqrt{\dfrac{8}{4}} = \sqrt{2}$

$$\frac{4\left(x - \frac{1}{2}\right)^2}{9} + \frac{y^2}{2} = 1$$

59.

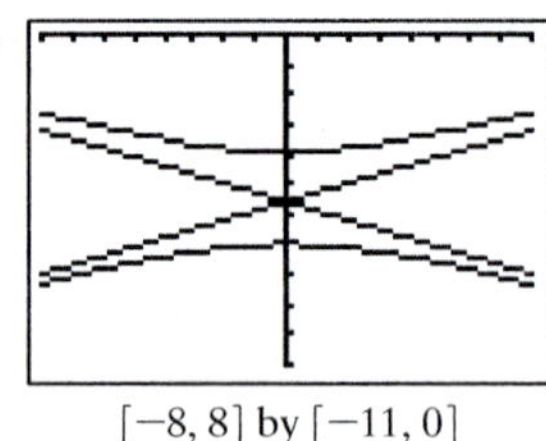

[−8, 8] by [−11, 0]

$e = \dfrac{7}{2}$, so a hyperbola. In polar coordinates the vertices are $\left(-7, \dfrac{\pi}{2}\right)$ and $\left(\dfrac{35}{9}, \dfrac{3\pi}{2}\right)$. Converting to Cartesian we have $(0, -7)$ and $\left(0, \dfrac{-35}{9}\right)$, so $2a = \dfrac{28}{9}$, $a = \dfrac{14}{9}$,

$c = ea = \dfrac{7}{2} \cdot \dfrac{14}{9} = \dfrac{49}{9}$ the center (h, k) $= \left(0, \dfrac{-35}{9} - \dfrac{14}{9}\right) = \left(0, \dfrac{-49}{9}\right)$ (since it's symmetric on the y-axis). Solving for

$b = \sqrt{c^2 - a^2} = \sqrt{\left(\dfrac{49}{9}\right)^2 - \left(\dfrac{14}{9}\right)^2} = \dfrac{21\sqrt{5}}{9} = \dfrac{7\sqrt{5}}{3}$

$$\frac{81\left(y + \frac{49}{9}\right)^2}{196} - \frac{9x^2}{245} = 1$$

61.

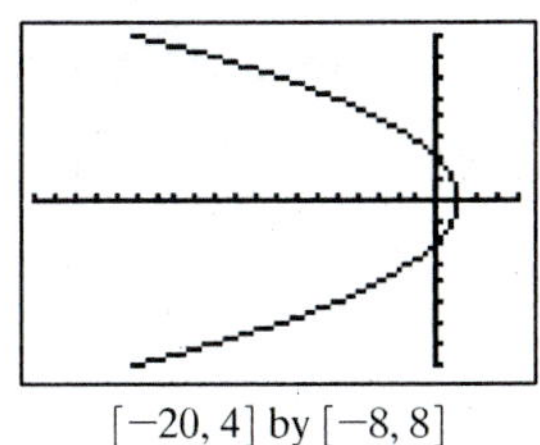

[−20, 4] by [−8, 8]

$e = 1$, so a parabola. In polar coordinates, the vertex is $(1, 0)$ and the parabola crosses the y-axis at $\left(2, \frac{\pi}{2}\right)$. Converting to Cartesian form, we have the vertex $(h, k) = (1, 0)$ and a point on the parabola is $(0, 2)$. Since the parabola opens to the left, $y^2 = 4p(x - 1)$. Substituting $(0, 2)$, we have $4 = -4p$, $p = -1$
$y^2 = -4(x - 1)$

63. $\sqrt{(3 - (-1))^2 + (-2 - 0)^2 + (-4 - 3)^2}$
$= \sqrt{16 + 4 + 49} = \sqrt{69}$

65. $\mathbf{v} + \mathbf{w} = \langle -3, 1, -2 \rangle + \langle 3, -4, 0 \rangle = \langle 0, -3, -2 \rangle$

67. $\mathbf{v} \cdot \mathbf{w} = \langle -3, 1, -2 \rangle \cdot \langle 3, -4, 0 \rangle = -9 - 4 + 0 = -13$

69. $\dfrac{\mathbf{w}}{|\mathbf{w}|} = \dfrac{\langle 3, -4, 0 \rangle}{\sqrt{3^2 + (-4)^2 + 0^2}} = \left\langle \dfrac{3}{5}, -\dfrac{4}{5}, 0 \right\rangle$

71. $(x + 1)^2 + y^2 + (z - 3)^2 = 16$

73. The direction vector is $\langle -3, 1, -2 \rangle$ so the vector equation of a line in the direction of $\mathbf{v}$ through P is
$\mathbf{r} = \langle -1, 0, 3 \rangle + t\langle -3, 1 -2 \rangle$

75. $4p = 18$, so $p = 4.5$; the focus is at $(0, 4.5)$.

77. (a) The "shark" should aim for the other spot on the table, since a ball that passes through one focus will end up passing through the other focus if nothing gets in the way.

(b) Let $a = 3$, $b = 2$, and $c = \sqrt{5}$. Then the foci are at $(-\sqrt{5}, 0)$ and $(\sqrt{5}, 0)$. These are the points at which to aim.

79. The major axis length is 18,000 km, plus 170 km, plus the diameter of the earth, so $a \approx 15{,}465$ km $= 15{,}465{,}000$ m. At apogee, $r = 18{,}000 + 6380 = 24{,}380$ km, so $v \approx 2633$ m/sec. At perigee, $r = 6380 + 170 = 6550$ km, so $v \approx 9800$ m/sec.

Chapter 8 Project

Answers are based on the sample data provided in the table.

1.

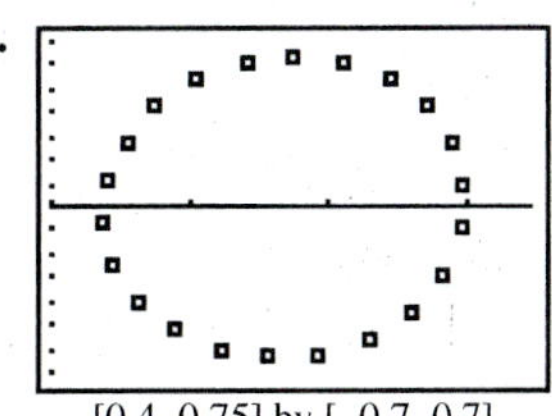

[0.4, 0.75] by [−0.7, 0.7]

2. The endpoints of the major and minor axes lie at approximately $(0.438, 0)$, $(0.700, 0)$, $(0.569, 0.640)$ and $(0.569, -0.640)$. The ellipse is taller than it is wide, even though the reverse appears to be true on the graphing calculator screen. The semimajor axis length is 0.640, and the semiminor axis length is $(0.700 - 0.438)/2 = 0.131$. The equation is
$$\frac{(y - 0)^2}{(0.640)^2} + \frac{(x - 0.569)^2}{(0.131)^2} = 1$$

3. With respect to the graph of the ellipse, the point (h, k) represents the center of the ellipse. The value a is the length of the semimajor axis, and b is the length of the semiminor axis.

4. Physically, $h = 0.569$ m is the pendulum's average distance from the CBR, and $k = 0$ m/sec is the pendulum's average velocity. The value $a = 0.64$ m/sec is the maximum velocity, and $b = 0.131$ m is the maximum displacement of the pendulum from its average position.

5. The parametric equations for the sample data set (using sinusoidal regression) are
$x_{1T} \approx 0.131 \sin(4.80T + 2.10) + 0.569$ and
$y_{1T} \approx 0.639 \sin(4.80T - 2.65)$.

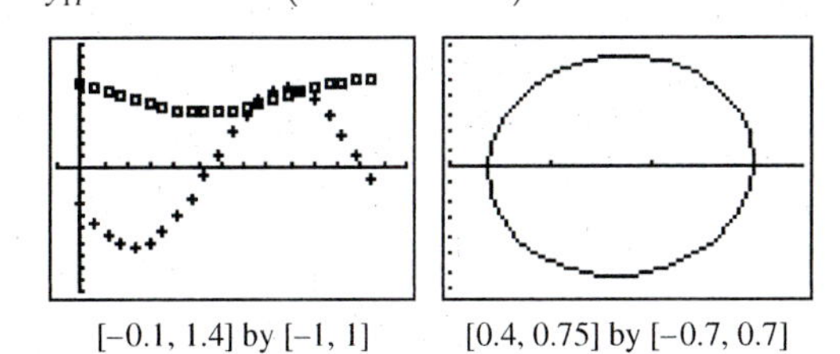

[−0.1, 1.4] by [−1, 1]   [0.4, 0.75] by [−0.7, 0.7]

Chapter 9 Discrete Mathematics

Section 9.1 Basic Combinatorics

Exploration 1

1. Six: $ABC, ACB, BAC, BCA, CAB, CBA$.
2. Approximately 1 person out of 6, which would mean 10 people out of 60.
3. No. If they all looked the same, we would expect approximately 10 people to get the order right simply by chance. The fact that this did not happen leads us to reject the "look-alike" conclusion.
4. It is likely that the salesman rigged the test to mislead the office workers. He might have put the copy from the more expensive machine on high-quality bond paper to make it look more like an original, or he might have put a tiny ink smudge on the original to make it look like a copy. You can offer your own alternate scenarios.

Quick Review 9.1

1. 52
3. 6
5. 10
7. 11
9. 64

Section 9.1 Exercises

1. There are three possibilities for who stands on the left, and then two remaining possibilities for who stands in the middle, and then one remaining possibility for who stands on the right: $3 \cdot 2 \cdot 1 = 6$.
3. Any of the five books could be placed on the left, and then any of the four remaining books could be placed next to it, and so on: $5 \cdot 4 \cdot 3 \cdot 2 \cdot 1 = 120$.
5. There are $3 \cdot 4 = 12$ possible pairs: K_1Q_1, K_1Q_2, K_1Q_3, K_1Q_4, K_2Q_1, K_2Q_2, K_2Q_3, K_2Q_4, K_3Q_1, K_3Q_2, K_3Q_3, and K_3Q_4.
7. $9! = 362{,}880$ (ALGORITHM)
9. There are 11 letters, where S and I each appear 4 times and P appears 2 times. The number of distinguishable permutations is $\frac{11!}{4!4!2!} = 34{,}650$.
11. The number of ways to fill 3 distinguishable offices from a pool of 13 candidates is ${}_{13}P_3 = \frac{13!}{10!} = 1716$.
13. $4 \cdot 3 \cdot 2 \cdot 1 = 24$
15. $\frac{6!}{(6-2)!} = \frac{6 \cdot 5 \cdot 4!}{4!} = 30$
17. $\frac{10!}{7!(10-7)!} = \frac{10 \cdot 9 \cdot 8 \cdot 7!}{7! \cdot 3 \cdot 2 \cdot 1} = 120$
19. combinations
21. combinations
23. There are 10 choices for the first character, 9 for the second, 26 for the third, then 25, then 8, then 7, then 6: $10 \cdot 9 \cdot 26 \cdot 25 \cdot 8 \cdot 7 \cdot 6 = 19{,}656{,}000$.
25. There are 6 possibilities for the red die, and 6 for the green die: $6 \cdot 6 = 36$.
27. ${}_{25}C_3 = \frac{25!}{3!(25-3)!} = \frac{25!}{3!22!} = 2300$
29. ${}_{48}C_3 = \frac{48!}{3!(48-3)!} = \frac{48!}{3!45!} = 17{,}296$
31. Choose A♠ and K♠, and 11 cards from the other 50: ${}_2C_2 \cdot {}_{50}C_{11} = 1 \cdot {}_{50}C_{11} = \frac{50!}{11!(50-11)!} = \frac{50!}{11!39!}$ $= 37{,}353{,}738{,}800$
33. We either have 3, 2, or 1 student(s) nominated: ${}_6C_3 + {}_6C_2 + {}_6C_1 = 20 + 15 + 6 = 41$
35. Each of the 5 dice have 6 possible outcomes: $6^5 = 7776$
37. $2^9 - 1 = 511$
39. Since each topping can be included or left off, the total number of possibilities with n toppings is 2^n. Since $2^{11} = 2048$ is less than 4000 but $2^{12} = 4096$ is greater than 4000, Luigi offers at least 12 toppings.
41. $2^{10} = 1024$
43. True. $\binom{n}{a} = \frac{n!}{a!(n-a)!} = \frac{n!}{a!b!} = \frac{n!}{(n-b)!b!} = \binom{n}{b}$
45. There are $\binom{6}{2} = 15$ different combinations of vegetables. The total number of entreé-vegetable-dessert variations is $4 \cdot 15 \cdot 6 = 360$. The answer is D.
47. ${}_nP_n = \frac{n!}{(n-n)!} = n!$ The answer is B.
49. Answers will vary. Here are some possible answers:
    (a) Number of 3-card hands that can be dealt from a deck of 52 cards
    (b) Number of ways to choose 3 chocolates from a box of 12 chocolates
    (c) Number of ways to choose a starting soccer team from a roster of 25 players (where position matters)
    (d) Number of 5-digit numbers that can be formed using only the digits 1 and 2
    (e) Number of possible pizzas that can be ordered at a place that offers 3 different sizes and up to 10 different toppings.

51. **(a)** Twelve

(b) Every 0 represents a factor of 10, or a factor of 5 multiplied by a factor of 2. In the product $50 \cdot 49 \cdot 48 \cdot \ldots \cdot 2 \cdot 1$, the factors 5, 10, 15, 20, 30, 35, 40, and 45 each contain 5 as a factor once, and 25 and 50 each contain 5 twice, for a total of twelve occurrences. Since there are 47 factors of 2 to pair up with the twelve factors of 5, 10 is a factor of 50! twelve times.

53. In the nth week, 5^n copies of the letter are sent. In the last week of the year, that's $5^{52} \approx 2.22 \times 10^{36}$ copies of the letter. This exceeds the population of the world, which is about 6×10^9, so someone (several people, actually) has had to receive a second copy of the letter.

55. Three. This is equivalent to the round table problem (Exercise 42), except that the necklace can be *turned upside-down*. Thus, each different necklace accounts for two of the six different orderings.

57. There are ${}_{52}C_{13} = 635{,}013{,}559{,}600$ distinct bridge hands. Every day has $60 \cdot 60 \cdot 24 = 86{,}400$ seconds; a year has 365.24 days, which is 31,556,736 seconds. Therefore it will take about $\dfrac{635{,}013{,}559{,}600}{31{,}556{,}736} \approx 20{,}123$ years. (Using 365 days per year, the computation gives about 20,136 years.)

Section 9.2 The Binomial Theorem

Exploration 1

1. ${}_3C_0 = \dfrac{3!}{0!3!} = 1$, ${}_3C_1 = \dfrac{3!}{1!2!} = 3$, ${}_3C_2 = \dfrac{3!}{2!1!} = 3$, ${}_3C_3 = \dfrac{3!}{3!0!} = 1$. These are (in order) the coefficients in the expansion of $(a + b)^3$.

2. $\{1\ 4\ 6\ 4\ 1\}$. These are (in order) the coefficients in the expansion of $(a + b)^4$.

3. $\{1\ 5\ 10\ 10\ 5\ 1\}$. These are (in order) the coefficients in the expansion of $(a + b)^5$.

Quick Review 9.2

1. $x^2 + 2xy + y^2$

3. $25x^2 - 10xy + y^2$

5. $9s^2 + 12st + 4t^2$

7. $u^3 + 3u^2v + 3uv^2 + v^3$

9. $8x^3 - 36x^2y + 54xy^2 - 27y^3$

Section 9.2 Exercises

1. $$\begin{aligned}(a + b)^4 &= \binom{4}{0}a^4b^0 + \binom{4}{1}a^3b^1 + \binom{4}{2}a^2b^2 + \binom{4}{3}a^1b^3 + \binom{4}{4}a^0b^4 \\ &= a^4 + 4a^3b + 6a^2b^2 + 4ab^3 + b^4\end{aligned}$$

3. $$\begin{aligned}(x + y)^7 &= \binom{7}{0}x^7y^0 + \binom{7}{1}x^6y^1 + \binom{7}{2}x^5y^2 + \binom{7}{3}x^4y^3 + \binom{7}{4}x^3y^4 + \binom{7}{5}x^2y^5 + \binom{7}{6}x^1y^6 + \binom{7}{7}x^0y^7 \\ &= x^7 + 7x^6y + 21x^5y^2 + 35x^4y^3 + 35x^3y^4 + 21x^2y^5 + 7xy^6 + y^7\end{aligned}$$

5. Use the entries in row 3 as coefficients: $(x + y)^3 = x^3 + 3x^2y + 3xy^2 + y^3$

7. Use the entries in row 8 as coefficients: $(p + q)^8 = p^8 + 8p^7q + 28p^6q^2 + 56p^5q^3 + 70p^4q^4 + 56p^3q^5 + 28p^2q^6 + 8pq^7 + q^8$

9. $\dbinom{9}{2} = \dfrac{9!}{2!7!} = \dfrac{9 \cdot 8}{2 \cdot 1} = 36$

11. $\dbinom{166}{166} = \dfrac{166!}{166!0!} = 1$

13. $\dbinom{14}{3} = \dbinom{14}{11} = 364$

15. $(-2)^8\dbinom{12}{8} = (-2)^8\dbinom{12}{4} = 126{,}720$

17. $$\begin{aligned}f(x) &= (x - 2)^5 \\ &= x^5 + 5x^4(-2) + 10x^3(-2)^2 + 10x^2(-2)^3 + 5x(-2)^4 + (-2)^5 \\ &= x^5 - 10x^4 + 40x^3 - 80x^2 + 80x - 32\end{aligned}$$

19. $$\begin{aligned}h(x) &= (2x - 1)^7 \\ &= (2x)^7 + 7(2x)^6(-1) + 21(2x)^5(-1)^2 + 35(2x)^4(-1)^3 + 35(2x)^3(-1)^4 + 21(2x)^2(-1)^5 + 7(2x)(-1)^6 + (-1)^7 \\ &= 128x^7 - 448x^6 + 672x^5 - 560x^4 + 280x^3 - 84x^2 + 14x - 1\end{aligned}$$

21. $$\begin{aligned}(2x + y)^4 &= (2x)^4 + 4(2x)^3y + 6(2x)^2y^2 + 4(2x)y^3 + y^4 \\ &= 16x^4 + 32x^3y + 24x^2y^2 + 8xy^3 + y^4\end{aligned}$$

23. $$\begin{aligned}(\sqrt{x} - \sqrt{y})^6 &= (\sqrt{x})^6 + 6(\sqrt{x})^5(-\sqrt{y}) + 15(\sqrt{x})^4 \cdot (-\sqrt{y})^2 + 20(\sqrt{x})^3(-\sqrt{y})^3 + 15(\sqrt{x})^2(-\sqrt{y})^4 + 6(\sqrt{x})(-\sqrt{y})^5 + (-\sqrt{y})^6 \\ &= x^3 - 6x^{5/2}y^{1/2} + 15x^2y - 20x^{3/2}y^{3/2} + 15xy^2 - 6x^{1/2}y^{5/2} + y^3\end{aligned}$$

25. $$\begin{aligned}(x^{-2} + 3)^5 &= (x^{-2})^5 + 5(x^{-2})^4 \cdot 3 + 10(x^{-2})^3 \cdot 3^2 + 10(x^{-2})^2 \cdot 3^3 + 5(x^{-2}) \cdot 3^4 + 3^5 \\ &= x^{-10} + 15x^{-8} + 90x^{-6} + 270x^{-4} + 405x^{-2} + 243\end{aligned}$$

27. Answers will vary.

29. $$\binom{n}{1} = \frac{n!}{1!(n-1)!} = n = \frac{n!}{(n-1)!1!} = \frac{n!}{(n-1)![n-(n-1)]!} = \binom{n}{n-1}$$

31. $$\begin{aligned}&\binom{n-1}{r-1} + \binom{n-1}{r} \\ &= \frac{(n-1)!}{(r-1)![(n-1)-(r-1)]!} + \frac{(n-1)!}{r!(n-1-r)!} \\ &= \frac{r(n-1)!}{r(r-1)!(n-r)!} + \frac{(n-1)!(n-r)}{r!(n-r)(n-r-1)!} \\ &= \frac{r(n-1)!}{r!(n-r)!} + \frac{(n-r)(n-1)!}{r!(n-r)!} \\ &= \frac{(r+n-r)(n-1)!}{r!(n-r)!} \\ &= \frac{n!}{r!(n-r)!} \\ &= \binom{n}{r}\end{aligned}$$

33. $\binom{n}{2} + \binom{n+1}{2} = \dfrac{n!}{2!(n-2)!} + \dfrac{(n+1)!}{2!(n-1)!}$

$= \dfrac{n(n-1)}{2} + \dfrac{(n+1)(n)}{2}$

$= \dfrac{n^2 - n + n^2 + n}{2} = n^2$

35. True. The signs of the coefficients are determined by the powers of the $(-y)$ terms, which alternate between odd and even.

37. The fifth term of the expansion is $\binom{8}{4}(2x)^4(1)^4 = 1120x^4$. The answer is C.

39. The sum of the coefficients of $(3x - 2y)^{10}$ is the same as the value of $(3x - 2y)^{10}$ when $x = 1$ and $y = 1$. The answer is A.

41. **(a)** 1, 3, 6, 10, 15, 21, 28, 36, 45, 55.

(b) They appear diagonally down the triangle, starting with either of the 1's in row 2.

(c) Since n and $n + 1$ represent the sides of the given rectangle, then $n(n + 1)$ represents its area. The triangular number is $1/2$ of the given area. Therefore, the triangular number is $\dfrac{n(n+1)}{2}$.

(d) From part (c), we observe that the nth triangular number can be written as $\dfrac{n(n+1)}{2}$. We know that binomial coefficients are the values of $\binom{n}{r}$ for $r = 0, 1, 2, 3, \ldots, n$. We can show that $\dfrac{n(n+1)}{2} = \binom{n+1}{2}$ as follows:

$\dfrac{n(n+1)}{2} = \dfrac{(n+1)n(n-1)!}{2(n-1)!}$

$= \dfrac{(n+1)!}{2!(n-1)!}$

$= \dfrac{(n+1)!}{2!((n+1)-2)!}$

$= \binom{n+1}{2}$

So, to find the fourth triangular number, for example, compute $\binom{4+1}{2} = \binom{5}{2} = \dfrac{5!}{2!(5-2)!} = \dfrac{5 \cdot 4 \cdot 3!}{2!3!}$

$= \dfrac{5 \cdot 4}{2} = 10.$

43. The sum of the entries in the nth row equals the sum of the coefficients in the expansion of $(x + y)^n$. But this sum, in turn, is equal to the value of $(x + y)^n$ when $x = 1$ and $y = 1$:

$2^n = (1 + 1)^n$

$= \binom{n}{0}1^n1^0 + \binom{n}{1}1^{n-1}1^1 + \binom{n}{2}1^{n-2}1^2 + \cdots + \binom{n}{n}1^01^n$

$= \binom{n}{0} + \binom{n}{1} + \binom{n}{2} + \cdots + \binom{n}{n}$

45. $3^n = (1 + 2)^n$

$= \binom{n}{0}1^n + \binom{n}{1}1^{n-1}2 + \binom{n}{2}1^{n-2}2^2 + \cdots + \binom{n}{n}2^n$

$= \binom{n}{0} + 2\binom{n}{1} + 4\binom{n}{2} + \cdots + 2^n\binom{n}{n}$

■ Section 9.3 Probability

Exploration 1

1.

Antibodies present

$p = 0.006$

$p = 0.994$

Antibodies absent

2.

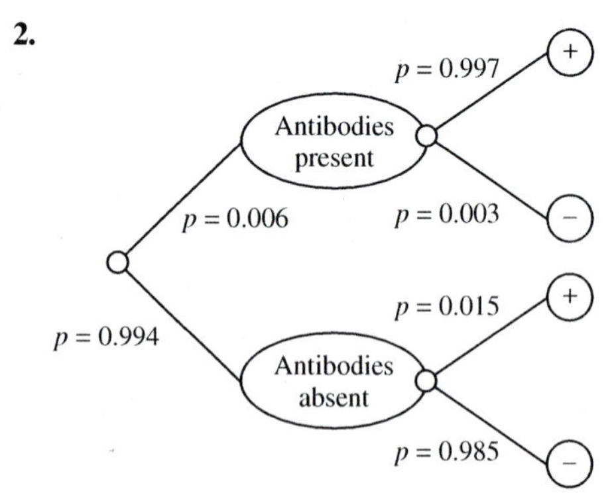

3.

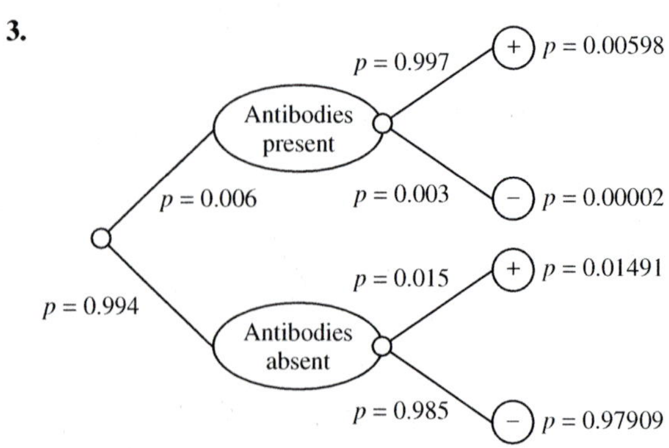

4. $P(+) = 0.00598 + 0.01491 = 0.02089$

5. $P(\text{antibody present} \mid +) = \dfrac{P(\text{antibody present and } +)}{P(+)}$

$\dfrac{0.00598}{0.02089} \approx 0.286$ (A little more than 1 chance in 4.)

Quick Review 9.3

1. 2

3. $2^3 = 8$

5. ${}_{52}C_5 = 2{,}598{,}960$

7. $5! = 120$

9. $\dfrac{{}_5C_3}{{}_{10}C_3} = \dfrac{\frac{5!}{3!2!}}{\frac{10!}{3!7!}} = \dfrac{1}{12}$

Section 9.3 Exercises

For #1–7, consider ordered pairs (a, b) where a is the value of the red die and b is the value of the green die.

1. $E = \{(3, 6), (4, 5), (5, 4), (6, 3)\}$: $P(E) = \dfrac{4}{36} = \dfrac{1}{9}$

3. $E = \{(2, 1), (3, 1), (3, 2), (4, 1), (4, 2), (4, 3), (5, 1), (5, 2), (5, 3), (5, 4), (6, 1), (6, 2), (6, 3), (6, 4), (6, 5)\}$:
$P(E) = \dfrac{15}{36} = \dfrac{5}{12}$

5. $P(E) = \dfrac{3 \cdot 3}{36} = \dfrac{1}{4}$

7. $E = \{(1, 1), (1, 2), (1, 4), (1, 6), (2, 1), (2, 3), (2, 5), (3, 2), (3, 4), (4, 1), (4, 3), (5, 2), (5, 6), (6, 1), (6, 5)\}$
$P(E) = \dfrac{15}{36} = \dfrac{5}{12}$

9. (a) No. $0.25 + 0.20 + 0.35 + 0.30 = 1.1$. The numbers do not add up to 1.

(b) There is a problem with Alrik's reasoning. Since the gerbil must always be in exactly one of the four rooms, the proportions must add up to 1, just like a probability function.

11. $P(B \text{ or } T) = P(B) + P(T) = 0.3 + 0.1 = 0.4$

13. $P(R) = 0.2$

15. $P[\text{not } (O \text{ or } Y)] = 1 - P(O \text{ or } Y)$
$= 1 - (0.2 + 0.1) = 0.7$

17. $P(B_1 \text{ and } B_2) = P(B_1) \cdot P(B_2) = (0.3)(0.3) = 0.09$

19. $P[(R_1 \text{ and } G_2) \text{ or } (G_1 \text{ and } R_2)] = P(R_1) \cdot P(G_2) + P(G_1) \cdot P(R_2) = (0.2)(0.2) + (0.2)(0.2) = 0.08$

21. P (neither is yellow) $= P(\text{not } Y_1 \text{ and not } Y_2)$
$= P(\text{not } Y_1) \cdot P(\text{not } Y_2) = (0.8)(0.8) = 0.64$

23. There are ${}_{24}C_6 = 134{,}596$ possible hands; of these, only 1 consists of all spades, so the probability is $\dfrac{1}{134{,}596}$.

25. Of the ${}_{24}C_6 = 134{,}596$ possible hands, there are ${}_4C_4 \cdot {}_{20}C_2 = 190$ hands with all the aces, so the probability is $\dfrac{190}{134{,}596} = \dfrac{5}{3542}$.

27. (a)

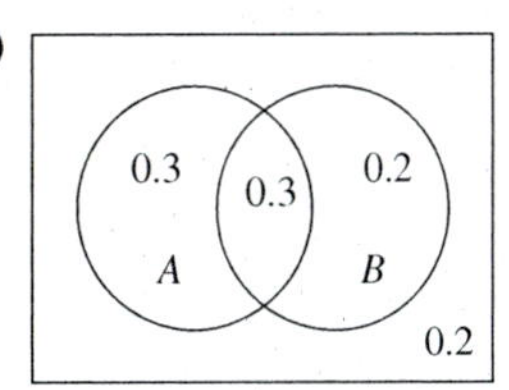

(b) 0.3

(c) 0.2

(d) 0.2

(e) Yes. $P(A \mid B) = \dfrac{P(A \text{ and } B)}{P(B)} = \dfrac{0.3}{0.5} = 0.6 = P(A)$

29. $P(\text{John will practice}) = (0.6)(0.8) + (0.4)(0.4) = 0.64$

31. If all precalculus students were put on a single list and a name then randomly chosen, the probability P(from Mr. Abel's class | girl) would be 12 | 22 = 6 | 11. But when one of the two classes is selected at random, and then a student from this class is selected,
P(from Mr. Abel's class | girl)

$$= \frac{P(\text{girl from Mr. Abel's class})}{P(\text{girl})} = \frac{\left(\frac{1}{2}\right)\left(\frac{12}{20}\right)}{\left(\frac{1}{2}\right)\left(\frac{12}{20}\right) + \left(\frac{1}{2}\right)\left(\frac{10}{25}\right)} = \frac{3}{5}$$

33. $\dfrac{{}_{20}C_2}{{}_{25}C_2} = \dfrac{190}{300} = \dfrac{19}{30}$

35. (a) $P(\text{cardiovascular disease or cancer}) = 0.45 + 0.22 = 0.67$

(b) $P(\text{other cause of death}) = 1 - 0.67 = 0.33$

37. The sum of the probabilities is greater than 1 – an impossibility, since the events are mutually exclusive.

39. (a) $P(\text{a woman}) = \dfrac{172}{254} = \dfrac{86}{127}$

(b) $P(\text{went to graduate school}) = \dfrac{124 + 58}{254} = \dfrac{91}{127}$

(c) $P(\text{a woman who went to graduate school}) = \dfrac{124}{254} = \dfrac{62}{127}$

41. $\dfrac{1}{{}_9C_2} = \dfrac{1}{36}$

43. $P(\text{HTTTTTTTTT}) = \left(\dfrac{1}{2}\right)^{10} = \dfrac{1}{2^{10}} = \dfrac{1}{1024}$

45. $P(\text{HHHHHHHHHH}) = \left(\dfrac{1}{2}\right)^{10} = \dfrac{1}{1024}$

47. $P(2\text{ H and } 8\text{ T}) = {}_{10}C_2 \cdot \left(\dfrac{1}{2}\right)^{10} = \dfrac{45}{1024}$

49. P (at least one H) $= 1 - P(0\text{ H}) = 1 - {}_{10}C_0 \cdot \left(\dfrac{1}{2}\right)^{10}$
$= 1 - \dfrac{1}{1024} = \dfrac{1023}{1024}$

51. False. A sample space consists of outcomes, which are not necessarily equally likely.

53. Of the 36 different, equally-likely ways the dice can land, 4 ways have a total of 5. So the probability is $4/36 = 1/9$. The answer is D.

55. $P(B \text{ and } A) = P(B)\,P(A|B)$, and for independent events, $P(B \text{ and } A) = P(B)P(A)$. It follows that the answer is A.

57. (a)

Type of Bagel	Probability
Plain	0.37
Onion	0.12
Rye	0.11
Cinnamon Raisin	0.25
Sourdough	0.15

(b) $(0.37)(0.37)(0.37) \approx 0.051$

(c) No. They are more apt to share bagel preferences if they arrive at the store together.

59. (a) We expect $(8)(0.23) = 1.84$ (about 2) to be married.

(b) Yes, this would be an unusual sample.

(c) $P(5 \text{ or more are married}) = {}_8C_5 \cdot (0.23)^5(0.77)^3 + {}_8C_6 \cdot (0.23)^6(0.77)^2 + {}_8C_7 \cdot (0.23)^7(0.77)^1 + {}_8C_8 \cdot (0.23)^8(0.77)^0 \approx 0.01913 = 1.913\%$.

61. (a) \$1.50

(b) $3 \cdot \frac{2}{6} + (-1) \cdot \frac{4}{6} = \frac{6}{6} - \frac{4}{6} = \frac{1}{3}$

Section 9.4 Sequences

Quick Review 9.4

1. $3 + (5 - 1)4 = 3 + 16 = 19$

3. $5 \cdot 4^2 = 80$

5. $a_{10} = \frac{10}{11}$

7. $a_{10} = 5 \cdot 2^9 = 2560$

9. $a_{10} = 32 - 17 = 15$

Section 9.4 Exercises

For #1–3, substitute $n = 1, n = 2, \ldots, n = 6$, and $n = 100$.

1. $2, \frac{3}{2}, \frac{4}{3}, \frac{5}{4}, \frac{6}{5}, \frac{7}{6}; \frac{101}{100}$

3. 0, 6, 24, 60, 120, 210; 999,900

For #5–9, use previously computed values of the sequence to find the next term in the sequence.

5. 8, 4, 0, −4; −20

7. 2, 6, 18, 54; 4374

9. 2, −1, 1, 0; 3

11. $\lim_{n\to\infty} n^2 = \infty$, so the sequence diverges.

13. $\frac{1}{1}, \frac{1}{4}, \frac{1}{9}, \frac{1}{16}, \ldots, \frac{1}{n^2}, \ldots$

$\lim_{n\to\infty} \frac{1}{n^2} = 0$, so the sequence converges to 0.

15. Since the degree of the numerator is the same as the degree of the denominator, the limit is the ratio of the leading coefficients. Thus $\lim_{n\to\infty} \frac{3n - 1}{2 - 3n} = -1$. The sequence converges to -1.

17. $\lim_{n\to\infty} (0.5)^n = \lim_{n\to\infty} \left(\frac{1}{2}\right)^n = 0$, so the sequence converges to 0.

19. $a_1 = 1$ and $a_{n+1} = a_n + 3$ for $n \ge 1$ yields $1, 4, 7, \ldots, (3n - 2), \ldots$

$\lim_{n\to\infty} (3n - 2) = \infty$, so the sequence diverges.

For #21–23, subtract the first term from the second to find the common difference d. Use the formula $a_n = a_1 + (n - 1)d$ with $n = 10$ to find the tenth term. The recursive rule for the nth term is $a_n = a_{n-1} + d$, and the explicit rule is the one given above.

21. (a) $d = 4$

(b) $a_{10} = 6 + 9(4) = 42$

(c) Recursive rule: $a_1 = 6$; $a_n = a_{n-1} + 4$ for $n \ge 2$

(d) Explicit rule: $a_n = 6 + 4(n - 1)$

23. (a) $d = 3$

(b) $a_{10} = -5 + 9(3) = 22$

(c) Recursive rule: $a_1 = -5$; $a_n = a_{n-1} + 3$ for $n \ge 2$

(d) Explicit rule: $a_n = -5 + 3(n - 1)$

For #25–27, divide the second term by the first to find the common ratio r. Use the formula $a_n = a_1 \cdot r^{n-1}$ with $n = 8$ to find the eighth term. The recursive rule for the nth term is $a_n = a_{n-1} \cdot r$, and the explicit rule is the one given above.

25. (a) $r = 3$

(b) $a_8 = 2 \cdot 3^7 = 4374$

(c) Recursive rule: $a_1 = 2$; $a_n = 3a_{n-1}$ for $n \ge 2$

(d) Explicit rule: $a_n = 2 \cdot 3^{n-1}$

27. (a) $r = -2$

(b) $a_8 = (-2)^7 = -128$

(c) Recursive rule: $a_1 = 1$; $a_n = -2a_{n-1}$ for $n \ge 2$

(d) Explicit rule: $a_n = (-2)^{n-1}$

29. $a_4 = -8 = a_1 + 3d$ and $a_7 = 4 = a_1 + 6d$, so $a_7 - a_4 = 12 = 3d$. Therefore $d = 4$, so $a_1 = -8 - 3d = -20$ and $a_n = a_{n-1} + 4$ for $n \ge 2$.

31. $a_2 = 3 = a_1 \cdot r^1$ and $a_8 = 192 = a_1 \cdot r^7$, so $a_8/a_2 = 64 = r^6$. Therefore $r = \pm 2$, so $a_1 = 3/(\pm 2) = \pm\frac{3}{2}$ and $a_n = -\frac{3}{2} \cdot (-2)^{n-1} = 3 \cdot (-2)^{n-2}$ or $a_n = \frac{3}{2} \cdot 2^{n-1} = 3 \cdot 2^{n-2}$.

33.

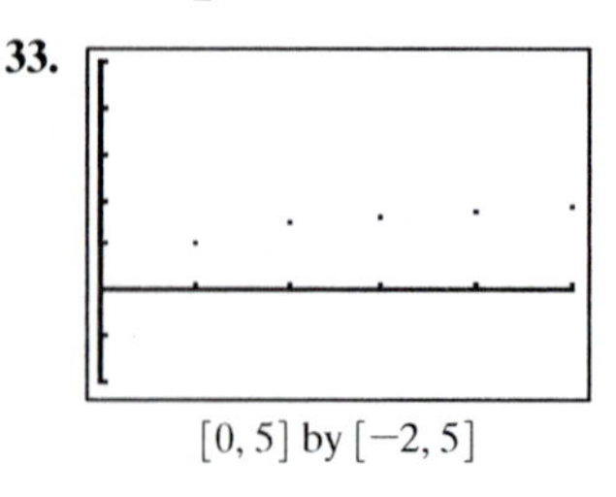

[0, 5] by [−2, 5]

35.

[0, 10] by [−10, 100]

37. The height (in cm) will be an arithmetic sequence with common difference $d = 2.3$ cm, so the height in week n is $700 + 2.3(n - 1)$: 700, 702.3, 704.6, 706.9, ..., 815, 817.3.

39. The numbers of seats in each row form a finite arithmetic sequence with $a_1 = 7, d = 2$, and $n = 25$. The total number of seats is

$$\frac{25}{2}[2(7) + (25 - 1)(2)] = 775.$$

41. The ten-digit numbers will vary; thus the sequences will vary. The end result will, however, be the same. Each limit will be 9. One example is:

Five random digits: 1, 4, 6, 8, 9
Five random digits: 2, 3, 4, 5, 6
List: 1, 2, 3, 4, 4, 5, 6, 6, 8, 9
Ten-digit number: 2, 416, 345, 689
Ten-digit number: 9, 643, 128, 564
a_1 = positive difference of the ten-digit numbers
= 7, 226, 782, 875
a_{n+1} = sum of the digits of a_n, so
a_2 = sum of the digits of $a_1 = 54$
a_3 = sum of the digits of $a_2 = 9$
All successive sums of digits will be 9, so the sequence converges and the limit is 9.

43. True. Since two successive terms are negative, the common ratio r must be positive, and so the sign of the first term determines the sign of every number in the sequence.

45. $a_1 = 2$ and $a_2 = 8$ implies $d = 8 - 2 = 6$
$c = a_1 - d$ so $c = 2 - 6 = -4$
$a_4 = 6 \cdot 4 + (-4) = 20$
The answer is A.

47. $r = \dfrac{a_2}{a_1} = \dfrac{6}{2} = 3$
$a_6 = a_1 r^5 = 2 \cdot 3^5 = 486$ and $a_2 = 6$, so
$\dfrac{a_6}{a_2} = \dfrac{486}{6} = 81.$
The answer is E.

49. **(a)** $a_1 = 1$ because there is initially one male-female pair (this is the number of pairs after 0 months). $a_2 = 1$ because after one month, the original pair has only just become fertile. $a_3 = 2$ because after two months, the original pair produces a new male-female pair.

(b) Notice that after $n - 2$ months, there are a_{n-1} pairs, of which a_{n-2} (the number of pairs present one month earlier) are fertile. Therefore, after $n - 1$ months, the number of pairs will be $a_n = a_{n-1} + a_{n-2}$: to last month's total, we add the number of new pairs born. Thus $a_4 = 3, a_5 = 5, a_6 = 8, a_7 = 13, a_8 = 21, a_9 = 34, a_{10} = 55, a_{11} = 89, a_{12} = 144, a_{13} = 233$.

(c) Since a_1 is the initial number of pairs, and a_2 is the number of pairs after one month, we see that a_{13} is the number of pairs after 12 months.

51. **(a)** For a polygon with n sides, let A be the vertex in quadrant I at the top of the vertical segment, and let B be the point on the x-axis directly below A. Together with (0, 0), these two points form a right triangle; the acute angle at the origin has measure $\theta = \dfrac{2\pi}{2n} = \dfrac{\pi}{n}$, since there are $2n$ such triangles making up the polygon. The length of the side opposite this angle is

$\sin\theta = \sin\dfrac{\pi}{n}$, and there are $2n$ such sides making up

the perimeter of the polygon, so $\sin\dfrac{\pi}{n} = \dfrac{a_n}{2n}$, or

$a_n = 2n\sin(\pi/n)$.

(b) $a_{10} \approx 6.1803$, $a_{100} \approx 6.2822$, $a_{1000} \approx 6.2832$, $a_{10,000} \approx 6.2832 \approx 2\pi$. It appears that $a_n \to 2\pi$ as $n \to \infty$, which makes sense since the perimeter of the polygon should approach the circumference of the circle.

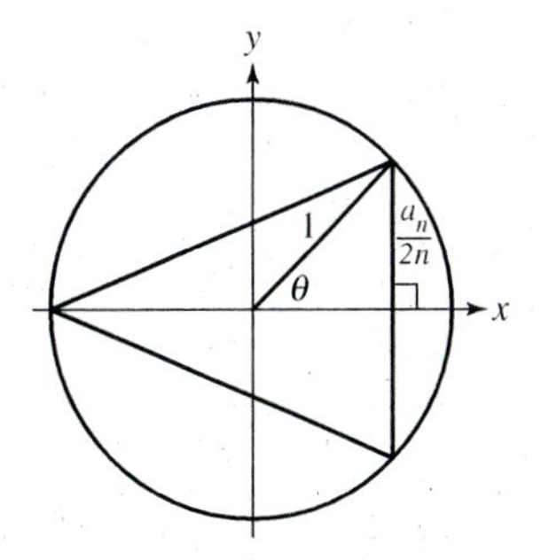

53. The difference of successive terms in $\{\log(a_n)\}$ will be of the form $\log(a_{n+1}) - \log(a_n) = \log\left(\dfrac{a_{n+1}}{a_n}\right)$. Since $\{a_n\}$ is geometric, $\dfrac{a_{n+1}}{a_n}$ is constant. This makes $\log\left(\dfrac{a_{n+1}}{a_n}\right)$ constant, so $\{\log(a_n)\}$ is a sequence with a constant difference (arithmetic).

55. $a_1 = [1\ 1], a_2 = [1\ 2], a_3 = [2\ 3], a_4 = [3\ 5], a_5 = [5\ 8], a_6 = [8\ 13], a_7 = [13\ 21]$. The entries in the terms of this sequence are successive pairs of terms from the Fibonacci sequence.

Section 9.5 Series

Exploration 1

1. $3 + 6 + 9 + 12 + 15 = 45$

2. $5^2 + 6^2 + 7^2 + 8^2 = 25 + 36 + 49 + 64 = 174$

3. $\cos(0) + \cos(\pi) + \cdots + \cos(11\pi) + \cos(12\pi)$
$= 1 - 1 + 1 + \cdots - 1 + 1 = 1$

4. $\sin(0) + \sin(\pi) + \cdots + \sin(k\pi)$
$+ \cdots = 0 + 0 + \cdots + 0 + \cdots = 0$

5. $\dfrac{3}{10} + \dfrac{3}{100} + \dfrac{3}{1{,}000} + \cdots + \dfrac{3}{1{,}000{,}000} + \cdots = \dfrac{1}{3}$

Exploration 2

1. $1 + 2 + 3 + \cdots + 99 + 100$

2. $100 + 99 + 98 + \cdots + 2 + 1$

3. $101 + 101 + 101 + \cdots + 101 + 101$

4. $100(101) = 10{,}100$
5. The sum in 4 involves two copies of the same progression, so it doubles the sum of the progression. The answer that Gauss gave was 5050.

Quick Review 9.5

1. $a_1 = 4;\ d = 2$ so $a_{10} = a_1 + (n-1)d$
$a_{10} = 4 + (10-1)2 = 4 + 18 = 22$
$a_{10} = 22$
3. $a_3 = 6$ and $a_8 = 21$
$a_3 = a_1 + 2d$ and $a_8 = a_1 + 7d$
$(a_1 + 7d) - (a_1 + 2d) = 21 - 6$ so $5d = 15 \Rightarrow d = 3.$
$6 = a_1 + 2(3)$ so $a_1 = 0$
$a_{10} = 0 + 9(3) = 27$
$a_{10} = 27$
5. $a_1 = 1$ and $a_2 = 2$ yields $r = \frac{2}{1} = 2$
$a_{10} = 1 \cdot 2^9 = 512$
$a_{10} = 512$
7. $a_7 = 5$ and $r = -2 \Rightarrow 5 = a_1(-2)^6$
$a_1 = \frac{5}{64};\ a_{10} = \frac{5}{24}(-2)^9 = \frac{-2560}{24} = -40$
$a_{10} = -40$
9. $\sum_{n=1}^{5} n^2 = 1 + 4 + 9 + 16 + 25 = 55$

Section 9.5 Exercises

1. $\sum_{k=1}^{11} (6k - 13)$
3. $\sum_{k=1}^{n+1} k^2$
5. $\sum_{k=0}^{\infty} 6(-2)^k$

For #7–11, use one of the formulas $S_n = n\left(\frac{a_1 + a_n}{2}\right)$ or $S_n = \frac{n}{2}[2a_1 + (n-1)d]$. In most cases, the first of these is easier (since the last term a_n is given); note that $n = \frac{a_n - a_1}{d} + 1.$

7. $6 \cdot \left(\frac{-7 + 13}{2}\right) = 6 \cdot 3 = 18$
9. $80 \cdot \left(\frac{1 + 80}{2}\right) = 40 \cdot 81 = 3240$
11. $13 \cdot \left(\frac{117 + 33}{2}\right) = 13 \cdot 75 = 975$

For #13–15, use the formula $S_n = \frac{a_1(1 - r^n)}{1 - r}$. Note that $n = 1 + \log_{|r|}\left|\frac{a_n}{a_1}\right| = 1 + \frac{\ln|a_n/a_1|}{\ln|r|}.$

13. $\frac{3(1 - 2^{13})}{1 - 2} = 24{,}573$
15. $\frac{42[1 - (1/6)^9]}{1 - (1/6)} = 50.4(1 - 6^{-9}) \approx 50.4$

For #17–21, use one of the formulas $S_n = \frac{n}{2}[2a_1 + (n-1)d]$ or $S_n = \frac{a_1(1 - r^n)}{1 - r}$.

17. Arithmetic with $d = 3$: $\frac{10}{2}[2 \cdot 3 + (10 - 1)(3)]$
$= 5 \cdot 31 = 155$
19. Geometric with $r = -\frac{1}{2}$: $\frac{4[1 - (-1/2)^{12}]}{1 - (-1/2)}$
$= \frac{8}{3} \cdot (1 - 2^{-12}) \approx 2.666$
21. Geometric with $r = -11$: $\frac{-1[1 - (-11)^9]}{1 - (-11)}$
$= -\frac{1}{12} \cdot (1 + 11^9) = -196{,}495{,}641$
23. (a) The first six partial sums are {0.3, 0.33, 0.333, 0.3333, 0.33333, 0.333333}. The numbers appear to be approaching a limit of $0.\overline{3} = 1/3$. The series is convergent.
(b) The first six partial sums are {1, −1, 2, −2, 3, −3}. The numbers approach no limit. The series is divergent.
25. $r = \frac{1}{2}$, so it converges to $S = \frac{6}{1 - (1/2)} = 12.$
27. $r = 2$, so it diverges.
29. $r = \frac{1}{4}$, so it converges to $S = \frac{3/4}{1 - (1/4)} = 1.$
31. $7 + \frac{14}{99} = \frac{693}{99} + \frac{14}{99} = \frac{707}{99}$
33. $-17 - \frac{268}{999} = -\frac{17{,}251}{999}$
35. (a) The ratio of any two successive account balances is $r = 1.1$. That is,
$\frac{\$22{,}000}{\$20{,}000} = \frac{\$24{,}200}{\$22{,}000} = \frac{\$26{,}620}{\$24{,}200} = \frac{\$29{,}282}{\$26{,}620} = 1.1.$
(b) Each year, the balance is 1.1 times as large as the year before. So, n years after the balance is \$20,000, it will be \$20,000 $(1.1)^n$.
(c) The sum of the eleven terms of the geometric sequence is $\frac{\$20{,}000(1 - 1.1^{11})}{1 - 1.1} = \$370{,}623.34$
37. (a) The first term, $120(1 + 0.07/12)^0$, simplifies to 120. The common ratio of terms, r, equals $1 + 0.07/12$.
(b) The sum of the 120 terms is
$\frac{120\,[1 - (1 + 0.07/12)^{120}]}{1 - (1 + 0.07/12)} = \$20{,}770.18$
39. $2 + 2 \cdot [2(0.9) + (2(0.9)^2 + 2(0.9)^3 + \ldots + 2(0.9)^9]$
$= 2 + \sum_{k=1}^{9} 4(0.9)^k = 2 + \frac{3.6(1 - 0.9^9)}{1 - 0.9} =$
$38 - 36(0.9^9) \approx 24.05$ m: The ball travels down 2 m. Between the first and second bounces, it travels 2(0.9) m up, then the same distance back down. Between the second and third bounces, it travels $2(0.9)^2$ m up, then the same distance back down, etc.

41. False. The series might diverge. For example, examine the series $1 + 2 + 3 + 4 + 5$ where all of the terms are positive. Consider the limit of the sequence of partial sums. The first five partial sums are $\{1, 3, 6, 10, 15\}$. These numbers increase without bound and do not approach a limit. Therefore, the series diverges and has no sum.

43. $3^{-1} + 3^{-2} + 3^{-3} + \cdots + 3^{-n} + \cdots =$

$\frac{1}{3} + \frac{1}{9} + \frac{1}{27} + \frac{1}{81} + \frac{1}{243} + \cdots + \frac{1}{3^n} + \cdots$

The first five partial sums are $\left\{\frac{1}{3}, \frac{4}{9}, \frac{13}{27}, \frac{40}{81}, \frac{121}{243}\right\}$. These appear to be approaching a limit of $1/2$, which would suggest that the series converges to $1/2$. The answer is A.

45. The common ratio is $0.75/3 = 0.25$, so the sum of the infinite series is $3/(1 - 0.25) = 4$. The answer is C.

47. **(a)** Heartland: 19,237,759 people.

Southeast: 42,614,977 people.

(b) Heartland: 517,825 mi^2.

Southeast: 348,999 mi^2.

(c) Heartland: $\frac{19{,}237{,}759}{517{,}825} \approx 37.15$ people/mi^2.

Southeast: $\frac{42{,}614{,}977}{348{,}999} \approx 122.11$ people/mi^2.

Heartland:		Southeast:	
Iowa	≈ 52.00	Alabama	≈ 86.01
Kansas	≈ 32.68	Arkansas	≈ 50.26
Minnesota	≈ 58.29	Florida	≈ 272.53
Missouri	≈ 80.28	Georgia	≈ 138.97
Nebraska	≈ 22.12	Louisiana	≈ 93.59
N. Dakota	≈ 9.08	Mississippi	≈ 59.65
S. Dakota	≈ 9.79	S.Carolina	≈ 128.95
Total	≈ 264.24	Total	≈ 829.96
Average	≈ 37.75	Average	≈ 118.57

(d) The table is shown on the right; the answer differs because the overall population density $\frac{\sum \text{population}}{\sum \text{area}}$ is generally not the same as the average of the population densities, $\frac{1}{n}\sum\left(\frac{\text{population}}{\text{area}}\right)$. The larger states within each group have a greater effect on the overall mean density. In a similar way, if a student's grades are based on a 100-point test and four 10-point quizzes, her overall average grade depends more on the test grade than on the four quiz grades.

49. The table suggests that $S_n = \sum_{k=1}^{n} F_k = F_{n+2} - 1$.

n	F_n	S_n	$F_{n+2} - 1$
1	1	1	1
2	1	2	2
3	2	4	4
4	3	7	7
5	5	12	12
6	8	20	20
7	13	33	33
8	21	54	54
9	34	88	88

51. Algebraically: $T_{n-1} + T_n = \frac{(n-1)n}{2} + \frac{n(n+1)}{2}$

$= \frac{n^2 - n + n^2 + n}{2} = n^2$.

Geometrically: the array of black dots in the figure represents $T_n = 1 + 2 + 3 + \cdots + n$ (that is, there are T_n dots in the array). The array of gray dots represents $T_{n-1} = 1 + 2 + 3 + \cdots + (n - 1)$. The two triangular arrays fit together to form an $n \times n$ square array, which has n^2 dots.

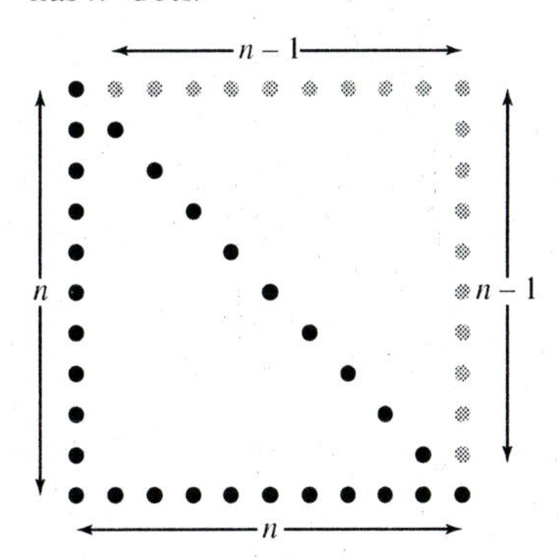

Section 9.6 Mathematical Induction

Exploration 1

1. Start with the rightmost peg if n is odd and the middle peg if n is even. From that point on, the first move for moving any smaller stack to a destination peg should be directly to the destination peg if the smaller stack's size n is odd and to the other available peg if n is even. The fact that the winning strategy follows such predictable rules is what makes it so interesting to students of computer programming.

Exploration 2

1. 43, 47, 53, 61, 71, 83, 97, 113, 131, 151. Yes.
2. 173, 197, 223, 251, 281, 313, 347, 383, 421, 461. Yes.
3. 503, 547, 593, 641, 691, 743, 797, 853, 911, 971. Yes. Inductive thinking might lead to the conjecture that $n^2 + n + 41$ is prime for all n, but we have no proof as yet!
4. The next 9 numbers are all prime, but $40^2 + 40 + 41$ is not. Quite obviously, neither is the number $41^2 + 41 + 41$.

Quick Review 9.6

1. $n^2 + 5n$
3. $k^3 + 3k^2 + 2k$
5. $(k + 1)^3$
7. $f(1) = 1 + 4 = 5, f(t) = t + 4,$ $f(t + 1) = t + 1 + 4 = t + 5$
9. $P(1) = \dfrac{2\cdot 1}{3\cdot 1 + 1} = \dfrac{1}{2},$ $P(k) = \dfrac{2k}{3k + 1} = \dfrac{2(k+1)}{3(k+1)+1} = \dfrac{2k+2}{3k+4}$

Section 9.6 Exercises

1. P_n: $2 + 4 + 6 + \cdots + 2n = n^2 + n.$
P_1 is true: $2(1) = 1^2 + 1.$
Now assume P_k is true: $2 + 4 + 6 + \cdots + 2k = k^2 + k.$ Add $2(k + 1)$ to both sides:
$2 + 4 + 6 + \cdots + 2k + 2(k + 1)$
$= k^2 + k + 2(k + 1) = k^2 + 3k + 2$
$= k^2 + 2k + 1 + k + 1 = (k + 1)^2 + (k + 1)$, so P_{k+1} is true. Therefore, P_n is true for all $n \geq 1$.

3. P_n: $6 + 10 + 14 + \cdots + (4n + 2) = n(2n + 4).$
P_1 is true: $4(1) + 2 = 1(2 + 4).$
Now assume P_k is true:
$6 + 10 + 14 + \cdots + (4k + 2) = k(2k + 4).$
Add $4(k + 1) + 2 = 4k + 6$ to both sides:
$6 + 10 + 14 + \cdots + (4k + 2) + [4(k + 1) + 2]$
$= k(2k + 4) + 4k + 6 = 2k^2 + 8k + 6$
$= (k + 1)(2k + 6) = (k + 1)[2(k + 1) + 4]$, so P_{k+1} is true. Therefore, P_n is true for all $n \geq 1$.

5. P_n: $5n - 2$. P_1 is true: $a_1 = 5\cdot 1 - 2 = 3.$
Now assume P_k is true: $a_k = 5k - 2.$
To get a_{k+1}, add 5 to a_k; that is,
$a_{k+1} = (5k - 2) + 5 = 5(k + 1) - 2$ This shows that P_{k+1} is true. Therefore, P_n is true for all $n \geq 1$.

7. P_n: $a_n = 2\cdot 3^{n-1}.$
P_1 is true: $a_1 = 2\cdot 3^{1-1} = 2\cdot 3^0 = 2.$
Now assume P_k is true: $a_k = 2\cdot 3^{k-1}.$
To get a_{k+1}, multiply a_k by 3; that is,
$a_{k+1} = 3\cdot 2\cdot 3^{k-1} = 2\cdot 3k = 2\cdot 3^{(k+1)-1}$. This shows that P_{k+1} is true. Therefore, P_n is true for all $n \geq 1$.

9. P_1: $1 = \dfrac{1(1 + 1)}{2}.$
P_k: $1 + 2 + \cdots + k = \dfrac{k(k + 1)}{2}$
P_{k+1}: $1 + 2 + \cdots + k + (k + 1) = \dfrac{(k + 1)(k + 2)}{2}.$

11. P_1: $\dfrac{1}{1\cdot 2} = \dfrac{1}{1 + 1}.$
P_k: $\dfrac{1}{1\cdot 2} + \dfrac{1}{2\cdot 3} + \cdots + \dfrac{1}{k(k + 1)} = \dfrac{k}{k + 1}.$
P_{k+1}: $\dfrac{1}{1\cdot 2} + \dfrac{1}{2\cdot 3} + \cdots + \dfrac{1}{k(k + 1)} + \dfrac{1}{(k + 1)(k + 2)} = \dfrac{k + 1}{k + 2}.$

13. P_n: $1 + 5 + 9 + \cdots + (4n - 3) = n(2n - 1).$
P_1 is true: $4(1) - 3 = 1\cdot(2\cdot 1 - 1).$
Now assume P_k is true:
$1 + 5 + 9 + \cdots + (4k - 3) = k(2k - 1).$
Add $4(k + 1) - 3 = 4k + 1$ to both sides:
$1 + 5 + 9 + \cdots + (4k - 3) + [4(k + 1) - 3]$
$= k(2k - 1) = 4k + 1 = 2k^2 + 3k + 1$
$= (k + 1)(2k + 1) = (k + 1)[2(k + 1) - 1]$,
so P_{k+1} is true.
Therefore, P_n is true for all $n \geq 1$.

15. P_n: $\dfrac{1}{1\cdot 2} + \dfrac{1}{2\cdot 3} + \cdots + \dfrac{1}{n(n + 1)} = \dfrac{n}{n + 1}.$
P_1 is true: $\dfrac{1}{1\cdot 2} = \dfrac{1}{1 + 1}.$
Now assume P_k is true:
$\dfrac{1}{1\cdot 2} + \dfrac{1}{2\cdot 3} + \cdots + \dfrac{1}{k(k + 1)}.$
Add $\dfrac{1}{(k + 1)(k + 2)}$ to both sides:
$\dfrac{1}{1\cdot 2} + \dfrac{1}{2\cdot 3} + \cdots + \dfrac{1}{k(k + 1)} + \dfrac{1}{(k + 1)(k + 2)}$
$= \dfrac{k}{k + 1} + \dfrac{1}{(k + 1)(k + 2)} = \dfrac{k(k + 2) + 1}{(k + 1)(k + 2)}$
$= \dfrac{(k + 1)(k + 1)}{(k + 1)(k + 2)} = \dfrac{k + 1}{k + 2} = \dfrac{k + 1}{(k + 1) + 1},$
so P_{k+1} is true.
Therefore, P_n is true for all $n \geq 1$.

17. P_n: $2^n \geq 2n.$
P_1 is true: $2^1 \geq 2\cdot 1$ (in fact, they are equal). Now assume P_k is true: $2^k \geq 2k.$
Then $2^{k+1} = 2\cdot 2^k \geq 2\cdot 2k$
$= 2\cdot(k + k) \geq 2(k + 1)$, so P_{k+1} is true. Therefore, P_n is true for all $n \geq 1$.

19. P_n: 3 is a factor of $n^3 + 2n$.
P_1 is true: 3 is a factor of $1^3 + 2 \cdot 1 = 3$.
Now assume P_k is true: 3 is a factor of $k^3 + 2k$.
Then $(k + 1)^3 + 2(k + 1)$
$= (k^3 + 3k^2 + 3k + 1) + (2k + 2)$
$= (k^3 + 2k) + 3(k^2 + k + 1)$.
Since 3 is a factor of both terms, it is a factor of the sum, so P_{k+1} is true. Therefore, P_n is true for all $n \geq 1$.

21. P_n: The sum of the first n terms of a geometric sequence with first term a_1 and common ratio $r \neq 1$ is $\dfrac{a_1(1 - r^n)}{1 - r}$.

P_1 is true: $a_1 = \dfrac{a_1(1 - r^1)}{1 - r}$.
Now assume P_k is true so that
$a_1 + a_1 r + \cdots = a_1 r^{k-1} = \dfrac{a_1(1 - r^k)}{(1 - r)}$.
Add $a_1 r^k$ to both sides: $a_1 + a_1 r + \cdots = a_1 r^{k-1}$
$+ a_1 r^k = \dfrac{a_1(1 - r^k)}{(1 - r)} + a_1 r^k$
$= \dfrac{a_1(1 - r^k) + ar^k(1 - r)}{1 - r}$
$= \dfrac{a_1 - a_1 r^k - a_1 r^{k+1}}{1 - r} = \dfrac{a_1 - a_1 r^{k+1}}{1 - r}$,
so P_{k+1} is true. Therefore, P_n is true for all positive integers n.

23. P_n: $\sum_{k=1}^{n} k = \dfrac{n(n + 1)}{2}$.

P_1 is true: $\sum_{k=1}^{1} k = 1 = \dfrac{1 \cdot 2}{2}$.

Now assume P_k is true: $\sum_{i=1}^{k} i = \dfrac{k(k + 1)}{2}$.
Add $(k + 1)$ to both sides, and we have
$\sum_{i=1}^{k+1} i = \dfrac{k(k + 1)}{2} + (k + 1)$
$= \dfrac{k(k + 1)}{2} + \dfrac{2(k + 1)}{2} = \dfrac{(k + 1)(k + 2)}{2}$
$= \dfrac{(k + 1)(k + 1 + 1)}{2}$, so P_{k+1} is true.
Therefore, P_n is true for all $n \geq 1$.

25. Use the formula in 23: $\sum_{k=1}^{500} k = \dfrac{(500)(501)}{2} = 125{,}250$

27. Use the formula in 23: $\sum_{k=4}^{n} k = \sum_{k=1}^{n} k - \sum_{k=1}^{3} k$
$= \dfrac{n(n + 1)}{2} - \dfrac{3 \cdot 4}{2} = \dfrac{n^2 + n - 12}{2} = \dfrac{(n - 3)(n + 4)}{2}$

29. Use the formula in 14: $\sum_{k=1}^{35} 2^{k-1}$
$= 2^{35} - 1 \approx 3.44 \times 10^{10}$

31. $\sum_{k=1}^{n} (k^2 - 3k + 4) = \sum_{k=1}^{n} k^2 - \sum_{k=1}^{n} 3k + \sum_{k=1}^{n} 4$
$= \dfrac{n(n + 1)(2n + 1)}{6} - 3\left[\dfrac{n(n + 1)}{2}\right] + 4n$
$= \dfrac{n(n^2 - 3n + 8)}{3}$

33. $\sum_{k=1}^{n} (k^3 - 1) = \sum_{k=1}^{n} k^3 - \sum_{k=1}^{n} 1 = \dfrac{n^2(n + 1)^2}{4} - n$
$= \dfrac{n(n^3 + 2n^2 + n - 4)}{4} = \dfrac{n(n - 1)(n^2 + 3n + 4)}{4}$

35. The inductive step does not work for 2 people. Sending them alternately out of the room leaves 1 person (and one blood type) each time, but we cannot conclude that their blood types will match *each other.*

37. False. Mathematical induction is used to show that a statement P_n is true for all positive integers.

39. The inductive step assumes that the statement is true for some positive integer k. The answer is E.

41. Mathematical induction could be used, but the formula for a finite arithmetic sequence with $a_1 = 1, d = 2$ would also work. The answer is B.

43. P_n: 2 is a factor of $(n + 1)(n + 2)$. P_1 is true because 2 is a factor of $(2)(3)$. Now assume P_k is true so that 2 is a factor of $(k + 1)(k + 2)$. Then
$[(k + 1) + 1][(k + 2) + 2]$
$= (k + 2)(k + 3) = k^2 + 5k + 6$
$= k^2 + 3k + 2 + 2k + 4$
$= (k + 1)(k + 2) + 2(x + 2)$. Since 2 is a factor of both terms of this sum, it is a factor of the sum, and so P_{k+1} is true. Therefore, P_n is true for all positive integers n.

45. Given any two consecutive integers, one of them must be even. Therefore, their product is even. Since $n + 1$ and $n + 2$ are consecutive integers, their product is even. Therefore, 2 is a factor of $(n + 1)(n + 2)$.

47. P_n: $F_{n+2} - 1 = \sum_{k=1}^{n} F_k$. P_1 is true since
$F_{1+2} = 1 = F_3 - 1 = 2 - 1 = 1$, which equals
$\sum_{k=1}^{1} F_k = 1$. Now assume that P_k is true:

$F_{k+2} - 1 = \sum_{i=1}^{k} F_i$. Then $F_{(k+1)+2} - 1$
$= F_{k+3} - 1 = F_{k+1} + F_{k+2} - 1$
$= (F_{k+2} - 1) + F_{k+1} = \left(\sum_{i=1}^{k} F_i\right) + F_{k+1}$
$= \sum_{i=1}^{k+1} F_i$, so P_{k+1} is true. Therefore, P_n is true for all $n \geq 1$.

49. P_n: $a - 1$ is a factor of $a^n - 1$. P_1 is true because $a - 1$ is a factor of $a - 1$. Now assume P_k is true so that $a - 1$ is a factor of $a^k - 1$. Then $a^{k+1} - 1 = a \cdot ak - 1$
$= a(a^k - 1) + (a - 1)$. Since $a - 1$ is a factor of both terms in the sum, it is a factor of the sum, and so P_{k+1} is true. Therefore, P_n is true for all positive integers n.

51. P_n: $3n - 4 \geq n$ for $n \geq 2$. P_2 is true since $3 \cdot 2 - 4 \leq 2$. Now assume that P_k is true: $3k - 4 \geq 2$. Then $3(k + 1) - 4 = 3k + 3 - 4$
$= (3k - 4) + 3k \geq k + 3 \geq k + 1$, so P_{k+1} is true. Therefore, P_n is true for all $n \geq 2$.

53. Use P_3 as the anchor and obtain the inductive step by representing any n-gon as the union of a triangle and an $(n - 1)$-gon.

■ Section 9.7 Statistics and Data (Graphical)

Exploration 1

1. We observe that the numbers seem to be centered a bit below 13. We would need to take into account the different state populations (not given in the table) in order to compute the national average exactly; but, just for the record, it was about 12.5 percent.
2. We observe in the stemplot that 3 states have percentages above 15.
3. We observe in the stemplot that the bottom five states are all below 10%. Returning to the table, we pick these out as Alaska, Colorado, Georgia, Texas, and Utah.
4. The low outlier is Alaska, where older people would be less willing or able to cope with the harsh winter conditions. The high outlier is Florida, where the mild weather and abundant retirement communities attract older residents.

Quick Review 9.7

1. $\approx 15.48\%$

3. $\approx 14.44\%$

5. ≈ 1723

7. 235 thousand

9. 1 million

Section 9.7 Exercises

1.

Stem	Leaf
0	5 8 9
1	3 4 6
2	3 6 8
3	3 9
4	
5	
6	1

61 is an outlier.

3. Males

Stem	Leaf
6	0 3
6	7 8 8 8
7	1 2 2 3 3 3
7	

This stemplot shows the life expectancies of males in the nations of South America are clustered near 70, with two lower values clustered near 60.

5.

Males	Stem	Females
3 0	6	
8 8 8 7	6	5 8
3 3 3 2 2 1	7	1 2
	7	5 6 7 7 9 9
	8	0 0

This stemplot shows that the life expectancies of the women in the nations of South America are about 5–6 years higher than that of the men in the nations of South America.

7.

Life expectancy (years)	Frequency
60.0–64.9	2
65.0–69.9	4
70.0–74.9	6

9.

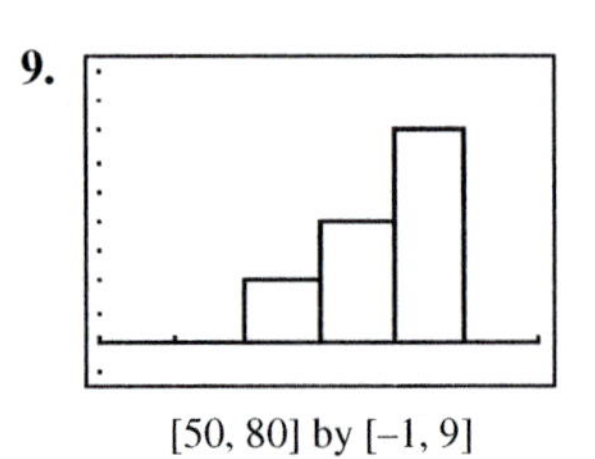

[50, 80] by [−1, 9]

11.

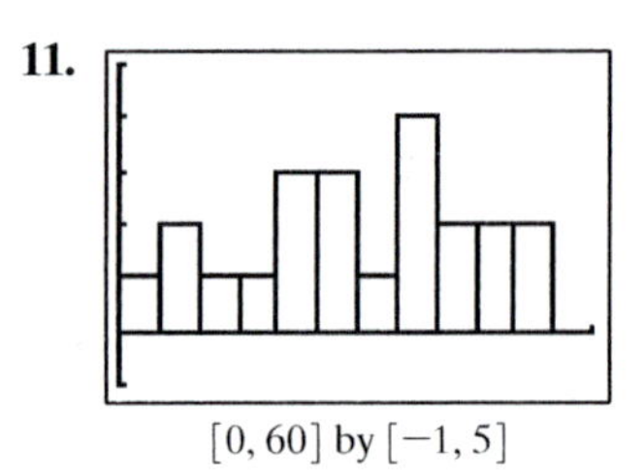

[0, 60] by [−1, 5]

13.

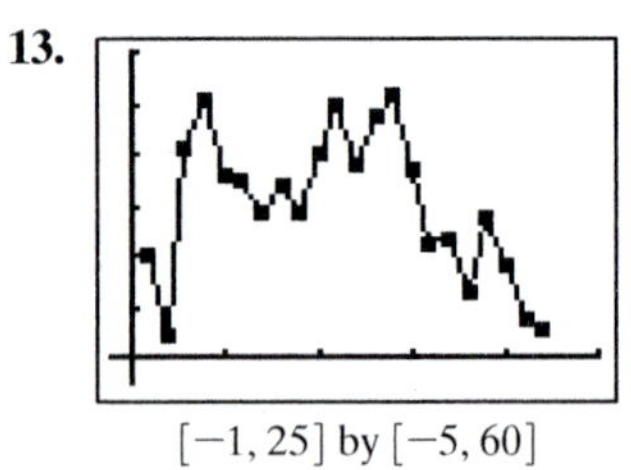

[−1, 25] by [−5, 60]

15.

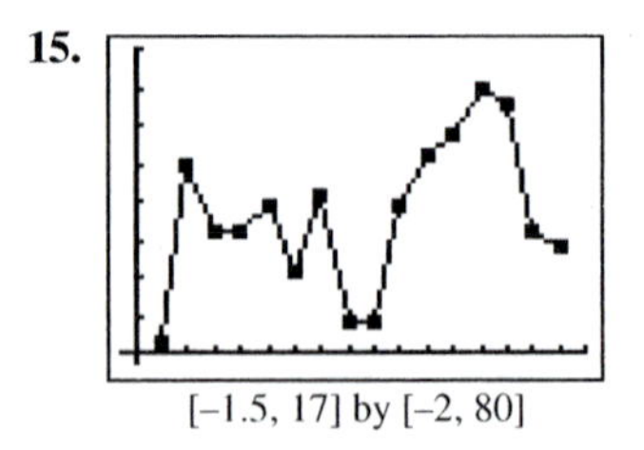

[−1.5, 17] by [−2, 80]

17.

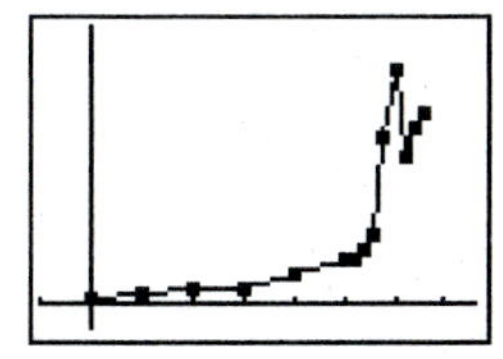

[1965, 2008] by [−1000, 11000]

The top male's earnings appear to be growing exponentially, with unusually high earnings in 1999 and 2000. Since the graph shows the earnings of only the top player (as opposed to a mean or median for all players), it can behave strangely if the top player has a very good year — as Tiger Woods did in 1999 and 2000.

19.

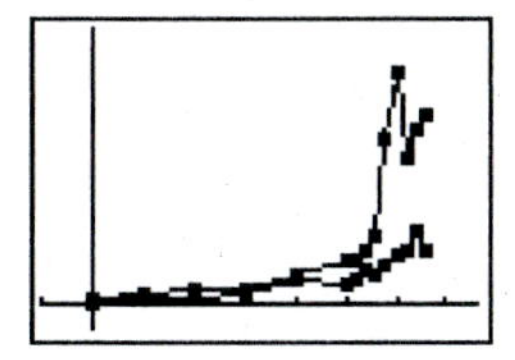

[1965, 2008] by [−1000, 11000]

After approaching parity in 1985, the top PGA player's earnings have grown much faster than the top LPGA player's earnings, even if the unusually good years for Tiger Woods (1999 and 2000) are not considered part of the trend.

21.

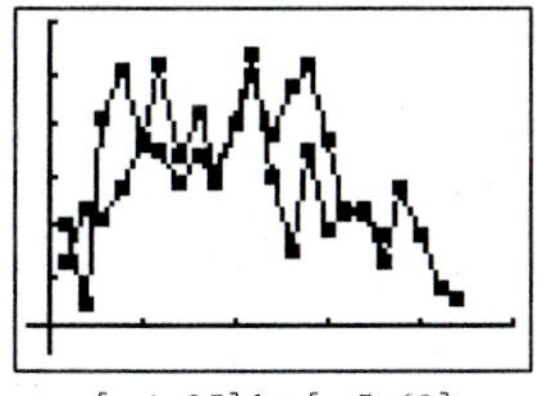

[−1, 25] by [−5, 60]

The two home run hitters enjoyed similar success, with Mays enjoying a bit of an edge in the earlier and later years of his career, and Mantle enjoying an edge in the middle years.

23. (a)

Stem	Leaf
28	2
29	3 7
30	
31	6 7
32	7 8
33	5 5 5 8
34	2 8 8
35	3 3 4
36	3 7
37	
38	5

(b)

Interval	Frequency
25.0–29.9	3
30.0–34.9	11
35.0–39.9	6

(c)

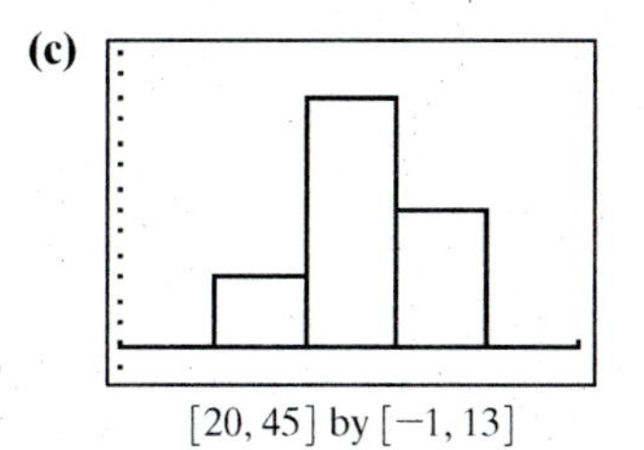

[20, 45] by [−1, 13]

(d) Time is not a variable in the data.

25.

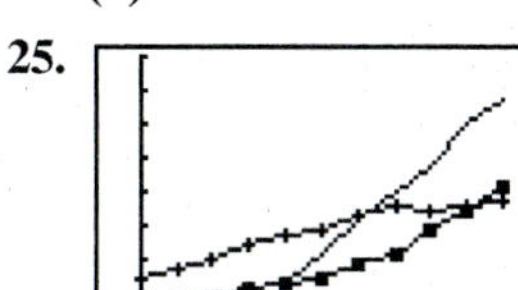

[1890, 2010] by [−4, 40]

— = CA   + = NY   ■ = TX

27. False. The empty branches are important for visualizing the true distribution of the data values.

29. A time plot uses a continuous line. The answer is C.

31. The histogram suggests data values clustered near an upper limit — such as the maximum possible score on an easy test. The answer is A.

33. Answers will vary. Possible outliers could be the pulse rates of long-distance runners and swimmers, which are often unusually low. Students who have had to run to class from across campus might have pulse rates that are unusually high.

35.

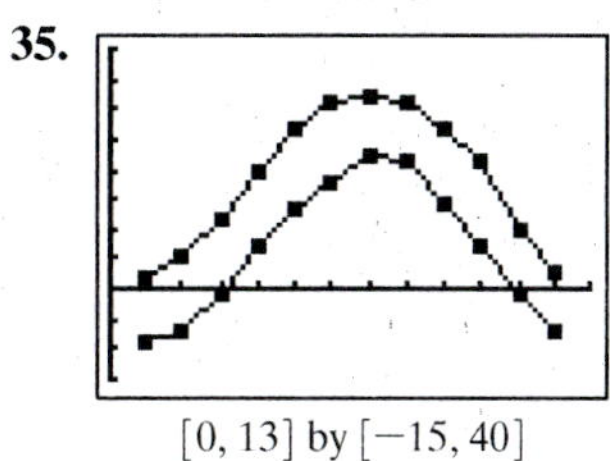

[0, 13] by [−15, 40]

■ Section 9.8 Statistics and Data (Algebraic)

Exploration 1

1. In Figure 9.19 (a) the extreme values will cause the range to be big, but the compact distribution of the rest of the data indicate a small interquartile range. The data in Figure 9.19 (b) exhibit high variability.
2. Figure 9.20 (b) has a longer "tail" to the right (skewed right), so the values in the tail pull the mean to the right of the (resistant) median. Figure 9.20 (c) is skewed left, so the values in the tail pull the mean to the left of the median. Figure 9.20 (a) is symmetric about a vertical line, so the median and the mean are close together.

Quick Review 9.8

1. $\sum_{i=1}^{7} x_i = x_1 + x_2 + x_3 + x_4 + x_5 + x_6 + x_7$

3. $\frac{1}{7}\sum_{i=1}^{7} x_i = \frac{1}{7}(x_1 + x_2 + x_3 + x_4 + x_5 + x_6 + x_7)$

5. The expression at the end of the first line is a simple expansion of the sum (and is a reasonable answer to the given question). By expanding further, we can also arrive at the final expression below, which is somewhat simpler.

$$\frac{1}{5}\sum_{i=1}^{5}(x_i - \bar{x})^2 = \frac{1}{5}[(x_1 - \bar{x})^2 + (x_2 - \bar{x})^2 + \cdots + (x_5 - \bar{x})^2]$$

$$= \frac{1}{5}[(x_1^2 - 2x_1\bar{x} + \bar{x}^2) + (x_2^2 - 2x_2\bar{x} + \bar{x}^2) + \cdots + (x_5^2 - 2x_5\bar{x} + \bar{x}^2)]$$

$$= \frac{1}{5}[x_1^2 + x_2^2 + \cdots + x_5^2 - 2\bar{x}(x_1 + x_2 + \cdots + x_5)] + \bar{x}^2$$

$$= \frac{1}{5}(x_1^2 + x_2^2 + \cdots + x_5^2) - 2\bar{x}^2 + \bar{x}^2 = \frac{1}{5}(x_1^2 + x_2^2 + \cdots + x_5^2) - \bar{x}^2$$

7. $\sum_{i=1}^{8} x_i f_i$

9. $\frac{1}{50}\sum_{i=1}^{50}(x_i - \bar{x})^2$

Section 9.8 Exercises

1. **(a)** Statistic. The number characterizes a set of known data values.

(b) Parameter. The number describes an entire population, on the basis of some statistical inference.

(c) Statistic. The number is calculated from information about all rats in a small, experimental population.

3. $\bar{x} = \frac{1}{5}(12 + 23 + 15 + 48 + 36) = \frac{134}{5} = 26.8$

5. $\bar{x} = \frac{1}{6}(32.4 + 48.1 + 85.3 + 67.2 + 72.4 + 55.3) = \frac{360.7}{6} = \frac{3607}{60} \approx 60.12$

7. $\bar{x} = \frac{1}{6}(1.5 + 0.5 + 4.8 + 7.3 + 6.3 + 3.0) = 3.9$ million, or 3,900,000

9. $\bar{x} = \frac{1}{9}(0 + 0 + 1 + 2 + 61 + 33 + 26 + 13 + 1) = \frac{137}{9} \approx 15.2$ satellites

11. There are 9 data values, which is an odd number, so the median is the middle data value when they are arranged in order. In order, the data values are {0, 0, 1, 1, 2, 8, 21, 28, 30}. The median is 2.

13. Mays: $\frac{660}{22} = 30$ home runs/year. Mantle: $\frac{536}{18} \approx 29.8$ home runs/year. Mays had the greater production rate.

15. No computations needed — Hip-Hop House: 1147 skirts in 4 weeks. What-Next Fashions: 1516 skirts in 4 weeks. What-Next Fashions had the greater production rate.

17. For {79.5, 82.1, 82.3, 82.9, 84.0, 84.5, 84.8, 85.4, 85.7, 85.7, 86.0, 86.4, 87.0, 87.5, 87.7, 88.0, 88.0, 88.2, 88.2, 88.5, 89.2, 89.2, 89.8, 89.8, 91.1, 91.3, 91.6, 91.8, 94.1, 94.4}, the median is $\frac{87.7 + 88.0}{2} = 87.85$, and there is no mode.

19. $\bar{x} = \frac{5(9879) + 4(5119) + 3(6143) + 2(2616) + 1(3027)}{9879 + 5119 + 6143 + 2616 + 3027}$

≈ 3.61

21. **(a)** Non-weighted: $\bar{x} = \frac{1}{12}(-9 - 7 - 1 + 7 + \cdots - 1 - 7) = \frac{77}{12} \approx 6.42°\text{C}$

(b) Weighted: $\bar{x} = \frac{(-9)(31) + (-7)(28) + (-1)(31) + (7)(30) + \cdots + (-1)(30) + (-7)(31)}{31 + 28 + 31 + 30 + \cdots + 30 + 31} = \frac{2370}{365} \approx 6.49°\text{C}$

(c) The weighted average is the better indicator.

23. The ordered data for Willie Mays is: {4, 6, 8, 13, 18, 20, 22, 23, 28, 29, 29, 34, 35, 36, 37, 38, 40, 41, 47, 49, 51, 52} so the median is

$\frac{29 + 34}{2} = 31.5, Q_3 = 40$, and $Q_1 = 20$

Five-number summary: {4, 20, 31.5, 40, 52}
Range: $52 - 4 = 48$
IQR: $40 - 20 = 20$
No outliers
The ordered data for Mickey Mantle is:
{13, 15, 18, 19, 21, 22, 23, 23, 27, 30, 31, 34, 35, 37, 40, 42, 52, 54} so the median is

$\frac{27 + 30}{2} = 28.5, Q_3 = 37$, and $Q_1 = 21$

Five-number summary: {13, 21, 28.5, 37, 54}
Range: $54 - 13 = 41$
IQR: $37 - 21 = 16$
No outliers

25. The sorted list is {28.2, 29.3, 29.7, 31.6, 31.7, 32.7, 32.8, 33.5, 33.5, 33.5, 33.8, 34.2, 34.8, 34.8, 35.3, 35.4, 36.7, 37.3, 38.5}. Since there are 19 numbers, the median is the 10th, Q_1 is the 5th, and Q_3 is the 15th; no additional computations are needed. All these numbers are underlined above, and are summarized below:

Min	Q_1	Median	Q_3	Max
28.2	31.7	33.5	35.3	38.5

The range is $38.5 - 28.2 = 10.3$, and
IQR $= 35.3 - 31.7 = 3.6$. There are no outliers, since none of the numbers fall below $Q_1 - 1.5 \times \text{IQR} = 26.3$ or above $Q_3 + 1.5 \times \text{IQR} = 40.7$.
Note that some computer software may return

$Q_1 = 32.2 = \frac{31.7 + 32.7}{2}$ and $Q_3 = 35.05 = \frac{34.8 + 35.3}{2}$

(The TI calculator which computes the five-number summary produces the results shown in the table).

27. (a)

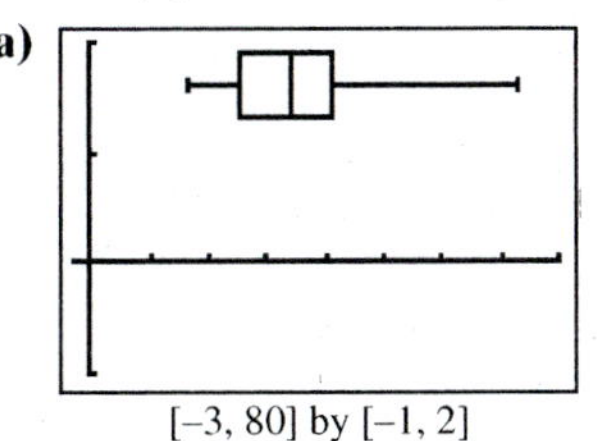

[–3, 80] by [–1, 2]

(b)

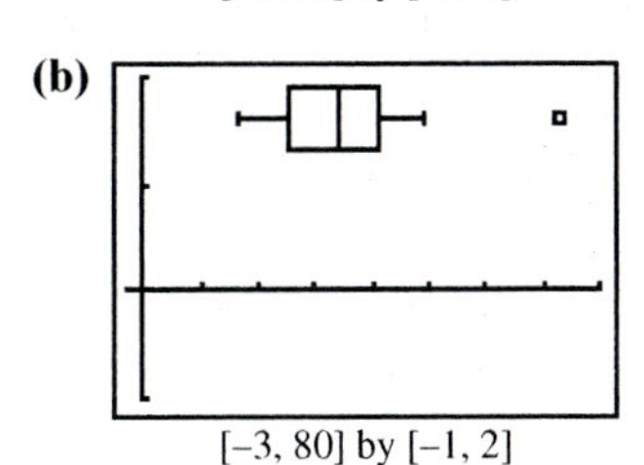

[–3, 80] by [–1, 2]

29. 12 of the 44 numbers are more than 10.5: $\frac{12}{44} = \frac{3}{11}$

31. The five-number summaries are:
Mays: 4, 20, 31.5, 40, 52 (top of graph)
Mantle: 13, 21, 28.5, 37, 54
(bottom of graph)

(a) Mays's data set has the greater range.

(b) Mays's data set also has the greater *IQR*.

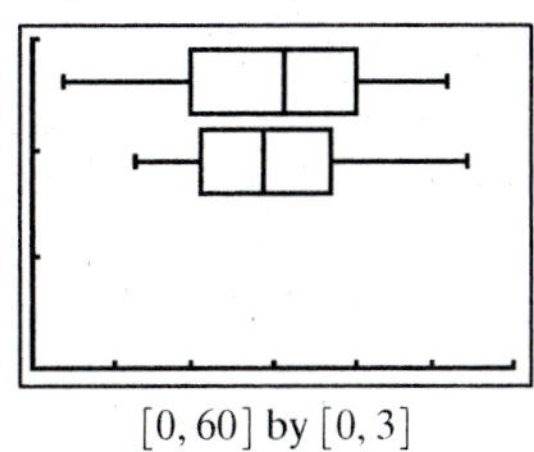

[0, 60] by [0, 3]

For #33–37, the best way to do the computation is with the statistics features of a calcuator.

33. $\sigma \approx 9.08, \sigma^2 = 82.5$

35. $\sigma \approx 186.36; \sigma^2 \approx 34{,}828.12$

37. $\sigma \approx 1.53, \sigma^2 \approx 2.34$

39. No. An outlier would need to be less than $25.5 - 1.5(25) = -12$ or greater than $50.5 + 1.5(25) = 88$.

41. (a) 68%

(b) 2.5%

(c) A parameter, since it applies to the entire population.

43. False. The median is a resistant measure. The *mean* is strongly affected by outliers.

45. The plot of an ideal normal distribution is a symmetric "bell curve." The answer is A.

47. The total number of points from all 30 exams combined is $30 \times 81.3 = 2439$. Adding 9 more points and recalculating produces a new mean of $(2439 + 9)/30 = 81.6$. The median will be unaffected by an adjustment in the top score. The answer is B.

49. There are many possible answers; examples are given.

(a) {2, 2, 2, 3, 6, 8, 20} — mode = 2, median = 3, and $\bar{x} = \frac{43}{7} \approx 6.14$.

(b) {1, 2, 3, 4, 6, 48, 48} — median = 4, $\bar{x} = 16$, and mode = 48

(c) {−20, 1, 1, 1, 2, 3, 4, 5, 6} — $\bar{x} = \frac{1}{3}$, mode = 1, and median = 2.

51. No: $(x_i - \bar{x}) \le (\max - \min)^2 = (\text{range})^2$, so that $\sigma^2 = \frac{1}{n}\sum_{i=1}^{n}(x_i - \bar{x})^2 \le \frac{1}{n}\sum_{i=1}^{n}(\text{range})^2 = (\text{range})^2$. Then $\sigma = \sqrt{\sigma^2} \le \sqrt{(\text{range})^2} = \text{range}$.

53. There are many possible answers; example data sets are given.

(a) $\{1, 1, 2, 6, 7\}$ — median $= 2$ and $\bar{x} = 3.4$.

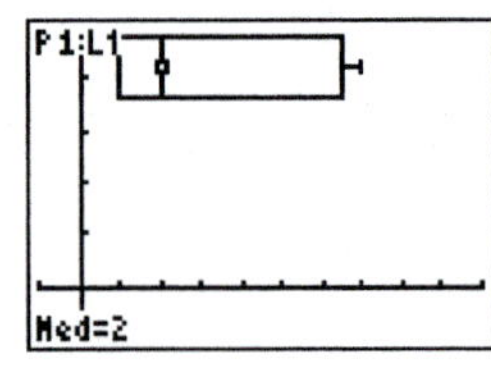

$[-1, 10]$ by $[-1, 5]$

(b) $\{1, 6, 6, 6, 6, 10\}$ — $2 \times \text{IQR} = 2(6 - 6) = 0$ and range $= 10 - 1 = 9$.

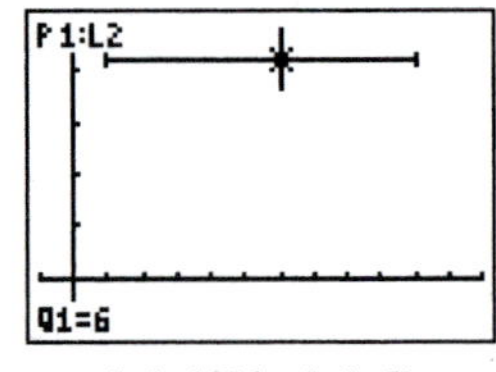

$[-1, 12]$ by $[-1, 5]$

(c) $\{1, 1, 2, 6, 7\}$ — range $= 7 - 1 = 6$ and $2 \times \text{IQR} = 2(6 - 1) = 10$.

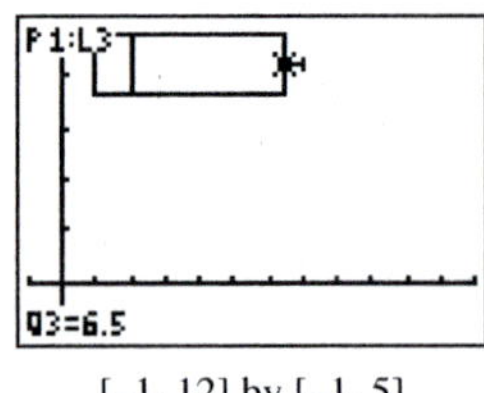

$[-1, 12]$ by $[-1, 5]$

55. For women living in South American nations, the mean life expectancy is

$$\bar{x} = \frac{(79.7)(39.1) + (67.9)(8.7) + \cdots + (79.2)(3.4) + (77.3)(25.0)}{39.1 + 8.7 + \cdots + 3.4 + 25.0} = \frac{27{,}825.56}{366.4} \approx 75.9 \text{ years.}$$

57. Since $\sigma = 0.05$ mm, we have $2\sigma = 0.1$ mm, so 95% of the ball bearings will be acceptable. Therefore, 5% will be rejected.

■ Chapter 9 Review

1. $\binom{12}{5} = \frac{12!}{5!(12-5)!} = \frac{12!}{5!7!} = 792$

3. ${}_{18}C_{12} = \frac{18!}{12!(18-12)!} = \frac{18!}{12!6!} = 18{,}564$

5. ${}_{12}P_7 = \frac{12!}{(12-7)!} = \frac{12!}{5!} = 3{,}991{,}680$

7. $26 \cdot 36^4 = 43{,}670{,}016$ code words

9. ${}_{26}P_2 \cdot {}_{10}P_4 + {}_{10}P_3 \cdot {}_{26}P_3 = 14{,}508{,}000$ license plates

11. Choose 10 more cards from the other 49: ${}_3C_3 \cdot {}_{49}C_{10} = {}_{49}C_{10} = 8{,}217{,}822{,}536$ hands

13. ${}_5C_2 + {}_5C_3 + {}_5C_4 + {}_5C_5 = 2^5 - {}_5C_0 - {}_5C_1 = 26$ outcomes

15. ${}_5P_1 + {}_5P_2 + {}_5P_3 + {}_5P_4 + {}_5P_5 = 325$

17. (a) There are 7 letters, all different. The number of distinguishable permutations is $7! = 5040$. (GERMANY can be rearranged to spell MEG RYAN.)

(b) There are 13 letters, where E, R, and S each appear twice. The number of distinguishable permutations is $\frac{13!}{2!2!2!} = 778{,}377{,}600$ (PRESBYTERIANS can be rearranged to spell BRITNEY SPEARS.)

19.
$$\begin{aligned}(2x + y)^5 &= (2x)^5 + 5(2x)^4y + 10(2x)^3y^2 \\ &\quad + 10(2x)^2y^3 + 5(2x)y^4 + y^5 \\ &= 32x^5 + 80x^4y + 80x^3y^2 + 40x^2y^3 \\ &\quad + 10xy^4 + y^5\end{aligned}$$

21.
$$\begin{aligned}(3x^2 + y^3)^5 &= (3x^2)^5 + 5(3x^2)^4(y^3) \\ &\quad + 10(3x^2)^3(y^3)^2 + 10(3x^2)^2(y^3)^3 \\ &\quad + 5(3x^2)(y^3)^4 + (y^3)^5 \\ &= 243x^{10} + 405x^8y^3 + 270x^6y^6 \\ &\quad + 90x^4y^9 + 15x^2y^{12} + y^{15}\end{aligned}$$

23.
$$\begin{aligned}(2a^3 - b^2)^9 &= (2a^3)^9 + 9(2a^3)^8(-b^2) + 36(2a^3)^7(-b^2)^2 \\ &\quad + 84(2a^3)^6(-b^2)^3 + 126(2a^3)^5(-b^2)^4 \\ &\quad + 126(2a^3)^4(-b^2)^5 + 84(2a^3)^3(-b^2)^6 \\ &\quad + 36(2a^3)^2(-b^2)^7 + 9(2a^3)(-b^2)^8 \\ &\quad + (-b^2)^9 \\ &= 512a^{27} - 2304a^{24}b^2 + 4608a^{21}b^4 \\ &\quad - 5376a^{18}b^6 + 4032a^{15}b^8 \\ &\quad - 2016a^{12}b^{10} + 672a^9b^{12} \\ &\quad - 144a^6b^{14} + 18a^3b^{16} - b^{18}\end{aligned}$$

25. $\binom{11}{8}(1)^8(-2)^3 = -\frac{11!8}{8!3!} = -\frac{11 \cdot 10 \cdot 9 \cdot 8}{3 \cdot 2 \cdot 1} = -1320$

27. $\{1, 2, 3, 4, 5, 6\}$

29. $\{13, 16, 31, 36, 61, 63\}$

31. {HHH, HHT, HTH, HTT, THH, THT, TTH, TTT}

33. {HHH, TTT}

35. $P(\text{HHTHTT}) = \left(\frac{1}{2}\right)^6 = \frac{1}{2^6} = \frac{1}{64}$

37. $P(1 \text{ H and } 3 \text{ T}) = {}_4C_1 \cdot \left(\frac{1}{2}\right)^4 = \frac{4}{2^4} = \frac{1}{4}$

39. $P(3 \text{ successes and } 1 \text{ failure}) = {}_4C_1 \cdot \left(\frac{1}{2}\right)^4 = \frac{4}{2^4} = \frac{1}{4} = 0.25$

41. $P(\text{SF}) = (0.4)(0.6) = 0.24$

43. $P(\text{at least 1 success}) = 1 - P(\text{no successes}) = 1 - (0.6)^2 = 0.64$

45. (a) $P(\text{brand } A) = 0.5$

(b) $P(\text{cashews from brand } A) = (0.5)(0.3) = 0.15$

(c) $P(\text{cashew}) = (0.5)(0.3) + (0.5)(0.4) = 0.35$

(d) $P(\text{brand } A/\text{cashew}) = \frac{0.15}{0.35} \approx 0.43$

For #47, substitute $n = 1, n = 2, \ldots, n = 6$, and $n = 40$.

47. 0, 1, 2, 3, 4, 5; 39

For #49–53, use previously computed values of the sequence to find the next term in the sequence.

49. $-1, 2, 5, 8, 11, 14; 32$

51. $-5, -3.5, -2, -0.5, 1, 2.5; 11.5$

53. $-3, 1, -2, -1, -3, -4; -76$

For #55–61, check for common difference or ratios between successive terms.

55. Arithmetic with $d = -2.5$;
$a_n = 12 + (-2.5)(n-1) = 14.5 - 2.5n$

57. Geometric with $r = 1.2$;
$a_n = 10 \cdot (1.2)^{n-1}$

59. Arithmetic with $d = 4.5$;
$a_n = -11 + 4.5(n-1) = 4.5n - 15.5$

61. $a_n = a_1 r^{n-1}$, so $-192 = a_1 r^3$ and $196{,}608 = a_1 r^8$. Then $r^5 = -1024$, so $r = -4$, and $a_1 = \dfrac{-192}{(-4)^3} = 3$;
$a_n = 3(-4)^{n-1}$

For #63–65, use one of the formulas $S_n = n\left(\dfrac{a_1 + a_n}{2}\right)$ or $S_n = \dfrac{n}{2}[2a_1 + (n-1)d]$. In most cases, the first of these is easier (since the last term a_n is given); note that $n = \dfrac{a_n - a_1}{d} + 1$.

63. $8 \cdot \left(\dfrac{-11 + 10}{2}\right) = 4 \cdot (-1) = -4$

65. $27 \cdot \left(\dfrac{2.5 - 75.5}{2}\right) = \dfrac{1}{2} \cdot 27 \cdot (-73) = -985.5$

For #67–69, use the formula $S_n = \dfrac{a_1(1 - r^n)}{1 - r}$. Note that $n = 1 + \log_{|r|}\left|\dfrac{a_n}{a_1}\right| = 1 + \dfrac{\ln |a_n/a_1|}{\ln |r|}$.

67. $\dfrac{4(1 - (-1/2)^6)}{1 - (-1/2)} = \dfrac{21}{8}$

69. $\dfrac{2(1 - 3^{10})}{1 - 3} = 59{,}048$

71. Geometric with $r = \dfrac{1}{3}$:
$$S_{10} = \frac{2187(1 - (1/3)^{10})}{1 - (1/3)} = \frac{29{,}524}{9} = 3280.\overline{4}$$

73.

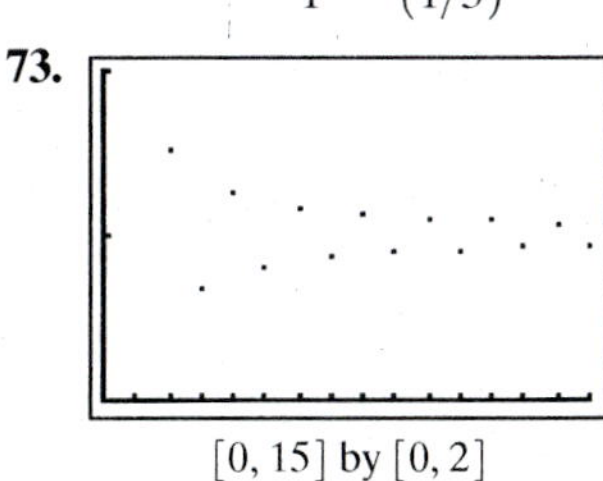

$[0, 15]$ by $[0, 2]$

75. With $a_1 = \$150$, $r = 1 + 0.08/12$, and $n = 120$, the sum becomes
$$\frac{\$150\,[1 - (1 + 0.08/12)^{120}]}{1 - (1 + 0.08/12)} = \$27{,}441.91$$

77. Converges: geometric with $a_1 = \dfrac{3}{2}$ and $r = \dfrac{3}{4}$, so
$$S = \frac{3/2}{1 - (3/4)} = \frac{3/2}{1/4} = 6$$

79. Diverges: geometric with $r = -\dfrac{4}{3}$

81. Converges: geometric with $a_1 = 1.5$ and $r = 0.5$, so
$$S = \frac{1.5}{1 - 0.5} = \frac{1.5}{0.5} = 3$$

83. $\displaystyle\sum_{k=1}^{21} [-8 + 5(k-1)] = \sum_{k=1}^{21} (5k - 13)$

85. $\displaystyle\sum_{k=0}^{\infty} (2k+1)^2$ or $\displaystyle\sum_{k=1}^{\infty} (2k-1)^2$

87. $\displaystyle\sum_{k=1}^{n} (3k+1) = 3\sum_{k=1}^{n} k + \sum_{k=1}^{n} 1$
$$= 3 \cdot \frac{n(n+1)}{2} + n = \frac{3n^2 + 5n}{2} = \frac{n(3n+5)}{2}$$

89. $\displaystyle\sum_{k=1}^{25} (k^2 - 3k + 4) = \frac{25 \cdot 26 \cdot 51}{6} - 3 \cdot \frac{25 \cdot 26}{2} + 4 \cdot 25 = 4650$

91. P_n: $1 + 3 + 6 + \cdots + \dfrac{n(n+1)}{2} = \dfrac{n(n+1)(n+2)}{6}$.

P_1 is true: $\dfrac{1(1+1)}{2} = \dfrac{1(1+1)(1+2)}{6}$.

Now assume P_k is true: $1 + 3 + 6 + \cdots + \dfrac{k(k+1)}{2} = \dfrac{k(k+1)(k+2)}{6}$. Add $\dfrac{(k+1)(k+2)}{2}$ to both sides:
$$1 + 3 + 6 + \cdots + \frac{k(k+1)}{2} + \frac{(k+1)(k+2)}{2}$$
$$= \frac{k(k+1)(k+2)}{6} + \frac{(k+1)(k+2)}{2}$$
$$= (k+1)(k+2)\left(\frac{k}{6} + \frac{1}{2}\right)$$
$$= (k+1)(k+2)\left(\frac{k+3}{6}\right)$$
$$= \frac{(k+1)((k+1)+1)((k+1)+2)}{6},$$
so P_{k+1} is true. Therefore, P_n is true for all $n \geq 1$.

93. P_n: $2^{n-1} \leq n!$. P_1 is true: it says that $2^{1-1} \leq 1!$ (they are equal). Now assume P_k is true: $2^{k-1} \leq k!$. Then $2^{k+1-1} = 2 \cdot 2^{k-1} \leq 2 \cdot k! \leq (k+1)k! = (k+1)!$, so P_{k+1} is true. Therefore, P_n is true for all $n \geq 1$.

95. (a)

Stem	Leaf
9	1 2
10	6 7
11	4 5 5 7 7
12	0 2 4 6 7 7
13	5 6
14	1 6 7 7 8
15	4 8
16	1 4
17	0 6
18	
19	
20	
21	9
22	
23	4

(b)

Price	Frequency
90,000–99,999	2
100,000–109,999	2
110,000–119,999	5
120,000–129,999	6
130,000–139,999	2
140,000–149,999	5
150,000–159,999	2
160,000–169,999	2
170,000–179,999	2
210,000–219,999	1
230,000–239,999	1

(c)

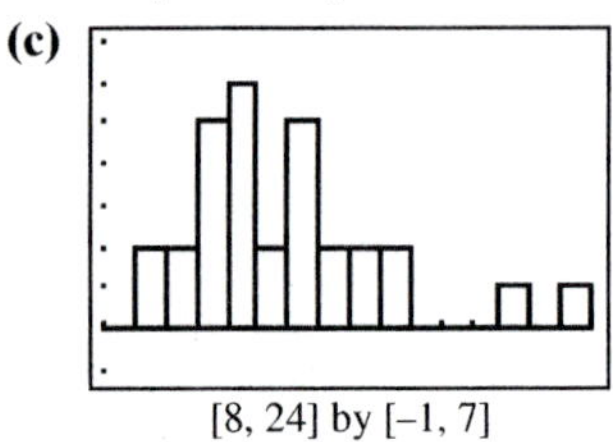

[8, 24] by [–1, 7]

97. (a)

Stem	Leaf
12	0 0 4 4
13	1 1 2 6 7 9
14	0 3 4 8
15	6
16	3
17	7 9
18	0
19	0 1 7
20	2
21	
22	
23	0

(b)

Length (in seconds)	Frequency
120—129	4
130—139	6
140—149	4
150—159	1
160—169	1
170—179	2
180—189	1
190—199	3
200—209	1
210—219	0
220—229	0
230—239	1

(c)

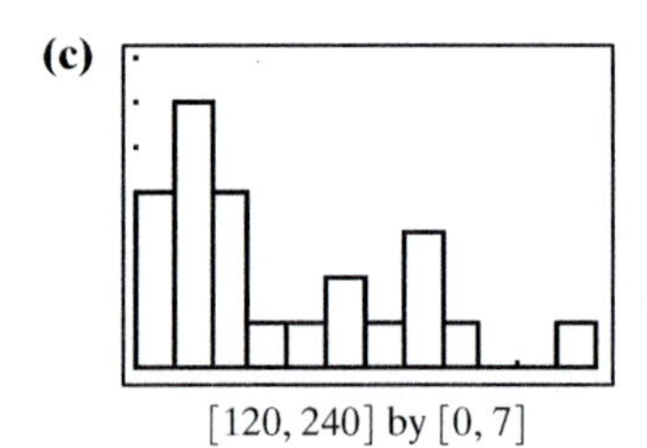

[120, 240] by [0, 7]

99. The ordered data is {9.1, 9.2, 10.6, 10.7, 11.4, 11.5, 11.5, 11.7, 11.7, 12, 12.2, 12.4, 12.6, 12.7, 12.7, 13.5, 13.6, 14.1, 14.6, 14.7, 14.7, 14.8, 15.4, 15.8, 16.1, 16.4, 17.0, 17.6, 21.9, 23.4} so the

median is $\frac{12.7 + 13.5}{2} = 13.1$, $Q_3 = 15.4$, and $Q_1 = 11.7$.

Five-number summary: {9.1, 11.7, 13.1, 15.4, 23.4}
Range: $23.4 - 9.1 = 14.3$ ($143,000)
IQR: $15.4 - 11.7 = 3.7$ ($37,000)
$\sigma \approx 3.19, \sigma^2 \approx 10.14$
Outliers: 21.9 and 23.4 are greater than $15.4 + 1.5(3.7) = 20.95$.

101. The ordered data is {120, 120, 124, 124, 131, 131, 132, 136, 137, 139, 140, 143, 144, 148, 156, 163, 177, 179, 180, 190, 191, 197, 202, 230} so the median is $\frac{143 + 144}{2} = 143.5$, $Q_3 = \frac{179 + 180}{2} = 179.5$, and

$Q_1 = \frac{131 + 132}{2} = 131.5$.

Five-number summary: {120, 131.5, 143.5, 179.5, 230}
Range: $230 - 120 = 110$
IQR: $179.5 - 131.5 = 48$
$\sigma = 29.9, \sigma^2 = 891.4$
No outliers.

103. (a)

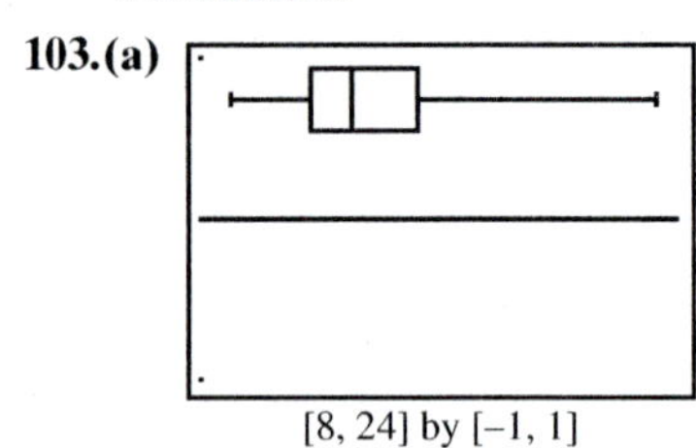

[8, 24] by [–1, 1]

(b)

[8, 4] by [–1, 1]

105. (a)

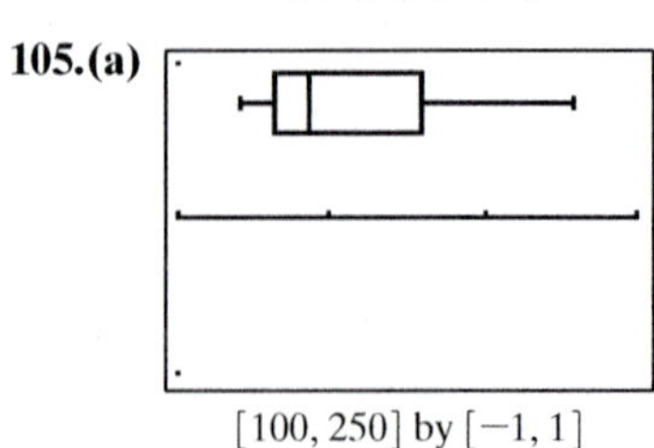

[100, 250] by [−1, 1]

(b)

[100, 250] by [−1, 1]

107.

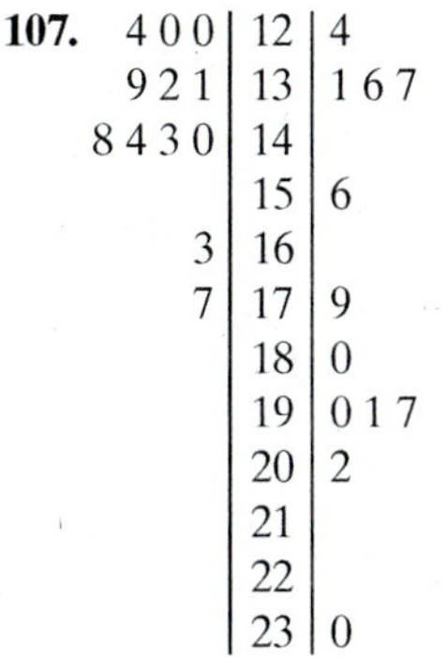

4 0 0	12	4
9 2 1	13	1 6 7
8 4 3 0	14	
	15	6
3	16	
7	17	9
	18	0
	19	0 1 7
	20	2
	21	
	22	
	23	0

The songs released in the earlier years tended to be shorter.

109. 

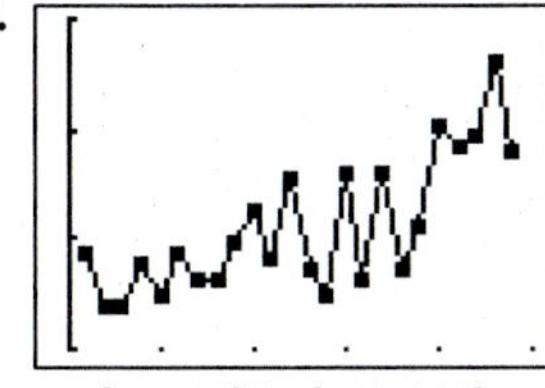

[−1, 25] by [100, 250]

Again, the data demonstrates that songs appearing later tended to be longer in length.

111. 1 9 36 84 126 84 36 9 1

113. **(a)** $P(\text{no defective bats}) = (0.98)^4 \approx 0.922$

(b) $P(\text{one defective bat}) = {}_4C_1 \cdot (0.98)^3(0.02)$
$= 4(0.98)^3(0.02) \approx 0.075$

Chapter 9 Project

Answers are based on the sample data shown in the table.

1. Stem Leaf

Stem	Leaf
5	
5	9
6	1 1 2 3 3 3 4 4 4 4
6	5 6 6 6 7 8 8 9 9
7	0 0 1 1 1 2 2 3
7	5

The average is about 66 or 67 inches.

2. A large number of students are between 63 and 64 inches and also between 69 and 72 inches.

Height	Frequency
59–60	1
61–62	3
63–64	7
65–66	4
67–68	3
69–70	5
71–72	5
73–74	1
75–76	1

3.

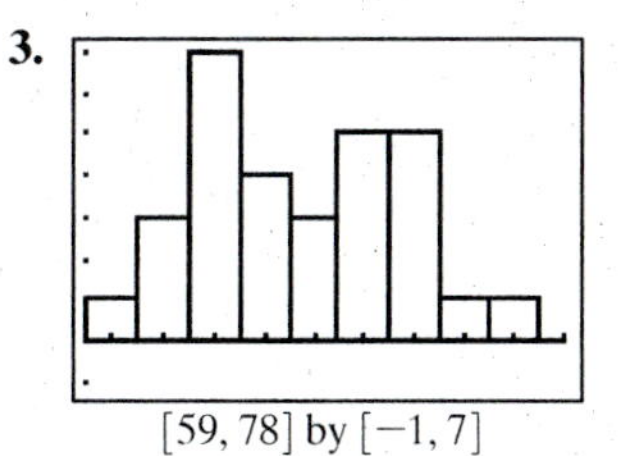

[59, 78] by [−1, 7]

Again, the average appears to be about 66 or 67 inches. Since the data are not broken out by gender, one can only speculate about average heights for males and females separately. Possibly the two peaks within the distribution represent an average height of 63–64 inches for females and 70–71 inches for males.

4. Mean = 66.9 in.; median = 66.5 in.; mode = 64 in. The mean and median both appear to be good measures of the average, but the mode is too low. Still, the mode might well be similar in other classes.

5. The data set is well distributed and probably does not have outliers.

6. The stem and leaf plot puts the data in order.
Minimum value: 59
Maximum value: 75
Median: $\dfrac{66 + 67}{2} = 66.5$
Q_1: 64
Q_3: 70
The five-number summary is $\{59, 64, 66.5, 70, 75\}$.

7.

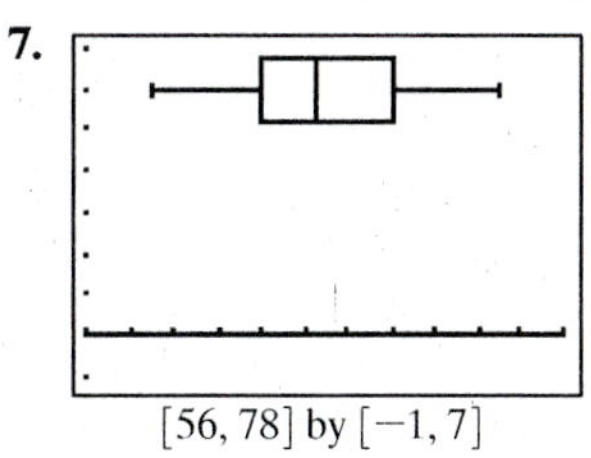

[56, 78] by [−1, 7]

The boxplot visually represents the five-number summary. The whisker-to-whisker size of the boxplot represents the range of the data, while the width of the box represents the interquartile range.

8. Mean = 67.5; median = 67; The new five-number summary is $\{59, 64, 67, 71, 86\}$.

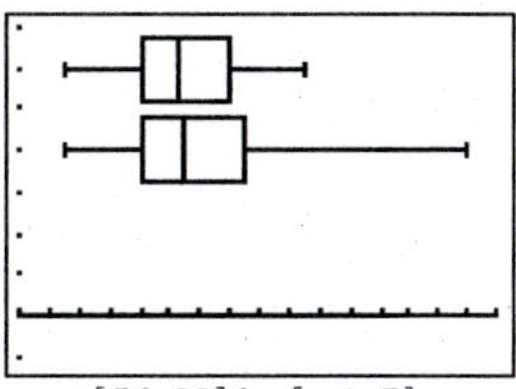

[56, 88] by [−1, 7]

The minimum and first quartile are unaffected, but the median, third quartile, and maximum are shifted upon to varying degrees.

9. The new student's height, 86 inches, lies 15 inches away from Q_3, and that is more than $1.5(Q_3 - Q_1) = 10.5$. The height of the new student should probably be tossed out during prediction calculations.

10. Mean = 68.9; median = 68; The new five-number summary is $\{59, 64, 68, 71, 86\}$.
All three of the new students are outliers by the $1.5 \times IQR$ test. Their heights should be left out during prediction-making.

Chapter 10
An Introduction to Calculus: Limits, Derivatives, and Integrals

Section 10.1 Limits and Motion: The Tangent Problem

Exploration 1

1. $m = \frac{4-1}{2-1} = \frac{3}{1} = 3$

2. $v_{ave} = \frac{\Delta s}{\Delta t} = \frac{3 \text{ ft}}{1 \text{ sec}} = 3 \text{ ft/sec}$

3. They are the same.

4. As the slope of the line joining $(a, s(a))$ and $(b, s(b))$

Quick Review 10.1

1. $m = \frac{-1-3}{5-(-2)} = \frac{-4}{7} = -\frac{4}{7}$

3. $y - 3 = \frac{3}{2}(x+2)$ or $y = \frac{3}{2}x + 6$

5. $y - 4 = \frac{3}{4}(x-1)$

7. $\frac{4 + 4h + h^2 - 4}{h} = \frac{4h + h^2}{h} = h + 4$

9. $\frac{\frac{1}{2+h} - \frac{1}{2}}{h} = \frac{2-(2+h)}{2(2+h)} \cdot \frac{1}{h}$
$= \frac{-h}{h} \cdot \frac{1}{2(2+h)} = -\frac{1}{2(h+2)}$

Section 10.1 Exercises

1. $v_{ave} = \frac{\Delta s}{\Delta t} = \frac{21 \text{ miles}}{1.75 \text{ hours}} = 12 \text{ mi per hour}$

3. $s'(4) = \lim_{h \to 0} \frac{s(4+h) - s(4)}{h}$
$= \lim_{h \to 0} \frac{3(h+4) - 5 - 7}{h}$
$= \lim_{h \to 0} 3 = 3$

5. $s'(2) = \lim_{h \to 0} \frac{s(2+h) - s(2)}{h}$
$= \lim_{h \to 0} \frac{a(h+2)^2 + 5 - (4a+5)}{h}$
$= \lim_{h \to 0} \frac{ah^2 + 4ah}{h}$
$= \lim_{h \to 0} (ah + 4a) = 4a$

7. Try $\frac{f(1) - f(0)}{1 - 0} = \frac{3-2}{1} = 1$

9. No tangent

11. 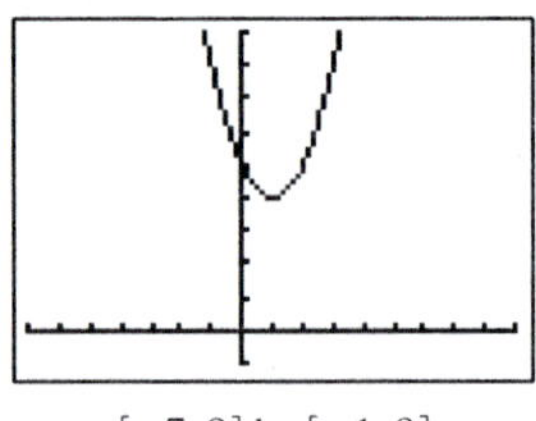

$[-7, 9]$ by $[-1, 9]$

$m = 4$

13. 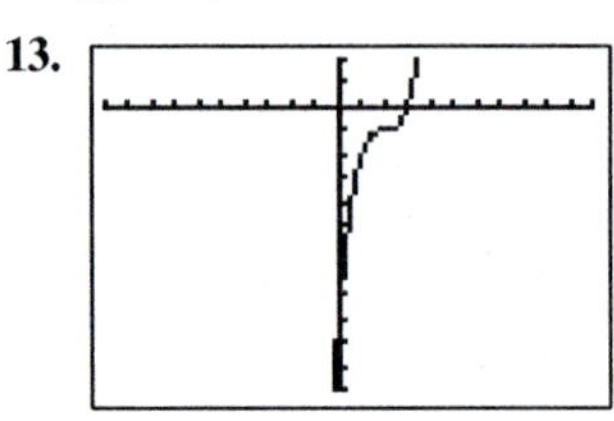

$[-10, 11]$ by $[-12, 2]$

$m = 12$

15. **(a)** $f'(0) = \lim_{h \to 0} \frac{f(0+h) - f(0)}{h}$
$= \lim_{h \to 0} \frac{3 + 48(0+h) - 16(0+h)^2 - 3}{h}$
$= \lim_{h \to 0} \frac{48h - 16h^2}{h} = 48$

(b) The initial velocity of the rock is $f'(0) = 48$ ft/sec.

17. **(a)** $m = \lim_{h \to 0} \frac{f(-1+h) - f(-1)}{h}$
$= \lim_{h \to 0} \frac{2(h-1)^2 - 2}{h} = \lim_{h \to 0} \frac{2h^2 - 4h + 2 - 2}{h}$
$= \lim_{h \to 0} (2h - 4) = -4$

(b) Since $(-1, f(-1)) = (-1, 2)$ the equation of the tangent line is $y - 2 = -4(x+1)$.

(c)

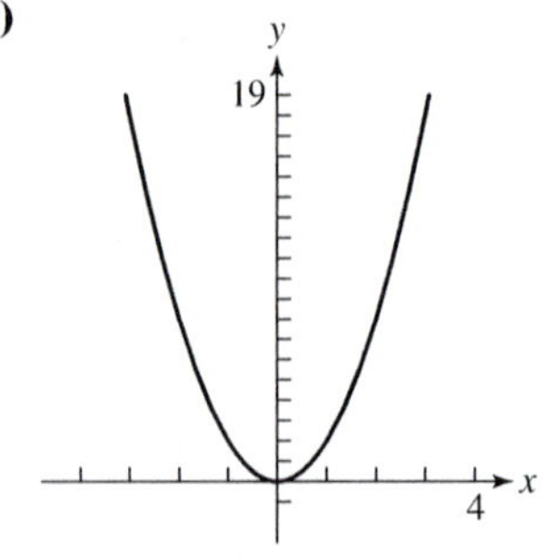

19. **(a)** $m = \lim_{h \to 0} \frac{f(2+h) - f(2)}{h}$
$= \lim_{h \to 0} \frac{2(h+2)^2 - 7(h+2) + 3 - (-3)}{h}$
$= \lim_{h \to 0} \frac{2h^2 + 8h + 8 - 7h - 14 + 6}{h}$
$= \lim_{h \to 0} (2h + 1) = 1$

(b) Since $(2, f(2)) = (2, -3)$ the equation of the tangent line is $y + 3 = 1(x - 2)$, or $y = x - 5$.

(c)

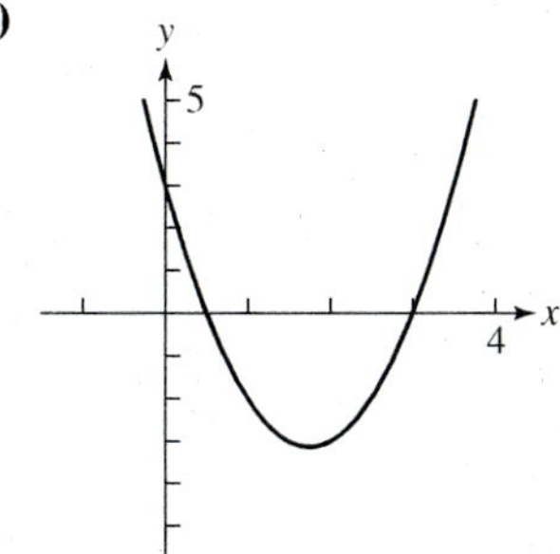

21.

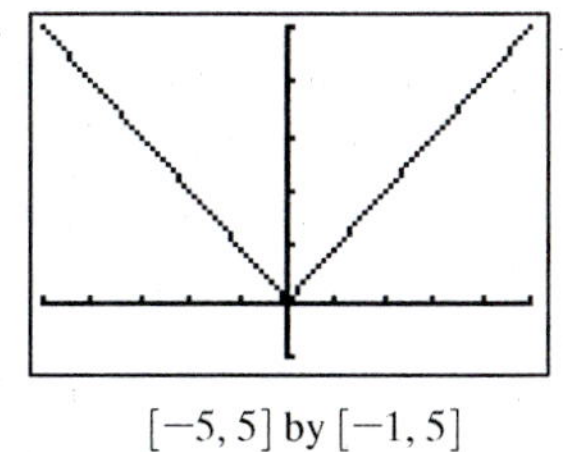

$[-5, 5]$ by $[-1, 5]$

At $x = -2$: $m = -1$, at $x = 2$: $m = 1$, at $x = 0$, m does not exist.

23. $\lim_{h \to 0} \dfrac{f(2 + h) - f(2)}{h} = \lim_{h \to 0} \dfrac{1 - (2 + h)^2 - (1 - 4)}{h}$

$= \lim_{h \to 0} \dfrac{-h^2 - 4h - 4 + 4}{h} = \lim_{h \to 0} (-h - 4) = -4$

25. $\lim_{h \to 0} \dfrac{f(-2 + h) - f(-2)}{h}$

$= \lim_{h \to 0} \dfrac{3(h - 2)^2 + 2 - (14)}{h}$

$= \lim_{h \to 0} \dfrac{3h^2 - 12h + 12 - 12}{h} = \lim_{h \to 0} (3h - 12) = -12$

27. $\lim_{h \to 0} \dfrac{f(-2 + h) - f(-2)}{h} = \lim_{h \to 0} \dfrac{|h - 2 + 2| - 0}{h}$

$= \lim_{h \to 0} \dfrac{|h|}{h}$. When $h > 0$, $\dfrac{|h|}{h} = 1$ while when $h < 0$, $\dfrac{|h|}{h} = -1$. The limit does not exist. The derivative does not exist.

29. $f'(x) = \lim_{h \to 0} \dfrac{2 - 3(x + h) - (2 - 3x)}{h}$

$= \lim_{h \to 0} \dfrac{2 - 3x - 3h - 2 + 3x}{h} = \lim_{h \to 0} \dfrac{-3h}{h} = -3$

31. $f'(x)$

$= \lim_{h \to 0} \dfrac{3(x + h)^2 + 2(x + h) - 1 - (3x^2 + 2x - 1)}{h}$

$= \lim_{h \to 0} \dfrac{3x^2 + 6xh + 3h^2 + 2x + 2h - 1 - 3x^2 - 2x + 1}{h}$

$= \lim_{h \to 0} \dfrac{6xh + 3h^2 + 2h}{h} = \lim_{h \to 0} (6x + 3h + 2) = 6x + 2$

33. (a) Between 0.5 and 0.6 seconds: $\dfrac{3.2 - 2.3}{0.6 - 0.5} = 9$ ft/sec

Between 0.8 and 0.9 seconds: $\dfrac{7.3 - 5.8}{0.9 - 0.8} = 15$ ft/sec

(b) $f(x) = 8.94x^2 + 0.05x + 0.01$, $x =$ time in seconds

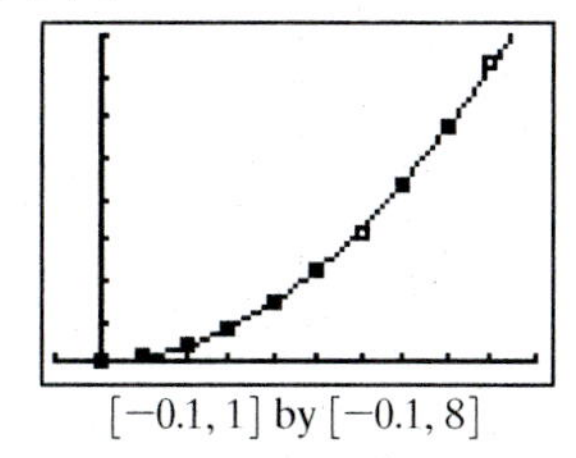

$[-0.1, 1]$ by $[-0.1, 8]$

(c) $f(2) \approx 35.9$ ft

35. (a)

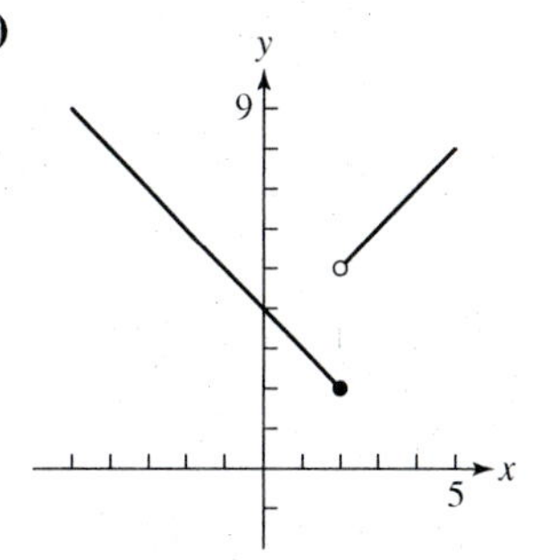

(b) Since the graph of the function does not have a definable slope at $x = 2$, the derivative of f does not exist at $x = 2$. The function is not continuous at $x = 2$.

(c) Derivatives do not exist at points where functions have discontinuities.

37. (a)

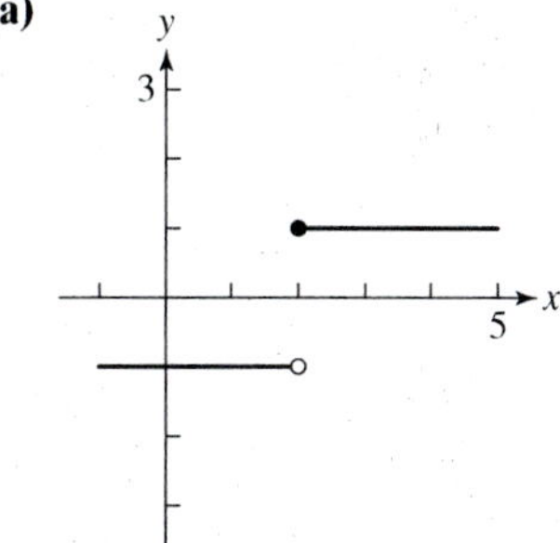

(b) Since the graph of the function does not have a definable slope at $x = 2$, the derivative of f does not exist at $x = 2$. The function is not continuous at $x = 2$.

(c) Derivatives do not exist at points where functions have discontinuities.

39. Answers will vary. One possibility:

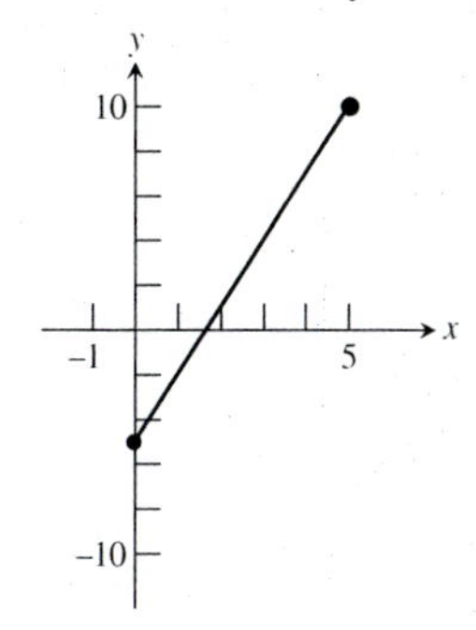

41. Answers will vary. One possibility:

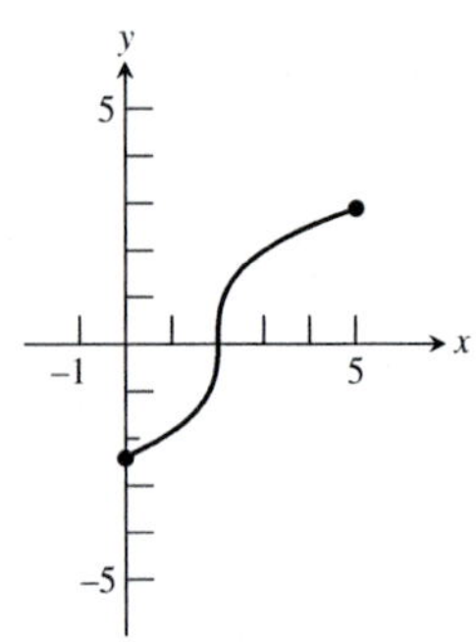

43. Since $f(x) = ax + b$ is a linear function, the rate of change for any x is exactly the slope of the line. No calculations are necessary since it is known that the slope $a = f'(x)$.

45. False. The instantaneous velocity is a limit of average velocities. It is nonzero when the ball is moving.

47. For $Y_1 = x^2 + 3x - 4$, at $x = 0$ the calculator shows $dy/dx = 3$. The answer is D.

49. For $Y_1 = x^3$, at $x = 2$ the calculator shows $dy/dx = 12$. The answer is C.

51. (a)

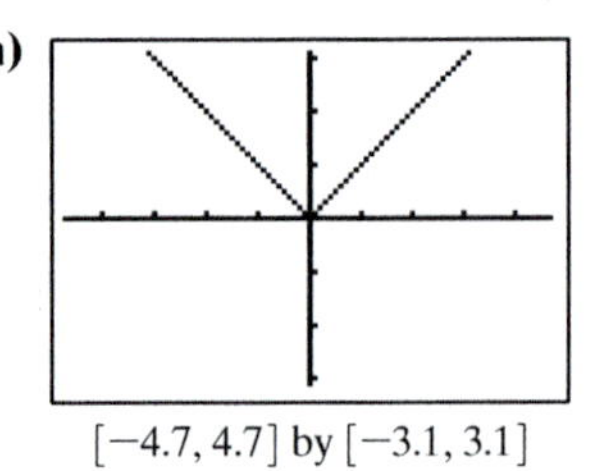

$[-4.7, 4.7]$ by $[-3.1, 3.1]$

No, there is no derivative because the graph has a corner at $x = 0$.

(b) No

53. (a)

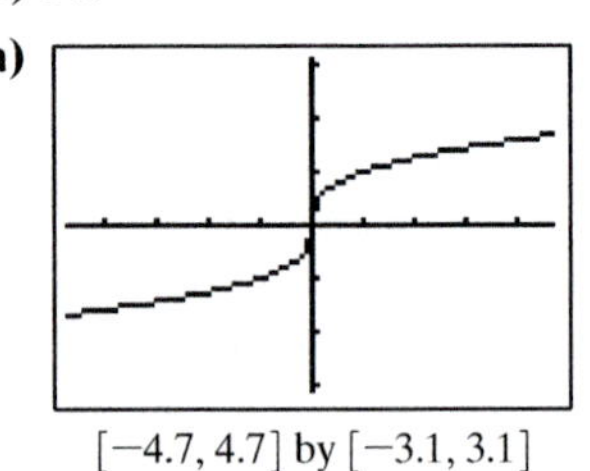

$[-4.7, 4.7]$ by $[-3.1, 3.1]$

No, there is no derivative because the graph has a vertical tangent (no slope) at $x = 0$.

(b) Yes, the tangent line is $x = 0$.

55. (a) The average velocity is

$$\frac{\Delta s}{\Delta t} = \frac{16(3)^2 - 16(0)^2}{3 - 0} = 48 \text{ ft/sec.}$$

(b) The instantaneous velocity is

$$\lim_{h\to 0} \frac{16(3 + h)^2 - 144}{h} = \lim_{h\to 0} \frac{96h + h^2}{h} = \lim_{h\to 0} (96 + h) = 96 \text{ ft/sec}$$

57.

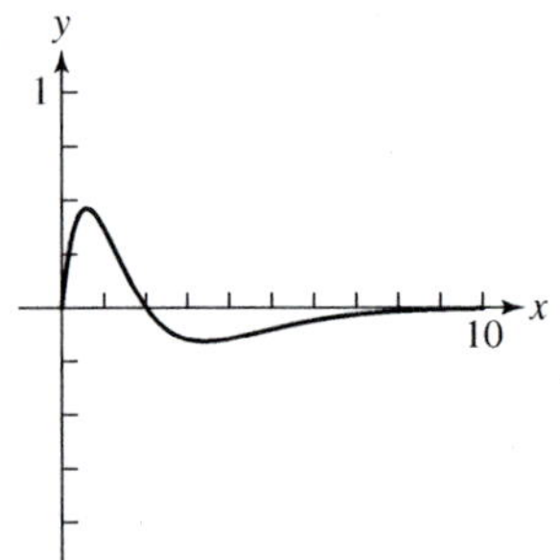

Section 10.2 Limits and Motion: The Area Problem

Exploration 1

1. The total amount of water remains 1 gallon. Each of the 10 teacups holds $\frac{1 \text{ gal}}{10} = 0.1$ gallon of water.
2. The total amount of water remains 1 gallon. Each of the 100 teacups holds $\frac{1 \text{ gal}}{100} = 0.01$ gallon of water.
3. The total amount of water remains 1 gallon. Each of the 1,000,000,000 teacups holds $\frac{1 \text{ gal}}{1{,}000{,}000{,}000} = 0.000\,000\,001$ gallon of water.
4. The total amount of water remains 1 gallon. Each of the teacups holds an amount of water that is less than what was in each of the 1 billion teacups in step 3. Thus each teacup holds about 0 gallons of water.

Quick Review 10.2

1. $\frac{1}{8}, \frac{1}{2}, \frac{9}{8}, 2, \frac{25}{8}, \frac{9}{2}, \frac{49}{8}, 8, \frac{81}{8}, \frac{25}{2}$

3. $\frac{1}{2}[2 + 3 + 4 + 5 + 6 + 7 + 8 + 9 + 10 + 11] = \frac{65}{2}$

5. $\frac{1}{2}[4 + 9 + \ldots + 121] = \frac{505}{2}$

7. (57 mph)(4 hours) = 228 miles

9. $\left(\frac{200 \text{ ft}^3}{\text{sec}}\right)(6 \text{ hours})\left(\frac{60 \text{ minutes}}{\text{hour}}\right)\left(\frac{60 \text{ seconds}}{\text{minute}}\right) = 4{,}320{,}000 \text{ ft}^3$

Section 10.2 Exercises

1. Let the line $y = 65$ represent the situation. The area under the line is the distance traveled, a rectangle, $(65)(3) = 195$ miles.

3. Let the line $y = 150$ represent the situation. The area under the line is the total number of cubic feet of water pumped, a rectangle, $(150)(3600) = 540{,}000 \text{ ft}^3$.

5. $\Delta s = \frac{\Delta s}{\Delta t} \cdot \Delta t = (640 \text{ km/h})(3.4 \text{ h}) = 2176 \text{ km}$

7. $\sum_{k=1}^{5} 1 \cdot f(k) = f(1) + f(2) + f(3) + f(4) + f(5)$

$= 3\frac{1}{2} + 4\frac{1}{4} + 3\frac{1}{2} + 1\frac{3}{4} + 0 = 13$ (answers will vary)

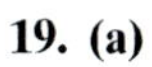

9. $\sum_{k=1}^{5} 1 \cdot f(k) = f(0.5) + f(1.5) + f(2.5) + f(3.5) + f(4.5)$
$= 3.5 + 5.25 + 2.75 + 0.25 + 1.25 = 13$ (answers will vary)

11. $\sum_{i=1}^{8} (10 - x_i^2)\Delta x_i$
$= (9 + 9.75 + 10 + 9.75 + 9 + 7.75 + 6 + 3.75)(0.5)$
$= 32.5$ square units

13. $\left[0, \frac{1}{2}\right], \left[\frac{1}{2}, 1\right], \left[1, \frac{3}{2}\right], \left[\frac{3}{2}, 2\right]$

15. $\left[1, \frac{3}{2}\right], \left[\frac{3}{2}, 2\right], \left[2, \frac{5}{2}\right], \left[\frac{5}{2}, 3\right], \left[3, \frac{7}{2}\right], \left[\frac{7}{2}, 4\right]$

For #17–19, the intervals are of width 1, so the area of each rectangle is $1 \cdot f(k) = f(k)$.

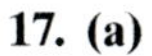

17. (a)

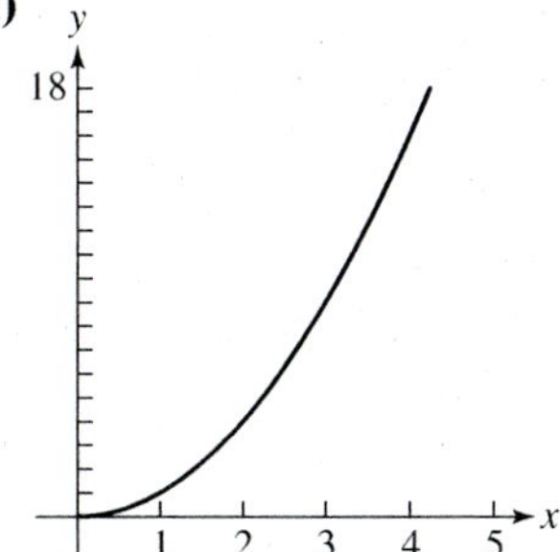

(b)

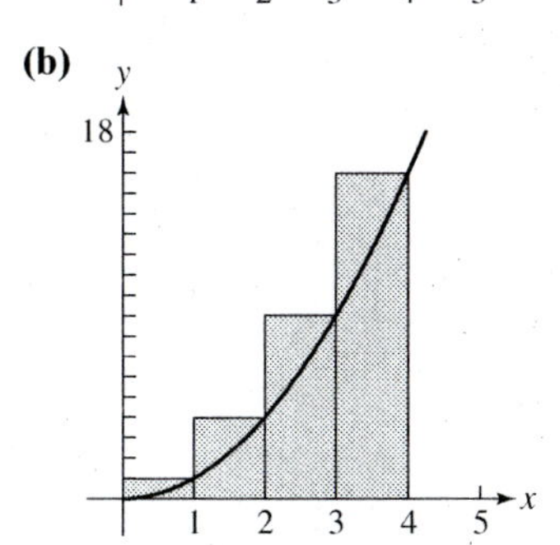

RRAM: $f(1) + f(2) + f(3) + f(4)$
$= 1 + 4 + 9 + 16 = 30$

(c)

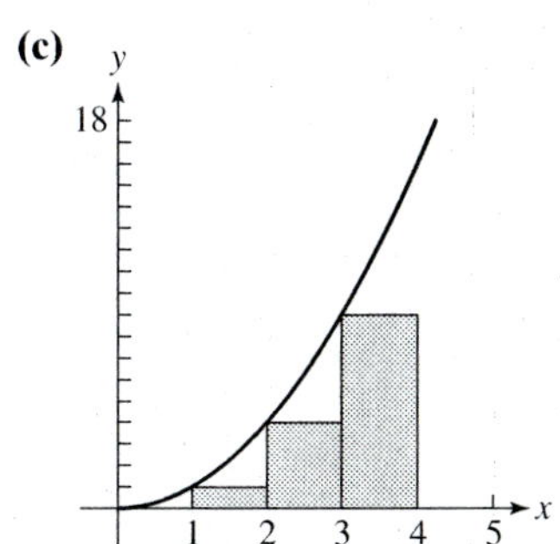

LRAM: $f(0) + f(1) + f(2) + f(3)$
$= 0 + 1 + 4 + 9 = 14$

(d) Average: $\frac{14 + 30}{2} = 22$

19. (a)

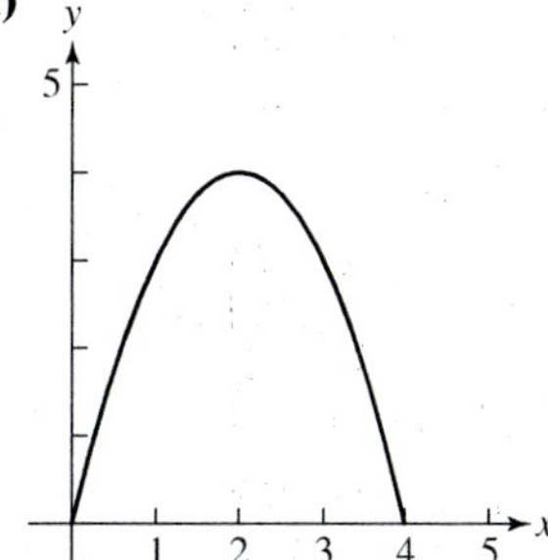

(b)

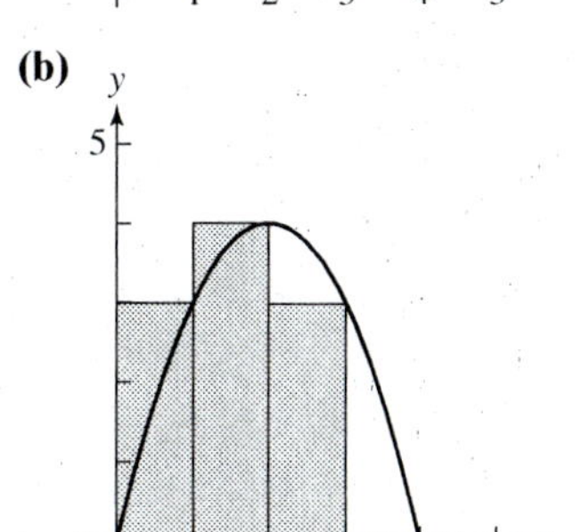

RRAM: $f(1) + f(2) + f(3) + f(4)$
$= 3 + 4 + 3 + 0 = 10$

(c)

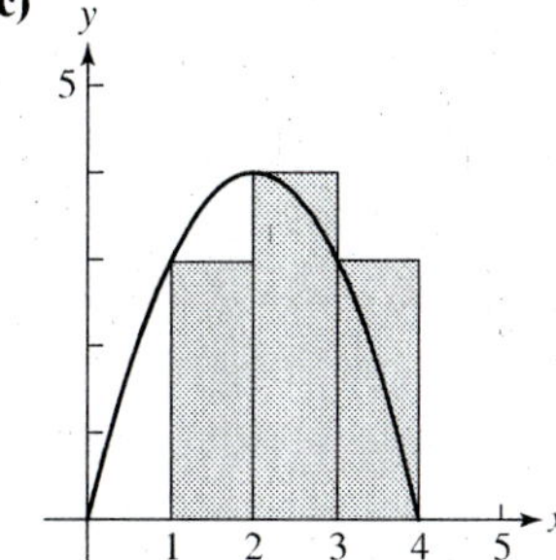

LRAM: $f(0) + f(1) + f(2) + f(3)$
$= 0 + 3 + 4 + 3 = 10$

(d) Average: $\frac{10 + 10}{2} = 10$

21. $\int_3^7 5\,dx = 20$ (Rectangle with base 4 and height 5)

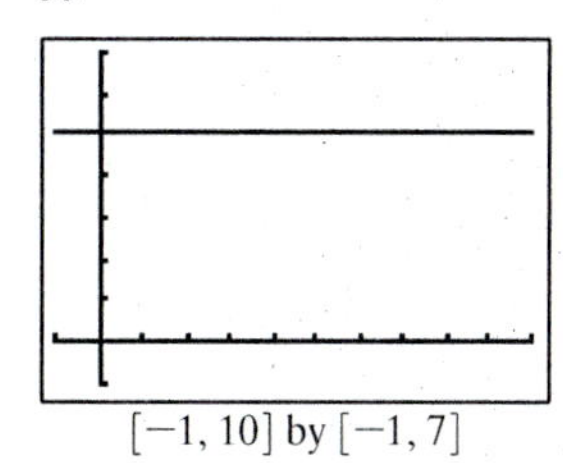

$[-1, 10]$ by $[-1, 7]$

23. $\int_0^5 3x\,dx = 37.5$ (Triangle with base 5 and altitude 15)

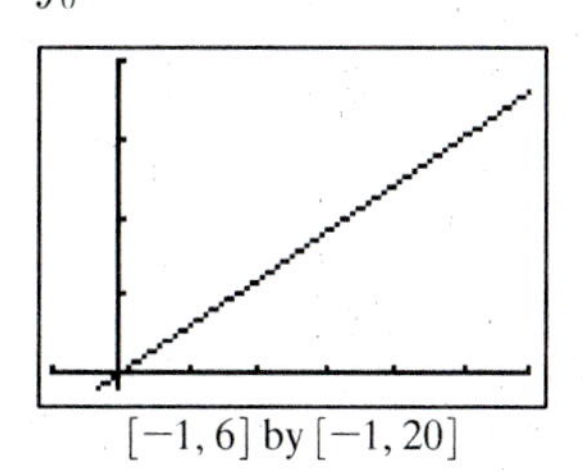

$[-1, 6]$ by $[-1, 20]$

25. $\int_1^4 (x + 3)\, dx = 16.5$ (Trapezoid with bases 4 and 7 and altitude 3)

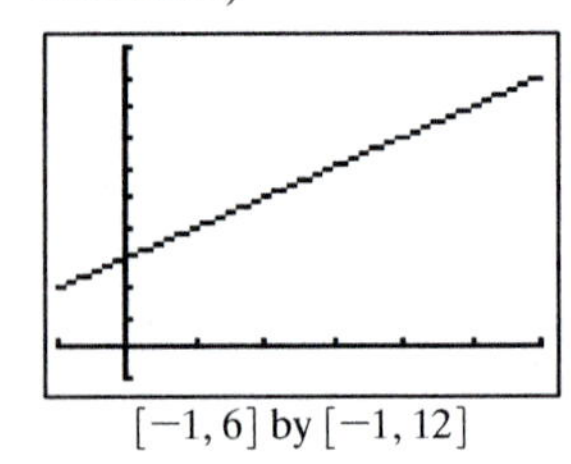

$[-1, 6]$ by $[-1, 12]$

27. $\int_{-2}^2 \sqrt{4 - x^2}\, dx = 2\pi$ (Semicircle with radius 2)

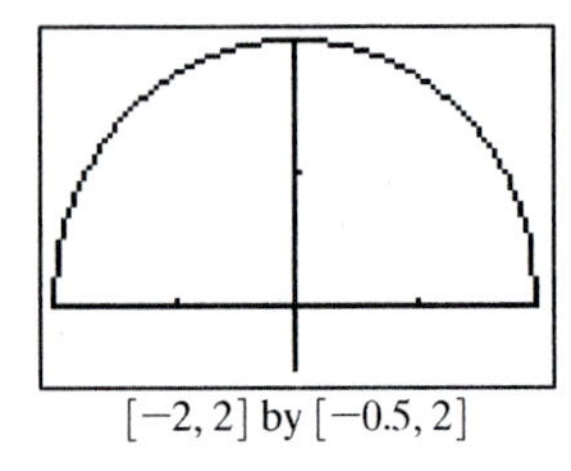

$[-2, 2]$ by $[-0.5, 2]$

29. $\int_0^\pi \sin x\, dx = 2$ (One arch of sine curve)

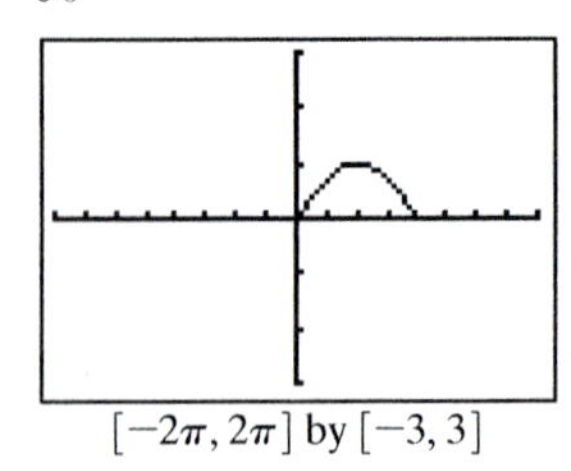

$[-2\pi, 2\pi]$ by $[-3, 3]$

31. $\int_2^{2+\pi} \sin (x - 2)\, dx = 2$ (One arch of sine curve translated 2 units right)

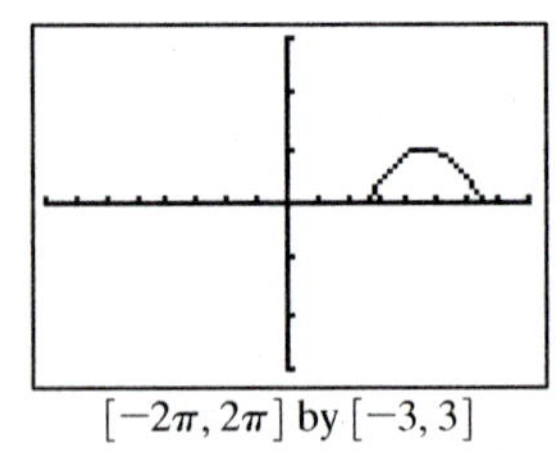

$[-2\pi, 2\pi]$ by $[-3, 3]$

33. $\int_0^{\pi/2} \sin x\, dx = 1$ (Half-arch of sine curve)

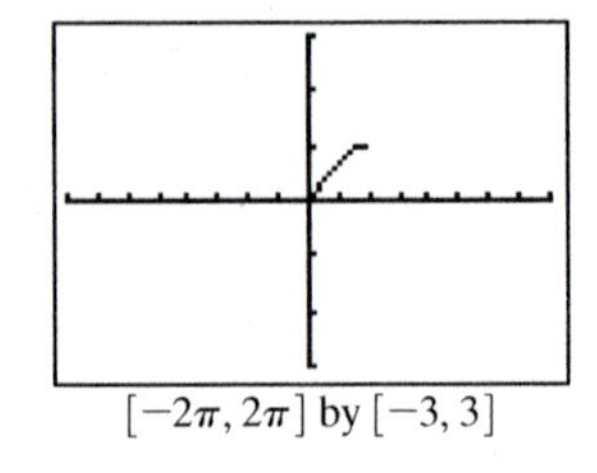

$[-2\pi, 2\pi]$ by $[-3, 3]$

35. $\int_0^\pi 2 \sin x\, dx = 4$ (Rectangles in sum are twice as tall, yielding twice the sum)

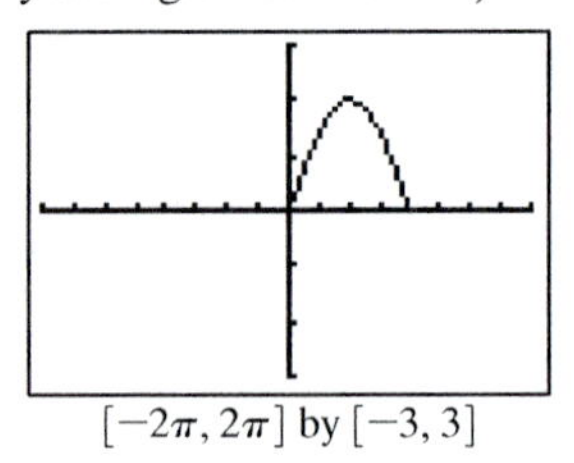

$[-2\pi, 2\pi]$ by $[-3, 3]$

37. $\int_0^{2\pi} |\sin x|\, dx = 4$ (Two arches of the sine curve)

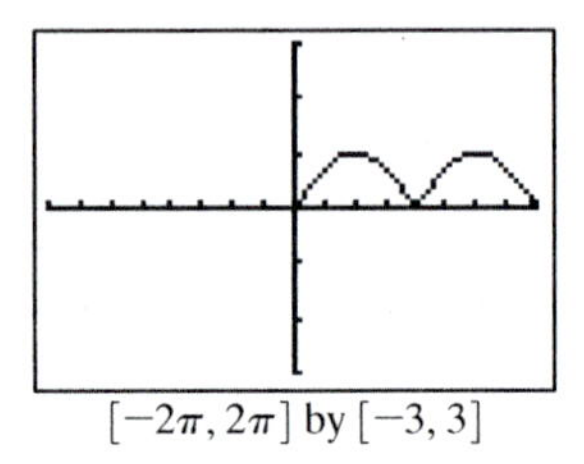

$[-2\pi, 2\pi]$ by $[-3, 3]$

39. The graph of $f(x) = kx + 3$ is a line. If k is a number between 0 and 4, the integral is the area of a trapezoid with bases of $0k + 3 = 3$ and $4k + 3$ and height of $4 - 0 = 4$. The area is $\frac{1}{2}(4)(3 + 4k + 3) = 2(4k + 6)$ $= 8k + 12$, so $\int_0^4 (kx + 3)dx = 8k + 12$.

41. The graph of $f(x) = 3x + k$ is a line. The integral is the area of a trapezoid with bases of $3 \cdot 0 + k = k$ and $3 \cdot 4 + k = 12 + k$ and height of $4 - 0 = 4$. The area is $\frac{1}{2}(4)(k + 12 + k) = 2(12 + 2k) = 24 + 4k$, so $\int_0^4 (3x + k)dx = 24 + 4k$.

43. Since $g(x) = -f(x)$, we consider g to be symmetric with f about the x-axis. For every value of x in the interval, $|f(x)|$ is the distance to the x-axis and similarly, $|g(x)|$ is the distance to the x-axis; $f(x)$ and $g(x)$ are equidistant from the x-axis. As a result, the area under $f(x)$ must be exactly equal to the area above $g(x)$.

45. The distance traveled will be the same as the area under the velocity graph, $v(t) = 32t$, over the interval $[0, 2]$. That triangular region has an area of $A = (1/2)(2)(64) = 64$. The ball falls 64 feet during the first 2 seconds.

47. (a)

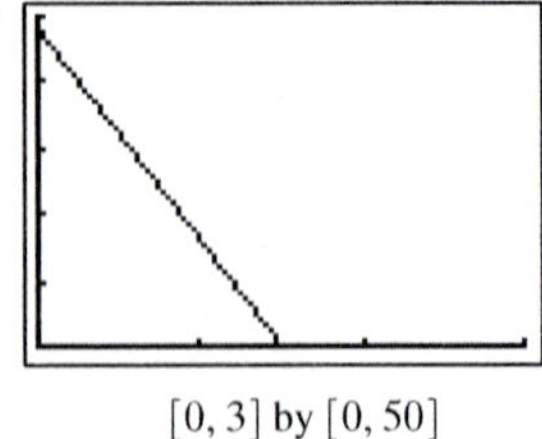

$[0, 3]$ by $[0, 50]$

(b) The ball reaches its maximum height when the velocity function is zero; this is the point where the ball changes direction and starts its descent. Solving for t when $48 - 32t = 0$, we find $t = 1.5$ sec.

(c) The distance the ball has traveled is the area under the curve, a triangle with base 1.5 and height 48 thus, $d = 0.5(1.5)(48) = 36$ units.

49. (a)

[0, 2] by [−50, 0]

(b) Each RRAM rectangle will have width 0.2. The heights (using the absolute value of the velocity) are 5.05, 11.43, 17.46, 24.21, 30.62, 37.06, and 43.47. The height of the building is approximately $0.2[5.05 + 11.43 + 17.46 + 24.21 + 30.62 + 37.06 + 43.47] = 33.86$ feet.

51. True. The exact area under a curve is given by the limit as n approaches infinity. This is true whether LRAM or RRAM is used.

53. Since $y = 2\sqrt{x}$ represents a vertical stretch, by a factor of 2, of $y = \sqrt{x}$, the area under the curve between $x = 0$ and $x = 9$ is doubled. The answer is A.

55. $y = \sqrt{x - 5}$ is shifted right 5 units compared to $y = \sqrt{x}$, but the limits of integration are shifted right 5 units also, so the area is unchanged. The answer is C.

57. In the definition of the definite integral, if $f(x)$ is negative, then $\sum_{i=1}^{n} f(x_i)\Delta x$ is negative, so the definite integral is negative. For $f(x) = \sin x$ on $[0, 2\pi]$, the "positive area" (from 0 to π) cancels the "negative area" (from π to 1), so the definite integral is 0.

Since $g(x) = x - 1$ forms a triangle with area $\frac{1}{2}$ below the x-axis on $[0, 1]$, $\int_0^1 (x - 1)dx = -\frac{1}{2}$.

59. True

$$\int_a^b f(x)dx + \int_a^b g(x)dx$$
$$= \left[\lim_{n\to\infty} \sum_{i=1}^{n} f(x_i)\Delta x\right] + \left[\lim_{n\to\infty} \sum_{i=1}^{n} g(x_i)\Delta x\right]$$
$$= \lim_{n\to\infty}\left[\sum_{i=1}^{n} f(x_i)\Delta x + \sum_{i=1}^{n} g(x_i)\Delta x\right]$$
$$= \lim_{n\to\infty} \sum_{i=1}^{n} [f(x_i) + g(x_i)]\Delta x$$
$$= \int_a^b (f(x) + g(x))dx$$

Note: There are some subtleties here, because the x_i that are chosen for $f(x)$ may be different from the x_i that are chosen for $g(x)$; however, the result is true, provided the limits exist.

61. False. Counterexample: Let $f(x) = 1$, $g(x) = 1$. Then $\int_0^2 f(x)g(x)dx = 2$ but $\int_0^2 f(x)dx \cdot \int_0^2 g(x)dx = 4$

63. False. Interchanging a and b reverses the sign of $\Delta x = \frac{b - a}{n}$, which reverses the sign of the integral.

Section 10.3 More on Limits

Exploration 1

1. Answers will vary. Possible answers include: Solving graphically or algebraically shows that $7x = 14$ when $x = 2$, so we know that 14 is the limit.

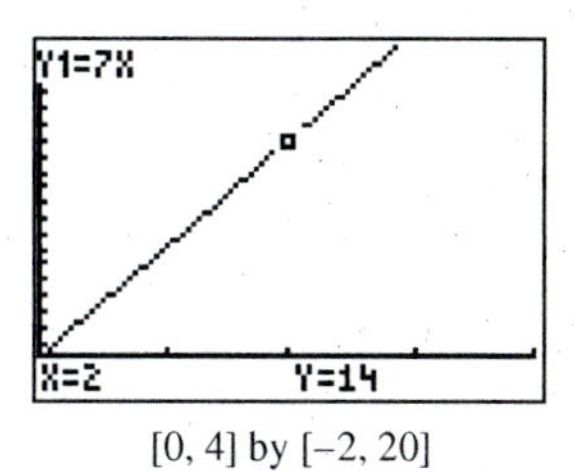

[0, 4] by [−2, 20]

A table of values also shows that the value of the function approaches 14 as x approaches 2 from either direction.

X	Y1
1.997	13.979
1.998	13.986
1.999	13.993
2	14
2.001	14.007
2.002	14.014
2.003	14.021
X=2	

2. Answers will vary. Possible answers include: The graphs suggest that the limit exists and is 2. Because the graph is a line with a discontinuity at $x = 0$, there is no asymptote at $x = 0$.

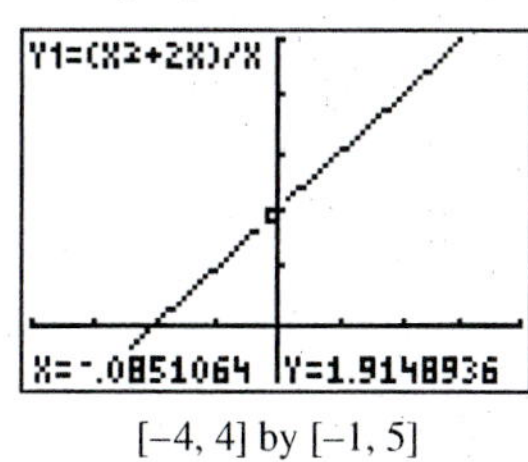

[−4, 4] by [−1, 5]

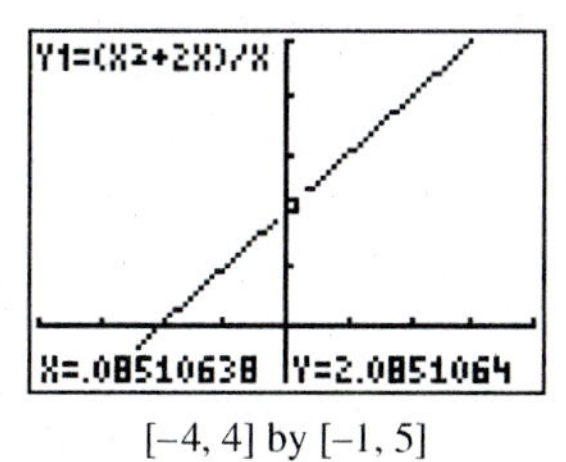

[−4, 4] by [−1, 5]

A table of values also suggests that the limit is 2.

X	Y1
-.003	1.997
-.002	1.998
-.001	1.999
0	ERROR
.001	2.001
.002	2.002
.003	2.003
X=0	

To show that 2 is the limit and $1.999\overline{9}$ is not, we can solve algebraically.

$$\lim_{x\to 0} \frac{x^2 + 2x}{x} = \lim_{x\to 0} \frac{x(x + 2)}{x} = \lim_{x\to 0} (x + 2) = 2$$

Exploration 2

1.

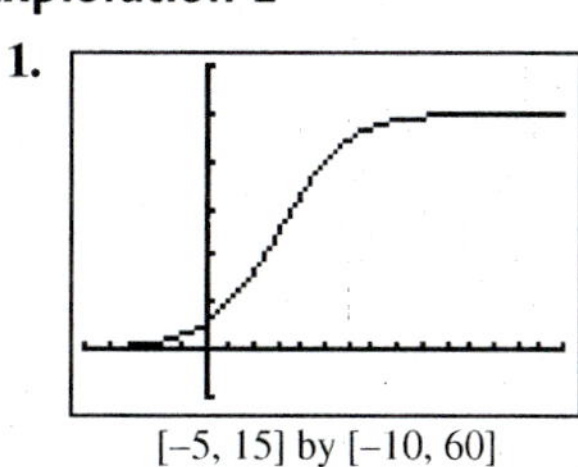

[−5, 15] by [−10, 60]

$\lim_{x\to\infty} f(x) = 50$, $\lim_{x\to -\infty} f(x) = 0$

2. The two horizontal asymptotes are $y = 50$ and $y = 0$.

3. As $x \to \infty$, $2^{3-x} \to 0$ and $1 + 2^{3-x} \to 1$.
As $x \to -\infty$, $2^{3-x} \to \infty$ and $1 + 2^{3-x} \to \infty$.

Quick Review 10.3

1. (a) $f(-2) = \dfrac{-4+1}{(-4-4)^2} = -\dfrac{3}{64}$,

(b) $f(0) = \dfrac{0+1}{(0-4)^2} = \dfrac{1}{16}$,

(c) $f(2) = \dfrac{4+1}{(4-4)^2}$ is undefined.

3. (a) Since $x^2 - 4 = 0$ and $x = \pm 2$, the graph of f has vertical asymptotes at $x = -2$ and $x = 2$.

(b) Since $\lim_{x \to -\infty} f(x) = 2$ and $\lim_{x \to \infty} f(x) = 2$, the graph of f has a horizontal asymptote of $y = 2$.

5. Since $\dfrac{2x^3}{-x} = -2x^2$, the end behavior asymptote is

(b) $y = -2x^2$.

7. (a) $[-2, \infty)$

(b) None

9.

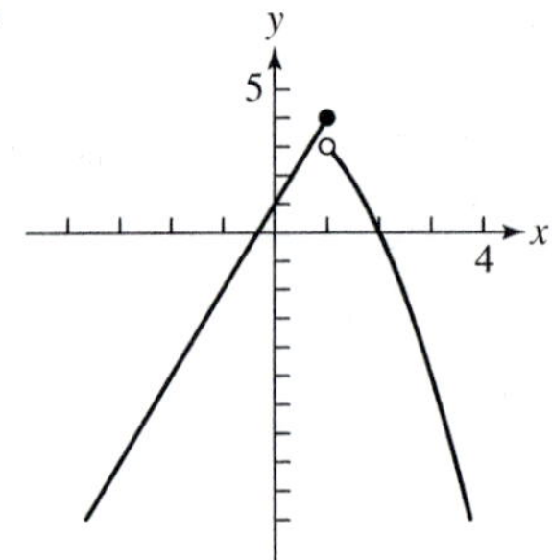

Section 10.3 Exercises

1. $(-1)(-2)^2 = -4$

3. $8 - 4 + 3 = 7$

5. $\sqrt{7}$

7. $\lim_{x \to 0} (e^x \sin(x)) = \lim_{x \to 0} e^x \lim_{x \to 0} \sin(x) = 1 \cdot 0 = 0$

9. $a^2 - 2$

11. (a) division by zero

(b) $\lim_{x \to -3} \dfrac{x^2 + 7x + 12}{x^2 - 9} = \lim_{x \to -3} \dfrac{(x+4)(x+3)}{(x+3)(x-3)}$
$= \lim_{x \to -3} \dfrac{x+4}{x-3} = -\dfrac{1}{6}$

13. (a) division by zero

(b) $\lim_{x \to -1} \dfrac{(x+1)(x^2 - x + 1)}{x+1} = \lim_{x \to -1} (x^2 - x + 1) = 3$

15. (a) division by zero

(b) $\lim_{x \to -2} \dfrac{(x+2)(x-2)}{x+2} = \lim_{x \to -2} (x - 2) = -4$

17. (a) The square root of negative numbers is not defined in the real plane.

(b) The limit does not exist.

19. $\lim_{x \to 0} \dfrac{\sin x}{2x^2 - x} = \lim_{x \to 0} \dfrac{\sin x}{x(2x-1)} = \lim_{x \to 0} \dfrac{\sin x}{x} \cdot \lim_{x \to 0} \dfrac{1}{2x - 1}$
$= 1 \cdot -1 = -1$ (Recall example 11 and the product rule)

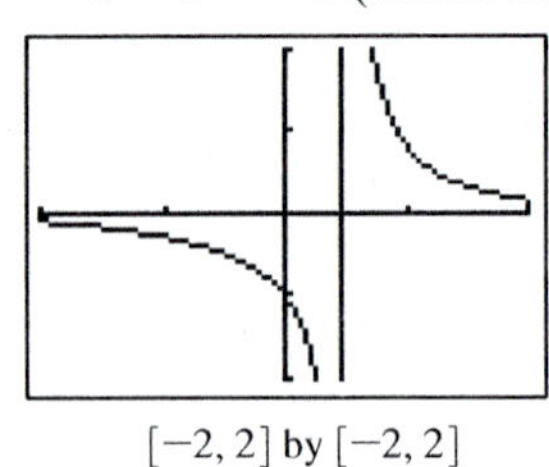

$[-2, 2]$ by $[-2, 2]$

21. $\lim_{x \to 0} \dfrac{\sin^2 x}{x} = \lim_{x \to 0} \dfrac{\sin x}{x} \cdot \sin x = \lim_{x \to 0} \sin x \cdot \lim_{x \to 0} \dfrac{\sin x}{x}$
$= 0 \cdot 1 = 0$

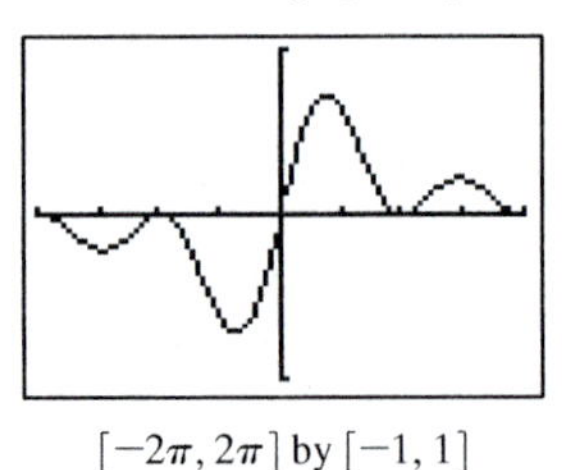

$[-2\pi, 2\pi]$ by $[-1, 1]$

In Exercises #23–25, the function is defined and continuous at the value approached by x, and so the limit is simply the function evaluated at that value.

23. $\lim_{x \to 0} \dfrac{e^x - \sqrt{x}}{\log_4(x+2)} = \dfrac{e^0 - \sqrt{0}}{\log_4(0+2)} = \dfrac{1}{1/2} = 2$

25. $\lim_{x \to \pi/2} \dfrac{\ln(2x)}{\sin^2 x} = \dfrac{\ln \pi}{\sin^2(\pi/2)} = \dfrac{\ln \pi}{1} = \ln \pi$

27. (a) $\lim_{x \to 2^-} f(x) = 3$

(b) $\lim_{x \to 2^+} f(x) = 1$

(c) $3 \neq 1$, so the limit does not exist.

29. (a) $\lim_{x \to 3^-} f(x) = 4$

(b) $\lim_{x \to 3^+} f(x) = 4$

(c) $\lim_{x \to 3} f(x) = 4$

31. (a) True

(b) True

(c) False

(d) False

(e) False

(f) False

(g) False

(h) True

(i) False

(j) True

33.

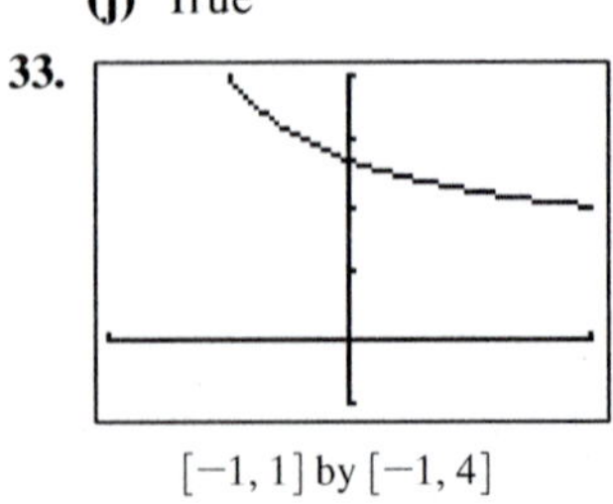

$[-1, 1]$ by $[-1, 4]$

(a) $\lim_{x\to 0^-} f(x) \approx 2.72$

(b) $\lim_{x\to 0^+} f(x) \approx 2.72$

(c) $\lim_{x\to 0} f(x) \approx 2.72$

35. (a) $\lim_{x\to 4} (g(x) + 2) = 4 + 2 = 6$ (sum rule)

(b) $\lim_{x\to 4} x \cdot f(x) = \lim_{x\to 4} x \cdot \lim_{x\to 4} f(x) = 4(-1) = -4$ (product rule)

(c) $\lim_{x\to 4} g^2(x) = \lim_{x\to 4} g(x) \cdot \lim_{x\to 4} g(x) = 4 \cdot 4 = 16$ (product rule)

(d) $\lim_{x\to 4} \dfrac{g(x)}{f(x) - 1} = \dfrac{\lim_{x\to 4} g(x)}{\lim_{x\to 4} f(x) - \lim_{x\to 4} 1} = \dfrac{4}{-1 - 1} = -2$ (quotient rule)

37. (a)

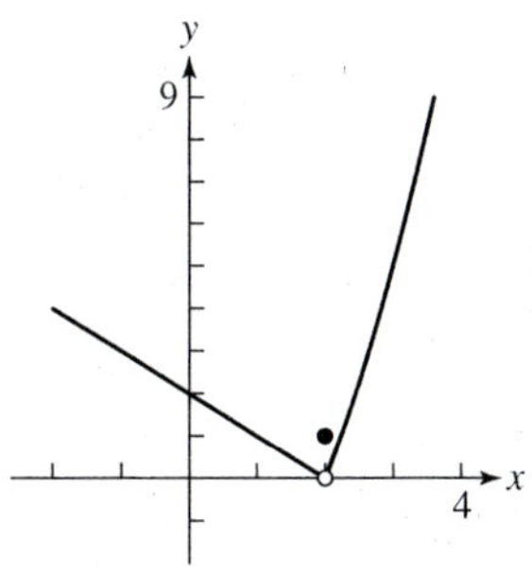

(b) $\lim_{x\to 2^+} f(x) = 0$, $\lim_{x\to 2^-} f(x) = 0$

(c) $\lim_{x\to 2} f(x) = 0$

39. (a)

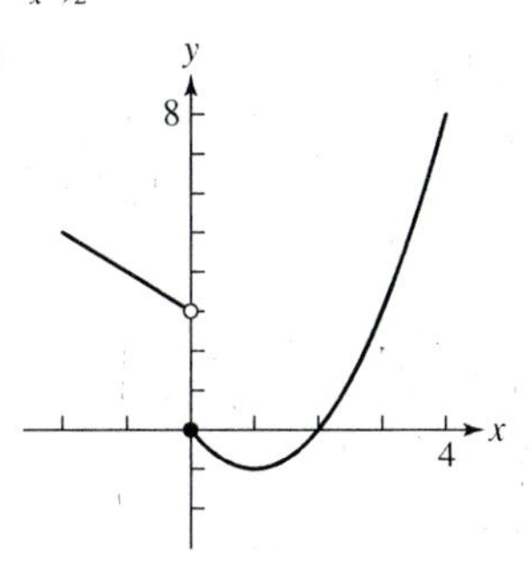

(b) $\lim_{x\to 0^+} f(x) = 0$, $\lim_{x\to 0^-} f(x) = 3$

(c) Limit does not exist because $\lim_{x\to 0^+} f(x) \neq \lim_{x\to 0^-} f(x)$.

For Exercises #41–43, use Figure 10.14.

41. $\lim_{x\to 2^+} \text{int } x = 2$

43. $\lim_{x\to 0.0001} \text{int } x = 0$

45. $\lim_{x\to -3^+} \dfrac{x + 3}{|x + 3|} = \lim_{x\to -3^+} \dfrac{x + 3}{x + 3} = 1$

47. (a) $\lim_{x\to\infty} \dfrac{\cos x}{1 + x} = 0$

(b) $\lim_{x\to -\infty} \dfrac{\cos x}{x} = 0$

49. (a) $\lim_{x\to\infty} (1 + 2^x) = \infty$

(b) $\lim_{x\to -\infty} (1 + 2^x) = 1$

51. (a) $\lim_{x\to\infty} (x + \sin x) = \infty$

(b) $\lim_{x\to -\infty} (x + \sin x) = -\infty$

53. (a) $\lim_{x\to\infty} (-e^x \sin x)$ is undefined, because $\sin x$ oscillates between positive and negative values.

(b) $\lim_{x\to -\infty} (-e^x \sin x) = 0$

55. $\lim_{x\to 3^-} \dfrac{1}{x - 3} = -\infty$; $x = 3$

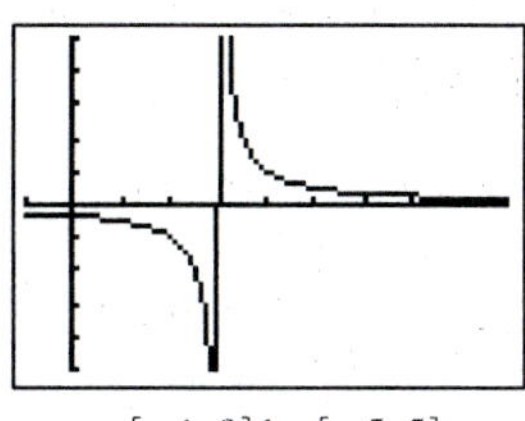

$[-1, 9]$ by $[-5, 5]$

57. $\lim_{x\to -2^+} \dfrac{1}{x + 2} = \infty$; $x = -2$

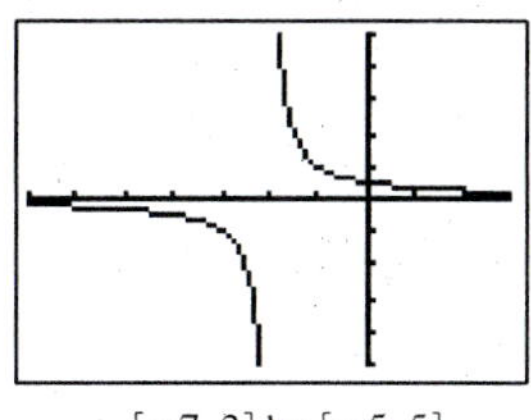

$[-7, 3]$ by $[-5, 5]$

59. $\lim_{x\to 2} \dfrac{1}{(x - 5)^5} = \infty$; $x = 5$

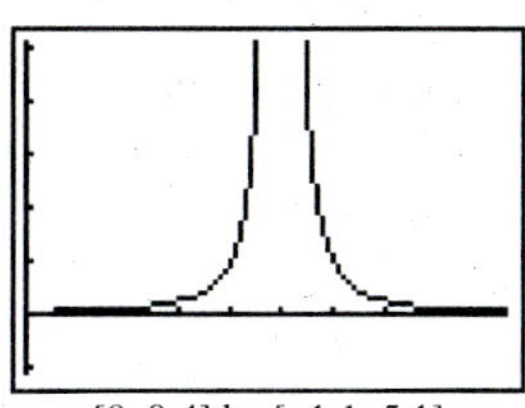

$[0, 9.4]$ by $[-1.1, 5.1]$

61. $\lim_{x\to 0} \dfrac{(1 + x)^3 - 1}{x} = \lim_{x\to 0} \dfrac{x^3 + 3x^2 + 3x + 1 - 1}{x}$

$= \lim_{x\to 0} \dfrac{x(x^2 + 3x + 3)}{x}$

$= \lim_{x\to 0} (x^2 + 3x + 3) = 3$

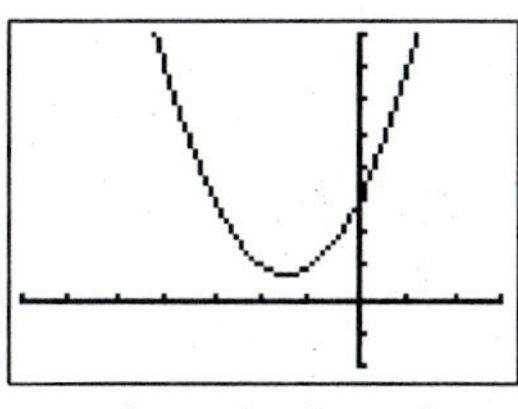

$[-7, 3]$ by $[-2, 8]$

63. $\lim_{x\to 0} \dfrac{\tan x}{x} = \lim_{x\to 0} \dfrac{\sin x}{x \cos x} = \lim_{x\to 0} \dfrac{\sin x}{x} \cdot \lim_{x\to 0} \dfrac{1}{\cos x}$

$= 1 \cdot 1 = 1$

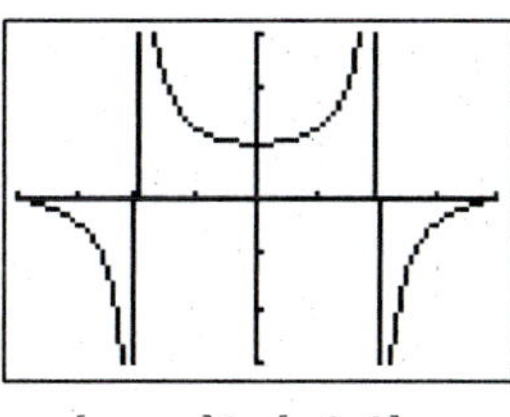

$[-\pi, \pi]$ by $[-3, 3]$

65. $\lim_{x \to 0} \frac{|x|}{x^2} = \lim_{x \to 0} \left|\frac{x}{x^2}\right| = \lim_{x \to 0} \left|\frac{1}{x}\right| = \infty$

67. $\lim_{x \to 0} \left[x \sin\left(\frac{1}{x}\right)\right] = 0$ because $x \to 0$

and $-1 \le \sin\left(\frac{1}{x}\right) \le 1$.

69. $\lim_{x \to 1} \frac{x^2+1}{x-1}$ is undefined, since $\lim_{x \to 1^-} \frac{x^2+1}{x-1} = -\infty$

and $\lim_{x \to 1^+} \frac{x^2+1}{x-1} = \infty$

71. $\lim_{x \to \infty} \frac{\ln x}{\ln x^2} = \lim_{x \to \infty} \frac{\ln x}{2 \ln x} = \frac{1}{2}$

73. False. $\lim_{x \to 3} f(x) = \lim_{x \to 3^-} f(x) = \lim_{x \to 3^+} f(x) = 5$

75. $\lim_{x \to 3} \frac{x^2 - 2x - 3}{x - 3} = \lim_{x \to 3} \frac{(x+1)(x-3)}{x-3}$

$= \lim_{x \to 3} (x + 1) = 4$. The answer is B.

77. $\lim_{x \to 3^-} \frac{x^2 - 2x - 9}{x - 3} = \infty$, $\lim_{x \to 3^+} \frac{x^2 - 2x - 9}{x - 3} = -\infty$.
The answer is C.

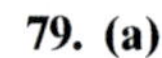

79. (a)

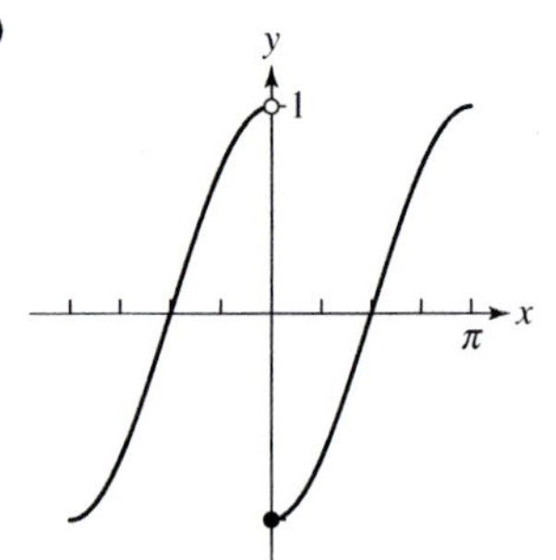

(b) $(-\pi, 0) \cup (0, \pi)$

(c) $x = \pi$

(d) $x = -\pi$

81. (a)

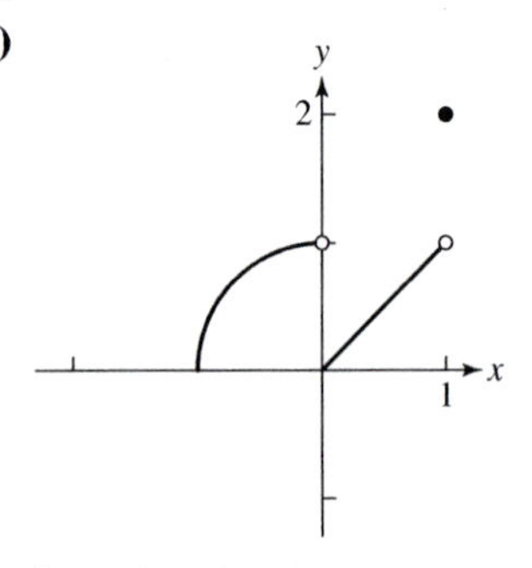

(b) $(-1, 0) \cup (0, 1)$

(c) $x = 1$

(d) $x = -1$

83. (a)

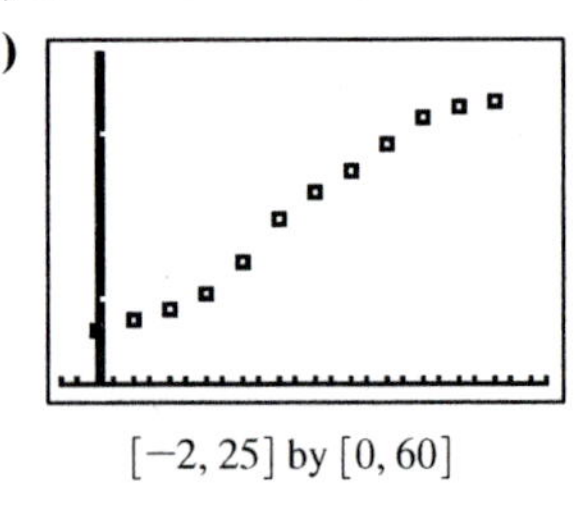

$[-2, 25]$ by $[0, 60]$

(b) $f(x) \approx \frac{57.71}{1 + 6.39e^{-0.19x}}$, where x = the number of months $\lim_{x \to \infty} f(x) \approx 57.71$

(c) The rabbit population will stabilize at a little less than 58,000.

(d) One possible answer: As populations burgeon, resources such as food, water, and safe havens from predators become more scarce and the population tends to stabilize based on the resources available to it—this is what is often call a maximum sustainable population.

85.

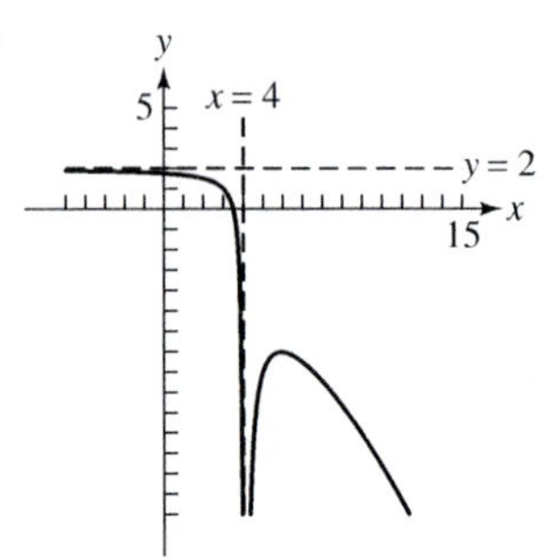

87.

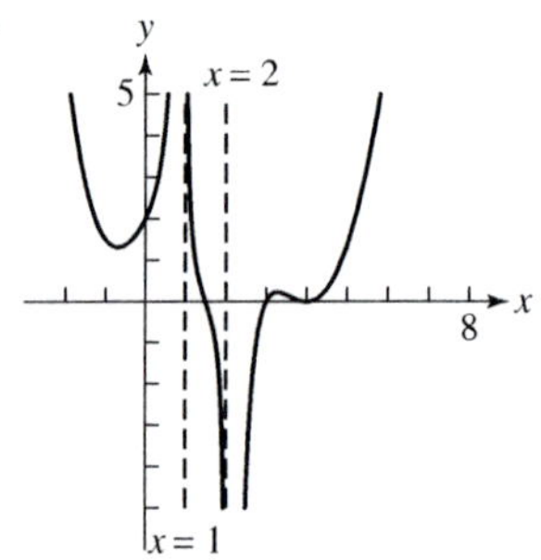

89. (a) For an 8-sided polygon, we have 8 isosceles triangles of area $\frac{1}{2}bh$. Thus, $A = 8 \cdot \frac{1}{2}bh = 4bh$. Similarly, for an n-sided polygon, we have n triangles of area $\frac{1}{2}bh$.

Thus $A = n \cdot \frac{1}{2} \cdot bh = \frac{1}{2}nhb$.

(b) Consider an n-sided polygon inscribed in a circle of radius r. Since a circle always is 360°, we see that each angle extending from the center of the circle to two consecutive vertices is an angle of $\frac{360°}{n}$. Dropping a perpendicular from the center of the circle to the midpoint of the base of the triangle (which is also one of the n sides) results in an angle of $\frac{360°}{2n}$. Since $\tan\left(\frac{360°}{2n}\right) = \frac{(b/2)}{h}$, we have $\frac{b}{2} = h \tan\left(\frac{360°}{2n}\right)$ and finally $b = 2h \tan\left(\frac{360°}{2n}\right)$.

(c) Since $A = \frac{1}{2}nhb$ and $b = 2h \tan\left(\frac{360°}{2n}\right)$, we have

$A = \frac{1}{2}nh\left(2h \tan\left(\frac{360°}{2n}\right)\right) = nh^2 \tan\left(\frac{360°}{2n}\right)$.

(d)

n	A
4	4
8	3.3137
16	3.1826
100	3.1426
500	3.1416
1,000	3.1416
5,000	3.1416
10,000	3.1416
100,000	3.1416

Yes, $A \to \pi$ as $n \to \infty$.

(e)

n	A
4	36
8	29.823
16	28.643
100	28.284
500	28.275
1,000	28.274
5,000	28.274
10,000	28.274
100,000	28.274

Yes, $n \to \infty$, $A \to 9\pi$.

(f) One possible answer:

$$\lim_{n\to\infty} A = \lim_{n\to\infty} nh^2 \tan\left(\frac{180°}{n}\right) = h^2 \lim_{n\to\infty} n \tan\left(\frac{180°}{n}\right)$$
$$= h^2\pi = \pi h^2$$

As the number of sides of the polygon increases, the distance between h and the edge of the circle becomes progressively smaller. As $n \to \infty$, $h \to$ radius of the circle.

91. (a)

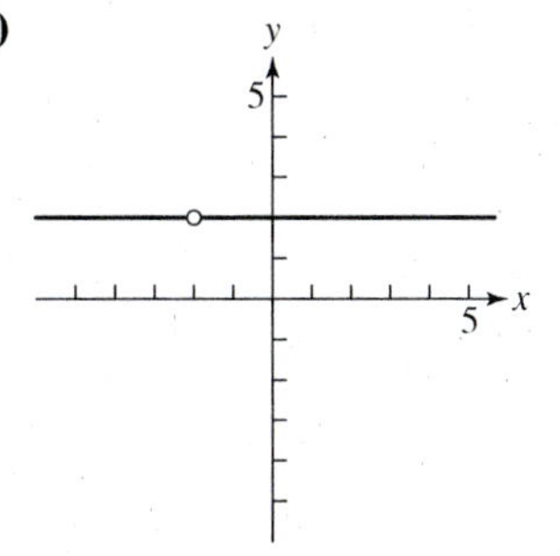

(b) $f(x) = \dfrac{2x + 4}{x + 2} = \dfrac{2(x + 2)}{x + 2} = 2$

(c) $g(x) = 2$

93. (a)

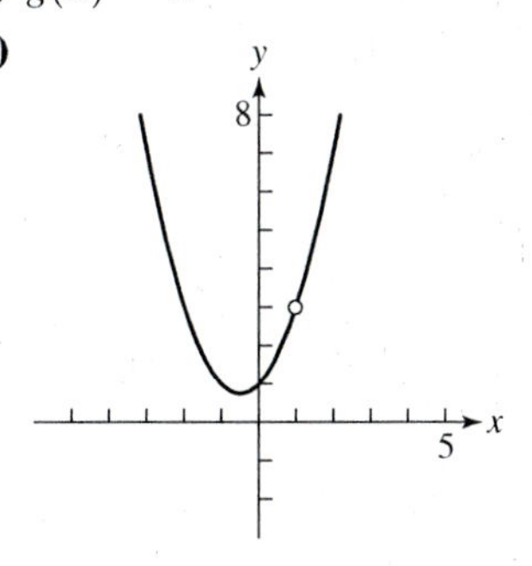

(b) $\dfrac{x^3 - 1}{x - 1} = \dfrac{(x - 1)(x^2 + x + 1)}{x - 1} = x^2 + x + 1$

(c) $g(x) = x^2 + x + 1$

Section 10.4 Numerical Derivatives and Integrals

Exploration 1

1. RRAM value $\approx$ 1.364075504 and the NINT value $\approx$ 1.386294361.
2. The new command is sum(seq(1/(1 + $K \cdot$ 3/100) $\cdot$ 3/100, K, 1, 100)). The calculated value $\approx$ 1.3751, which is a better approximation than for 50 rectangles.
3. The integral is $\int_0^{\pi} \sin x \, dx$. The RRAM value is $\approx$1.999342 and the NINT value is 2.
4. The command is sum(seq($\sqrt{}$(4 + $K \cdot$ 5/50) $\cdot$ 5/50, K, 1, 50)). The calculated value $\approx$ 12.7166 and the NINT value is 12.666667.

Quick Review 10.4

1. $\dfrac{\Delta y}{\Delta x} = \dfrac{4^2 - 1^2}{4 - 1} = \dfrac{15}{3} = 5$

3. $\dfrac{\Delta y}{\Delta x} = \dfrac{\log_2 4 - \log_2 1}{4 - 1} = \dfrac{2 - 0}{3} = \dfrac{2}{3}$

5. $\dfrac{\Delta y}{\Delta x} = \dfrac{11 - 2}{4 - 1} = \dfrac{9}{3} = 3$

7. $\dfrac{\sin(1.01) - \sin(0.99)}{2(0.01)} \approx 0.5403$

9. $\dfrac{\ln 1.001 - \ln 0.999}{2(0.001)} \approx 1.000$

Exercises 10.4

In #1–9, use NDER on a calculator to find the numerical derivative of the function at the specific point.

1. −4

3. −12

5. 0

7. $\approx$ 1.0000

9. $\approx$ −3.0000

In #11–21, use NINT on a calculator to find the numerical integral of the function over the specified interval.

11. $\dfrac{64}{3}$

13. 2

15. $\approx$ 0

17. 1

19. $\approx$ 3.1416

21. $\approx$ 106.61 mi

23. (a) $v_{ave} = \dfrac{435 - 485}{2 - 1} = \dfrac{-50}{1} = -50$ ft/sec

(b)

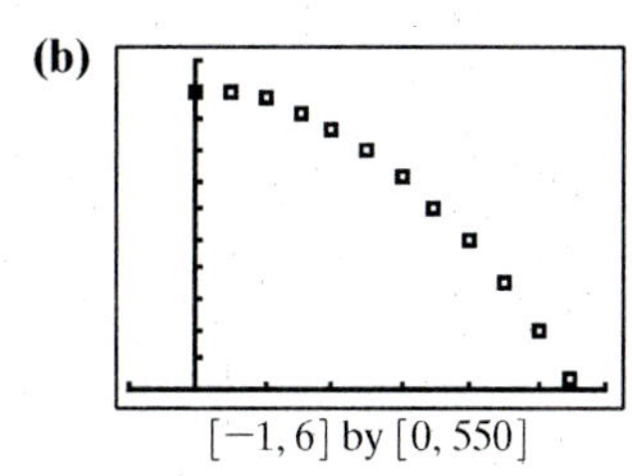

[−1, 6] by [0, 550]

(c) $s(t) \approx -16.08t^2 + 0.36t + 499.77$

(d) $v(1.5 \text{ sec}) \approx -47.88$ ft/sec

(e) Set $s(t)$ equal to zero and solve for t using the quadratic equation.

$$t = \frac{-0.36 - \sqrt{0.36^2 - 4(-16.08)(499.77)}}{2(-16.08)}$$

≈ 5.586 sec (The minus sign was chosen to give $t \geq 0$.) Using NDER at $t = 5.586$ sec gives $v \approx -179.28$ ft/sec.

25. (a) The midpoints of the subintervals will be 0.25, 0.75, 1.25, etc. The average velocities will be the successive height differences divided by 0.5—that is, times 2.

Midpoint	$\Delta s/\Delta t$
0.25	−10 ft/sec
0.75	−20
1.25	−40
1.75	−60
2.25	−70
2.75	−90
3.25	−100
3.75	−120
4.25	−140
4.75	−150
5.25	−170

(b)

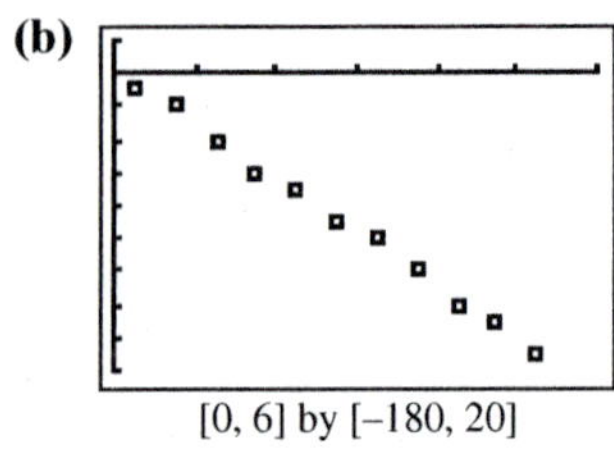

[0, 6] by [−180, 20]

$y \approx -32.18t + 0.32$

(c) Substituting $t = 1.5$ leads to $y \approx -47.95$ ft/sec. This is close to the value of -47.88 ft/sec found in Exercise 23d.

27. The average velocities, $\Delta s/\Delta t$, for the successive 0.5-second intervals are 8, 24, 40, 56, and 72 ft/sec. Multiplying each by 0.5 sec and then summing them gives the estimated distance: 100 ft.

29. The program accepts inputs which determine the width and number of approximating rectangles. These rectangles are summed and the result is output to the screen.

For #31–41, verify the function is non-negative by graphing it over the interval.

31. (b)

N	LRAM	RRAM	Average
10	15.04	19.84	17.44
20	16.16	18.56	17.36
50	16.86	17.82	17.34
100	17.09	17.57	17.33

(c) fnlnt gives 17.33; at N_{100}, the average is 17.3344.

33. (b)

N	LRAM	RRAM	Average
10	7.84	11.04	9.44
20	8.56	10.16	9.36
50	9.02	9.66	9.34
100	9.17	9.49	9.33

(c) fnlnt gives 9.33; at N_{100}, the average is 9.3344.

35. (b)

N	LRAM	RRAM	Average
10	98.24	112.64	105.44
20	101.76	108.96	105.36
50	103.90	106.78	105.34
100	104.61	106.05	105.33

(c) fnlnt gives 105.33, at N_{100}, the average is 105.3344.

37. (b)

N	LRAM	RRAM	Average
10	7.70	8.12	7.91
20	7.81	8.02	7.91
50	7.87	7.95	7.91
100	7.89	7.93	7.91

(c) fnlnt gives 7.91, the same result as N_{100}.

39. (b)

N	LRAM	RRAM	Average
10	1.08	0.92	1.00
20	1.04	0.96	1.00
50	1.02	0.98	1.00
100	1.01	0.99	1.00

(c) fnlnt gives 1, the same result as N_{100}.

41. (b)

N	LRAM	RRAM	Average
10	0.56	0.62	0.59
20	0.58	0.61	0.593
50	0.59	0.60	0.594
100	0.59	0.60	0.594

(c) fnlnt $=$ 0.594, the same result as N_{100}.

43. True. The notation NDER refers to a symmetric difference quotient using $\Delta x = h = 0.001$.

45. NINT will use as many rectangles as are needed to obtain an accurate estimate. The answer is B. (Note: NDER estimates the derivative, not the integral.)

47. Instantaneous velocity is the derivative, not an integral, of the position function. The answer is C.

49. (a) $f'(x)$

$$= \frac{2(x+h)^2 + 3(x+h) + 1 - 2x^2 - 3x - 1}{h}$$

$$= \lim_{h \to 0} \frac{2(x^2 + 2xh + h^2) + 3x + 3h - 2x^2 - 3x}{h}$$

$$= \lim_{h \to 0} \frac{4xh + 2h^2 + 3h}{h} = \lim_{h \to 0} 4x + 2h + 3$$

$$= 4x + 3$$

(b) $$g'(x) = \lim_{h \to 0} \frac{(x+h)^3 + 1 - x^3 - 1}{h}$$

$$= \lim_{h \to 0} \frac{3x^2h + 3xh^2 + h^3}{h}$$

$$= \lim_{h \to 0} 3x^2 + 3xh + h^2 = 3x^2$$

(c) Standard: $$\frac{f(2.001) - f(2)}{0.001} = \frac{15.011002 - 15}{0.001} = 11.002$$

Symmetric: $$\frac{f(2.001) - f(1.999)}{0.002}$$

$$= \frac{15.011002 - 14.989002}{0.002} = 11$$

(d) The symmetric method provides a closer approximation to $f'(2) = 11$.

(e) Standard: $\dfrac{g(2.001) - g(2)}{0.001} \approx \dfrac{9.012006 - 9}{0.001}$

$= 12.006001$

Symmetric: $\dfrac{g(2.001) - g(1.999)}{2(0.001)} \approx 12.000001$

The symmetric method provides a closer approximation to $g'(2) = 12$.

51. $f'(0) = \lim\limits_{h \to 0} \dfrac{f(0 + h) - f(0)}{h} = \lim\limits_{h \to 0} \dfrac{|h| - 0}{h}$

$= \lim\limits_{h \to 0} \dfrac{|h|}{h} = 1$

if 0 is approached from the right, and -1 if 0 is approached from the left. This occurs because calculators tend to take average values for derivatives instead of applying the definition. For example, a calculator may calculate the derivative of $f(0)$ by taking

$\dfrac{f(0.0001) - f(-0.0001)}{0.0001 - (-0.0001)} = 0.$

53. (a) Let $y_1 = \text{abs}(\sin(x))$, which is $|f(x)|$. Then NINT $(Y_1, X, 0, 2\pi)$ gives 4.

(b) Let $y_1 = \text{abs}(x^2 - 2x - 3)$, which is $|f(x)|$. Then NINT $(Y_1, X, 0, 5)$ gives ≈ 19.67.

55. Since $f(x) \geq g(x)$ for all values of x on the interval,

$$A = \lim_{N \to \infty} \sum_{k=1}^{n} \left\{ \left[\left(\frac{b - a}{N} \right) f\left(a + \frac{k(b - a)}{N} \right) \right] - \left[\left(\frac{b - a}{N} \right) g\left(a + \frac{k(b - a)}{N} \right) \right] \right\}$$

$$= \lim_{N \to \infty} \sum_{k=1}^{n} \left(\frac{b - a}{N} \right) \left[f\left(a + \frac{k(b - a)}{N} \right) - g\left(a + \frac{k(b - a)}{N} \right) \right].$$

If the area of both curves is already known and $f(x) \geq g(x)$ for all values of x, the area between the curves is simply the area under f minus the area under g.

57. (b)

x	$A(x)$
0.25	0.0156
0.5	0.125
1	1
1.5	3.375
2	8
2.5	15.625
3	27

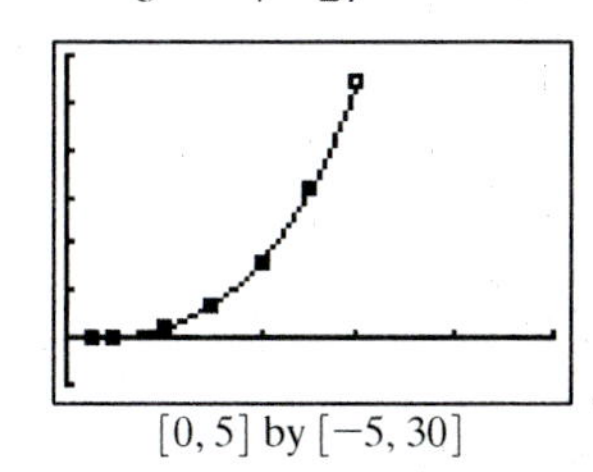

$[0, 5]$ by $[-5, 30]$

(c) $f(x) \approx x^3$

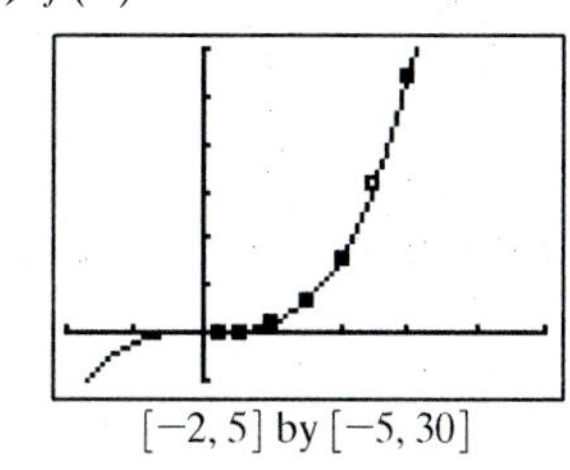

$[-2, 5]$ by $[-5, 30]$

(d) The exact value of $A(x)$ for any x greater than zero appears to be x^3.

(e) $A'(x) = \lim\limits_{h \to 0} \dfrac{f(x + h) - f(x)}{h} = \lim\limits_{h \to 0} \dfrac{(x + h)^3 - x^3}{h}$

$= \lim\limits_{h \to 0} \dfrac{3x^2h + 3xh^2}{h} = 3x^2.$

The functions are exactly the same.

Chapter 10 Review

1. (a) 2

(b) Does not exist.

3. (a) 2

(b) 2

5. $\lim\limits_{x \to -1} \dfrac{x - 1}{x^2 + 1} = \dfrac{-1 - 1}{2} = -1$

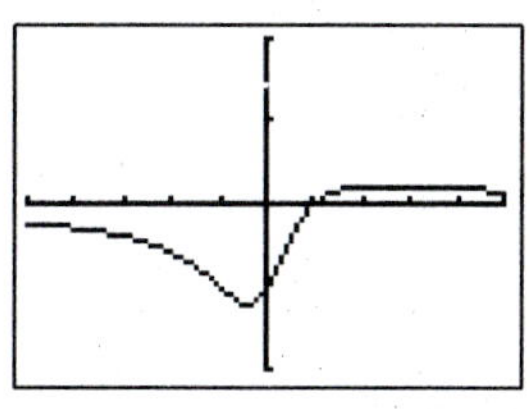

$[-5, 5]$ by $[-2, 2]$

7. $\lim\limits_{x \to -2} \dfrac{(x - 5)(x + 2)}{x + 2} = \lim\limits_{x \to -2} x - 5 = -7$

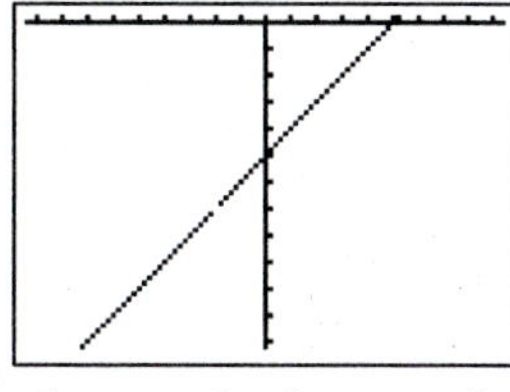

$[-9.4, 9.4]$ by $[-12.2, 0.2]$

9. $\lim\limits_{x \to 0} 2 \tan^{-1}(x) = \lim\limits_{x \to 0} 2\dfrac{\sin^{-1}(x)}{\cos^{-1}(x)} = \dfrac{2 \cdot 0}{1} = 0$

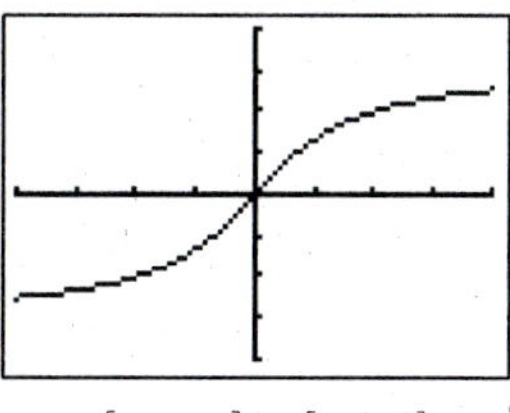

$[-\pi, \pi]$ by $[-4, 4]$

11.

x	$x(f)$
-3	-1
-10	$-\frac{1}{64}$
-100	$-\frac{1}{9604}$
-1000	$-\frac{1}{996{,}004}$

$\Rightarrow \lim_{x \to -\infty} f(x) = 0$

13.

x	$f(x)$
1	2
2	-1
5	-4.6
10	-9.8
100	-99.98
1000	-999.998

$\Rightarrow \lim_{x \to \infty} f(x) = -\infty$

15. $\lim_{x \to 2^+} \frac{1}{x-2} = \infty$

17. $\lim_{x \to 0} \frac{\frac{1}{2+x} - \frac{1}{2}}{x} = \lim_{x \to 0} \frac{2-(2+x)}{2(2+x)} \cdot \frac{1}{x}$

$= \lim_{x \to 0} \frac{-x}{x} \cdot \frac{1}{2(2+x)} = -\frac{1}{4}$

19. $f(x) = \frac{x-5}{(x+5)(x+1)}$, so f has vertical asymptotes at $x = -1$ and $x = -5$. Since $\lim_{x \to \infty} f(x) = 0$ and $\lim_{x \to -\infty} f(x) = 0$, f also has a horizontal asymptote at $y = 0$.

21. $\lim_{x \to 3} \frac{(x+5)(x-3)}{3-x} = \lim_{x \to 3} -x - 5 = -8.$

23. $\lim_{x \to 0} \frac{\frac{1}{x-3} + \frac{1}{3}}{x} = \lim_{x \to 0} \frac{3 + (x-3)}{3(x-3)} \cdot \frac{1}{x}$

$= \lim_{x \to 0} \frac{x}{x} \cdot \frac{1}{3(x-3)} = -\frac{1}{9}$

25. $\lim_{x \to 2} \frac{(x-2)(x-3)}{(x-2)(x-1)} = \lim_{x \to 2} \frac{x-3}{x-1} = -\frac{1}{1} = -1$

27. $\lim_{x \to 1} \frac{(x-1)(x^2+x+1)}{x-1} = \lim_{x \to 1} (x^2+x+1) = 3$

$$F(x) = \begin{cases} \frac{x^3-1}{x-1} & x \neq 1 \\ 3 & x = 1 \end{cases}$$

29. $f'(x) = \lim_{h \to 0} \frac{f(2+h) - f(2)}{h}$

$= \lim_{h \to 0} \frac{1 - (h+2) - 2(h+2)^2 - (-9)}{h}$

$= \lim_{h \to 0} \frac{-2h^2 - 8h - 8 - h + 8}{h} = \lim_{h \to 0} \frac{-2h^2 - 9h}{h}$

$= \lim_{h \to 0} (-2h - 9) = -9$

31. (a) $\frac{f(3.01) - f(3)}{3.01 - 3} = \frac{12.0801 - 12}{0.01} = 8.01$

(b) $\lim_{h \to 0} \frac{f(3+h) - f(3)}{h}$

$= \lim_{h \to 0} \frac{(h+3)^2 + 2(h+3) - 3 - 12}{h}$

$= \lim_{h \to 0} \frac{h^2 + 6h + 9 + 2h + 6 - 15}{h}$

$= \lim_{h \to 0} \frac{h^2 + 8h}{h} = 8$

33. (a) $m = \lim_{h \to 0} \frac{f(h+1) - f(1)}{h}$

$= \lim_{h \to 0} \frac{(h+1)^3 - 2(h+1) + 1 - 0}{h}$

$= \lim_{h \to 0} \frac{h^3 + 3h^2 + h}{h} = \lim_{h \to 0} h^2 + 3h + 1 = 1$

(b) $(1, f(1)) = (1, 0)$ so the equation of the tangent line at $x = 1$ is $y = x - 1$.

35. $\lim_{h \to 0} \frac{f(x+h) - f(x)}{h}$

$= \lim_{h \to 0} \frac{5(x+h)^2 + 7(x+h) - 1 - 5x^2 - 7x + 1}{h}$

$= \lim_{h \to 0} \frac{5x^2 + 10xh + 5h^2 + 7x + 7h - 5x^2 - 7x}{h}$

$= \lim_{h \to 0} \frac{10xh + 5h^2 + 7h}{h} = 10x + 7$

For #37, verify the function is non-negative through graphical or numerical analysis.

37. (b) LRAM: 42.2976
RRAM: 40.3776
Average: $\frac{42.2976 + 40.3776}{2} = 41.3376$

39. (a) Using $x = 0$ for 1990, the scatter plot of the data is:

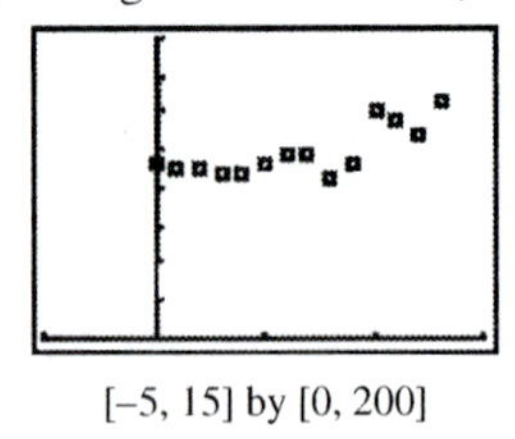

[–5, 15] by [0, 200]

(b) The average rate of change for 1990 to 1991 is found by examining $\frac{\Delta x}{\Delta y}$.

$\frac{\Delta y}{\Delta x} = \frac{114.0 - 116.4}{1 - 0} = -2.4$ cents per year.

The average rate of change for 1997 to 1998 is found by examining $\frac{\Delta x}{\Delta y}$.

$\frac{\Delta y}{\Delta x} = \frac{105.9 - 123.4}{1 - 0} = -17.5$ cents per year.

(c) The average rate exhibits the greatest increase from one year to the next consecutive year in the interval from 1999–2000.

(d) The average rate exhibits the greatest decrease from one year to the next consecutive year in the interval from 1997–1998.

(e) The linear regression model for the data is:
$y = 3.0270x + 104.6600$.

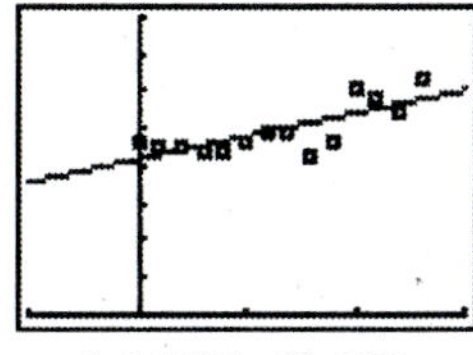

[−5, 15] by [0, 200]

(f) The cubic regression model for the data is:
$y = 0.0048x^3 + 0.3659x^2 - 2.4795x + 116.2006$.

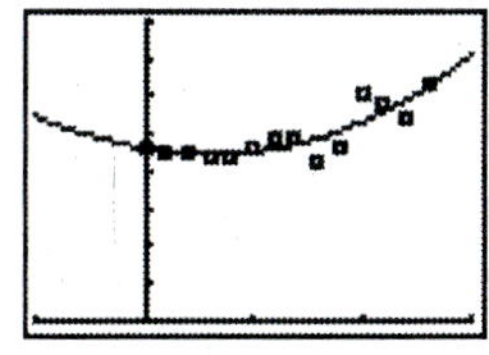

[−5, 15] by [0, 200]

Most of the data points touch the regression curve, which would suggest that this is a fairly good model. One possible answer: The cubic regression model is the best. It shows a larger rate of increase in fuel prices, which is what is happening in today's market.

(g) The cubic regression model for the data is:
$y = 0.0048x^3 + 0.3659x^2 - 2.4795x + 116.2006$.
Using NDER with this regression model yields the following instantaneous rates of change for:

1997	3.4 cents per gallon
1998	4.3 cents per gallon
1999	5.3 cents per gallon
2000	6.3 cents per gallon

The cubic regression model is showing a rate of increase. The true rate of change, however, may be higher than what the model indicates.

(h) If we use the given cubic regression model, the average price of a gallon of regular gas in 2007 will be 203.4 cents per gallon. Based on what is happening with the current fuel prices, this prediction may not be high enough.

Chapter 10 Project

1. The scatter plot of the population data for Clark County, NV, is as follows. The year 1970 is represented by $t = 0$.

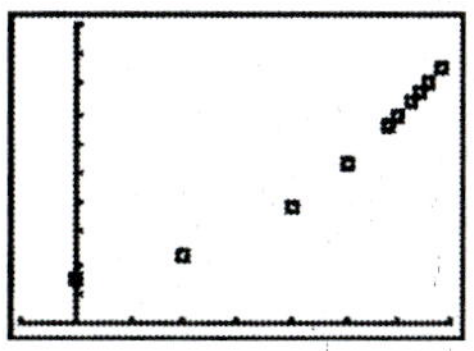

[−5, 35] by [0, 2000000]

2. The average population growth rate from 1970–2004 is
$$\frac{1{,}715{,}337 - 277{,}230}{2004 - 1970} = \frac{1{,}438{,}107}{34} = 42{,}297 \text{ people/year.}$$
The average population growth rate from 1980–2004 is
$$\frac{1{,}715{,}337 - 463{,}087}{2004 - 1980} = \frac{1{,}252{,}250}{24} = 52{,}177 \text{ people/year.}$$
The average population growth rate from 1990–2004 is
$$\frac{1{,}715{,}337 - 770{,}280}{2004 - 1990} = \frac{945{,}057}{14} = 67{,}504 \text{ people/year.}$$
The average population growth rate from 1995–2004 is
$$\frac{1{,}715{,}337 - 1{,}055{,}435}{2004 - 1995} = \frac{659{,}902}{9} = 73{,}322 \text{ people/year.}$$

3. The exponential regression model for the population data is $y = 271{,}661.8371 \cdot 1.0557797^t$.

4. If we use NDER with the given exponential regression model, the instantaneous population growth rate in 2004 is 93,359 people/year. The average growth rate from 1995–2004 most closely matches the instantaneous growth rate of 2004.

5. Using the exponential regression model $y = 271{,}661.8371 \cdot 1.0557797^t$ to predict the population of Clark County for the years 2010, 2020, and 2030 yields:
$y = 271{,}661.8371 \cdot 1.0557797^{40} = 2{,}382{,}105$ people
$y = 271{,}661.8371 \cdot 1.0557797^{50} = 4{,}099{,}152$ people
$y = 271{,}661.8371 \cdot 1.0557797^{60} = 7{,}053{,}864$ people
The web site predictions are probably more reasonable, since the given data and scatter plot suggest that growth in recent years has been fairly linear.

Appendix A

Section 1 Radicals and Rational Exponents

Section 1 Exercises

For Exercises 1–5, recall that there are two real nth roots if n is even, and only one if n is odd.

1. $\sqrt{81} = 9$ or -9, since $81 = (\pm 9)^2$

3. $\sqrt[3]{64} = 4$, since $64 = 4^3$

5. $\sqrt{\frac{16}{9}} = \frac{\sqrt{16}}{\sqrt{9}} = \frac{4}{3}$ or $-\frac{4}{3}$, since $\frac{16}{9} = \left(\pm\frac{4}{3}\right)^2$

7. $\sqrt{144} = 12$ since $12 \cdot 12 = 144$

9. $\sqrt[3]{-216} = -6$, since $(-6)^3 = -216$

11. $\sqrt[3]{-\frac{64}{27}} = -\frac{4}{3}$, since $\left(-\frac{4}{3}\right)^3 = -\frac{64}{27}$

13. 4

15. $\frac{5}{2}$ or 2.5

17. 729

19. $\frac{1}{4}$ or 0.25

21. -2

23. $\sqrt{1.69} = 1.3$ since $1.3^2 = 1.69$

25. $\sqrt[4]{19.4481} = 2.1$ since $2.1^4 = 19.4481$

27. $\sqrt{288} = \sqrt{12^2 \cdot 2} = \sqrt{12^2} \cdot \sqrt{2} = 12\sqrt{2}$

29. $\sqrt[3]{-250} = \sqrt[3]{(-5)^3 \cdot 2} = \sqrt[3]{(-5)^3} \cdot \sqrt[3]{2} = -5\sqrt[3]{2}$

31. $\sqrt{2x^3y^4} = \sqrt{(xy^2)^2 \cdot 2x}$
$= \sqrt{(xy^2)^2} \cdot \sqrt{2x} = |x|y^2\sqrt{2x}$

33. $\sqrt[4]{3x^8y^6} = \sqrt[4]{(x^2y)^4 \cdot 3y^2}$
$= \sqrt[4]{(x^2y)^4} \cdot \sqrt[4]{3y^2} = |x^2y|\sqrt[4]{3y^2} = x^2|y|\sqrt[4]{3y^2}$

35. $\sqrt[5]{96x^{10}} = \sqrt[5]{(2x^2)^5 \cdot 3} = \sqrt[5]{(2x^2)^5} \cdot \sqrt[5]{3} = 2x^2\sqrt[5]{3}$

37. $\frac{4}{\sqrt[3]{2}} \cdot \frac{\sqrt[3]{4}}{\sqrt[3]{4}} = \frac{4\sqrt[3]{4}}{\sqrt[3]{8}} = \frac{4\sqrt[3]{4}}{2} = 2\sqrt[3]{4}$

39. $\frac{1}{\sqrt[5]{x^2}} \cdot \frac{\sqrt[5]{x^3}}{\sqrt[5]{x^3}} = \frac{\sqrt[5]{x^3}}{\sqrt[5]{x^5}} = \frac{\sqrt[5]{x^3}}{x}$

41. $\sqrt[3]{\frac{x^2}{y}} = \frac{\sqrt[3]{x^2}}{\sqrt[3]{y}} \cdot \frac{\sqrt[3]{y^2}}{\sqrt[3]{y^2}} = \frac{\sqrt[3]{x^2y^2}}{\sqrt[3]{y^3}} = \frac{\sqrt[3]{x^2y^2}}{y}$

43. $[(a + 2b)^2]^{1/3} = (a + 2b)^{2/3}$

45. $2x(x^2y)^{1/3} = 2x(x^2)^{1/3}y^{1/3} = 2x^{3/3}x^{2/3}y^{1/3} = 2x^{5/3}y^{1/3}$

47. $a^{3/4}b^{1/4} = \sqrt[4]{a^3} \cdot \sqrt[4]{b} = \sqrt[4]{a^3b}$

49. $x^{-5/3} = \sqrt[3]{x^{-5}} = \frac{1}{\sqrt[3]{x^5}}$

51. $\sqrt{\sqrt{2x}} = [(2x)^{1/2}]^{1/2} = (2x)^{1/4} = \sqrt[4]{2x}$

53. $\sqrt[4]{\sqrt{xy}} = [(xy)^{1/2}]^{1/4} = (xy)^{1/8} = \sqrt[8]{xy}$

55. $\frac{\sqrt[5]{a^2}}{\sqrt[3]{a}} = \frac{a^{2/5}}{a^{1/3}} = a^{2/5 - 1/3} = a^{1/15} = \sqrt[15]{a}$

57. $a^{3/5}a^{1/3}a^{-3/2} = a^{3/5 + 1/3 - 3/2} = a^{-17/30} = \frac{1}{a^{17/30}}$

59. $(a^{5/3}b^{3/4})(3a^{1/3}b^{5/4}) = 3 \cdot a^{5/3}a^{1/3} \cdot b^{3/4}b^{5/4} = 3 \cdot a^{6/3} \cdot b^{8/4} = 3a^2b^2$ $(b \geq 0)$

61. $\left(\frac{-8x^6}{y^{-3}}\right)^{2/3} = (-8x^6y^3)^{2/3} = (-8)^{2/3}(x^6)^{2/3}(y^3)^{2/3}$
$= [(-8)^2]^{1/3}x^{12/3}y^{6/3} = 64^{1/3}x^4y^2 = 4x^4y^2$

63. $\frac{(x^9y^6)^{-1/3}}{(x^6y^2)^{-1/2}} = \frac{(x^6y^2)^{1/2}}{(x^9y^6)^{1/3}} = \frac{\sqrt{x^6y^2}}{\sqrt[3]{x^9y^6}} = \frac{|x^3y|}{x^3y^2} = \frac{|x^3| \cdot |y|}{x^3y^2}$
$= \frac{1}{|y|} \cdot \frac{|x|}{x} = \frac{|x|}{x|y|}$

65. $\sqrt{9x^{-6}y^4} = |3x^{-3}y^2| = 3y^2|x^{-3}| = \frac{3y^2}{|x^3|}$

67. $\sqrt[4]{\frac{3x^8y^2}{8x^2}} = \sqrt[4]{\frac{2 \cdot 3x^8y^2}{2 \cdot 8x^2}} = \frac{\sqrt[4]{6x^6y^2}}{2} = \frac{\sqrt[4]{6x^4x^2y^2}}{2}$
$= \frac{|x|\sqrt[4]{6x^2y^2}}{2}$

69. $\sqrt[3]{\frac{4x^2}{y^2}} \cdot \sqrt[3]{\frac{2x^2}{y}} = \sqrt[3]{\frac{(4x^2)(2x^2)}{(y^2)(y)}} = \sqrt[3]{\frac{8x^4}{y^3}} = \frac{2\sqrt[3]{x^4}}{y}$
$= \frac{2x\sqrt[3]{x}}{y}$

71. $3\sqrt{4^2 \cdot 3} - 2\sqrt{6^2 \cdot 3} = 3 \cdot 4\sqrt{3} - 2 \cdot 6\sqrt{3}$
$= 12\sqrt{3} - 12\sqrt{3} = 0$

73. $\sqrt{x^2 \cdot x} - \sqrt{(2y)^2 \cdot x} = |x|\sqrt{x} - 2|y| \cdot \sqrt{x}$
$= (|x| - 2|y|)\sqrt{x} = (x - 2|y|)\sqrt{x}$ (since the square root is undefined when $x < 0$)

For #75–81, evaluate each side using a calculator or paper and pencil.

75. $\sqrt{2 + 6} < \sqrt{2} + \sqrt{6}$ $(2.828\ldots < 3.863\ldots)$

77. $(3^{-2})^{-1/2} = 3$

79. $\sqrt[4]{(-2)^4} > -2$ $(2 > -2)$

81. $2^{2/3} < 3^{3/4}$ $(1.587\ldots < 2.279\ldots)$

83. $t = 1.1\sqrt{10} \approx 3.48$ sec

85. If n is even, then there are two real nth roots of a (when $a > 0$): $\sqrt[n]{a}$ and $-\sqrt[n]{a}$.

Section 2 Polynomials and Factoring

Section 2 Exercises

1. $3x^2 + 2x - 1$; degree 2

3. $-x^7 + 1$; degree 7

5. No — cannot have a negative exponent like x^{-1}

7. Yes

9. $(x^2 - 3x + 7) + (3x^2 + 5x - 3) = (x^2 + 3x^2) + (-3x + 5x) + (7 - 3) = 4x^2 + 2x + 4$

11. $(4x^3 - x^2 + 3x) + (-x^3 - 12x + 3)$
$= (4x^3 - x^3) - x^2 + (3x - 12x) + 3$
$= 3x^3 - x^2 - 9x + 3$

13. $2x(x^2) - 2x(x) + 2x(3) = 2x^3 - 2x^2 + 6x$

15. $(-3u)(4u) + (-3u)(-1) = -12u^2 + 3u$

17. $2(5x) - x(5x) - 3x^2(5x) = 10x - 5x^2 - 15x^3$
$= -15x^3 - 5x^2 + 10x$

19. $x(x + 5) - 2(x + 5) = (x)(x) + (x)(5) - (2)(x)$
$- (2)(5) = x^2 + 5x - 2x - 10 = x^2 + 3x - 10$

21. $3x(x + 2) - 5(x + 2) = (3x)(x) + (3x)(2)$
$- (5)(x) - (5)(2) = 3x^2 + 6x - 5x - 10$
$= 3x^2 + x - 10$

23. $(3x)^2 - (y)^2 = 9x^2 - y^2$

25. $(3x)^2 + 2(3x)(4y) + (4y)^2 = 9x^2 + 24xy + 16y^2$

27. $(2u)^3 - 3(2u)^2(v) + 3(2u)(v)^2 - (v)^3$
$= 8u^3 - 3v(4u^2) + 6uv^2 - v^3$
$= 8u^3 - 12u^2v + 6uv^2 - v^3$

29. $(2x^3)^2 - (3y)^2 = 4x^6 - 9y^2$

31. $x^2(x + 4) - 2x(x + 4) + 3(x + 4)$
$= (x^2)(x) + (x^2)(4) - (2x)(x) - (2x)(4)$
$+ (3)(x) + (3)(4)$
$= x^3 + 4x^2 - 2x^2 - 8x + 3x + 12$
$= x^3 + 2x^2 - 5x + 12$

33. $x^2(x^2 + x + 1) + x(x^2 + x + 1) - 3(x^2 + x + 1)$
$= (x^2)(x^2) + (x^2)(x) + (x^2)(1) + (x)(x^2) + (x)(x)$
$+ (x)(1) - (3)(x^2) - (3)(x) - (3)(1)$
$= x^4 + x^3 + x^2 + x^3 + x^2 + x - 3x^2 - 3x - 3$
$= x^4 + 2x^3 - x^2 - 2x - 3$

35. $(x)^2 - (\sqrt{2})^2 = x^2 - 2$

37. $(\sqrt{u})^2 - (\sqrt{v})^2 = u - v, u \geq 0$ and $v \geq 0$

39. $x(x^2 + 2x + 4) - 2(x^2 + 2x + 4) = (x)(x^2)$
$+ (x)(2x) + (x)(4) - (2)(x^2) - (2)(2x) - (2)(4)$
$= x^3 + 2x^2 + 4x - 2x^2 - 4x - 8 = x^3 - 8$

41. $5(x - 3)$

43. $yz(z^2 - 3z + 2)$

45. $z^2 - 7^2 = (z + 7)(z - 7)$

47. $8^2 - (5y)^2 = (8 + 5y)(8 - 5y)$

49. $y^2 + 2(y)(4) + 4^2 = (y + 4)^2$

51. $(2z)^2 - 2(2z)(1) + 1^2 = (2z - 1)^2$

53. $y^3 - 2^3 = (y - 2)[y^2 + (y)(2) + 2^2]$
$= (y - 2)(y^2 + 2y + 4)$

55. $(3y)^3 - 2^3 = (3y - 2)[(3y)^2 + (3y)(2) + 2^2]$
$= (3y - 2)(9y^2 + 6y + 4)$

57. $1^3 - x^3 = (1 - x)[1^2 + (1)(x) + x^2]$
$= (1 - x)(1 + x + x^2) = (1 - x)(1 + x + x^2)$

59. $(x + 2)(x + 7)$

61. $(z - 8)(z + 3)$

63. $(2u - 5)(7u + 1)$

65. $(3x + 5)(4x - 3)$

67. $(2x + 5y)(3x - 2y)$

69. $(x^3 - 4x^2) + (5x - 20) = x^2(x - 4) + 5(x - 4)$
$= (x - 4)(x^2 + 5)$

71. $(x^6 - 3x^4) + (x^2 - 3) = x^4(x^2 - 3) + 1(x^2 - 3)$
$= (x^2 - 3)(x^4 + 1)$

73. $(2ac + 6ad) - (bc + 3bd) = 2a(c + 3d) - b(c + 3d)$
$= (c + 3d)(2a - b)$

75. $x(x^2 + 1)$

77. $2y(9y^2 + 24y + 16) = 2y[(3y)^2 + 2(3y)(4)$
$+ 4^2] = 2y(3y + 4)^2$

79. $y(16 - y^2) = y(4^2 - y^2) = y(4 + y)(4 - y)$

81. $y(5 + 3y - 2y^2) = y(1 + y)(5 - 2y)$

83. $2[(5x + 1)^2 - 9] = 2[(5x + 1)^2 - 3^2]$
$= 2[(5x + 1) + 3][(5x + 1) - 3]$
$= 2(5x + 4)(5x - 2)$

85. $2(6x^2 + 11x - 10) = 2(2x + 5)(3x - 2)$

87. $(2ac + 4ad) - (2bd + bc) = 2a(c + 2d) - b(2d + c)$
$= (c + 2d)(2a - b) = (2a - b)(c + 2d)$

89. $(x^3 - 3x^2) - (4x - 12) = x^2(x - 3) - 4(x - 3)$
$= (x - 3)(x^2 - 4) = (x - 3)(x + 2)(x - 2)$

91. $(2ac + bc) - (2ad + bd)$
$= c(2a + b) - d(2a + b) = (c - d)(2a + b)$

Neither of the groupings $(2ac - bd)$ and $(-2ad + bc)$ has a common factor to remove.

Section 3 Fractional Expressions

Section 3 Exercises

1. $\frac{5}{9} + \frac{10}{9} = \frac{5 + 10}{9} = \frac{15}{9} = \frac{5}{3}$

3. $\frac{20}{21} \cdot \frac{9}{22} = \frac{20 \cdot 9}{21 \cdot 22} = \frac{180}{462} = \frac{30}{77}$

5. $\frac{2}{3} \div \frac{4}{5} = \frac{2}{3} \cdot \frac{5}{4} = \frac{2 \cdot 5}{3 \cdot 4} = \frac{10}{12} = \frac{5}{6}$

7. The LCD of the denominators is $2 \cdot 7 \cdot 3 \cdot 5 = 210$:
$\frac{1}{14} + \frac{4}{15} - \frac{5}{21} = \frac{15}{210} + \frac{56}{210} - \frac{50}{210}$
$= \frac{15 + 56 - 50}{210} = \frac{21}{210} = \frac{1}{10}$

9. No values are restricted, so the domain is all real numbers.

11. The value under the radical must be nonnegative, so $x - 4 \geq 0$: $x \geq 4$ or $[4, \infty)$.

13. The denominator cannot be 0, so $x^2 + 3x \neq 0$ or $x(x + 3) \neq 0$. Then $x \neq 0$ and $x + 3 \neq 0$: $x \neq 0$ and $x \neq -3$.

15. The denominator cannot be 0, so $x - 1 \neq 0$, or $x \neq 1$. Then $x \neq 2$ and $x \neq 1$.

17. $x^{-1} = \frac{1}{x}$ and the denominator cannot be 0, so $x \neq 0$.

19. The denominator is $12x^3 = (3x)(4x^2)$, so the new numerator is $2(4x^2) = 8x^2$.

21. The numerator is $x^2 - 4x = (x - 4)(x)$, so the new denominator is $(x)(x) = x^2$.

23. The denominator is $x^2 + 2x - 8 = (x + 4)(x - 2)$, so the new numerator is $(x + 3)(x + 4) = x^2 + 7x + 12$.

25. The numerator is $x^2 - 3x = x(x - 3)$, so the new denominator is $x(x^2 + 2x)$ or $x^3 + 2x^2$.

27. $(x - 2)(x + 7)$ cancels out during simplification; the restriction indicates that the values 2 and -7 were not valid in the original expression.

29. No factors were removed from the expression; we can see by inspection that $\frac{2}{3}$ and 5 are not valid.

31. $(x - 3)$ ends up in the numerator of the simplified expression; the restriction reminds us that it began in the denominator so that 3 is not allowed.

33. $\frac{3x(6x^2)}{3x(5)} = \frac{6x^2}{5}, x \neq 0$

35. $\frac{x(x^2)}{x(x-2)} = \frac{x^2}{x-2}, x \neq 0$

37. $\frac{z(z-3)}{(3-z)(3+z)} = -\frac{z}{z+3}, z \neq 3$

39. $\frac{(y+5)(y-6)}{(y+3)(y-6)} = \frac{y+5}{y+3}, y \neq 6$

41. $\frac{(2z)^3 - 1^3}{(z+3)(2z-1)} = \frac{(2z-1)[(2z)^2 + (2z)(1) + 1^2]}{(z+3)(2z-1)}$
$= \frac{4z^2 + 2z + 1}{z+3}, z \neq \frac{1}{2}$

43. $\frac{(x^3 + 2x^2) - (3x + 6)}{x^2(x+2)} = \frac{x^2(x+2) - 3(x+2)}{x^2(x+2)}$
$= \frac{(x+2)(x^2-3)}{x^2(x+2)} = \frac{x^2-3}{x^2}, x \neq -2$

45. $\frac{1}{x-1} \cdot \frac{(x+1)(x-1)}{3} = \frac{x+1}{3}, x \neq 1$

47. $\frac{x+3}{x-1} \cdot \frac{-(x-1)}{(x+3)(x-3)} = -\frac{1}{x-3}, x \neq 1 \text{ and } x \neq -3$

49. $\frac{(x-1)(x^2+x+1)}{2x^2} \cdot \frac{4x}{x^2+x+1} = \frac{2(x-1)}{x}$

51. $\frac{(y+5)(2y-1)}{(y+5)(y-5)} \cdot \frac{y-5}{y(2y-1)} = \frac{1}{y}, y \neq 5, y \neq -5$ and $y \neq \frac{1}{2}$

53. $\frac{1}{2x} \cdot \frac{4}{1} = \frac{2}{x}$

55. $\frac{x(x-3)}{14y} \cdot \frac{3y^2}{2xy} = \frac{3(x-3)}{28}, x \neq 0 \text{ and } y \neq 0$

57. $\frac{2x^2y}{(x-3)^2} \cdot \frac{x-3}{8xy} = \frac{x}{4(x-3)}, x \neq 0 \text{ and } y \neq 0$

59. $\frac{2x+1-3}{x+5} = \frac{2x-2}{x+5}$

61. $\frac{3}{x(x+3)} - \frac{1}{x} - \frac{6}{(x+3)(x-3)}$
$= \frac{3(x-3)}{x(x+3)(x-3)} - \frac{1(x+3)(x-3)}{x(x+3)(x-3)}$
$- \frac{6x}{x(x+3)(x-3)}$
$= \frac{(3x-9) - (x^2-9) - (6x)}{x(x+3)(x-3)}$
$= \frac{-x^2-3x}{x(x+3)(x-3)} = -\frac{x(x+3)}{x(x+3)(x-3)}$
$= -\frac{1}{x-3} = \frac{1}{3-x}, x \neq 0 \text{ and } x \neq -3$

63. $\frac{\frac{x^3-y^3}{x^2y^2}}{\frac{x^2-y^2}{x^2y^2}} = \frac{x^3-y^3}{x^2y^2} \cdot \frac{x^2y^2}{x^2-y^2}$
$= \frac{(x-y)(x^2+xy+y^2)}{(x-y)(x+y)} = \frac{x^2+xy+y^2}{x+y},$
$x \neq y, x \neq 0, \text{ and } y \neq 0$

65. $\frac{\frac{2x(x-4)+13x-3}{x-4}}{\frac{2x(x-4)+x+3}{x-4}} = \frac{2x^2+5x-3}{x-4} \cdot \frac{x-4}{2x^2-7x+3}$
$= \frac{(2x-1)(x+3)}{(2x-1)(x-3)} = \frac{x+3}{x-3}, x \neq 4 \text{ and } x \neq \frac{1}{2}$

67. $\frac{\frac{x^2-(x+h)^2}{x^2(x+h)^2}}{h} = \frac{x^2-(x^2+2xh+h^2)}{x^2(x+h)^2} \cdot \frac{1}{h}$
$= \frac{-2xh-h^2}{hx^2(x+h)^2} = \frac{-h(2x+h)}{hx^2(x+h)^2}$
$= -\frac{2x+h}{x^2(x+h)^2}, h \neq 0$

69. $\frac{\frac{b^2-a^2}{ab}}{\frac{b-a}{ab}} = \frac{(b+a)(b-a)}{ab} \cdot \frac{ab}{b-a} = b + a$
$= a + b, a \neq 0, b \neq 0, \text{ and } a \neq b$

71. $\left(\frac{x+y}{xy}\right)\left(\frac{1}{x+y}\right) = \frac{1}{xy}, x \neq -y$

73. $\frac{1}{x} + \frac{1}{y} = \frac{y}{xy} + \frac{x}{xy} = \frac{x+y}{xy}$

Appendix C

Section 1 Logic: An Introduction

Section 1 Exercises

1. **(a)** $2 + 4 = 8$ is a false statement.

(b) "Shut the window" is an instruction, which is neither true nor false. It is not a statement.

(c) "Los Angeles is a state" is a false statement.

(d) "He is in town" is neither true nor false when "he" is unspecified. It is not a statement.

(e) "What time is it?" is a question, which is neither true nor false. Is it not a statement.

(f) $5x = 15$ is neither true nor false when x is unspecified. It is not a statement.

(g) $3 \cdot 2 = 6$ is a true statement.

(h) $2x^2 > x$ is neither true nor false when x is unspecified. It is not a statement.

(i) "This statement is false" is true if it is false, and false if it is true. So it fails to be either true or false but not both. It is not a statement.

(j) "Stay put!" is a command, which is neither true nor false. It is not a statement.

3. In each case, negate the corresponding quantified statement from Exercise 2.

(a) There is no natural number x such that $x + 8 = 11$.

(b) It is not true that for all natural numbers x, $x + 0 = x$. That is: There exists a natural number x such that $x + 0 \neq x$.

(c) There is no natural number x such that $x^2 = 4$.

(d) There exists a natural number x such that $x + 1 = x + 2$.

5. **(a)** The book does not have 500 pages.

(b) Six is not less than eight.

(c) $3 \cdot 5 \neq 15$

(d) It is not true that some people have blond hair. In other words: No people have blond hair.

(e) Not all dogs have four legs. (Or, in other words: Some dogs do not have four legs.)

(f) It is not true that some cats do not have nine lives. In other words: All cats have nine lives.

(g) Not all squares are rectangles. (Or, in other words: Some squares are not rectangles.)

(h) All rectangles are squares.

(i) It is not true that for all natural numbers x, $x + 3 = 3 + x$. In other words: There exists a natural number x such that $x + 3 \neq 3 + x$.

(j) There does not exist a natural number x such that $3 \cdot (x + 2) = 12$. In other words: For all natural numbers x, $3 \cdot (x + 2) \neq 12$.

(k) Not every counting number is divisible by itself and 1. (Or, in other words: Some natural counting number is not divisible by itself and 1.)

(l) All natural numbers are divisible by 2.

(m) It is not true that for all natural numbers x, $5x + 4x = 9x$. In other words: For some natural number x, $5x + 4x \neq 9x$.

7. Use the truth tables for the connectives.

(a) $p \wedge q$ is false because p is false (and a true conjunction requires both sides true).

(b) $p \vee q$ is true because q is true (and a true disjunction requires only one side true).

(c) $\sim p$ is true because p is false.

(d) $\sim q$ is false because q is true.

(e) $\sim(\sim p)$ is false because $\sim p$ is true [part (c)].

(f) $\sim p \vee q$ is true because $\sim p$ and q are both true. (Either one would suffice.)

(g) $p \wedge \sim q$ is false because p and $\sim q$ are both false. (Either one false would suffice to make the conjunction false.)

(h) $\sim(p \vee q)$ is false because $p \vee q$ is true [part (b)].

(i) $\sim(\sim p \wedge q)$ is false because $\sim p$ and q both true makes $\sim p \wedge q$ true.

(j) $\sim q \wedge \sim p$ is false because $\sim q$ is false [part (d)].

9. **(a)** r, $\vee$, and s are analogous to R, $\cup$, and S, respectively, so $r \vee s$ corresponds to $R \cup S$.

(b) q, $\wedge$, and $\sim q$ are analogous to Q, $\cap$, and $\overline{Q}$, respectively, so $q \wedge \sim q$ corresponds to $Q \cap \overline{Q}$.

(c) r, $\vee$, and q are analogous to R, $\cup$, and Q, respectively, so $\sim(r \vee q)$ corresponds to $\overline{R \cup Q}$.

(d) p, $\wedge$, r, $\vee$, and s are analogous to P, $\cap$, R, $\cup$, and S, respectively, so $p \wedge (r \vee s)$ corresponds to $P \cap (R \cup S)$.

11. **(a)** The statements $\sim(p \vee q)$ and $\sim p \wedge \sim q$ are equivalent, and the statements $\sim(p \wedge q)$ and $\sim p \vee \sim q$ are equivalent.

(b) The corresponding De Morgan Laws for sets are $\overline{P \cup Q} = \overline{P} \cap \overline{Q}$ and $\overline{P \cap Q} = \overline{P} \cup \overline{Q}$. The analogy comes from letting p mean "x is a member of P" and letting q mean "x is a member of Q." Then, for the first law, $\sim(p \vee q)$ means "x is a member of $\overline{P \cup Q}$," which is equivalent to "x is a member of $\overline{P} \cap \overline{Q}$," which translates into $\sim p \wedge \sim q$. Similar reasoning holds for the second law.

Section 2 Conditionals and Biconditionals

Section 2 Exercises

1. Use p: "It is raining" and q: "The grass is wet."

(a) The conditional "If p, then q" is $p \rightarrow q$.

(b) The conditional "If not-p, then q" is $\sim p \rightarrow q$.

(c) The conditional "If p, then not-q" is $p \to \sim q$.

(d) The conditional "q if p" is $p \to q$.

(e) The conditional "Not-q implies not-p" is $\sim q \to \sim p$.

(f) The biconditional "q if, and only if, p" is $q \leftrightarrow p$.

3. If the implication is $p \to q$, then the converse is $q \to p$, the inverse is $\sim p \to \sim q$, and the contrapositive is $\sim q \to \sim p$.

(a) Converse: If you are good in sports, then you eat Meaties; Inverse: If you do not eat Meaties, then you are not good in sports; Contrapositive: If you are not good in sports, then you do not eat Meaties.

(b) Converse: If you do not like mathematics, then you do not like this book; Inverse: If like this book, then you like mathematics; Contrapositive: If you like mathematics, then you like this book.

(c) Converse: If you have cavities, then you do not use Ultra Brush toothpaste; Inverse: If you use Ultra Brush toothpaste, then you do not have cavities; Contrapositive: If you do not have cavities, then you use Ultra Brush toothpaste.

(d) Converse: If your grades are high, then you are good at logic; Inverse: If you are not good at logic, then your grades are not high; Contrapositive: If your grades are not high, then you are not good at logic.

5. Use the truth tables for the connectives.

(a) $\sim p \to \sim q$ is true because $\sim p$ is false and $\sim q$ is true.

(Either one would suffice to make the implication true.)

(b) $\sim(p \to q)$ is true because p true and q false makes $p \to q$ false.

(c) $(p \vee q) \to (p \wedge q)$ is false because p true and q false makes $p \vee q$ true and $p \wedge q$ false.

(d) $p \to \sim p$ is false because p is true and $\sim p$ is false.

(e) $(p \vee \sim p) \to p$ is true because p is true. ($p \vee \sim p$ is always true.)

(f) $(p \vee q) \leftrightarrow (p \wedge q)$ is false because $(p \vee q) \to (p \wedge q)$ is false [part (c)].

7. No. If it does not rain and Iris goes to the movies, then the first statement is true, but the second statement is false.

9. The contrapositive is logically equivalent: "If a number is not a multiple of 4, it is not a multiple of 8."

11. **(a)** For $r \to s$ to be true when s is false, r must be false also. By the same reasoning, q must be false, and so must p.

(b) p must be false, because if p were true, $p \wedge q$ would be true, and then $(p \wedge q) \to r$ could not be true while r was false.

(c) Not only can q be true, but it has to be true, since q true and p false makes $p \to q$ true and is the only way for $q \to p$ to be false.

13. **(a)** All college students are poor.
Helen is a college student.
Helen is poor.
(A Venn diagram will confirm that this is valid.)

(b) Some freshmen like mathematics.
All people who like mathematics are intelligent.
Some freshmen are intelligent.
(A Venn diagram will confirm that this is valid.)

(c) If I study for the final, then I will pass the final.
If I pass the final, then I will pass the course.
If I pass the course, then I will look for a teaching job.
If I study for the final, then I will look for a teaching job.
(This involves two successive applications of the chain rule.)

(d) Every equilateral triangle is isosceles.
There exist triangles that are equilateral.
There exist triangles that are isosceles.
(A Venn diagram will confirm that this is valid.)

15. **(a)** If a figure is a square, then it is a rectangle.

(b) If a number is an integer, then it is a rational number.

(c) If a figure has exactly three sides, then it may be a triangle.

(d) If it rains, it is cloudy.

Solutions
to
Section 1.2 Exercise
for the Media Update
and
Florida Edition

Extending the Ideas

Section 1.2

87. Looking Ahead to Calculus A Key theorem in calculus, **The Extreme Value Theorem**, states, if a function f is continuous on a closed interval $[a, b]$, then f has both a maximum value and a minimum value on the interval. For each of the following functions, verify that the function is continuous on the given interval and find the maximum and minimum values of the function and the x values at which these extrema occur.

(a) $f(x) = x^2 - 3, [-2, 4]$

(b) $f(x) = 1/x, [1, 5]$

(c) $f(x) = |x + 1| + 2, [-4, 1]$

(d) $f(x) = \sqrt{x^2 + 9}, [-4, 4]$

Solutions

(a) f is continuous on $[-2, 4]$; the maximum value is 13, which occurs at $x = 4$, and the minimum value is -3, which occurs at $x = 0$.

(b) f is continuous on $[1, 5]$; the maximum value is 1, which occurs at $x = 1$, and the minimum value is 0.2, which occurs at $x = 5$.

(c) f is continuous on $[-4, 1]$; the maximum value is 5, which occcurs at $x = -4$, and the minimum value is 2, which occurs at $x = -1$.

(d) f is continuous on $[-4, 4]$; the maximum value is 5, which occurs at both $x = -4$ and $x = 4$, and the minimum value is 3, which occurs at $x = 0$.